Diesel Fundamentals

Diesel Fundamentals
Principles and Service

Second Edition

Frank J. Thiessen
Davis N. Dales

A RESTON BOOK
PRENTICE HALL, Englewood Cliffs, New Jersey 07632

Library of Congress Cataloging in Publication Data

Thiessen, F. J.
  Diesel Fundamentals.
  "A Reston book."
  Bibliography: P.
  Includes index.
  1. Diesel motor. I. Dales, D. N.
II. Title.
TJ795.D29 1986   621.43'6   85-19121
ISBN 0-8359-1286-8

A Reston Book
Published by Prentice Hall
A Division of Simon & Schuster, Inc.
Englewood Cliffs, NJ 07632

10 9 8 7 6 5

Printed in the United States of America

Contents

Preface

The application and use of the diesel engine as a source of power is widespread and still increasing. Diesel engines are used in transportation in cars, trucks, buses, railroads, ships, and boats. In agriculture they are used in tractors, harvesters, and other self-propelled equipment. In construction and the mining industry they power earth movers, ore carriers, bulldozers, backhoes, cranes, and trenchers. Diesel engines are also used to drive generators for electric power, irrigation pumps, compressors, and other stationary equipment.

This widespread and increasing use of the diesel engine as a power source provides ample opportunity for a number of rewarding career opportunities in such areas as:

- service sales
- service management
- shop supervision
- diesel machine shop
- fleet maintenance management
- parts sales and management
- new and used equipment sales
- diesel service instruction

Good pay is available in all of these career areas, depending on the ability and efforts of the individual. The challenge of continuing technological change makes these careers personally rewarding as well.

This text provides a sound basis upon which the above careers can be built. The subject matter is presented in a straightforward and concise manner. Information on each system and each component of each system is divided into the following categories:

- Purpose of the system or unit
- Construction of the system or unit
- Operation of the system or unit
- Service of the system or unit

The operating principles of the various systems of the engine are essentially the same. It follows, then, that if the principles of each system or component are understood, the differences among the various makes will be understood as well. Service procedures follow the generally accepted sequence common to the trade.

The text is generously illustrated to help in explaining systems and components, as well as service. Each illustration is accompanied by a brief explanation tying the illustration to the text.

The authors' experience in the industry as mechanic, shop foreman, service manager, garage owner–operator, instructor, and program coordinator is reflected in the design and content of this text, and it represents a thoroughly practical approach to the subject.

IMPORTANT SAFETY NOTICE

Proper service and repair is important to the safe, reliable operation of all equipment. The procedures recommended and described in this text are effective methods of performing service operations. Some of these service operations require the use of tools specially designed for the purpose. The special tools should be used as recommended in the manufacturers' service manuals.

It is important to note that this text contains various general precautions, which should be read carefully in order to minimize the risk of personal injury or damage to the unit resulting from improper service methods. It also is important to understand that these general precautions are not exhaustive. The authors could not possibly know, evaluate, and advise the service trade of all conceivable ways in which service might be carried out, or of the possible hazardous consequences of each method. Accordingly, anyone who uses any given service procedure or tool must first satisfy himself thoroughly that neither his safety nor the safety of the unit will be jeopardized by the service method he selects.

CAUTION

Equipment contains many parts dimensioned in the metric system as well as in the customary system. Many fasteners are metric and are very close in dimension to familiar customary fasteners in the inch system. It is important to assure that, during any maintenance procedures, replacement fasteners have the same measurements and strength as those removed, whether metric or customary. (Numbers on the heads of metric bolts and on surfaces of metric nuts indicate their strength. Customary bolts use radial lines for this purpose, while most customary nuts do not have strength markings.) Mismatched or incorrect fasteners can result in damage or malfunction, and possibly in personal injury. Therefore, fasteners removed should be saved for re-use in the same locations whenever possible. In cases where the fasteners are not satisfactory for re-use, care should be taken to select a replacement that matches the original. For information on any specific make or model, refer to the appropriate service manual.

Acknowledgments

The authors have made every effort to give proper credit to the sources of illustrations and other materials, and would be grateful for information from readers on errors or omissions in this regard. We wish also to thank the following companies for their assistance in developing this second edition of *Diesel Fundamentals, Principles and Service:*

Cummins Engine Company

Caterpillar Tractor Company

Detroit Diesel Allison Division, General Motors Corporation

Mack Trucks Inc.

J.I. Case, a Tenneco Company

Ford Tractor Operations, Ford Motor Company

International Harvester Company

Deere and Company

Robert Bosch Canada Ltd.

American Bosch, United Technologies Corporation

Diesel Systems Group, Stanadyne Inc.

California Air Resources Board

Garrett Corporation

Wallace Murray Corporation

Donaldson Company Inc.

Chrysler Corporation

Imperial Oil Limited

Grey Goose Bus Lines Limited

Sioux Tools Inc.

Hastings Manufacturing Company

Sunnen Products Company

Weatherhead Company of Canada

National Seal

H. Pauling and Company Ltd.

L. S. Starrett Company

Mac Tools Inc.

Owatonna Tool Company

Snap-On Tools Corporation

Proto Canada Division, Ingersoll Rand Canada Inc.

Diesel Fundamentals

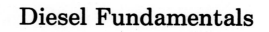

Introduction

PART 1 DIESEL SERVICE CAREERS

The diesel service industry is one of the larger industries worldwide. In industrialized countries, employment in this industry is very high.

The size and diversity of the industry offer many opportunities for interesting, challenging, and rewarding careers. Continuing technological change in the industry demands that the diesel technician continue to learn and update his knowledge and skills.

Career choices include becoming a specialist in one or more specific subject areas such as the following:

- Light service and lubrication
- Tires
- Brakes
- Wheel alignment and balance
- Manual transmissions and differentials
- Automatic transmissions
- Engines
- Tune up
- Air conditioning
- Electrical
- Diesel injection
- Trim
- Diagnostics

In addition, there are career opportunities in rebuilding plants as engine rebuilders, transmission rebuilders, fuel injection pump and injector rebuilders and calibrators, etc.

Service career opportunities are also available in the frame repair and alignment fields.

Supervisory Careers

Larger diesel shops, dealers, vehicle manufacturers, fleet owners, and public and private technical schools offer opportunities for supervisory careers in the diesel service industry. Among these are:

- Service sales
- Service control operator
- Shop foreman
- Service manager
- Garage manager or proprietor
- Manufacturer's district service representative
- Manufacturer's technical training instructor
- Manufacturer's factory quality control inspector
- Manufacturer's regional service manager
- Diesel instructor in public or private schools, high schools, vocational-technical schools, and community colleges
- Fleet maintenance manager in industry, federal or local governments

Some of these positions require additional experience and training.

Related Career Opportunities

Many related career opportunities exist both inside and outside of the diesel service industry. Some of these are:

- Parts sales
- Parts manager
- New and used sales
- Used parts sales
- Recreational vehicle sales and service
- Agricultural equipment sales and service
- Highway transport sales and service
- Off-road industrial and construction equipment sales and service
- Government vehicle safety inspection
- Government emissions control inspection

PART 2   PREPARING FOR A DIESEL SERVICE CAREER

The degree of success achieved in the chosen career is dependent on a number of factors, not the least of which is desire—the kind that results in determination, effort, and achievement.

It is a good idea to "stay with it" in high school to gain the necessary skills in communications, mathematics, and science. Knowledge of the basic principles of the following subject matter should be acquired either prior to taking a diesel engine course, or during the course. Some diesel engine courses may require a knowledge of these principles as prerequisites, while others may provide related class instruction in science, mathematics, and communication.

Some of the major areas in which knowledge of basic principles is required are:

- Matter
- Mass
- Atmospheric Pressure
- Weight
- Gravity
- Absolute Pressure
- Vacuum
- Density
- Hydraulics
- Pneumatics
- Friction
- Work
- Force
- Force Multiplication
- Energy
- Power
- Torque
- Heat
- Electricity
- Areas
- Volumes
- Ratios
- Speed

Also required is the ability to calculate these as they relate to the diesel engine by means of addition, subtraction, multiplication, and division using whole numbers, fractions, decimals, and percentages.

This text deals with each of these topics in their respective chapters where necessary. However, it is assumed that the needed skills in mathematics have already been acquired.

Understanding the Measurement Systems

The modern diesel engine consists of many individual parts and components held together by a variety of fasteners of different shapes and sizes.

Some diesel engine components are dimensioned in the customary U.S. system of measurement while others are dimensioned in the SI metric system. The modern technician and the modern diesel shop must, therefore, be equipped with tools and measuring devices designed to be used on U.S. dimensioned components and on SI metric components as well. The technician must also be able to distinguish between the two. The importance of this is obvious when fasteners (bolts, nuts, screws, etc.) used in the industry are considered. It is critical to the safety of the vehicle and its passengers that only replacement fasteners of original equipment size, type, and quality be used. Metric dimensioned fasteners should never be used when U.S. dimensioned fasteners were originally used. Nor should U.S. dimensioned fasteners be used where metric fasteners were originally used. Even though they may appear to be similar, they do not fit properly and are not interchangeable.

Current manufacturer's service repair manuals, as well as other types of service manuals, are not confined to using one system of measurement for vehicle specifications. U.S. measurements, metric measurements, or both, may appear in any given service manual.

This text gives most measurements and specifications in the customary units, with metric equivalents in brackets following. Metric equivalents are stated in figures that are rounded off to the most appropriate value. This avoids the cumbersome figures with three or four decimals which result from stating exact equivalents.

The U.S. system of measurement is the same as the Imperial system used in Canada, with the exception of liquid volume measurements such as the gallon, quart, and pint. Reference charts in the Appendix for this text provide comparative equivalent measurements in the U.S. system, the SI metric system, and the Imperial system, as well as the appropriate conversion factors, should conversion be necessary.

Schools and Courses

Many different types of schools offer a variety of diesel service training programs and courses.

High schools in many areas offer courses which allow the student to major in diesel engine service. Courses may be of two or three years duration and may offer an in-school service shop as well as some industry experience through cooperative training programs. These types of courses are of the job entry or pre-employment type, meaning that sufficient skills training is provided to allow a student to become employable in the industry. Additional experience and/or training are required to become a certified technician or mechanic.

Vocational-technical schools, community colleges, and universities may offer similar job entry type courses, usually of one or two years duration, and may offer more advanced, more technical courses as well.

The job entry type courses normally offer a certificate after successful completion of the course. The more technical courses offer a diploma or a degree.

Community colleges and vocational-technical schools may also offer short courses for apprentices leading to eligibility for certification as a licensed journeyman heavy duty mechanic.

High schools, community colleges, and vocational-technical schools also usually offer short courses in theory only for those experienced in the trade but not yet certified. These courses are designed to help the student in taking certification exams. Equipment manufacturers also offer a variety of courses on basic principles as well as the more technical and product improvement courses.

PART 3   BECOMING CERTIFIED

Reasons for Certification

Doctors, lawyers, accountants, nurses, plumbers, electricians, dentists, and other professions are licensed or certified in order to practice their professions.

In order to become certified, certain strict uniform requirements must be met. This protects the general public and the practitioner or profession. Licensing or certification tells the general public and the prospective employer that certain minimum standards of performance have been met. Standards for knowledge and skill are established. Usually the certified technician receives higher pay than the noncertified operator. The certified technician is recognized as a professional by the public, by the employer, and by his peers.

Kinds of Certification

The National Institute for Automotive Service Excellence (ASE), with headquarters in Washington, D.C., has offered a program of voluntary certification since 1972.

Most departments of labor include an apprenticeship division responsible for apprenticeship training and certification.

PART 4   HOW TO PASS A TEST

Be Confident

If you didn't think you could pass the test, you would not have spent the time studying. You know you can pass the test because you have done your work and have studied the subject.

Review Your Notes

You should have made a set of notes during your studies listing key points. Review these a day or two before the test. Don't cram the night before. This will only cause you mental and physical fatigue and perhaps rob you of needed sleep. Cramming can also upset you emotionally.

Get a Good Night's Rest

Get to bed at a respectable time the night before the exam. Don't go out on the town or entertain until all hours. You need the rest in order to be calm and relaxed.

Eat Properly

Don't skip supper, breakfast, or lunch before a test. Keep your normal, regular eating routine. Don't attempt to fortify your courage with stimulants—you want to be clear headed and able to think.

Bring Pencils

Pencils are often not provided for the examination so bring two or three so you won't have to sharpen a broken pencil.

Be on Time

Allow yourself enough time to eat and arrive calm and relaxed. If you don't, you may be emotionally upset before you start. If the test is to be written at a location unfamiliar to you, be sure you know how to get there. It may be worthwhile to check it out a day or two before test time by going there.

Listen to Instructions

Be sure you pay attention to any verbal instructions that may be given by the examiner. If you fail to hear or understand, be sure to ask for clarification.

Read Test Instructions Carefully

Make sure you understand all written instructions for the test. If you do not understand the written instructions fully, raise your hand and ask the test supervisor to explain them to you.

Write the Test

If there is a time limit on the test, allocate your time. For example, if the test has 200 questions and you are allowed two hours, you need to pace yourself to about 50 questions every 30 minutes. However, don't rush yourself. If necessary, read the question twice. If you are still baffled, you should go on to the next question and come back to this one later. By then, some of the other questions may have helped you to understand. Don't assume that there are trick questions. Trying to figure out what you may think is a trick question is a waste of time. Test questions are not intentionally tricky. If you think the question is tricky, perhaps you did not fully understand the instructions. Read them again.

Try to answer every question. Most tests have a choice of one answer out of four, so you have a 25% chance of being correct before you start.

When you have finished, review your questions and answers briefly. Do not try to read unintended meanings into them. Correct any errors or missed questions.

Test Types

Most examination questions you will face in certification tests are of the multiple choice type. There may also be a few true-or-false questions and some fill-in-the-blank-to-complete-the-statement types.

There are several types of multiple choice questions to be considered. First there is the "simple choice" question in which there is only one correct answer among the four choices. Based on your knowledge of the subject, you must be able to select that one correct answer.

Another type is the "best choice" or the "one best answer" type. In this type of question, you must be able to decide which of the answers given is the most correct or the most complete. This requires careful consideration of and discrimination between each of the answers given, to choose the one best answer based on your knowledge of the subject.

A third type of question is the "all except one is correct" type. In this type of question all the answers given are correct except for one which is wrong. You must select the one wrong answer. Make sure you understand that, to answer the question correctly, you must select the one wrong answer.

A fourth type of question is the one in which "comparative statements" are given. A lead statement is given describing some operation or repair procedure. Then two more statements are given which comment on the first statement. This is usually in the form of "Mechanic A says . . . and Mechanic B says. . . ." These statements must be evaluated to determine the correct answer to a question about the statements made by Mechanic A and Mechanic B. The question is usually "Who is correct?"

(a) Mechanic A

(b) Mechanic B

(c) Both Mechanic A and Mechanic B are correct

(d) Neither A nor B

If you understand the type of question you are faced with and you have read the question correctly, then you should be able to come up with the correct answer based on your knowledge of the subject.

PART 5   HOW TO STUDY

Decide on a definite study plan. When you are faced with a repair job in the shop, you follow a definite plan of procedure with the proper tools and a specific allocation of time. Then you proceed to do one thing at a time in the right sequence until the job is successfully completed. The same methods must be applied to study habits in order to be successful. Here are some suggestions that have proved to work well.

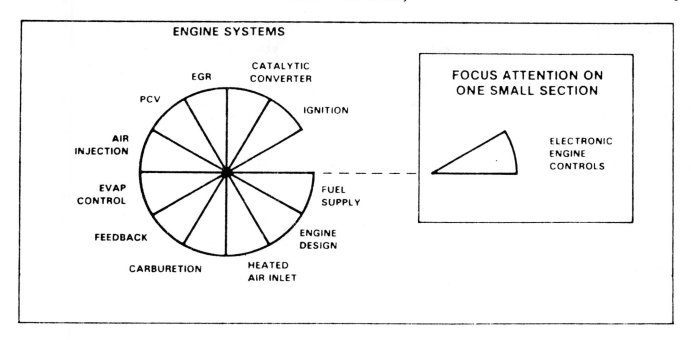

FIGURE I-1. Learn to concentrate on one small section of subject matter at a time. When you have all the individual sections mastered, they will fit together to form a complete unit. All professionals acquire their expertise in this manner. *(Courtesy of Chrysler Corporation)*

1. Establish a definite time period for study. Set aside at least one hour for study, but don't try to overdo it. More than two hours at one sitting is too much.

2. Find a place for study that will allow you to concentrate on what you are doing. Distractions caused by children, the TV or hi-fi, the telephone, or disturbing noise should be avoided.

3. Decide on a definite time of day for your study. You may function best in your studies sometime in the afternoon or early evening, or you may be a morning person. Make sure it is a time during which you are alert and during which there will be minimum interruptions or distractions.

4. Study and learn in small chunks. Like eating an elephant, it must be done one bite at a time; it cannot be swallowed whole. Concentrate on the specific subject matter at hand until you have grasped it fully. When you have learned all the smaller chunks well, the larger concept or subject matter will easily fall into place without gaps or gray areas.

5. Underline key points in your text as you read.

This helps you to remember them better and allows you to review material more easily and quickly.

6. Take notes during class of key points made by your instructor. Don't try to write down everything the instructor says, a few words to remind you of what was said about each point is enough.

7. If you were given a course outline, use it. It is a summary of the subject matter being covered and is very useful in developing a good set of notes.

8. Make sure you understand the diagrams and schematics, such as wiring diagrams and test instrument connections. Don't guess—you must understand the reason why connections are made a certain way.

9. Keep fit physically and mentally. Poor physical health can lower mental ability and efficiency. Be on time for all classes and be prepared to pay attention and participate in class activities, such as discussions and projects. Complete assignments on time.

Good study habits, a healthy body and mind, a keen desire to succeed, and confidence in yourself and your ability are key elements to success.

PART 6   WORKING CONDITIONS

Diesel shops today usually are modern facilities which provide clean, well-lighted working conditions. Good equipment is provided for the type of service work done by the shop.

Employees often work eight hours a day for five and a half days a week, though this varies somewhat in the different shops. Some shift work and some outdoor work may be required.

Most of the larger diesel shops provide cafeteria facilities or lunch rooms, showers, and specific types of clothing.

In many of the larger shops, technicians work "flat rate." This means that there is a given time, obtained from a "flat rate manual," allotted to each repair job. If, for example, the flat rate time for a starter overhaul is two hours, the technician would be paid for two hours of work for doing that job. This would be so whether the job took more or less than the two hour flat rate time.

Other technicians work on "straight time"—an hourly, weekly, or monthly rate. However, productivity remains a factor in maintaining job security.

The diesel service industry offers many rewarding and challenging career opportunities as well as the opportunity for advancement and a good standard of living.

Chapter 1

Safety

Performance Objectives

After thorough study of this chapter and your school shop or service shop, you should be able to do the following:

1. Recognize and practice safety in selecting and using proper clothing for work in a diesel engine shop.
2. Follow the required procedures in case of fire in the shop.
3. Use proper ventilation and shop exhaust equipment whenever needed.
4. Follow the first-aid procedures given for the shop in which you are working.
5. Complete the self-check questions with at least 80 percent accuracy.

If you are the type of person who likes a variety of work, you will find it in shop work, where a large number of different service jobs and procedures are carried out. The variety of jobs and procedures, however, requires a high degree of awareness of the importance of safety. Safety is your job, everyone's job.

Safety in the shop includes avoiding injury to yourself and to others working near you. It also includes avoiding damage to vehicles in the shop and damage to shop equipment and parts. The following are some of the factors to consider in practicing shop safety.

PART 1   PERSONAL SAFETY

1. Wear proper clothing. Loose clothing, ties, uncontrolled long hair, rings, etc., can get caught in rotating parts or equipment and cause injury. Wear the kind of shoes that provide protection for your feet; steel-capped work boots with nonskid soles are best. Keep clothing clean.

2. Use protective clothing and equipment where needed. Use rubber gloves and apron as well as a face mask for handling batteries. Protective goggles or safety glasses are recommended at all times.

3. Keep hands and tools clean to avoid injury to hands and to avoid falling due to slipping when pulling on a wrench.

4. Do not use compressed air to clean your clothes. This can cause dirt particles to be embedded in your skin and cause infection. Do not point the compressed air hose at anyone. Compressed shop air, used for cleaning, should not exceed 30psi.

5. Be careful when using compressed air to blow away dirt from parts. You should not use compressed air to blow dirt from clutch parts since cancer-causing asbestos dust may be inhaled as a result.

6. Do not carry screwdrivers, punches, or other sharp objects in your pockets. You could injure yourself or damage the equipment you are working on.

7. Never get involved in horseplay or other practical jokes. They can lead to injury.

8. Make sure you use the proper tool for the job and use it the right way. The wrong tool or its incorrect use can damage the part you are working on or cause injury or both.

FIGURE 1-1. Safety goggles shield the eyes from injury. (Courtesy of Mac Tools Inc.)

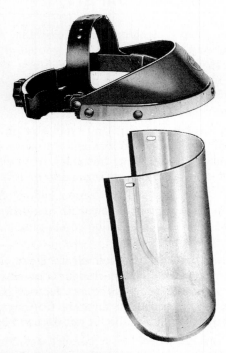

FIGURE 1-2. Face shields protect the face from possible injury resulting from acid, Freon, flying particles from grinding wheels, chipping metal, and the like. *(Courtesy of Mac Tools Inc.)*

9. Never work under anything that is not properly supported. Use safety stands properly placed to work under equipment. Use a creeper.

10. Do not jack any equipment while someone is under it.

11. Never run an engine without proper ventilation and adequate means of getting rid of exhaust gases. Exhaust gas contains deadly carbon monoxide. It can and does kill.

FIGURE 1-3. Those requiring glasses to correct a visual defect should use glasses with safety lenses. *(Courtesy of Mac Tools Inc.)*

12. Keep your work area clean at all times. Your safety and the quality of work you do depend on it.

13. Lifting and carrying should be done properly to avoid injury. Heavy objects should be lifted and moved with the right equipment for the job.

14. Do not stand in the plane of rotating parts such as fans, etc.

15. Never smoke while working on any vehicle.

16. When working with others, note any unsafe practices and report them.

PART 2   SHOP SAFETY

1. Familiarize yourself with the way the shop is laid out. Find out where things are in the shop. You will need to know where the shop manuals are kept in order to obtain specifications and service procedures. Make sure you know the route to the exit in case of fire.

2. Find out whether there are certain stalls that are reserved for special jobs. Abide by these rules.

3. Take note of all the warning signs around the shop. No smoking signs, special instructions for some shop tools and equipment, danger zones, etc., are all there to help the shop run smoothly and safely.

FIGURE 1-4. Dirty shop floors can cause major injury.

FIGURE 1-5. Proper lifting methods are a must to avoid back injury.

4. Note the location of fire extinguishers. Take time to read their operating instructions and the type of fire they are meant to be used on.

5. Follow local regulations with regard to storing fuel and other flammable liquids. Fuel should be stored only in approved containers and locations.

6. Never use gasoline to clean parts.

7. Always immediately wipe up any fuel that has been spilled.

8. Fuel vapors are highly explosive. If vapors are present in the shop, have the doors open and the ventilating system turned on to get rid of these dangerous vapors.

9. Repair any fuel leak immediately. The potential fire hazard is very high.

10. Dirty and oily rags should be stored in closed metal containers to avoid catching fire.

11. Keep the shop floor and work benches clean and tidy. Oil on the floor can cause serious personal injury.

12. Do not operate shop tools or equipment that are in unsafe condition. Electrical cords and connectors must be in good condition. Bench grinding wheels and wire brushes should be replaced if defective. Floor jacks and hoist must be in safe operating condition and should not be used above their rated capacity. The same applies to mechanical and hydraulic presses, drills, and drill presses. Draw the attention of your instructor or shop foreman to any unsafe equipment or conditions.

13. Extension cords should not pose a hazard by being strung across walkways.

14. Do not leave jack handles in the down position across the floor. Someone could trip over them.

FIGURE 1-6. Avoid getting clothing caught in rotating parts such as fans, pulleys, grinders, and drills. Do not stand in the plane of rotating parts that are not shielded.

FIGURE 1-7. Electrical cords and connectors must be in good condition to avoid injury.

15. Do not drive equipment over electrical cords. This could cause short circuits.

16. Do not steam clean fuel injection parts. Expansion could cause them to seize.

PART 3   FIRE SAFETY

Several classes of fires may occur in a diesel shop. Classifications of fires are determined by the type of combustible material involved. Fire extinguishers are classified in a similar manner, depending on their effectiveness on a particular class of fire. In some cases the use of an improper fire extinguisher may, in fact, increase the intensity of the flames. The use of water on a fuel oil fire is one example of this hazard.

Following is a list of fire classifications and the appropriate fire extinguishers that should be used for each.

1. Class A fires. These fires occur in normal combustible materials such as paper, wood, rags, and rubbish.

2. Class B fires. These fires result from flammable liquids such as diesel fuel, gasoline, oil, grease, paint, paint thinners, and similar substances being ignited.

3. Class C fires. These fires occur in electrical equipment and usually involve insulating materials and overheated electrical wiring due to electrical overload in switch panels, electric motors, and the like.

4. Class D fires. Although not common in diesel shops, these fires can occur where combustible metals such as lithium, sodium potassium, magnesium, titanium, and zirconium are present.

Pressurized Water. Usually a hose with a hand-squeezed, trigger-operated valve. Use on class A fires.

Dry Chemical. Usually a portable hand-held tank with a valve operated by squeezing a handle or lever and aiming the chemical at the base of the fire. Use on class B and C fires. Dry chemical fire extinguishers that are suitable for three classes of fires, classes A, B and C, are also available.

Carbon Dioxide—CO_2. Usually a portable hand-held tank with a valve operated by squeezing a trigger or handle and aiming the CO_2 at the fire. Use on class B and C fires.

PART 4   FIRST AID

1. Make sure you are aware of the location and contents of the first-aid kit in your shop.

2. Find out if there is a resident nurse in your shop or school, and find out where the nurse's office is.

3. If there are any specific first-aid rules in your school or shop, make sure you are aware of them and follow them. You should be able to locate emergency telephone numbers quickly, such as ambulance, doctor, and police.

4. There should be an eye-wash station in the shop to thoroughly rinse your eye should you get acid or some other irritant into it.

5. Burns should be cooled immediately by rinsing with water and then treating as recommended.

6. If someone is overcome by carbon monoxide, immediately get him or her to fresh air.

7. In case of severe bleeding, try to stop blood loss by applying pressure with clean gauze on or around the wound, and summon medical aid.

8. Do not move someone who may have broken bones unless life is further endangered. Moving a person may cause additional injury. Call for medical assistance.

Safety is the responsibility of everyone. The following is a good procedure to use:

- Study safety regulations
- Set up a safe working area
- Report any unsafe working conditions
- Be safety conscious
- Practice safety on every job

PART 5   SELF CHECK

1. What type of shoes are best for shop use?

2. Why is adequate ventilation of such critical importance in a diesel shop?

3. List as many safety rules as you can think of regarding the use of shop tools and equipment.

4. List three safety rules that should be practiced when using power tools.

5. What is the purpose of jack stands?

6. List three types of fire extinguishers and the types of fires on which each should be used.

Chapter 2

Tools and Fasteners

Performance Objectives

Study this chapter and the proper use of the various tools and shop equipment used in the diesel shop. After having had enough opportunity to practice using the tools and equipment in a safe and efficient manner, you should be able to accomplish the following:

1. Select and use the correct hand tools and power equipment in a safe and efficient manner.

2. Select the appropriate shop manual, and locate and use the required information for the job being done.

3. Interpret repair order information correctly.

4. Prepare a parts list with all the required information necessary for correct replacement.

5. Select and use the correct type of fastener.

6. Complete the self-check questions with at least 80 percent accuracy.

PART 1 SHOP MANUALS AND FORMS

Shop Manuals

Shop manuals are a necessary part of the diesel service shop. They are needed to obtain the desired specifications and for specific service procedures. Mistakes and comebacks can be almost eliminated by the proper use of the correct shop manual.

Manufacturers' shop manuals are the most reliable source of information. Other shop manuals are also available that often provide helpful hints and suggestions.

At the front is the index of different sections in a manual. This directs the reader to the desired section, where there is another index. The section index leads to the particular area for which information is being sought.

Work Orders

The work order has room for the following information:

- Place of business
- Name of customer
- Date of work order
- Work order number
- Equipment identification
- Type of service required
- Customer's signature
- List of parts used and their cost
- Labor costs
- Tax
- Responsibilities and liabilities of place of business and customer

The work order serves as a means of communication between the various parties involved in the repair procedure, such as the service writer, customer, technician, shop foreman, parts department, cashier, and accounting department. The technician usually gets the hard copy, on which he records the time used to repair the vehicle and the parts required. The original copy is given to the customer on receipt of payment, and the remaining copies stay with the place of business for its records.

Parts Lists

Whether the hard copy of the work order is used for a parts list or a separate requisition is used, the parts department requires at least the following information to be able to provide the correct parts for the unit being serviced: equipment type, make, year, and model must be provided in every case.

Further information is included depending on which component of the equipment is being serviced. For example:

- Engine: displacement, engine number, fuel system, with or without turbocharger, etc.
- Transmission—type (standard or automatic transmission) number, and model.

11

• Clutch—diameter and number of springs in pressure plate.

Naturally, the correct names of the parts required must be used when ordering or requisitioning parts. When the correct name is not known, it usually becomes quite difficult to communicate with the parts department. Learning the correct name for each part is one of the first things a prospective technician should do if he expects to function in the service industry.

PART 2   HAND TOOLS

The technician's job is made easier by a good selection of quality tools and equipment. The quality and speed of work are also increased. Fast and efficient work is necessary to satisfy the customer and the employer. An efficient, productive technician also experiences greater job satisfaction and earns more money as a result. The technician should not jeopardize his ability by selecting tools that are inadequate or of poor quality. Good tools are easier to keep clean and last longer than tools of inferior quality.

Good tools deserve good care. Select a good roll cabinet and a good tool box to properly store your tools. They represent a fairly large investment and should be treated accordingly. Measuring instruments and other precision tools require extra care in handling and storage to prevent damage. Keeping tools clean and orderly is time well spent. It increases your speed and efficiency on each job you do.

Wrenches

Open end, box end, combination, and Allen wrenches are used to turn bolts, nuts, and screws. (See Fig. 2–1 to 2–10.) The open end wrench holds the nut or bolt only on two flat sides. They slip or round off the nut more readily than do box end wrenches. It is better to use box end wrenches wherever possible. Various offsets are available to make it easier to get at tight places. Both six- and twelve-point box end wrenches are available.

Sizes range from 3/8 to 1¼ inches in the average set and larger in 1/16-inch steps. Electrical wrench sets come in a smaller range of sizes. Metric wrench sets range from 6 to 32 millimeters. Allen or hex wrench sizes generally range from 2 to 20 millimeters and from 1/8 to 7/16 inch. Other sizes are also available.

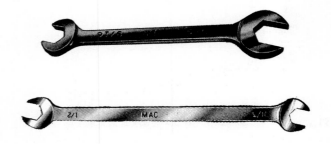

FIGURE 2–1. Heavy open-end wrench (above) and thinner open-end wrench (below). *(Courtesy of Mac Tools Inc.)*

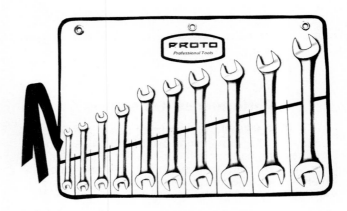

FIGURE 2–2. Open-end wrench set of various sizes in storage pouch. *(Courtesy of Proto Canada Division of Ingersoll-Rand Canada Inc.)*

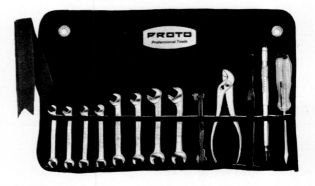

FIGURE 2–3. Electrical wrench set includes smaller open-end wrenches and small pliers. *(Courtesy of Proto Canada Division of Ingersoll-Rand Canada Inc.)*

FIGURE 2-4. Box-end wrenches are available in many sizes in both 6- and 12-point styles. Various offset styles are also available. *(Courtesy of Mac Tools Inc.)*

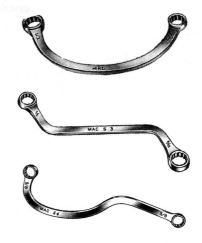

FIGURE 2-7. A number of special wrenches of various shapes for hard-to-get-at places are essential. *(Courtesy of Mac Tools Inc.)*

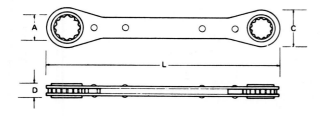

FIGURE 2-5. Ratcheting box wrenches can be a handy addition to any tool kit. *(Courtesy of Mac Tools Inc.)*

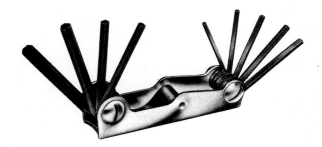

FIGURE 2-8. Hex or Allen wrenches are a must in the technician's tool kit.

FIGURE 2-6. Combination wrenches have an open-end wrench on one end and a box-end on the other.

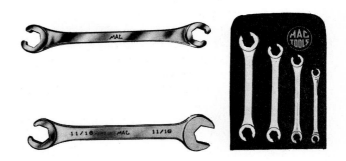

FIGURE 2-9. Flare nut or tubing wrenches should be used on tubing fittings to avoid rounding off the fittings.

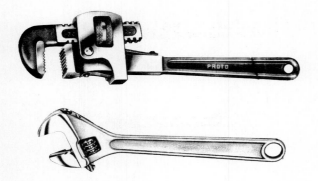

FIGURE 2-10. The adjustable wrench and the pipe wrench are necessary parts of a tool kit but should not be used in place of wrenches or sockets. *(Courtesy of Proto Canada Division of Ingersoll-Rand Canada Inc.)*

Ratcheting box end wrenches are very handy and come in similar size ranges as open end and box end wrenches. For tubing fittings, flare nut wrenches should be used rather than open end wrenches. The technician should also have an adjustable wrench, but this wrench should not be used in place of the proper open end or box end wrench.

Striking Wrenches

Striking wrenches (Fig. 2-11) are designed to be used with a hammer to break loose nuts or bolts that are particularly tight or seized. The wrench is placed on the fastener so that it will not slip off when struck with the hammer. A hammer is used to strike the striking face of the wrench in a direction to loosen the fastener. The sudden impact of the hammer blow applies a tremendous torque load to the wrench and fastener. Only wrenches designed for that purpose should be struck with a hammer.

FIGURE 2-11. Striking wrench. *(Courtesy of Proto Canada Division of Ingersoll-Rand Canada Inc.)*

Sockets and Drives

The well-equipped technician should have a 1/4-inch drive socket set, a 3/8-inch drive set, and a 1/2-inch drive set with standard and metric sockets (Fig. 2-12 to 2-21). Socket wrenches are fast and convenient to use. Both six- and twelve-point sockets should be included in the well-equipped tool kit, as well as deep sockets and flex sockets. Socket sizes are similar to wrench sizes and metric sizes. Other drive sizes such as 3/4 or 1 inch are used for heavy duty work.

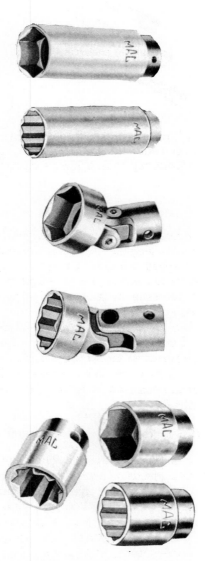

FIGURE 2-12. Standard, flex, and deep sockets are available in both 6- and 12-point types, as well as 8-point. *(Courtesy of Mac Tools Inc.)*

FIGURE 2-13. A ratchet allows turning the fastener without repeated removal of the wrench, thereby speeding up the work. Common drive sizes are 1/4, 3/8, 1/2, and 3/4 inch. *(Courtesy of Mac Tools Inc.)*

FIGURE 2-14. Drive extensions allow access to hard-to-get-at bolts and nuts. Extensions are available in various lengths. *(Courtesy of Mac Tools Inc.)*

Socket drives include universal joints, extensions of different lengths, ratchets, flex handles, T-handles, and speed handles. Drive sizes are available, as mentioned above. Drive adapters are also available to increase or reduce drive sizes to fit available sockets.

A number of other socket attachments are available and are handy to have. Both electric and air-operated impact wrenches are used to drive sockets to speed up the work; however, the sockets used are of a special design to withstand the continuous impacts of the driver.

Torque Wrenches

To tighten a bolt or nut to specifications, it is necessary to use a torque wrench. (See Fig. 2-22 to 2-26.) If a bolt or nut is overtightened, it puts excessive strain on the parts and damages the bolt or nut as well. On the other hand, if not tightened enough, the unit may come apart during use. Torque can be defined as twisting force. A 1-pound pull on a 1-foot wrench (center of bolt to point of pull) equals 1 pound-foot. Force times distance equals torque *(F × D = T)* or 1 pound × 1 foot = 1 pound-foot of

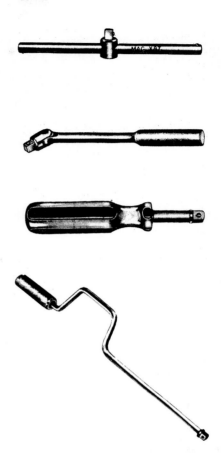

FIGURE 2-15. Other socket drives include (from top to bottom) T-bar handle, flex handle, spinner, and speed handle.

FIGURE 2-16. A universal joint converts the standard socket drive to a flex drive for hard-to-get-at places.

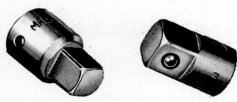

FIGURE 2-17. Drive size adapters convert socket drives up or down in size. Care must be exercised when using a large driver on a small socket. Breakage can result. *(Courtesy of Mac Tools Inc.)*

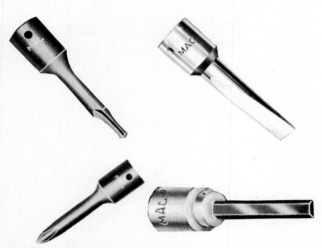

FIGURE 2-18. Other socket attachments include a standard screwdriver tip, Phillips screwdriver, clutch screwdriver, and hex wrench. *(Courtesy of Mac Tools Inc.)*

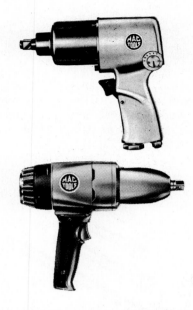

FIGURE 2-20. Impact tools speed the technician's work. Both air-operated (top) and electric impact tools (bottom) are available. *(Courtesy of Mac Tools Inc.)*

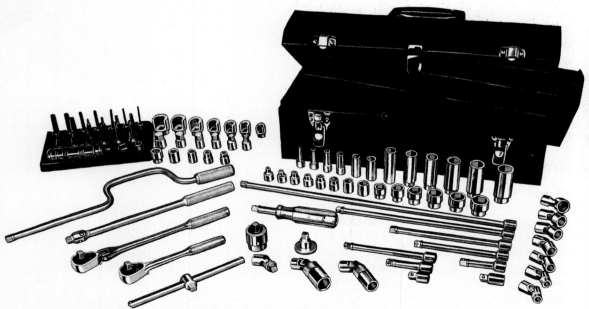

FIGURE 2-19. A complete 3/8-inch drive socket set including a swivel head ratchet and other special attachments.

FIGURE 2-21. Special sockets such as these should be used with impact tools. They are able to withstand the heavier use and continuous impacts. *(Courtesy of Mac Tools Inc.)*

torque. One hundred pounds of force applied to a torque wrench three feet long results in 300 lb. ft. of torque.

When an adapter is used on a torque wrench, this increases the effective length of the torque wrench thereby changing the actual torque from that which is read on the dial. To calculate actual torque, the following formula may be used:

A = Torque reading on dial
B = Length of torque wrench without adapter
C = Length of adapter

$$\frac{A \times B + C}{B} = \text{Actual torque}$$

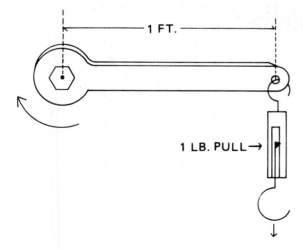

FIGURE 2-22. Torque is twisting force. Amount of torque is calculated by multiplying force times distance. *F × D = T.*

FIGURE 2-23. The dial-type torque wrench is available in various drive sizes. Dials are read directly and must be closely observed to torque fasteners correctly.

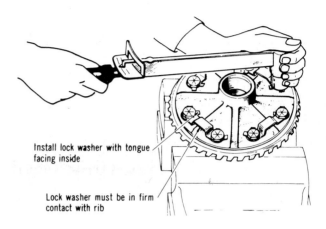

Install lock washer with tongue facing inside

Lock washer must be in firm contact with rib

FIGURE 2-24. The scale-type torque wrench is read directly just as the dial-type. A feelable-audible click attachment is provided on some models that signals when predetermined preset torque has been reached.

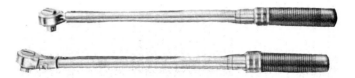

FIGURE 2-25. Another click-type torque wrench has a micrometer-type adjustment. There is no direct reading scale. Desired torque must be set on the micrometer adjustment. Preset torque is reached when the wrench clicks. *(Courtesy of Mac Tools Inc.)*

Other values for force are used, such as kilograms (kiloponds) or newtons. In this case, the values used for distance are centimeter and meter. The amount of torque, depending on the type of torque wrench used, is measured in inch-pounds or foot-pounds for domestic equipment. For imports, the centimeter-kilogram and meter-kilogram torque values are used. Another term for kilogram is kilopond when used in connection with torque values. The

FIGURE 2-26. Gear type torque multiplier. *(Courtesy of Proto Canada Division of Ingersoll Rand Canada Inc.)*

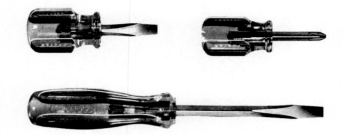

FIGURE 2-27. Common screwdriver types. The square socket type is known as the *Robertson screwdriver* (not shown).

current SI torque values are given in newton-meters. This system is also currently being used by many domestic manufacturers.

Use the appropriate type of torque wrench for the specifications given. If specifications are given in centimeter-kilograms, then a torque wrench with a centimeter-kilogram scale should be used. If torque values stated do not match the values on your torque wrench, use the metric conversion chart in the Appendix to convert the values given.

Abbreviations for the various torque values given are as follows:

- Pound-inches: lb-in
- Pound-feet: lb-ft
- Centimeter kilograms: cm kg or cm kp
- Meter kilograms: m kg or m kp
- Newton meters: N • m

Torque Multipliers

Torque multipliers provide a means for increasing the applied torque through a gear drive mechanism. Actual torque applied is calculated by multiplying the torque wrench reading (setting) by the gear ratio reduction of the torque multiplier. For example: using a 10-to-1 torque multiplier and applying 100 pound-feet of torque to it would result in an output torque of 10 times 100, or 1000 lb.-ft. being applied to the fastener. Torque multipliers are available in a range of gear ratios.

Screwdrivers

Screwdrivers (Fig. 2-27 and 2-28) are probably abused more than any other tool. Use them only for the purpose for which they were intended. There is no all-purpose screwdriver. Use the right type and size of screwdriver for the job. Slotted screws require flat-blade screwdrivers. Select a screwdriver with a tip that is as wide as the screw slot is long.

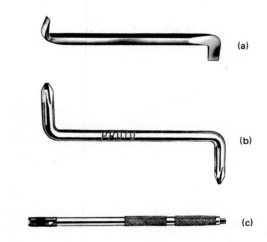

FIGURE 2-28. (A) Offset tip. (B) Phillips screwdrivers. (C) Screw-holding screwdriver for starting screws. *(Courtesy of Proto Canada Division of Ingersoll-Rand Canada Inc.)*

Use the correct size of Phillips, Reed and Prince, Robertson, or clutch-type screwdriver. Never make do with the wrong size.

Pliers

There are two groups of pliers (Fig. 2-29). One is used for gripping and the other for cutting. Diagonal cutting pliers are sometimes called side-cutting pliers or side cutters. Gripping pliers should not be used in place of wrenches or sockets since this damages nuts and bolts. Do not grip machined or hardened surface parts with pliers; it will damage the surface.

Hammers

Ball-peen hammers and soft hammers are the types used by diesel technicians (Fig. 2–30). Soft hammers such as plastic, rawhide, lead, or brass types are used on easily damaged surfaces.

A hammer should be held at the end of the handle. The hammer should land flat on the surface being struck.

Handles should be kept secure in the hammer head to avoid injury and damage. Select the right size (weight) of hammer for the job.

Punches

Pins and rivets are removed with punches (Fig. 2–31). A tapered starting punch is used to start rivet removal after the rivet head has been chiseled or ground off. The rivet is then driven out the rest of the way with the pin punch. A long tapered punch is used for aligning parts. A center punch is used to mark parts before disassembly and to mark the spot where a hole is to be drilled.

Punches should be kept in good condition. Do not allow mushrooming to take place.

Chisels

Chisels (Fig. 2–32 to 2–35) are used to cut rivet heads and other metal. A chisel holder can be used for heavy work. Chisels should be kept sharp. Sharpen at approximately a 60° included angle.

Any sign of mushrooming should be ground off. Use safety goggles or a face mask when using a chisel.

Files

Files are cutting tools used to remove metal, to smooth metal surfaces, etc. Many types of files are available for different jobs, with different sizes, shapes, and coarse or fine cutting edges. Size is determined as shown in Figure 2–36, which also shows some different shapes. The range of coarseness in order from coarse to fine is rough, coarse, bastard, second cut, smooth, and dead smooth. Never use a file without a handle. Do not use a file as a pry. Files are brittle and break easily.

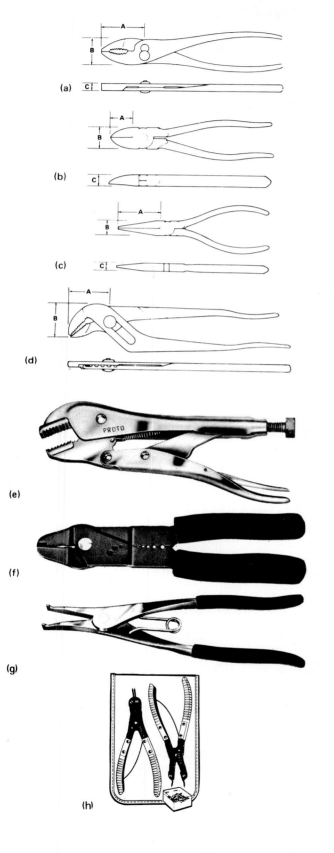

FIGURE 2–29. (A) Combination pliers. (B) Diagonal or sidecutting pliers. (C) Needle-nose pliers. (D) Channel lock pliers. (E) Lock grip pliers. (F) Wire-cutting and stripping tool. (G) External snap ring pliers. (H) Lock ring plier set with interchangeable tips.

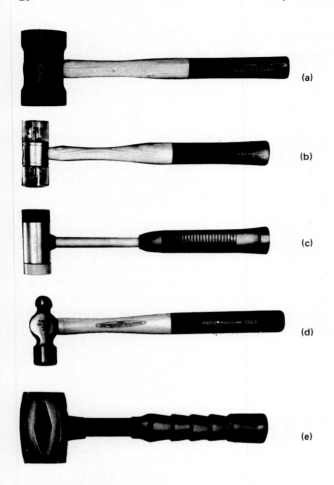

FIGURE 2-32. Punch or chisel holder avoids injury to hands in heavy work.

Hacksaws

A hacksaw consists of an adjustable frame with a handle and a replaceable hacksaw blade. Blades are commonly available in 10- and 12-inch lengths. The coarseness of the cutting teeth is stated in number of teeth per inch, usually 14, 18, 24, or 32 teeth per inch. The finer blades are used for materials of thin cross section. Select a blade for the job that will have at least two teeth cutting at all times. (See Fig. 2-39 to 2-41.)

Apply light pressure on the forward stroke and release the pressure on the return stroke. When replacing a hacksaw blade, install it with the teeth pointing away from the handle.

Cleaning Tools

Hand-held scrapers are used to scrape gasket or other material from flat surfaces. This should be followed by a light sanding.

The wire brush is used to clean rough surfaces. A soft bristle brush is used to help clean parts being washed in solvent. Rotary wire carbon brushes are used to remove carbon and the like.

FIGURE 2-30. (A) Rubber mallet. (B) Plastic-faced hammer. (C) Rubber- or plastic-faced hammer. (D) Ball-peen hammer. (E) Sledge hammer. *(Courtesy of Proto Canada Division of Ingersoll-Rand Canada Inc.)*

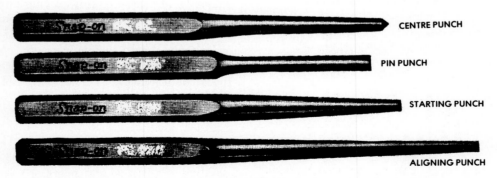

FIGURE 2-31. Common types of punches used by diesel technicians. *(Courtesy of Snap-on Tools Corporation)*

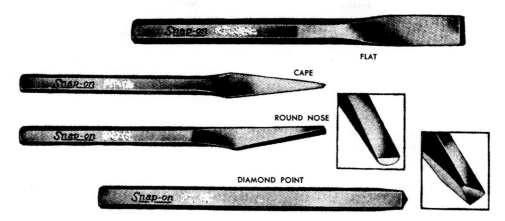

FIGURE 2-33. Common chisel types that should be included in the technician's tool kit. *(Courtesy of Snap-on Tools Corporation)*

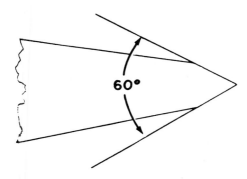

FIGURE 2-34. Chisel sharpening angle.

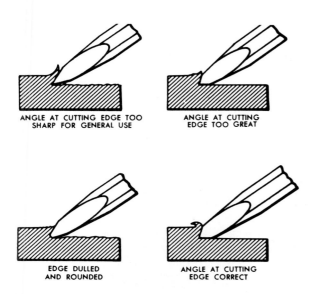

ANGLE AT CUTTING EDGE TOO SHARP FOR GENERAL USE

ANGLE AT CUTTING EDGE TOO GREAT

EDGE DULLED AND ROUNDED

ANGLE AT CUTTING EDGE CORRECT

FIGURE 2-35. Incorrectly and correctly sharpened chisel angles. *(Courtesy of Ford Motor Company of Canada Ltd.)*

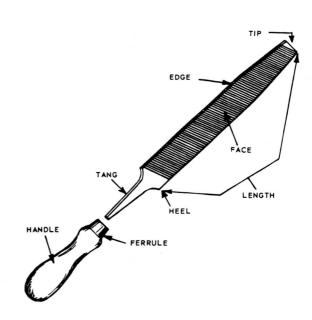

FIGURE 2-36. Parts of a file. Size is determined by length.

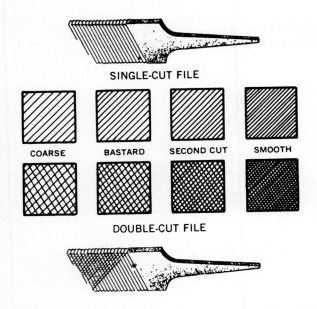

FIGURE 2-37. Coarse to fine files, left to right. Single-cut (upper) and double-cut (lower) file types.

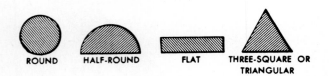

FIGURE 2-38. Cross-sectional shapes of files. (Courtesy of Ford Motor Company of Canada Ltd.)

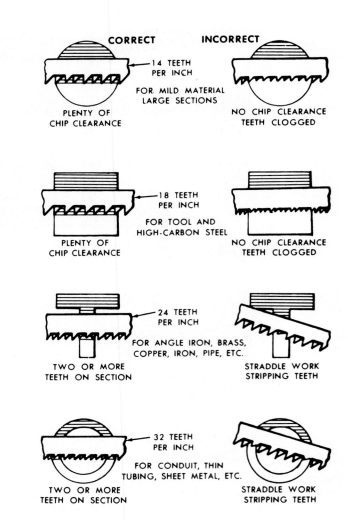

CORRECT   INCORRECT

14 TEETH PER INCH
FOR MILD MATERIAL LARGE SECTIONS
PLENTY OF CHIP CLEARANCE
NO CHIP CLEARANCE TEETH CLOGGED

18 TEETH PER INCH
FOR TOOL AND HIGH-CARBON STEEL
PLENTY OF CHIP CLEARANCE
NO CHIP CLEARANCE TEETH CLOGGED

24 TEETH PER INCH
FOR ANGLE IRON, BRASS, COPPER, IRON, PIPE, ETC.
TWO OR MORE TEETH ON SECTION
STRADDLE WORK STRIPPING TEETH

32 TEETH PER INCH
FOR CONDUIT, THIN TUBING, SHEET METAL, ETC.
TWO OR MORE TEETH ON SECTION
STRADDLE WORK STRIPPING TEETH

FIGURE 2-40. Correct and incorrect use of different hacksaw blades. (Courtesy of Ford Motor Company of Canada Ltd.)

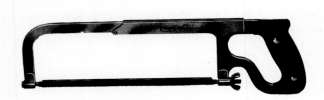

FIGURE 2-39. Adjustable frame hacksaw.

FIGURE 2-41. Power-driven hole saw. (Courtesy of Ford Motor Company of Canada Ltd.)

(a)

(b)

FIGURE 2-42. Parts cleaning scrapers: (A) for irregular surfaces; (B) for flat surfaces. *(Courtesy of Proto Canada Division of Ingersoll-Rand Canada Inc.)*

(a)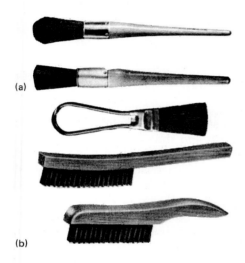

(b)

FIGURE 2-43. (A) Brushes for cleaning parts in solvents. (B) Wire brushes. *(Courtesy of Proto Canada Division of Ingersoll-Rand Canada Inc.)*

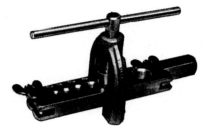

FIGURE 2-45. Flaring tool kit required to flare tubular lines. *(Courtesy of Proto Canada Division of Ingersoll-Rand Canada Inc.)*

FIGURE 2-46. Tubing cutter with reamer attached *(Courtesy of Proto Canada Division of Ingersoll-Rand Canada Inc.)*

FIGURE 2-44. Power-driven rotary wire brushes. *(Courtesy of Proto Canada Division of Ingersoll-Rand Canada Inc.)*

Tubing Tools

Steel and copper tubing should be cut with a tubing cutter. Avoid applying too much pressure during cutting. This can collapse the tube. Low-pressure lines need only a single flare, whereas high-pressure lines require double-lap flaring.

After cutting, ream the tubing and then flare as required. The reamer is usually part of the tubing cutter. Make sure there are no metal chips left in the tubing.

Threading Tools

Tools used for cutting threads are called taps and dies. Taps are used to cut inside threads and dies for outside threads. The most common thread types are (1) coarse, known as UNC, NC, or Unified National Coarse, (2) fine, known as UNF, NF, or Unified Na-

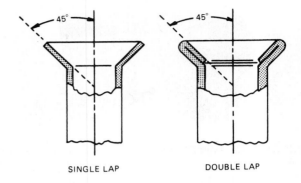

SINGLE LAP   DOUBLE LAP

FIGURE 2-48. Completed single-lap flare for low-pressure lines and double-lap flare for high-pressure lines.

tional Fine, and (3) extra fine, known as UNEF, NEF, or Unified National Extra Fine. In addition, there are machine-thread taps and dies and metric taps and dies.

Always make sure that the correct taps and dies are being used to fit the fastener threads. Mistakes can be costly and time consuming. The taper tap is used to thread a hole through a piece of metal. The bottoming tap is used to thread a hole that does not go all the way through. In this case the taper tap should be used first, followed by the bottoming tap to complete the job. The plug tap is used to thread a hole partway through for a plug. To ensure that smooth, undamaged threads are produced, use a good lubricant and frequently back up the threading tool to remove metal chips. Chamfer rods or other round stock to be threaded. This makes it easier to start the die.

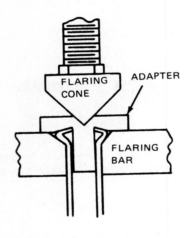

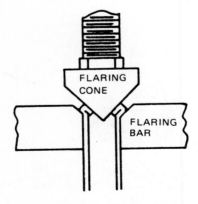

FIGURE 2-47. Example of double-lap flaring procedure required for high-pressure lines.

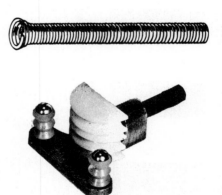

FIGURE 2-49. Two types of tubing benders. Tubing benders must be used when bending tubing to avoid collapsing or kinking of tubing. A collapsed tube reduces flow. *(Courtesy of Proto Canada Division of Ingersoll-Rand Canada Inc.)*

FIGURE 2-50. Adjustable round dies for threading bolts and studs.

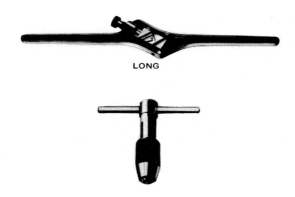

FIGURE 2-53. Handle types for taps.

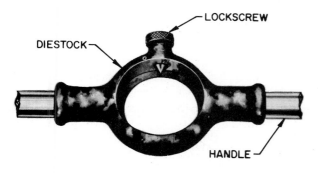

FIGURE 2-51. Die holder required to use dies.

FIGURE 2-54. A good drill index showing drill sizes and a good selection of drill bits. *(Courtesy of Mac Tools Inc.)*

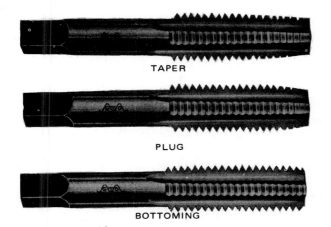

FIGURE 2-52. Three types of taps for threading holes: taper tap for threading right through a hole, plug tap to thread holes for plugs such as drain and fill plug holes, and bottoming tap for threading holes that do not go all the way through.

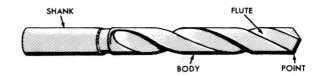

FIGURE 2-55. Parts of a drill bit.

Drills and Reamers

Drills used by diesel technicians are known as twist drills. Drill sizes or diameters are identified in five ways: by number, by letter, in fractions, in decimals, and in millimeters. Refer to the Appendix for cross reference.

Fractional sizes range from 1/64 inch and up in 1/64-inch steps. There are 80 numbered sizes from 0.0135 to 0.228 inch. Sizes identified by letter range from 0.234 to 0.413 inch. Metric drill sizes range from less than 1 millimeter and up.

The point to be drilled should first be marked with a center punch. This prevents the drill from wandering. Drill speed varies with drill size; in general, the larger the drill, the slower the drill speed should be. The type of material being drilled also affects drill speed.

A good drilling lubricant should be used on the material being drilled. High-speed carbon drills should be used on harder metals.

Reamers are used to produce a smooth, perfectly round hole. First use a drill 1/64 inch smaller than the desired finished size; then finish by reaming. Turn the reamer slowly and only in a forward direction until the desired size is reached.

Screw Extractors

Several methods are used to remove broken screws or studs. If the stud is not very tight, it can sometimes be removed by turning it with a center punch and hammer. If enough of the stud extends above the surface, saw a slot in it and turn it out with a screwdriver. Another method is to drill the stud in the exact center with a drill that leaves only a thin shell of the stud in the hole. Then turn the extractor (Fig. 2–56) into the hole in a counterclockwise direction. Penetrating oil helps to loosen the threads. Be careful not to break the screw extractor in the hole. It is extremely hard and cannot be drilled out.

Bushing and Seal Tools

There are many different types of bushing and seal drivers and pullers (Fig. 2–57 and 2–58). Some jobs require very special types of pullers or drivers. Pullers are either of the threaded type or the slide hammer type. Care must be exercised in selecting the correct type for the job. Bushings and seals are very easily damaged if an improper tool or method is used. Refer to the tool maker's instructions and the shop manual for specific information.

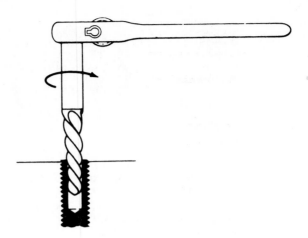

FIGURE 2–56. Using a screw extractor to remove a broken stud after it has been properly drilled.

Miscellaneous Tools

A great variety of miscellaneous tools is available to the service industry. Many such tools can lighten and speed up the task of the technician.

A magnetic retrieving tool is very handy to have. A stethoscope helps to localize and identify abnormal knocks and other noises. Essential miscellaneous tools include extension lights and cords, and a good creeper.

A good technician should be aware of any new developments in tools and equipment that will improve the quality and productivity of his work.

PART 3   SOLDERING TOOLS

The technician may be required to solder such items as radiators and wiring connections. To do a good job of soldering, the surfaces must be smooth and absolutely clean. A good soldering iron of the proper size should be used: heavier irons for heavier work, a soldering gun for electrical work, etc. The soldering iron should be thoroughly cleaned and tinned (coated with solder). Solder is available in bar or wire form. The correct flux must also be used. Rosin flux must be used for electrical work to prevent corrosion. Acid flux is used on other work. Acid-core and rosin-core wire solder are most frequently used.

Soldering irons should be placed in holders during heating and cooling.

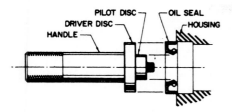

Installing oil seal in housing
is easily accomplished here.

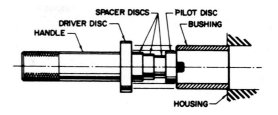

Pilot disc prevents cocking as
bushing is installed with tool.

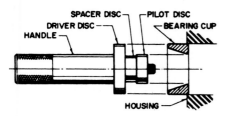

Bearing cup driven into housing.
Alignment is maintained.

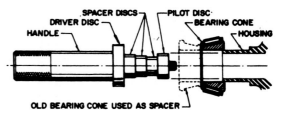

Here, a bearing cone is being
installed on a shaft correctly.

FIGURE 2-57. A universal bushing, bearing, and seal tool kit for removing and installing bushings, bearings, and seals. A variety of disc sizes and combinations can be used for different jobs as illustrated. *(Courtesy of Owatonna Tool Company)*

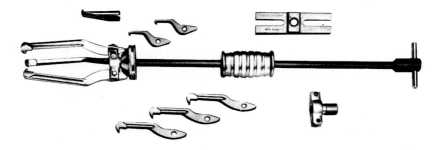

FIGURE 2-58. A slide hammer puller kit with attachments. Inside and outside jaws are interchangeable. *(Courtesy of Owatonna Tool Company)*

FIGURE 2-59. A long-handle, flex-head magnet for retrieving parts.

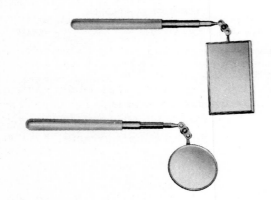

FIGURE 2-60. Mirrors with flex heads and long handles allow the technician to look at otherwise inaccessible places.

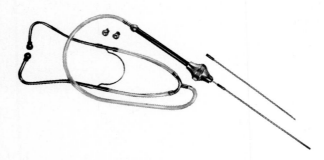

FIGURE 2-61. The stethoscope is used to determine the exact location of knocks and other abnormal noises.

FIGURE 2-62. Electric soldering iron. *(Courtesy of Snap-on Tools Corporation)*

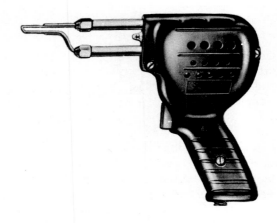

FIGURE 2-63. Electric soldering gun used for soldering electrical connections. *(Courtesy of Snap-on Tools Corporation)*

FIGURE 2-64. Bar- and wire-type solders.

FIGURE 2-65. Standard English feeler gauge with ten blades from 0.002 to 0.015 inch. *(Courtesy of Mac Tools Inc.)*

PART 4 MEASURING TOOLS

All measuring tools should be treated as precision instruments. They should be properly used, cared for, and stored. Tools provided with special storage cases should be put back in the case when no longer being used. Be careful not to drop precision tools. Do not have them lying around among or under other tools. Precision surfaces, such as straight edges, micrometer anvils, and caliper jaws, should not be marred, scratched, or dented. The accuracy of measuring instruments should be checked if any doubt exists. They can be checked against other tools known to be accurate. Gauge blocks for micrometers determine micrometer accuracy. If no means of checking are available in the shop, they should be sent out to be checked.

Feeler Gauges

Feeler gauges (Fig. 2-66 and 2-67) are precision measuring tools used to measure small clearances. Various blade lengths are available. Flat feeler gauges have their thicknesses marked in thousandths of an inch or in millimeters or both. Nonmagnetic feeler gauges are used for checking clearances where magnetic force exists. Stepped feeler gauges have the tip of the blade two thousandths of an inch thinner than the rest of the blade. These are used for "go-no go" quick measurements.

Never bend, twist, force, or wedge feeler gauges since this destroys their accuracy. Wipe blades clean with an oily cloth to prevent rust.

FIGURE 2-67. Nonmagnetic feeler gauge with brass blades in both English and metric sizes. *(Courtesy of Mac Tools Inc.)*

Calipers

Inside and outside calipers are useful in measuring to an accuracy of approximately 1/100 inch. Vernier calipers measure inside and outside dimensions. English, metric, and combination types are available. English vernier calipers measure to an accuracy of 0.001 inch, while metric calipers measure to an accuracy of 0.02 millimeters. Most vernier calipers also have a depth measuring rod attached.

Straight Edges

Straight edges (Fig. 2-69) are used to measure surface irregularities. A feeler gauge is needed to determine the amount of warpage shown by the straight edge. Cylinder heads, valve bodies, and other machined surfaces are checked with a straight edge.

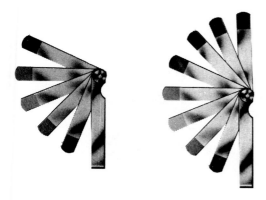

FIGURE 2-66. Go/no-go gauge with stepped feeler blades.

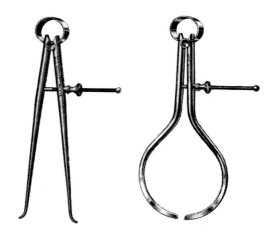

FIGURE 2-68. Inside and outside calipers. *(Courtesy of The L. S. Starrett Company)*

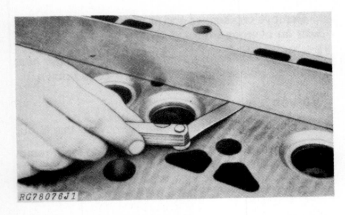

FIGURE 2-69. The straightedge is used to determine whether a flat surface is warped. Here it is used with a feeler gauge to check the amount of head warpage. If head is warped beyond manufacturer's specifications, it must be machined or ground to restore the surface.

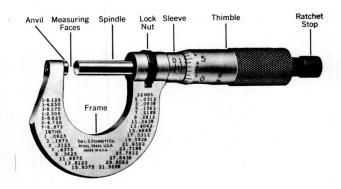

FIGURE 2-70. Outside micrometer with major parts identified. A piece of work is measured by placing it between the anvil and the spindle faces and turning the spindle by means of the thimble until both faces contact the work. The micrometer reading is then taken. *(Courtesy of The L. S. Starrett Company)*

Micrometers

Outside micrometers are used to measure the size of parts, such as diameter and thickness. The size of a micrometer is determined by the distance between the face of the anvil and the face of the spindle when the micrometer is adjusted to its minimum and maximum adjustments. A 1- to 2-inch micrometer would not be able to measure anything less than 1 inch or anything more than 2 inches. A 25- to 50-millimeter micrometer would not be able to measure anything less than 25 millimeters or anything more than 50 millimeters.

Digital micrometers measure inside and outside diameters, are available in sizes and types similar to conventional micrometers, and are read directly. (See Fig. 2-70 to 2-78.)

English Micrometer Scale

Each division on the sleeve represents 0.025 or 25/1000 inch. From 0 to 1 represents four such divisions (4 × 0.025 inch) or 0.100 inch (100/1000 inch). Each division on the thimble represents 0.001 or 1/1000 inch. There are 25 such divisions on the thimble. Therefore, one turn of the thimble moves the spindle 0.025 or 25/1000 inch. To read the micrometer in thousandths, multiply the number of divisions visible on the sleeve by 0.025, then add the number of thousandths indicated by the line on the thimble that coincides with the long reading line on the sleeve.

Example: Refer to Figure 2-71.

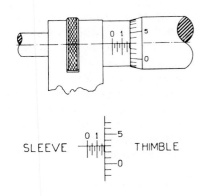

READING .178"

FIGURE 2-71. Micrometer reading (English scale).

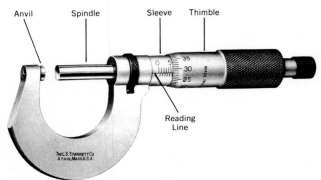

FIGURE 2-72. The metric micrometer is similar in appearance to the English micrometer. *(Courtesy of The L. S. Starrett Company)*

The 1 line on sleeve is visible, representing 0.100

There are three additional lines visible, each representing 0.025 inch 0.075

Line 3 on the thimble coincides with the reading line on the sleeve, each line representing 0.001 inch 0.003

The micrometer reading is 0.178 inch

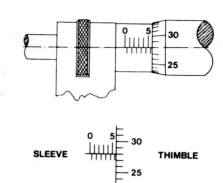

Reading 5.78 mm

FIGURE 2-73. Metric micrometer scale.

Metric Micrometer Scale

One revolution of the thimble advances the spindle toward or away from the anvil 0.5 millimeter distance (Fig. 2-73).

The reading line on the sleeve is graduated in millimeters (1.0 millimeter) from 0 to 25. Each millimeter is also divided in half (0.5 millimeter). It requires two revolutions of the thimble to advance the spindle 1.0 millimeter. The beveled edge of the thimble is graduated in 50 divisions from 0 to 50. One revolution of the thimble moves the spindle 0.5 millimeter. Each thimble graduation equals 1/50 of 0.5 millimeter or 0.01 millimeter.

To read the micrometer, add the number of millimeters and half-millimeters visible on the sleeve to the number of hundredths of a millimeter indicated by the thimble graduation that coincides with the reading line on the sleeve.

Example: Refer to Figure 2-73.

The 5-millimeter sleeve graduation is visible 5.00

One additional 0.5-millimeter line is visible on the sleeve 0.50

Line 28 on the thimble coincides with the reading line on the sleeve; 28 × 0.01 0.28

The micrometer reading is 5.78 millimeters

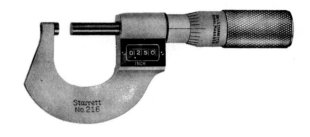

FIGURE 2-74. Digital readout micrometers are quicker to read but are more expensive.

Telescoping Gauges

Telescoping gauges can be used, in conjunction with outside micrometers, for measuring inside diameters, instead of using inside micrometers.

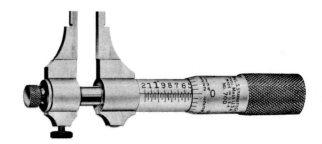

FIGURE 2-75. Inside micrometers in English and metric are read the same as outside micrometers. *(Courtesy of The L. S. Starrett Company)*

FIGURE 2-76. Depth-gauge micrometer is used to measure hole depth or the difference in the levels of two surfaces. English scale is shown here.

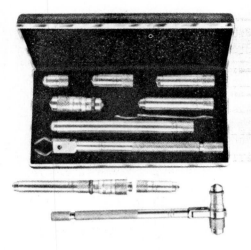

FIGURE 2-78. Inside micrometers are used to measure inside diameters.

FIGURE 2-77. Metric depth-gauge micrometer.

Dial Indicators

The dial indicator is needed to perform many of the measurements required in the shop. Graduations on the scale are in either 0.001 inch (1/1000 inch) or in 0.01 millimeter (1/100 millimeter) for general shop use. (See Fig. 2-80.)

The measurement range of the dial indicator is determined by the amount of plunger travel provided. Special dial indicators are available for measuring cylinder taper and out of round, as shown in Chapter 9.

Manometers and Pressure Gauges

A measuring instrument known as a *manometer* is used in diesel engine performance diagnosis to measure certain engine pressures and vacuum. Both water and mercury types are used. Manometers are available in the rigid U tube or the flexible clear plastic tube design. The plastic type, also called a *slack tube manometer,* is more convenient to store in a tool box since it is more compact and not as easily broken as the rigid type. Both are operated similarly in a U shaped form and are, therefore, known as the *U tube manometers.* The fixed type is usually mounted on a stand for protection.

A sliding scale is provided on the manometer, and must be adjusted before its use. The zero line must coincide with the height of the liquid in both sides of the tube before being connected to the engine. The scale is calibrated in inches, millimeters, or both. Water manometers are more easily read if a small amount of color dye is added to the water.

A water manometer is used to measure very low pressures such as inlet air pressure and crankcase pressure. Since mercury is 13.6 times as heavy as water, a mercury manometer is used to measure higher pressures such as turbo boost pressure, air box pressure, and exhaust back pressure.

Specially calibrated dial or digital type gauges have replaced the use of the manometer in many shops.

Range, English	Range, Metric
5/16 to 1/2 in.	7.9 to 12.7 mm
1/2 to 3/4 in.	12.7 to 19 mm
3/4 to 1-1/4 in.	19 to 31.7 mm
1-1/4 to 2-1/8 in.	31.7 to 54 mm
2-1/8 to 3-1/2 in.	54 to 89 mm
3-1/2 to 6 in.	89 to 152.4 mm

FIGURE 2–79. Telescoping gauges are used to measure inside diameters when used in conjunction with an outside micrometer. The knurled handle releases the spring-loaded plungers to diameter size; then they are locked by turning the knurled handle. The telescoping gauge is then carefully removed and measured with an outside micrometer. *(Courtesy of The L. S. Starrett Company)*

Pyrometers

A pyrometer is used to measure the exhaust temperature of a diesel engine and to compare exhaust temperatures between cylinders. There are two types of pyrometers: the portable type, which is hand held; and the fixed type, which is engine- and instrument panel-mounted.

Both pyrometers use the thermocouple type of sensor at the engine exhaust manifold. The thermocouple consists of two dissimilar metals covered by an insulator. The bimetal sensor element in the fixed type is located in the exhaust manifold near the cylinder head in such a way that it exposes it to exhaust gases. The fixed type is automatically compensated, whereas the hand held unit must be adjusted to zero on the scale before taking a reading.

The pyrometer is used to check overall engine exhaust temperature as well as to isolate any cylinder operating at temperatures either above or below normal.

PART 5   POWER TOOLS

Bench Grinder

The bench grinder is an indispensable item in any shop. Usually one end of the motor shaft has a grinding wheel and the other has a wire wheel. The grinder can then be used for sharpening tools and cleaning parts. Other jobs for the grinder are grinding rivets or removing stock from metal parts. Grinding wheels of different types and sizes are available for specific jobs.

Grinding and cleaning require skill and careful handling to avoid injury to the operator or the tools and parts being reworked.

FIGURE 2-80. This dial indicator set takes care of many measuring requirements in the shop. Dial indicators are used to measure differential gear backlash, disc and flywheel run-out, crankshaft and transmission shaft end play, and so on. (A) Hole attachment or wiggle bar. (B) Clamps. (C) Toolpost holder. (D) Upright spindle. (E) Universal clamp. *(Courtesy of The L. S. Starrett Company)*

Electric Drills

Hand-held electric drills perform a variety of jobs, as listed in Figure 2-84. When using a hand-held drill, a few general rules will make the job easier. Do not apply any side pressure when drilling since this can break drill bits. Apply only enough pressure for good drilling; too much pressure can cause overheating and destroy the drill bit. Ease up on the pressure just before the drill breaks through to prevent grabbing. Keep a firm grasp on the drill at all times. Use a small amount of cutting oil when drilling steel. Make sure the piece of work being drilled is held securely. These general rules apply to the drill press as well. The drill press has a movable table which can be raised, lowered, and turned sideways. A drilling block should be used on top of the table to avoid drilling into the table.

Many other power tools designed for special procedures are used in the diesel shop. These tools and their uses are discussed in the particular chapters to which they apply.

PART 6   SHOP TOOLS

Hydraulic Jacks

Hydraulic floor jacks, transmission jacks, and portable jacks are used in the diesel shop. The floor jack is used to raise equipment at the front, rear, or sides. The jack should be properly placed so that parts underneath are not damaged. Do not work under equipment supported only on a jack. Always use jack stands properly placed.

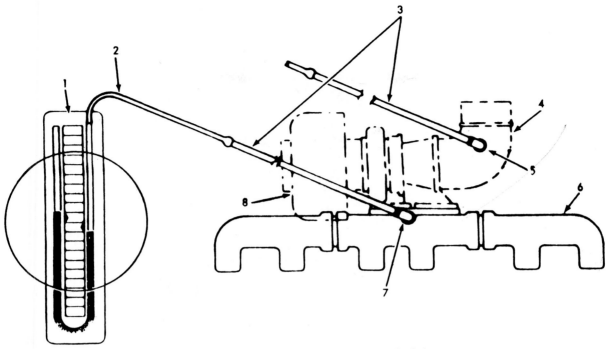

NOTE: MANOMETER AS SHOWN INDICATES
2" MERCURY BACK-PRESSURE

1 U-TUBE MANOMETER
2 RUBBER TUBING
3 COPPER TUBING
4 EXHAUST ELBOW
5 LOCATION OF FITTING
  FOR TURBOCHARGED
  ENGINES

6 EXHAUST MANIFOLD
7 LOCATION OF FITTING
  FOR NATURALLY
  ASPIRATED ENGINES
8 TURBOCHARGER

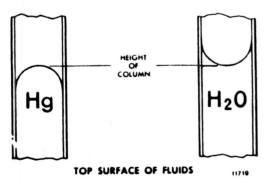

HEIGHT
OF
COLUMN

Hg

H₂O

TOP SURFACE OF FLUIDS

11719

CONVEX FOR MERCURY   CONCAVE FOR WATER

PRESSURE CONVERSION CHART		
1" water	=	.0735" mercury
1" water	-	.0361 psi
1" mercury	=	13.6000" water
1" mercury	=	.4910 psi
1 psi	=	27.7000" water
1 psi	=	2.0360" mercury
1 psi	=	6.895 kPa
1 kPa	=	.145 psi

FIGURE 2–81. (a) The U-tube manometer is a primary measuring device indicating pressure or vacuum by the difference in the height of the two columns of fluid. Connect the manometer to the source of pressure, vacuum, or differential pressure. When the pressure is imposed, add the number of inches one column of fluid travels up to the amount the other column travels down to obtain the pressure (or vacuum) reading. *(Courtesy of Allis-Chalmers Manufacturing Co.)* (b) Comparison of column height for mercury and water manometers (left) and pressure conversion chart (right). *(Courtesy of Detroit Diesel Allison)*

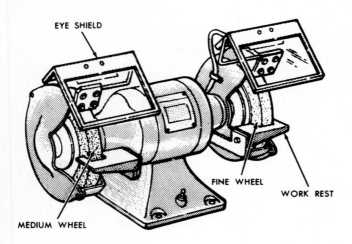

FIGURE 2-82. Bench grinder with tool supports and protective shields in place. *(Courtesy of Ford Motor Company of Canada Ltd.)*

FIGURE 2-83. Wire wheel and grinding wheel attachments for the bench grinder.

FIGURE 2-84. Hand-held electric drill. Common types include 1/4, 3/8, and 1/2 inch. Available in low-speed, high-speed, and variable-speed models. Common uses are for drilling, honing, driving, cleaning brushes, hole saws, and rotary files. *(Courtesy of Sioux Tools Inc.)*

The transmission jack is indispensable for transmission removal and installation. The platform adjusts by tilting forward, back, or sideways, and can be raised or lowered hydraulically. The hold-down chain should always be fastened securely to prevent the transmission from falling off the jack. Use of the transmission jack allows proper alignment of transmission and bell housing for removal and installation.

Hoists and Lifts

Different types of hoists and lifts are used to raise vehicles for under-vehicle work. Transmission removal, drive line work, and exhaust system repairs are usually done on a hoist. A good hoist is one that allows easy vehicle positioning and has minimal under-vehicle obstruction. Twin-post and drive-on hoists are commonly used in diesel shops. Design capacity of lifts and hoists should not be exceeded.

Chassis Lubrication Equipment

Chassis lubrication equipment is located in the lube bay and provides a motor oil dispenser, gear oil dispenser, grease gun, and air pressure hose. Different equipment manufacturers recommend lubrication at different intervals. The number of points requiring lubrication varies among different equipment makes. Some equipment has lubrication fittings already installed; others have plugs which must be removed and lube fittings installed before lubrication can be done.

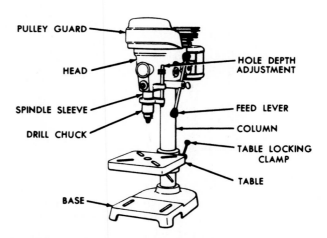

FIGURE 2-85. The drill press provides for a number of drilling speeds by changing the drive belt to different pulley positions. The drill press is used for precision drilling of parts that can be carried by hand. *(Courtesy of Ford Motor Company of Canada Ltd.)*

FIGURE 2-86. Portable hydraulic shop crane used for a variety of lifting jobs. Sling adapter at right allows tilting of load as required for engine removal and installation.

FIGURE 2-87. Chain hoist used for lifting heavy objects. Mechanical gear reduction makes lifting easier.

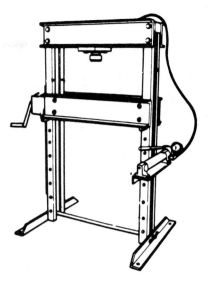

FIGURE 2-88. Floor-type and bench-top hydraulic shop presses. Floor model has an adjustable press bed for different working positions. Presses are used to remove and install bearings, gears, bushing, piston pins, and so on.

Follow manufacturer's recommendations and specifications for periodic lubrication and types of lubricants required.

Portable Cranes

Cranes are used to lift and transport heavy components. The crane should not be adjusted beyond its designed adjustment limits. For transporting a component with a crane, the load should be lowered to avoid upsetting. Never exceed the lift capacity of the crane.

FIGURE 2-89. The bench vise is used to hold items on which work is being done. The soft jaw (right) must be used when clamping easily damaged surfaces. *(Courtesy of The L. S. Starrett Company)*

Hydraulic Presses

The hydraulic press should not be used beyond its rated capacity. All work should be properly positioned and supported, and all recommended shields and protective equipment should be used. The hydraulic press exerts tremendous pressure and can cause parts under pressure to literally explode, causing serious injury.

PART 7   FASTENERS

A great variety of types and sizes of fasteners is used in the diesel industry. (See Fig. 2–90 to 2–101.) Each fastener is designed for a specific purpose and for specific conditions that are encountered in equipment operation.

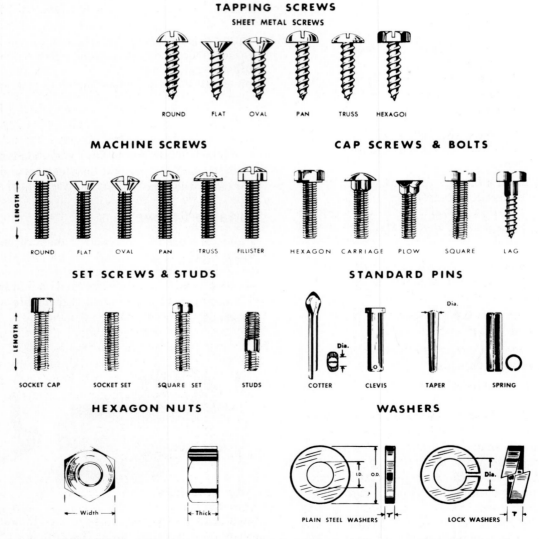

FIGURE 2–90. Many of the common types of fasteners used in the industry. *(Courtesy of H. Paulin & Co. Limited)*

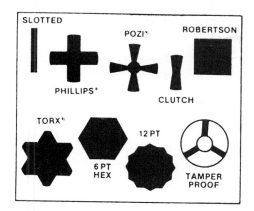

FIGURE 2-91. Various types of fastener head drive designs. *(Courtesy of Mac Tools Inc.)*

Using an incorrect fastener or a fastener of inferior quality for the job can result in early failure and even injury to the operator and others.

Some precautions to observe when replacing fasteners are the following:

• Always use the same diameter, length, and type of fasteners as were used originally by the vehicle manufacturer.

• Never thread a fastener of one thread type to a fastener of a different thread type.

• Always use the same number of fasteners as were originally used by the manufacturer of the equipment.

• Always observe the equipment manufacturer's recommendations for tightening sequence, tightening steps (increments), and torque values.

• Always use the correct washers, pins, and locks as specified by the manufacturer.

• Always replace stretched fasteners or fasteners with damaged threads.

• Never use a cotter pin more than once.

Damaged threads in threaded parts can be restored by the use of helically coiled thread inserts. Replace damaged snap rings and keys with new ones. The completed work is only as good as the technician's desire and ability to do a professional job with the use of correct parts and fasteners.

A number of terms have been used over the years to identify the various types of threads. Some of these have been replaced with new terms. The terms most commonly used in the trade are as follows:

The United States Standard (USS), the American National Standard (ANS), and the Society of Automotive Engineers Standard (SAE) have all been replaced by the Unified National Series. The Unified National Series consists of four basic classifications:

1. Unified National Coarse (UNC or NC).
2. Unified National Fine (UNF or NF).
3. Unified National Extrafine (UNEF or NEF).
4. Unified National Pipe Thread (UNPT or NPT).

The two common metric classifications are coarse and fine and can be identified by the letters SI (Système International d'Unités or International System of Units) or ISO (International Standards Organization).

Bolt Size

Bolt size is determined by two measurements. This is true of both English and metric bolts. Bolt length is the distance measured (in inches for English bolts and in millimeters or centimeters for S.I. metric bolts) from the bottom of the head to the tip of the bolt. Bolt diameter is the measurement across (in inches or fractions of an inch in the English system and in millimeters (mm) in the metric system) the major diameter of the threaded area.

Thread Pitch

The thread pitch of a bolt in the English system is determined by the number of threads there are in one inch of threaded bolt length and is expressed in "number of threads per inch." The thread pitch in the metric system is determined by the distance in millimeters between two adjacent threads. To check the thread pitch of a bolt or stud, a thread pitch gauge is used. Gauges are available in both English and metric dimensions.

Tensile Strength

The tensile strength of a bolt is the amount of stress or stretch it is able to withstand. The type of bolt material and the diameter of the bolt determine its tensile strength. In the English system, the tensile strength of a bolt is identified by the number of radial lines on the bolt head. More lines means higher tensile strength. See Figures 2-92 to 2-94.

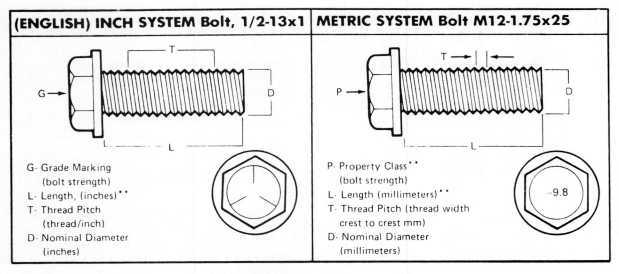

(ENGLISH) INCH SYSTEM Bolt, 1/2-13x1	METRIC SYSTEM Bolt M12-1.75x25

G- Grade Marking
   (bolt strength)
L- Length, (inches)**
T- Thread Pitch
   (thread/inch)
D- Nominal Diameter
   (inches)

P- Property Class**
   (bolt strength)
L- Length (millimeters)**
T- Thread Pitch (thread width
   crest to crest mm)
D- Nominal Diameter
   (millimeters)

*The property class is an Arabic numeral distinguishable from the slash SAE English grade system.
**The length of all bolts is measured from the underside of the head to the end.

BOLT STRENGTH IDENTIFICATION

(ENGLISH) INCH SYSTEM

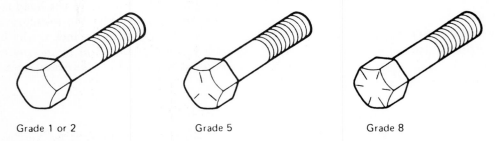

Grade 1 or 2   Grade 5   Grade 8

(English) Inch bolts - Identification marks correspond to bolt strength - increasing number of slashes represent increasing strength.

METRIC SYSTEM

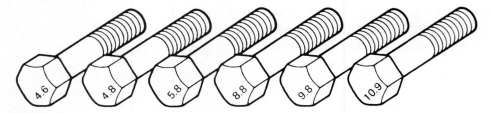

Metric bolts - Identification class numbers correspond to bolt strength - increasing numbers represent increasing strength. Common metric fastener bolt strength property are 9.8 and 10.9 with the class identification embossed on the bolt head.

FIGURE 2-92. Bolt teminology. *(Courtesy of Ford Motor Co. of Canada Ltd.)*

(ENGLISH) INCH SYSTEM		METRIC SYSTEM	
Grade	Identification	Class	Identification
Hex Nut Grade 5	3 Dots	Hex Nut Property Class 9	Arabic 9
Hex Nut Grade 8	6 Dots	Hex Nut Property Class 10	Arabic 10
Increasing dots represent increasing strength.		May also have blue finish or paint daub on hex flat. Increasing numbers represent increasing strength.	

OTHER TYPES OF PARTS

Metric identification schemes vary by type of part, most often a variation of that used of bolts and nuts. Note that many types of English and metric fasteners carry no special identification if they are otherwise unique.

—Stamped "U" Nuts

—Tapping, thread forming and certain other case hardened screws

—Studs, Large studs may carry the property class number. Smaller studs use a geometric code on the end.

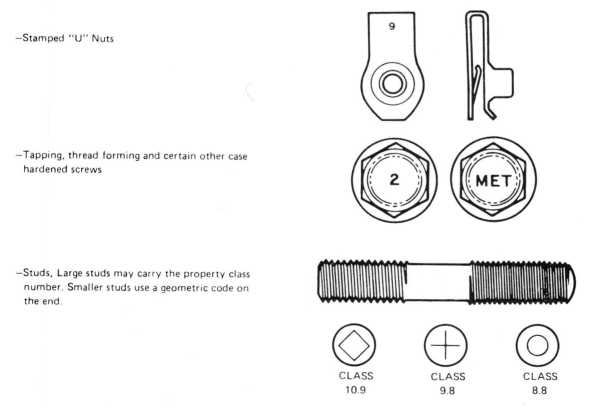

FIGURE 2-93. Nut and stud terminology. (Courtesy of Ford Motor Co. of Canada Ltd.)

Capscrew Markings and Torque Values

Current Usage	Much Used	Much Used	Used at Times	Used at Times
Minimum Tensile Strength — PSI [MPa]	To 1/2 — 69,000 [476] To 3/4 — 64,000 [421] To 1 — 55,000 [379]	To 3/4 — 120,000 [827] To 1 — 115,000 [793]	To 5/8 — 140,000 [965] To 3/4 — 133,000 [917]	150,000 [1 034]
Quality of Material	Indeterminate	Minimum Commercial	Medium Commercial	Best Commercial
SAE Grade Number	1 or 2	5	6 or 7	8

Capscrew Head Markings

Manufacturer's marks may vary

These are all SAE Grade 5 (3 line)

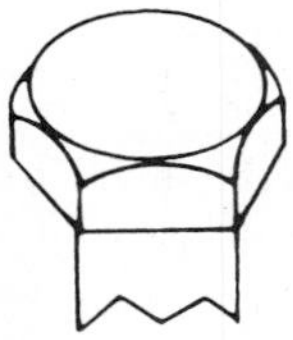

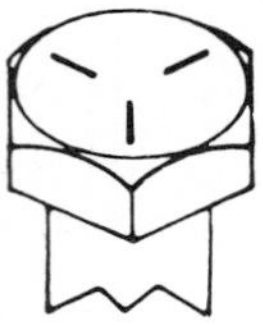

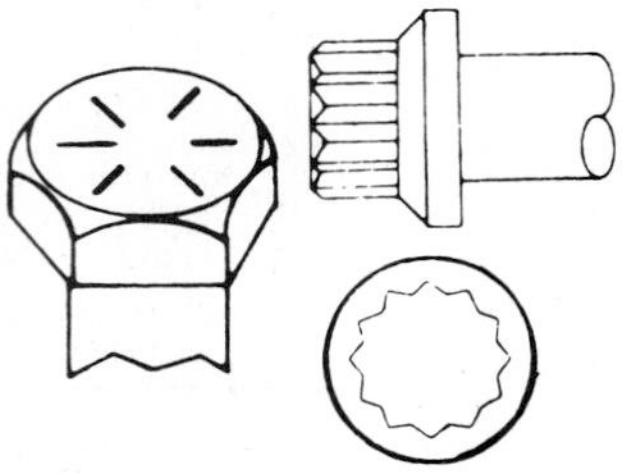

Capscrew Body Size (Inches) · (Thread)	Torque Ft.-Lbs. [N•m]	Torque Ft.-Lbs. [N•m]	Torque Ft.-Lbs. [N•m]	Torque Ft.-Lbs. [N•m]
1/4 — 20	5 [7]	8 [11]	10 [14]	12 [16]
— 28	6 [8]	10 [14]		14 [19]
5/16 — 18	11 [15]	17 [23]	19 [26]	24 [33]
— 24	13 [18]	19 [26]		27 [37]
3/8 — 16	18 [24]	31 [42]	34 [46]	44 [60]
— 24	20 [27]	35 [47]		49 [66]
7/16 — 14	28 [38]	49 [66]	55 [75]	70 [95]
— 20	30 [41]	55 [75]		78 [106]
1/2 — 13	39 [53]	75 [102]	85 [115]	105 [142]
— 20	41 [56]	85 [115]		120 [163]
9/16 — 12	51 [69]	110 [149]	120 [163]	155 [210]
— 18	55 [75]	120 [163]		170 [231]
5/8 — 11	83 [113]	150 [203]	167 [226]	210 [285]
— 18	95 [129]	170 [231]		240 [325]
3/4 — 10	105 [142]	270 [366]	280 [380]	375 [508]
— 16	115 [156]	295 [400]		420 [569]
7/8 — 9	160 [217]	395 [536]	440 [597]	605 [820]
— 14	175 [237]	435 [590]		675 [915]
1 — 8	235 [319]	590 [800]	660 [895]	910 [1234]
— 14	250 [339]	660 [895]		990 [1342]

Notes:
1. Always use the torque values listed above when specific torque values are not available.
2. Do not use above values in place of those specified in other sections of this manual; special attention should be observed when using SAE Grade 6, 7 and 8 capscrews.
3. The above is based on use of clean, dry threads.
4. Reduce torque by 10% when engine oil is used as a lubricant.
5. Reduce torque by 20% if new plated capscrews are used.
6. Capscrews threaded into aluminum may require reductions in torque of 30% or more of Grade 5 capscrews torque and must attain two capscrew diameters of thread engagement.

Caution: If replacement capscrews are of a higher grade than the original capscrew, tighten the replacement capscrew to the torque value used for the original capscrew.

FIGURE 2-94. Torque chart for typical fasteners. *(Courtesy of Cummins Engine Company, Inc.)*

REUSE OF PREVAILING TORQUE NUT(S) AND BOLT(S)

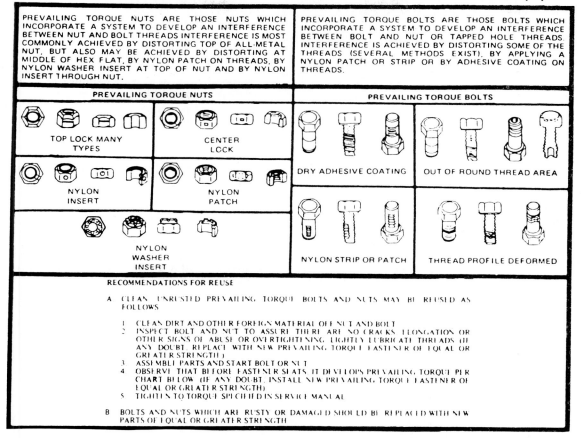

| PREVAILING TORQUE NUTS ARE THOSE NUTS WHICH INCORPORATE A SYSTEM TO DEVELOP AN INTERFERENCE BETWEEN NUT AND BOLT THREADS INTERFERENCE IS MOST COMMONLY ACHIEVED BY DISTORTING TOP OF ALL-METAL NUT, BUT ALSO MAY BE ACHIEVED BY DISTORTING AT MIDDLE OF HEX FLAT, BY NYLON PATCH ON THREADS, BY NYLON WASHER INSERT AT TOP OF NUT AND BY NYLON INSERT THROUGH NUT. | PREVAILING TORQUE BOLTS ARE THOSE BOLTS WHICH INCORPORATE A SYSTEM TO DEVELOP AN INTERFERENCE BETWEEN BOLT AND NUT OR TAPPED HOLE THREADS. INTERFERENCE IS ACHIEVED BY DISTORTING SOME OF THE THREADS (SEVERAL METHODS EXIST), BY APPLYING A NYLON PATCH OR STRIP OR BY ADHESIVE COATING ON THREADS. |

PREVAILING TORQUE NUTS

TOP LOCK MANY TYPES   CENTER LOCK

NYLON INSERT   NYLON PATCH

NYLON WASHER INSERT

PREVAILING TORQUE BOLTS

DRY ADHESIVE COATING   OUT OF ROUND THREAD AREA

NYLON STRIP OR PATCH   THREAD PROFILE DEFORMED

RECOMMENDATIONS FOR REUSE

A CLEAN UNRUSTED PREVAILING TORQUE BOLTS AND NUTS MAY BE REUSED AS FOLLOWS

   1 CLEAN DIRT AND OTHER FOREIGN MATERIAL OFF NUT AND BOLT
   2 INSPECT BOLT AND NUT TO ASSURE THERE ARE NO CRACKS, ELONGATION OR OTHER SIGNS OF ABUSE OR OVERTIGHTENING. LIGHTLY LUBRICATE THREADS (IF ANY DOUBT, REPLACE WITH NEW PREVAILING TORQUE FASTENER OF EQUAL OR GREATER STRENGTH)
   3 ASSEMBLE PARTS AND START BOLT OR NUT
   4 OBSERVE THAT BEFORE FASTENER SEATS IT DEVELOPS PREVAILING TORQUE PER CHART BELOW (IF ANY DOUBT, INSTALL NEW PREVAILING TORQUE FASTENER OF EQUAL OR GREATER STRENGTH)
   5 TIGHTEN TO TORQUE SPECIFIED IN SERVICE MANUAL

B BOLTS AND NUTS WHICH ARE RUSTY OR DAMAGED SHOULD BE REPLACED WITH NEW PARTS OF EQUAL OR GREATER STRENGTH

FIGURE 2-95. Reuse of prevailing torque nuts and bolts. *(Courtesy of General Motors Corporation)*

In the metric system tensile strength of a bolt can be identified by a number on the bolt head. See Figures 2-92 to 2-94. The higher the number, the greater the tensile strength.

Fastener Locks

Cotter pins are probably the most common type of fastener locking device. A split pin with a looped type head, it is available in many different lengths and diameters. Cotter pins should only be used once, since several bends cause breakage. The cotter pin should be large enough in diameter to fit the hole in the fastener and long enough to allow proper locking of the exposed split ends. Special cotter pins may be required for such applications as clutch pins and connecting rod nuts where applicable.

Flat metal locks are used in many cases to prevent nuts or cap screws from loosening. Be sure to install the correct type for the fastener being used. Install the lock in a manner that will prevent the nut or cap screw from loosening. Several designs are in use, samples of which are illustrated. Flat metal locks should not be used more than once, since repeated bends cause breakage.

HEXAGON

HEXAGON WASHER FACED

SQUARE (CHAMFER)

PLAIN

SELF LOCKING NUT

WHEEL NUT

WING NUT

HEX. CAP

CONTRACTING

EXPANDING

FIGURE 2-99. Several types of commonly used snap rings. Snap rings are used to prevent endwise movement of shafts and bearings. Damaged or distorted snap rings must be replaced. *(Courtesy of Ford Motor Company of Canada Ltd.)*

LOCKING ACTION:

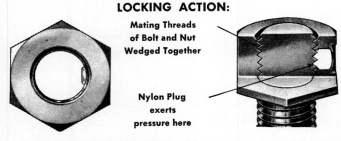

Mating Threads of Bolt and Nut Wedged Together

Nylon Plug exerts pressure here

FIGURE 2-96. Several types of nuts. Cutaway view of one type of self-locking nut action.

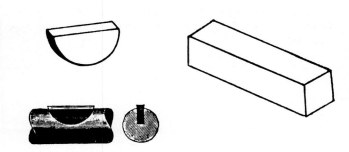

FIGURE 2-100. Woodruff keys (left) and straight keys are used to hold parts to shafts so that part and shaft rotate as a unit. In other cases splined shaft and parts are used to accomplish the same result.

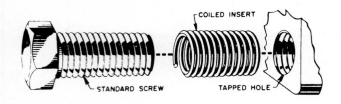

COILED INSERT

STANDARD SCREW

TAPPED HOLE

FIGURE 2-97. Damaged threads can be restored in a threaded part by the use of a helically coiled insert. The damaged hole is drilled to a precise oversize, tapped, and a coiled insert installed. This provides new threads of the original diameter and type.

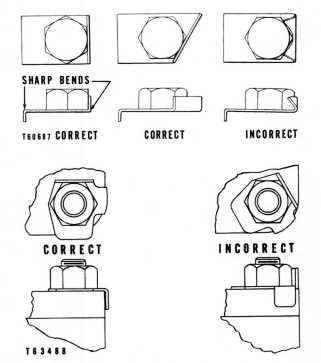

SHARP BENDS

T60687 CORRECT

CORRECT

INCORRECT

CORRECT

INCORRECT

T63488

FIGURE 2-101. Typical flat metal locking procedure. *(Courtesy of Caterpillar Tractor Company)*

FLAT

COMMON LOCK

EXTERNAL LOCK

INTERNAL LOCK

EXTERNAL-INTERNAL LOCK

COUNTER-SUNK LOCK

FIGURE 2-98. Common types of washers are used in diesel equipment. Lock washers prevent loosening of fasteners and should be used wherever original equipment was so equipped.

PART 8   SELF-CHECK

1. Why are shop manuals necessary in the diesel shop?

2. What information is necessary about equipment in order to obtain correct replacement parts for it?

3. What is torque?

4. Why is a torque wrench necessary?

5. A well-equipped technician will have a good range of English or U.S. size wrenches as well as ____sizes.

6. To measure with an outside micrometer, the part being measured must be placed between the ____and the ____.

7. The straight edge is not a precision measuring tool. True or false?

8. A lift used to raise equipment may be positioned to contact any point underneath. True or false?

9. Why should the load on a portable floor crane be lowered for movement from one place to another?

10. What pressure tests are made with a manometer on a diesel engine?

Chapter 3

Gaskets, Seals, and Bearings

Performance Objectives

After thorough study of this chapter and the appropriate gaskets, seals, and bearings, you should be able to do the following:

1. State the purpose of gaskets and sealants.
2. Identify the different types of gaskets and sealants and state their use.
3. State the purposes of static and dynamic seals.
4. Identify different types of seals and state their use.
5. State the purposes of friction and antifriction bearings.
6. Describe the basic construction and design differences of various bearing types.
7. Complete the self-check questions with at least 80 percent accuracy.

PART 1 GASKETS

Gaskets are used to prevent leakage of gases or liquids between two parts bolted together. The gasket is placed between the two mating machined surfaces; the bolts or fasteners are then tightened in the recommended sequence, and to the specified torque. The gasket is designed for the particular job it is required to do. Different materials, such as cork, synthetic rubber, Vellumoid®, steel, copper, and asbestos are used. In some cases special sealant is used with the gasket.

Gasket Requirements

Gaskets are required to function without deterioration, over long periods of time, and in various conditions. The type of gasket that is used in any given application is carefully chosen to achieve the proper results. Determining factors include:

• High and low temperature extremes that will be encountered
• Type of material to be sealed: gases, liquids, etc.
• Pressure of confined material
• Finish or smoothness of mating parts
• Clearance required between assembled mating parts

Different applications require a variety of different materials and designs.

Gasket Qualities

Good gasket qualities include the following:

• *Compressibility* and *extrudability*—required to allow the gasket to "flow" slightly to conform to the minor irregularities of the mating surfaces.
• *Resilience*—the ability of the gasket material to contract and expand with the changes in temperature and still provide good sealing.
• *Permeability*—the degree to which the gasket material is able to prevent liquids or gases from passing through the material itself.

These qualities are required to a greater or lesser degree depending on the particular application for which the gasket is designed.

Gasket Storage

Many gaskets are encapsulated in plastic and cardboard packages to prevent deterioration and breakage. These gaskets are more easily stored than single gaskets that are not packaged. Loose gaskets should be stored lying flat. Other parts should not be laid on top of them. Nevertheless, cork and paper or fiber may still deteriorate and shrink due to drying out. They may be restored, however, by being soaked in warm water until original size has returned.

PART 2   SEALANTS

There are two types of sealants used in diesel equipment service. One type is used with gaskets and is applied directly to the gasket or to the mating surfaces to be sealed by the gasket. A great variety of materials is available for this purpose. The manufacturer's service manual recommendations should be followed as to whether to use a sealer with the gasket, and if so, which type of sealer to use (i.e., hardening or nonhardening, compatibility, and the like). Another type of sealant used is the "form-in-place gasket material." There are two basic types of this material available. They are not interchangeable and should not be so used.

Aerobic Sealant (RTV Sealant)

Aerobic or Room Temperature Sealant cures when exposed to moisture in the air. It has a shelf storage life of one year and should not be used if the expiration date on the package has passed. Always inspect the date on the package before use. This material is normally used on flexible metal flanges.

Anaerobic Sealant

This type of sealant cures in the absence of air (as when squeezed between two metal surfaces). It will not cure if left in an uncovered tube. This material is used between two smooth machined surfaces. It should not be used on flexible metal flanges.

To use either of these form-in-place gasket materials, be sure to follow the material supplier and equipment manufacturer's recommendations, including preparation of mating surfaces, application methods, and maximum exposure time allowed before assembly.

Anti-Seize Compound

An anti-seize compound must be used on the threads of some fasteners during component assembly when specified by the equipment manufacturer. The material is used to coat the threaded portion of the bolt or stud lightly before installation and tightening. Use of the compound prevents the fasteners from seizing during normal operation. This allows the fasteners to be removed later for service and prevents damage to components that can result from removal of seized fasteners.

The compound should not be used indiscriminately, since such use could result in fasteners becoming loose during equipment operation.

Locking Compounds

Some fasteners on certain applications require the use of a thread-locking compound to prevent loosening of the fasteners during normal operation due to vibration. The locking compound is also used on the inner or outer race of certain bearings for the same reason. The locking compound is applied to the threaded portion of the fastener, the bearing race, or the bearing bore as specified in the appropriate service repair manual.

Locking compounds should be used only when and where specified by the equipment manufacturer.

Teflon tape is another type of locking and sealing material used in some applications. The tape is wrapped around the threaded portion of the bolt, cap screw, or stud before installation. Teflon tape should also be used only where specified by the equipment manufacturer.

PART 3   SEALS

Oil seals are classified as static or dynamic. The static seal is used between two stationary parts. The dynamic seal provides a seal between a stationary and a moving part. One example of a static seal is the O ring seal between a transmission hydraulic pump and the transmission case. The rear main bearing crankshaft oil seal is an example of a dynamic seal.

FIGURE 3-1. A variety of seals of various shapes and designs is required in the engine. Some examples are shown above. *(Courtesy of National Seal)*

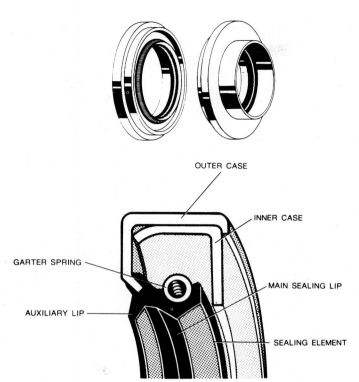

Concentric Wall Bearing
(uniform wall thickness)

Eccentric Wall Bearing
(wall heavier at crown than at parting faces)

FIGURE 3-2. A typical seal (above) and common seal terminology (below). Seals are used to keep in oils, fluids, and grease or to exclude dirt, or both. The main sealing element can be synthetic rubber, leather, or felt. Some seals have both an inner and outer case; others have only an outer case. Some seals include a bolt on flange.

FIGURE 3-3. Several types of friction bearings and bushings. Two-piece bearings are needed on applications such as an engine crankshaft. Eccentric bearing design improves the ability to maintain an oil film between the bearing and bearing journal. One-piece bushings are used on camshafts and other parts. Some bushings of a porous material are permanently lubricated with a special lubricant that saturates the bushing.

Some seals are designed to withstand high pressures. Piston rings, for instance, are designed to withstand high combustion pressures and seal both gases and liquids. Other seals use felt, synthetic, rubber, fiber, or leather. Many seals have a metal case and a tension spring. Both single- and double-lip seals are used.

Seals should always be installed according to manufacturer's specifications. In general, though, the sealing lip should be installed toward the fluid or gas being contained. Felt dust seals should be installed with the felt toward the outside.

PART 4   BEARINGS

Bearings (Fig. 3-3 to 3-7) reduce the friction of rotating parts. Overhauling and reconditioning are easier and less expensive when bearings can be replaced.

There are two basic classifications of bearings: friction and antifriction bearings.

Friction Bearings

Friction bearings rely on sliding friction with a film of oil between the bearing and bearing journal. A friction bearing is a thin-walled cylindrical part located between the rotating part and the supporting part. It can be of one-piece construction or it may consist of several pieces. The one-piece friction bearing is often called a bushing; the terms bearing in-

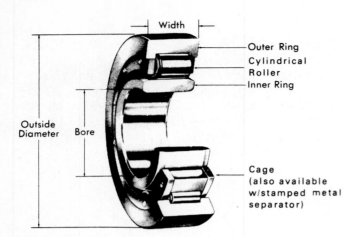

FIGURE 3-4. Cutaway view of a straight roller bearing with parts identified. This bearing is designed to carry a radial load (90° to the shaft).

serts or bearing shells are used for multiple-piece friction bearings. Various types of combinations of metals are used for friction bearings. Sintered bronze, brass, steel, Babbitt metal, lead, tin, nylon, and aluminum are examples. There are no balls or rollers in friction-type bearings.

Antifriction Bearings

Antifriction bearings rely on rolling friction for operation. Rolling friction requires less effort to rotate a part than does sliding friction. The greater the load, the more evident this fact becomes.

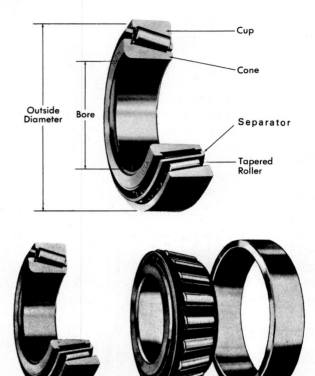

FIGURE 3-6. Tapered roller bearing and component part names. This bearing is able to carry a radial load as well as an axial load in one direction. *(Courtesy of General Motors Corporation)*

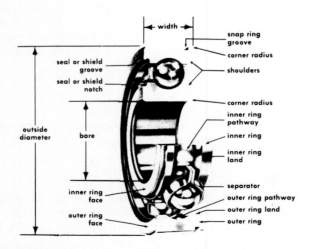

FIGURE 3-5. Cutaway view of a ball bearing with parts identified. This bearing is also designed to carry a radial load.

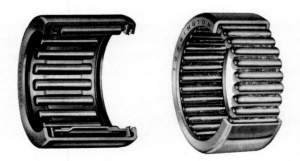

FIGURE 3-7. Common types of needle bearings designed to carry radial loads.

Antifriction bearings include various designs of ball bearings and roller bearings.

Thrust Washers

Thrust washers are also a type of friction bearing. They are designed to absorb an axial load or thrust load. They may be designed to rotate with a rotating part against a stationary part, they may be "lugged" to the stationary part, or they may "float"—be free to rotate between two parts, not being lugged to either part.

Thrust washers are designed to reduce wear on adjacent parts and to maintain specified clearances between parts. Various metals, metal combinations, and plastics are used in thrust washer design including steel, bronze, steel-backed bronze, aluminized steel, nylon, and teflon. Thrust washer surfaces are often grooved for improved lubrication.

Shielded Bearings

Some ball bearing designs include a shield on one side. The shield is attached to the outer race and shields or hides the balls with only a small opening around the inner race. This restricts the amount of lubrication that can pass through the bearing. In this way it is possible to prevent overloading a seal with lubricant, which reduces the chance of leakage past the seal.

Sealed Bearings

Sealed bearings are not to be confused with shielded bearings. Sealed bearings appear similar but are completely sealed to prevent any lubricant from escaping from the bearing or past the bearing. Some designs are sealed on one side only allowing bearing lubrication from the open side of the bearing. The open side of the bearing is installed toward the lubricant being confined. Other designs are sealed on both sides; they are pre-lubricated for life during the manufacturing process.

Needle Thrust Bearings

The needle thrust bearing or Torrington type bearing is designed to absorb thrust loads. Compared to a thrust washer, it provides reduced friction. It consists of a series of small diameter rollers or "needle" rollers arranged radially and held in position in a washer-shaped cage. The assembly may include separate washer type thrust surfaces—one on each side—or the thrust surfaces may be incorporated into the adjacent parts.

Bearing Loads

Bearings are designed to carry radial loads, axial loads, or both. A radial load is a load imposed at right angles to the shaft. An axial load is a load imposed parallel to the shaft. (See Fig. 3–8.)

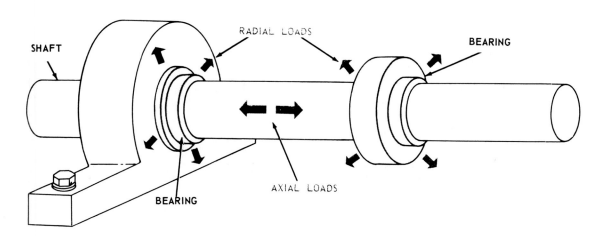

FIGURE 3-8. Load forces acting on bearings. *(Courtesy of Deere and Company)*

A straight roller bearing is able to carry a radial load, whereas a thrust bearing is able to absorb an axial load. A tapered roller bearing can carry both a radial load and a unidirectional axial load. A double row opposed tapered roller bearing is able to carry a radial load as well as axial loads in both directions.

Specific types of friction and antifriction bearings are dealt with in different chapters as they apply.

PART 5   SELF-CHECK

1. Why are gaskets required?

2. List four factors that determine the type of gasket required for a particular application.

3. How should gaskets be stored to prevent damage?

4. Explain the use of aerobic and anaerobic sealants.

5. List two classifications of seals.

6. The two basic classifications of bearings are the_____type and the_____type.

7. Define radial loads and axial loads on bearings.

8. What are locking compounds and anti-seize compounds used for?

Chapter 4

Lines, Fittings, and Belts

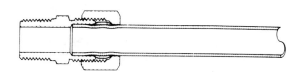

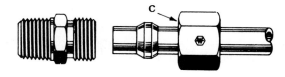

Use long nut when excessive vibration may be encountered.

FIGURE 4-1. Compression-type brass fittings used with copper and steel lines. *(Courtesy of The Weatherhead Company of Canada Ltd.)*

Performance Objectives

After a thorough study of this chapter and the appropriate training aids, you should be able to do the following:

1. Identify various types of lines, tubing, and fittings, and state their general uses.

2. Properly cut tubing and perform satisfactory single- and double-lap flares with fittings properly installed.

3. Bend tubing using a tubing bender to fit a given application, without cracking or kinking the tubing.

4. Identify various types of belt drives, belts, and pulleys, and state their general applications.

5. Properly adjust belt tension as specified by the service manual.

6. Complete the self-check questions with at least 80 percent accuracy.

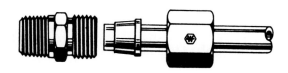

Use long nut when excessive vibration may be encountered.

FIGURE 4-2. One type of plastic tubing and fittings uses a sleeve type of seal as shown here. *(Courtesy of The Weatherhead Company of Canada Ltd.)*

PART 1   LINES AND FITTINGS

Lines and fittings (Fig. 4-1 to 4-10) are used to carry liquids and gases in such engine systems as the cooling system, the lubrication system, the fuel system, the exhaust system, and so on. Some lines are subjected to relatively low pressures while other lines are required to withstand very high pressures.

FIGURE 4-3. Slip-on type of plastic tubing and fitting often used with a clamp. *(Courtesy of The Weatherhead Company of Canada Ltd.)*

UNION   COUPLING

FIGURE 4-4. The union shown connects two lines, using compression sleeves, while the coupling on the right connects two threaded lines. *(Courtesy of The Weatherhead Company of Canada Ltd.)*

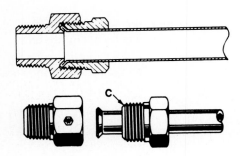

FIGURE 4-5. Inverted flare type of fittings. *(Courtesy of The Weatherhead Company of Canada Ltd.)*

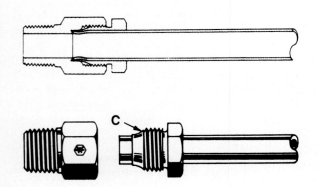

FIGURE 4-6. Threaded sleeve type of fitting. *(Courtesy of The Weatherhead Company of Canada Ltd.)*

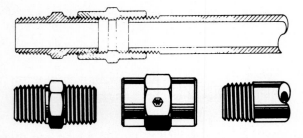

FIGURE 4-7. Pipe thread type of fittings. *(Courtesy of The Weatherhead Company of Canada Ltd.)*

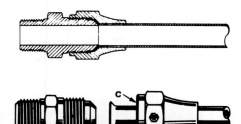

Use long nut when excessive vibration may be encountered.

FIGURE 4-8. Flare nut type of fittings. *(Courtesy of The Weatherhead Company of Canada Ltd.)*

It is essential to use only the recommended type and size of line for any particular application in order to avoid trouble and failure. These lines should be used only with the recommended fittings installed in the correct manner.

Lines include relatively thick walled pipes, which do not lend themselves readily to bending; thin walled tubing, which is easier to bend; and flexible hoses.

Copper, aluminum, plastic, and steel tubing are used. Only the seamless steel tubing is suitable for high-pressure applications such as fuel injection lines and brakes.

Care must be taken not to kink or flatten tubing when it is necessary to bend it to required shape. Kinks or flattened sections restrict flow in the tubing. Use a good type of tubing bender to avoid damage to the tubing.

It is essential that the different thread types in fittings be recognized. Never mix thread types or seal types in fittings. Do not cross thread fittings. Route lines and fittings in such a way as to avoid abrasion, which can cause a line to leak. Use all original brackets and clamps to ensure that lines are properly supported. Do not allow lines to twist during removal or installation. Route lines to avoid damage from exhaust heat.

Flexible lines and hoses are used for various pressure ranges. Hose construction and flexible line construction is different for low- or high-pressure use. Always use the recommended hose for a particular application.

Flaring Tubing

Steel and copper tubing should be cut with a tubing cutter. Avoid applying too much pressure during

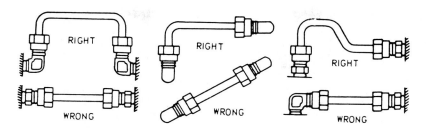

FIGURE 4-9. Right and wrong methods of tubing installation. Sufficient tubing length must always be allowed to prevent the tubing from cracking due to expansion and contraction as well as vibration.

cutting. This can collapse the tube. Low-pressure lines need only a single flare, whereas brake lines require double-lap flaring or International Standards Organization (ISO) flaring. Some vehicles have the double-lap flared brake lines while others have the ISO-type flared lines and fittings. Fittings and lines of different designs should never be mixed, used together, or interchanged.

After cutting, ream the tubing and then flare as required. The reamer is usually part of the tubing cutter. Make sure that no metal chips are left in the tubing.

HOW TO TIGHTEN FLARE-TYPE TUBE FITTINGS

Line Size (Outside Diameter)	Flare Nut Size (Across Flats)	Tightness (Ft-lbs)	Recommended Turns of Tightness (After Finger Tightening)	
			Original Assembly	Re-assembly
3/16″	7/16″	10	1/3 Turn	1/6 Turn
1/4″	9/16″	10	1/4 Turn	1/12 Turn
5/16″	5/8″	10-15	1/4 Turn	1/6 Turn
3/8″	11/16″	20	1/4 Turn	1/6 Turn
1/2″	7/8″	30-40	1/6 to 1/4 Turn	1/12 Turn
5/8″	1″	80-110	1/4 Turn	1/6 Turn
3/4″	1 1/4″	100-120	1/4 Turn	1/6 Turn

FIGURE 4-10. General tightening specifications, which may be used if manufacturer's specifications for tightening tubing fittings are not available.

PART 2   FUEL INJECTION TUBING

High-pressure fuel injection tubing (Fig 4-11 and 4-12) is used to connect the injection pump to the fuel injectors. The tubing is of thick-walled, seamless design with precise inside diameter dimensions to ensure equal fuel delivery to each engine cylinder. All the lines on an engine are of exactly the same length for the same reason.

Injection tubing has a tensile strength of about 55,000 pounds (24,948 kg) to prevent tubing expansion during injection. Inside diameters are accurately sized and free of any irregularities to ensure even injection to all cylinders. This requires that outside dimensions also be precise and free of irregularities.

Injection tubing bends are formed with a mechanical tubing bender that prevents flattening and kinks. Internal dimensions must not be altered in bends. Tubing bends should have a radius of at least one inch (25.4 cm) or more to ensure that flow characteristics are maintained.

The most common injection tubing size is one-quarter inch outside diameter. The tubing is available in several inside diameters to suit individual injection system requirements.

Common inside dimensions and their exterior color code markings are as follows:

0.063 inch (1.6002 mm) Red
0.067 inch (1.7018 mm) Black
0.078 inch (1.9812 mm) Yellow
0.084 inch (2.1336 mm) Blue
0.093 inch (2.3622 mm) White

Color code markings are in the form of intermittent striping on the tubing exterior. Other sizes of injec-

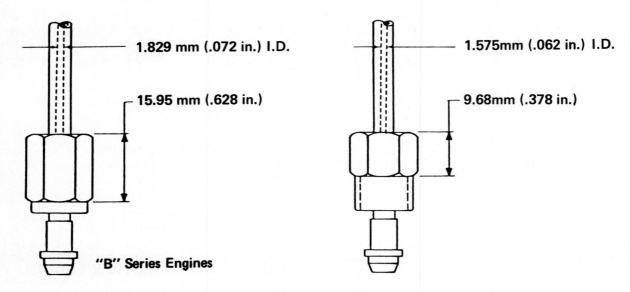

FIGURE 4-11. Typical fuel injection line and fitting dimensions. *(Courtesy of International Harvester Company)*

tion tubing for larger diesel engines are also used and include 5/16 inch (7.9375 mm), 3/8 inch (9.525 mm), 7/16 inch (11.1125 mm), and 1/2 inch (12.70 mm) outside diameters.

Several different types of injection tubing fittings are used to attach the tubing to the pump and injectors. These include the tapered seat type, the flared type, and the ferrule type.

Absolute cleanliness is essential to ensure that fittings are leak free after assembly. It is also important that sealing surfaces not be nicked or damaged in any way during handling. Open lines and fittings must be capped to prevent the entry of dirt or moisture during service.

When assembling the ferrule type injection line, the tubing must be held in place against the fitting

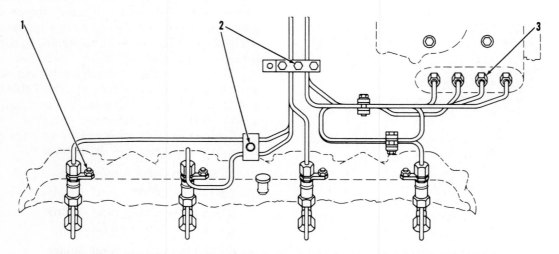

FIGURE 4-12. Fuel injection lines and brackets. (1) Injector adapter nut. (2) Fuel injection line clamps. (3) Injection line fitting nuts. *(Courtesy of Caterpillar Tractor Company)*

counterbore bottom while the fitting nut is tightened. Otherwise the one end of the ferrule may crimp over the end of the tubing. A new seal must be used with the flared type connector to ensure against leakage.

Many fuel injection systems use a series of injection line support brackets and insulators. These are designed to reduce injection line vibration and prevent breakage. It is important that these items be in good condition and be properly installed and tightened to ensure long injection line life.

Injection lines must be handled with care so that they are not bent or distorted during removal or installation. Injection lines that are in good condition need not be bent or distorted to facilitate starting the fitting nuts. Fitting nuts should start easily by hand if the lines have been properly positioned. Care

must be exercised that fittings are not crossthreaded during installation. Fittings damaged by cross threading must be replaced and this, of course, means replacing the entire injection line assembly.

PART 3   HYDRAULIC HOSES

Hydraulic hoses (Fig. 4–13 to 4–16) are very convenient for transmitting fluids and pressures. Hoses are flexible and can be routed in many different ways. They withstand vibrations and pressure surges easily and they are easily replaced.

Hydraulic hoses consist of three basic components: an inner tube-type liner, layers of reinforcing material, and outer cover. The inner tube is smooth,

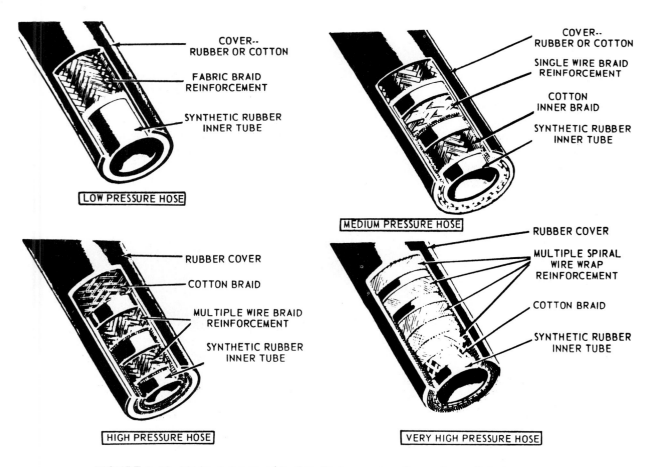

FIGURE 4–13. Various types of hydraulic hoses are shown here. It is important to use only the correct type of hose as specified by the equipment manufacturer. *(Courtesy of Deere and Company)*

Hose Size in Inches	1. Use MEDIUM Pressure Single Wire Braid Hose if System Working Pressure Equals . . .	2. Use HIGH Pressure Multiple Wire Braid Hose if System Working Pressure Equals . . .	3. Use VERY HIGH Pressure Spiral Wire Wrap Hose if System Working Pressure Equals . . .
¼"	3000 psi	5000 psi	— — —
⅜"	2250 psi	4000 psi	5000 psi
½"	2000 psi	3500 psi	4000 psi
⅝"	1750 psi	2750 psi	— — —
¾"	1500 psi	2250 psi	3000 psi
1"	800 psi	1875 psi	3000 psi
1¼"	600 psi	1625 psi	3000 psi
1½"	500 psi	1250 psi	3000 psi
2"	350 psi	1125 psi	2500 psi

FIGURE 4-14. Hydraulic hose application chart. *(Courtesy of Deere and Company)*

oil-, heat- and corrosion-resistant synthetic rubber material. The reinforcing material can be fiber or wire braid, or a combination of these. The type of material and the number of plies used is determined by the hydraulic pressures the hose will be required to withstand. The outer cover, usually a special rubber compound, is designed to protect the hose assembly. It resists deterioration caused by weather, oil, and abrasion.

Hydraulic hose includes the following:

Fabric braid

Single wire braid

Double wire braid

Spiral wire braid

Hoses subject to higher pressures use the stronger wire braid design and extra plies depending on application. Larger diameter hoses withstand less

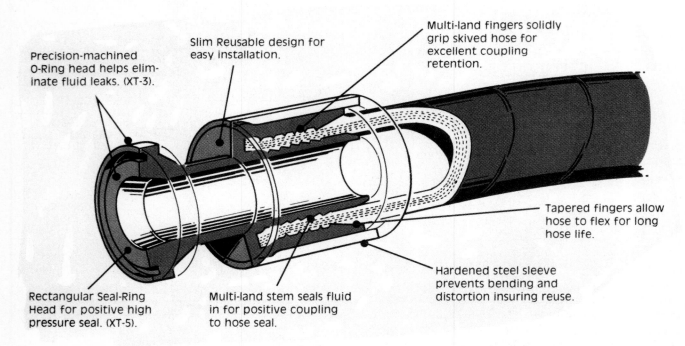

Precision-machined O-Ring head helps eliminate fluid leaks. (XT-3).

Slim Reusable design for easy installation.

Multi-land fingers solidly grip skived hose for excellent coupling retention.

Tapered fingers allow hose to flex for long hose life.

Rectangular Seal-Ring Head for positive high pressure seal. (XT-5).

Multi-land stem seals fluid in for positive coupling to hose seal.

Hardened steel sleeve prevents bending and distortion insuring reuse.

FIGURE 4-15. Reusable hydraulic hose fitting. *(Courtesy of Caterpillar Tractor Company)*

HOSE INSTALLATION

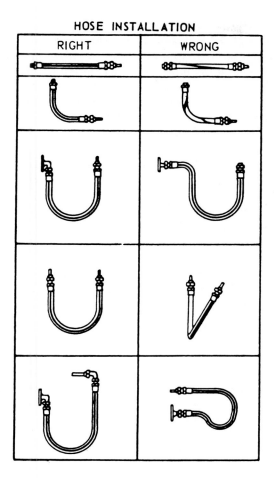

FIGURE 4-16. Right and wrong methods of hose installation. Installation should not result in twisted hose after fittings are tightened.

pressure than smaller hoses of the same design. This results from the fact that the larger hose has a greater surface area on which the pressure is applied.

Hydraulic hoses use threaded fittings or connectors at each end. These may be straight or angled. Materials used for fittings and connectors are steel, brass, and in some cases stainless steel or even plastic. Two types of fittings are in use: the permanent type and the reusable type. The permanent type is discarded with the hose when replacement is required. A new hose with new fittings already attached must be used for replacement. Some shops stock hydraulic hose in bulk as well as a variety of replacement fittings. A crimping machine is used to attach the fittings to the desired length of new hose.

Reusable hose fittings are removed from the failed hose and installed on the new hose cut from bulk hose in stock. The fittings are screwed on or clamped on depending on fitting design.

PART 4   REPLACING HOSES

Hydraulic hose failure can be caused for any of a variety of reasons, including abrasion, twisting, overheating, incorrect hose application, incorrect routing, incorrect hose length, improper use of or type of fittings, cracking, bursting, or pin hole leaks.

When hose replacement is required it is good practice to determine the cause of failure and to avoid that problem when installing the replacement hose. Here are some important guidelines to follow when replacing hydraulic hose:

1. Be sure to use the correct replacement hose and fittings for the application. Use only the type of hose and fittings designed to handle the pressure requirements of the system.

2. Be sure the replacement hose is the correct length: not too long and not so short as to create tension on the hose when assembled.

3. Be sure to route the hose in a manner that will prevent loops, tension, rubbing, heating, sharp bends, or twisting of the hose.

4. Be sure to install all required hose support brackets or clamps as originally equipped.

PART 5   BELTS AND PULLEYS

A system of belts and pulleys (Fig. 4-17) is used to drive such engine accessories as the alternator, water pump, fan, power steering pump, and air compressor. None of these needs to be timed to rotate in a precise relationship to the engine crankshaft.

However, such items as overhead camshafts and diesel fuel injection pumps require precise timing to the engine crankshaft on a continuing basis and, therefore, require more positive drive mechanisms. Among these is the toothed or cog belt and sprocket drive. Gear drive systems for engines are discussed in Chapter 9 of this text.

"V" Belts

Several different designs of V belts are used to drive engine accessories. These are the conventional V

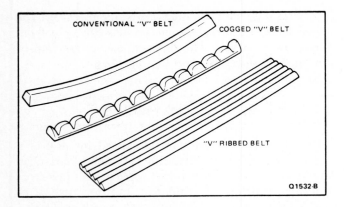

FIGURE 4-17. Three types of belts used to drive engine accessories. *(Courtesy of Ford Motor Co. of Canada Ltd.)*

belt, the cogged V belt, and the poly V or ribbed V belt, also known as a serpentine belt.

All of these belts are constructed of a combination of rubber, fabric, and rubber-impregnated fabric. Reinforcing fabric or steel cords are used in some belts that are required to draw heavier loads, and to reduce stretch and slippage (such as air conditioning compressors). Conventional and cogged V belt drives may be of the single belt or dual matched belt type.

In some cases a single, poly V, serpentine belt is used to drive several accessories while in many other applications two, three, or more belt drive systems may be used.

All V belt drives rely on proper belt tension to provide the necessary wedging action of the belt in the pulleys to keep belt slippage at a minimum. However, excessive belt tension will cause early bearing failure in engine driven accessories, as well as excessive belt and pulley wear. Belts that are too loose will allow slippage, causing engine accessories to be driven too slowly. Loose belts will also cause overheating and rapid wear of belts and pulleys, as well as engine overheating due to insufficient fan and water pump speeds.

Belt tension adjustment is provided by means of an adjustable idler pulley or by one or more of the accessories (usually the alternator and power steering pump) being on a sliding adjustable mounting.

"V" Belt Size

Efficient belt drive operation is dependent on proper belt size to match pulley V width. A belt that is too narrow will not result in good side gripping action

since it will bottom out in the pulleys. A belt that is too wide will ride too high in the pulleys resulting in slippage, belt damage, and possibly in the belt jumping off the pulleys during operation.

Belts that are worn, glazed, oil or grease contaminated, cracked or torn, should be replaced.

Belt squeal is usually the result of glazed and slipping belts. Adjusting belt tension may not eliminate the squeal since the belt may be glazed or worn, in which case the belt should be replaced. Dual matched belts should always be replaced in matched pairs.

"V" Pulleys

The crankshaft drive pulley may be a single, dual, triple, or quadruple V design, depending on how many belt drive systems the engine has. Pulleys on accessories may be of single or multiple V design.

Pulley construction varieties include stamped steel, cast iron, steel alloy, and die cast aluminum.

Pulley diameters determine the speed ratio between engine crankshaft and driven accessories. A crankshaft drive pulley and a driven accessory pulley of the same size would drive the accessory at crankshaft speed minus some minor slippage. An accessory pulley that is smaller than the drive pulley would result in the accessory being driven at greater than crankshaft speed.

Proper pulley size is determined by the manufacturer to insure the best of a wide range of operating speeds for engine accessories.

Pulleys may be mounted in several different ways: press fit on the shaft, keyed to the shaft, splined to the shaft, tapered shaft, and hub. Any one of these mountings may also include a retaining bolt and washer.

Pulleys must run straight and true and be in alignment with each other for efficient belt drive operation. Bent, damaged, cracked, worn, or broken pulleys must be replaced with pulleys equivalent to original equipment type and size.

Poly V Belts

Poly V, ribbed V, or serpentine belts are more flexible than conventional V belts since they are of smaller cross sectional dimension. They are also constructed of a combination of rubber and fabric. The poly V belt consists of a series of small inner surface Vs that grip corresponding V grooves in poly V pulleys.

A unique feature of this type of belt is that it can be routed in a variety of ways including looping the back of the belt over a flat surfaced pulley. This is not possible with the conventional V belt since it would crack and tear if routed in this manner.

Similar operating principles regarding belt tension, belt condition, pulley condition, and pulley ratios apply as they do to conventional V belt drives discussed earlier.

Cog Belt Drives

Cog belt drives, in some cases, are used to drive overhead camshafts, auxiliary shafts, and diesel fuel injection pumps on small diesel engines.

In all of these cases, a precise drive relationship must be maintained between the engine crankshaft and the driven component.

On a small automotive engine for example, the camshaft and the diesel injection pump must be driven at exactly half the speed of the crankshaft. They must also be precisely timed to the engine crankshaft and piston position. This timing and speed ratio relationship must be continuously maintained during all phases of the operating life of the engine. This places more rigorous requirements on the cog belt drive than are required for other V belt drives.

The cog belt must not stretch or lose its tension. Belt construction such as fiber glass reinforcement provides this characteristic. The cog belt must not slip. Teeth or cogs on the inner circumference of the belt, and corresponding teeth on the drive and driven sprockets, prevent slippage. The cog belt must not deteriorate over long periods of time from slight oil or water contamination. Synthetic rubber compounds assure long life under these conditions. The cog belt must not encounter foreign objects such as twigs, stones, ice, or snow during operation which could cause the drive to fail. A shield almost completely enclosing the cog belt drive prevents entry of such foreign objects.

Proper cog belt tension is provided by a belt tensioner adjustment. Proper cog belt operation (and engine operation) requires that precise belt tension specifications be followed when making adjustments.

PART 6   SELF-CHECK

1. What materials are used for diesel engine lines and fittings?
2. Kinked or flattened sections in tubing _____ flow.
3. Name three types of diesel engine drive belts.
4. All belt drives rely on proper belt _____ to prevent slippage.
5. Cogged or toothed drive belts are used where no _____ can be tolerated.
6. What level of pressure must fuel injection tubing be able to withstand?

Chapter 5

Diesel Engine Operation, Classification, and Measurements

Performance Objectives

After thorough study of this chapter, the appropriate training models, and manufacturer's manuals, you should be able to do the following:

1. Complete the self-check questions with at least 80 percent accuracy.
2. Define energy conversion as it applies to the diesel engine.
3. List the basic components of a diesel engine.
4. Describe basic four-stroke-cycle and two-stroke-cycle diesel engine operation and compare the two.
5. Describe and chart basic valve and port timing.
6. Describe the relationship between piston and crank pin travel.
7. List eight diesel engine classifications.
8. Define and calculate cylinder bore, piston stroke, displacement, and compression ratios.
9. Define diesel engine speed classifications.
10. List eight diesel engine support systems.

The diesel engine must produce the power needed to drive the vehicle or tractor under varying conditions of speed, load, and environment.

The size of the engine required is determined by the size and type of the vehicle and the load and performance requirements of the unit.

Both two-stroke-cycle and four-stroke-cycle engines are in common use. Multi-cylinder reciprocating piston diesel engines with two, three, four, five, six, eight, or more cylinders are produced.

Regardless of the type of engine being used, the demands placed on the engine to perform are rigorous. Speed and load requirements in many types of service are constantly changing during operation. Seasonal and regional temperature extremes are also encountered. All of these factors are considered in engine design, operation, and service.

The diesel engine is easily recognized by the absence of such components as spark plugs, ignition wires, coil, and distributor, common to gasoline engines.

You will find diesel engines on such units as buses, trucks, tractors, mining equipment, construction equipment, agricultural equipment, railroad locomotives, ships and boats, power-generating units, and pumping and irrigation equipment.

To diagnose engine problems or service engines effectively, it is necessary to have a thorough understanding of the operating principles and construction features of the internal combustion diesel engine.

PART 1   ENERGY CONVERSION

The internal combustion diesel engine is a device used to convert the chemical energy of the fuel into heat energy and then convert this heat energy into usable mechanical energy. This is achieved by combining the appropriate amounts of air and fuel and burning them in an enclosed cylinder at a controlled rate. A movable piston in the cylinder is forced down by the expanding gases of combustion.

The movable piston in the cylinder is connected to the top of a connecting rod. The bottom of the connecting rod is attached to the offset portion of a crankshaft. As the piston is forced down, this force is transferred to the crankshaft, causing the crankshaft to rotate. The reciprocating (back and forth or up and down) movement of the piston is converted to rotary (turning) motion of the crankshaft, which supplies the power to drive the vehicle.

In general an average air-fuel ratio for good combustion is about 15 parts of air to 1 part of fuel by weight. However, the diesel engine always takes in a full charge of air (since there is no throttle plate in most systems), but only a small part of this air is used at low or idle engine speeds. Air consists of about 20 percent oxygen while the remaining 80 percent is mostly nitrogen. This means that, for every gallon of fuel burned, the oxygen in 9,000 to 10,000 gallons of air is required.

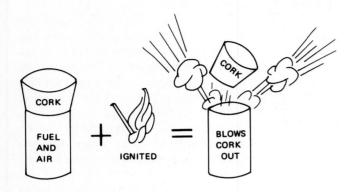

FIGURE 5-1. When fuel and air are burned in a confined area, heat and pressure are created. This process is the conversion of the chemical energy of the fuel and oxygen to heat energy.

A large amount of the heat energy produced by the engine is lost through the exhaust, cooling, and lubrication systems, as well as through radiation, and is therefore not available to produce power. A further loss occurs through the friction of moving engine and drive train parts. Consequently, of the total potential heat energy in the fuel, available power at the drive wheels could be as low as 15 percent.

PART 2   BASIC ENGINE COMPONENTS

The basic single-cylinder engine consists of a cylinder (engine block), a movable piston inside this cylinder, a connecting rod attached at the top end to the piston and at the bottom to the offset portion of a crankshaft, a camshaft to operate the two valves (intake and exhaust), and a cylinder head. A flywheel is attached to one end of the crankshaft. The other end of the crankshaft has a gear to drive the camshaft gear. The camshaft gear is twice as large as the crankshaft gear. This drives the camshaft at half the speed of the crankshaft on four-stroke-cycle engines. On two-stroke-cycle engines, the crankshaft and camshaft run at the same speed. (See Fig. 5-3.)

These are all described in more detail in Chapters 7 to 12.

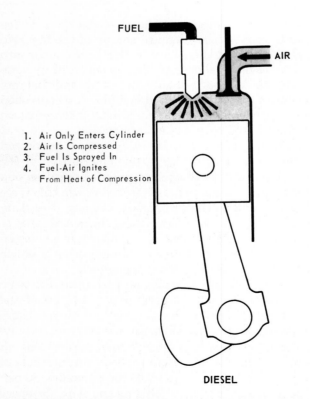

1. Air Only Enters Cylinder
2. Air Is Compressed
3. Fuel Is Sprayed In
4. Fuel-Air Ignites
   From Heat of Compression

FIGURE 5-2. Basic principle of four-stroke diesel engine operation. *(Courtesy of Deere and Company)*

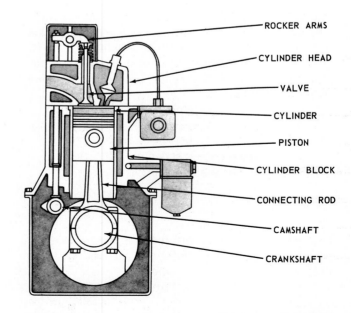

FIGURE 5-3. One of the most common engine block designs showing basic engine components. *(Courtesy of Deere and Company)*

PART 3 FOUR-STROKE-CYCLE ENGINE OPERATION

The movement of the piston from its uppermost position (TDC, top dead center) to its lowest (BDC, bottom dead center) position is called a stroke. Many engines operate on the four-stroke-cycle principle. A series of events involving four strokes of the piston completes one cycle. These events are: (1) the intake stroke, (2) the compression stroke, (3) the power stroke, and (4) the exhaust stroke. Two revolutions of the crankshaft and one revolution of the camshaft are required to complete one cycle. (See Fig. 5–4 to 5–7.)

On the intake stroke the piston is pulled down in the cylinder by the crankshaft and connecting rod. During this time the intake valve is held open by the camshaft. Since the piston has moved down in the cylinder, creating a low-pressure area (vacuum), atmospheric pressure forces air past the intake valve into the cylinder. Atmospheric pressure is approximately 14.7 pounds per square inch (about 101.35 kilopascals) at sea level. Pressure in the cylinder during the intake stroke is considerably less than this. The pressure difference is the force that causes the air to flow into the cylinder, since a liquid or a gas (vapor) will always flow from a high- to a low-pressure area.

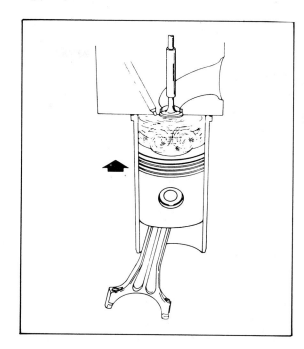

FIGURE 5–5. Compression stroke.

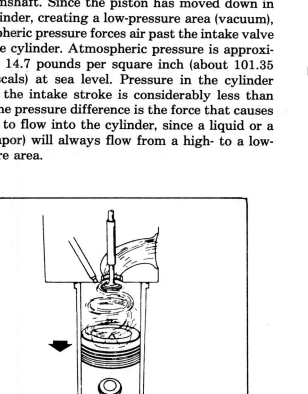

FIGURE 5–4. Intake stroke.

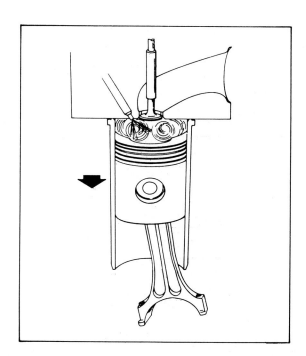

FIGURE 5–6. Power stroke.

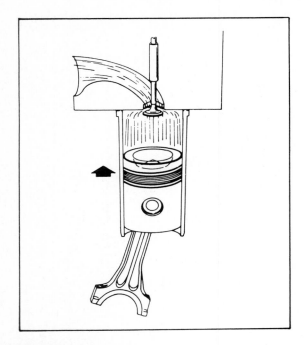

FIGURE 5-7. Exhaust stroke.

As the piston is moved up by the crankshaft from BDC, the intake valve closes. The air is trapped in the cylinder above the piston. Further piston travel compresses the air to approximately 1/20 of its original volume (approximately 20:1 compression ratio) when the piston has reached TDC. This completes the compression stroke. (Compression ratios vary from about 13:1 to about 22:1.)

Compressing the air to this extent creates a great deal of friction between the air molecules, which creates sufficient heat to ignite the fuel when it is injected. This temperature could be in the area of 800°F (426.6°C) to 1200°F (648.8°C). For this reason a spark ignition system is not required on a diesel engine.

The burning of the air-fuel mixture (combustion) occurs at a controlled rate. Expansion of the burning mixture increases pressure. This increased pressure forces the piston down on the power stroke, causing the crankshaft to rotate.

At the end of the power stroke, the camshaft opens the exhaust valve, and the exhaust stroke begins. Remaining pressure in the cylinder and upward movement of the piston force the exhaust gases out of the cylinder. At the end of the exhaust stroke, the exhaust valve closes and the intake valve opens, continually repeating the entire cycle of events.

To start the engine, some method of cranking the engine is required to turn the crankshaft and cause piston movement. This is done by the starting motor when the key is turned to the start position. When sufficient air has entered the cylinders and fuel injected, the power strokes create enough energy to continue the crankshaft rotation. At this point the key is released and the starter is disengaged.

Sufficient energy is stored in the flywheel and other rotating parts on the power strokes to move the pistons and related parts through the other three strokes (exhaust, intake, and compression). The amount of fuel allowed to enter the cylinders will determine the power and speed developed by the engine.

PART 4   TWO-STROKE-CYCLE ENGINE OPERATION

The two-stroke-cycle diesel engine completes all four events (intake, compression, power, and exhaust) in one revolution of the crankshaft or two strokes of the piston. (See Fig. 5-8.)

A series of ports or openings is arranged around the cylinder in such a position that the ports are open when the piston is at the bottom of its stroke. A blower forces air into the cylinder through the open ports, expelling all remaining exhaust gases past the open exhaust valves and filling the cylinder with air. This is called *scavenging*.

As the piston moves up, the exhaust valves close and the piston covers the ports. The air trapped above the piston is compressed since the exhaust valve is closed. Just before the piston reaches top dead center, the required amount of fuel is injected into the cylinder. The heat generated by compressing the air ignites the fuel almost immediately. Combustion continues until the fuel injected has been burned. The pressure resulting from combustion forces the piston downward on the power stroke. When the piston is approximately halfway down, the exhaust valves are opened, allowing the exhaust gases to escape. Further downward movement uncovers the inlet ports, causing fresh air to enter the cylinder and expel the exhaust gases. The entire procedure is then repeated, as the engine continues to run.

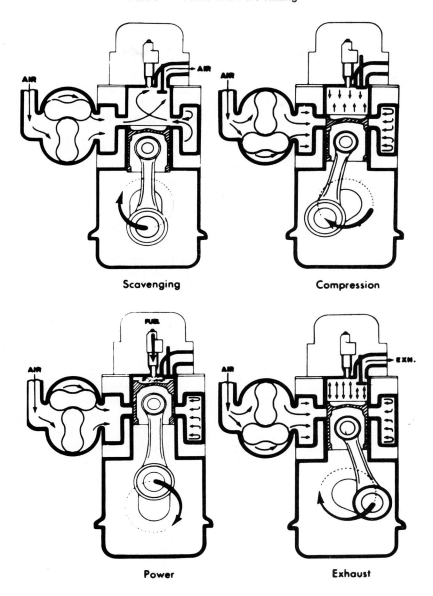

FIGURE 5–8. Two-stroke-cycle engine operation. All four events—intake (scavenging), compression, power, and exhaust—are completed in only two strokes of the piston (one turn of the crankshaft). *(Courtesy of Detroit Diesel Allison Division of General Motors Corporation)*

PART 5   VALVE AND PORT TIMING

Intake Valve Timing, Four-Stroke-Cycle Engine

It is advantageous to have the intake valve begin to open before the piston reaches the TDC position on the exhaust stroke for two reasons. First, this ensures that the valve will be fully open to take advantage of the rapid downward movement of the piston during the first 90 degrees of crank rotation when the greatest pressure drop in the cylinder occurs. Second, it helps to overcome the static inertia of the air in the intake manifold and, as a result, starts air flow into the cylinder sooner.

It also helps engine breathing to keep the intake valve open until the piston has passed the BDC po-

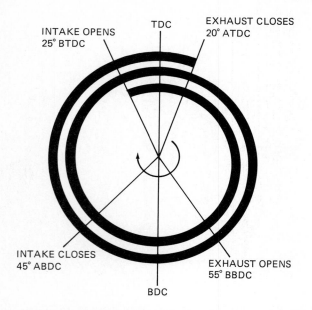

FIGURE 5-9. Typical valve timing diagram for four cylinder engine. Diagram represents two crankshaft revolutions. Stroke duration in degrees of crankshaft rotation: intake—250°; compression—135°; power—125°; exhaust—255°; valve overlap—45°.

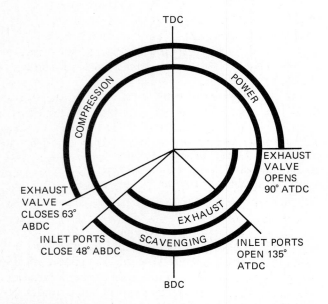

FIGURE 5-10. Typical two-cycle engine valve and port timing. Diagram represents one crankshaft revolution. Duration of events in degrees of crankshaft rotation: power—90°; exhaust—153°; scavenging—93°; compression—117°.

sition. This allows the kinetic inertia of the incoming air to fill the cylinder more completely.

Intake valve timing varies somewhat among the various makes and models of diesel engines. Not uncommon is an intake valve opening at 25 to 30 degrees BTDC and an intake valve closing at 40 to 45 degrees ABDC.

Exhaust Valve Timing

To ensure expulsion of exhaust gases, the exhaust valve opens at about 50 to 55 degrees BBDC. Exhaust gas pressure in the cylinder begins to force the gases past the exhaust valve into the exhaust system. As the piston passes BDC and moves up on the exhaust stroke, it forces the exhaust gases out of the cylinder. As the piston nears the top, the velocity of the escaping exhaust gases creates a pressure in the cylinder that is slightly lower than atmospheric pressure and, since the intake valve opens at about 30 degrees BTDC, a scavenging effect is created. At about 20 to 25 degrees ATDC, the exhaust valve is closed and a new cycle of events begins all over again.

The period when the piston is near the TDC position and the intake and exhaust valves are both open at the same time is known as *valve overlap*. See Chapters 8, 9, and 10 for more information on valves and valve timing.

Intake Port And Exhaust Valve Timing, Two-Stroke-Cycle Engine

As the piston is forced downward on the power stroke to about 90 degrees ATDC, the exhaust valve begins to open. Exhaust gases escape past the open exhaust valve causing pressure in the cylinder to drop. This continues until the piston reaches a point about 130 to 135 degrees ATDC. At this point the inlet ports are uncovered by the downward-moving piston. Air pressurized by the engine blower forces fresh air into the cylinder through the ports. This causes any remaining exhaust gases in the cylinder to be forced out past the open exhaust valve.

During this time the piston continues past the BDC position and starts to move upward. When it reaches a point about 48 degrees ABDC, the inlet ports are closed by the upward moving piston. A little later, about 63 degrees ABDC, the exhaust valve is closed and the compression stroke begins. Just before the piston reaches TDC, about 23 degrees BTDC, fuel is injected into the cylinder, causing ignition and expansion to take place. The cycle then repeats itself.

PART 6   COMPARISON OF TWO-CYCLE AND FOUR-CYCLE ENGINES

It could be assumed that a two-cycle engine with the same number of cylinders, the same displacement, compression ratio, and speed as a four-cycle engine would have twice the power since it has twice as many power strokes. However, this is not the case, since both the power and compression strokes are shortened to allow scavenging to take place. The two-cycle engine also requires a blower, which takes engine power to drive.

About 160 degrees out of each 360 degrees of crankshaft rotation are required for exhaust gas expulsion and fresh air intake (scavenging) in a two-cycle engine. About 415 degrees of each 720 degrees of crankshaft rotation in a four-cycle engine are required for intake and exhaust. These figures indicate that about 44.5% of crank rotation is used for the power producing events in the two-cycle engine, while about 59% of crank rotation is used for these purposes in the four-cycle engine. Friction losses are consequently greater in the four-cycle engine. Heat losses, however, are greater in the two-cycle engine through both the exhaust and the cooling systems. In spite of these differences, both engine types enjoy prominent use worldwide.

PART 7   PISTON AND CRANK PIN TRAVEL

In general, the speed and distance traveled by the crank pin can be considered to be constant and in a uniform path at any given engine speed. This is not the case, however, with piston speed and travel. (See Fig. 5–11.)

When the movement of the crank pin is uniform and at a constant velocity, the speed and distance traveled by the piston connected to it varies due to crank pin and connecting rod angle. When the piston reaches the TDC position, its speed is zero. As it begins to move downward, its speed increases rapidly. At a point where the crank pin is about 63 degrees ATDC, the piston has reached its maximum speed. This is the point at which the crank throw centerline and the connecting rod centerline form an angle of 90 degrees. After this point, the piston speed decreases until it reaches zero at the BDC position. As the piston moves upward from this point, its speed increases until it reaches its maximum at about 63 degrees BTDC. From this point on the pis-

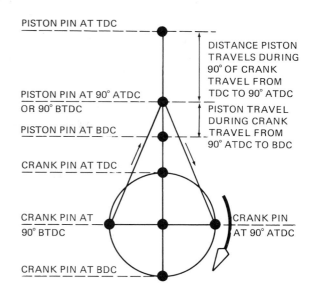

FIGURE 5–11. Piston and crank pin travel diagram. Note that the distance traveled by the piston is greater during the first and last quarter revolution than during the second and third quarter revolution.

ton slows down until it reaches TDC again, where it once more comes to a stop.

Piston speed is normally stated as an average speed in feet per minute (ft/min) and can be calculated as follows: Piston Speed = Stroke (in feet) × RPM ÷ 2. You must divide by two since the piston travels the stroke distance twice (up and down) during each revolution.

It should also be noted that the distance traveled by the piston varies with the crank angle. Starting with the TDC position of the crank pin, the piston travels a greater distance during the first 90 degrees of crank rotation than it does during the second 90 degrees. As the crank pin continues past the BDC position through the third 90 degrees of rotation, the piston travel is less than it is for the final 90 degrees when it reaches the TDC position.

The force of combustion pressure has no rotating effect on the crankshaft when the piston is at the TDC position. As the crank pin passes the TDC position on the power stroke, the mechanical advantage through the angle of the connecting rod increases until it reaches its maximum at about 63 degrees ATDC. This is the point where the crank pin and connecting rod form a 90 degree angle. Thereafter, the force advantage decreases rapidly while combustion pressure also decreases.

PART 8   ENGINE CLASSIFICATION

Diesel engines can be classified in a number of different ways, depending on engine design.

By Cycles. Two-stroke- and four-stroke-cycle engines are being used.

By Cooling Systems. Liquid-cooled engines and air-cooled engines are being used. Liquid-cooled engines are the most common in the diesel industry.

By Fuel System. Gasoline, diesel and propane fuel systems are currently used in a wide variety of engines.

By Ignition Method. Gas engines use the spark (electrical) ignition system. The electrical ignition system causes a spark across the spark plug electrodes in the cylinder at the end of the compression stroke, which ignites the vaporized fuel and air mixture.

Diesel engines use the heat from compressing the air to ignite the fuel when it is injected into the cylinder at the end of the compression stroke. Since diesel engine compression ratios are much higher than gasoline engine compression ratios, sufficient heat is generated by compressing the air to ignite the fuel upon injection.

By Valve Arrangement. Four types of valve arrangements have been used in gasoline and diesel engines as shown in Figure 5-12. Of the four types (L, T, F, and I heads), the I head is commonly used on diesel engines.

By Cylinder Arrangement. Engine block configuration or cylinder arrangement depends on cylinder block design. Cylinders may be arranged in a straight line one behind the other. The most common *in-line* designs are the four- and six-cylinder engines.

The V type of cylinder arrangement uses two *banks* of cylinders arranged in a 60° to 90° V design. The most common examples are those with two banks of three to eight cylinders each. The opposed engine uses two banks of cylinders opposite each other with the crankshaft in between.

By Displacement. Engine displacement is the amount of air displaced by the piston when it moves from BDC to TDC; it varies with cylinder bore size, length of piston stroke, and number of cylinders.

By Engine Speed. Engines are classified as low, medium, high, and super high speed.

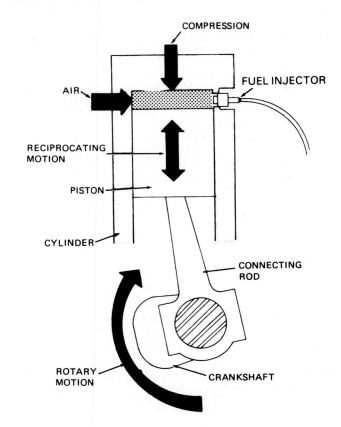

FIGURE 5-12. Pressure from the expanding gases of combustion in the cylinder forces the piston down on the power stroke. Downward movement of the piston is transferred to the crankshaft, causing it to turn (rotary motion). Heat energy is converted to mechanical energy, and reciprocating motion of the piston is converted to rotary motion of the crankshaft.

By Application. Industrial, highway, rail, agricultural, marine, or stationary use.

PART 9   BORE, STROKE, AND DISPLACEMENT

A number of factors determine the ability of an engine to produce usable power. Some of these factors are determined by the manufacturer, such as cylinder bore, stroke, displacement, and compression ratio. Other factors such as force, pressure, vacuum, and atmospheric pressure also affect the power output of an engine. The amount of power an engine is able to produce is measured in several ways, as is the efficiency of an engine. All of these factors,

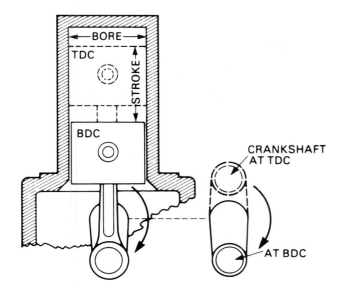

FIGURE 5–13. The distance the piston travels from its highest position in the cylinder (TDC, top dead center) to its lowest position in the cylinder (BDC, bottom dead center) is known as the *stroke.* The length of the stroke is determined by crankshaft design. The bore dimension is the cylinder diameter.

terms, and conditions must be properly understood to gain an understanding of their individual and combined effects on engine performance.

The following definitions are intended to help the student gain competence in the understanding of an engine's operating principles, its ability to produce power, and some of its limitations.

Cylinder Bore

Cylinder bore is the diameter of the engine's cylinder measured in inches or millimeters. Cylinder bore size is a major factor in determining engine displacement.

Piston Stroke

The engine's stroke is the distance traveled by the piston from its bottom dead center position to its top dead center position measured in inches or millimeters. The distance traveled by the piston is determined by crankshaft design and is exactly twice the crank throw measurement.

The engine's stroke is also a major factor in engine displacement.

Displacement

The displacement of an engine is determined by cylinder bore diameter, length of stroke, and number of cylinders. Simply stated, it is the amount or volume of air pushed out of the cylinder (displaced) by one piston as it moves from BDC to TDC, multiplied by the number of cylinders in the engine.

Displacement is calculated as follows:

$$\pi\, r^2 \times \text{stroke} \times \text{no. of cylinders}$$

$$\pi = \frac{22}{7}$$

$$r^2 = \text{radius} \times \text{radius}$$

$$(\text{radius} = 1/2 \text{ of cylinder bore})$$

Therefore, a six-cylinder engine with a 3.800-inch bore and a 3.4-inch stroke would have a displacement of:

$$\frac{22}{7} \times (1.9 \times 1.9) \times (3.4 \times 6) = \frac{22}{7} \times 3.61 \times 20.4$$

$$\frac{22}{7} \times 73.644 = \frac{1620.168}{7} = 231.45 \text{ CID}$$

(cubic inches of displacement)

Another method is to proceed as follows: multiply $0.7854 \times$ stroke $\times$ no. of cylinders. The proportionate area of a circle is 0.7854 of a square with the same dimensions as the circle diameter.

In metric terms, the displacement of a six-cylinder engine with a 100.0-millimeter bore and a stroke of 80 millimeters would be calculated as follows. First, since metric displacement is stated in cubic centimeters, it is necessary to convert the bore and stroke dimensions to centimeters.

100 mm = 10-cm bore   80 mm = 8-cm stroke

$$\frac{22}{7} \times (5 \times 5) \times (8 \times 6) = \frac{22}{7} \times 25 \times 48$$

or

$$\frac{22}{7} \times 1200 = \frac{26,400}{7} = 3771 \text{ cubic centimeters}$$
of displacement or
3.771 liters (L)

The power an engine is able to produce depends very much on its displacement. Engines with more displacement are able to take in a greater amount of air and fuel on each intake stroke and can therefore produce more power. Engine displacement can be increased by engine design in three ways: (1) increasing cylinder bore diameter, (2) lengthening the stroke, and (3) increasing the number of cylinders.

PART 10   COMPRESSION RATIOS

Compression Ratio—Gas Engine

The compression ratio (Fig. 5-14) of an engine is a comparison of the total volume of one cylinder (cylinder displacement plus combustion chamber volume) to combustion chamber volume. To calculate compression ratio, *divide the combustion chamber volume into the total cylinder volume.* For example, if the combustion chamber is 5 cubic inches and the cylinder displacement is 36 cubic inches, the total cylinder volume is 41 cubic inches.

$$41 \div 5 = 8.2:1$$

Metric Example. If combustion chamber volume is 90 cubic centimeters and cylinder displacement is 650 cubic centimeters, the total cylinder volume is 740 cubic centimeters.

$$740 \div 90 = 8.22:1$$

These are typical compression ratios for gasoline engines.

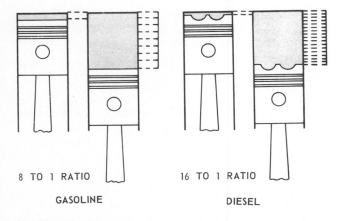

8 TO 1 RATIO
GASOLINE

16 TO 1 RATIO
DIESEL

FIGURE 5-14. Typical gasoline engine and diesel engine compression ratios. *(Courtesy of Deere and Company)*

An engine with an 8.2 compression ratio will produce approximately 150 pounds per square inch (psi) of compression pressure (1034.25 kilopascals, kPa). This rises to approximately 600 psi of combustion pressure (4137 kPa).

The compression ratio of an engine can be changed in several ways. To raise the compression ratio, the combustion chamber volume can be reduced and the cylinder volume left unchanged, or the cylinder volume can be increased (increased bore or stroke) and the combustion chamber volume left unchanged.

To lower the compression ratio, increase combustion chamber volume and leave cylinder volume unchanged or reduce cylinder volume (decrease bore or stroke) and leave combustion chamber volume unchanged.

The approximate maximum compression ratio for gasoline engines is about 14:1. Detonation and serious engine damage are the result of too high a compression ratio in gasoline engines. Combustion chamber design and type of gasoline (octane or antiknock rating) also affect the point at which detonation will occur. Detonation is the ignition of the fuel due to the high temperature caused by the high pressure in the combustion chamber. Fuel may be ignited before the spark occurs at the spark plug, and burning is rapid and uncontrolled. This causes parts to be subjected to excessive heat and stress.

Compression ratios must be high enough to produce adequate pressures and power. Compression ratios in passenger car engines prior to emission-control legislation were at approximately 12:1. This resulted in good power production but caused high exhaust emissions. One method of reducing exhaust emissions includes lowering the compression ratio to reduce combustion temperatures. High combustion temperatures cause excessive nitric oxide exhaust emissions.

Compression Ratio—Diesel Engines

The compression ratio of a diesel engine is much higher than that of a gasoline engine. This is possible since air only is compressed. Compressing air at diesel compression ratios causes air molecules to collide rapidly with each other. The friction caused by these collisions creates heat. Temperatures of 1000° Fahrenheit (540°C) or higher can be reached depending on the compression ratio. This is hot enough to ignite the fuel when it is injected near the top of the compression stroke.

Compression ratios are determined by cylinder displacement and combustion chamber volume. To calculate compression ratio, divide the combustion chamber volume into the total cylinder volume. For example, if the combustion chamber volume is 2 cubic inches and the cylinder displacement is 36 cubic inches, the total cylinder volume is 38 cubic inches.

$$38 \div 2 = 19:1$$

Metric Example. If combustion chamber volume is 36 cubic centimeters and cylinder displacement is 650 cubic centimeters, the total cylinder volume is 686 cubic centimeters.

$$686 \div 36 = 19:1$$

These are typical compression ratios for diesel engines.

An engine with a 19:1 compression ratio will produce approximately 500 pounds per square inch (psi) of compression pressure (3447.5 kilopascals [kPa]). This rises to approximately 2000 psi of combustion pressure (13,790 kPa) when the piston reaches the top dead center position.

The compression ratio of an engine can be changed in several ways. To raise the compression ratio, the combustion chamber volume can be reduced and the cylinder volume left unchanged, or the cylinder volume can be increased (increased bore or stroke) and the combustion chamber volume left unchanged.

To lower the compression ratio, the combustion chamber volume can be increased and the cylinder volume left unchanged, or the cylinder volume can be reduced (decreased bore or stroke) and the combustion chamber volume left unchanged.

Compression ratios in diesel engines must be high enough to create sufficient heat for good ignition and adequate power. Ratios range from about 11:1 to as high as 22:1 in some small high-speed diesel engines.

PART 11   ENGINE SPEED

Diesel engines are often classified according to their speed capability. Low, medium, high, and super high speed classifications are used. Both piston speed and rotating speed have been used as measures of engine speed capability.

Piston speed is expressed in feet per minute (or meters per minute), while rotating speed refers to crankshaft speed expressed in revolutions per min-ute (rpm). Both methods leave something to be desired. Rotating speed alone does not take into consideration the size and piston stroke of an engine. A long-stroke engine with a low rotating speed not exceeding 750 rpm may have quite a high piston speed. A high-speed short-stroke engine, on the other hand, has a relatively low piston speed. For these reasons an engine speed characteristic known as the *speed factor* is used.

The speed factor of an engine is determined as follows: Speed Factor = crankshaft rpm × piston speed (in feet per minute) ÷ 100,000. The division by 100,000 is an arbitrary figure to produce a less cumbersome final figure. This allows all diesel engines to be assigned a speed factor between the figures 1 and 81. This results in the following classifications:

1. Engines with a speed factor of 1 to 3 are designated as low-speed engines.

2. Speed factors of 3 to 9 are medium-speed engines.

3. Speed factors of 9 to 27 are high-speed engines.

4. Speed factors of 27 to 81 are super-high-speed engines.

It should be noted that these classifications have been arbitrarily determined by multiplying each successively higher speed classification by a factor of three times the next lower speed category.

To find the speed factor of a six-cylinder engine with a five-inch bore and a six-inch stroke with a rated speed of 2500 rpm, proceed as follows. First find the piston speed of the engine. We need the stroke distance and the time it takes the piston to travel the stroke distance—to determine the mean piston speed. The piston travels up once and down once during each revolution or two strokes of the piston. The distance traveled by the piston during one revolution is twice the stroke distance. Since piston speed is expressed in feet per minute and piston stroke is expressed in inches, we must convert inches to feet by dividing by 12. From this information we can calculate mean piston speed as follows: Piston Speed = rpm ÷ 2 × stroke ÷ 12. Using the engine figures mentioned earlier, we calculate thus: 2500 ÷ 2 × 6 ÷ 12 = 625 ft/min. To obtain the speed factor, we calculate thus: rpm × piston speed − 100,000 or 2500 × 625 ÷ 100,000 = 15.625.

PART 12   DIESEL ENGINE SYSTEMS

For a diesel engine to operate, a number of systems must perform their functions in an efficient manner. These are as follows:

• The *fuel injection system* provides fuel to the cylinders.

• The *air induction system* provides the air to the cylinders.

• The *compression system* sufficiently compresses the air to provide a controlled rate of combustion and expansion.

• The *exhaust system* provides the means to efficiently dispose of the burned gases.

• The *lubrication system* reduces friction and wear and helps in cooling, sealing, and cleaning.

• The *cooling system* maintains the engine's most efficient operating temperature.

• The *ventilation system* removes harmful crankcase vapors.

• The *emission control systems* control the amount of harmful pollutants emitted into the atmosphere by the engine.

PART 13   SELF-CHECK

1. The internal combustion engine converts _____ energy of the fuel into _____ energy and to usable _____ energy.

2. What is the approximate air/fuel ratio of a diesel engine?

3. List the major components that make up the internal combustion engine.

4. List the four strokes of a four-stroke-cycle engine and describe the basic operation of each.

5. Describe the basic operating principle of a two-stroke-cycle engine.

6. The diesel engine compresses the air/fuel mixture on the compression stroke. True or false.

7. List the 8 engine systems required for engine operation.

8. List ten methods of classifying engine types and sizes.

9. Define engine bore, stroke, and displacement.

10. How is the compression ratio of an engine calculated?

Chapter 6

Engine Performance Factors

Performance Objectives

After thorough study of this chapter, the appropriate training models and manufacturer's manuals, you should be able to do the following:

1. Complete the self-check questions with at least 80 percent accuracy.
2. Define the following terms:
   - force
   - pressure
   - vacuum
   - energy
   - work
   - power
   - friction
   - inertia
3. Describe atmospheric pressure.
4. Define the following types of engine efficiency:
   - mechanical
   - thermal
   - fuel
   - cycle
   - volumetric
   - scavenge
5. Calculate engine power, torque, mechanical efficiency, thermal efficiency, fuel efficiency, and volumetric efficiency.
6. Define engine and cylinder balance.
7. List three types of dynamometers and describe basic hydraulic dynamometer operation.

PART 1   FORCE AND PRESSURE

Force and Force Measurement

A force is a push, pull, or twist that acts on something. The push or pull attempts to change the state of motion of that thing or the at rest state of that thing. In other words, a force is something that can try to stop something that is moving or it can try to keep it moving. It can try to move something that is stopped, or it can try to prevent its movement.

The force of expanding gases in a cylinder causes the piston to move down on the power stroke. The force of a brake pad against a brake disc tries to stop the disc from turning.

Diesel equipment uses the principle of force in a great variety of applications. The degree or magnitude of a force can be measured and is stated in pounds in the English system and in newtons in the metric system. For example, an engine weighing 300 pounds exerts a pulling force of 300 pounds on the hook of a portable engine lift on which it is suspended. The same engine, in metric terms, exerts a pulling force of 1334.4 newtons on that hook.

A twisting force (called *torque*) exerted on a drive shaft by an engine might be in the area of 50 pound feet under certain driving conditions. In metric terms this would be 67.79 newton meters (N m).

Pressure and Pressure Measurement

Pressure can be defined as a force applied over a specific area. For example, 100 pounds of metal resting on a 10 square inch area exerts a pressure of 100 pounds over the 10 square inches of area. The pressure exerted on one square inch is, therefore, 100 pounds divided by 10 square inches, (100 ÷ 10), which is 10 pounds per square inch of pressure.

Compression pressure in the engine's cylinder is measured in pounds per square inch. A typical example could be a compression pressure of 400 pounds per square inch.

In metric terms pressure is stated in kilopascals (kPa). For purposes of comparison, one pound per square inch is equal to 6.895 kPa. A typical compression pressure could be 2700 kPa.

To understand the metric term *kilopascals*, we know that the prefix *kilo* means one thousand. Therefore, the term *kilopascals* means 1000 pascals. A pascal is a force of one newton over an area of one square meter (1 N/m^2).

PART 2   ATMOSPHERIC PRESSURE

The atmosphere is a layer of air surrounding the earth's surface. This layer of air exerts a force against the earth's surface because of the earth's force of gravity. This force or pressure of the atmosphere against the earth's surface is called *atmospheric pressure* (Figure 6–1).

Atmospheric pressure is greatest at sea level, since there is more atmosphere above a given point at sea level than there is at a given point on a high mountain. The air is, therefore, also less dense (air molecules not packed together as tightly) at higher altitudes.

A one square inch column of atmosphere at sea level weighs 14.7 pounds. Atmospheric pressure at sea level is, therefore, 14.7 pounds per square inch. However, at the top of a 10,000-foot-high mountain, a one square inch column of air weighs only 12.2 pounds; therefore, atmospheric pressure is 12.2 pounds per square inch at that altitude. It is important to recognize this fact since the air intake of an engine is affected adversely by increased altitude.

It should also be remembered that the atmosphere consists of approximately 20 percent oxygen and nearly 80 percent nitrogen, with a small percentage of other elements. It is the oxygen in the air that is needed to combine with fuel to support combustion.

The air at sea level is more dense, or more tightly compacted, than it is at higher altitudes. For this reason, an engine is able to produce more power at sea level than at higher altitudes.

The temperature of air also has a bearing on an engine's ability to produce power. When air is heated, it expands and becomes less dense. The engine is not able to take in as much air on the intake stroke because of this and will, therefore, produce less power. Today's engines, however, are equipped with intake air temperature control systems which more closely control inlet air temperatures. Air density is stated as lb/ft^3 or as kg/m^3.

The humidity of the air is the percentage of moisture the air is able to keep in suspension at a given temperature. At 100 percent humidity, the air cannot support any additional moisture. At 50 percent humidity, there is half as much moisture in the air as it is able to support at that temperature. Moisture in the air improves engine performance since it has a cooling effect. Engines do not perform as well in hot dry air.

Atmospheric Pressure Measurement

Atmospheric pressure is measured with a barometer and is expressed in inches or millimeters of mercury (or hg).

A simple barometer is a glass tube closed at one end and open at the other. The tube is filled with mercury. The open end is then held shut while the tube is inverted and the open end is submerged in an open top dish filled with mercury. With the mercury-filled tube fixed in this position, the mercury level in the tube will drop slightly, leaving part of the tube empty at the top. As atmospheric pressure is able to act on the mercury in the open dish, the mercury level in the tube will rise or fall as atmospheric pressure rises or falls. The height of the mercury in the tube above the surface of the mercury in the dish is measured in inches; this is the barometric pressure. At sea level, 14.7 pounds per square inch of atmospheric pressure results in 29.92 inches of mercury in the barometer. In metric terms, barometric pressure at sea level is expressed as 101.35

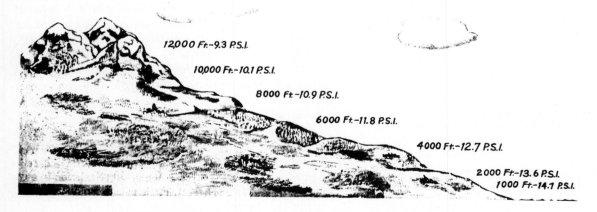

FIGURE 6–1. The effects of altitude on atmospheric pressure.

kilopascals (kPa) since one inch of mercury is equal to 3.38 kPa. Or to put it another way, atmospheric pressure at sea level is 14.7 pounds per square inch, or 101.35 kilopascals, since one pound per square inch of pressure is equal to 6.895 kPa. Another unit of pressure measurement is the bar. One bar is equal to 0.986923 atmosphere.

A kilopascal is 1000 pascals since the prefix kilo represents 1000. A pascal is equal to a newton of force applied over one square meter.

PART 3   MEASURING OTHER PRESSURES

Pressure testing gauges used in the diesel industry register zero at atmospheric pressure. All pressure measurements taken are, therefore, actually pressures above atmospheric pressure, except for vacuum measurements.

Vacuum Measurements

Any pressure which is less than atmospheric pressure is called a vacuum. Actually, it is a partial vacuum.

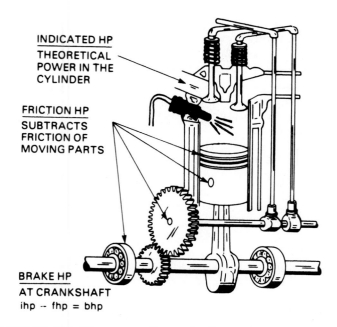

INDICATED HP
THEORETICAL
POWER IN THE
CYLINDER

FRICTION HP
SUBTRACTS
FRICTION OF
MOVING PARTS

BRAKE HP
AT CRANKSHAFT
ihp -- fhp = bhp

FIGURE 6-2. The power theoretically available in an engine is reduced by the friction of moving parts. Actual usable power at the crankshaft is calculated as follows: IHP - FHP = BHP.

Vacuum is measured as a difference in pressure. The difference in pressure between atmospheric pressure and the pressure being measured (intake manifold vacuum for instance) is expressed in inches of water or millimeters of water. This is so since the measurement is taken with a "U" tube water manometer.

A "U" tube manometer is a measuring device that indicates a pressure difference by the difference in the height of fluid in the two columns of the "U" tube. A mercury manometer is used to measure a relatively large pressure difference from atmospheric pressure. A water manometer is used to measure minor differences, and it can measure more accurately since water is lighter than mercury. Manometers are used to measure such items as intake manifold pressure (or vacuum), exhaust back pressure, crankcase pressure, and boost pressure (turbocharger output pressure).

Dial type vacuum gauges are commonly used to measure intake manifold vacuum. These gauges are calibrated to indicate inches or millimeters of water on the dial and can be read directly.

PART 4   ENERGY

Energy can be defined as the potential or ability to do work. The fuel in the tank has the potential for doing work if it is placed in the engine's cylinders, combined with the heat, compressed air, and ignited. When these conditions are met, the potential energy of the raw fuel becomes the kinetic energy of rapidly expanding gases caused by combustion.

This kinetic energy of the burning fuel forces the piston downward in the cylinder, which results in usable crankshaft rotation.

From this we can see that there are two basic forms of energy—potential energy, which does not result in any action or motion until proper conditions are met, and kinetic energy, which is the ability to do work because of motion.

Another example of kinetic energy is the energy of a vehicle moving at road speed. The vehicle continues to move even after the source of power (engine) is disconnected from the drive train.

The vehicle continues to move because of the kinetic energy stored in the moving vehicle. When the brakes are applied, the kinetic energy of the moving vehicle is converted, by friction, to heat energy.

PART 5   WORK

To do work requires energy. Work is said to be done when an applied force overcomes a resistance and moves through a distance. Work produces measurable results. When sufficient energy is expended through an application of force (push, pull, or twist) to overcome the resistance to motion of any particular object, movement is the result.

Pulling a one pound object a distance of one foot results in one foot pound of work being done (if friction is ignored). In other words, force times distance equals work ($W = F \times D$).

Units of measurement for work in the English system include foot pounds and inch pounds—both being the most commonly used in the diesel industry.

In metric terms, the formula $W = F \times D$ also applies. The unit of force measurement is the newton (N), and the unit of distance measurement is the meter (m). Therefore, when a force of one newton is required to move an object a distance of one meter, the unit measurement of work done is one newton meter (1 N m), which is equal to one joule (J). The relationship of the joule to the kilowatt is discussed under engine power.

PART 6   ENGINE POWER

Power is the rate at which work is being done. Power can also be defined as the ability to do a specific amount of work in a specific amount of time.

Engine power (Figure 6-3) in the English system is stated in horsepower and in the metric system in kilowatts. Both systems are explained here. The formula for calculating power is $P = F \times D \div T$, where F = force, D = distance, and T = time.

Engine Power—Brake Power

A man named Watt, observing the ability of a horse to do work in a mine, decided arbitrarily that this ability to do work was the equivalent of raising 33,000 pounds of coal a distance of one foot in one minute. This became the standard measurement of a unit of power called horsepower (HP).

This can be expressed as a formula.

$$1\text{HP} = 33{,}000 \text{ pounds} \times 1 \text{ foot} \times 1 \text{ minute}$$

This formula allows the horsepower of an engine to be calculated if certain factors are known. These

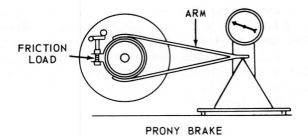

FIGURE 6-3. The prony brake uses a friction device to apply a load to an engine. An arm connected to the friction device deflects in proportion to the load applied and actuates the gauge, which indicates load in pounds or kilograms. Engine speed and load figures are then used to calculate brake horsepower.

factors are: the force produced by an engine and the distance through which that force moves in one minute.

A device known as the prony brake can be used to obtain these factors. Since the prony brake is a braking device, the output of an engine is stated in terms of brake power.

To determine the brake horsepower of an engine, we need to calculate the engine's force times the distance through which that force travels in one minute, and divide Watt's formula for one horsepower into the result of that calculation.

A prony brake uses a drum attached to the engine flywheel. A contracting brake band surrounds the drum. The band can be tightened to increase the load on the engine. An arm is attached at one end to the band. The other end of the arm is connected to a scale through a knife edge device. This assures accuracy of arm length from the center of crankshaft rotation to the scale.

With the engine running, the band is slowly tightened. This causes the arm to exert pressure on the scale. The brake horsepower output of an engine can be calculated using Watt's formula for one horsepower and the prony brake test results. Simply divide Watt's formula into the prony brake test results.

Let's calculate the brake horsepower of a theoretical engine, assuming the following conditions.

• Engine speed: 2000 revolutions per minute (rpm).

• Arm length: 3 feet (radius of circle arm would make if allowed to turn)

• Reading on scale: 100 pounds

Don't forget the formula: F × D × T divided by 33,000 pounds × 1 ft. × 1 minute = Brake Horsepower (BHP).

And, don't forget that, to determine the circumference of a circle, we calculate 2π × radius or $2\pi r$, and we know that π is 22/7.

Therefore:

$$F = 100 \text{ pounds}$$
$$D = 3 \text{ ft.} \times 2\pi \times 2000$$
$$T = 1 \text{ minute}$$

Using this information produces this formula.

$$\frac{2\pi \times 3 \times 2000 \times 100 \times 1}{33,000 \times 1 \times 1} =$$

$$\frac{3 \times 2000 \times 100 \times 1}{5250} = \frac{600,000}{5250} = 114.28 \text{ BHP}$$

Dividing 2π into 33,000 gives us the denominator of 5,250. Multiplying the remaining numerator figures results in 600,000. One horsepower is equal to 0.746 kilowatts (kW). Therefore, 114.28 BHP = 85.25 kW.

Indicated Power

Indicated power is the theoretical power an engine is able to produce. It is calculated by using the following factors:

• P = mean effective pressure in the cylinder in pounds per square inch
• L = length of piston stroke in feet
• A = area of cylinder cross section in square inches
• N = number of power strokes per minute for one cylinder
• K = number of cylinders in the engine.

The formula for calculating indicated horsepower is, therefore,

$$\frac{\text{PLANK}}{33,000} = \text{IP}$$

Using this formula, it is possible to calculate the IP of an engine if the number of cylinders in the engine, the engine's bore and stroke, the engine's speed, and the mean effective pressure in the cylinder are known.

Friction Power

Friction power is the power required to overcome the friction of the various moving parts of the engine as it runs. Friction power increases as engine size and speed are increased.

The friction power of an engine can be calculated (if the indicated power and brake power are known) by subtracting the BP from the IP (IP − BP = FP).

SAE Power

SAE power is the power of an engine as determined by the Society of Automotive Engineers. Tests are performed under rigorously controlled conditions including the inlet air temperature, ambient temperature, humidity, and the like. A number of specific conditions, such as inlet air restriction and exhaust restriction are also stated, since these are determining factors. Other factors and conditions must also be met.

Aside from all these factors and conditions, SAE power is measured at the transmission output shaft, with all normal engine accessories mounted and operating. This includes the air cleaner and exhaust system.

Since a particular engine model may be used by a vehicle manufacturer for several different applications and may be equipped differently on different models, the SAE power for a particular engine varies depending on how it is equipped.

Engine Power—Kilowatts (kW)

In the metric system engine power is stated in kilowatts.

The power output of an engine is calculated, as stated earlier, by using the formula P = W ÷ T, where P = power, W = work, and T = time. We also know that work represents force × distance.

Force in the metric system is measured in newtons or N. Distance, for our purposes here, is measured in meters or m. Therefore, work can be expressed in terms of newton meters (Nm).

Time is stated in minutes. To determine the power of an engine, we calculate force times distance divided by time (F × D ÷ T).

The electrical unit for measuring work is the joule. One joule is the equivalent of one ampere of current under one volt of pressure for one second. One joule is also the equivalent of one newton of force moving one meter of distance in one second.

The watt is the unit of electrical power and is the equivalent of one joule per second. One kilowatt is 1000 watts.

To summarize:

$$1 \text{ N m} = 1 \text{ joule}$$
$$1 \text{ joule per second} = 1 \text{ watt (W)}$$
$$1000 \text{ watts} = 1 \text{ kilowatt (kW)}$$

It would be possible to determine the power output in kilowatts of an engine by using a prony brake described earlier.

If, for example, the prony brake had a torque arm length of one meter, and the scale upon which it acted measured the applied force in newtons, the resultant power would be stated in newton meters. The newton meter output, at a given engine speed measured in revolutions per minute, could then be used to calculate engine power in kilowatts.

Assuming an engine speed of 2000 rpm, we would obtain the engine speed per second by dividing 2000 by 60, since kilowatts is joules per second.

$$2000 \div 60 = 33.33$$

Assuming further that the applied force at the end of the torque arm is 1000 newtons and the length of the torque arm is one meter, we calculate as follows:

$$2\pi \times 1 \times 33.33 = 209.5 \text{ kW}$$
or
$$\frac{2000}{60} \times 2\pi \times 1 = 209.5 \text{ kW}$$
or
$$2\pi \times 1 \times \frac{2000}{60} = 209.5 \text{ kW}$$

One kilowatt is 1.341 horsepower. Therefore, 209.5 kW is equal to 280 horsepower.

Friction

A certain amount of force (push or pull) is required to slide one object over the surface of another. This resistance to motion between two objects in contact with each other is called friction. Friction increases with load. It requires more effort to slide a heavy object across a surface than it does to slide a lighter object over the same surface.

The condition of the two surfaces in contact also affects the degree of friction. Smooth surfaces produce less friction than rough surfaces. Dry surfaces cause more friction than surfaces that are lubricated or wet.

Residual oil clinging to the cylinder walls, rings, and pistons of an engine that has been stopped for some time will produce a greasy friction when the engine is started. Of course, as soon as the engine starts, the lubrication system supplies increased lubrication, which results in viscous friction.

Viscous comes from the word *viscosity*, which is a measure of an oil's ability to flow or its resistance to flow. Some energy is still required to slide a well-lubricated object over the surface of another. Although the layer of lubricant separates the two surfaces, the lubricant itself provides some resistance to motion. This is called viscous friction. Friction bearings provide a sliding friction action, while ball and roller bearings provide rolling friction. Rolling friction offers less resistance to motion than sliding friction.

Inertia

Inertia is the tendency of an object in motion to stay in motion or the tendency of an object at rest to stay at rest. The first can be called *kinetic inertia* and the latter *static inertia*. The moving parts of an engine are affected by kinetic inertia. A piston moving in one direction tries to keep moving in that direction because of kinetic inertia. The crankshaft and connecting rod must overcome this kinetic inertia by stopping the piston at its travel limit and reversing its direction. The static inertia of a vehicle that is stopped must be overcome by engine power to cause it to move.

PART 7   ENGINE TORQUE AND BRAKE POWER

As an engines runs, the crankshaft is forced to turn by the series of pushes or power impulses imposed on the crank pins by the pistons and rods. This twisting force is called *torque*. (Fig. 6–4 and 6–5).

Engine torque and engine power are closely related. For instance, as we learned earlier, if we know the torque and speed of an engine, we can calculate its power.

Torque is equal to $F \times R$ where F is the force applied to the end of a lever and R is the length of the lever from the center of the turning shaft to the point on the lever at which force is being applied. R represents the radius of a circle through which the

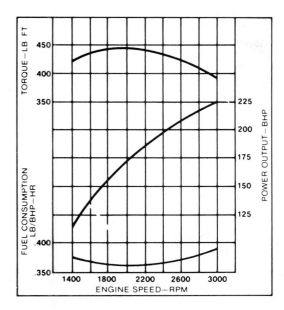

FIGURE 6-4. This chart shows the relationship between torque in lb.-ft, brake horsepower, and fuel consumption over the operating speed range of one typical diesel engine. *(Courtesy of Cummins Engine Company Inc.)*

applied force would move, if moved through a complete revolution. Therefore, $T = F \times R$.

Engine torque is expressed in terms of pound feet in the English system and in newton meters in the metric system.

Maximum torque is produced in an engine when there is maximum pressure in the cylinders. Peak torque is, therefore, reached when there is maximum air and fuel delivery to the engine. This normally occurs at a somewhat lower engine speed than that at which maximum brake power is produced.

Engine torque drops off as engine speed increases to the point where cylinders take in less air. At this point, engine brake power is still increasing due to the increased number of power impulses per minute. Engine brake power drops off when the effect of an increased number of power impulses is offset by the reduced air intake of the cylinders.

As can be seen from this, engine torque and engine brake power are closely related to the volumetric efficiency of the engine.

PART 8   ENGINE EFFICIENCY

The degree of engine efficiency is expressed in percentage figures resulting from a comparison of the

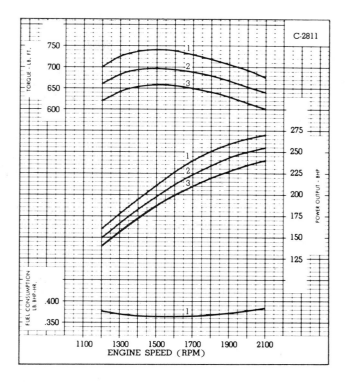

FIGURE 6-5. This chart shows an example of custom rating choices of one particular engine to suit torque and power requirements for particular applications. Rate of fuel delivery and engine governed speeds are tailored to provide desired performance. *(Courtesy of Cummins Engine Company Inc.)*

theoretical power of an engine without any power losses, to the actual power available from the engine.

Mechanical efficiency and thermal efficiency are two ways used to express engine efficiency.

Mechanical Efficiency

Indicated power is the theoretical power an engine is able to produce as discussed earlier. It is expressed as horsepower or as kilowatts. Brake power is the actual power delivered by an engine expressed in horsepower or kilowatts.

The formula for calculating the mechanical efficiency of an engine compares the brake power to the indicated power and is calculated by dividing IP into BP.

$$BP \div IP = \text{mechanical efficiency}$$

For example, an engine that produces 72 brake horsepower and has an indicated power of 90 hp, would have a mechanical efficiency of 72/90 or 80 percent. This means that, although 90 BHP is produced in the cylinders, only 72 BHP is being delivered at the flywheel.

Or, using the metric units, if an engine has an indicated power of 120 kW but delivers only 102 kW, the mechanical efficiency would be 102/120 or 85 percent.

Thermal Efficiency

The thermal efficiency of an engine is the degree to which the engine is successful in converting the energy of the fuel into usable heat energy or power. It is the heat energy in the cylinder that forces the pistons to move which results in crankshaft rotation.

Volumetric Efficiency

The amount of air an engine is able to take into the cylinder on the intake stroke, compared to filling the cylinder completely with air at atmospheric pressure, is known as the volumetric efficiency of an engine.

Another way to describe it would be to say that volumetric efficiency is the engine's ability to get rid of exhaust gases and take in the air as compared to the displacement of the engine.

The engine is not able to take in a 100 percent fill on each intake stroke because of design limitations. Such factors as valve and port diameters, manifold runner configuration, valve timing, engine speed, and atmospheric pressure all affect volumetric efficiency.

An engine running at 3000 rpm will have only half the time to fill the cylinder on each intake stroke as it would have at 1500 rpm. Since this is the case, volumetric efficiency drops as engine speed increases. As a result, engine torque also decreases (when engine speed exceeds a certain range).

An engine operating in an area that is 5000 feet above sea level will have less volumetric efficiency than the same engine at sea level because atmospheric pressure is lower at 5000 feet above sea level than at sea level. Since it is atmospheric pressure that forces the air into the cylinder, it is easy to see that there will be a corresponding decrease in volumetric efficiency as the altitude (at which an engine operates) increases.

To achieve 100 percent volumetric efficiency in an engine, it is necessary to use a blower to force the air into the cylinders. A turbocharger or a supercharger can be used for this. The turbocharger uses exhaust gas pressure to drive a turbine, and the turbine drives the blower. A supercharger uses a mechanically driven blower to do the job. Turbochargers and superchargers must have precise controls to prevent overcharging of the cylinders and consequent engine damage. Turbochargers are described later in this text.

Scavenge Efficiency

The term *scavenge efficiency* is used to describe how thoroughly the burned gases are removed from the cylinder and how well it is filled with fresh air. The term is used for two-stroke-cycle engines, whereas the term *volumetric efficiency* is used for four-stroke-cycle engines.

It is desirable that the exhaust gases be removed as thoroughly as possible and that the cylinder be as completely filled with fresh air as possible.

Scavenge efficiency depends primarily on the precise location of the inlet ports, the design and arrangement of the exhaust valves and ports, and the efficiency of the intake and exhaust systems. Scavenge efficiency is reduced when intake or exhaust systems are restricted.

Fuel Efficiency

Fuel efficiency is actually the rate of fuel consumption over a distance travelled for highway vehicles or pounds of fuel consumed per brake horsepower hour. Fuel efficiency is dependent on all the foregoing factors as well as vehicle weight, size, and load.

Diesel engine fuel efficiency is considerably better than that of the gasoline engine for several reasons. Diesel fuel has a higher heat value than gasoline. Also the compression ratio of the diesel engine is considerably higher than that of the gasoline engine. See Chapter 19 for more on diesel fuels.

Cycle Efficiency

Cycle efficiency is equal to engine output divided by fuel input, where output is expressed in units of power and input is expressed in units of heat value of the fuel consumed.

The diesel cycle is considerably more efficient than the gasoline engine Otto cycle. Since diesel fuel has a higher heat value than gasoline and since diesel engine compression ratios are also higher, the combustion temperature of the diesel is considerably higher.

PART 9   ENGINE BALANCE

Balance of rotating engine parts is achieved by equal distribution of mass radially around the axis of the rotating part. However the reciprocating masses such as the pistons and connecting rods create an unbalanced condition that must be compensated for in engine design. (See Fig. 6-6 and 6-7.)

These unbalanced forces are classified as primary and secondary forces. Primary unbalanced forces are equal in frequency to engine speed, whereas secondary force frequencies are at twice the engine speed. The primary forces are of greater consequence. The arrangement of the pistons and rods on the crankshaft determine the effect of these forces on the engine. These forces, known as *unbalanced couples*, tend to move the ends of the engine in an elliptical path.

Balance weights attached to the outer ends of camshafts, special weighted balance shafts, acces-

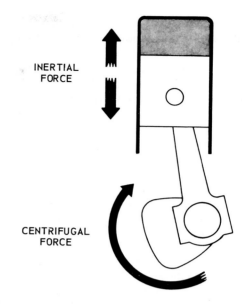

FIGURE 6-6. Unbalancing engine forces. *(Courtesy of Deere and Company)*

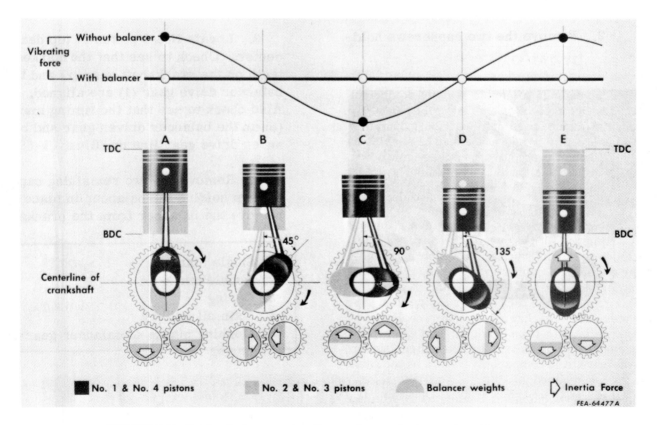

FIGURE 6-7. Engine balancer operation on four cylinder engine. *(Courtesy of Deere and Company)*

sory gears, pulleys, or special balance gears are methods used to cancel the effects of unbalanced couples. The number, size, and position of the weights are designed to produce a couple that is equal in magnitude and opposite in direction to the primary couple. This produces forces of the same magnitude and frequency, thereby canceling the effect of the unbalanced couple.

PART 10   CYLINDER BALANCE

The efficiency and smooth operation of an engine depend to a great degree on the balance of power output between all engine cylinders. Any conditions that result in reduced power output of one or more cylinders as compared to the remaining cylinders results in cylinder imbalance. The resulting uneven power impulses transmitted to the crankshaft create torsional vibrations as well as overall reduced power and torque.

Factors contributing to cylinder imbalance include incorrect valve adjustment, incorrect injector timing, unequal injector output, worn piston rings, leaking valves, and the like.

PART 11   DYNAMOMETERS

The dynamometer is a reliable piece of equipment used to test all aspects of engine performance. By varying the engine speed and load, most operating conditions except for the weather can be simulated. The dynamometer is used to determine engine power output as well as to allow testing of engine systems under various speed and load conditions. (See Fig. 6-8 and 6-9.)

1. *Engine Dynamometer:* Used in a room especially equipped for the purpose in the diesel shop. It includes the dynamometer, provision for mounting the engine and connecting it to the dynamometer, an exhaust system, batteries for the starting and charging systems, a clean air supply, a fuel supply system, a heat exchanger for engine cooling, and the appropriate test instruments and gauges to monitor engine performance. In this system friction from the drive train components is not a factor, since they are not involved.

2. *P.T.O. Dynamometer:* Usually a two-wheeled portable unit used to test the performance of power take-off-equipped tractors in agriculture and industry. The unit is hitched to the tractor and the power

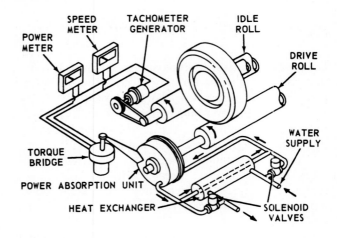

FIGURE 6-8. Major components of dynamometer used to test the road horsepower of a vehicle. Road horsepower is the available horsepower at the drive wheels.

FIGURE 6-9. Testing engine power with a power take off type dynamometer. *(Courtesy of Deere and Company)*

shaft is connected to the tractor power take-off shaft. Test gauges are mounted in a panel on the unit. A water supply is needed for heat removal from the dynamometer.

3. *Chassis Dynamometer:* Equipped with floor-mounted drive rollers to accommodate highway tractor single- or tandem-drive axles. This unit is also usually installed in a special room for the purpose. No engine support systems are required with this unit, since only mobile units are tested. A set of test instruments and gauges are used to indicate vehicle performance. In this system drive train friction is a factor and output is called *road horsepower*.

Steep grades, level roads, stop-and-go city driving, acceleration and deceleration, and a wide range of load conditions all can be simulated on a chassis dynamometer in the shop. With the proper diagnostic equipment connected to the vehicle during these tests, engine condition and performance can be accurately determined in a very few minutes. This kind of diagnosis and testing cannot be done on the road nor can it be done in the shop without a dynamometer. In addition to the test results obtained from the auxiliary test equipment, the dynamometer indicates vehicle or engine speed and power.

A simple type of test equipment called a *prony brake* can be used to measure engine brake power. This device uses a slipping friction type of brake with an arm attached to it. The other end of the arm is attached to a scale. If the length of this arm is 4 feet from the center of the brake drum shaft to the point at which the scale is attached, and the brake is applied to provide a reading on the scale of 30 pounds, the torque would be: 30 pounds × 4 feet = 120 pound-feet of torque.

If engine speed under these conditions is 1500 revolutions per minute, the brake horsepower can be calculated as follows: torque (in lb.-ft) × rpm ÷ 5250 = BHP or (120 × 1500) ÷ 5250 = 34.28 brake power. This can be converted to kilowatts (kW) by multiplying 34.28 × 0.746 = 25.57 kilowatts.

Most dynamometers, whether of the engine type or the chassis type, convert the torque and speed factors automatically to a brake power or road power reading on a dial.

The engine converts the heat energy of combustion to mechanical energy to drive a drive shaft or drive wheels. The drive shaft or drive wheels transfer this mechanical energy to the dynamometer by means of a shaft in the case of an engine dynamometer, and by means of rollers mounted in the shop floor in the case of a chassis dynamometer. This mechanical energy is transmitted to the dynamometer's power absorption unit, which converts the mechanical energy back to heat energy.

The power absorption unit, the torque bridge, and the connecting arm serve the same function as the prony brake. The torque bridge, however, converts the applied force to an electrical signal, which varies with the amount of force applied. This electrical signal provides a reading on the brake horsepower dial. The dynamometer also measures speed, which is indicated on a second dial.

Chassis, P.T.O. and engine dynamometers used in service shops are generally of the hydraulic type. The power absorption unit consists basically of two units: a drive unit and a driven unit. The drive unit is a drum with vanes attached to it internally. The driven unit is also a drum with vanes attached to its interior. The driven unit has an arm attached to it. The other end of the arm is connected to the torque bridge. The drive unit and driven unit are enclosed in a sealed housing, which can be filled and emptied of fluid.

The fluid used in some dynamometers is water; others may use oil. In either case, the amount of load applied is in direct proportion to the amount of fluid permitted to enter the power absorption unit. This is controlled by electrically operated solenoid valves. The solenoids are operated by a hand-held control device.

As fluid is allowed to enter the power absorption unit, the rotating drive member throws the fluid against the driven member, which is held by the connecting arm. As more fluid is allowed to enter the unit (more load applied), the force against the driven member is increased, causing the arm to move slightly. This arm movement is converted to an electrical signal by the torque bridge, which then is indicated on the dial in brake horsepower. In this manner any combination of vehicle or engine speed and load can be observed.

As a load is applied to the dynamometer, the fluid in the power absorption unit heats up. This heat must be dissipated in order to prevent overheating. In the open water-type dynamometers, the absorption unit is connected to a cool-water pressure source that is constant (city main water) in order to keep the unit cool and dissipate the heat. The heated water is directed to the floor drain.

In the closed hydraulic-type of absorption unit, the oil is circulated through a heat exchanger, which is water cooled. This type requires less water.

Many dynamometers are equipped with an inertia flywheel, usually belt-driven from the rollers in the floor. The inertia flywheel can be used to simulate vehicle inertia during acceleration, deceleration, and coasting modes. It is useful in diagnosing engine and drive train problems. The flywheel can be engaged or disengaged by a manually operated lever.

When doing any type of dynamometer testing, make sure that the engine or vehicle is in a safe condition to be tested. Serious damage to the engine or vehicle can result from improper testing methods.

Be sure to follow all procedures recommended by the vehicle manufacturer and by the manufacturer of the equipment (dynamometer) being used.

It must be remembered that all test results observed during dynamometer testing are valid only for the conditions that existed at the time of the test, including engine and vehicle condition.

PART 12   SELF-CHECK

1. Describe atmospheric pressure.
2. What is a vacuum?
3. Define energy and work.
4. What is the formula for calculating power?
5. Engine power is measured in units of _____ or _____.
6. How would you calculate engine brake power when indicated power and friction power are known?
7. What is kinetic inertia?
8. Define engine torque.
9. What is the mechanical efficiency of an engine with 72 BHP and 90 IHP?
10. Which engine has the higher thermal efficiency, the gasoline engine or the diesel engine?
11. Volumetric efficiency increases with engine speed. True or false.
12. What are the advantages of using a dynamometer for engine testing?

Chapter 7

Engine Block Components

Performance Objectives

After adequate study of this chapter and the appropriate engine components you should be able to do the following:

1. Complete the self-check questions with at least 80 percent accuracy.

2. State the purposes of engine components (a) to (c) below.

3. Describe the construction and design differences of engine components (a) to (c) below.

4. Describe the operation, wear, and deterioration of engine components (a) to (c) below.

(a) cylinder block and all its components

(b) crankshaft assembly and balance shafts

(c) piston and connecting rod assembly

The basic construction of diesel engines is generally quite similar except for the opposed piston types. The major components of a diesel engine (excluding accessories) may be divided into two groups. One group includes all the parts that do not involve any motion with respect to engine operation; these are called stationary engine parts. The other group includes all the parts that move in order for the engine to operate; these are called the moving parts of an engine. The major function of the stationary parts is to maintain all the moving parts in proper relative position to each other. The job of the moving parts is to produce power and to maintain that power-producing capability.

PART 1 CYLINDER BLOCK CONSTRUCTION

The cylinder block is the main supporting structure to which all other engine parts are directly or indirectly attached. Two basic types of block construction are used: the en-bloc and the fabricated, welded steel types. (See Fig. 7-1 to 7-4.)

Most large diesel engines as used in large power-generating plants and large marine applications are fabricated by welding steel components together to form the block. Reinforcing plates are welded in place at points of heavy load or stress. Deck plates are designed to hold the cylinder liners in place. These plates are welded to the uprights to form a single strong unit.

Other diesel engines use the en-block method of construction. In this design the block is manufactured as a one-piece cast iron alloy unit.

The block is formed by pouring the molten metal into a mold with a sand-based core. When the metal has cooled and hardened, the mold and sand core are removed. Removing the sand core leaves openings for cylinders, water jackets, crankcase, and bearing bores. Rough holes in the side of the block through which the sand core was supported are machined and closed with soft metal plugs. These core hole plugs are also known as expansion plugs or frost plugs. They are either of the dished type or the cup type.

Oil galleries are either cast in the block or drilled and plugged during the machining process.

Oil galleries are required to provide lubrication for bearings and cam followers as well as other engine parts. Two-cycle engines have air passages, called *air boxes*, in the block to provide air for induction.

Block mating surfaces are carefully machined to provide good sealing surfaces for attaching cylinder heads, gear covers, oil pans, and flywheel housings. Main bearing caps are installed and align bored, as are the camshaft bearing bores. This ensures that the shafts can rotate freely without any binding.

Particular attention is paid to such items as the block deck surface, cylinder liner counterbores, and crankshaft centerline to deck surface dimensions. The cylinder centerline must also be at exactly 90 degrees to the crankshaft centerline to ensure that pistons can move freely without creating side stresses on the pistons, rods, and cylinders.

Cam follower bores are machined at right angles to the camshaft bearing bores. Separate cam follower housings bolted to the block are used in many engines. Holes are drilled and threaded to allow parts to be attached.

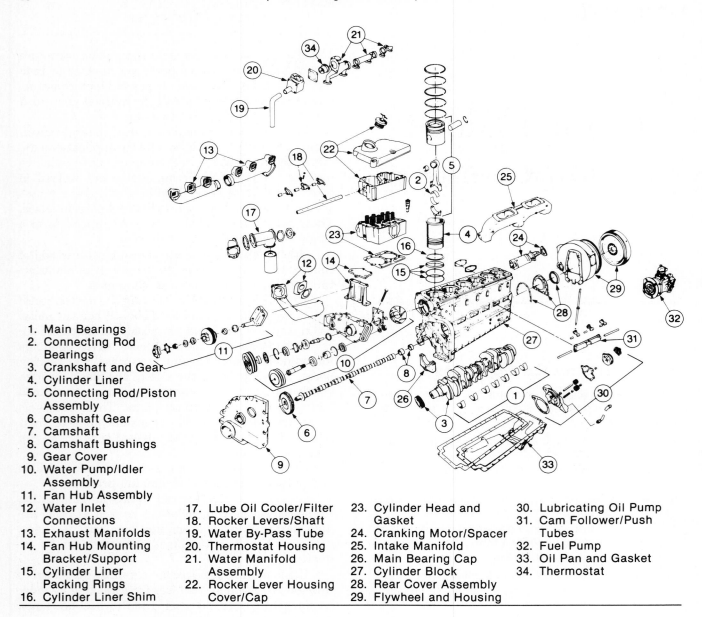

1. Main Bearings
2. Connecting Rod
   Bearings
3. Crankshaft and Gear
4. Cylinder Liner
5. Connecting Rod/Piston
   Assembly
6. Camshaft Gear
7. Camshaft
8. Camshaft Bushings
9. Gear Cover
10. Water Pump/Idler
    Assembly
11. Fan Hub Assembly
12. Water Inlet
    Connections
13. Exhaust Manifolds
14. Fan Hub Mounting
    Bracket/Support
15. Cylinder Liner
    Packing Rings
16. Cylinder Liner Shim

17. Lube Oil Cooler/Filter
18. Rocker Levers/Shaft
19. Water By-Pass Tube
20. Thermostat Housing
21. Water Manifold
    Assembly
22. Rocker Lever Housing
    Cover/Cap

23. Cylinder Head and
    Gasket
24. Cranking Motor/Spacer
25. Intake Manifold
26. Main Bearing Cap
27. Cylinder Block
28. Rear Cover Assembly
29. Flywheel and Housing

30. Lubricating Oil Pump
31. Cam Follower/Push
    Tubes
32. Fuel Pump
33. Oil Pan and Gasket
34. Thermostat

FIGURE 7-1. Engine components of Cummins NH855 engine. *(Courtesy of Cummins Engine Company Inc.)*

Some engines use flat steel end plates bolted to the ends of the block. Gaskets are used to prevent fluid leakage. Gears and other accessories are mounted to these plates.

Many engines are provided with inspection holes in the sides of the block. This allows inspection of internal parts such as cylinder liners, bearings, pistons, rods, and the like. These inspection holes are closed with a gasketed cover plate. Some engines have a safety feature built into the inspection plate. A spring-loaded cover keeps the opening sealed during normal operation. In the event of an explosion in the crankcase, the explosive force opens the cover against spring pressure. This relieves pressure in the crankcase and prevents more serious damage to the engine.

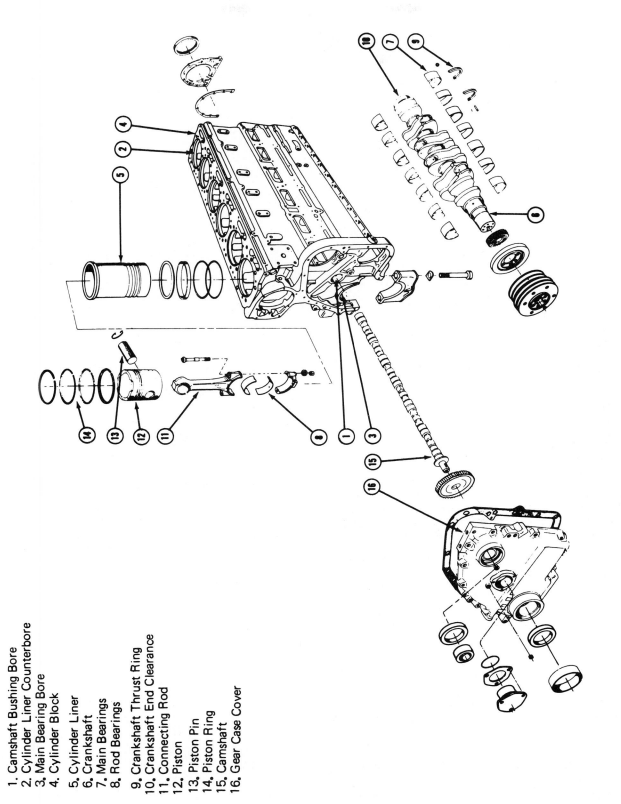

1. Camshaft Bushing Bore
2. Cylinder Liner Counterbore
3. Main Bearing Bore
4. Cylinder Block

5. Cylinder Liner
6. Crankshaft
7. Main Bearings
8. Rod Bearings

9. Crankshaft Thrust Ring
10. Crankshaft End Clearance
11. Connecting Rod
12. Piston
13. Piston Pin
14. Piston Ring
15. Camshaft
16. Gear Case Cover

FIGURE 7-2. Major components of the engine block assembly. (*Courtesy of Cummins Engine Company Inc.*)

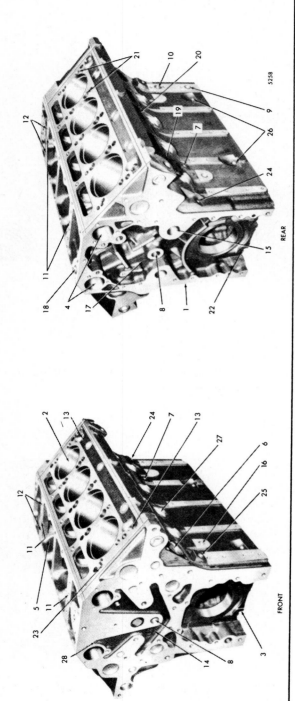

FRONT

REAR

5258

1. Cylinder Block
2. Bore for Cylinder Liner
3. Front Main Bearing Cap
4. Bores for Camshafts
5. Air Box
6. Water into Block
7. Water Drain
8. Main Oil Gallery
9. Oil Passage from Pump to Oil Filter
10. Oil Passage from Oil Cooler

11. Oil Drain Passages to Blower
12. Oil Drain Passages from Blower
13. Oil Drain from Cylinder Head
14. Oil Drain Passages
15. Oil Passage (Idler Gear Bearing)
16. Oil Return Passages to Crankcase

17. Crankcase Breather Cavity
18. Crankcase Breather Outlet
19. Air Box Drain
20. Inspection Hole Opening
21. Water Passages to Cylinder Head
22. Rear Main Bearing Cap
23. Oil Passage to Cylinder Head

24. Oil Gallery
25. Oil Gallery
26. Dipstick Hole or Oil Return Passage to Crankcase
27. Core Hole for Water Jacket
28. Oil Tube

FIGURE 7-3. Cylinder block components of a V8 diesel. (*Courtesy of Detroit Diesel Allison*)

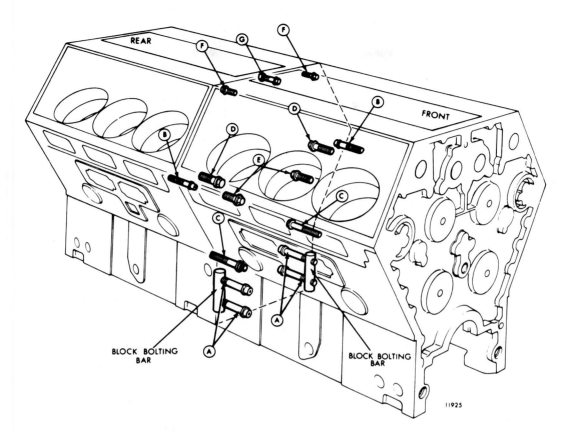

FIGURE 7-4. V12 engine block consisting of two V6 engine blocks bolted together. *(Courtesy of Detroit Diesel Allison)*

PART 2   BLOCK DESIGN

In-line cylinder block designs have all cylinders arranged in line—one behind the other. V-type blocks have cylinders arranged in a V shape. A variation of the V block is the Y block. On a Y block, the sides of the block extend down, well past the crankshaft center line. This type of block is heavier and requires only a shallow oil pan. The V block oil pan mounting surfaces are at the same level as the crankshaft center line. This requires a deeper oil pan but reduces engine weight.

Cylinder Numbering and Firing Order

Cylinder numbering (Fig. 7-5 and 7-6) is usually done from the front to the back, or from the end opposite to the flywheel, on in-line engines. (This is not always the case.)

On 8-cylinder V-type engines, the farthest forward cylinder is usually the number one position.

Remaining cylinders may be numbered from 1 to 4 down the right bank, and from 5 to 8 down the left bank, or they may be numbered alternately from side to side in sequence, as the connecting rods are mounted on the crankshaft.

Some engine manufacturers use a numbering system where each bank of cylinders is numbered separately. Cylinders on the left bank are numbered L1, L2, L3 etc. while cylinders on the right bank are numbered R1, R2, R3, etc.

The manufacturer decides the type of cylinder firing order the engine will have. The cylinder numbering system used and the crankshaft design will determine the firing order of an engine.

PART 3   CYLINDER LINERS

Cylinder liners (Fig. 7-7 to 7-11) are used to permit engine rebuilding without the expense of cylinder block replacement. If the cylinder block is in good

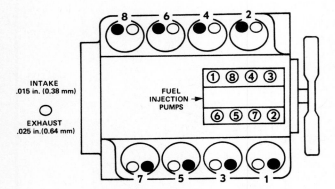

**CYLINDER, VALVE AND INJECTION
PUMP LOCATION**

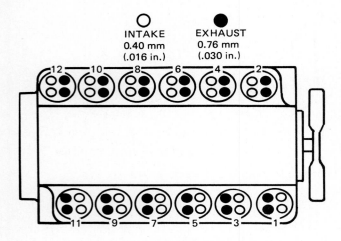

CYLINDER AND VALVE LOCATION

FIGURE 7-5. Caterpillar 3208 (top) and 3512 (bottom) cylinder numbering. 3208 firing order is 1,2,7,3,4,5,6,8. The firing orders (injection sequence) for the 3512 are:
SAE standard rotation 1-12-9-4-5-8-11-2-3-10-7-6
SAE opposite rotation................1-4-9-8-5-2-11-10-3-6-7-12
(Courtesy of Caterpillar Tractor Company)

condition, the cylinders can be restored to like-new condition by replacing the cylinder liners. Cylinder liners are also known as *cylinder sleeves*.

Several different types of cylinder liners are used in diesel engines. These include the dry, wet, water-jacketed, and ported types. Some small diesel engine designs do not use cylinder liners. In these cases cylinder blocks are reconditioned by reboring cylinders to oversize and installing new oversize pistons, or by replacing the block.

Dry Cylinder Liners

The dry cylinder liner does not come in contact with the engine coolant. Dry liners are relatively thin walled steel and are supported full-length by the liner bore of the cylinder block. Some designs are a press fit in the cylinder block, while others are a looser slide-in fit. Press fit liners are usually replaced when worn with the loose fit type. This requires resizing the bores in the block to a slightly larger diameter. All cylinder liners in the block must be of the same type and fit. Dry liners usually have a flanged top that fits into the counterbore in the block deck. The liners are held in place tightly by the cylinder head and are sealed at that point by a gasket or sealing ring.

Wet Cylinder Liners

In this design the liner forms part of the water jacket. The liner is of thick wall design to withstand all combustion pressures and is usually of cast iron alloy construction. Liner-to-block seals are provided at both top and bottom to prevent coolant leakage into the combustion chamber and crankcase. Wet liners also have a flanged top that fits into the counterbore in the block deck. Wet liners are held in place by the cylinder head and are sealed at the top by a gasket or seal ring similar to the dry liner.

Water-Jacketed Cylinder Liners

This type of liner is used in many of the larger diesel engines with fabricated engine frames or blocks. They are of double-wall construction with space between for coolant circulation. Coolant enters the liner jacket at one end and exits at the other.

Jacketed liners may be cast integrally or they may be a two-piece design in which the liner is installed in a separate water jacket. Seals are required at the top and bottom. Each jacket is provided with a coolant drain plug at the bottom.

Some cylinder liners are designed with a counterbore at the top inside circumference of the liner just above piston ring travel. This prevents the formation of a cylinder ridge as the liner wears due to piston ring friction. The counterbore is slightly larger in diameter than the rest of the liner.

Ported Cylinder Liners

Two-cycle diesel engines are equipped with ported cylinder liners. Ports are required to provide the cylinder with air for scavenging and combustion. These ports are precisely designed and positioned to pro-

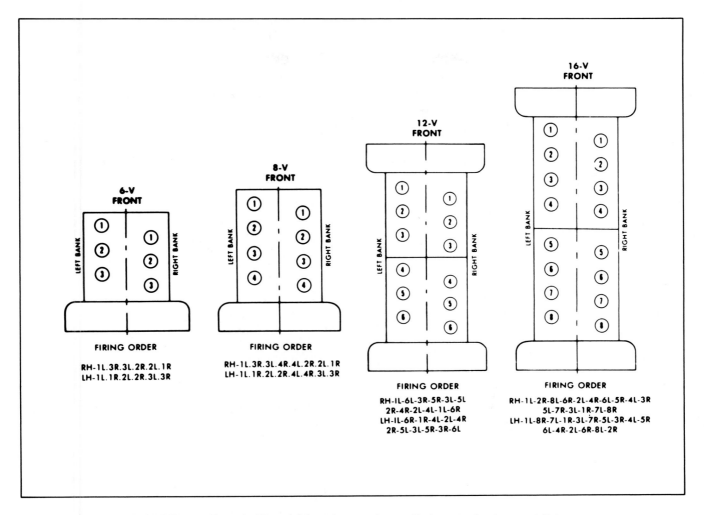

FIGURE 7-6. Detroit Diesel 92 series engine cylinder numbering and firing orders. *(Courtesy of Detroit Diesel Allison Division of General Motors Corporation)*

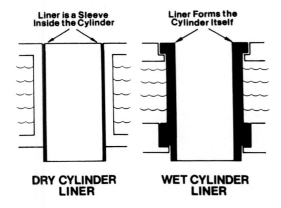

FIGURE 7-7. Cylinder liner types.

vide the desired port timing and air flow characteristics. Seals are required above and below the ported area to seal the air passages from engine coolant and lubricant.

Cylinder Liner Material

Cylinder liner materials must be able to withstand the extreme heat and pressures developed by the engine. At the same time friction and wear must be kept at a minimum. Fine-grained cast iron or steel are commonly used. The wear surface of many liners is plated with finely porous chrome or treated chemically to improve wear resistance and lubrication.

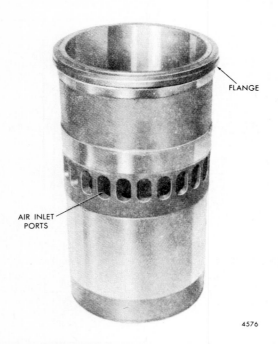

FIGURE 7-8. Wet-type ported cylinder liner for two-stroke-cycle diesel engine. *(Courtesy of Detroit Diesel Allison)*

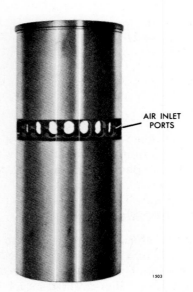

FIGURE 7-9. Dry-type ported cylinder liner for two-stroke-cycle diesel engine. *(Courtesy of Detroit Diesel Allison)*

FIGURE 7-10. Cylinder becomes glazed (hard-polished surface) during operation as shown on left. New or reconditioned cylinder liner has slight crosshatch pattern. *(Courtesy of Detroit Diesel Allison)*

Molybdenum is used in some liners for this purpose. Chrome plating provides excellent wear resistance and low friction.

The cylinder wear surface is a 25- to 30-microinch finish. It is produced with a fine hone to provide a visible 25 to 35 degree cross hatch pattern (50 to 70 degree included angle). This is necessary to help piston rings to seat properly and provide effective sealing against combustion pressures and lubricant leakage past the rings into the combustion chamber. The oil suspended on the cylinder walls as a result of this surface finish facilitates the lapping process of the rings sliding over the cylinder walls and results in a very smooth, close fit between the rings and the cylinder walls.

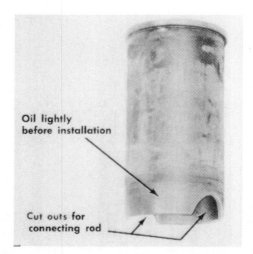

FIGURE 7-11. Some cylinder liners are designed with cutouts for connecting rod clearance when crank pin is at 90° ATDC or 90° BTDC positions. *(Courtesy of International Harvester Company)*

PART 4   CYLINDER BLOCK OPERATION

Cylinder Liner Wear and Stresses

In operation, the cylinder block components are subject to great changes in temperature, pressure of combustion, stress from expansion and contraction, cylinder liner wear from piston thrust, ring pressure, abrasives (possible scoring), and distortion. (See Fig. 7–12 to 7–17.)

Major thrust forces of the piston against the cylinder wall occur as a result of combustion pressures against the piston and the angle of the connecting rod during the power stroke. See Figure 7–12 and 7–13. Piston thrust is considerably less during the other strokes. Piston thrust contributes to cylinder wear.

Piston rings push and slide against cylinder walls as the piston moves up and down. Small particles of carbon and other abrasives that may enter the lubricating oil can cause wear.

Heat and pressure are most severe when the piston is near the top on the power stroke. Lubrication at this point and under these conditions is also least effective. Consequently, most cylinder wear resulting in cylinder taper takes place at the very top of ring travel in the cylinder. This wear results in a cylinder ridge developing at the top of the cylinder.

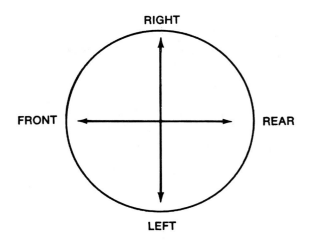

FIGURE 7–13. Major and minor thrust results in cylinders wearing out-of-round. Cylinders can also become distorted due to thermal shock, hot spots, and incorrect tightening of cylinder head bolts.

Hot spots can develop as a result of rust and scale buildup in the water jacket. This prevents good heat transfer to the engine coolant and causes this area of the cylinder to overheat. Distortion and a wavy cylinder surface can result. If severe, metal from the piston ring can be transferred to the cylinder wall and cause piston, cylinder, and ring scoring.

A *loose piston pin, broken rings, dirt,* and *carbon* can also damage cylinder walls. An *excessively rich fuel delivery* can cause lubrication to be washed away and increase cylinder wear.

Incorrectly tightened cylinder head bolts or *thermal shock* (sudden and extreme temperature change) can cause cylinder distortion. Coolant seepage into the combustion chamber can cause corrosion, as can combustion by product acids, especially if the engine is not serviced frequently enough.

Cylinder Liner Problems

Since cylinder liners are subjected to such extreme conditions, problems can develop that can result in liner failure. Some of the more common problems and their causes are listed here.

1. *Scoring:* Surface damage such as severe scratches and abrasions. This results in rapid wear of rings and cylinder walls with resulting combustion pressure loss into the crankcase and oil consumption. Scoring can be caused by scuffed rings, scored pistons, improper cold starting procedure, residual abrasives from honing cylinders, dust in in-

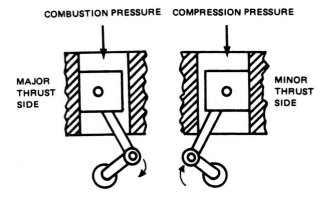

FIGURE 7–12. Combustion pressures force the piston against the major thrust side due to connecting rod angle and the crankshaft's resistance to rotation during the power stroke (above left). Compression pressures resisting piston movement during the compression stroke result in less thrust on the opposite side of the cylinder.

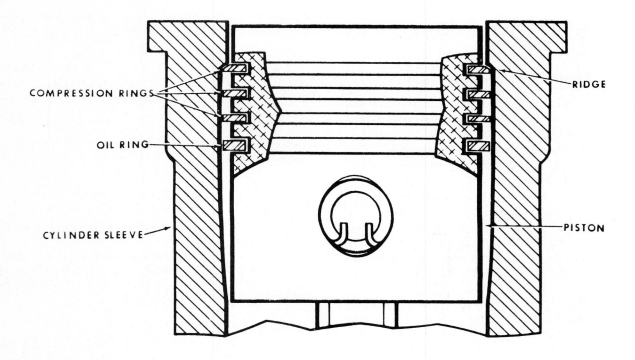

FIGURE 7-14. Cylinder wear is greatest at top of ring travel. At this point there is the least lubrication, the greatest ring pressure against the cylinder, and the highest temperature. *(Courtesy of J. I. Case, a Tenneco company)*

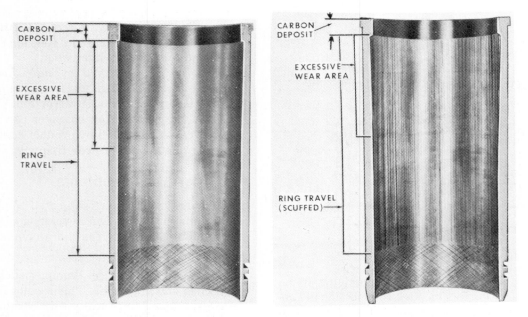

FIGURE 7-15. Normal cylinder wear after extended engine operating life (left). Scuffed and scored cylinder caused by lack of lubrication, excessive heat, or foreign matter such as dirt or carbon (right). *(Courtesy of J. I. Case, a Tenneco company)*

FIGURE 7-16. Cylinder liner erosion. *(Courtesy of Deere and Company)*

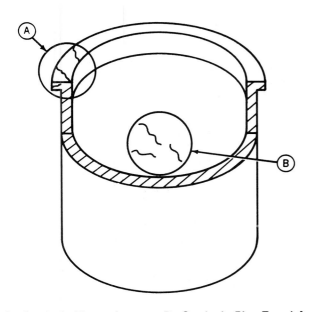

A—Cracks in Flange Area   B—Cracks in Ring Travel Area

FIGURE 7-17. Areas of liner where cracks may occur. *(Courtesy of Deere and Company)*

take air, broken piston pin snap ring, broken ring land, incorrectly installed liner or liner seal rings, or a distorted cylinder block.

2. *Corrosion:* Damage to the metal surface resulting in actual loss of metal and leaving a rough pitted surface. On cylinder sleeves in contact with engine coolant the causes include lack of proper chemical treatment of coolant, a long period of engine storage without draining the coolant, incorrectly maintained corrosion inhibitor, high mineral or chemical content in the coolant.

3. *Erosion:* Loss of metal through the effects of abrasives in the coolant. Causes include residual casting core sand or residual grit and sand from block cleaning procedures.

4. *Cavitation:* Pitting and wearing away of surface metal. Causes include aerated engine coolant, excessively high coolant temperature, inadequate coolant flow, and inadequate coolant treatment.

5. *Scale:* Deposits of minerals on metal surfaces. Causes include high chemical or mineral content of coolant and inadequate coolant treatment.

6. *Flange cracking:* Cylinder liner flange cracked from liner body. Caused by incorrect counterbore machining or shimming, overtightened head bolts, incorrect liner protrusion, or a distorted block deck.

7. *Liner cracking:* Cracks in liner body. Caused by overheating, hot spots, corrosion, scoring, improper seal or packing installation, severe cavitation, piston seizure, or improper liner handling.

8. *Discoloration:* Outside surface of liner discolored. A light or medium gray wet liner outside surface is normal and is caused by coolant additives. Brownish or beige colored liners are usually the result of overheating.

Main Bearing Bores

Main bearing bores support the crankshaft and main bearings. Main bearing caps must absorb all the force imposed by all the power impulses of the engine. These loads are quite high and can cause out of round and misalignment in time. Radial loads of approximately three tons per power impulse are imposed on the crankshaft and main bearing caps. At a speed of 3000 rpm, an eight-cylinder engine would subject the main bearing caps to 12,000 such impacts per minute. To withstand such punishment, engine tolerances and torque values must be very precise. Larger engines use larger-diameter main bearings and four-bolt main bearing caps. (See Fig. 7–18 to 7–20.)

Camshaft Bearing Bores

Camshaft bores support the camshaft and are subjected to loads imposed by valve springs during valve operation. Bearing bores must be in alignment to allow proper camshaft rotation and even wear distribution. In-block camshaft bearing bores are not distorted due to normal operation, but may distort due to block distortion.

Lifter Bores

Lifter bores are subject to wear and scoring due to abrasives such as carbon and varnish, particularly if engine oil is not changed at regular intervals.

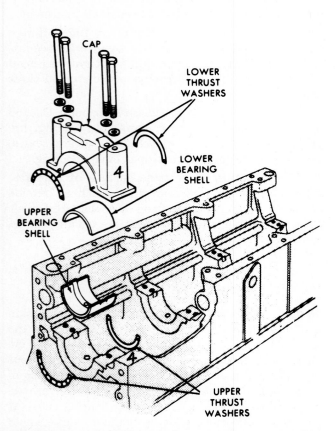

FIGURE 7–18. Four-bolt main bearing cap and crankshaft thrust washers. Note main bearing cap number and corresponding number on block. *(Courtesy of Detroit Diesel Allison)*

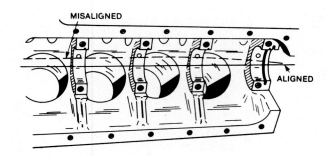

FIGURE 7–19. Main bearing bores should be in perfect alignment. Misalignment can result from overloading, thermal shock, incorrect tightening of main bearing cap bolts, overheating, and so on.

Oil Galleries

Oil galleries can become restricted due to sludge buildup. Reduced lubrication of moving parts can result from partially plugged passages.

Water Jackets

Water jackets surrounding the cylinders can collect sludge, scale, and rust, which reduces heat-transfer ability and coolant circulation, resulting in overheating and block distortion. Core hole plugs in water jackets may look good on the outside, but may be corroded and nearly rusted through from the inside.

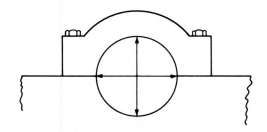

FIGURE 7–20. Main bearing bores can become out-of-round due to normal operation and incorrect tightening of bolts and can wear due to a bearing becoming loose and moving in the bore.

PART 5   CRANKSHAFT, BEARINGS AND SEALS

Purpose of Crankshaft

The crankshaft is one of the major moving parts of the engine. It is the only connection between the power-producing parts of the engine and the drive train. It is designed to transmit all the power the engine is able to produce and must be able to do so trouble free for many thousands of hours of operation. (See Fig. 7–21)

As the name implies, the crankshaft consists of a series of cranks offset from the crankshaft center-line. Inline engine crankshafts usually have one crank for each cylinder, while V-type engines nor-

mally have one crank for each pair of cylinders. The crankshaft converts the reciprocating motion of the pistons and connecting rods to rotary motion of the crankshaft.

Crankshaft Construction

Modern diesel engine crankshafts are usually constructed of one-piece forged alloy steel. Forged crankshafts are much stronger than the cast steel crankshafts usually used in automobile engines. The forging method forms the crankshaft in a process where very high pressure is applied to the metal to create the desired shape and stress direction. Some large engines use a crankshaft built up from two forgings by bolting them together at the two flanged ends. Bearing surfaces or even the entire crankshaft

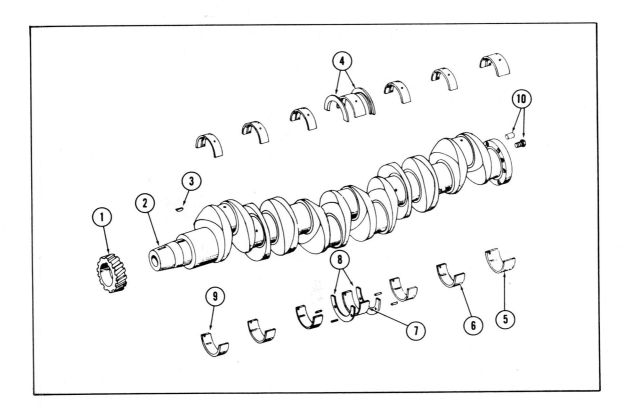

1. **Crankshaft Gear**	6. **Lower Intermediate Main Bearing Shell**
2. **Crankshaft**	7. **Lower Center Main Bearing Shell**
3. **Woodruff Key**	8. **Lower Thrust Flanges**
4. **Upper Thrust Flanges**	9. **Lower Front Main Bearing Shell**
5. **Lower Rear Main Bearing Shell**	

FIGURE 7–21. Crankshaft and main bearings. *(Courtesy of Allis-Chalmers)*

are induction hardened to reduce the wear rate. (See Fig. 7–22 to 7–31.)

The major portion of the crankshaft consists of the main and connecting rod journals. Connecting rod journals are also known as crank pins. The rod journals are connected to the main journals by two webs also called cheeks or arms. The crank pin and its two connecting arms are known as a *crank throw.*

The entire crankshaft is supported in the cylinder block by means of friction bearings at each of the main bearing journals. In some large engines the crank pins and main journals are hollow to reduce weight. Drilled lubrication passages connect the main and rod journals. These passages are usually chamfered at the journal surface to aid in oil distribution and prevent bearing damage from sharp edges. Endwise movement of the crankshaft is controlled by a flanged main bearing or by thrust washer-type bearing inserts that bear against machined surfaces on the crankshaft at one of the main bearing journals. Machined surfaces are also provided for oil seals to prevent oil leakage past the ends of the crankshaft.

Many crankshafts are equipped with counterweights designed to offset the weight of the crank pin and connecting rod during rotation. Counterweights may be an integral part of the crankshaft or they may be attached to it by means of capscrews.

A machined section at one end of the shaft provides the means for mounting accessory drive gears, a vibration damper, and at the other end a flywheel or torque converter mounted to the crankshaft flange. Keyways provide the means for maintaining a positive drive connection to the gears, vibration damper, and pulley. The crankshaft may also have a threaded hole at the end for a retaining bolt.

Smooth engine operation depends to a large extent on evenly spaced power impulses being transmitted to the crankshaft. Uniform crankshaft rotation can result only if power impulses are evenly spaced at alternating ends of the shaft as much as possible. The precise positioning of the crank throws is designed to achieve these results. In some cases, however, successive firing of cylinders cannot be avoided, especially in 2-, 3- and 4-cylinder engines.

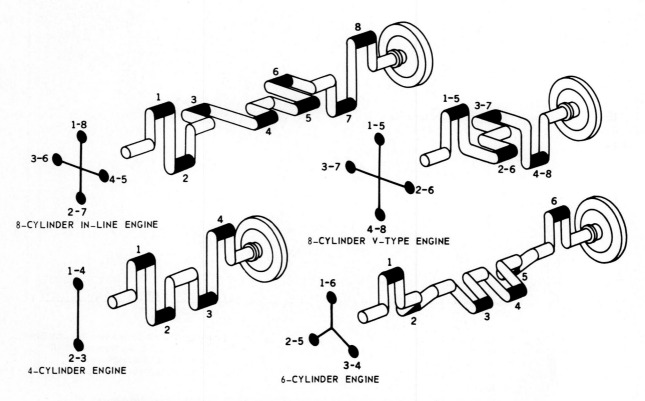

FIGURE 7-22. Crankshaft design variations showing crank throw positions by numbers. *(Courtesy of Deere and Company)*

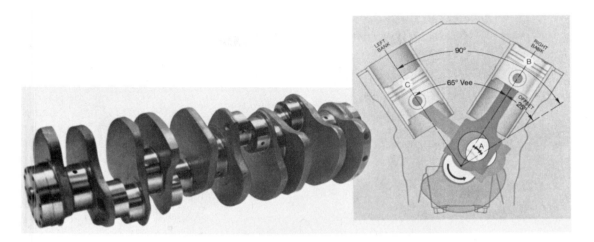

FIGURE 7-23. Crankshaft for 65° V-type engine, showing 25° offset connecting rod journals to provide evenly spaced power impulses. *(Courtesy of Detroit Diesel Allison)*

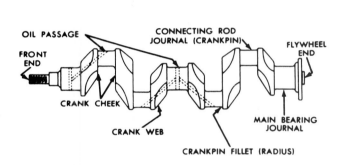

FIGURE 7-24. Crankshaft terminology and phantom view of oil passages. *(Courtesy of Ford Motor Company of Canada Ltd.)*

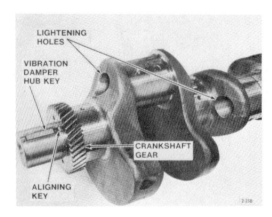

FIGURE 7-25. Lightening holes resulting from crankshaft balancing during manufacture. *(Courtesy of Mack Trucks Inc.)*

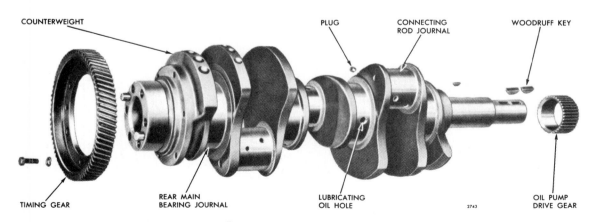

FIGURE 7-26. Crankshaft for V6 engine. Note timing gear drive at rear and plugs for drilled lubrication passages. *(Courtesy of Detroit Diesel Allison)*

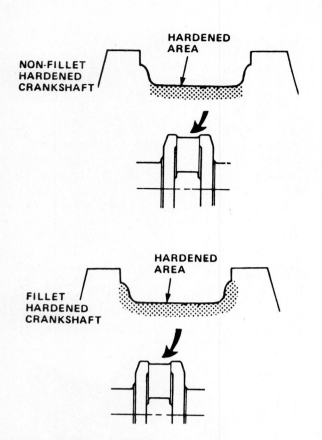

FIGURE 7-27. Surface hardening of crankshaft bearing journals.

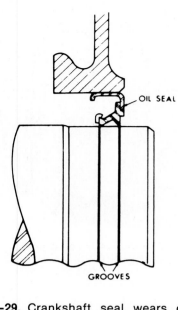

FIGURE 7-29. Crankshaft seal wears grooves into crankshaft seal journal. *(Courtesy of Detroit Diesel Allison)*

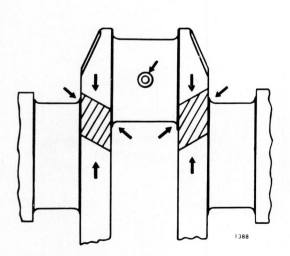

FIGURE 7-28. Arrows indicate critical crankshaft stress areas. *(Courtesy of Detroit Diesel Allison)*

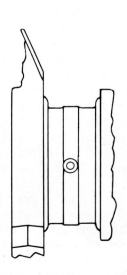

FIGURE 7-30. Crankshaft bearing journal showing ridging of journal from grooved bearing wear. *(Courtesy of Detroit Diesel Allison)*

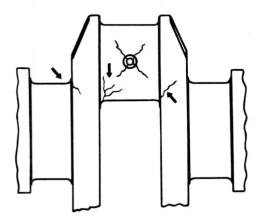

FIGURE 7-31. Fatigue cracks on crankshaft. *(Courtesy of Detroit Diesel Allison)*

Crankshaft Bearings

Crankshaft main bearings and connecting rod bearings (Fig. 7-32 to 7-50) are of the precision insert split type. Bearings are steel-backed and have laminated layers of softer bearing liner materials. Alloys, including such materials as babbitt, copper, aluminum, and lead, are used as bearing liners.

Bearings have a number of characteristics and design features that extend their service life. One of the main bearings is flanged (has thrust surfaces) to control crankshaft end play. Lubrication holes and grooves facilitate maintaining a good oil film between the bearing and the journal. Locating lugs align the bearing location in the bore. Crush height ensures good radial pressure of the bearing against the bore for good heat transfer. Bearing spread keeps the bearing in position during assembly. Good bearing characteristics include conformability, embedability, fatigue resistance, corrosion resistance, and long life.

On inline engine design, the crank throws are spaced radially an equal number of degrees apart. A six-cylinder, four-cycle, in-line engine has the crank throws spaced exactly 120 degrees apart. This results from dividing 720 degrees by 6, the number of cylinders. This means that cylinders are paired with respect to their radial position. In a firing order of 1-5-3-6-2-4, for example, the pairs are 1 and 6, 2 and 5, 3 and 4. When pistons 1 and 6 are at TDC and number 1 is beginning the power stroke, then number 6 is beginning the intake stroke. In other words, each pair of cylinders is 360 degrees apart in the four-stroke cycle of events.

In V-type engine design the concept is not quite as simple, since the angle formed between the cylinder banks must be considered. This is so since cylinders fire in opposite banks. When angles used do not divide evenly into 720 degrees, radial positioning of crank throws must be staggered to achieve even cylinder firing. In the 60 degree V6 design crank throws can be evenly spaced, since 60 divides evenly into 720 degrees. In the 65 degree V6 design this is not the case, since 65 does not divide evenly into 720. Three of the crank pins must be offset 5 degrees from the other three to provide even firing.

The number and arrangement of crank throws on a crankshaft depends on engine design factors such as the operating cycle (two- or four-stroke), cylinder arrangement (in line, V, or opposed), the angle formed between banks of cylinders on V engines, and the number of cylinders.

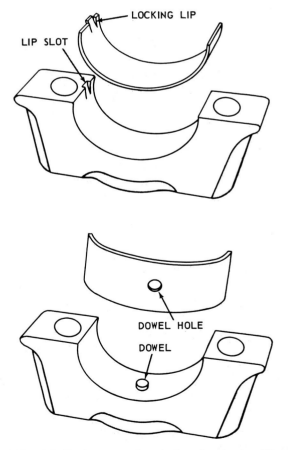

FIGURE 7-32. Two types of main bearing locks. *(Courtesy of Deere and Company)*

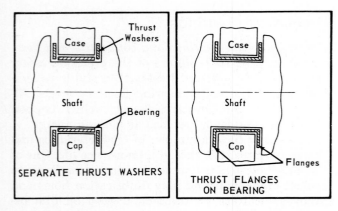

FIGURE 7-33. Two types of crankshaft thrust bearings. *(Courtesy of Deere and Company)*

FIGURE 7-35. Bearing spread makes it necessary to snap the bearing into place when installing. The bearing is slightly out-of-round before installation but becomes round when snapped into place in the connecting rod or main bearing bore. This helps keep bearings in place during assembly. *(Courtesy of Sunnen Products Company Ltd.)*

The clearance between bearing journals and bearings must be of precise dimension to be able to maintain a good oil film and prevent metal-to-metal contact. As soon as this oil film breaks down, rapid wear on the crankshaft and bearings results.

Bearing clearances on journals of 3.5 inches in diameter (88.9 millimeters) are generally from 0.003 to 0.006 inch (0.075 to 0.152 millimeter). The rotating journal forces the oil between the journal and the bearing on the loaded side in the direction of journal rotation when bearing clearances are correct. This is konwn as *hydrodynamic lubrication.* The ability to maintain hydrodynamic lubrication decreases rapidly with increased bearing clearances.

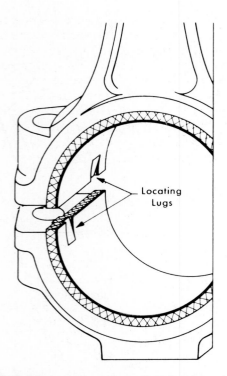

FIGURE 7-34. Locating lugs (bearing-tangs) fit into notches in the bearing bore to prevent fore-and-aft movement of the bearing inserts. Rotation of the bearing in its bore is also prevented by these lugs. *(Courtesy of Sunnen Products Company Ltd.)*

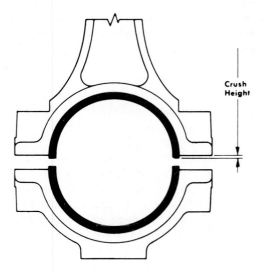

FIGURE 7-36. With the bearings in place and before tightening, the bearing extends slightly past the mating surfaces of the connecting rod or main bearing bores and caps. This is called *crush height. (Courtesy of Sunnen Products Company Ltd.)*

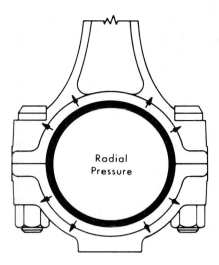

FIGURE 7-37. Proper bearing crush creates a radial pressure of the bearing against its bore for good heat transfer and load-carrying ability. Any dirt or foreign matter between the bearing insert and bearing bore would destroy these qualities and reduce bearing inside diameter. *(Courtesy of Sunnen Products Company Ltd.)*

Excessive bearing clearances also increase the amount of oil throw-off from the bearings. A certain amount of oil throw-off is needed for oil circulation and for lubrication of other internal engine parts. However, if oil throw-off is excessive, the amount of oil thrown up on the cylinder walls is more than the rings are able to control. As a result of this, oil passes by the rings into the combustion chamber, where it is burned.

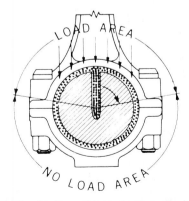

FIGURE 7-38. As the oil hole in the crankshaft sweeps past the load area, no oil is being fed into the bearing. Oil already there is forced between bearing and journal by journal rotation (hydrodynamic lubrication). *(Courtesy of Sunnen Products Company Ltd.)*

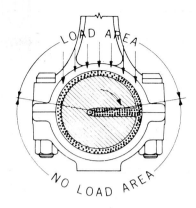

FIGURE 7-39. As the oil hole reaches the no-load area, oil is again fed into the bearing to replace oil being thrown off at the edges. As the shaft rotates to the loaded area, the process in Figure 7-38 is repeated. *(Courtesy of Sunnen Products Company Ltd.)*

Close examination of bearings will indicate reasons for bearing failure, as shown in Figures 7-46 through 7-50.

Crankshaft Oil Seals

Crankshaft oil seals are of the single- or double-lip synthetic rubber design. The sealing element is attached to the metal case, which fits tightly in its bore in the block. The sealing lip applies sufficient static pressure against the rotating journal to prevent oil leakage with minimum wear. Some seals have a garter spring behind the sealing element to provide added static pressure against the journal. Seals are always installed with the sealing lip facing the fluid or pressure to be contained.

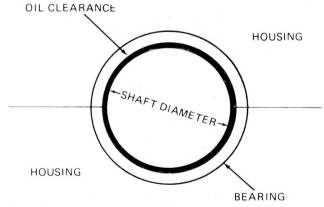

FIGURE 7-40. A specified amount of clearance between the bearing journal and the bearing surface must be provided to maintain a film of oil between these two parts to reduce friction and wear.

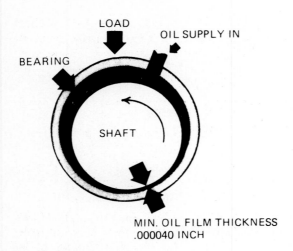

MIN. OIL FILM THICKNESS
.000040 INCH

FIGURE 7-41. Oil is fed to the lightly loaded side of the bearing. Shaft rotation forces oil around the bearing to maintain an oil film around the entire bearing journal. This is called *hydrodynamic lubrication. (Courtesy of Sunnen Products Company Ltd.)*

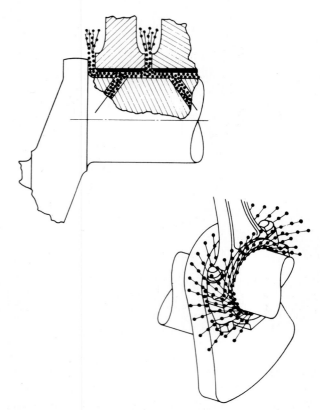

FIGURE 7-43. A controlled amount of oil throw-off is normal and is dependent on bearing clearances being correct. Oil throw-off helps lubricate cylinder walls, piston pins, and cam lobes. *(Courtesy of Sunnen Products Company Ltd.)*

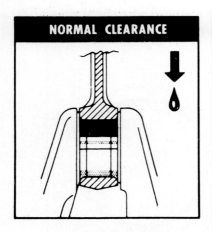
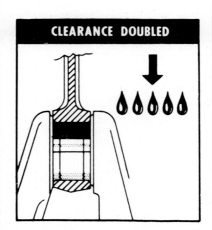

FIGURE 7-42. With correct bearing clearance, the amount of oil throw-off from the rotating shaft is minimal. When clearance is doubled, oil throw-off is five times greater; if clearance is four times as great, oil throw-off is 25 times as great. Piston rings are unable to scrape this excessive oil from the cylinder walls, and oil then enters the combustion chamber and is burned. *(Courtesy of Hastings Ltd.)*

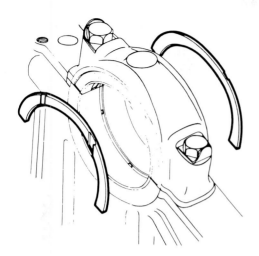

FIGURE 7-44. Some engines use separate thrust washers instead of a flanged main bearing to control crankshaft end play. (Main bearing cap halves only shown.)

Dirt, rough journal surfaces, and overheating are the major causes of seal failure. Excessive main bearing clearance or crankshaft end play also contribute to premature seal failure.

An oil slinger or oil deflector is often used behind main bearing seals. The oil slinger directs excess oil from the main bearing away from the seal, thereby reducing the workload of the seal. Some crankshafts have smooth-edged spiral grooves machined into the seal journal surface. As the shaft turns, the spiral grooves direct oil back into the crankcase through the wiping action of the seal.

PART 6 THE FLYWHEEL AND HARMONIC BALANCER

A flywheel is required to stabilize the speed fluctuations of the crankshaft resulting from power impulses of the engine cylinders. (See Fig. 7-51 to 7-56.) The flywheel stores energy during the power stroke and releases it during the non-power-producing strokes of the engine. The flywheel has an important speed-governing effect since it limits the speed increase or decrease during sudden changes in engine load. In many cases the flywheel provides a convenient means for mounting a large ring gear used for starting purposes. It also provides the frictional drive surface for a clutch type mechanical drive. Engine timing marks are often located on the flywheel. These degree markings indicate the TDC piston position usually for No. 1 cylinder and are used to determine fuel injection timing.

Flywheels are made from cast iron alloys, cast, or rolled steel. Provision for mounting is provided by the flywheel web in the center area. The precisely machined opening in the center ensures centering of the flywheel with the rotating axis of the crankshaft. The radial positioning of the flywheel is main-

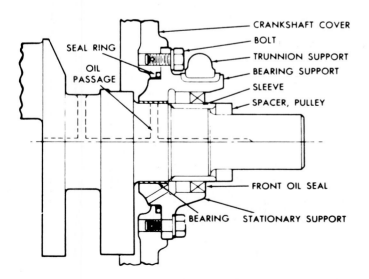

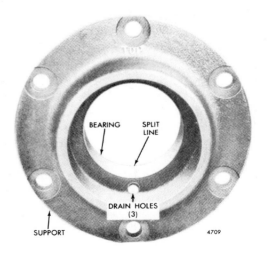

FIGURE 7-45. Outboard bearing and support on front of crankshaft used on some engines. *(Courtesy of Detroit Diesel Allison)*

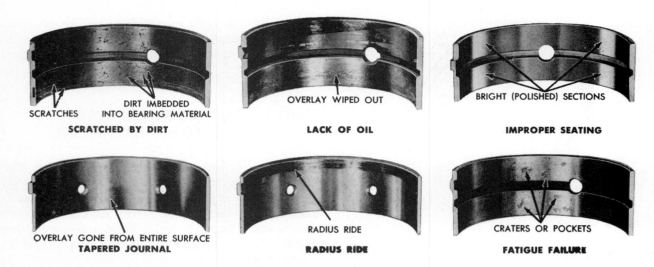

SCRATCHES   DIRT IMBEDDED INTO BEARING MATERIAL
SCRATCHED BY DIRT

OVERLAY WIPED OUT
LACK OF OIL

BRIGHT (POLISHED) SECTIONS
IMPROPER SEATING

OVERLAY GONE FROM ENTIRE SURFACE
TAPERED JOURNAL

RADIUS RIDE
RADIUS RIDE

CRATERS OR POCKETS
FATIGUE FAILURE

FIGURE 7-46. Several types and causes of bearing failures. *(Courtesy of Ford Motor Company of Canada Ltd.)*

FIGURE 7-47. Type of bearing failure caused by antifreeze leakage into crankcase.

tained by unevenly spaced mounting bolt holes or by a dowel pin. This relationship must be maintained for timing marks to be valid.

Flywheel designs include the flat and the recessed types. The recessed type reduces the overall size of the clutch and flywheel assembly. On engines equipped with hydraulic drive mechanisms, the torque converter serves the purpose of the flywheel and clutch assembly.

The starter ring gear is a shrink fit on the flywheel, and it may also be welded to it as well. The ring gear is heated prior to installation. Heating the ring gear increases its diameter, allowing it to slip easily over the flywheel mounting surface. As the

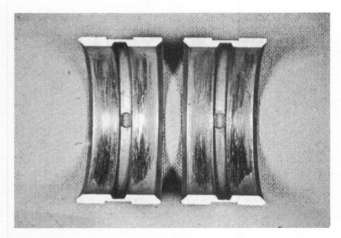

FIGURE 7-48. Oil diluted by fuel caused this bearing condition.

FIGURE 7-49. Incorrect crankshaft end play caused thrust flange failure on this main bearing.

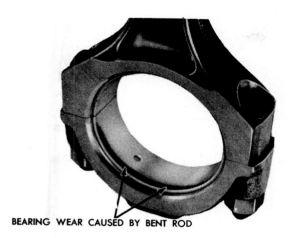

BEARING WEAR CAUSED BY BENT ROD

FIGURE 7-50. Upper half of bearing would show similar wear on opposite side. Other results of this condition are increased oil throw-off, bearing journal wear, and connecting rod big end out of round. *(Courtesy of Ford Motor Company of Canada Ltd.)*

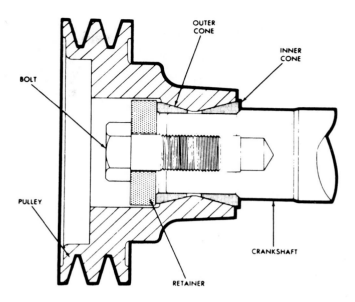

FIGURE 7-52. Cone-mounted crankshaft pulley. *(Courtesy of Detroit Diesel Allison)*

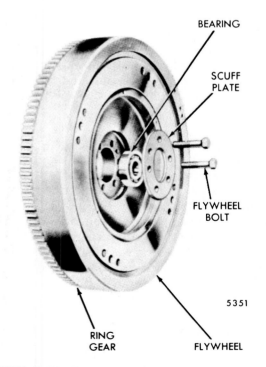

5351

FIGURE 7-51. Typical flywheel and pilot bearing. *(Courtesy of Detroit Diesel Allison)*

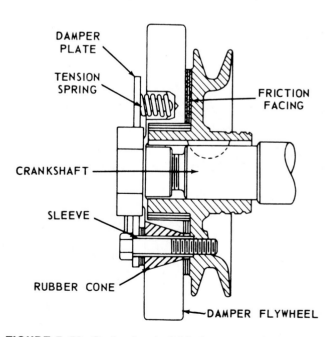

FIGURE 7-53. Spring-loaded friction type of vibration damper and pulley. *(Courtesy of Deere and Company)*

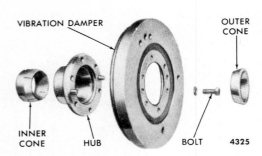

FIGURE 7-54. Crankshaft vibration damper and mounting hub and cones. *(Courtesy of Detroit Diesel Allison)*

ring gear cools, it shrinks to a very tight fit. The ring gear is supported by a flange on the side of the flywheel opposite the starter drive gear.

Flywheel Problems

A number of problems can develop, all of which adversely affect engine operation. Some of the more common problems and their effects are as follows.

1. Damaged or worn ring gear teeth can prevent starter drive pinion engagement and cause drive pinion damage.

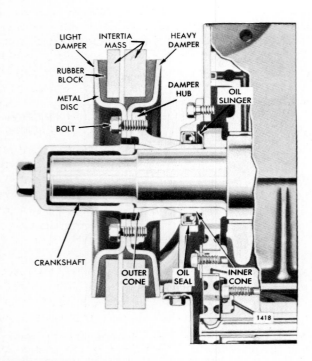

FIGURE 7-55. Double damper assembly detail. *(Courtesy of Detroit Diesel Allison)*

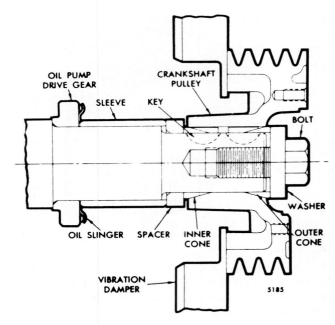

FIGURE 7-56. Woodruff key-mounted crankshaft pulley. *(Courtesy of Detroit Diesel Allison)*

2. Worn or overheated clutch friction surface can cause clutch slippage and further damage to the clutch and flywheel.

3. Flywheel runout can cause vibration, poor clutch release, and clutch slippage.

4. Loose starter ring gear can cause ring gear to turn on flywheel while attempting to start engine, preventing engine starting and wearing flywheel.

Harmonic Balancer

The harmonic balancer or vibration damper is needed to dampen normal torsional vibrations of the engine crankshaft. As each cylinder fires, it causes the crank throw to speed up. The inertia of the rest of the shaft causes it to stay slightly behind, resulting in a twisting action on the crankshaft. The torsional pulsations of successive cylinder firings create vibrational frequencies that vary with engine speed and with the number of engine cylinders. The vibration damper reduces the effects of these vibrations.

The vibration damper consists primarily of a hub and an inertia ring. The inertia ring is either bonded to the hub through a flexible elastomer (rubber compound) insert or it is enclosed in a housing filled with a viscous, grease-like fluid. The inertia ring in the latter arrangement is free to float in the fluid. The clearance between the inertia ring and the housing is only about 0.010 inch (0.25 mm) on all sides.

In both designs the inertia ring moves slightly in relation to crankshaft rotation as each cylinder fires, thereby dampening the torsional vibrations of the crankshaft over a wide range of engine speeds. Some dampers are designed with two inertia rings of different sizes for more effective control over a wider range of vibrational frequencies.

Over an extended period of time, the elastomer can deteriorate or the bonding can let go, rendering the damper ineffective or causing vibrations itself as a result. As long as the viscous type damper is not damaged physically, it remains effective. Dents in the viscous housing prevent the inertia ring from operating as intended. A damaged damper must be replaced. On damper designs where the hub is also a seal journal, the seal can wear a groove in the hub, resulting in oil leakage. A sleeve-type repair can restore the damper if otherwise in good condition. The hub may require machining in some cases to accommodate the repair sleeve.

PART 7   ENGINE BALANCER

Even though individual engine components such as crankshafts and flywheels are balanced, the entire rotating assembly must also be balanced. (See Fig. 6-6, 6-7, and 7-57.) There are two forces at work that tend to create an unbalanced condition. These are:

1. *Centrifugal Force.* This results from rotary motion of the crankshaft creating a force outward from the axis of rotation. The magnitude of this force changes with engine speed.

2. *Inertia Forces.* These are caused by the reciprocating movement of pistons and connecting rods. The magnitude of these forces also changes with engine speed.

The centrifugal forces of the rotating crankshaft and the inertia forces of the reciprocating pistons can be offset by the use of crankshaft counterweights in four-cycle engines with more than four cylinders. This is possible since crank throws can be positioned to achieve the desired result. On engines with four cylinders or less, this is not possible. Even though the crankshaft as a whole can be balanced, individual crank throws remain unbalanced.

To reduce the stress caused by these forces on higher speed engines, gear-driven balance weights or balance shafts are used. Balance weights are positioned to offset the inertial forces of the pistons.

Two gears with balance weights are used in one balancer design. The gears are timed to the crankshaft gear in a manner that results in the upward force of the piston being offset by the downward force of the counterweight on the blancer. Gear driven balance shafts with counterweights operate in a similar manner.

PART 8   PISTONS, RINGS, AND CONNECTING RODS

The piston-and-rod assembly (in conjunction with the cylinder and valves) acts as a pump on the intake and exhaust strokes. On the power stroke, it transmits the pressure of expanding gases to the crankshaft, forcing it to turn.

Connecting Rods

The *connecting rod* is attached to the crankshaft at one end (big end) and to the piston at the other end (small end). The tapered I-beam type of connecting rod is the most common. The big end of the rod is split so that it can be connected to the crankshaft. The cap and yoke are a matched pair, and their relationship must not be altered. Both rod and cap are numbered. Special precision connecting rod bolts and nuts keep the cap in proper alignment with the rod. Some connecting rods have an oil spurt hole in the yoke or at the cap mating surface to provide cylinder wall lubrication. Notches in the yoke and cap provide proper bearing positioning. The small end of the connecting rod is attached to the piston by a piston pin. In some cases the small end of the rod is clamped to the pin or has a bushing in it to allow pin and rod oscillation. In other designs the pin is bolted to the rod. Connecting rods are usually drilled to provide lubrication to the piston pin and also to spray oil into the bottom of the piston for piston cooling on some designs. (See Fig. 7-58 to 7-62.)

The connecting rod undergoes high loads and speeds as well as constant change of direction. Overloading the engine can cause bent connecting rods and bearing failure. The big end of the connecting rod can also become out-of-round as a result of high loads and high mileage or extended hours.

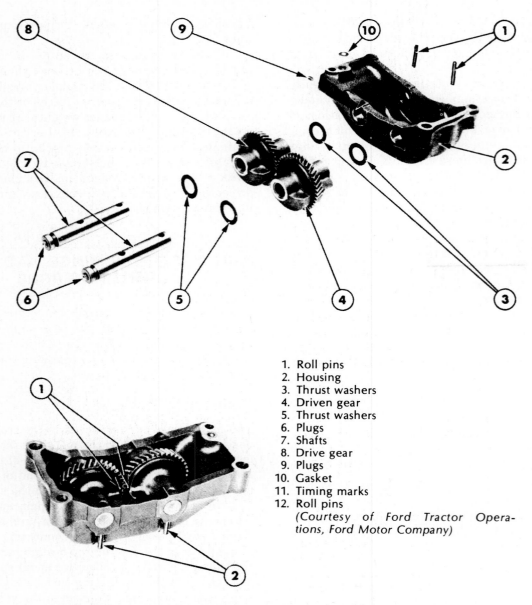

1. Roll pins
2. Housing
3. Thrust washers
4. Driven gear
5. Thrust washers
6. Plugs
7. Shafts
8. Drive gear
9. Plugs
10. Gasket
11. Timing marks
12. Roll pins
    (Courtesy of Ford Tractor Operations, Ford Motor Company)

FIGURE 7-57. Crankshaft balancer assembly. Weighted gears are timed to crankshaft drive gear to balance crankshaft operation.

Pistons

The piston forms the movable bottom of the cylinder and combustion chamber. It is designed to withstand normal loads and temperatures and provides long service life under these conditions. The piston must absorb all the thrust that a cylinder is able to produce. Most pistons are made of a special alloy of aluminum because of its lighter weight. Some pistons are made of cast iron. (See Fig. 7-63 to 7-76.)

Two or more grooves are provided for the piston rings. Lands separate the ring grooves. Round or slotted holes are located in or just below the bottom ring groove to allow oil scraped from the cylinder walls by the rings to return to the crankcase.

The piston head diameter (which includes the ring area) is usually approximately 0.030 inch smaller (0.76 millimeter) than the skirt diameter. This is necessary since the top of the piston is sub-

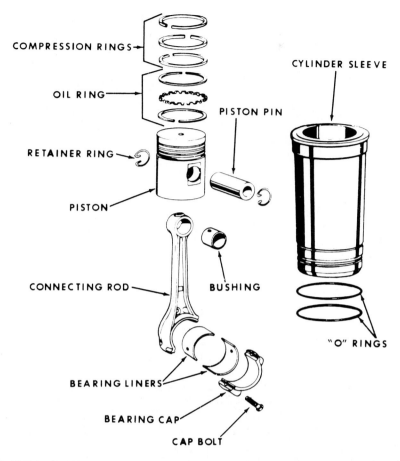

FIGURE 7-58. Piston and cylinder liner assembly. Note offset rod design at big end of rod. *(Courtesy of J. I. Case, a Tenneco company)*

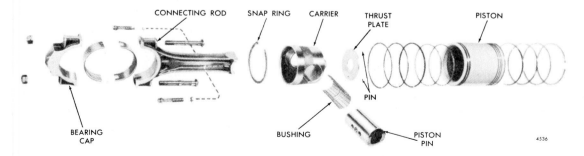

FIGURE 7-59. Floating piston skirt design allows piston to rotate for even wear distribution. Note piston pin is bolted to the connecting rod. *(Courtesy of Detroit Diesel Allison)*

113

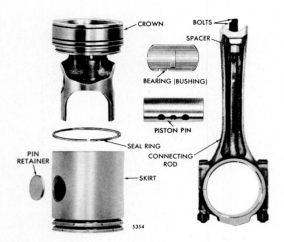

FIGURE 7-60. Cross head type of piston design prevents vertical load distortion and reduces thermal distortion of the piston skirt. Note semicircular pin bushing (150°). *(Courtesy of Detroit Diesel Allison)*

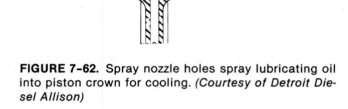

FIGURE 7-62. Spray nozzle holes spray lubricating oil into piston crown for cooling. *(Courtesy of Detroit Diesel Allison)*

jected to the hottest temperatures, and the most expansion. The piston head is round, whereas the piston skirt is cam ground (slight oval shape) in some designs.

Clearance between the piston skirt and cylinder wall must be maintained at precise low tolerances.

The top of the piston head or crown usually also forms the combustion chamber. A bowl-shaped depression with a raised center area is used in many direct injection engines. This is sometimes called the mexican hat design.

A stainless steel heat plug is used in the top of the piston in some engines with a precombustion chamber. The plug protects the aluminum piston

from being damaged by the very high pressure and temperature of the burning gases escaping from the prechamber.

Two general classifications of pistons are used in diesel engines: the one-piece trunk-type piston and the two-piece crosshead design. In the crosshead design, the crown that includes the skirt carrier and the upper ring area forms one piece, while the skirt with the remaining rings forms the other piece. The

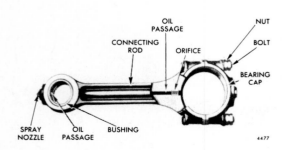

FIGURE 7-61. Rifle-drilled connecting rod with metered orifice to control amount of oil provided for pin lubrication and piston cooling. *(Courtesy of Detroit Diesel Allison)*

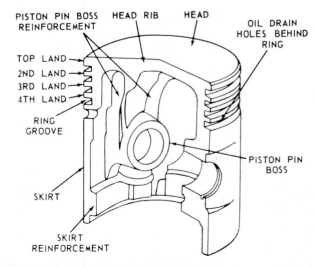

FIGURE 7-63. Construction details of piston and piston terminology. *(Courtesy of Deere and Company)*

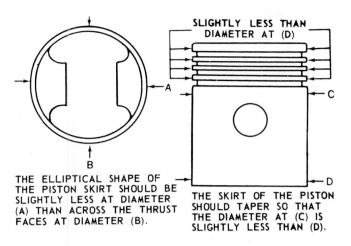

THE ELLIPTICAL SHAPE OF THE PISTON SKIRT SHOULD BE SLIGHTLY LESS AT DIAMETER (A) THAN ACROSS THE THRUST FACES AT DIAMETER (B).

THE SKIRT OF THE PISTON SHOULD TAPER SO THAT THE DIAMETER AT (C) IS SLIGHTLY LESS THAN (D).

FIGURE 7-64. Many pistons are cam ground (slightly oval-shaped skirt) for better piston fit. Piston is designed to expand parallel to the pin as it warms up, thus allowing closer piston to cylinder fit when cold to reduce piston slap. *(Courtesy of Deere and Company)*

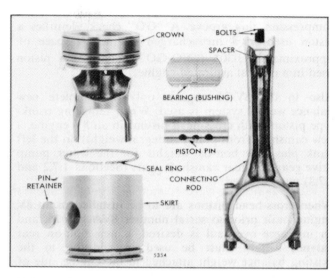

FIGURE 7-66. Cross head piston components and connecting rod. *(Courtesy of Detroit Diesel Allison)*

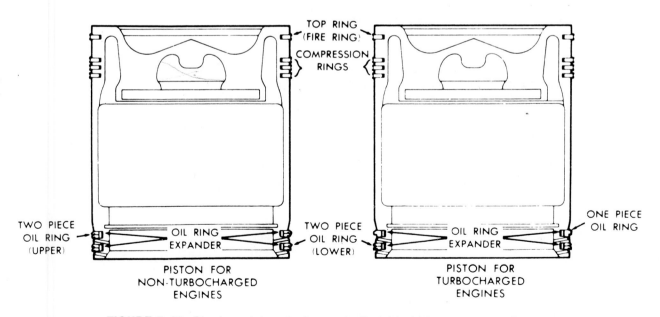

FIGURE 7-65. Six-ring piston design uses fire ring in top groove and uses different oil rings for turbocharged engine from those for nonturbocharged engine. *(Courtesy of Detroit Diesel Allison)*

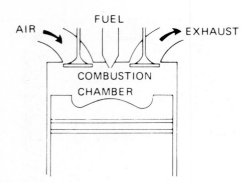

FIGURE 7-67. One type of piston head design used with open-type combustion chamber and direct fuel injection. *(Courtesy of Cummins Engine Company Inc.)*

pieces are held together by the piston pin, which passes through both. The crosshead design has reduced thermal distortion and the skirt is not subjected to vertical load distortion.

Several piston designs are used in diesel engines. Piston head designs include dished, flat, and irregular types. The shape of a combustion chamber depends on the shape of the cylinder head and the shape of the piston head. Combustion is affected by the shape of the combustion chamber to a large extent.

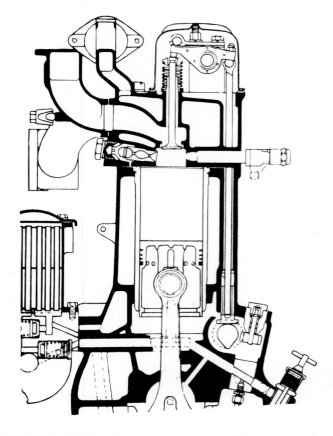

FIGURE 7-69. Flat head piston used with energy cell and combustion chamber in cylinder head. Energy cell is at left of valve and injector. *(Courtesy of J. I. Case, a Tenneco company)*

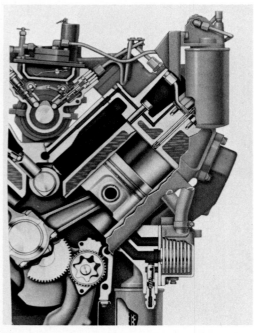

FIGURE 7-68. Two-ring piston used in four-stroke diesel with direct injection. *(Courtesy of Caterpillar Tractor Company)*

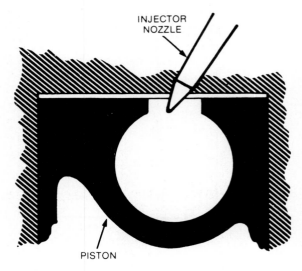

FIGURE 7-70. Piston head with swirl chamber known as the M type of combustion chamber reduces diesel knock at low speeds.

FIGURE 7-71. Turbulence-type combustion chamber uses flat head piston.

Cam-ground piston skirts (slightly oval-shaped) are larger in diameter across the thrust surfaces as compared to the diameter parallel to the piston pin. This allows the piston to fit the cylinder better when cold as well as at operating temperatures. In other words, clearance between the piston skirt and the cylinder on the thrust side remains relatively constant whether the engine is cold or at operating temperature. Since the slot above the skirt directs heat away from the skirt area and to the pin boss area, expansion takes place parallel to the pin. As the piston expands and becomes more round, a wider area of the thrust sides (skirts) of the piston comes in contact with the cylinder wall.

FIGURE 7-72. Corroded piston has gray pitted appearance resulting from moisture or antifreeze. *(Courtesy of Deere and Company)*

Floating skirts are not cam-ground since they are less subject to heat and are free to rotate in the cylinder.

A steel strut is used in many piston designs to help control the direction of piston expansion as well as to strengthen the piston.

THIS PISTON SUITABLE FOR
INSTALLATION AS IS

SLIGHTLY SCORED, USE ONLY
AFTER REMOVING SCORE MARKS
BY POLISHING WITH CROCUS
CLOTH OR HARD INDIA STONE

BADLY SCORED—UNFIT
FOR USE

4094

FIGURE 7-73. Various stages of piston wear. *(Courtesy of Detroit Diesel Allison)*

FIGURE 7-74. Piston damaged from overheating and broken rings (top). Clogged and seized oil ring and scored piston resulting from below normal temperature operation (bottom). *(Courtesy of General Motors Corporation)*

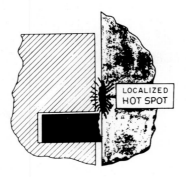

FIGURE 7-75. Localized hot spot can result from rust or scale buildup in water jacket. Metal is actually transferred from the ring to the cylinder wall, similar to a weld, causing ring and piston damage. *(Courtesy of General Motors Corporation)*

A tapered skirt piston design is also used for expansion control. Since the top of the skirt runs at a higher temperature than the bottom of the skirt, some pistons are designed to be slightly smaller in diameter at the top of the skirt than at the bottom of the skirt.

The piston pin is of tempered-steel construction. It may be mounted in one of several different manners, as illustrated.

In operation the piston is subjected to wide temperature variations, high combustion pressures, rapid changes in speed and direction, friction, and the effects of improper cooling, improper combustion, and incorrect injection timing if engine service is neglected.

Detonation and preignition can result in serious piston damage. Excessive carbon buildup can cause both preignition and detonation. Carbon can also cause the rings to become seized in their grooves and cause cylinder and piston scoring.

Carbon is a soft or hard deposit that builds up on engine parts such as rings pistons, injectors, combustion chambers, and valves. It is the result of

FIGURE 7-76. Excessive groove wear can cause excessive oil consumption and ring breakage. *(Courtesy of General Motors Corporation)*

the residues left after combustion. Some residue comes from the fuel being burned, but most of it is from oil getting into the combustion chamber and burning.

A piston pin that is too tight in the piston can prevent proper connecting rod oscillation and the piston from expanding or contracting properly. This could result in piston slap and piston damage.

Piston Rings

Piston rings provide a dynamic seal between the piston and the cylinder wall. Their purpose is to prevent combustion pressures from entering the crankcase and crankcase oil from entering the combustion chamber. They also control the degree of cylinder wall lubrication. (See Fig. 7–77 to 7–79.)

The general classification of piston rings includes compression rings and oil control rings. Some manufacturers refer to the top compression ring as a fire ring. Some diesel engines use only two piston rings, one compression ring at the top and one oil control ring just below the compression ring. Other designs use three, four, five, or even more rings per piston, as shown in the illustrations. Rings may be chrome-faced, copper oxide-coated or plasma-faced.

Compression rings are required to seal against combustion and compression pressures. Oil control rings are required to control cylinder wall lubrication. Both types of rings may be designed to scrape excess oil from the cylinder walls. Piston ring design therefore requires that rings be installed correctly. Lower rings on Detroit Diesel two-cycle engines also serve to prevent scavenging air from entering the crankcase.

Some combustion gases and unburned fuel (hydrocarbons) will inevitably pass the rings and reach the crankcase. If the rings, pistons, cylinders, and fuel and lubrication systems are in good condition, this will be minimal.

Since engine oil from the lubrication system helps seal the rings, any decrease in lubrication system efficiency will affect the rings' ability to seal and will increase ring and cylinder wear.

Counterbores and chamfers on compression rings assist the rings to slide over the oil on the cylinder walls during upward movement of the piston and scrape the oil off the cylinder walls on downward movement. Tapered-face and barrel-face ring designs are also used for this purpose.

Ring Expanders

Expanders used behind specially designed compression rings increase ring pressure against the cylinder walls for increased sealing ability. Rings without expanders rely on ring tension alone for static pressure against the cylinder wall. Static tension is caused because the piston ring in its relaxed state is larger in diameter than the cylinder and must be compressed to be installed. In its relaxed state the ring is slightly out-of-round; however, when it is installed, it becomes round and seals against the surface of the cylinder wall. The ring face and the 60° crosshatch pattern of the cylinder wall wear in during the first period of engine operation to form a good seal. On the power stroke the dynamic pressure of combustion increases ring pressure against the cylinder wall. This is the result of combustion pressure getting between the top compression ring and the piston.

PART 9   RING SIDE CLEARANCE AND WEAR

Ring side clearance is provided in the ring grooves to prevent rings from sticking or binding in the grooves due to expansion. A gap at the ring ends is provided to prevent ring ends from butting and causing the ring to become tight in the cylinder due to expansion.

During operation, piston rings are subjected to dynamic pressures, friction, heat, constant change of direction and speed, and inertia. Since there is some side clearance between the ring and the land, the piston ring moves up and down in the ring groove on the different strokes of the engine. Due to ring pressure against the cylinder wall and the inertia of the piston ring, the ring tends to stay behind when the piston changes direction. This causes the ring to move up and down in the groove and eventually causes ring groove wear. The ring itself also wears, increasing ring side clearance even further. When clearance is excessive, ring breakage can occur.

Another factor concerning ring and groove wear is cylinder condition. When a cylinder wears, it becomes tapered (larger diameter at the top of ring travel than at the bottom of ring travel). As a result, the piston rings are forced deeper into the ring grooves as the piston moves down in the cylinder. As the piston moves up in the cylinder, ring tension causes the ring to expand to fit the worn part of the cylinder. The rings continuously expand and contract in the ring grooves as the piston moves up and down in the worn cylinder.

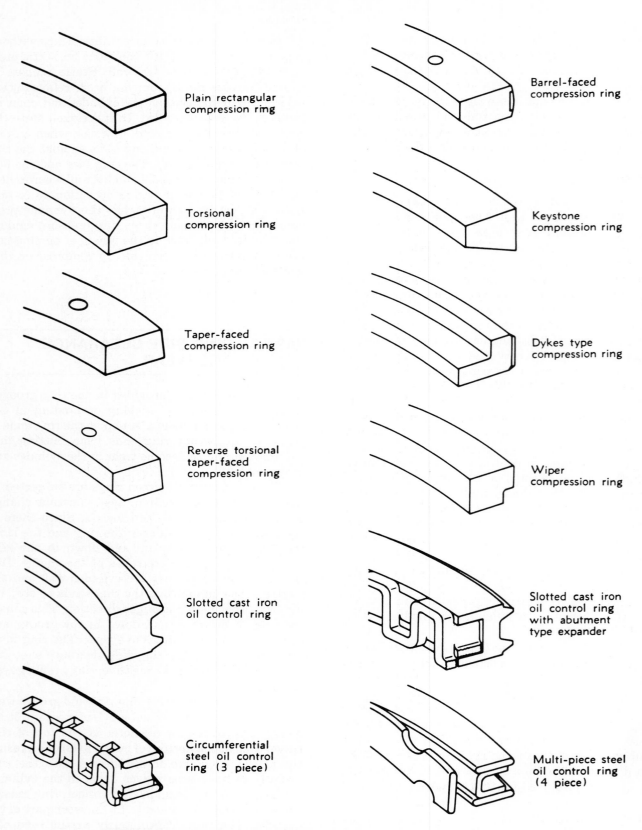

FIGURE 7-77. Various piston ring designs. *(Courtesy of Hastings Manufacturing Company Ltd.)*

Plain rectangular compression ring

Torsional compression ring

Taper-faced compression ring

Reverse torsional taper-faced compression ring

Slotted cast iron oil control ring

Circumferential steel oil control ring (3 piece)

Barrel-faced compression ring

Keystone compression ring

Dykes type compression ring

Wiper compression ring

Slotted cast iron oil control ring with abutment type expander

Multi-piece steel oil control ring (4 piece)

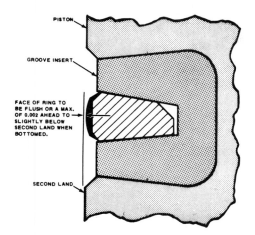

FIGURE 7-78. Keystone piston ring cross section and piston groove insert. *(Courtesy of Mack Trucks Inc.)*

PART 10   SELF-CHECK

1. Engine blocks are manufactured from _____ or _____.

2. Name four types of engine block design.

3. The minor thrust side of an engine is on the opposite side to the crank throw when the piston is on the _____ stroke.

4. List the three major reasons for cylinders wearing to a taper.

5. State the purpose of the crankshaft.

6. What three functions are performed by the flywheel?

7. The 90 degree V8-crank throws are indexed every _____ degrees.

8. The purpose of the connecting rod is to _____.

9. List three methods of controlling piston expansion.

10. Three methods used to mount piston pins are _____.

11. Why are piston rings needed?

12. Why are ring side clearance and ring gap clearance required?

13. Describe the difference between a cross head piston and a floating skirt piston.

14. Which engine component has the greatest effect on combustion chamber design?

15. Describe the proper cylinder surface finish.

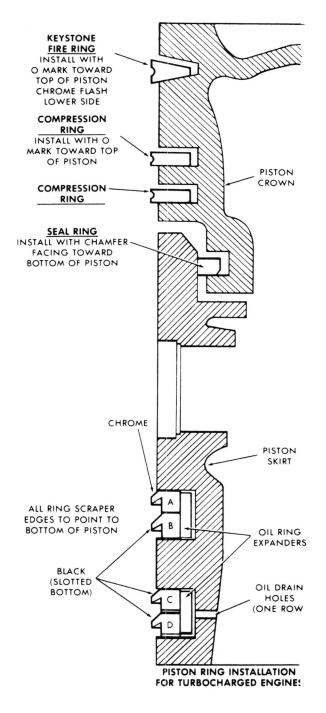

PISTON RING INSTALLATION FOR TURBOCHARGED ENGINES

FIGURE 7-79. Details of piston rings and grooves for cross-head type of piston. *(Courtesy of Detroit Diesel Allison)*

Chapter 8

Camshafts and Camshaft Drives

Performance Objectives

After thorough study of this chapter, the appropriate training models and manufacturer's service manuals you should be able to do the following:

1. Complete the self-check questions with at least 80 percent accuracy.
2. State the purpose of engine components (a) to (d) below.
3. Describe the basic construction differences of engine components (a) to (d) below.
4. Describe the operation, wear, and deterioration of engine components (a) to (d) below.

    (a) camshaft
    (b) camshaft bearings
    (c) camshaft drive mechanisms
    (d) accessory drive gears

The camshaft assembly includes the camshaft, bearings, and drive mechanism. The camshaft and drive assembly are required to control the opening and closing of the valves. In some cases the camshaft also operates the injectors.

The design of the camshaft and drive results in valves being opened and closed at a controlled rate of speed as well as at a precise time in relation to piston position. This also applies to injector operation on engines where injectors are engine camshaft-operated (injector timing). Camshafts are gear driven at half crankshaft speed on four-stroke-cycle engines and at crankshaft speed on two-stroke-cycle engines.

PART 1   CAMSHAFTS

Camshafts (Fig. 8–1) are of forged steel construction for strength and are induction-hardened or heat-treated for wear resistance and long life.

Four-stroke-cycle engines have two cam lobes per cylinder when injectors are not engine camshaft-operated. On four-stroke-cycle engines with engine camshaft-operated injectors, the camshaft is designed with three lobes per cylinder, one intake cam lobe, one exhaust cam lobe, and a third lobe operating the injector.

The two-stroke-cycle diesel engine has three lobes per cylinder: one lobe to operate the unit injector, and two lobes to operate the two or four exhaust valves per cylinder. (There are no intake valves).

The cam lobe converts the rotary motion of the camshaft to reciprocating motion of the cam follower.

Endwise movement of the camshaft is limited by a thrust plate located between the bearing journal and the drive gear or sprocket. The thrust plate is bolted to the engine. On some engines there is no thrust plate to hold the camshaft in place. This design uses a thrust pin installed in the block, which engages a groove in the camshaft.

PART 2   CAM LOBE DESIGN AND VALVE TIMING

The design of the cam lobe contour has a major effect on engine performance. The amount of valve opening, the amount of time that the valve remains open (duration), the time when the valves open and close (valve timing), and the speed at which valves open and close are determined by cam lobe design (Fig. 8–2).

The difference between the diameter of the lobe base circle and the nose-to-heel diameter plus rocker arm ratio determines the amount of valve opening. In other words, lobe lift times rocker arm ratio equals valve lift.

The time at which valves open and close (valve timing) and the duration of valve opening are stated in degrees of crankshaft rotation. For example, on a four-stroke-cycle engine, the intake valve normally begins to open just before the piston has reached

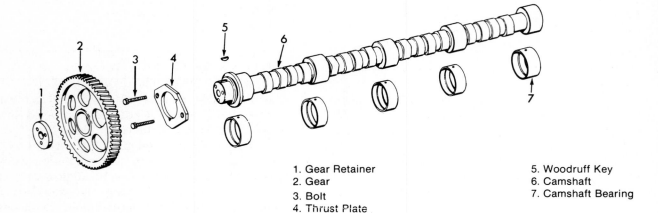

1. Gear Retainer
2. Gear
3. Bolt
4. Thrust Plate
5. Woodruff Key
6. Camshaft
7. Camshaft Bearing

FIGURE 8-1. Typical camshaft assembly components. *(Courtesy of International Harvester Company).*

top dead center. The valve remains open as the piston travels down to BDC and even past BDC. This is intake valve duration. An example of this could be stated as follows: IO at 20° BTDC, IC at 50° ABDC, or intake opens 20° before top dead center, intake closes 50° after bottom dead center. Intake valve duration in this case is 250° of crankshaft rotation.

This leaves 130° duration for the compression stroke since compression ends when the piston reaches TDC. At this point the power stroke begins. The power stroke ends when the exhaust valve be-

gins to open at approximately 50° before bottom dead center. The duration of the power stroke in this case is also 130°.

The exhaust valve opening at 50° BBDC begins the exhaust stroke. The exhaust stroke continues as the piston passes BDC and moves upward to and past TDC. With the exhaust valve closing at 20° ATDC, the duration of the exhaust stroke is 250°.

It is apparent from this description that the exhaust valve stays open for a short period of time during which the intake valve is also open. In other words, the end of the exhaust stroke and the beginning of the intake stroke overlap for a short period of time. This is called *valve overlap.*

Opening the intake valve before TDC and closing it after BDC increases the fill of air in the cylinder. Opening the intake valve early helps overcome the static inertia of the air at the beginning of the intake stroke, while leaving the intake valve open after BDC takes advantage of the kinetic inertia of the moving air. This increases volumetric efficiency.

As the piston moves down on the power stroke past the 90° ATDC position, pressure in the cylinder has dropped, and the leverage to the crankshaft has decreased due to connecting rod angle and crankshaft position. This ends the effective length of the power stroke, and the exhaust valve can then be opened to begin expelling the burned gases. The exhaust valve remains open until the piston has moved up past the TDC position. This helps to remove as much of the burned gases as necessary and increases volumetric efficiency.

On the two-stroke-cycle diesel engine, inlet port location in the cylinder in relation to piston position determine intake timing and duration.

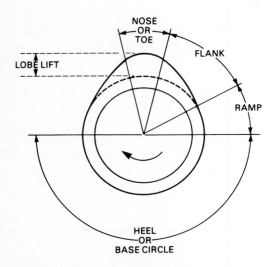

FIGURE 8-2. Cam lobe terminology.

PART 3   CAMSHAFT BEARINGS

Camshaft bearings are of the bushing type or split type and are similar in construction to crankshaft bearings. On some engines the camshaft bearings are all of equal diameter. On other engines the front camshaft bearings and journal are the largest, and the remaining bearings and journals are of progressively smaller diameter, with the smallest at the rear of the engine. This sizing makes it easier to remove and install the camshaft. Camshaft bearings are lubricated from oil galleries in the block or cylinder head.

PART 4   CAMSHAFT DRIVES

Camshafts are usually driven by gears marked for correct timing with the crankshaft. Some smaller diesel engines have a chain- or cog belt-driven camshaft. A V-type engine with two camshafts may have one camshaft turning clockwise and the other turning counterclockwise. (See Fig. 8-3 to 8-5.)

Camshaft drive wear results from torsional loads of driving the camshaft. Opening the valves against spring pressure causes a reverse twist on the camshaft, which has to be overcome by the camshaft drive. When the cam lobe passes the high lift

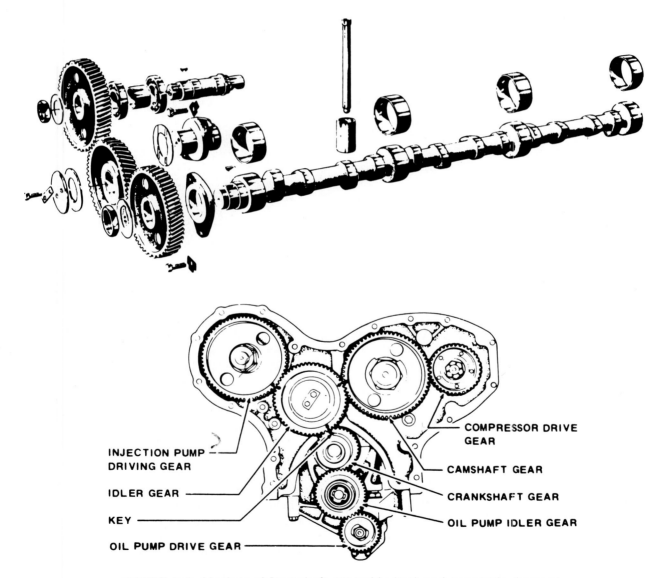

FIGURE 8-3. Mack truck camshaft assembly (top) and gear train (bottom). *(Courtesy of Mack Trucks Inc.)*

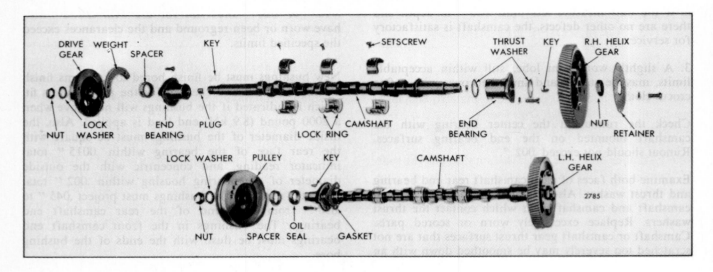

FIGURE 8-4. Camshafts and related parts for Detroit Diesel V92 engine. Note split-type camshaft bearings. *(Courtesy of Detroit Diesel Allison)*

point, torque is reversed. This torque reversal eventually results in camshaft drive mechanism wear. As the drive mechanism wears, valve timing is retarded, which reduces volumetric efficiency and therefore power. When wear is excessive on the chain and sprocket type of drive, it can cause the drive to jump one or more teeth and change valve timing. If valve timing is sufficiently affected, the valves will not close in time, and the piston will hit the valves, possibly causing valve and piston damage. This is also the case when cam drive gear teeth are broken or when a cam drive belt jumps a cog or more.

PART 5   CAM LOBE WEAR

Camshaft lobe wear is caused by friction between the cam lobe and the lifter or cam follower. Insufficient lubrication, excessive valve spring tension, excessive valve lash, hydraulic lifter failure, and dirty oil will contribute to early and rapid wear. Worn cam lobes retard valve timing, reduce valve opening, and decrease duration. This reduces volumetric efficiency and therefore engine power.

PART 6   ACCESSORY DRIVES, GEARS, AND COVERS

Engine gear drives include gears to drive such components as camshafts, balance shafts, crankshaft balancers, blowers, oil pumps, water pumps, injection pumps, and compressors. One or more idler gears may be required to transmit rotary motion from the crankshaft gear to the driven component. Most of these components operate in a definite sequence and relationship to the crankshaft and are therefore accurately timed to crankshaft position. All such gears have timing marks or indicators to assure correct assembly.

Gear covers are of cast-iron or die-cast aluminum construction. They keep the gears housed in a dirt-free and properly lubricated environment. Covers are attached with cap screws. A gasket type of seal prevents oil leakage and entry of dirt. Aligning dowel pins assures proper gear cover alignment with the engine block. A lip type of seal located in the gear cover provides sealing between the stationary gear cover and rotating crankshaft. The gear cover may incorporate outboard bearings for some of the gears.

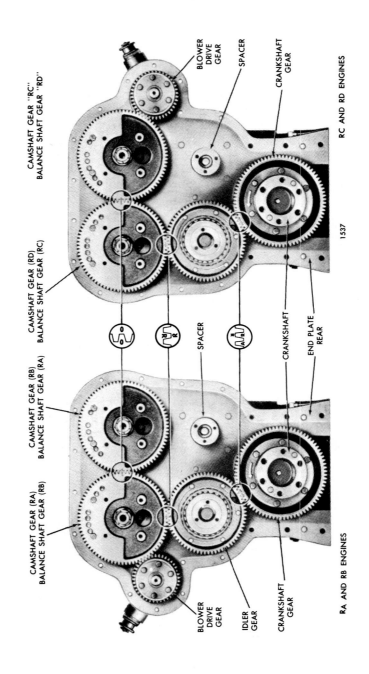

CAMSHAFT GEAR "RC"
BALANCE SHAFT GEAR "RD"

CAMSHAFT GEAR (RD)
BALANCE SHAFT GEAR (RC)

BLOWER DRIVE GEAR

SPACER

CRANKSHAFT GEAR

CAMSHAFT GEAR (RB)
BALANCE SHAFT GEAR (RA)

CAMSHAFT GEAR (RA)
BALANCE SHAFT GEAR (RB)

SPACER

CRANKSHAFT

END PLATE REAR

RC AND RD ENGINES

1537

BLOWER DRIVE GEAR

IDLER GEAR

CRANKSHAFT GEAR

RA AND RB ENGINES

FIGURE 8-5. Detroit Diesel single-camshaft gear train, showing blower drives on opposite sides and balance shaft drive gears. (*Courtesy of Detroit Diesel Allison*)

127

PART 7   SELF-CHECK

1. Why is a camshaft needed in a two-cycle engine? In a four-cycle engine?

2. The cam lobe converts _____ motion to _____ motion.

3. Why is it advantageous to open the exhaust valve before the piston reaches BDC?

4. Define valve overlap.

5. List three effects of worn cam lobes.

6. What is the camshaft-to-crankshaft speed ratio on a two-cycle engine? On a four-cycle engine?

7. Camshafts are made from what?

8. The camshaft controls the _____ and the _____, and in some engine designs also the _____.

9. List three types of camshaft drive mechanisms.

Chapter 9

Valve Operating Mechanisms

Performance Objectives

After thorough study of this chapter, the appropriate training models and manufacturer's service manuals you should be able to do the following:

1. Complete the self-check questions with at least 80 percent accuracy.

2. State the purpose of each of the engine components (a) to (g) below.

3. Describe the construction and various design differences of each of the engine components (a) to (g) below.

4. Describe the operation, wear, and deterioration of each of the engine components (a) to (g) below.

    (a) cam followers
    (b) hydraulic valve lifters
    (c) push rods and tubes
    (d) rocker arms
    (e) valve bridge or crosshead
    (f) decompressor
    (g) vehicle retarder

PART 1 CAM FOLLOWERS AND VALVE LIFTERS

Cam followers and valve lifters are camshaft-lobe operated. They are required to reduce wear on the camshaft and valve or injector operating push rods or tubes. (See Fig. 9-1 to 9-4.)

There are several types of cam followers and valve lifters. Included are mechanical and hydraulic valve lifters. Some are equipped with rollers at the base, while others simply slide over the cam lobe. Rollers reduce friction and wear. Hinged roller cam followers reduce these factors even more. Hinged roller lifters are used for injector operation as well as for valve operation. Many engine designs use springs to ensure that the cam follower remains in constant contact with the cam lobe. This increases cam and cam follower life.

The cam follower, as the name implies, is designed to follow the cam lobe surface or profile. This action converts rotary motion of the camshaft to reciprocating motion of the valve train and valves or injector plunger.

Cam follower types include the straight roller type; the rocker type, which also includes a roller; the mushroom type; and the hydraulic type. The hydraulic type is used only on smaller diesel engines.

The roller type of cam follower reduces friction and wear on both the camshaft and the cam follower since it provides for rolling friction rather than sliding friction.

Springs are used on some engines to keep the cam follower in proper contact with the cam.

PART 2 HYDRAULIC LIFTER OPERATION

Hydraulic lifter operation (Fig. 9-5) relies on engine oil pressure. When the intake or exhaust valve is closed and the lifter is on the base circle or heel of the cam lobe, engine oil pressure is fed into the lifter body and the lifter plunger. Oil flows through the

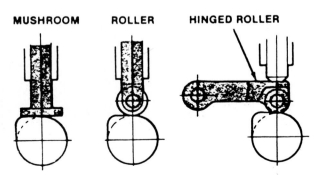

MUSHROOM   **ROLLER**   **HINGED ROLLER**

FIGURE 9-1. Typical mechanical valve lifter designs. Roller lifters have less friction. *(Courtesy of Ford Motor Co. of Canada Ltd.)*

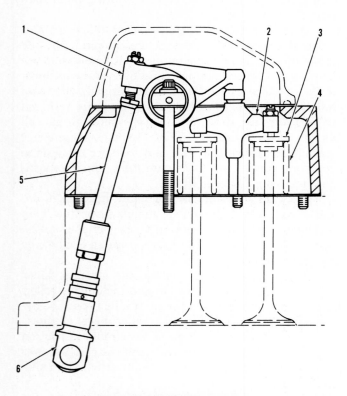

VALVE SYSTEM COMPONENTS

1. Rocker arm. 2. Bridge. 3. Rotocoil. 4. Valve spring. 5. Push rod. 6. Lifter.

FIGURE 9-2. Caterpillar 3512 engine valve train. *(Courtesy of Caterpillar Tractor Co.)*

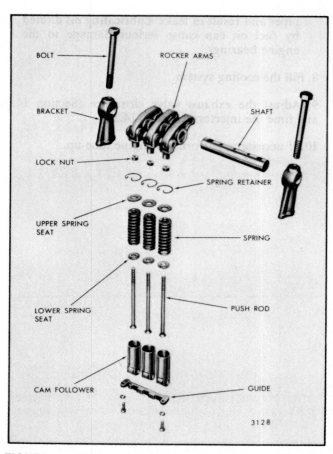

FIGURE 9-3. Detroit Diesel V-71 cam followers and related parts. *(Courtesy of Detroit Diesel Allison)*

FIGURE 9-4. Cummins Big Cam engine cam followers are mounted in separate housing. Note larger roller center cam follower for injector operation. *(Courtesy of Cummins Engine Company Inc.)*

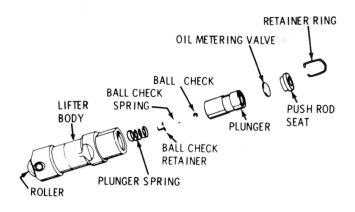

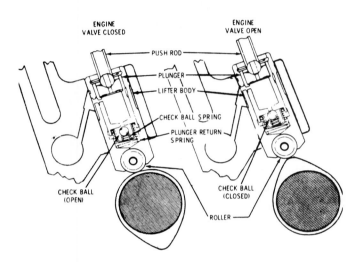

FIGURE 9–5. Exploded view of hydraulic roller type valve lifter (top) and operation (bottom). *(Courtesy of Ford Motor Co. of Canada Ltd.)*

check valve and fills the area below the lifter plunger. This takes up all valve lash. As the camshaft rotates, the cam lobe begins to push against the bottom of the lifter body. Valve-spring pressure through the rocker arm and the pushrod attempt to keep the lifter plunger down. This causes pressure in the lifter below the plunger to increase and close the check valve. The oil is trapped below the plunger, and the lifter in effect becomes solid since oil is not compressible.

A small amount of oil leakage past the plunger allows the lifter to leak down should the lifters pump up. Lifter pump-up is caused by anything that causes even momentary clearance anywhere in the valve-operating train. Valves that are sticky in the valve guides, weak valve springs, and overspeeding of an engine can cause lifter pump-up. Another reason for this slight leakage past the plunger is to al-

low oil to escape as parts expand, as their temperature increases. If oil could not escape, the valves would not be able to seal. A metering disc just below the pushrod seat meters the amount of oil delivered through the pushrod to the rocker arm for lubrication.

Lifter and Cam Follower Wear

Lifter wear is most evident on the lifter base, which is subjected to high loads and the friction of the cam lobe. In hydraulic lifters excessive leakage past the plunger or a leaking check valve can cause lifters to bottom out and cause a noisy valve train and retarded valve opening. Varnish buildup inside the lifter can cause the hydraulic lifter to stick, causing the same problems. Cam follower wear takes place on the rollers, roller pin, and bushing and pushrod seat. Stress cracks may also develop. Cracked cam followers must be replaced. Cam followers that are not cracked can usually be rebuilt by replacing the worn parts.

PART 3   PUSH RODS AND TUBES

Push rods and tubes transfer cam follower motion to the rocker arms and valve bridge. Push rods may be solid or tubular, hence the two names. Push tubes have reduced weight, which reduces valve train or injector train inertia. The top and bottom are usually designed with a ball or cup type seat to engage the cam follower or rocker arm.

Push rods can become bent as the result of incorrect valve or injector adjustment, sticking valves, or incorrect valve timing. Bent push rods reduce valve opening or injector plunger travel and increase lash with the resulting increase in valve train wear. Balls and seats may also wear, causing the same problems. Excessively worn or bent push rods must be replaced.

PART 4   ROCKER ARMS AND VALVE BRIDGE

Rocker arms or levers transmit and reverse push rod movement to the valves or injectors. Rocker arms that are center pivoted cause movement at each end of the arm to be equal but in opposite directions. If the pivot point is off center toward the push rod, valve opening (or injector plunger travel) is increased proportionally. In some engine designs the camshaft bears directly on the rocker arms, thereby eliminating the need for push rods and their built-in delay factor due to deflection. (See Fig. 9-6 to 9-8.)

A valve bridge is used on engines with four valves per cylinder. This allows one cam to operate two valves. The valve bridge is also known as a crosshead. A bridge guide is located between the two valves it operates to keep the bridge in alignment.

Rocker arms are usually mounted on shafts, which act as the pivot.

Rocker arms, shafts, and pivots wear, owing to friction from valve-train operation. Rocker shaft wear is easily detected at the bottom of the shaft. Rocker arms wear at the valve steam end as well as at the pushrod end. The bottom of the rocker shaft hole in the rocker arm will also wear. The valve bridge will wear where it contacts the valve, the pushrod or pin, and the guide.

PART 5   DECOMPRESSOR— COMPRESSION RELEASE SYSTEM

Many diesel engines are provided with a method of releasing cylinder compression in order to make it easier to crank the engine for starting. The system is designed to hold all the intake valves or all the exhaust valves open during cranking until cranking speed is high enough to ensure that the engine will start when compression is restored (Fig. 9-10).

The mechanism utilizes a shaft equipped with pins that act on the valves when the shaft is turned. Another method is to use a shaft that has flat sections on one side. The round side pushes against the rocker arms, thereby holding the valves open when engaged.

A system of linkage allows the operator to engage the decompressor from the operator's seat. Once the engine is cranking over fast enough for starting, the decompressor is disengaged, allowing the engine to start and restoring normal valve operation.

PART 6   VEHICLE RETARD SYSTEMS

Retard systems (Fig. 9-11 to 9-15) are designed to retard vehicle speed when descending a hill. This reduces the need to use the vehicle brakes, thereby extending their life and reducing the potential for overheated brakes. Several design types are used by different vehicle manufacturers. They can be classified as follows:

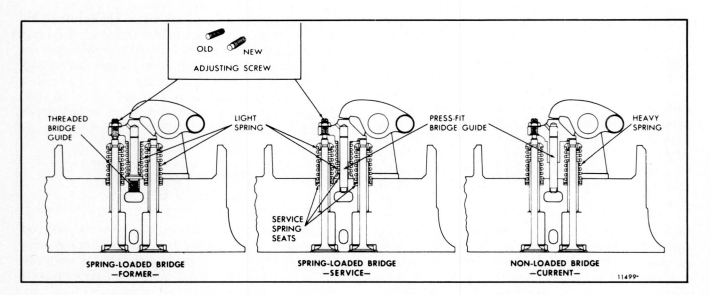

FIGURE 9-6. Different valve bridge designs for Detroit Diesel V-71 engine with four exhaust valves per cylinder. *(Courtesy of Detroit Diesel Allison)*

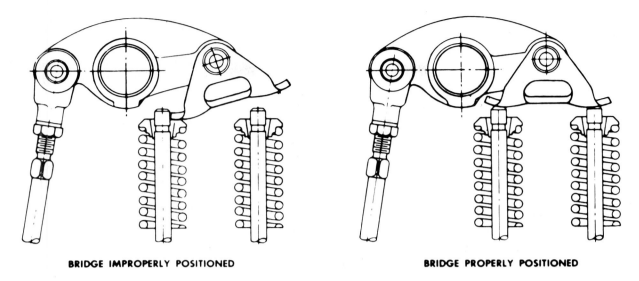

BRIDGE IMPROPERLY POSITIONED   BRIDGE PROPERLY POSITIONED

FIGURE 9-7. One type of valve bridge, showing incorrect and correct operating position. *(Courtesy of Detroit Diesel Allison)*

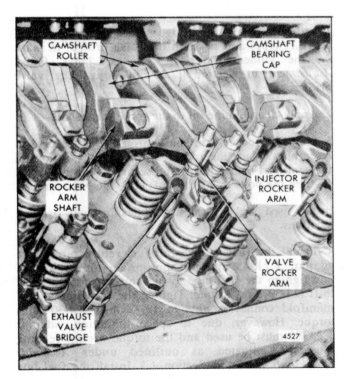

FIGURE 9-8. In this Detroit Diesel series 149 engine the camshaft acts directly on the combination cam follower and rocker arm. *(Courtesy of Detroit Diesel Allison)*

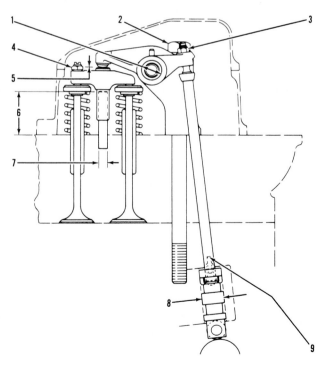

FIGURE 9-9. Valve train components of 3408 Caterpillar engine. *(Courtesy of Caterpillar Tractor Company)*

133

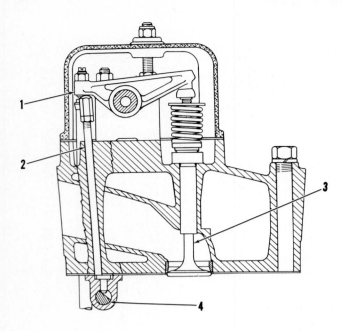

FIGURE 9-10. Compression release mechanism (earlier engines). 1. Inlet valve rocker arm. 2. Compression release push rod. 3. Inlet valve. 4. Compression release shaft. *(Courtesy of Caterpillar Tractor Company)*

1. Those that act on the engine, using it as an air compressor that makes the engine harder to turn.

2. Those that act on the vehicle drive train by electrical or hydraulic means in a manner that retards vehicle movement without mechanical friction such as that used in wheel brakes.

Engine Braking Systems

Vehicle retard systems that act on the engine include the Jacobs brake used by several engine manufacturers, the Mack Dynatard system used on Mack trucks, and the older butterfly exhaust brake used on some older vehicles.

The Jacobs brake converts the engine cylinders into power-absorbing air compressors. On engines such as Cummins and Detroit Diesel, the injector push rod is used to actuate a master piston which is connected to a slave piston positioned to open the exhaust valve at the end of the compression stroke. This can happen only when the engine throttle is closed, the clutch is engaged, and a switch on the dash panel is in the on position. The dash switch activates a solenoid that opens a valve to allow engine oil pressure to activate the brake. Jacob's brake oil is instantly dumped through a separate passage when the dash switch is turned off.

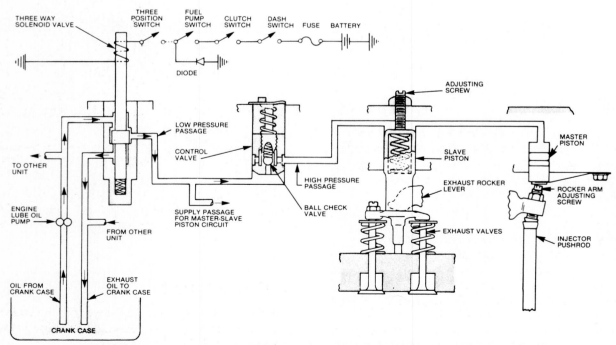

FIGURE 9-11. Typical Jacobs Brake operation. *(Courtesy of Cummins Engine Company Inc.)*

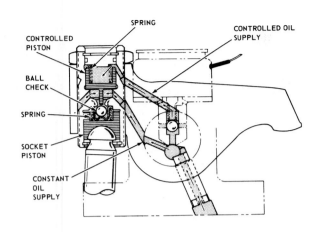

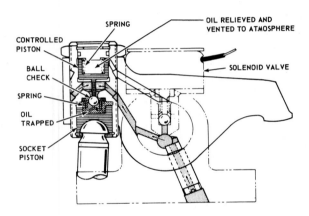

FIGURE 9-12. Mack Dynatard lash adjuster operation brake off (top) and brake on (bottom). *(Courtesy of Mack Trucks Inc.)*

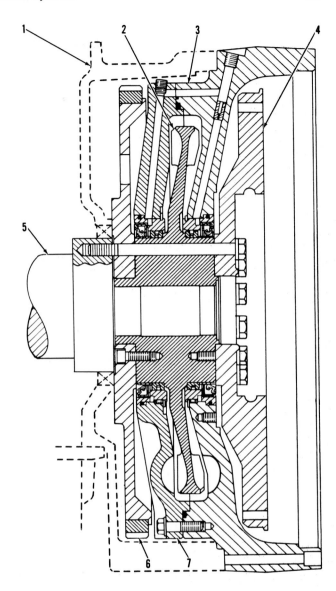

1. Flywheel housing. 2. Rotor. 3. BrakeSaver housing. 4. Flywheel. 5. Crankshaft flange. 6. Ring gear plate. 7. Stator.

FIGURE 9-13. Hydraulic engine retarder, BrakeSaver, used on Caterpillar truck engines. *(Courtesy of Caterpillar Tractor Company)*

Compressed air from the engine cylinders is vented through the exhaust valves near the end of the compression stroke, thereby preventing the stored energy of the compressed air from acting on the pistons and crankshaft on the downstroke of the piston. The Mack Dynatard system operates on the same principle.

The butterfly type exhaust brake utilizes a valve to close the exhaust pipe, which created backpressure. As the pressure built up, the engine had to work harder to try to push more exhaust out of the cylinders, thereby retarding the vehicle. Maximum pressure in the system was prevented from exceeding safe limits by the use of an automatic venting system. An air-operated cylinder controlled the position of the brake valve in response to a driver-operated dash control.

Hydraulic Retarder

Hydraulic retarders are used in the drive train and can be mounted at the rear of the engine between the engine and the transmission.

This unit consists of a vaned drive member (rotor) and a vaned stationary member (stator) en-

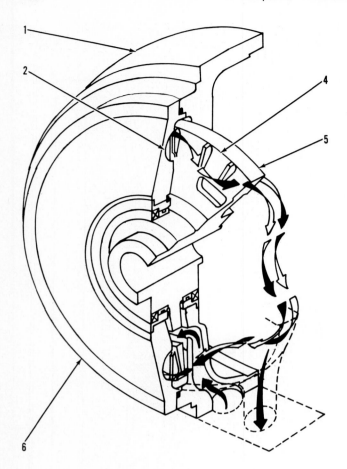

**1. BrakeSaver housing. 2. Pocket. 4. Pocket. 5. Rotor.
6. Stator.**

FIGURE 9-14. Oil flow through BrakeSaver. *(Courtesy of Caterpillar Tractor Company)*

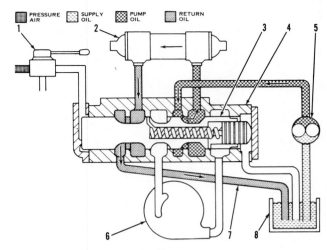

BRAKESAVER OIL FLOW (OFF)

**1. BrakeSaver control lever. 2. Oil cooler. 3. Valve spool.
4. BrakeSaver control valve. 5. Oil pump. 6. BrakeSaver.
7. Line. 8. Oil pan.**

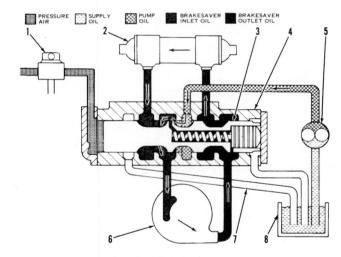

BRAKESAVER OIL FLOW (FILL)

**1. BrakeSaver control lever. 2. Oil cooler. 3. Valve spool.
4. BrakeSaver control valve. 5. Oil pump. 6. BrakeSaver.
7. Line. 8. Oil pan.**

FIGURE 9-15. BrakeSaver oil flow. *(Courtesy of Caterpillar Tractor Company)*

closed in a housing. The engine crankshaft drives the rotor. The stator is mounted so that it cannot turn. With no oil in the unit, it is ineffective. When activated, engine oil is pumped into the unit, which creates a hydraulic coupling. Oil is picked up by the vanes of the spinning rotor and thrown against the vanes of the stator. The turbulent churning action of the oil creates enough friction to create a braking effect sufficient to retard the vehicle. As more oil is pumped into the unit, the braking effect is increased. When the vehicle operator disengages the unit, the oil supply is cut off and the rotor quickly pumps out the remaining oil. The system requires no maintenance or periodic adjustment.

Electrical Retarder

Electric retarders are used on some large off-road equipment. In these units diesel-generated electric motors are used to drive the wheels. These motors are electrically connected in a manner that converts them to generators when the retard mode is selected. This system is beyond the scope of this text.

PART 7   SELF-CHECK

1. What is the purpose of the cam follower and rocker lever?

2. Are push rods or tubes required on all engines? Explain your answer.

3. How can rocker lever design be used to increase valve opening?

4. What is the purpose of the valve bridge?

5. Why are some engines equipped with a compression release system?

6. Briefly describe how a vehicle retard system using the engine cylinders as air compressors works.

Chapter 10

Cylinder Heads and Valves

Performance Objectives

After thorough study of this chapter, the appropriate training models and manufacturer's service manuals you should be able to do the following:

1. Complete the self-check questions with at least 80 percent accuracy.

2. State the purpose of each of the engine components (a) to (e) below.

3. Describe the construction and various design differences of engine components (a) to (e) below.

4. Describe the operation, wear, and deterioration of engine components (a) to (e) below.

    (a) cylinder head

    (b) cylinder head gasket

    (c) intake and exhaust valves, guides, seats, seals, and ports

    (d) valve springs, retainers, and rotators

    (e) valve bridge or crosshead

PART 1   CYLINDER HEADS

The cylinder head forms the top or lid for the engine cylinders. It contains the valves that open and close the intake and exhaust ports that lead from the combustion chamber to the exhaust manifold and from the intake manifold to the combustion chamber. It also contains the valve guides that keep the valves in position, the valve seats, water passages, provision for mounting fuel injectors, threaded holes for attaching valve operating parts, and other accessories as well as a number of non-threaded holes for attaching the cylinder head to the block. Machined surfaces allow for positive sealing of combustion chambers and other openings with the proper gaskets in between. (See Fig. 10–1 to 10–4.)

Some in-line diesel engines use a single-cylinder head assembly that serves all cylinders, usually up to six. A cylinder head for each pair of cylinders is also used on some engines. Large engines normally use a separate cylinder head for each cylinder.

Cylinder heads are usually of cast iron alloy construction. The casting process is similar to that used for casting engine blocks. Internal passages are created by a sand core during casting, after which the sand is removed. Other openings and surfaces are machined. Particular attention is given to provide adequate coolant flow over the exhaust seat areas and may include coolant nozzles that direct coolant at the hottest areas.

Cylinder head problems include cracking that causes fluid or pressure leaks either internally or externally, rust and scale buildup in the water jackets resulting in poor heat conduction, warpage, valve guide and bridge guide wear, and valve seat wear or damage.

A stamped steel, cast iron, or aluminum cylinder head cover and gasket are used to cover the valve and injector operating parts and to prevent oil leakage and entry of dirt.

Cylinder Head Gaskets

Cylinder head gaskets are designed to provide a positive seal between the head and engine block. One of the primary jobs of the gasket is to prevent any escape of combustion pressures from the cylinders. Very high pressures occur in the cylinder and no leakage must be allowed. Other jobs include sealing and metering coolant flow, sealing lubrication passages, and the like.

Materials used in head gaskets include steel, copper, asbestos, fiber, synthetic rubber, and silicone. Gasket materials must be compatible with the fluids and gases they are required to seal and must not deteriorate from contact with them. Gaskets are also subject to a wide range of temperature and pressure extremes, as well as to metal expansion and contraction. In spite of all these hazardous conditions, properly installed gaskets last a very long time.

Gasket design includes the single-piece and multiple-piece types. The multi-piece types may include a separate fire ring for each cylinder as well

Two Cylinders

One Cylinder

Six Cylinders

Four Cylinders

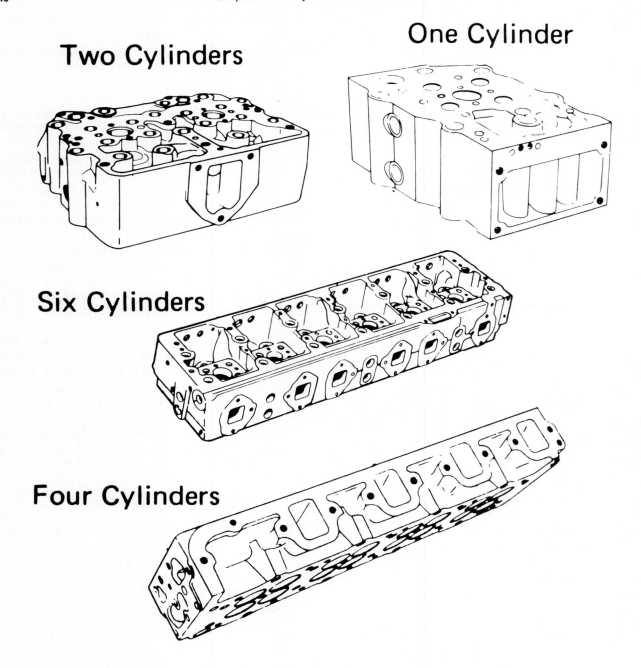

FIGURE 10-1. Various cylinder head designs. *(Courtesy of Cummins Engine Company, Inc.)*

as any number of O rings and seals to seal individual openings. Some of these fit into recesses to ensure proper positioning and compression.

Some engine manufacturers may require a sealant to be used on certain applications. These should be used only as and where specified.

PART 2   INTAKE AND EXHAUST VALVES

Depending on engine size and design, there may be one, two, or four exhaust valves for each cylinder

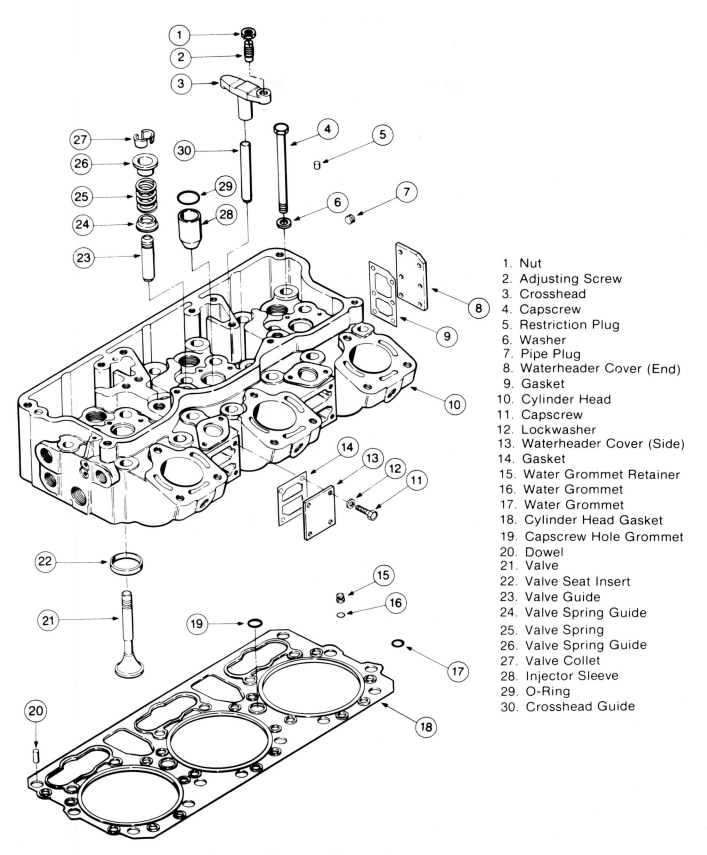

1. Nut
2. Adjusting Screw
3. Crosshead
4. Capscrew
5. Restriction Plug
6. Washer
7. Pipe Plug
8. Waterheader Cover (End)
9. Gasket
10. Cylinder Head
11. Capscrew
12. Lockwasher
13. Waterheader Cover (Side)
14. Gasket
15. Water Grommet Retainer
16. Water Grommet
17. Water Grommet
18. Cylinder Head Gasket
19. Capscrew Hole Grommet
20. Dowel
21. Valve
22. Valve Seat Insert
23. Valve Guide
24. Valve Spring Guide
25. Valve Spring
26. Valve Spring Guide
27. Valve Collet
28. Injector Sleeve
29. O-Ring
30. Crosshead Guide

FIGURE 10–2. Exploded view of one of four heads for a V12 Cummins engine. *(Courtesy of Cummins Engine Company Inc.)*

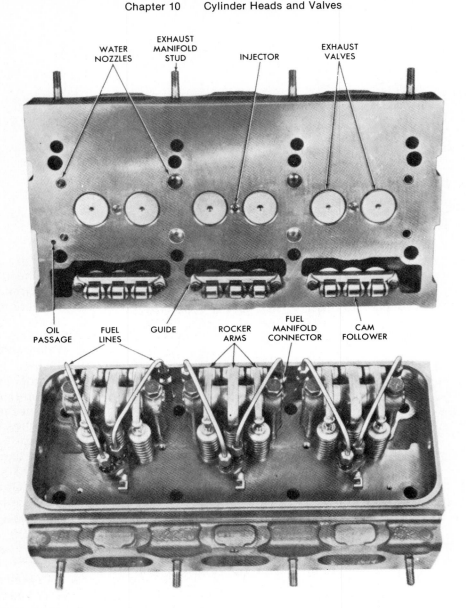

FIGURE 10-3. Detroit Diesel two-exhaust-valve V6 cylinder head. *(Courtesy of Detroit Diesel Allison)*

and one, two, or no intake valves per cylinder. Two-stroke-cycle engines do not have any intake valves, while four-stroke-cycle engines have both intake and exhaust valves. The four-valve-per-cylinder design provides better engine breathing than the two-valve design since it provides more total ported area for any cylinder diameter than the two-valve design. (Fig. 10-5 to 10-8.)

The *valve assemblies* open and close the intake and exhaust ports that connect the intake and exhaust manifolds to the combustion chamber. The valve assembly includes the valve, valve seat, spring, spring retainer, locks, valve guide, and seal. A rotator is sometimes used instead of a retainer.

Poppet-type intake and exhaust valves are used in diesel engines. The intake valve head diameter is

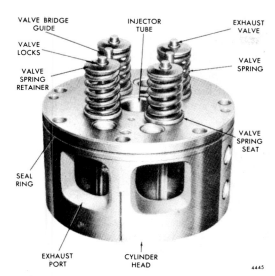

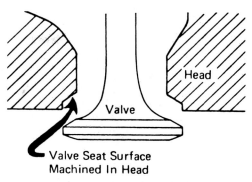

FIGURE 10-6. Integral valve seat is simply machined into cylinder head material. *(Courtesy of Cummins Engine Company Inc.)*

FIGURE 10-4. Pot-type four-exhaust-valve cylinder head for two-stroke diesel engine. *(Courtesy of Detroit Diesel Allison)*

usually larger than the exhaust valve head, because the intake valve and port handle the slow-moving air (on naturally aspirated engines), whereas the piston moves the exhaust gases out with more positive force.

The valve stem passes through the valve guide in the port, and the valve head closes the port when the valve is seated. The valve seat is located at the combustion chamber end of the ports. The valve spring is located on the spring seat and keeps the

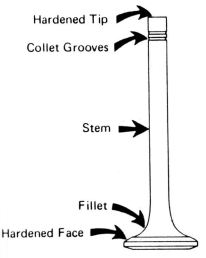

FIGURE 10-5. Valve terminology. *(Courtesy of Cummins Engine Company Inc.)*

valve seated until the camshaft causes the valve to open against spring pressure. The spring is held in place by a spring retainer that is locked to the valve stem by two split locks. The valve stem seal is located on the valve stem to prevent engine oil from entering the combustion chamber or exhaust manifold through the valve guide.

Intake and exhaust valves are designed to operate over a long period of time with relatively little attention or problems. Valve heads are sufficiently heavy for good heat capacity, yet light enough not to cause valve float very readily. Valve face and seat angles can be 30°, 45°, or 60°. Seating angles are required to provide a positive seal. The wiping action of angled seating surfaces helps clear minor carbon particles that could prevent proper sealing. The wedging action of the angled surfaces also contributes to better sealing. Valves are made of high-grade steel alloy for long service life.

Intake and exhaust valve materials include alloy steel valves with an aluminized face and chrome stem, Silchrome valve with an aluminized face, stainless steel, austenitic steel with aluminized face and chrome stem, and SAE 21-2 steel with a nickel-plated face. Exhaust valves with sodium-filled stems are sometimes used for better valve cooling. At operating temperature, the sodium is liquefied. Valve movement causes the sodium to transfer heat from the head of the valve to the valve stem and then to the valve guide.

Exhaust valve temperature may reach approximately 1300° to 1500°F (704° to 815°C). This means that they are in fact running red-hot. Good heat transfer therefore is essential.

It is important that valves be fully seated when they are closed. The exhaust valve is closed approximately two-thirds of the time on four-stroke en-

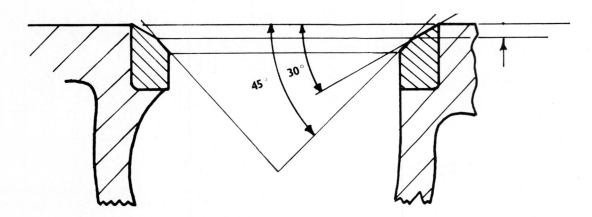

FIGURE 10-7. Valve seat insert and seat angles. A 30° overcut has been used to narrow the 45° valve seat to its proper width. Common valve seat angles are 30°, 45°, and 60°.

gines while the engine is running. It is during this time that a large part of the heat transfer from the valve head to the seat takes place. The heat from the seat is transferred to the engine coolant. The remaining heat transfer takes place from the valve stem to the valve guide to the engine coolant. The sodium-filled valve increases the amount of heat transfer from the valve head to the stem, which helps the valve head to run cooler.

Intake valve temperatures are considerably less than exhaust valve temperatures. Incoming air cools the intake valve while the valve is open.

The valve seat has a great deal to do with good heat transfer. A valve seat that is too narrow will

not absorb sufficient heat from the valve head and will wear both seat and valve more rapidly. A seat that is too wide is not able to clear carbon particles as readily, will therefore not seal as well, and can cause valve and seat burning. Some engines have integral valve seats that are induction-hardened. Most diesel engines have replaceable alloy-steel valve seat inserts.

Valve guides keep the valves in proper alignment with the valve seat during operation. Valve guides are either integral or replaceable.

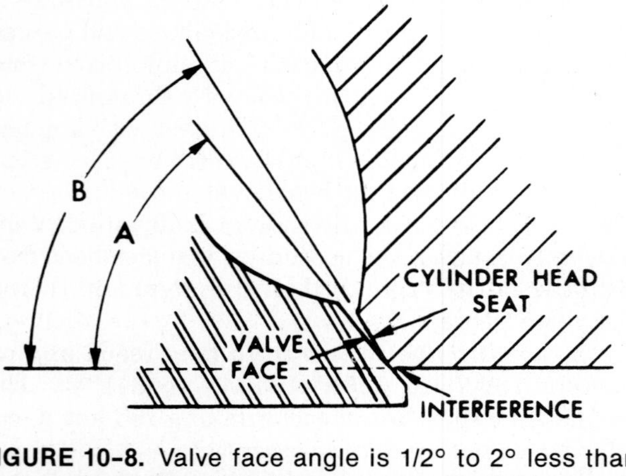

PART 3   VALVE, SEAT, AND GUIDE WEAR

Valves, seats, and guides will wear from normal friction, heat, and pressures. Valve stems and valve guides will wear, resulting in increased valve stem-to-guide clearance. Valve seats and valve faces will also wear eventually from continued opening and closing of the valves. Rocker arm action causes both rocker arms and valve stem tips to wear. The rocking action of the rocker arm also creates a side pressure on the valve stem, which contributes to valve stem and guide wear. Byproducts from normal combustion, such as carbon, acids, and moisture, also have a bearing on the wear of these parts. (See Fig. 10-9 and 10-10.)

Even in normal use the intake and the exhaust valves absorb a fantastic amount of punishment. Extreme temperatures scorch it many times each second, and powerful spring tension pounds the red-hot valve head. Hot gases under tremendous pres-

FIGURE 10-8. Valve face angle is 1/2° to 2° less than seat angle B. This interference angle provides a positive seal on the combustion chamber side of the valve and seat. *(Courtesy of General Motors Corporation)*

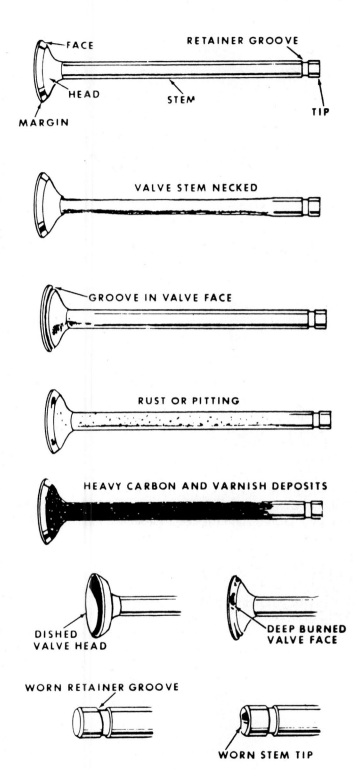

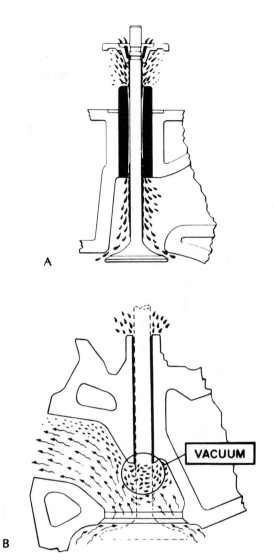

FIGURE 10-10. (A) Vacuum during the intake stroke can draw oil into the combustion chamber, past the valve guide, if clearance is excessive or stem seals are faulty or missing. (B) As exhaust gases are expelled past the valve guide area, a low-pressure area is created, causing oil to pass by the valve guide into the exhaust gases.

FIGURE 10-9. Most of the common valve problems are shown here. *(Courtesy of J. I. Case, a Tenneco company)*

sures swirl past it. Carbon deposits form on the face, preventing the valve from seating properly or cooling efficiently. As a result, the valves, particularly the exhaust valves, become pitted, burned, warped, and grooved. No longer concentric with the valve seat, they leak compression and fail to dissipate heat. Engine efficiency and economy nosedive.

The valve seat also wears. Hot gases burn it. Carbon particles that retain heat pit it. The valve guide wears in a corresponding position to the valve stem. Between stem and guide, carbon residues form, which cause the valves to stick. Excessive valve lash or excessive valve-spring pressures can cause high-velocity valve seating and rapid valve and seat wear. Valves can stick in the guides as a result of bent valves, insufficient stem-to-guide clearance, incorrect cylinder head tightening, and resulting guide distortion and varnish and carbon deposits.

Valve float refers to the condition when a valve is not closing fast enough. Weak or broken valve springs can cause valve float, which reduces the time a valve is seated and therefore increases valve temperatures. Valve float can also result in bent valves from pistons hitting slow-moving valves. If the valve is not out of the way by the time the piston comes up, it may hit the valve and bend it. This condition is usually experienced at high engine speeds or when maximum engine speed is exceeded during downhill engine braking.

Malfunctioning valve rotators can also contribute to increased valve and seat wear.

PART 4   CRACKED, BURNED, AND BROKEN VALVES

High engine speeds and loads, poor valve seating due to carbon deposits, warpage, misalignment, or insufficient lash, preignition, and detonation are factors that cause excessive valve temperature and cracked or burned or broken valves. High temperatures combined with combustion byproducts (acids and moisture) and high-velocity gases can erode the valve stem sufficiently to weaken the stem, resulting in valve breakage. (See Fig. 10–11 to 10–13.)

Carbon deposits on the valves are a result of fuel residues and engine oil being burned. Engine oil can enter the combustion chamber past the valve stems if valve stem and guide wear are excessive. Engine oil can also get past worn cylinders and piston rings. Carbon deposits in the combustion chamber can cause preignition and detonation, which increase combustion and valve temperatures.

Poorly seated valves allow hot-burning gases to escape past the valve with a cutting torch effect and can cause valves to be burned and cracked. Severely cracked valves can have pieces break away and cause piston and cylinder damage. A valve head that has separated from the stem also causes severe piston, cylinder, and cylinder head damage. Stem erosion and acid etching can cause this kind of valve breakage.

FIGURE 10–11. Valve cracked from excessive heat. *(Courtesy of Deere and Company)*

Excessive oil consumption can be caused by excessive valve stem-to-guide clearance or hardened and cracked valve stem seals. Excessive guide and valve stem wear will allow too much heat to get to the seals and cause them to harden and crack.

Uneven wear patterns develop at the valve stem tip if rotators are not functioning.

Some engine designs mount the rocker assemblies directly to the cylinder head. Other engines may have a separate cast-iron rocker housing, which is bolted to the cylinder head. Both designs use a rocker cover. Several types of these rocker covers are used, such as the stamped-steel, die-cast aluminum, and cast-iron covers. In many cases the rocker cover is provided with an oil filler cap, and in some cases the crankcase breather may be connected to the cover. The cover is sealed against entry of dirt and leakage of oil by a gasket or by a self-curing silicone sealer.

FIGURE 10–12. Valve burned from too little tappet clearance. *(Courtesy of Deere and Company)*

FIGURE 10-13. Valve breakage may be caused by fatigue or by impact. *(Courtesy of Deere and Company)*

PART 5   VALVE SPRINGS, ROTATORS, AND CROSSHEADS

Springs

Valve springs are needed to close the valves. A spiral winding of high-grade spring steel is ground flat at each end for square seating and even pressure distribution. Valve-spring rates are designed to cause the valve-operating mechanism to follow the cam lobe profile for controlled opening and closing of the valve. A valve spring that is not square (tilted) will cause a valve to close only on one side of the seat.

Valve-spring action is dampened by spring design or by a separate spring dampener to reduce spring vibrations. A variable-rate valve spring (coned spring or unequally spaced coils) is one method of spring dampening.

A reverse-wound secondary spring inside the main valve spring reduces spring vibration by friction as well as different vibration frequencies.

Rotators

Valve rotators are used to cause the valve to rotate slightly each time that the valve is operated. Some engines use rotators only on the exhaust valves; others have them on both intake and exhaust valves. Slight rotation of the valve promotes even wear distribution on the seat, valve, and tip and helps in removing small particles between the valve and the seat. Better valve seating and longer valve life are the result. (See Fig. 10-14.)

Several types of rotators are used. Positive rotator types include the spring-loaded ball and ramp type and the single circular coil spring type. These two types ensure some valve rotation each time that the valve is operated. The spring-loaded ball and ramp type will cause the valve to be rotated in one direction only. The single circular coil spring type can cause the valve to rotate in either direction.

The free type of valve rotator releases spring tension from the valve momentarily during opening so that the valve is free to rotate. This type does not cause positive valve rotation. Engine vibration and turbulence of gases contribute to valve rotation during the time the valve is free to rotate.

Some positive rotators are designed to operate between the valve spring and the cylinder head. Others are designed to operate at the valve tip in place of the spring retainer. The free-type rotator also operates at the valve tip.

Valve Crosshead or Bridge

Engines that have four valves per cylinder use a valve bridge or crosshead to allow each rocker arm to operate two valves at once. The bridge is mounted on a machined guide pin which is installed in the cylinder head between the two valves. The pin is usually a press fit in the cylinder head. One end of the bridge contacts one valve stem tip with a hardened wear surface. The other end has an adjusting screw and locknut for setting crosshead adjustments. A hardened contact pad in the top center is provided for rocker arm contact. As the rocker arm bears down on this pad, the two valves are opened at the same time. The crosshead slides down and up on the guide pin as the rocker arm operates the valves. (See Fig. 10-16.)

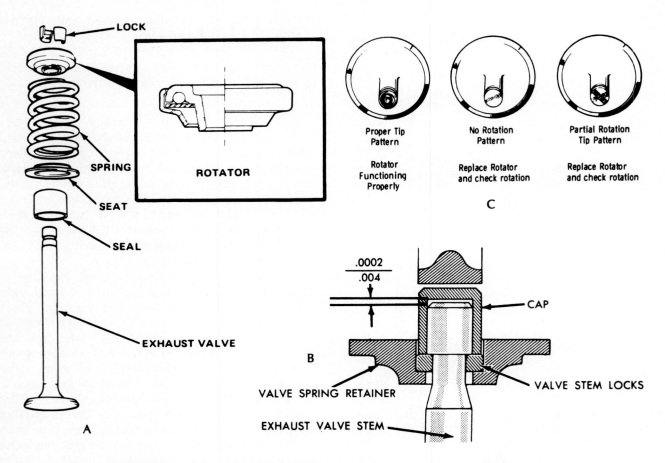

FIGURE 10-14. (A) Positive type of valve rotator. Valve rotation helps the valve and seat to maintain a positive seal. (B) Free type of valve rotator. (C) Wear patterns on the valve tip are good indicators of rotator malfunction. *(Courtesy of General Motors Corporation)*

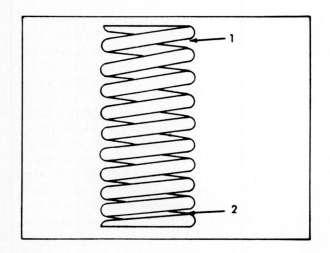

INTAKE AND EXHAUST VALVE SPRING

1. **Retainer End (top)**
2. **Damper End**

FIGURE 10-15. Note spacing of spring coils to provide dampening. *(Courtesy of Allis-Chalmers)*

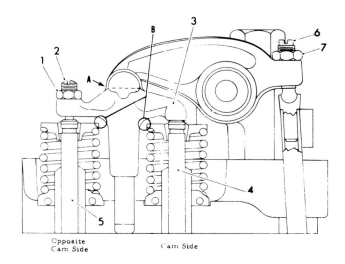

Opposite
Cam Side   Cam Side

1. Bridge Adjusting Screw Nut
2. Adjusting Screw
3. Bridge
4. Valve (cam side)
5. Valve (opposite cam side)
6. Rocker Lever Screw
7. Rocker Lever Screw Nut

FIGURE 10–16. Valve bridge components. Note critical clearance at circled areas at B. *(Courtesy of Deere and Company)*

PART 6   SELF-CHECK

1. What is the purpose of the intake and exhaust valves?
2. Describe how the exhaust valves are cooled.
3. An engine with four valves per cylinder could have two _____ and two _____ valves or four _____ valves.
4. What is the purpose of the crosshead or bridge?
5. Name three types of cylinder head design.
6. How can valve rotator malfunction be detected?
7. What methods of valve spring dampening are used?
8. What causes valve guide wear?
9. What causes exhaust valves to become burned or cracked?

Chapter 11

Lubrication Systems

Performance Objectives

After thorough study of this chapter and sufficient practice on the appropriate engine components, you should be able to:

1. Complete the self-check questions with at least 80 percent accuracy.

2. Describe the purpose, construction, and operation of the lubrication system and its components.

3. Diagnose lubrication system problems.

4. Recondition or replace faulty components.

5. Test the reconditioned system to determine the success of the service performed.

PART 1   LUBRICATION PRINCIPLES

Engine oils are the lifeblood of the internal combustion engine. The quality and performance are often taken for granted, without full appreciation of the severe service conditions under which the engine oil must function. Lubrication must be accomplished despite highly oxidizing conditions, very high temperatures, and the presence of large amounts of contaminants. The high output of today's engines, coupled with extended drain intervals, has increased the severity under which the oil must perform. (See Fig. 11-1 and 11-2.)

Extensive research and in-service field testing are required to develop oils meeting the increasing severity demanded by today's engines.

Engine oils are formulated from selected lube basestocks and fortified with the right additives to provide the performance level required.

Engine Oil Functions

The engine oil must perform many direct functions, while not causing any debits in other areas of engine performance.

Lubrication

The oils must provide a fluid film between all moving engine parts to reduce friction, heat, and wear. Friction and wear are caused by metal-to-metal contact of the moving parts. Wear is also caused by acidic corrosion, rusting, and the abrasion from the contaminants carried in the oil.

Sealing

High combustion pressures are encountered. Piston rings require an oil film between ring and liner and between ring and piston groove to seal against these high pressures and to prevent blowby.

Cooling

The engine oil is largely responsible for piston cooling. This is done by direct heat transfer through the oil film to the cylinder walls and on to the cooling system. Heat is also carried by the oil from the undercrown and skirt to the crankcase. Thermally stable oils are required to withstand the high temperatures encountered.

Deposit Control

Rings must remain free to function properly and to maintain a good seal. Deposit buildup in the ring grooves and on piston lands must therefore be controlled.

Varnish Control

Engine parts, particularly the piston, must be kept free of varnish to ensure performance and proper cooling.

Sludge Control

High- and low-temperature sludge-forming contaminants must be held in suspension and not allowed to drop out and accumulate. The larger particles are removed by the filter. Sludge and abrasives are removed with the oil when drained.

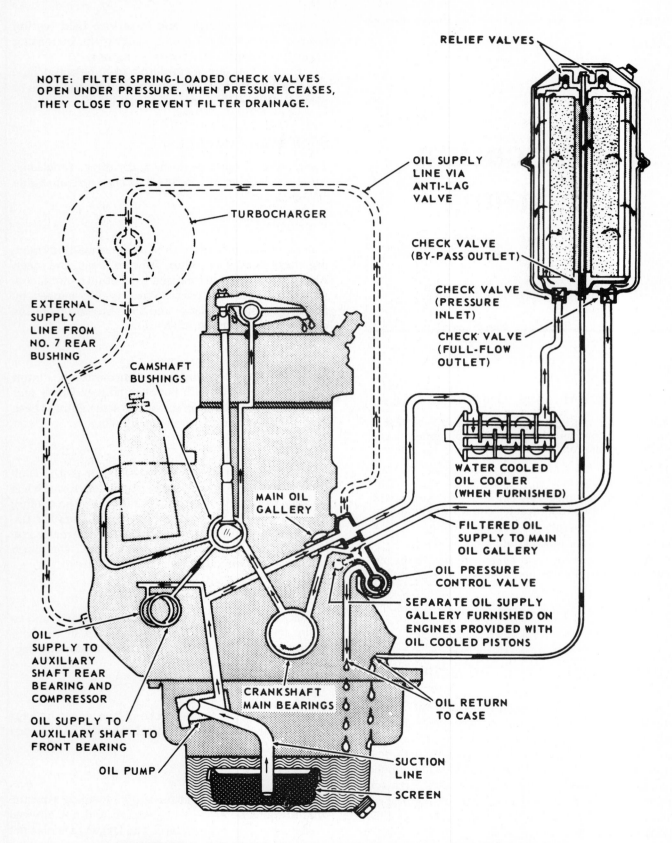

NOTE: FILTER SPRING-LOADED CHECK VALVES OPEN UNDER PRESSURE. WHEN PRESSURE CEASES, THEY CLOSE TO PREVENT FILTER DRAINAGE.

RELIEF VALVES

OIL SUPPLY LINE VIA ANTI-LAG VALVE

TURBOCHARGER

CHECK VALVE (BY-PASS OUTLET)

CHECK VALVE (PRESSURE INLET)

CHECK VALVE (FULL-FLOW OUTLET)

EXTERNAL SUPPLY LINE FROM NO. 7 REAR BUSHING

CAMSHAFT BUSHINGS

WATER COOLED OIL COOLER (WHEN FURNISHED)

MAIN OIL GALLERY

FILTERED OIL SUPPLY TO MAIN OIL GALLERY

OIL PRESSURE CONTROL VALVE

SEPARATE OIL SUPPLY GALLERY FURNISHED ON ENGINES PROVIDED WITH OIL COOLED PISTONS

OIL SUPPLY TO AUXILIARY SHAFT REAR BEARING AND COMPRESSOR

OIL RETURN TO CASE

CRANKSHAFT MAIN BEARINGS

OIL SUPPLY TO AUXILIARY SHAFT TO FRONT BEARING

OIL PUMP

SUCTION LINE

SCREEN

FIGURE 11-1. Lubrication system and oil circulation in a Mack truck engine. *(Courtesy of Mack Trucks Inc.)*

152

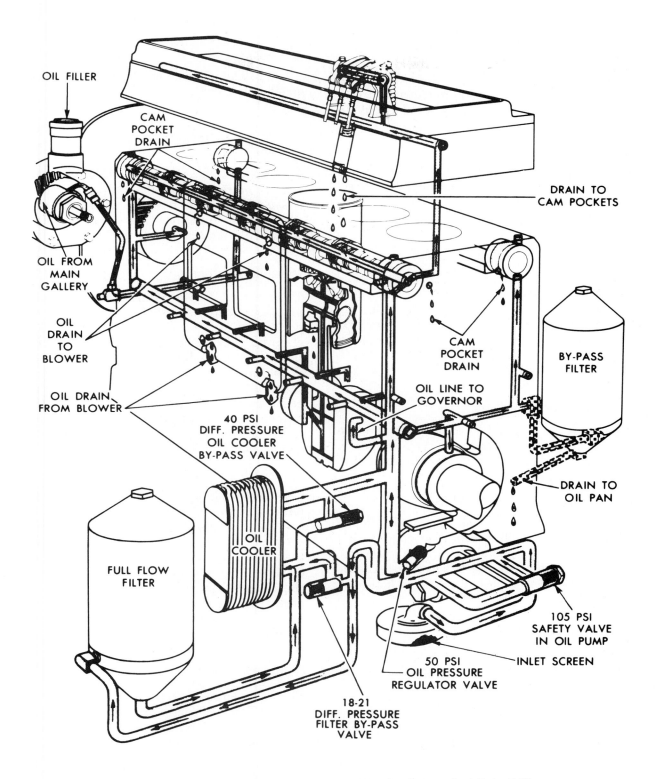

FIGURE 11-2. Schematic diagram and oil circulation in a typical Detroit Diesel engine. *(Courtesy of Detroit Diesel Allison)*

OIL FILLER

CAM POCKET DRAIN

OIL FROM MAIN GALLERY

OIL DRAIN TO BLOWER

OIL DRAIN FROM BLOWER

40 PSI DIFF. PRESSURE OIL COOLER BY-PASS VALVE

OIL COOLER

FULL FLOW FILTER

18-21 DIFF. PRESSURE FILTER BY-PASS VALVE

50 PSI OIL PRESSURE REGULATOR VALVE

INLET SCREEN

105 PSI SAFETY VALVE IN OIL PUMP

OIL LINE TO GOVERNOR

CAM POCKET DRAIN

BY-PASS FILTER

DRAIN TO OIL PAN

DRAIN TO CAM POCKETS

Bearing Protection

Oil breakdown and corrosive products of combustion can cause bearing corrosion. The additives in the oil counteract this action by minimizing breakdown, neutralizing blowby products, and helping to form a protective film.

Rust Control

Engine components including valve lifters, valve stems, rings, cylinder wall, etc., are subjected to severe rust-promoting conditions, particularly in winter stop-and-go driving. Rusting is controlled by oil formulation.

Wear Control

Wear occurs through metal-to-metal contact, acidic corrosion, rusting, and abrasive action of the oil's contaminant load. Metal-to-metal contact is controlled by proper viscosity selection and use of film-forming compounds. Acidic corrosion and rusting are controlled by formulation of the oil while the abrasive wear is controlled by air and oil filtration and oil drain intervals.

Scuff Protection

High-peak pressures occur in such areas as valve train mechanisms, particularly the camshaft lobes. Antiweld- or antiseize-type additives are required to minimize this type of wear.

Control of Combustion Chamber Deposits

Deposits, including oil-derived, accumulate in the combustion chamber, increasing the compression ratio and creating hot spots.

Combustion chamber deposits increase exhaust emissions. Oils must be formulated to reduce such deposits.

Control of Valve Deposits

Some higher ash oils tend to create deposits on exhaust valves in some severe type services. This tendency must be minimized.

How successful the lubrication system is in performing all these functions depends on a number of factors and conditions. There must be an adequate supply of good-quality lubricant delivered to all moving engine parts under sufficient pressure to provide hydrodynamic lubrication for rotating parts and oil adhesion to surfaces subject to sliding friction.

- The oil and filter must be changed at regular intervals.

- The enigne must operate at its most efficient temperature.

- Engine oil temperatures must not be excessively hot or cold.

PART 2   LUBRICATING OILS

Motor oils for diesel engines fall into two basic categories: petroleum-based oils and synthetic oils. Petroleum-based oils, however, contain a variety of additives; so in fact they, too, are partly synthetic. Some of the major additives include those described here.

Metallic Detergents

These are metallic ash-containing compounds having a detergent/dispersant action in controlling deposits and keeping engine parts clean. They will clean up existing deposits as well as disperse particulate contaminants in the oil. As the performance of ashless dispersants improves, they are replacing the metallic detergents, thus reducing the ash level of the finished oil. Metallic detergents are the major contributor of the base number required for acidic corrosion control. They are also good antiwear, antiscuff, and antirust additives.

Ashless Dispersants

These are ashless organic compounds having a detergent/dispersant action in controlling deposits and keeping the engine parts clean. Their cleanup action on existing deposits is much slower and less effective than the metallic detergents. However, their dispersancy is much more effective in suspending in the oil potential carbon-forming deposits and, in particular, low-temperature sludge. Improvements in their diesel performance over the past few years have increased their use in combination with the metallic detergents in lower ash formulations.

Oxidation Inhibitors

These prevent oxygen from attacking the lubricant base oil. Without inhibitors the oil would react with oxygen, eventually solidifying in service or turning acidic and causing bearing corrosion.

Bearing Corrosion Inhibitors

Bearing corrision is the result of acid attack on the oxides of the bearing metals. The acids involved originate either from the blowby combustion gases or from oxidation of the crankcase oil. Acidic corrosion is controlled by the addition of inhibitors, which form protective barrier films on the bearing surfaces.

Rust Inhibitors

Rusting results from the oxygen attack on the metal surface and usually occurs in thin film areas such as hydraulic lifters, push rods, etc. It is controlled by the addition of an inhibitor in the oil formulation.

Antiwear

Wear results from metal-to-metal contact, acidic corrosion, and contaminant or dirt load. Metal-to-metal contact is overcome by the use of film-forming compounds. The acidic corrosion, originating mainly from acidic blowby gases, is neutralized by the use of alkaline additives.

Reserve alkalinity is required to control corrosive wear. The type and sulfur level of the fuel will determine the alkalinity reserve required. The use of highly alkaline or "overbased" additives provides high alkaline reserve where needed.

Foam Depressants

Detergent/dispersant-type oils tend to entrain air, which, when rapidly released, causes foaming. Foam depressants are added to control release of entrained air, thus eliminating this problem.

Pour Point Depressants

Base oils contain hydrocarbons that tend to solidify or crystallize into waxy materials at lower temperatures. Use of pour point depressants in the oil formulation modifies the wax crystal structure, resulting in a lower pour point and, in some instances, improved low-temperature fluidity. (Pour point is the lowest temperature at which the oil will flow when tested under prescribed conditions.)

Viscosity Index Improvers

Petroleum oils thin out with increasing temperature. Viscosity index is a measure of this rate of viscosity change. The addition of a viscosity index improver will slow down the rate of "thinning"; thus the oil will remain thicker at the engine-operating temperature. Viscosity index improvers are used extensively to formulate multigrade oils.

PART 3   ENGINE OIL CLASSIFICATION

Engine oils are classified by viscosity and performance. (See Fig. 11–3 to 11–6.)

SAE Viscosity Classification

The Society of Automotive Engineers (SAE) classifies engine oils by SAE viscosity numbers, commonly called *viscosity grades* or *SAE grades*.

The W grades are defined by cold cranking simulator viscosity (in centipoises) at 0°F or −18°C, a value related to ease of cranking an engine in cold weather. The other grades are defined by kinematic viscosity (in centistokes) at 212°F or 100°C, to define high-temperature wear protection.

The SAE system is based solely on viscosity; other factors of oil quality or performance are not considered.

In the viscosity test a measured amount of oil is brought to the specified temperature (usually 100°C or 212°F). The length of time in seconds required for a specified volume of oil to flow through a small orifice in an instrument such as a saybolt or kinematic viscometer is recorded. The SAE grade is determined by referring to the SAE chart.

Multigrade Oils

The viscosity of an oil changes with temperature. At low temperatures the oil is thick, its viscosity high. As temperature increases, the oil becomes thinner and its viscosity decreases. A sluggish oil makes engine starting difficult and delays warm-up lubrication, while excessively thin oils give poor lubrication and high oil consumption.

Figure 11–5 represents the temperature viscosity relationship of typical SAE 10W, 40, and 10W-40 grades. The 10W grade is controlled at 0°F or −18°C and is represented by the broken line. The 40 grade is controlled at 212°F or 100°C and is represented by the solid line. The dotted line passes through the specification for both the 10W and 40 grades; they therefore can be called 10W-40. The reduced rate of thinning out with increasing temperature is due to the viscosity index improver additive used.

The 15W-40 and 10W-30 multigrade oils provide the cranking ease and reliability required for some areas.

API Engine Service Classifications

The American Petroleum Institute (API) has classified and described engine oils in terms intended to aid equipment makers in recommending proper oils and consumers in selecting them.

The API engine service classification is divided into an S series covering engine oils generally sold

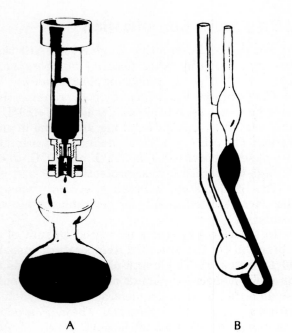

SAE CRANKCASE OIL CLASSIFICATION				
SAE Viscosity Number	Time of Flow Through Saybolt Viscometer in Seconds			
	at 0°F.		at 210°F.	
	Min.	Max.	Min.	Max.
5W	—	6,000	—	—
10W	6,000 (Note A)	less than 12,000	—	—
20W	12,000 (Note B)	48,000	—	—
20	—	—	45	less than 58
30	—	—	58	less than 70
40	—	—	70	less than 85
50	—	—	85	110

Note A. Minimum viscosity at 0°F may be waived provided viscosity at 210°F is not below 40 seconds, Saybolt Universal.
Note B. Minimum viscosity at 0°F may be waived provided viscosity at 210°F is not below 45 seconds, Saybolt Universal.

A   B   C

FIGURE 11-3. Saybolt viscometer (A) and kinematic viscometer (B) are used to determine viscosity of lubricating oils. Chart (C) shows time required in seconds for specified volume of various SAE grades of oil to flow through saybolt viscometer. *(Courtesy of Deere and Company)*

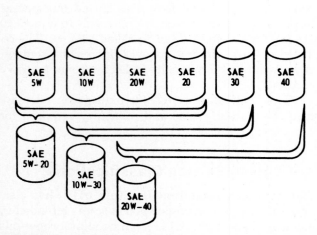

FIGURE 11-4. Multiviscosity oils can replace a number of single viscosity oils when recommended. *(Courtesy of Deere and Company)*

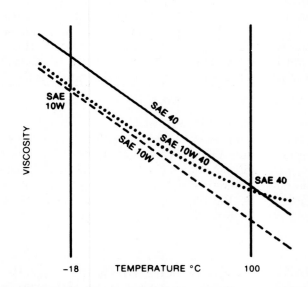

FIGURE 11-5. Temperature-viscosity relationship of single- and multiviscosity lubricating oils. *(Courtesy of Deere and Company)*

| API Engine Service Classification | | Select the equivalent API Engine Service Classification when one of the following is recommended: | | | |
Service	Commercial	API Classification*	U.S. Ordnance	Canadian Govt.	Industry
SA	—	ML	—	—	—
SB	—	MM	—	—	—
SC	—	MS ('64)	—	—	Ford M2C101A GM 4745M
SD	—	MS ('68)	—	—	Ford M2C101B GM 6041M
SE	—	—	—	—	Ford M2C101C Ford M2C144A GM 6136M
—	CA	DG	MIL-L-2104A	—	—
—	CB	—	MIL-L-2104A (Sup 1)	—	—
—	CC	DM	MIL-L-2104B	—	—
—	CD	DS	MIL-L-45199B	—	Cat. Series 3
—	SE-CC	—	MIL-L-46152	3-GP-302	—
—	SD-CD	—	MIL-L-2104C	3-GP-304	—
—	SD-CC	—	—	—	GM 6042M IHC # 1 (Gas)
—	CC-CD	—	—	—	Mack EO-G
—	SE-CD	—	—	—	IHC # 1 (Diesel) Mack EO-H
—	—	—	—	—	Mack EO-J

FIGURE 11-6. Crankcase oil API engine service classification and equivalent engine manufacturer's specifications. *(Courtesy of Deere and Company)*

157

in service stations for use in passenger cars and light trucks (mainly gasoline engines), and a C series for oils for use in commercial, farm, construction, and off-highway vehicles (mainly diesel engines). An oil can meet more than one classification, e.g., SE/CC. The system is open-ended so that new categories can be added in either series as needed.

However, the single classification of CC oil, for example, is so broad that it may not be suitable for certain applications. A CC/SC classification may be required for stop-and-go service to help reduce carbon and sludge formation while a CC/CD classification may be required for turbocharged engines.

Figure 11-6 compares API engine service classification levels of engine oils with equivalent specifications issued by U.S. Ordinance, the Canadian government, and industry. Thus a recommendation to use an oil meeting a MILL-2104B specification would be met with an API engine service classification CC oil.

Fuel Economy Oils

Engine oils will be expected to play a more significant role in providing improved fuel economy. Additives such as colloidal graphite, colloidal molybdenum disulfide, and soluble friction modifiers are being used. There is also a trend to lighter viscosity grades and multigrades for greater fuel efficiency. The extended use of lighter grades will demand improved wear performance.

Synthetic Oils

Synthetic oils are made from man-made synthetic basestocks and fortified with additives similar to conventional petroleum basestocks. The advantage of the synthetic basestock is that it can be made to meet the required performance characteristics.

Properly formulated synthetic oils can outperform conventional oils in oil life and the control of oil thickening in service. They also have extremely good low-temperature performance and should be used when extreme low-temperature starting might be anticipated. Some fuel economy advantage can be gained with the use of the lighter viscosity grades 5W-20 and 5W-30. Advantages of synthetic oils include:

- higher viscosity stability over a wider operating temperature range
- greatly reduced effects of oxidation (reduced oil thickening)

- reduced wear and increased load-carrying ability
- reduced loss through evaporation
- reduced crankcase oil temperatures
- considerably reduced oil consumption
- fewer engine deposits
- less frequent oil changes
- increased fuel economy

PART 4   OIL PUMPS AND PRESSURE REGULATION

The engine oil pump must provide a continuous supply of oil at sufficient pressure and in sufficient quantity to provide adequate lubrication at all times to the entire engine. The oil pump is mounted on the engine block on either the outside or inside. It picks up oil from the reserve in the oil pan through the inlet screen and pickup tube. The oil is forced out of the pump outlet to the pressure regulator valve, which limits maximum oil pressure, and to the main oil gallery. (See Fig. 11-7 to 11-10.)

Two types of oil pumps are used: the rotor type and the gear type. One of the rotors or gears is gear-driven by the crankshaft.

Since there is no direct path or opening for oil to flow through between the pump inlet and outlet, the pump is a positive displacement pump. Bearing clearances and metered oil holes restrict the flow of oil from the pump. This results in backpressure and pressure buildup. To limit maximum pressure, a pressure regulator valve is incorporated into the oil pump. Excess oil pressure is vented to the pump inlet or to the oil pan. It is therefore apparent that engine oil pressure is dependent on pressure regulator valve-spring tension and restriction to flow on the outlet side of the pump. It follows, then, that excessive bearing clearances will cause a drop in oil pressure. A sticking regulator valve or damaged regulator valve spring will also affect oil pressure.

Internal wear in the oil pump, allowing oil to bypass back to the inlet side, will cause oil pressure to drop correspondingly. Oil pumps are designed to have sufficient reserve capacity to allow for normal wear.

Scavenge Pump

Many diesel engines used in equipment required to operate at fairly steep grades use an additional oil pump known as a scavenge pump (Fig. 11-11). It is

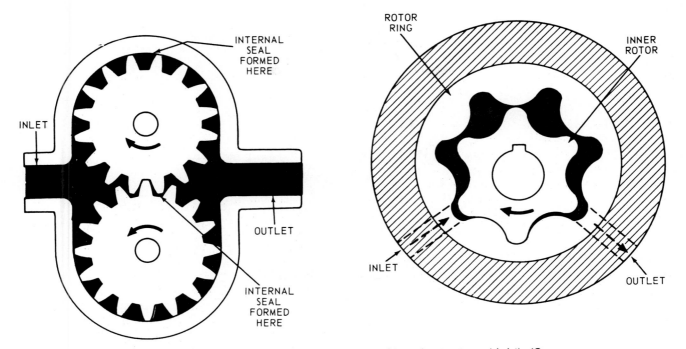

FIGURE 11-7. External gear-type oil pump (left) and rotor-type (right). *(Courtesy of Deere and Company)*

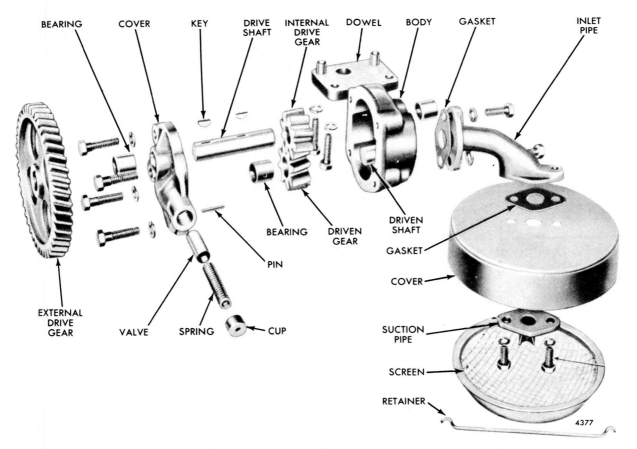

FIGURE 11-8. Exploded view of gear type of lubricating oil pump with external gear type of drive. *(Courtesy of Detroit Diesel Allison)*

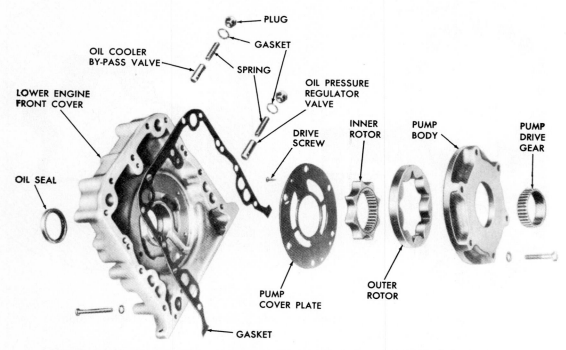

FIGURE 11-9. Rotor type of oil pump with internal pump drive. *(Courtesy of Detroit Diesel Allison)*

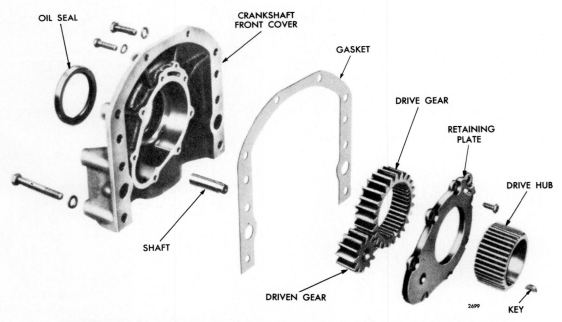

FIGURE 11-10. External gear type of oil pump with internal drive. *(Courtesy of Detroit Diesel Allison)*

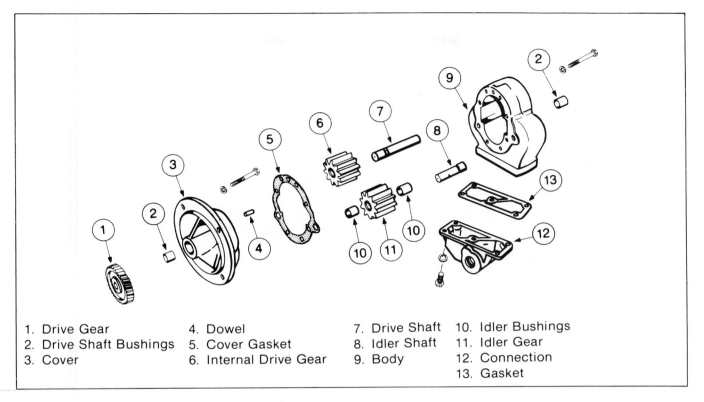

1. Drive Gear   4. Dowel   7. Drive Shaft   10. Idler Bushings
2. Drive Shaft Bushings   5. Cover Gasket   8. Idler Shaft   11. Idler Gear
3. Cover   6. Internal Drive Gear   9. Body   12. Connection
                                                                             13. Gasket

FIGURE 11-11. Scavenge oil pump for Cummins 1710 C.I.D. engine. *(Courtesy of Cummins Engine Company, Inc.)*

designed and positioned to ensure that the main oil pump is always supplied with adequate lubricating oil regardless of equipment attitude. Depending on oil pan design, oil may accumulate at the front or rear in the pan. Without a scavenge pump, the main oil pump would be starved of oil when the equipment is operated at certain steep angles. The scavenge pump ensures that oil is returned to the main sump and oil pump at all times. Scavenge oil pump design is similar to main oil pump design.

PART 5   OIL FILTERS, LINES, AND COOLERS

Oil filters are designed to trap foreign particles suspended in the oil and prevent them from getting to engine bearings and other parts. Some engines use the full-flow filtering system in which all the oil delivered by the pump must pass through the filter before reaching the engine bearings. (See Fig. 11-12 to 11-20.)

Should the filter be neglected and become completely clogged, a bypass valve allows oil to flow directly from the oil pump to the bearings. Another valve prevents the oil from draining out of the filter when the engine is stopped. This ensures oil delivery to engine parts immediately after the engine is started.

Other engines may use the shunt type or bypass type of filtering systems (see illustrations). Filters may be mounted directly on the engine block or on a special adapter with no external oil lines, or they may be remote mounted requiring external oil lines.

Some engines have only one lubricating oil filter; others may have two or more.

Filter elements come in three types: the surface type, the depth type, and the combination type. The surface type is usually of the paper bellows design; foreign matter collects on the surface of the paper element as the lubricating oil flows through the filter element. The depth type consists of a perforated canister filled with a fibrous material. Foreign matter is trapped in this element at various depths in the fibrous material as the oil flows through. The

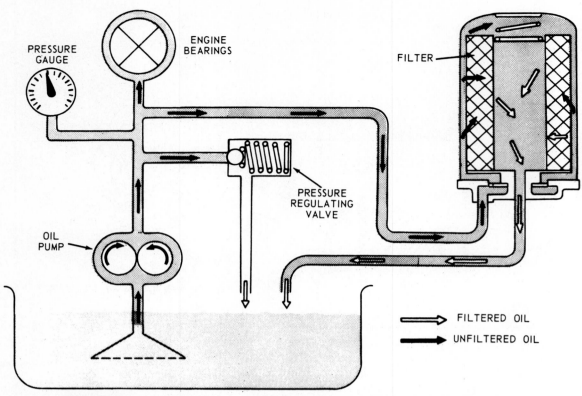

FIGURE 11-12. Bypass type of oil filtering system. Engine bearings are fed from oil pump. Some oil is directed from the main gallery to the filter. Filtered oil is returned to sump. *(Courtesy of Deere and Company)*

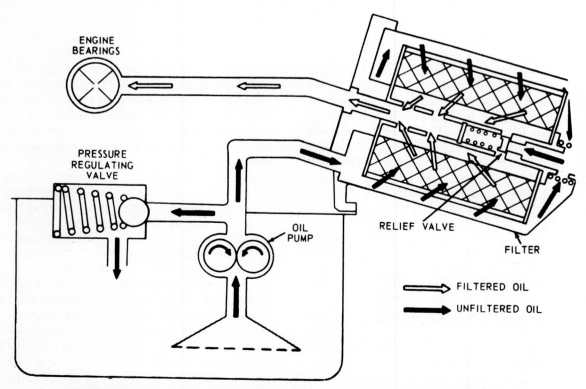

FIGURE 11-13. Full-flow filtering system. All the oil passes through the filter before it reaches the bearings. If filter is plugged, relief valve opens, allowing unfiltered oil to lubricate bearings. *(Courtesy of Deere and Company)*

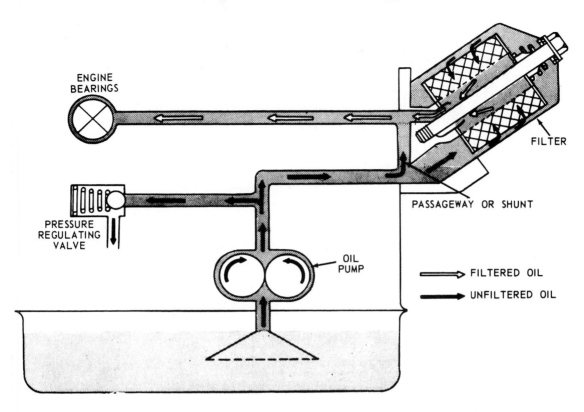

FIGURE 11-14. Shunt type of filtering system. Engine bearings are supplied with both filtered and unfiltered oil. When filter becomes plugged, bearings are supplied with unfiltered oil only through the shunt. *(Courtesy of Deere and Company)*

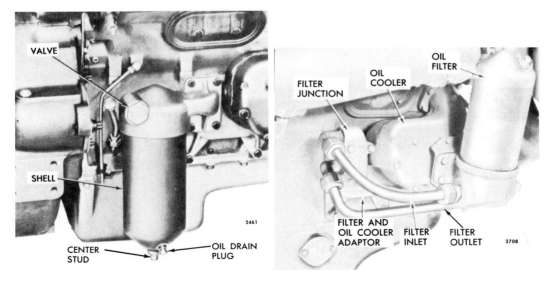

FIGURE 11-15. Typical filter mountings. Remote-mounted filter on right requires external oil lines. *(Courtesy of Detroit Diesel Allison)*

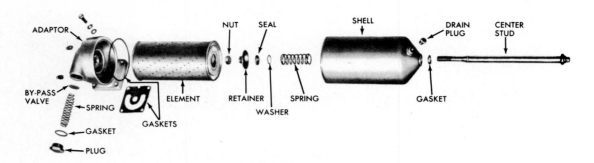

FIGURE 11-16. Exploded view of cartridge type of full-flow oil filter. *(Courtesy of Detroit Diesel Allison)*

combination type of filter element combines the surface and depth designs in one element. Another type of oil filter is the centrifugal type. Lubricating oil forced against a rotor causes the rotor to spin. Heavier foreign particles are separated from the oil in this manner and deposited on the filter wall while the oil drains back to the sump.

Lubricating oil lines include flexible synthetic rubber and braided lines as well as the tubular metal lines. Flexible lines usually have the fittings permanently attached. Metal lines allow replacement of fittings or lines separately. Flexible lines have the advantage of being able to withstand vibration without damage. Metal tubing requires adequate

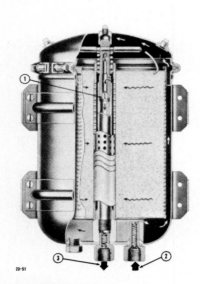

1. Metering Orifice
2. Oil Inlet
3. Oil Outlet

FIGURE 11-17. Cutaway view of cartridge type of bypass oil filter. Bypass type of filter requires a metered orifice to restrict flow into filter. Without this orifice, engine bearings would receive insufficient lubrication. *(Courtesy of Mack Trucks Inc.)*

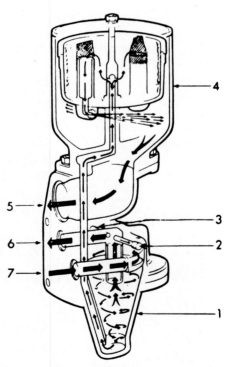

FIGURE 11-18. Phantom view of certrifugal type of oil cleaner: (1) Cyclone; (2) Oil outlet; (3) Oil outlet; (4) Cleaner housing and rotor; (5) Clean oil outlet to sump; (6) Clean oil outlet to engine bearings; (7) Oil inlet. *(Courtesy of Mack Trucks Inc.)*

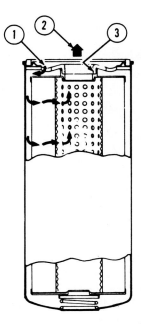

FIGURE 11-19. Spin-on type of full-flow filter: (1) Oil inlet holes. (2) Oil outlet. (3) Threaded cover. *(Courtesy of Mack Trucks Inc.)*

support by means of brackets on longer lines to prevent damage from vibration.

Lubricating Oil Cooler

In order to do its job satisfactorily, the engine lubricating oil must be kept within the proper temperature range. If the oil is too cold, it does not flow freely. If it is too hot, it is not able to support the bearing loads; it cannot carry away enough heat; and too great an oil flow could result. If this happened, oil pressure could drop below specified limits and oil consumption could become excessive.

During engine operation, the lubricating oil absorbs a considerable amount of heat that must be dissipated by an oil cooler.

The oil cooler consists of a housing containing a series of plates or tubes. The cooler is connected to the engine cooling system and to the lubricating system. This requires two inlets and two outlets. One inlet and one outlet are connected to the cooler housing, while the other inlet and outlet are connected to the oil cooler core plates or tubes. In the typical tube type cooler, coolant from the engine water pump flows through a passage in the oil cooler; it then passes through the tubes of each section of the oil cooler, back to the outlet passage, and into the water jackets in the engine block. Engine oil from the engine oil pump enters a passage in the oil cooler, passes through the oil filter, then around the tubes of the oil cooler, back through the outlet passage, and on to the oil galleries in the cylinder block.

A bypass valve is provided to permit engine oil to flow to the engine lubricating system should the oil cooler become plugged.

Oil Pressure Indicators

Oil pressure indicators inform the operator whether the lubrication system is functioning normally. Direct-reading gauges that indicate actual oil pressure are used in some instances. These are either pres-

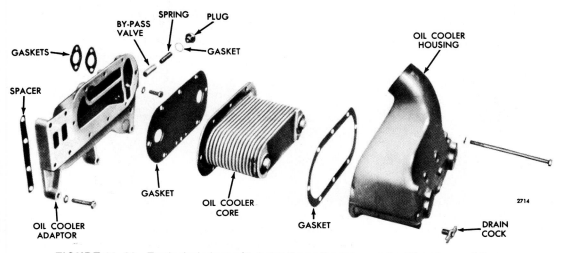

FIGURE 11-20. Exploded view of lubricating oil cooler parts. *(Courtesy of Detroit Diesel Allison)*

sure-sensitive direct-acting gauges or electric gauges that respond to a pressure-sensitive variable-resistance sending unit switch.

Oil warning lights that indicate inadequate oil pressure when they light are also used.

Some engines are equipped with automatic alarms or automatic shutdown devices, which are triggered when oil pressure drops below specified levels.

PART 6   CRANKCASE VENTILATION

Crankcase ventilation is required to remove harmful crankcase vapors and to prevent pressure buildup in the crankcase. Combustion byproducts, condensation, and contaminants are removed by the ventilation system. Clean air enters the crankcase from a breather cap usually equipped with a filter or in some cases from the air cleaner. (See Fig. 11-21 to 11-23.)

FIGURE 11-22. Crankcase ventilation valve controls flow of crankcase vapors from valve cover to intake manifold. *(Courtesy of Caterpillar Tractor Company)*

FIGURE 11-21. Road draft type of crankcase ventilation. Note angled end of tube at bottom. This provides a low pressure at the tube when the vehicle is driving down the road, causing crankcase vapors to flow out of tube. Fresh clean air enters filtered oil filler cap. *(Courtesy of Mack Trucks Inc.)*

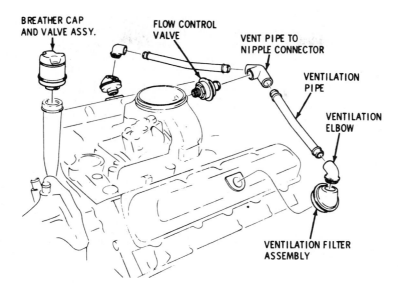

FIGURE 11-23. Crankcase ventilation system as used on automotive diesel engine. *(Courtesy of General Motors Corporation)*

A completely open system includes a vent pipe and an open filtered air inlet. Harmful vapors are simply vented to the atmosphere. Other systems route the crankcase vapors to the air induction system. One method used is an internal pipe or tube leading from the rocker cover to the air inlet.

Another system uses a valve mounted on top of the rocker cover. The valve regulates flow from the rocker chamber to the air manifold.

Crankcase pressure is caused by combustion pressures getting past the piston rings into the crankcase; this is called blowby. If the engine is in good condition, blowby will be minimal. Worn rings and cylinders will increase blowby.

Some crankcase pressure is desirable to help keep dust out and to help keep any dust in suspension so it will be vented through the vent system.

Lubrication System Service

For lubrication system problem diagnosis and corrective steps, refer to the diagnostic chart. The following service operations are common to lubrication system service.

Checking Oil Level and Condition

It is important that the engine oil level be checked in a consistent manner in order to be able to determine accurately the level of oil in an engine. It is equally important that the engine oil level not be allowed to fall below the recommended level and not be overfilled.

Several methods of checking crankcase oil levels are recommended by vehicle manufacturers. One manufacturer includes an oil level check with the engine running at idle. One side of the dipstick is marked for the proper oil level with the engine running and the other side indicates the oil level with the engine stopped. When checking the engine oil level, the vehicle should be level, the engine at operating temperature, and the engine stopped for at least 20 minutes to allow time for oil drainback. Oil levels should be kept between the add-oil and full marks on the dipstick. Overfilling may cause oil to be thrown around by the crankshaft, which can cause foaming and increased oil consumption from too much oil being thrown up onto the cylinder walls. Too low a lubricating oil level will cause higher oil temperatures and consequently reduced lubricating effectiveness.

Oil Consumption, Oil Leaks

A certain amount of oil consumption is normal. It is therefore normal that oil may have to be added between oil changes.

To determine whether a vehicle is using an excessive amount of oil, the method of checking the oil level must be applied consistently over a period of time and mileage.

PART 7 LUBRICATION SYSTEM DIAGNOSIS & SERVICE

PROBLEM	CAUSE	CORRECTION
Low lubricating oil pressure	1. Oil leak—line gasket, etc. Insufficient lubricating oil level.	1. Check oil level and add make-up oil as required. Oil must be to recommended specifications. Check for oil leaks.
	2. Wrong oil viscosity	2. Drain lubricating oil. Change oil filters and fill with oil meeting specifications.
	3. Defective oil pressure gauge	3. Check operation of oil gauge. If defective, replace gauge.
	4. Dirty oil filter(s)	4. Check operation of bypass valve for the filter. Install new oil filter elements. Clean or install new oil cooler core. Drain oil from engine and install oil meeting specifications.
	5. Lubricating oil diluted with fuel oil	5. Check fuel system for leaks. Make necessary repairs. Drain diluted lubricating oil. Install new filter elements and fill crankcase with recommended oil meeting specifications.
	6. Defective oil pump relief valve	6. Remove valve, check for seat condition and sticking relief valve-spring tension and cap. Check assembly parts. The use of incorrect parts will result in improper oil pressure. Make necessary repairs or install new relief valve.
	7. Incorrect meshing of oil pump gears	7. Check mounting arrangement. If engine has been rebuilt, check for proper gear ratio combination of the oil pump-driven gear and drive gear. Incorrect gear combinations will result in immediate gear failure and possible engine damage.
	8. Excessive clearance between crankshaft and bearings	8. Overhaul engine and replace worn defective parts.
Oil in cooling system	1. Defective oil cooler O ring(s)	1. Disassemble and replace O ring(s).
	2. Defective oil cooler core	2. Remove oil cooler. Disassemble and repair/replace oil cooler core.
	3. Blown head gasket	3. Replace head gasket.
Coolant in lubricating oil	1. Defective oil cooler core	1. Disassemble and repair/replace oil cooler core.
	2. Blown head gasket	2. Replace head gasket.
	3. Defective water pump oil seal(s)	3. Remove water pump; disassemble and replace defective parts.
	4. Cylinder sleeve seals failure	4. Replace cylinder sleeve seals.
Excessive oil consumption	1. External oil leaks	1. Check engine for visible signs of oil leakage. Look for loose, stripped oil drain plug, broken gaskets (cylinder head cover, etc.), front and rear oil seal leakage. Replace all defective parts.
	2. Clogged crankcase breather	2. Remove obstruction.
	3. Excessive exhaust back pressure	3. Check exhaust pressure and make necessary corrections.
	4. Worn valve guides	4. Replace valve guides.
	5. Air compressor passing oil	5. Repair or replace air compressor.
	6. Failure of seal rings in turbocharger	6. Check inlet manifold for oil and make necessary repairs.
	7. Internal engine wear	7. Overhaul engine.

Oil consumption may be the result of worn piston rings, worn valve guides, excessive bearing clearance, and oil leakage. If an oil consumption problem exists, any oil leakage must be corrected first before the condition of the engine is blamed. If no oil leakage exists, a thorough diagnosis of the engine's mechanical condition should be performed to determine the cause. A wet and dry compression test can help to decide whether the piston rings, valve guides, or seals may be at fault.

Oil and Filter Change Intervals

The oil and filter change intervals should be followed as outlined in the owner's manual or shop service manual. The SAE rating (viscosity) and the service rating of the oil used should comply with the manufacturer's recommendations.

The frequency of oil and filter changes may vary, depending on the type of service that the vehicle is required to provide. Under severe operating conditions such as dusty conditions, extensive idling, frequent short trips (especially in cold weather), and sustained high-speed driving in hot weather, more frequent oil changes may be required. Oil and filter change intervals may be based on number of miles or kilometers driven, number of hours of service, or analysis of engine oil samples established, over a period of time. Engine manufacturer's or lubricating oil supplier's recommendations should be followed to ensure long engine service life.

Oil Contamination

Excessive oil dilution requires changing the engine oil and filters and correcting the cause of oil dilution. This may require adjusting or correcting any fuel delivery system problems resulting in excessive fuel delivery to the cylinders, including injector operation.

Coolant leakage past cylinder liner seals or head gaskets may cause sludge buildup and bearing damage and must be corrected.

Engine Oil Analysis

Periodic analysis of the engine oil can provide valuable information about engine operation and condition. Chemical analysis includes such methods as spectrophotometry, atomic absorption, and infrared analysis. These tests can be performed only in a properly equipped chemical lab.

The results of these tests monitored over a period of time can provide valuable engine wear rate data from which the need for engine repair can be predicted with considerable accuracy. This, in turn,

can prevent costly on-the-job engine breakdown. Tests indicate wear of engine parts in the form of fine metallic particles found in the oil. Based on the amount and types of metal identified, degree of wear of bearings, rings, and cylinder walls can be determined.

Oil Pressure Testing

If for any reason an oil pressure problem is suspected, a master test gauge with a range of 0 to 100 psi (0 to 689.5 kilopascals) should be used to verify actual oil pressure produced by the engine (Fig. 11-24).

Generally, procedures include bringing the engine oil to operating temperature. The engine is then shut off and the test gauge installed at the point indicated in the service manual (usually a plug in the main oil gallery). The engine is restarted and readings are taken at specified engine speed. These readings are then compared to those provided in the shop manual.

If the oil pressure is too high, the problem is usually a stuck oil pump pressure relief valve. The relief valve should be removed and polished with crocus cloth to correct this condition. Relief valve-spring pressure should also be checked at this time.

If the oil pressure is too low, the cause may be any of the following:

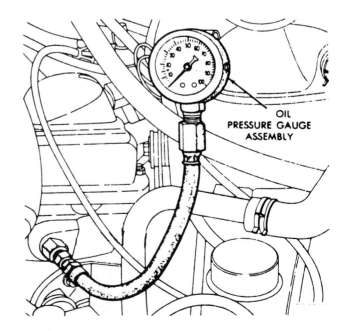

OIL PRESSURE GAUGE ASSEMBLY

FIGURE 11-24. Using master oil pressure gauge to check lubrication system-operating pressure.

- worn oil pump
- excessive bearing clearances (camshaft or crankshaft)
- weak or broken pressure relief valve spring
- relief valve stuck in the open position
- excessive oil dilution
- plugged oil pickup screen
- air leak into oil pump inlet

If the oil pump is worn, it is usually replaced. The oil pump must be removed and checked as outlined in the following section to determine its condition.

If the bearing clearances are excessive, they must be corrected to bring clearances to specified limits. This requires replacement of the bearings and may require replacing the camshaft and crankshaft as well. If the relief valve or spring is at fault, it should be cleaned or replaced.

If the oil pickup tube or screen is faulty, it should be cleaned or replaced.

Oil Pump Service

If the oil pump is suspected of being faulty, it must be removed and checked for wear. Both gear- and rotor-type pumps are measured similarly to determine if wear is excessive (See Fig. 11–25 and 11–26).

The relationship of the pump parts to each other must be maintained during this procedure if the pump is to be used again. Do not mark pump parts with a center punch. Use a felt pen or chalk to mark parts for correct reassembly. Some pumps have the gears or rotors marked during manufacture. Procedures for measuring pump clearances are illustrated, and results should be compared to limits specified in the service shop manual.

The oil pump should be assembled properly, lubricated, and primed to ensure immediate lubrication on engine start-up. The correct thickness of the cover plate gasket must be used (where applicable) to ensure correct gear or rotor tolerances and adequate oil pump pressure. Install the pump as outlined in the appropriate service manual and tighten all bolts to specified torque and proper sequence.

Crankcase Ventilation Service

A plugged or restricted crankcase ventilation system will contribute to oil leaks and oil consumption by causing the crankcase to be pressurized. The ventilation system should be inspected and cleaned to make sure that there is no restriction in either the air inlet or the vent outlet. Any filters in the system

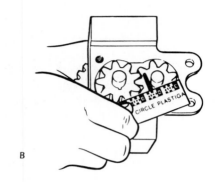

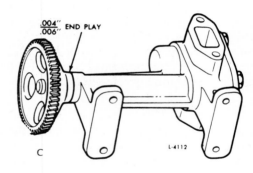

FIGURE 11–25. Measuring oil pump clearances. A. Measuring gear to housing clearance with feeler gauge. B. Measuring gear to cover plate clearance with Plastigage ® C. Oil pump drive shaft end play requirement, typical. *(Courtesy of Deere and Company[A], [B] and Detroit Diesel Allison [C])*

should be cleaned or replaced, as well as the valve, if so equipped.

Oil Cooler Service

The oil cooler core should be removed and cleaned according to the manufacturer's recommendations.

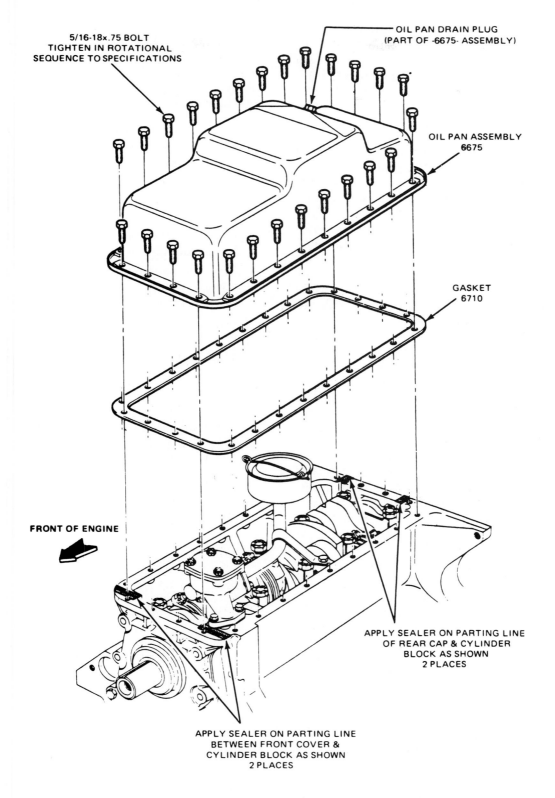

5/16-18x.75 BOLT
TIGHTEN IN ROTATIONAL
SEQUENCE TO SPECIFICATIONS

OIL PAN DRAIN PLUG
(PART OF -6675- ASSEMBLY)

OIL PAN ASSEMBLY
6675

GASKET
6710

FRONT OF ENGINE

APPLY SEALER ON PARTING LINE
OF REAR CAP & CYLINDER
BLOCK AS SHOWN
2 PLACES

APPLY SEALER ON PARTING LINE
BETWEEN FRONT COVER &
CYLINDER BLOCK AS SHOWN
2 PLACES

FIGURE 11-26. Oil pump and pan installation, typical. *(Courtesy of Ford Motor Company)*

This may include removing the cooler and immersing the core in the recommended chemical solution for cleaning. Thorough flushing with water is necessary after chemical cleaning or a chemical neutralizing flushing solution may be used. Pressure test the cooler as recommended in the service manual.

If the engine failure has resulted in metallic particles being circulated in the lubricating oil and oil cooler, the best procedure is to replace the cooler core. This will ensure that no metal particles will be recirculated from the oil cooler into the lubrication system after the engine overhaul.

All gaskets, lines, and fittings must be properly sealed and installed. Tighten all connections and bolts to the recommended torque values given in the service manual.

After completing all required lubrication system service, the crankcase should be filled with the recommended amount and type of lubricating oil. The entire lubrication system, all lines, filters, coolers, etc., can be filled and primed by using a pressure tank connected to the main oil gallery to charge the system. The pressure tank has a reservoir for lubricating oil, a connecting line and valve to the engine, and a valve for pressurizing with compressed air. After connecting the line to the engine, the valve is opened, allowing oil to enter the lubrication system. This process is continued until system pressure is reached (engine lubrication pressure). This procedure ensures immediate lubrication when the engine is started.

After the pressure tank is removed and the engine oil gallery plug replaced, the engine can be started, Lubricating oil pressure should be observed to make sure that it is within specifications. The entire system should be checked for leaks during engine warm-up and again after engine-operating temperature has been reached.

PART 8   SELF-CHECK

1. What four jobs are done by the lubrication system?
2. Name at least five common additives used in compounding motor oils.
3. Give one example of the SAE rating and the service rating of engine oils.
4. Two types of engine oil pumps are _____, and _____.
5. Why is the engine oil filter needed?
6. The positive crankcase ventilation system is required to _____.
7. Define the API classification of engine lubricating oils.
8. What advantages are claimed for synthetic oil?
9. Why is a scavenge pump needed on some engines?
10. List three causes of excessive oil consumption.
11. How often should engine oil be changed?

Chapter 12

Cooling Systems

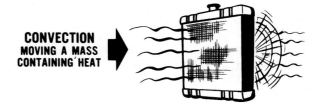

FIGURE 12-1. Heat transfer takes place in an engine by heat moving through objects such as engine parts, from hot areas to cooler areas.

CONDUCTION
HEAT MOVING THROUGH AN OBJECT

CONVECTION
MOVING A MASS CONTAINING HEAT

FIGURE 12-2. Heat is transferred from the hot radiator to atmosphere by air passing through the fins and tubes.

_____ **Performance Objectives** _____

After thorough study of this chapter and sufficient practice on the appropriate engine components, you should be able to:

1. Complete the self-check questions with at least 80 percent accuracy.

2. Describe the purpose, construction, and operation of the cooling system and its components.

3. Diagnose cooling system problems.

4. Recondition or replace the faulty components.

5. Test the reconditioned system to determine the success of the service performed.

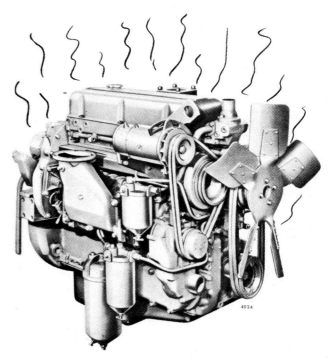

FIGURE 12-3. Heat is also transferred from the hot engine to the atmosphere by radiation.

PART 1   COOLING SYSTEM PRINCIPLES

The diesel engine cooling system is designed to bring the engine to its most efficient operating temperature (as soon as possible after starting) and to maintain that temperature through all operating conditions. Some of the heat absorbed by the cooling system is used to heat the cab interior in cold weather and to keep windows clear of moisture and frost. (See Fig. 12-1 to 12-3.)

The cooling system of an engine relies on the principles of conduction, convection, and radiation. Heat is conducted from the metal surrounding the cylinders, from valves, and from cylinder heads to the coolant in the water jackets of the block and head. The hot coolant is forced out of the block and cylinder heads by the water pump to the radiator, where the heat is removed by convection. Some cooling of the engine takes place through radiation. Air flow around the engine carries this heat away.

Since about 30 percent of the heat energy of the burning fuel in an engine, as well as heat from friction, is absorbed by the cooling system, all components involved must be of sufficient capacity, and they must be in good operating condition.

There are two general classifications of diesel engine cooling systems: the liquid system and the air cooling system. Of these the liquid cooling system is by far the most common.

The liquid system uses a liquid (water or antifreeze) as the medium to absorb engine heat and transfer it to the radiator or heat exchanger. A fan blows air across the radiator tubes and fins to dissipate the heat. A centrifugal water pump circulates the coolant through the engine and radiator. A flow control valve called a thermostat determines whether circulation is allowed through the radiator. The opening temperature range of the thermostat, approximately 160 to 185 degrees Fahrenheit (71 to 85 degrees Celsius), maintains operating temperature in that range. Engine coolant is also used to control lubricating oil temperature by means of a liquid-to-liquid heat exchanger or oil cooler.

Cooling systems for marine engines are often equipped with two water pumps. One pump circulates coolant as described above but uses a liquid-to-liquid heat exchanger instead of a radiator. The other pump, known as the raw water pump, circulates raw sea water through separate passages in the heat exchanger to cool the engine coolant with sea water.

In the keel cooling system the coolant is drawn by the water pump from the keel cooler and is forced through the engine oil cooler, cylinder block, cylinder heads, and exhaust manifold to the thermostat housings. A by-pass from the thermostat housings to the inlet side of the water pump permits circulation of coolant through the engine while the thermostats are closed. When the thermostats open the coolant can flow through the keel cooling coils and then to the suction side of the water pump for recirculation. The keel cooling coils are located at the bottom of the vessel where they are exposed to sea water. The heat of the engine coolant is transferred through the coils of the keel cooler to the surrounding water.

The air cooling system relies on controlled air flow across the engine's finned cylinders and cylinder heads. Metal shrouds direct the air flow from a blower over the engine, thereby transferring engine heat to the atmosphere. Thermostatically-controlled damper valves regulate air flow to maintain engine temperature within the desired range.

Coolant Circulation

In the liquid cooling system, the coolant circulates through the cylinder block, up through the cylinder head, and back through the bypass to the water pump when the thermostat is closed. When the temperature of the coolant reaches thermostat-opening temperature, coolant circulates from the water pump to the block, and the cylinder head, through the open thermostat to the radiator inlet, through the radiator and the radiator outlet, and back to the water pump. Coolant circulates through the heater core at all times on some vehicles, and only when the heater temperature control is turned on in others. (See Fig. 12-4 and 12-5.)

Temperature

The average temperatures of engine components are relatively high, when compared to the boiling point of water; pistons will run as high as 500°F (260°C), exhaust valves 1200°F (649°C), and the water side of the cylinder liners 250°F (121°C). These temperatures are high enough to cause the water to boil. The flow of coolant in these areas is critical to assure that the water velocity is high enough to prevent localized boiling, which would result in loss of heat transfer and hot spots. The engine design includes porting cylinder blocks in the water header to assure a high velocity swirl around the cylinder liner and in the head areas.

It is necessary to maintain a continuous coolant circulation throughout the engine under all operating conditions. When the coolant flow stops, even for a short interval of time, the engine is immediately placed in danger. The piston rings will scuff and pistons score or seize as quickly as 30 seconds after flow ceases, when operating at full throttle. Even though the coolant is present, local boiling at the point of maximum heat transfer will occur; the temperature is high enough to destroy the lubricating oil film. The piston also expands as its temperature is raised, and scoring will inevitably occur.

The engine may continue to operate after brief periods of overheating, but the residual damage due to overheating will begin to affect various compo-

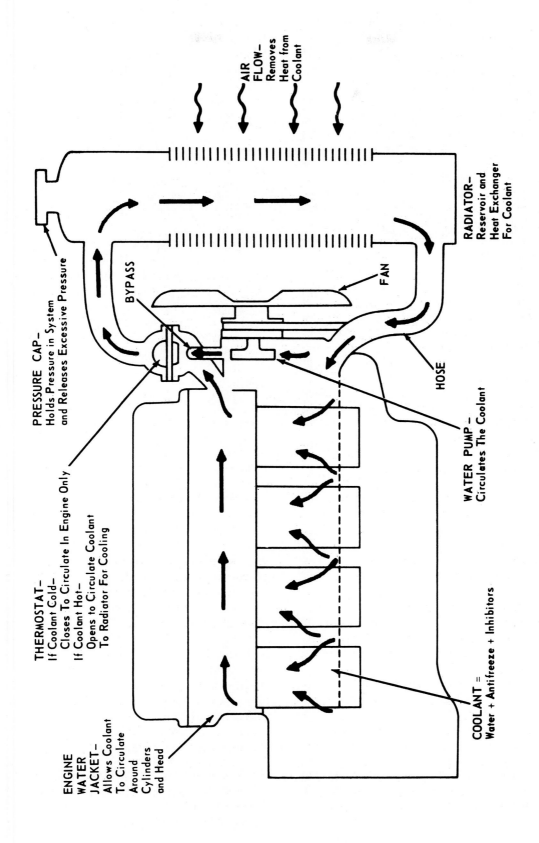

AIR FLOW—
Removes Heat from Coolant

RADIATOR—
Reservoir and Heat Exchanger For Coolant

BYPASS

FAN

HOSE

WATER PUMP—
Circulates The Coolant

PRESSURE CAP—
Holds Pressure in System and Releases Excessive Pressure

THERMOSTAT—
If Coolant Cold—
Closes To Circulate In Engine Only
If Coolant Hot—
Opens to Circulate Coolant To Radiator For Cooling

ENGINE WATER JACKET—
Allows Coolant To Circulate Around Cylinders and Head

COOLANT = Water + Antifreeze + Inhibitors

FIGURE 12-4. Typical coolant circulation in an engine. *(Courtesy of Deere and Company)*

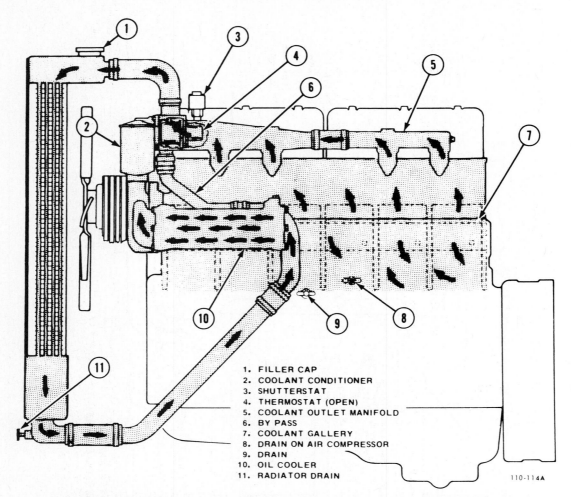

1. FILLER CAP
2. COOLANT CONDITIONER
3. SHUTTERSTAT
4. THERMOSTAT (OPEN)
5. COOLANT OUTLET MANIFOLD
6. BY PASS
7. COOLANT GALLERY
8. DRAIN ON AIR COMPRESSOR
9. DRAIN
10. OIL COOLER
11. RADIATOR DRAIN

110-114A

FIGURE 12-5. Mack truck six-cylinder diesel-engine coolant flow and major cooling system components. *(Courtesy of Mack Trucks Inc.)*

nents. Liner O-ring life may be shortened because of accelerated heat-age hardening. Scuff marks on the liner or piston may lead to high oil consumption and can lead to excessive blowby and piston seizure at later dates. In many cases, failures such as high oil consumption are caused by the cooling system malfunction.

PART 2   WATER PUMPS

The water pump is belt- or gear-driven from the crankshaft pulley. Water pump capacity must be sufficient to provide adequate coolant circulation. Centrifugal, vane-type, nonpositive displacement pumps are commonly used. The impeller, shaft, fan hub, and pulley are supported in the water pump

housing by one or more bearings. A water pump seal prevents coolant from leaking. (See Fig. 12-6 and 12-7.)

The water pump forces coolant into the engine block as the impeller rotates. Coolant enters the center area of the impeller from the radiator outlet and is thrown outward centrifugally to create a flow into the block. Coolant flow returns to the water pump through the bypass when the thermostat is closed, and through the radiator when the thermostat is open.

Raw (Sea) Water Pump

The raw water pump used on some marine diesel engines is similar in design to the conventional fresh water engine coolant pump in many ways (Fig. 12-8). The shaft, bearings, seal, and housing are very

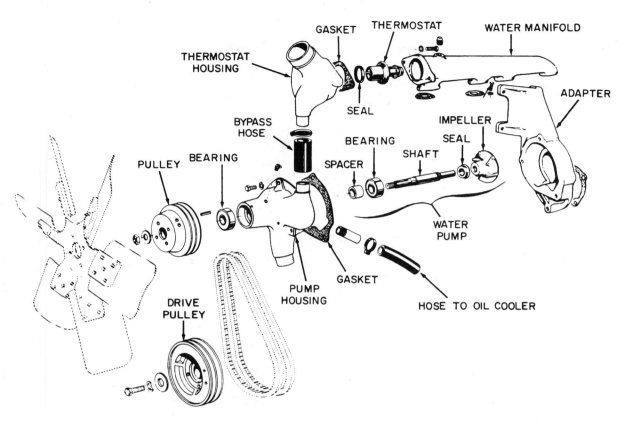

FIGURE 12-6. Exploded view of fan, water pump, and water distribution parts. *(Courtesy of J. I. Case, a Tenneco company)*

similar to those found in other water pumps. The major difference is in the type of impeller used. The raw water pump usually uses a flexible vane type impeller made from a special rubber compound and bonded to a steel hub which is mounted to the pump shaft. Since this type of pump is subjected to impurities such as salt and sand normally found in raw sea water, a conventional pump would not last in this type of application.

PART 3   THERMOSTATS

The thermostat is a temperature-sensitive flow control valve located in the thermostat housing at the front of the engine. The thermostat remains closed until the engine reaches operating temperature. As the temperature increases, the thermostat opens. This allows coolant to be circulated through the radiator for cooling. When the engine coolant falls below operating temperature, the thermostat closes once again. Coolant circulation is restricted to the

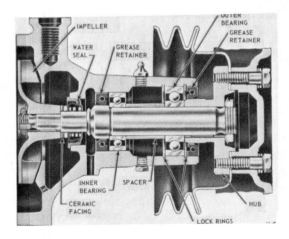

FIGURE 12-7. Cross-sectional view of belt-driven water pump. *(Courtesy of Mack Trucks Inc.)*

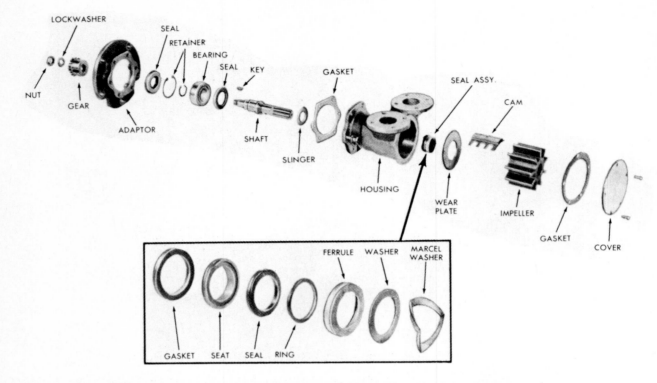

FIGURE 12-8. Raw water pump as used on marine diesel installations, gear driven. This type of pump uses a flexible vane synthetic or natural rubber impeller. *(Courtesy of Detroit Diesel Allison)*

engine block and cylinder heads and the cab's interior heater when the thermostat is closed. A bypass provides the passage for coolant return to the pump. (See Fig. 12-9 to 12-12.)

Many V-type engines have two or more thermostats, one or two for each bank of cylinders.

Semi-blocking thermostats are used in the rapid warm-up cooling system.

In this warm-up system enough coolant to vent the system is by-passed to the radiator top tank by means of a separate external deaeration line and then back to the water pump without going through the radiator cores. As the coolant temperature rises above 170°F., the thermostat valves start to open, restricting the by-pass system, and permit a portion of the coolant to circulate through the radiator. When the coolant temperature reaches approximately 185°F., the thermostat valves are fully open, the by-pass system is completely blocked off, and all of the coolant is directed through the radiator.

A defective thermostat which remains closed, or only partially open, will restrict the flow of coolant and cause the engine to overheat. A thermostat which is stuck in a full open position may not permit the engine to reach its normal operating temperature. The incomplete combustion of fuel due to cold engine operation will result in excessive carbon deposits on the pistons, rings, and valves.

Properly operating thermostats are essential for efficient operation of the engine. If the engine operating temperature deviates from the normal range the thermostats should be removed and checked.

PART 4   RADIATOR HOSES AND CLAMPS

Radiator hoses include the straight, molded, and flexible types. The straight and molded types must not be distorted when installed. The flexible hose can be bent as needed. The hose connecting the radiator to the inlet side of the water pump usually has a spiral wire support built in to prevent hose collapse due to low internal pressure. Hoses are required to make flexible connections between cooling system components. (See Fig. 12-13 to 12-15.)

Hoses are secured to their connections with hose clamps. A variety of designs are in use, the most

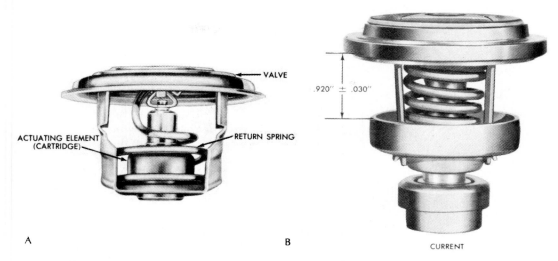

FIGURE 12-9. Typical thermostats: butterfly valve type (A) and poppet valve type (B). *(Courtesy of Detroit Diesel Allison)*

common being the geared type. Some engines use special silicone hoses that require hose clamps specially designed for them.

Radiator hoses can deteriorate both internally and externally. Hoses soften and disintegrate from contamination by lubricating or fuel oil. Silicone hoses are not affected in this way. Rubber hoses can swell and restrict coolant circulation. Material from damaged or deteriorated interior walls of rubber hoses can contaminate the coolant and impede thermostat operation.

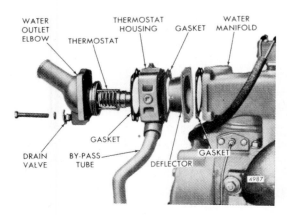

FIGURE 12-10. Typical thermostat installation. Bypass tube allows coolant circulation in the engine when thermostat is closed. *(Courtesy of Detroit Diesel Allison)*

Water Manifold

Many multi-cylinder diesel engines are equipped with water manifolds. The water manifold is designed to ensure more even temperature control throughout the engine. Coolant heated by the cylinders at the rear of the engine must flow forward through the other cylinders on its way to the radiator. Consequently the forward cylinders run somewhat hotter than the others when not equipped with a water manifold. The water manifold collects water from each cylinder or pair of cylinders individually and routes it to the thermostats and radiator or heat exchanger. This results in cylinder temperatures being more equal. Coolant outlets in the cylinder head conecting to the water manifold are usually located near the exhaust valve port where the hottest coolant is continually removed.

PART 5   RADIATORS AND CAPS

Radiators are designed to allow rapid heat dissipation and good air flow through the radiator core. A heat exchanger in the outlet tank of the radiator, connected to the transmission, regulates transmission fluid temperature in some applications. (See Fig. 12-16 to 12-18.)

A radiator consists of two metal tanks connected to each other by a core consisting of a series of thin tubes and fins. Coolant flows from the inlet tank through the tubes to the outlet tank whenever

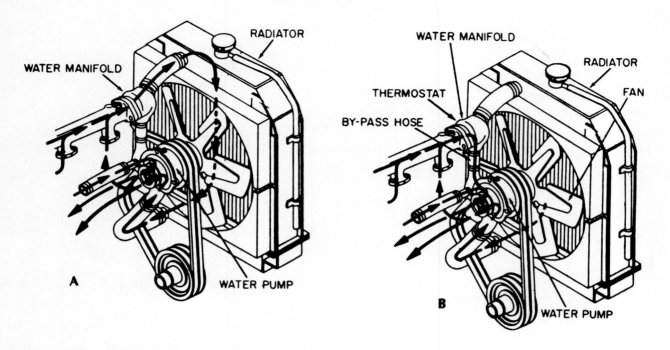

FIGURE 12-11. Coolant flow with thermostat open (A) and closed (B). *(Courtesy of J. I. Case, a Tenneco company)*

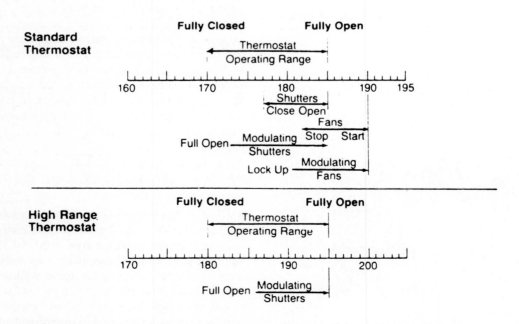

FIGURE 12-12. Operating range of typical thermostats, shutters, and fans in relation to each other. *(Courtesy of Cummins Engine Company Inc.)*

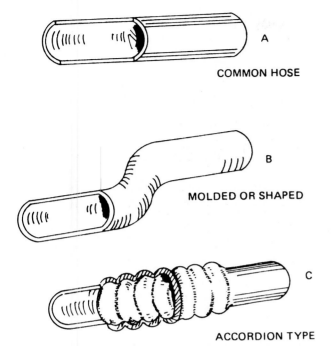

FIGURE 12-13. Cooling system hose types. *(Courtesy of Ford Motor Company)*

FIGURE 12-15. Typical hose clamps.

the thermostat in the engine is open. The tubes and fins radiate heat from the hot coolant, and the air flow created by the fan, or ram air, dissipates the heat to the atmosphere.

The inlet tank is equipped with a filler neck and radiator cap as well as an overflow tube. The overflow tube allows excess pressure to escape either to the ground or to the coolant reserve tank.

The radiator is made of aluminum or copper-brass. Both of these metals have good heat conductivity. Some radiators have the inlet tank at the top

and the outlet tank at the bottom. This is a vertical-flow radiator. The horizontal-flow radiator has one tank at each side. The inlet tank is connected to the thermostat housing, while the outlet tank is connected to the water pump inlet. Hoses are used to make these connections.

The *radiator cap* incorporates two valves. The pressure relief valve limits pressure in the cooling system to a predetermined level. By using a pressure cap, the system becomes pressurized as a result of coolant expansion. Engine heat causes the coolant to expand. Pressurizing the cooling system raises the boiling point of the coolant by approximately 3.25°F (1.8°C) for every pound per square inch (6.895 kilopascals) of pressure increase. This reduces the tendency for coolant to boil.

A 10-psi (68.95-kPa) radiator pressure cap would increase the boiling point of a 50:50 solution of antifreeze from 230°F (110°C) to 262.5°F (128°C).

If the coolant expands sufficiently to cause system pressure above radiator cap design relief pressure, the pressure valve opens and allows coolant to escape, via the overflow tube, to the reservoir, until pressure in the system is stabilized.

When the engine is shut off, the coolant cools down and contracts. This creates a low pressure in the cooling system and allows coolant to re-enter from the reservoir through the vacuum valve in the radiator pressure cap. This prevents radiator hoses from collapsing and allows the cooling system to remain full of coolant at all times. Reduced oxidation

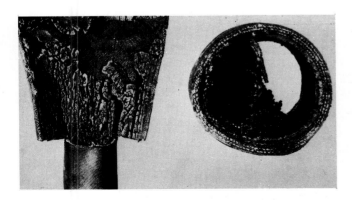

FIGURE 12-14. Radiator hose interior deterioration.

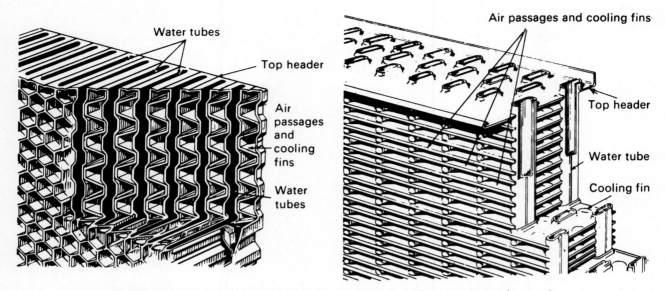

FIGURE 12-16. Two types of radiator core construction: honeycomb or cellular type (left) and tube and fin (right). *(Courtesy of Ford Motor Company)*

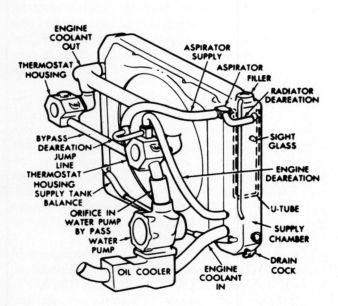

FIGURE 12-17. Cross-flow radiator design and connections. *(Courtesy of Detroit Diesel Allison)*

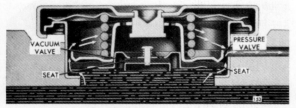
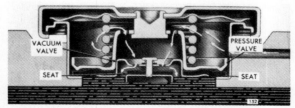

FIGURE 12-18. Radiator pressure cap operation: pressure relief valve open at left and vacuum valve open at right. *(Courtesy of Detroit Diesel Allison)*

and rust formation are benefits of the constant full cooling system.

Radiator capacity is determined by core size, thickness, and surface area. Engine size, number of accessories such as air conditioning, and type of service determine the radiator capacity of different equipment.

PART 6   FANS AND FAN DRIVES

Radiator Fan

The fan is designed to provide sufficient air flow through the radiator core to ensure adequate cooling at all engine speeds and loads. The fan is bolted to a drive hub, which may be mounted on the water pump shaft or mounted and driven separately. Fans mounted on the water pump rely on water pump bearings for support, while fans mounted separately are supported by a hub and bearings of their own. A fan clutch or viscous fluid drive may be used to control fan operation according to engine need. (See Fig. 12–19 and 12–20.)

Fan capacity is determined by the number of blades, total fan diameter, blade pitch, and fan speed. Fan efficiency can be increased by the use of a shroud around the circumference of the fan. This reduces air turbulence and air recirculation around the tips of the fan blades.

The pitch of the fan blades may be fixed or variable. Variable pitch fans have flexible curved blades that tend to flatten out as fan speed increases. On highway vehicles, ram air adds to the air flow through the radiator at highway speeds. Variable-pitch fans and clutch-controlled fans reduce the power required to drive the fan, thereby reducing fuel consumption.

Fans may push or pull air through the radiator and may move air forward, sideways, or rearward in relation to the engine depending on equipment design and the manufacturer's preference. The fan may be direct-drive from an engine gear or shaft, or may be driven by a belt from the crankshaft pulley.

Fan Drives

Fan drives are designed to achieve the following beneficial results.

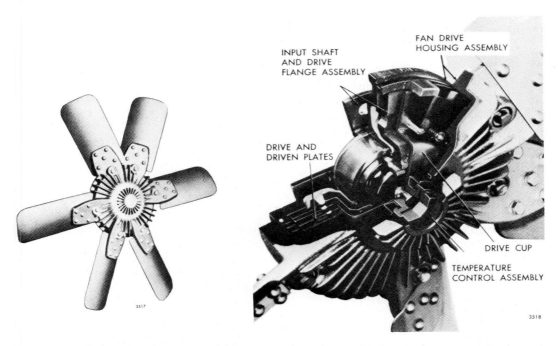

FIGURE 12-19. Typical thermomodulated fan drive assembly. Temperature control assembly controls clutch application to vary fan speed to provide required amount of air flow for cooling. *(Courtesy of Detroit Diesel Allison)*

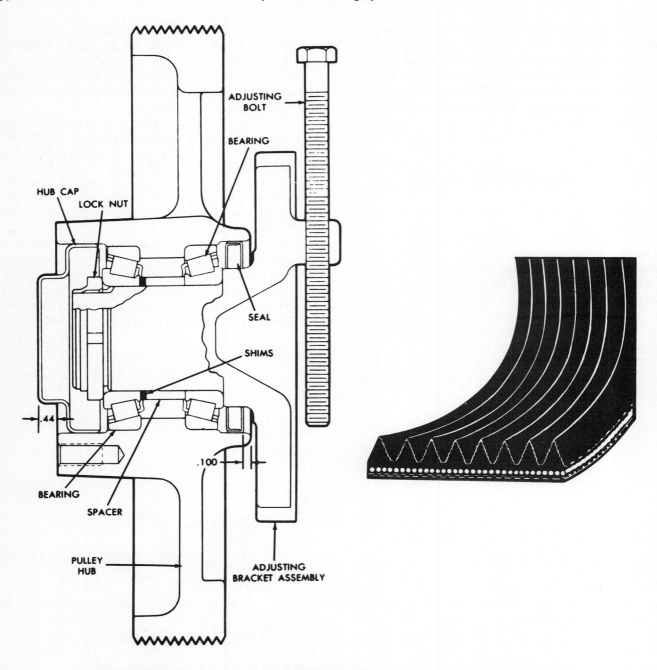

FIGURE 12–20. Poly-V fan drive cutaway view (left) and section of poly-V belt (right). *(Courtesy of Detroit Diesel Allison, left, and Deere and Company, right)*

1. Faster engine warm up.

2. Reduced fuel consumption.

3. Reduced fan noise.

4. Increased fan drive belt life.

5. Improved engine temperature control.

Two types of fan drives are commonly used: the viscous drive and the clutch drive. Variations in design and operation exist in each type, but the operating principles are similar.

The thermo-modulated viscous fan drive is an integral unit with no external controls or control lines. It operates on the principle of transmitting drive torque from the input shaft to the fan through the shearing of a viscous silicone fluid film between the input and output plates of the unit. The input plate is connected to the fan drive shaft and the output plate is connected to the fan. The plates and fluid are contained in a housing. An integral thermostatic control element reacts to changes in engine temperature and varies the thickness of the fluid film between the plates, thereby varying the fan speed. No periodic service is required by this unit.

Several makes of clutch type fan drives such as Horton, Rockford, and Bendix are used on diesel engines. A single plate clutch is released by spring pressure and applied by air pressure. Air pressure is controlled by a coolant-temperature-sensing switch, which in turn controls a solenoid-operated air control valve. As engine coolant temperature rises to about 185 degrees F (85 C), the coolant sensor closes its contacts, completing the electrical circuit to the air control solenoid. This opens the air valve, applying air pressure (from the vehicle reservoir) to engage the fan clutch against spring pressure to drive the fan. When coolant temperature drops below sensor switch closing temperature, the switch opens and the solenoid cuts off air pressure to the fan clutch, at the same time, venting air pressure from the clutch. The clutch spring disengages the clutch and the fan freewheels.

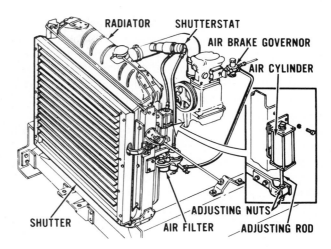

FIGURE 12-21. Air-operated type of radiator shutters. Automatically controlled shutters maintain relatively constant engine operating temperature. *(Courtesy of Ford Motor Company)*

control valve is reached, air is allowed to exhaust from the shutter cylinder, and the springs open the shutter.

The set temperature is maintained by the shutter remaining open and closed for varying periods of time, depending upon engine load and air temperature. This oscillation of the shutter between fully open and fully closed is accompanied by a lag of approximately 7°F in the control valve to prevent excessive cycling.

PART 7   SHUTTERS AND SHUTTER CONTROLS

The radiator shutters provide the means for controlling air flow through the radiator to maintain proper engine operating temperature. The shutter consists of a series of vanes mounted in a frame and attached in front of the radiator. The vanes are able to pivot in order to open and close. (See Fig. 12-21 and 12-22.)

The shutterstat in the air-operated system is mounted in the upper radiator hose and senses engine coolant temperature. The temperature-sensing element operates the needle-type air valve to control compressed air to the shutter cylinder.

When the engine is below operating temperature, the control valve directs air to the shutter cylinder to close the shutter. Engine temperature then rises quickly because air cannot be drawn through the radiator. When the operating temperature of the

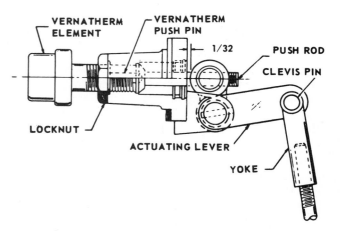

FIGURE 12-22. Vernatherm unit operates radiator shutters by expansion and contraction of material in vernatherm element. Element is mounted in lower radiator tank where it senses coolant temperatures. *(Courtesy of Mack Trucks Inc.)*

In the vernatherm shutter control system, the vernatherm temperature-sensing element is located in the bottom radiator tank. When the coolant temperature rises to the correct operating temperature, the material in the element expands against a diaphragm, which in turn is connected to the operating piston. Continued expansion causes the diaphragm and piston to move against the push rod, which is connected to the shutter control linkage. This movement opens the shutter, allowing air to pass through the radiator and cause the coolant temperature to drop. This cools the sensing element, causing the material in the element to contract. As a result, the piston returns to the original position once again to close the shutter.

The engine thermostat, fan, and shutter operate as a team to control engine-operating temperature. Each member of the team must operate at precise on and off coolant temperatures to maintain proper temperature control. Service manuals provide accurate and detailed specifications on temperature-sensing devices and controls.

PART 8   COOLANT FILTERS AND CONDITIONERS

Some engines are equipped with coolant filters and conditioners. They are designed to provide a cleaner cooling system, better heat dissipation, improved heat transfer, and thereby increased engine efficiency and service life. (See Fig. 12-23 and 12-24.)

Both spin-on and canister (replaceable element) types are used. The filter element removes any impurities such as sand or rust suspended in the coolant. The filter element includes corrosion inhibitors and water softeners. The corrosion inhibitors dissolve in the coolant and form a protective film on the metal parts of the cooling system to prevent rust. Softening the water reduces scale deposits and maintains the coolant in an acid-free condition.

Periodic replacement of the chemically treated filter element is required to ensure effective cooling system protection. If engine coolant is drained and discarded, the filter must be replaced since many of its protective elements will have been consumed in the discarded coolant.

PART 9   HEAT EXCHANGER (MARINE)

Some marine engines are equipped with a liquid-to-liquid heat exchanger. The heat exchanger takes the place of the radiator on land-based engines. (See Fig. 12-25 and 12-26)

FIGURE 12-23. Canister type of water filter and conditioner has removable element for replacement. *(Courtesy of Detroit Diesel Allison)*

The unit consists of a housing with an expansion tank and a core of cells or tubes. Hot engine coolant circulates through the expansion tank and around the core cells back to the engine water pump and the engine. The core mounted inside the housing has separate inlet and outlet passages. Raw sea water is circulated through the core by the raw water pump. As engine coolant flows over the outside of the core, it is cooled by the cooler sea water inside the core.

FIGURE 12-24. Spin-on type of water filter and conditioner. *(Courtesy of Detroit Diesel Allison)*

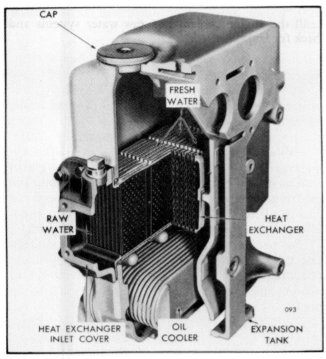

FIGURE 12-25. Cross section of typical heat exchanger for marine engine. *(Courtesy of Detroit Diesel Allison)*

Zinc electrodes are used to protect the heat exchanger from the corrosive effects of the electrolytic action of the raw sea water. One or more electrodes extending into the sea water in the heat exchanger may be used.

The expansion tank provides the means for filling the system and is provided with a filler cap. An overflow pipe connected to the expansion tank provides a vent to atmosphere or to a coolant reserve tank.

PART 10   ANTIFREEZE

Liquid coolant is the medium used to absorb heat while it is in the engine and transfer it to the radiator, where it is dissipated to the atmosphere. Although water is a satisfactory liquid to use for absorbing and transferring heat, it has several deficiencies. It has a relatively low boiling point and freezes readily. Also, inhibitors must be added to water to prevent rust and scale formation and for water pump seal lubrication. For these reasons, an ethylene glycol-based liquid is used for year-round service. (See Fig. 12-27 to 12-29.)

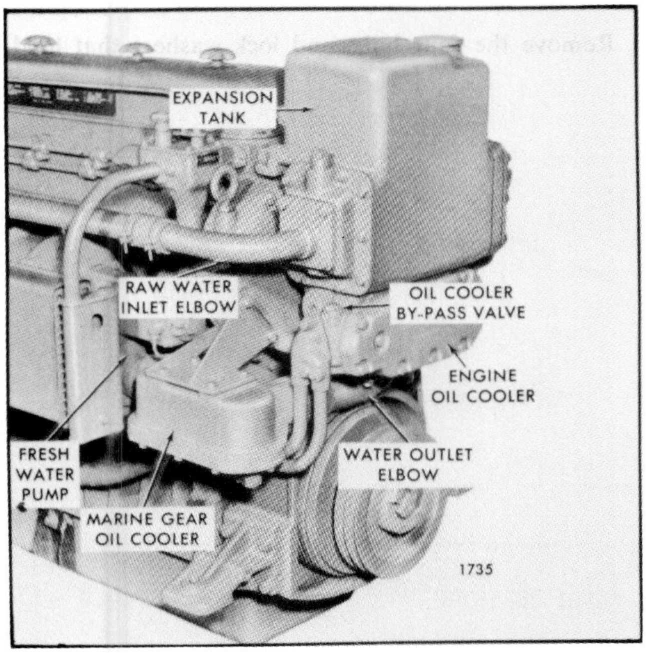

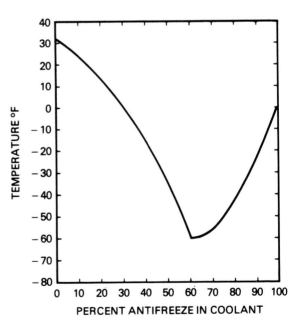

FIGURE 12-27. This graph shows the degree of freeze protection with various percentages of ethylene glycol in the coolant.

FIGURE 12-26. Heat exchanger installation on engine. *(Courtesy of Detroit Diesel Allison)*

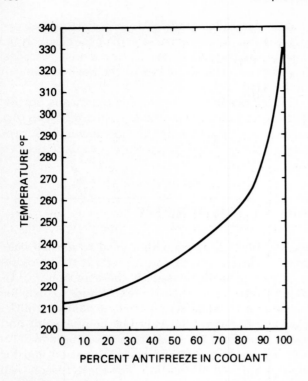

FIGURE 12-28. This graph shows the boiling point of coolant in relation to percentage of ethylene glycol in the coolant.

Ethylene glycol-based antifreeze coolant has a higher boiling point than water; it has the necessary inhibitors and additives required to retard the buildup of rust and scale, and it also has a water pump seal lubricant.

A mix of 50 percent undiluted ethylene glycol antifreeze and 50 percent water will provide freeze protection at approximately −34°F (−36°C) and will have a boiling point at approximately 230°F (110°C) at 14.7 psi atmospheric pressure (101.4 kilopascals). A greater than 60 percent ethylene glycol content is not practical, since further increasing the antifreeze content can cause the antifreeze to thicken at low temperatures, restrict coolant circulation, and cause antifreeze in the engine to boil.

In operation, antifreeze additives and inhibitors tend to lose their effectiveness. For this reason, engine manufacturers and antifreeze manufacturers recommend changing the coolant every 12 to 24 months.

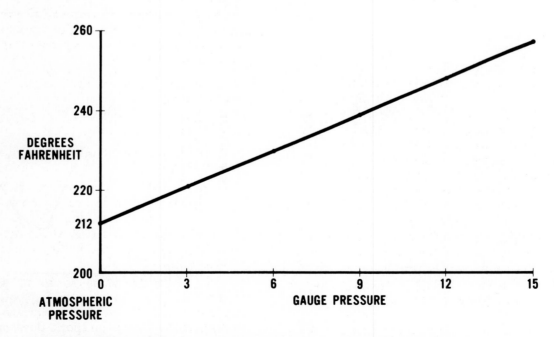

FIGURE 12-29. Effects of pressure in the cooling system on the boiling point of the coolant. *(Courtesy of Ford Motor Company)*

PART 11 COOLING SYSTEM DIAGNOSIS AND SERVICE

Visual Inspection

1. Coolant level
2. Cleanliness of coolant
3. Radiator cap gasket
4. Hose and connections for leaks or hose collapse
5. Engine lubricating oil level

Note: Oil level too low or too high will cause overheating and possible loss of the coolant.

6. Coolant in the oil and vice versa
7. Bent or damaged blades in the fan
8. Tension and condition of the belt
9. Bent core fins or accumulation of road debris in the radiator; damage or linkage wear in the shutters
10. Leaks at the cooler, water pump, truck heater, or other accessories

If there are any problems indicated by these checks, determine the cause and make the appropriate correction.

Most cooling system malfunctions result in one or the other of two conditions: overheating or overcooling.

Overheating

The obvious effects of severe overheating are seized pistons and resulting progressive damage. Intermittent loss of coolant flow can result in ring or piston scuffing and scoring, damage to the liner seals and the head gaskets, loosening of the injector sleeves, scuffed valve guides, and even cracked cylinder heads. This damage is often not obvious at the time of flow loss but later develops into an engine failure. In the case of piston and/or ring scuffing and scoring, this may show up as high oil consumption or excessive blowby several miles or hours later.

Continued operation with coolant temperatures above the recommended outlet temperatures can result in thermal fatigue of such parts as pistons, cylinder heads, and valves. High temperature causes the rubber parts to age prematurely, harden, and fail.

The resulting high lubricating oil temperatures will result in much thinner or lower viscosity oil and therefore higher oil consumption; higher than normal wear on the rings, bearings, and other parts; increased lacquer or varnish deposits; and reduced life of various assemblies.

Overcooling

Too much cooling may result from a faulty thermostat or shutterstat operation, leaking thermostat seal, too low temperature setting, poor installation, or no thermostat. As a result, many parts are operating at lower than normal temperature at greater than normal clearances. The lubricating oil temperature (even engines without oil cooler) is directly related to coolant temperature; thus low coolant temperatures will result in the oil temperature being low.

Low oil temperatures reduce the effectiveness of oil detergency and the ability of filters to clean and increase the possibility of water content and acidity. Several gases pass through the sump cavity containing water vapor (i.e., sulfur-dioxide, nitrous oxide). These gases combine with water to form acid. Low oil temperatures from overcooling will allow the water to condense in the crankcase, causing increased acidity and internal formation of sludge, lacquer, and rust. Operating with extreme overcooling may result in excessive wear and/or poor engine-operating performance. *Caution:* Never allow an engine to idle over 15 minutes at a time. Long periods of idling can be harmful for an engine because the operating temperatures drop so low that the fuel may not burn completely, resulting in deposit on the valves, clogged injector spray holes, and sticking piston rings.

1. Periodically check the coolant temperature and level indicator.
2. Listen for changes in the fan noise at all speeds.
3. Listen for belt squealing, particularly during acceleration.
4. Always allow the engine to cool before shutdown. Operate the engine for at least 3 minutes, preferably 5 minutes, at idle speed. Much can be done on the road by easing off at the end of the run. This will prevent localized boiling and loss of coolant at shutdown.

The cooling system diagnostic chart lists common problems and causes in the cooling system and related systems. The technician should have a thorough knowledge of the purpose, construction, and operation of the entire system and its components as well as the operating conditions to which the engine is subjected in order to make an intelligent diagnosis of the problem and its cause. If this is the case, the appropriate correction is obvious from the stated cause, and the necessary repair or service procedures should be made in accordance with service manual procedures and specifications.

COOLING SYSTEM DIAGNOSIS

PROBLEM	CAUSE
Loss of coolant due to external leakage	1. Leaking pipe plugs, such as core plugs, which seal off coolant passages. Core hole plugs are sometimes loosened by corrosion or vibration. 2. Loose clamps, faulty hose and piping 3. Leaking radiator or cab heater core. 4. Leaking radiator deaeration top tank or surge tank 5. Leaking gaskets due to improper tightening of capscrews, uneven gasket surfaces, or faulty gasket installation 6. Leaking drain cocks 7. Leaking water pump. Badly worn or deteriorated seals are the cause of leaks at the pump. Premature failure of the pump seals often results from suspended abrasive materials in the cooling system, excessive heat from lack of coolant or cavitation. 8. Leaks at engine or air compressor cylinder heat gasket 9. Leaks at upper cylinder liner counterbore 10. Leaking engine or auxiliary oil cooler 11. Leaking air intake aftercooler (intake air heater in some cases) 12. Leaking coolant-cooled exhaust manifold 13. Leaking water manifold and/or connections
Loss of coolant due to internal leakage	1. Leaking engine or air compressor cylinder head gasket. Coolant passes into the cylinders or the crankcase. 2. Cracked engine or air compressor cylinder head. Coolant passes into the cylinders and is blown out the exhaust or out the air compressor discharge. 3. Deteriorated, severed, or chafed liner packing; defective liner packing bore. Coolant passes into the engine crankcase. 4. Improperly seated or defective injector sleeves. Coolant can pass into the cylinder or crankcase and, in case of the heads having cylindrical injectors, coolant can enter the fuel system if it gets past the injector body O-ring. 5. Porous cylinder block or head casting. Coolant can pass into the crankcase via holes in the cooling system jacket. In case of the cylinder heads having internal fuel passages, coolant can enter the fuel system via holes in the wall of the fuel rifle, which adjoins coolant passages. 6. Cracked or porous water coolant exhaust and faulty manifold to head mounting gaskets. Coolant passes into the cylinder or out through the exhaust system. 7. Leaking engine air intake aftercooler or intake air heater. Coolant passes into the cylinders on naturally aspirated engines; also on turbocharged and supercharged engines when the intake manifold pressure is less than the coolant pressure.
Loss of coolant due to overflow	1. Slush freeze of frozen coolant resulting from insufficient antifreeze in the system or poor mixing of antifreeze and water 2. Dirt, scale, or sludge in the cooling system 3. Plugged radiator core 4. Combustion gas entering the cooling system and displacing the coolant, causing it to overflow

Engine overheating	1. Poor circulation of coolant caused by collapse of soft hose and restriction
	2. Overfueling the engine
	3. Radiator shutter malfunction or improper adjustment of the thermal controls
	4. Incorrect adjustment or malfunction of variable pitch or modulating fans
	5. Crankcase oil level too high. Crankshaft dips in oil and causes a corresponding increase in temperatures due to friction and parasitic load on the engine. Crankshaft oil dipping can also be encountered when operating an engine beyond the angles for which the oil pan was originally designed.
	6. Dirty engine exterior; heavy accumulations of dirt and grease can severely hinder normal heat dissipation through the exterior walls of all the engine components.
	7. Pressure of air in the cooling system. The following are the most common causes of air entrainment:
	  a. Low coolant level due to a leaking water pump or leaks at hose and coolant accessories
	  b. Insufficient venting of the coolant system or plugging of vent lines
	  c. Leaking air intake aftercooler
	  d. Inadequate cooling system deaerating top tank. Allows air entrainment of the coolant by failing to control the top tank coolant turbulence or by not maintaining the proper level of coolant

PROBLEM	CAUSE
	e. Leaking engine or air compressor cylinder head gasket
	f. Improperly seated injector sleeves (where applicable)
	8. Inadequate cooling capacity. This condition can be the result of the misapplication of any one or any combination of the following cooling system components:
	a. radiator
	b. top tank
	c. surge tank
	d. fan
	e. fan shroud
	f. water pump
	g. auxiliary coolers
	h. recirculation baffles
	i. fan speed
	9. Afterboil. Coolant boils and overflows after the engine is abruptly shut down following heavy loading.
	10. Improper fan belt tension
Coolant contaminated with combustion gases	1. Cracked cylinder head
	2. Blown head gasket
	3. Injector sleeves leaking at the bore (where applicable)
Coolant contaminated with fuel oil	1. Injector sleeves leaking at the bore (where applicable)
	2. Porous cylinder head casting; the fuel enters the cooler system through holes in the wall of the fuel rifle, which adjoins the coolant passages.
Coolant circulating at high pump speed only	1. Badly deteriorated water pump impeller; impeller damage is primarily caused by corrosion and cavitation erosion.
	2. Excessive impeller-to-body clearance
	3. Loose water pump drive belts
	4. Impeller slipping on the shaft
	5. Cracked impellers
Fan or water pump belts breaking prematurely	1. Foreign material falls in drive
	2. Shock or extreme overloads
	3. Belt damaged on installation by localized stretching
	4. Belts not properly matched
	5. Pulleys misaligned
	6. Pulleys nicked or rough
	7. Guard or shield interfering during operation
Engine operating too hot or overheating when loaded and coolant is known to be at the proper level	1. An altered horsepower rating of an engine or new engine installation, which exceeds the original design of the cooling system or the vehicle
	2. Clogged radiator air passages
	3. Damaged radiator core fins
	4. Heat exchanger element, which contains heavy lime and scale deposits

PROBLEM	CAUSE
	5. Thermostat not opening fully
	6. Fan shrouding missing, damaged, or improperly positioned with respect to the fan
	7. Recirculation baffles on the sides of the radiator missing or damaged
	8. Fan drive belts slip. Impeller slips on the water pump shaft.
	9. Excessive heat load from the torque converter operating below 0.3 speed ratio
	10. Clogged coolant passages, radiator, and engine
	11. Faulty automatic radiator shutters. Shutters only open partially or open too late
	12. Faulty thermatic or modulated fan drive. Fan engages too late or operates too slowly
	13. Faulty variable pitch fan. Fan operates with the blades at an insufficient pitch
Engine coolant temperature too low	1. Thermostat stuck in open position or malfunctions to allow premature opening
	2. Thermostat seal lip deteriorating and hardening, allowing the coolant to bypass the closed thermostat and enter the radiator, heat exchanger, or keel cooler
	3. Excessive bypassing of coolant to the radiator or heat exchanger, with the thermostat closed and properly sealed
	4. Engine exposed to very low temperatures and high wind with a low load factor
Lack of temperature control	1. Defective thermostat. Thermostat fails to open at the proper coolant temperature range and/or does not open completely.
	2. Defective thermostat seal. Coolant leaks by the seal and on to the water pump, therefore bypassing the radiator or heat exchanger as the case may be
	3. Operating engine having a pressurized cooling system without a pressure cap, with a defective pressure cap or a defective self-contained relief valve
Clogged water filter element	1. Contamination of the coolant caused by soluble oil, engine lube oil, gear oil or hydraulic transmission fluid, engine-fuel, or rust, scale, and lime
	2. Contamination of the cooling system with dirty fill or makeup coolant
	3. Use of antileak additive
Corrosion, rust and scale buildup occurring in the cooling system even though the water filter is serviced regularly	1. Insufficient water filter capacity
	2. Improper water filter installation
	3. Careless servicing of the water filter
	4. Aeration of the coolant. Free oxygen and carbon dioxide contained in the air are highly corrosive.

COOLING SYSTEM DIAGNOSIS (Cont'd.)

PROBLEM	CAUSE
Oil overheating even though the cooling system has adequate capacity and is free of defects	1. Clogged or restricted oil cooler oil passages 2. Restricted oil cooler oil supply and return lines. 3. Clogged or restricted oil cooler passages due to dirt, rust, and scale 4. Restricted oil cooler supply and discharge lines
Cab heater not producing heat at normal engine operating temperatures	1. Valves in the supply and/or return coolant lines shut off. Improperly sized heater for the compartment size. 2. Restricted heater cooler lines 3. Plugged heater core 4. Air entrainment of the heater core caused by operating the engine with a low coolant level 5. Heater lines not installed across a pressure drop or in the hot coolant portion of the circuit

Cooling System Service

Water Pump, Water Pump Idler, and Fan Hub

1. Inspect the water pump, idler pulley, and fan hub for wobble and evidence of grease or coolant leakage. Replace the damaged units.

2. Check the weep holes for evidence of continuous water leakage past the seal and make sure that the holes are clean and clear.

Normally a water pump impeller will not need replacement between pump overhauls. However, at overhaul be sure to check for any impeller deterioration or damage.

3. The pulley grooves should be checked for wear by the belt method. Push a new belt down in the groove. There should be 1/16-inch to 1/8-inch (0.06 to 0.13 mm) protrusion above the pulley groove. The belts should not bottom in the grooves.

V-Belt Drive

Selecting the V Belt

1. When using matched sets of belts, be sure that the belts in the set are from a single manufacturer.

Note: Do not install or try to make a matched set of belts by mixing old belts with new belts or belts from different manufacturers.

2. If one belt of a matched set fails, remove all the belts in the set and install a new set of matched belts.

Inspection of Belts

1. All the belts of a matched set should be riding at approximately the same height in the groove. Differences between the belts should not exceed 1/16 inch (0.06 mm).

2. No object should rub against the belts. Check the heat shields, guards, etc., for proper clearance and stability at all engine speeds.

3. If the side walls of the belt become frayed, remove and replace with a matched set of belts. Check to determine the reason for fraying.

4. If the belts become saturated with oil, they should be replaced. Loss of friction surface will cause slippage.

5. Small shallow cracks in the cushion or at the top of cog may be ignored. If one or more cracks are deep, the belts must be replaced.

6. Check the tensioning of the belt. If the belt squeals, it is a sign of insufficient tension. Squealing most often occurs during engine acceleration.

7. Check the pulleys for damage or nicked grooves. Replace the pulley if the nick cannot be removed with a file.

8. Check for loose pulley shafts at the bearing. Replace worn or damaged bearings or shafts, which cause pulley movement or wobble.

Belt Tension

In order for the belt to drive its intended accessory properly, it must be properly tensioned (Fig. 12-30). The belt must have sufficient tension to transmit

FIGURE 12-30. Checking fan drive belt tension with special gauge. *(Courtesy of Cummins Engine Company Inc.)*

the load of the accessory, or excessive slip will result. This slip glazes the sides of the belt, reducing its friction; more slippage results, causing premature belt failure.

Excessive tension will damage the accessory bearings, resulting in premature bearing failure and reduced belt life.

When tensioning the belts, a new set of belts must be tensioned at a higher value than the belts that have been in operation for at least 15 minutes. This higher value allows for the "seating-in" effect and tension loss.

Note: After the belts have seated in, check the tension to see that they are at the proper operating tension. The belt tension must be checked by using an approved belt tension gauge. This gauge must be properly calibrated for acceptable accuracy. Used belts (belts that have been run for at least 15 minutes) are tensioned at a lower value. Using the appropriate gauge, check and/or adjust belts to the tension as specified in the appropriate service manual.

Thermostat and Seal

Check the thermostat operation by immersing the thermostat and a thermometer in a container of water. Heat the water to the thermostat-operating temperature. Discard the thermostat if it does not operate within the range stamped on the body. It is recommended that the thermostat seal in the thermostat housing be replaced any time that the thermostat is removed. (See Fig. 12-31.)

Radiator

The cooling system should be cleaned and flushed when necessary. (See Fig. 12-32 to 12-34.)

Cleaning Procedure

1. Drain all the coolant from the cooling system.

2. Add the recommended commercial radiator cleaning compound and fill with water.

3. Run the engine at a high idle for 30 minutes. Keep constant watch on the temperature. The thermostat must be open for at least 15 minutes.

Note: Temporary covering of the radiator may be required to obtain the proper coolant temperature.

4. Drain the system completely and allow the engine to cool.

5. Add neutralizer and fill with water.

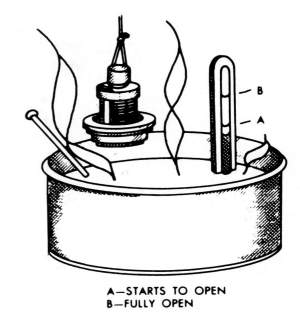

A—STARTS TO OPEN
B—FULLY OPEN

FIGURE 12-31. Testing thermostat operation. Thermostat is suspended in water being heated. Thermometer shows water temperature at which thermostat opens. *(Courtesy of Detroit Diesel Allison)*

6. Run the engine with the thermostat open for 5 minutes after the coolant temperature reaches 180°F (85°C).

7. Drain the complete cooling system and allow the engine to cool.

8. Flush with clean water and refill the system with antifreeze solution.

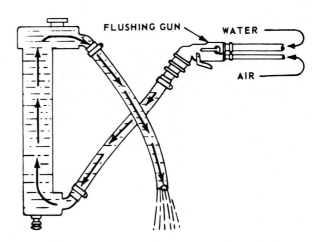

FIGURE 12-32. Reverse flow flushing the radiator. *(Courtesy of Mack Trucks Inc.)*

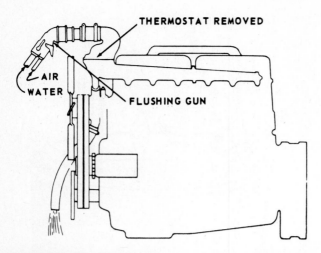

FIGURE 12-33. Reverse flow flushing the engine block. *(Courtesy of Mack Trucks Inc.)*

9. Run the engine at high idle until the thermostat opens, and check for leaks. If none are found, shut down the engine.

10. Blow compressed air through the core in the reverse direction from normal air flow until all the debris is blown out. Use care to avoid bending the fins of the radiator.

Backflushing

Backflushing of the radiator and block may be attempted if chemical cleaning does not satisfactorily clean the radiator. *Caution:* Backflushing can loosen

scale formation causing the cooling system to clog at a later date during the operation.

Since the cooling system may clog from the material loosened during flushing, it is usually better, if the radiator condition warrants, to take the radiator to a qualified radiator repair shop.

Radiator Shroud

Shrouds should be checked periodically for the proper clearance around the fan and for air leaks, cracks, and deterioration.

Pressure Cap

There are several different cap testers that can be used to check the cap. The cap is turned on to the end of the tester, and a pump built into the tester is used to apply pressure. A pressure gauge, which is a part of the tester, indicates the pressure as it is applied. A pressure cap that tests more than 2 psi above or below its required value should be replaced. (See Fig. 12-35 and 12-36.)

Adapters usually come with the cap pressure-tester for long or short caps, which permits the use of the tester during the cylinder block pressure checks.

Check the filler neck to make sure that the cap is sealing when installed. The vacuum relief valve in the cap should be checked for seal when seated and for adequate spring tension to hold a seal in position. A vacuum check should be made to verify reseating of the vacuum relief valve.

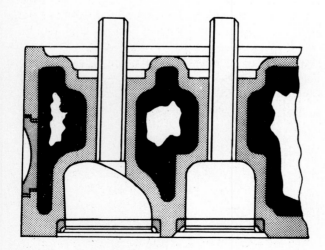

FIGURE 12-34. Typical areas of rust and scale buildup in small passages in head that are subject to greatest heat, the valve seat areas. *(Courtesy of Mack Trucks Inc.)*

FIGURE 12-35. Testing radiator pressure cap with cooling system tester. *(Courtesy of General Motors Corporation)*

FIGURE 12-36. Testing cooling system for internal and external leaks by pressurizing system to specified pressure with cooling system tester. *(Courtesy of Chrysler Corporation)*

Shutter

Periodically all moving parts of the shutters should be cleaned. The shutterstat, filter-lubricator, and operating cylinder must be cleaned and inspected following the manufacturer's instructions.

The shutter-blade pivot points require cleaning with compressed air. Little lubricant is required for these pivots; use a small amount of nongumming low-temperature lubricant. *Never use engine oil.*

If installed, replenish the filter-lubricator with the correct fluid as prescribed by the shutter manufacturer.

The temperature-sensing unit or shutterstat must be set to operate within the ranges indicated in the appropriate service manual.

Adjustments

1. Check to make sure that the shutters are completely sealed in the closed position.

2. Make sure that the shutters are completely open at the desired temperature setting.

Caution: Shutters must sense a temperature in the coolant flow stream. This requirement prohibits putting the sensing unit in the top tank above the baffle. If a point other than the engine water outlet is used, the temperature setting must be adjusted to allow for the temperature drop that occurs between the water outlet and the sensing point.

Change Coolant Filter

Canister-Type Element

1. Close the shut-off valves on inlet and drain lines. Unscrew the drain plug at the bottom of the housing.

2. Remove the cover capscrews and lift off the cover; discard the gasket and canister.

3. Remove the new canister from the package; install the canister in the housing.

4. Install the cover.

5. Reinstall the drain plug and open the shut-off valves in the inlet and drain lines.

Spin-On Element

1. Close the shut-off valve on the inlet and drain lines.

2. Unscrew the element and discard.

3. Install a new element and tighten until the seal touches the filter head. Tighten an additional one-half to three-fourths turn. Open the shut-off valves.

Heat Exchanger

Heat exchangers require inspections (at engine rebuild) of the interior tube bundles to ensure against build up on the tubes or clogging from oil sludge, which reduces heat transfer. Marine units and some industrial heat exchangers have electrolytic protection provided by a zinc plug. These plugs have a sacrificial anode that is meant to be consumed while protecting other metallic portions of the system.

These plugs should be removed and inspected at recommended intervals. If the zinc anode has deteriorated to less than approximately half the original diameter, it should be replaced. If these are neglected, the raw or sea water could react with and consume the heat exchanger or oil cooler elements.

Filling the Cooling System

After all the necessary flushing and cooling system repairs are made, the system should be filled with the recommended coolant as follows (general procedure):

1. Close drain cocks and install drain plugs.

2. Turn the heater temperature control to high.

3. Add coolant to the system until radiator remains full.

4. Add additional coolant to the required level in the coolant reserve tank.

5. Observe sight glass to make sure that the system is completely deaerated.

6. Install the radiator cap.

7. Start the engine and run until operating temperature is reached.

8. Switch the heater blower fan on. If the heater produces heat, the heater core is full; if not, accelerate the engine several times to remove air lock from the heater until heat is produced.

9. Switch the engine off.

10. Correct coolant level in the reserve tank.

11. Make sure that there are no coolant leaks.

PART 12   SELF-CHECK

1. The engine cooling system is designed to _____ _____ .

2. Describe the path of coolant flow (a) when the thermostat is closed and (b) when the thermostat is open.

3. The sole purpose of using antifreeze as a coolant is for freeze protection. True or false?

4. The radiator cap has two valves. What is the purpose of each of these?

5. Three factors that determine the capacity of a radiator fan are
(a) _____ ,
(b) _____ ,
and (c) _____ .

6. What is the purpose of the shutterstat?

7. The heat exchanger transfers engine heat to the radiator. True or false?

8. The raw water pump circulates engine coolant. True or false?

9. Why is a zinc plug used in a heat exchanger?

10. Why should a matched belt set not be mixed with other belts when installed on an engine?

Chapter 13

Engine Diagnosis, Removal, and Disassembly

is also true for the short block assembly. A short block assembly normally includes a rebuilt block with new pistons, rings, camshaft bearings, crankshaft bearings, a rebuilt camshaft, crankshaft, and connecting rods. A new oil pump is usually included, as well as a new camshaft drive.

Completely rebuilt engines, sometimes called remanufactured engines, are also available from engine rebuilding shops. These include the complete block assembly and cylinder heads as well as the oil pan, valve covers, and water pump.

A complete overhaul of an engine should be accompanied by cleaning and servicing of the engine support systems as well, to ensure troublefree operation.

Service shops that do not overhaul engines but do replace them usually base their decision for replacing cylinder heads on the results of a compression test and/or leak-down test. Cylinder wear is usually measured as soon as the heads have been removed. If liner wear is excessive, new liners or a completely rebuilt engine are installed.

Performance Objectives

After thorough study of this chapter and sufficient practical work on the appropriate components, and with the appropriate shop manuals, tools, and equipment, you should be able to do the following:

1. Complete the self-check questions with at least 80 percent accuracy.
2. Follow the accepted general precautions as outlined in this chapter.
3. Diagnose basic engine problems according to the diagnostic chart provided.
4. Safely remove and properly disassemble the engine.

PART 1   INTRODUCTION

Service shops differ considerably in their approach to engine service. Some fully equipped shops will perform most of the service operations required to overhaul an engine completely; others rely on rebuilt unit replacement.

Service procedures such as crack detection and repair, cylinder liner reboring, block deck machining, main bearing bore line boring, connecting rod straightening and resizing, and crankshaft and camshaft rebuilding are normally done by rebuilding shops. Many shops do their own work on valve guides, valve seats, and valves; other shops replace the complete cylinder head with a rebuilt unit. This

PART 2   ENGINE PROBLEM DIAGNOSIS

It is essential that a thorough understanding of the purpose, construction, and operation of the engine and all its components (as well as the engine support system) be acquired before any intelligent diagnosis and service can be performed.

The course of action to be taken in any repair work is decided by following a systematic approach to problem diagnosis. The following diagnostic chart will assist in the procedure of identifying the source of the problem and the extent of work required to correct the problem. Although this chart is courtesy of Mack Trucks Incorporated, most of the items listed will apply to other makes as well.

Where Mack-specified products or specifications are given, these of course should be followed for Mack engines. When servicing other makes such as Detroit Diesel, Caterpillar, Cummins, IHC, Deere & Company, or Allis, their recommendations must be followed.

In general, the problem must be isolated as follows:

a noise problem only

a performance problem only

a noise problem and a performance problem

an engine mechanical problem

an engine support system problem

A number of test procedures must be performed to determine where the problem exists:

fuel and induction system tests

electrical system tests

compression tests

exhaust system tests

cooling system tests

lubrication system tests

(Consult the appropriate chapters in this text for procedures.)

It must also be determined whether the service required can be done without removing the engine. If engine removal is not required, perform only those service procedures that apply to the unit being serviced. Should engine removal be required, proceed as outlined in this chapter and in the appropriate service manual.

PART 3   GENERAL PRECAUTIONS

Follow all safety precautions outlined in Chapter 1. The following general precautions should also be observed during engine service.

Servicing cab-over and cab-forward truck engines requires additional safety precautions as follows:

Before Tilting Cab

• Check clearance above and in front of truck cab.

• Keep tool chests and workbenches away from front of cab.

• Inspect sleeper and cab interior for loose luggage, tools, and liquid containers that could fall forward.

• Check "buddy" or "jump" seats on right side of cab to be sure that they are secured in place.

While Tilting Cab

• Engine must not be running.

• Check position of steering shaft U joint to prevent binding.

• Never work under a partially tilted cab, unless it is properly supported.

• When cab is tilted past the overcenter position, use cables or chains to prevent it from falling. Do not rely on the cab hydraulic lift mechanism to retain or break the fall of the cab in any position beyond the overcenter point.

While Cab is Tilted

• If it's necessary to open a door, take care to avoid damage to door hinges and/or window glass.

While Lowering Cab

• Engine must not be running.

• Be sure that the cab lowers properly on mounting and locating pins. Twisted or misaligned cabs may miss locating pins, and cab latch will not secure cab in locked position.

• Lock cab-latching mechanism when the cab is all the way down. Failure to lock the cab will allow it to swing forward when the truck is stopped suddenly.

Industrial Attachments

• Be sure that all hydraulic, mechanical, or air-operated equipment such as loaders, buckets, and the like are lowered to their at-rest position on the floor or are safely supported by other means before you attempt to do any engine work.

Other Precautions

• Be aware of the potential danger of disconnecting any pressurized systems such as hydraulics, fuel, air, or air conditioning, including refrigerant-handling precautions.

• Disconnect battery ground cable at battery to avoid electrical system damage.

ENGINE DIAGNOSIS
(COURTESY OF MACK TRUCKS INC.).

DIAGNOSIS GUIDE			
ITEM	SYMPTOM	ITEM	SYMPTOM
A	Engine not cranking	J	Excessive black or gray smoke
B	Engine cranking, not starting	K	Excessive blue or white smoke
C	Engine misfiring	L	Excessive fuel consumption
D	Engine stalling at low speed	M	Excessive oil consumption
E	Erratic engine speed	N	Engine overheating
F	Low power	O	High exhaust temperature
G	Engine not reaching no-load governed rpm	P	Low lubricating oil pressure
		Q	Oil in cooling system
H	Excessive engine vibration	R	Coolant in lubricating oil
I	Excessive smoke during acceleration	S	Low compression

DIAGNOSTIC CHART

SYMPTOM	PROBABLE CAUSE	REMEDY
A. Engine not cranking	1. Batteries with low output	1. Check batteries, charge, or replace if required.
	2. Loose or corroded battery connection	2. Clean and tighten battery connections.
	3. Broken or corroded wires	3. Check voltage at connections—switch to starter and battery to starter; replace defective parts.
	4. Faulty starter, solenoid/starter	4. Check operation of starter, solenoid/starter as outlined under electrical section repair. Replace defective parts.
	5. Faulty key switch	5. Replace key switch.
	6. Internal seizure	6. Bar the engine over at least one complete revolution. If the engine cannot be rotated a complete revolution, internal damage is indicated, and the engine must be disassembled.
B. Engine cranking, not starting	1. Slow cranking speed	1. Check items listed for engine will not crank.
	2. Chassis equipped with Mack puff limiter. Puff limiter air cylinder preventing full starting fuel to be delivered by the injection pump	2. Check emergency spring brakes. They must be set prior to starting. Also check all service lines for air leaks and operation of air cylinder and reversing defective parts.
	3. Emergency shut-off valve closed or partly closed	3. Check emergency shut-off system and make necessary correction.
	4. Low ambient air temperature	4. Use starting aid (20°F or below).
	5. No fuel to engine	5. Check for empty fuel tank, plugged fuel tank connections, obstructed or kinked fuel suction lines, fuel transfer pump failure, or plugged fuel filters.
	6. Governor throttle shaft linkage binding/improper setting of accelerator linkage	6. Check accelerator linkage; free all binding parts/make necessary adjustments.
	7. Defective fuel transfer pump	7. Check transfer pump for minimum output pressure. If low, change fuel filters. Look for air leaks and recheck pressure. If still below minimum, pump is defective. Replace fuel transfer pump.
	8. Poor quality fuel or water in fuel	8. Drain fuel from tank. Install new fuel filters and fill tank with recommended Mack-specified diesel fuel.
	9. Improper lubrication oil viscosity. Oil too thick for free crankshaft rotation.	9. Drain oil. Install new oil filters and fill crankcase with Mack-specified lubricating oil.
	10. Low compression	10. Check cylinder compression; if low, see symptom of low compression.
	11. Engine improperly timed	11. Check engine timing. The setting is protected by seals and therefore this item must be checked at a factory authorized service station.
C. Engine misfiring	1. Poor quality fuel/water or dirt in fuel	1. Drain fuel from tank. Install new fuel filters and fill tank with recommended Mack-specified diesel fuel.
	2. Air in fuel system	2. Check system for air leaks and correct. Air will generally get into the fuel system on the suction side of the fuel transfer pump.
	3. Broken or leaking high-pressure fuel lines	3. Check for fuel leaks and replace defective parts.
	4. Restrictions in fuel lines/drain lines	4. Check for fuel flow. If no flow, replace lines.

SYMPTOM	PROBABLE CAUSE	REMEDY
	5. Low fuel supply pressure	5. Check to be sure that there is fuel in fuel tank. Look for leaks or sharp bends or kinks in fuel line between fuel tank and fuel transfer pump, also clogged suction pipe in tank and plugged fuel suction hose. Look for air in the system. Check fuel pressure; if lower than specified, change filters and recheck. If still low, replace or repair transfer pump.
	6. Improper valve lash	6. Check and make necessary adjustment.
	7. Defective fuel injection nozzle or fuel pump	7. Run engine at speed that gives maximum misfiring or rough running. Loosen fuel line nuts, one at a time, on injection pump, cutting fuel flow to cylinder; when fuel is cut from a given cylinder, and running speed does not change, it is an indication that the cylinder is not firing. Remove nozzle and check; if defective, repair or replace. If nozzle checks ok, check for low compression. However, if no fuel is evident when nut is loosened, it is an indication pump is defective. Repair or replace defective parts.
	8. Engine improperly timed	8. Check engine timing.
	9. Cylinder head gasket leakage	9. Check for visible signs for leakage, coolant in the lubricating oil, also oil traces in the coolant. Use a compression tester to check each cylinder. Replace cylinder head gasket.
	10. Worn camshaft lobe	10. Replace camshaft.
D. Engine stalling at low speeds	1. Idle speed too low	1. Check idle setting and make the necessary adjustments
	2. Fuel tank vent plugged or partly plugged	2. Check vent arrangement and make necessary repairs.
	3. Low fuel supply	3. Check to be sure fuel is in fuel tank. Look for leaks, sharp bends, or dents in the fuel supply lines. Check for air in fuel system. Check fuel pressure; it must be within the recommended specification. If not, check filters, replace, and recheck pressure. If still low, repair or replace transfer pump.
	4. Injection pump overflow valve leaky, stuck open, or closed	4. Repair or replace valve.
	5. Defective fuel injection nozzle	5. Isolate defective fuel injection nozzle and replace (refer to C, item 7).
	6. Defective fuel injection pump	6. Remove injection pump. Test and make repairs as required. Reinstall pump.
	7. High parasitic load	7. Check for excessive loading due to engaged auxiliary attachments.
E. Erratic engine speed	1. Air leaks in fuel suction line	1. Check for air leaks and make necessary corrections.
	2. Throttle linkage loose or out of adjustment	2. Check all throttle linkage and make necessary adjustment.
	3. Injection pump governor failure	3. Remove injection pump and look for damaged or broken springs or other components. Check for free travel of fuel rack. Check for correct governor spring. Install new parts in place of those having damage or defects. Recalibrate injection pump and install.

SYMPTOM	PROBABLE CAUSE	REMEDY
F. Low power NOTE: When diagnosing low-power complaints, it is possible the trouble can be traced to chassis components other than the engine. Make sure the chassis rolls freely when the brakes are released.	1. Restrictions in the air intake system; clogged air filter(s), etc.	1. Check the air pressure in air intake manifold. Replace air filter and make necessary corrections to air system.
	2. Poor quality fuel	2. Drain fuel tank(s); clean and bleed fuel system. Replace fuel filters. Fill tank with the recommended Mack-specified diesel fuel.
	3. Damage or restrictions in the accelerator/shut-off cable linkage	3. Check linkage, adjust to obtain sufficient travel. Replace if damaged or bent.
	4. Low fuel pressure	4. Check fuel supply lines for leaks, kinks, restrictions, air in system, etc. Check fuel pressure. If low, replace filters and re-check; if still low, replace or repair fuel transfer pump. Also check for sticking, binding, or defective fuel overflow valve. Repair or replace valve.
	5. Excessive valve lash	5. Set valve lash to specified clearance.
	NOTE: In addition to the items already listed, the following probable causes (items 6–8) are for units equipped with Mack puff limiter.	
	6. Air cylinder improperly shimmed	6. Remove air cylinder. Check operation and PLE dimension. Make corrections and install on unit.
	7. Puff-limiter air cylinder stuck in puff limiting position	7. Check operation of air cylinder and replace if defective.
	8. Defective reversing relay, permitting excessive air pressure being directed to the air cylinder	8. Check air pressure at relay outlet; if over specified limit, replace relay.
	NOTE: In addition to the items listed, the following probable causes (items 9–15) relate to the turbo-charged inter-cooled diesel engine.	
	9. Restrictions in the tip turbine air cleaner/air induction system	9. Check air induction system. Replace air cleaner and make any necessary repairs.
	10. Air leak in tip turbine fan. Bleed air connection.	10. Check all air connections. They must be tight.
	11. Tip turbine fan exhaust port plugged	11. Remove blockage. Exhaust port must be open at all times.
	12. Excessive dirt buildup in tip turbine fan housing or wheel	12. Clean, repair, or replace unit.
	13. Tip turbine wheel rub	13. Repair or replace unit.
	14. Intercooler core blocked	14. Remove core. Clean or replace. Reinstall core.
	15. Intercooler gasket leakage	15. Disassemble intercooler and replace gasket.
	16. Incorrect fuel injection timing	16. Make adjustment to timing.
	17. Plugged fuel tank vents	17. Clean vents.
	18. Fuel injection nozzle failure	18. Isolate defective nozzle and replace.
	19. Turbocharger with carbon deposits or other causes of friction.	19. Make inspection of the turbocharger. Clean, repair, or replace unit.
	20. Internal fuel injection pump wear preventing full rack travel	20. Remove injection pump. Make necessary repairs, recalibrate pump, and install on engine.
	21. High altitude operation	21. Engine loses horsepower with increases in altitude. The percentage of power loss is governed by the altitude at which the engine is operated. Make the necessary adjustments.
	22. Low compression	22. Check items listed for low compression.

DIAGNOSTIC CHART (Cont'd.)

SYMPTOM	PROBABLE CAUSE	REMEDY
G. Engine not reaching no-load governed rpm.	1. Air in fuel system.	1. Check system for air leaks and correct. Air will generally get into the fuel system on suction side of fuel transfer pump.
	2. Accelerator linkage loose or out of adjustment	2. Check all linkage and make necessary adjustment.
	3. Restricted fuel lines/stuck overflow valve	3. Check flow in fuel lines and overflow valve for defective spring, poor valve, valve setting, or sticking. Make all necessary repairs.
	4. High-idle adjustment set too low	4. Check high-idle adjustment. Make necessary adjustment.
	5. Chassis equipped with Mack maxi miser: air leaks in the air supply line or defective control valve	5. Should the air supply to the transmission or air line from the transmission to the fuel injection pump governor leak, the maximum engine rpm would automatically be restricted in all gears and not just fifth gear. Check all connecting hoses and make necessary repairs. Replace control valve if required.
	6. Fuel injection pump calibration incorrect	6. Remove injection pump and nozzle assemblies from engine. Check calibration. Make necessary adjustments and install on engine.
	7. Internal fuel pump governor wear	7. Remove injection pump from engine. Make all necessary repairs, recalibrate, and install on engine.
H. Excessive smoke during acceleration NOTE: This item is geared to chassis equipped with Mack puff-limiter.	1. Broken outlet line to air cylinder	1. Check and replace broken/deteriorated air line.
	2. Plugged or broken pressure-sensing line	2. Check and replace broken/deteriorated air line.
	3. Inoperative air cylinder	3. Check operation of air cylinder; replace if defective.
	4. Air cylinder improperly shimmed	4. Remove air cylinder and check operation, PLE dimension, and shim pack. Make necessary corrections and install an injection pump.
	5. Defective or stuck reversing relay valve	5. Correct sticking or replace valve
I. Excessive black or gray smoke	1. Insufficient air for combustion	1. Check for air cleaner restrictions. Check inlet manifold pressure and inspect turbo-charger for correct operation. Make necessary repairs.
	2. High exhaust back pressure	2. Check for faulty exhaust piping or muffler obstruction. Repair/replace defective parts.
	3. Improper grade of fuel	3. Drain fuel from tank. Install new fuel filters and fill tank with recommended Mack specified diesel fuel.
	4. Faulty injection nozzle	4. Isolate faulty nozzle and replace. Refer to engine misfiring, item 7.
	5. Improper engine timing	5. Check timing and make necessary corrections.
J. Excessive black or white smoke	1. Engine lubricating oil level too high	1. Remove excess lubricating oil. If oil is contaminated with either fuel or coolant, completely drain system. Change oil filters, locate source of leak, and correct. Fill with Mack-specified lubricating oil.

SYMPTOM	PROBABLE CAUSE	REMEDY
		Check oil level with dipstick. *Do not over-fill.*
	2. Failure of turbo-charger oil seals	2. Check inlet manifold for and make repair to turbocharger as required.
	3. Worn piston rings	3. Check cylinder walls for scuffing. Clean up or replace as required. Install new piston rings.
	4. Engine misfiring or running rough	4. Check all items listed under engine misfiring.
	5. Engine to pump timing	5. Check and reset timing.
K. Excessive fuel consumption	1. Restrictions in air induction system	1. Inspect system and remove restrictions. Replace defective parts.
	2. External fuel system leakage	2. Check fuel system external piping for signs of fuel leakage. Make necessary corrections.
	3. Incorrect injection timing	3. Check engine timing and make corrections.
	4. Defective injection nozzle assembly	4. Isolate defective nozzle assembly. Remove and replace defective parts.
	5. Fuel injection pump calibration incorrect	5. Remove injection pump and nozzle assemblies from engine. Check calibration and make necesssary adjustments.
	6. Internal engine wear	6. Overhaul engine.
L. Excessive oil consumption	1. External oil leaks	1. Check engine for visible signs of oil leakage. Look for loose/stripped oil drain plug, broken gaskets (cylinder head cover, etc.), front and rear oil seal leakage. Replace all defective parts.
	2. Clogged crankcase breather/pipe	2. Remove obstruction.
	3. Excessive exhaust back pressure	3. Check exhaust pressure and make necessary corrections.
	4. Worn valve guides	4. Replace valve guides.
	5. Air compressor passing oil	5. Repair or replace air compressor.
	6. Failure of seal rings in turbocharger	6. Check inlet manifold for oil and make necessary repairs.
	7. Internal engine wear	7. Overhaul engine.
M. Engine overheating	1. Coolant level low	1. Determine cause. Replace leaking gaskets and hoses. Tighten connections and add coolant.
	2. Loose or worn fan belts	2. Adjust belt tension/replace belts.
	3. Air flow through radiator restricted	3. Remove all obstructions from outer surface of radiator.
	4. Radiator pressure cap defective	4. Check pressure release of radiator cap. Replace cap if defective.
	5. Defective coolant thermostat/water gauge	5. Check thermostat for proper opening temperature and correct installation. Check temperature gauge. Repair, if necessary.
	6. Fan improperly positioned/viscous fan drive not operating properly	6. Check operation of fan and make necessary adjustment.
	7. Chassis with shutters: shutters not opening properly	7. Check shutter operation and make necessary repairs.
	8. Combustion gases in coolant	8. Determine point at which gases enter the system. Repair or replace components as required.
	9. Plugged oil cooler	9. Remove oil cooler from engine. Disassemble, remove restrictions/replace defective parts.

DIAGNOSTIC CHART (Cont'd.)

SYMPTOM	PROBABLE CAUSE	REMEDY
	10. Defective coolant pump	10. Remove pump and make necessary repairs.
N. High exhaust temperature	1. Operating chassis in wrong gear ratio for load, grade, and altitude	1. Select the correct gear ratio for load and grade conditions.
	2. Restrictions in air induction system	2. Inspect system and remove restrictions. Replace defective parts.
	3. Air leaks in air induction system	3. Check pressure in the air intake manifold. Look for restrictions at the air cleaner/intercooler. Make necessary corrections.
	4. Leaks in exhaust system (pre-turbo)	4. Check exhaust system for leaks and make necessary repairs.
	5. Fuel injection timing incorrect	5. Make adjustment to injection timing.
	6. Restriction in the exhaust system	6. Inspect system and make necessary repairs.
	NOTE: In addition to the items listed above, the following (items 7–12) relate to chassis equipped with intercooler engine.	
	7. Plugged tip turbine fan air cleaner	7. Check air cleaner and remove restriction.
	8. Leaking tip turbine fan, bleed air connection	8. Check all air connections. They must be airtight for efficient operation.
	9. Tip turbine fan exhaust port plugged	9. Remove blockage. Exhaust port must be open at all times.
	10. Excessive dirt built up in fan housing or wheel	10. Clean, repair, or replace unit.
	11. Turbine wheel rub	11. Repair or replace the unit.
	12. Intercooler gasket leaking	12. Disassemble intercooler and replace gasket.
	13. Improper valve lash	13. Set valve lash to specified clearance.
	14. Defective injection nozzle assembly	14. Isolate defective nozzle assembly. Remove and replace defective parts.
	15. Fuel injection pump calibration incorrect	15. Remove injection pump and nozzle assemblies from engine. Check calibration and make necessary adjustment.
O. Low lubricating oil pressure	1. Oil leak—line gasket, etc. Insufficient lubricating oil level	1. Check oil level and add makeup oil as required. Oil must be to Mack—recommended specifications. Check for oil leaks.
	2. Wrong oil viscosity	2. Drain lubricating oil. Change oil filters and fill with oil meeting Mack specifications.
	3. Defective oil pressure gauge	3. Check operation of oil gauge. If defective, replace gauge.
	4. Dirty oil filter(s)	4. Check operation of bypass valve for the filter. Install new oil filter elements. Clean or install new oil cooler core. Drain oil from engine and install oil meeting Mack specifications.
	5. Lubricating oil diluted with fuel oil	5. Check fuel system for leaks. Make necessary repairs. Drain diluted lubricating oil. Install new filter elements and fill crankcase with recommended oil meeting Mack specifications.
	6. Defective oil pump relief valve	6. Remove valve check for seat condition and sticking relief valve-spring tension and cap. Check assembly parts. The use of incorrect parts will result in improper oil pressure. Make necessary repairs or install new relief valve.

SYMPTOM	PROBABLE CAUSE	REMEDY
	7. Incorrect meshing of oil pump gears	7. Check mounting arrangement. If engine has been rebuilt, check for proper gear ratio combination of the oil pump driven gear and drive gear. Incorrect gear combinations will result in immediate gear failure and possible engine damage.
	8. Excessive clearance between crankshaft and bearings	8. Overhaul engine and replace worn/defective parts.
P. Oil in cooling system	1. Defective oil cooler O ring(s)	1. Disassemble and replace O ring(s).
	2. Defective oil cooler core	2. Remove oil cooler. Disassemble and repair/replace oil cooler core.
	3. Blown head gasket	3. Replace head gasket.
Q. Coolant in lubricating oil	1. Defective oil cooler core	1. Disassemble and repair/replace oil cooler core.
	2. Blown head gasket	2. Replace head gasket.
	3. Defective water pump oil seal(s)	3. Remove water pump, disassemble, and replace defective parts.
	4. Mack V8 and Mack Scania engines only: cylinder sleeve (seals O rings) failure	4. Replace cylinder sleeve seals (O rings).
R. Low compression	1. Improper valve lash	1. Set valve lash to specified clearance.
	2. Blown head gasket	2. Replace head gasket.
	3. Broken or weak valve spring	3. Check for and replace defective parts.
	4. Burned valves or seat and parts	4. Remove head from engine and recondition head.
	5. Piston rings stuck, worn, broken/improperly seated	5. Overhaul engine.
	6. Camshaft or valve lifters worn	6. Replace camshaft and/or valve lifters. Overhaul engine if required.

• Be careful when handling hot cooling systems (the system is pressurized; radiator cap removal can cause system to boil and overflow, causing burns to hands and face). Turn cap to *first notch only* to relieve system pressure (use rag) and then remove cap.

• Be careful of hot engine parts.

• Engine oil and transmission fluid may be hot enough to cause burns; handle with care to avoid injury.

• Engine parts are heavy. Use proper methods and equipment for handling parts in order to avoid personal injury and parts damage.

• Engine parts have many precision-machined surfaces. They must be handled with extreme care to avoid damage.

• Absolute cleanliness of all parts and the entire area is essential during engine assembly. A small piece of dirt can cause complete engine failure. Abrasives damage friction surfaces severely. Extremely close tolerances are destroyed by small particles of dirt or foreign material.

• Store such components as the hood, radiator, and battery in a safe manner in which they will not be damaged during engine overhaul.

• Do not damage the radiator or AC condenser during engine service. They are easily dented or ruptured, causing a leak.

PART 4   ENGINE REMOVAL

The engine and the engine compartment should be thoroughly steam cleaned prior to removing any parts or disconnecting any systems. This will avoid getting dirt and water into engine components and systems. (See Fig. 13-1 to 13-3.)

Procedures for engine removal will vary from one application to another; however, the following steps will apply in general and should be followed.

1. Drain the cooling system (raw water and engine coolant).

FIGURE 13-1. Proper lift attachment (left) must be used when lifting heavy components such as engines. *(Courtesy of Caterpillar Tractor Company)*

FIGURE 13-3. Engine mounted on repair stand. Engine can be rotated to any position desired. *(Courtesy of Owatonna Tool Company)*

2. Drain the lubricating oil.

3. Disconnect the fuel lines.

4. Remove the air silencer or air cleaner and mounting bracket.

5. Remove the turbochargers, where applicable.

6. Disconnect the exhaust piping and remove the exhaust manifolds.

7. Disconnect the throttle controls.

8. Disconnect and remove the starting motor, battery-charging generator, and other electrical equipment.

9. Remove the air compressor, where applicable.

10. Remove the radiator and fan guard or the heat exchanger and other related cooling system parts.

11. Remove the air box drain tubes and fittings, where applicable.

12. Remove the air box covers, where applicable.

13. Disconnect any other lubricating oil lines, fuel lines, or electrical connections.

14. Attach engine lift brackets to the engine, one at the front and one at the rear. Use bolts of sufficient diameter and length to carry engine weight. Bolts used must tighten lift brackets firmly to engine. Connect the lift and apply some tension. Support the transmission properly if the transmission is not mounted to the frame.

15. Separate the engine from the transmission or other driven mechanism.

16. Remove the bolts that fasten the engine to the equipment that it powers.

17. Raise the engine from its mounting. Move the engine away from the mount and lower immediately to transport to the engine repair stand. Failure to lower the engine for transport could cause the engine and stand to upset.

18. Mount the engine on an overhaul stand.

FIGURE 13-2. Engine lift sling used for engine removal. *(Courtesy of Detroit Diesel Allison)*

Caution: Make sure that the engine is securely fastened to the overhaul stand before releasing the lifting sling. Severe injury to personnel and destruction of engine parts will result if the engine breaks away from the overhaul stand.

PART 5   ENGINE DISASSEMBLY AND CLEANING

It is important that engine disassembly be done systematically and in proper sequence. (See Fig. 13–4 to 13–28.) A number of checks and measurements must be made during disassembly to determine the condition of engine components. All measurements taken during disassembly should be compared to the manufacturer's specifications to determine serviceability of parts.

A thorough visual inspection of all parts should also accompany the disassembly procedure. The serviceability of parts is often determined by visual inspection when their condition is obvious. This eliminates cleaning and other service procedures on parts that require replacement.

The following general procedures are typical.

1. Remove externally mounted parts: turbocharger, air compressor, starter, water manifolds, water pumps, fuel pumps, flywheel, housing, etc.

2. Remove manifolds.

3. Remove valve covers (cylinder head covers).

4. Remove necessary valve and injector operating parts (rocker arms, push rods, lifters). Keep these parts in order for proper reassembly.

Caution: Loosen rocker shaft mounting bolts evenly, a little at a time, to prevent shaft distortion from valve-spring pressure. Procedure will vary depending on engine design: in-block camshaft, overhead camshaft, horizontally opposed, and so on. Remove cylinder heads.

Caution: Fuel injectors may protrude below head surface. Handle with care to avoid damage to injector or remove injectors first.

5. Remove the harmonic balancer attaching bolt if so equipped.

6. Remove the harmonic balancer, using a special puller attached to the balancer hub with two or three bolts. Be careful not to damage the crankshaft with puller screw. Never use a two- or three-leg puller hooked over the pulley or inertia ring for balancer removal. This would damage the balancer or the pulley.

7. Remove gear covers.

Caution: This may require removal of the oil pan first on some models and removal of the flywheel and housing.

8. Check gear drive condition to determine gear run-out, gear backlash, and crankshaft end play (some models); check chain deflection or backlash, sprocket wear, and camshaft end play (other models); check belt and sprocket condition and camshaft end play on belt drives.

9. Remove gear drives as outlined in service manual. Be sure to locate all gear timing marks. On some engines there may be no marks. In this case mark gears as specified in the service manual before removal. Remove cam followers.

10. Remove the camshaft (this may require removal of the thrust plate bolts first on some models). Support the camshaft during removal to avoid dragging lobes over bearing surfaces, which could damage bearings and lobes. Do not bump cam lobe edges, which can cause damage.

11. Remove balance shafts and auxiliary shafts where applicable.

12. Remove cylinder ridges if necessary to facilitate piston removal. Remove carbon with a scraper, then grind the ridge with a rotary grinder and an electric drill. A ridge reamer can only be used if the cylinder liner is prevented from turning in the block.

13. Remove the oil pan if not previously removed.

14. Remove the oil pump assembly.

15. Remove the crankshaft balancer, if so equipped.

16. Make sure that connecting rods and main bearing caps are properly identified to ensure proper reassembly. All connecting rods should have numbers on both yokes and caps. Main bearing caps should be identified for location and position. If rods and main bearing caps do not have proper identification, they must be stamped with number stamps. The rods should be numbered in accordance with the engine cylinder numbering system on both yoke and cap on the same side of the rod. The main bearing caps and engine block should also be properly numbered.

17. Remove the piston and rod assemblies as follows:
   a. Position the crankshaft throw at the bottom of its stroke.

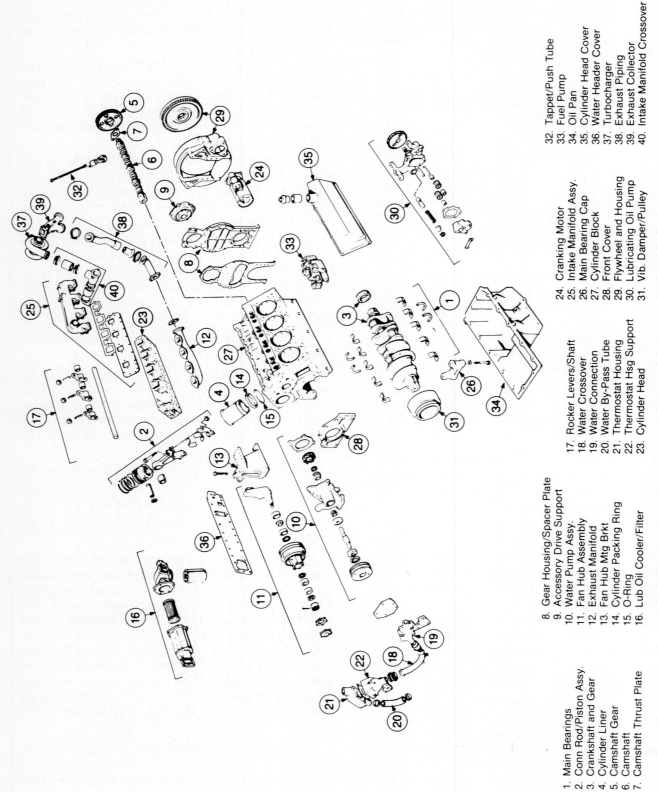

1. Main Bearings
2. Conn Rod/Piston Assy.
3. Crankshaft and Gear
4. Cylinder Liner
5. Camshaft Gear
6. Camshaft
7. Camshaft Thrust Plate
8. Gear Housing/Spacer Plate
9. Accessory Drive Support
10. Water Pump Assy.
11. Fan Hub Assembly
12. Exhaust Manifold
13. Fan Hub Mtg Brkt
14. Cylinder Packing Ring
15. O-Ring
16. Lub Oil Cooler/Filter

17. Rocker Levers/Shaft
18. Water Crossover
19. Water Connection
20. Water By-Pass Tube
21. Thermostat Housing
22. Thermostat Hsg Support
23. Cylinder Head

24. Cranking Motor
25. Intake Manifold Assy.
26. Main Bearing Cap
27. Cylinder Block
28. Front Cover
29. Flywheel and Housing
30. Lubricating Oil Pump
31. Vib. Damper/Pulley

32. Tappet/Push Tube
33. Fuel Pump
34. Oil Pan
35. Cylinder Head Cover
36. Water Header Cover
37. Turbocharger
38. Exhaust Piping
39. Exhaust Collector
40. Intake Manifold Crossover

FIGURE 13-4. Exploded view of Cummins VT-504 engine, showing parts relationship. *(Courtesy of Cummins Engine Company Inc.)*

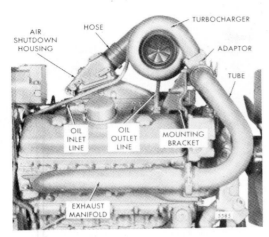

FIGURE 13-5. Turbocharger components must be removed for engine overhaul. *(Courtesy of Detroit Diesel Allison)*

FIGURE 13-8. Removing rocker housing assembly from Cummins diesel engine. *(Courtesy of Cummins Engine Company, Inc.)*

FIGURE 13-6. Removing injection pump and governor from Caterpillar V-type engine. *(Courtesy of Caterpillar Tractor Company)*

FIGURE 13-9. Removing fuel injector using a slide hammer type of puller. *(Courtesy of Cummins Engine Company Inc.)*

FIGURE 13-7. Removing rocker housing cover. *(Courtesy of Cummins Engine Company, Inc.)*

FIGURE 13-10. Removing cylinder head with the use of special T-handle lift attachments. *(Courtesy of Cummins Engine Company Inc.)*

FIGURE 13-11. Special lift attachment used to remove large cylinder heads. *(Courtesy of Detroit Diesel Allison)*

FIGURE 13-12. Removing rocker shaft assembly prior to cylinder head removal on Caterpillar engine. *(Courtesy of Caterpillar Tractor Company)*

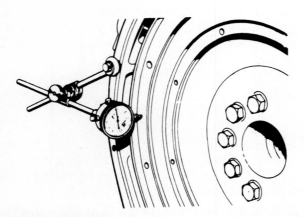

FIGURE 13-13. Measuring flywheel run-out using a dial indicator. *(Courtesy of Caterpillar Tractor Company)*

FIGURE 13-14. Flywheel removal using a crane and lift attachment. *(Courtesy of Detroit Diesel Allison)*

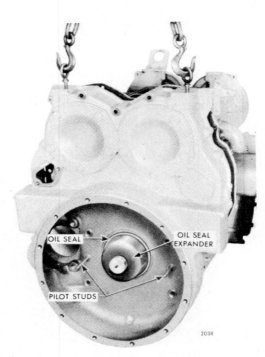

FIGURE 13-15. Removing flywheel housing and gear cover. *(Courtesy of Detroit Diesel Allison)*

FIGURE 13-16. Removing blower assembly from Detroit Diesel engine. *(Courtesy of Detroit Diesel Allison)*

FIGURE 13-19. Removing camshaft. Camshaft must be properly supported during removal and installation to prevent damage to camshaft and bearings. *(Courtesy of Caterpillar Tractor Company)*

FIGURE 13-17. Removing crankshaft pulley with special puller. *(Courtesy of Detroit Diesel Allison)*

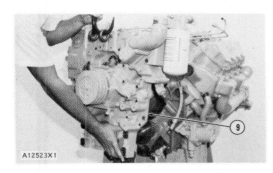

FIGURE 13-18. Removing front gear housing assembly. *(Courtesy of Caterpillar Tractor Company)*

b. Remove connecting rod nuts or cap screws and cap. Tap the cap lightly with a soft hammer to aid in cap removal.

c. Cover rod bolts to avoid damage to crankshaft journals. Pieces of rubber hose of suitable size may be used.

d. Carefully push out the piston-and-rod assembly with a wooden hammer handle or wooden drift and support the piston by hand as it comes out of the cylinder. Be sure that the connecting rod does not damage the cylinder during removal.

e. With bearing inserts in the rod and cap, replace the cap (numbers on the same side) and install nuts. (Store the piston-and-rod assembly properly). Repeat the procedure for all other piston-and-rod assemblies.

18. Check the flywheel or converter drive plate run-out with dial indicator to determine whether replacement or reconditioning is required.

19. Remove the flywheel or flex plate; scribe marking crankshaft and flywheel or plate aids in reassembly.

20. Remove main bearing cap bolts and main bearing caps.

21. Carefully remove the crankshaft by lifting both ends equally to avoid binding and damage. Heavy crankshafts in larger engines may require special precautions to ensure that accidents and damage to not ocur. Consult the appropriate manual for specific procedures.

22. Store the crankshaft in a vertical position to avoid damage, or support in enough positions to avoid sag.

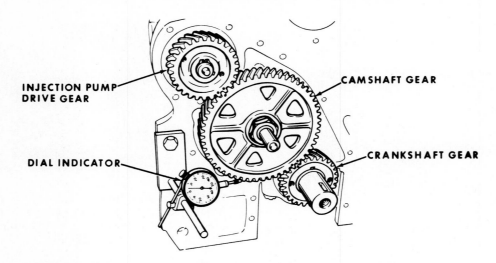

FIGURE 13-20. Gear backlash must be measured before removal to determine gear condition. *(Courtesy of J. I. Case, a Tenneco company)*

23. Remove main bearings from block and main bearing caps. Remove the rear main oil seal from the block and main bearing cap.

24. Install main bearing caps loosely in preparation for degreaser or hot tank cleaning.

25. Remove all core hole plugs in preparation for hot tank cleaning. This allows the cleaning solution to clean all water jackets thoroughly.

26. Remove oil gallery plugs and cam bearing bore plug for cleaning.

27. Remove cylinder liners, using the appropriate liner puller for the job.

28. Carefully place the engine block (and all other parts that are not subject to corrosion from the cleaning fluid being used) into the cleaning tank. Some cleaning solutions attack aluminum and other alloys. (Follow recommendations provided by the cleaning fluid manufacturer and engine manufacturer.) Some engines have the water jackets coated with oxide, which would be destroyed in a cleaning solution.

FIGURE 13-21. Removing thrust washer retaining bolts before camshaft or balance shaft removal. *(Courtesy of Detroit Diesel Allison)*

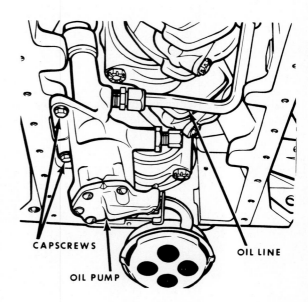

FIGURE 13-22. Removing oil pump requires disconnecting oil lines and removing mounting capscrews. *(Courtesy J. I. Case, a Tenneco company)*

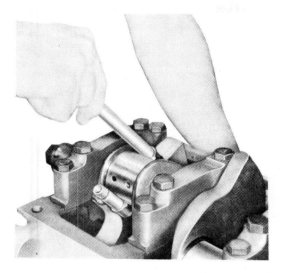

FIGURE 13-23. Connecting rod bolts should be covered to prevent damage to bearing journals of crankshaft during piston and rod removal. *(Courtesy of General Motors Corporation)*

FIGURE 13-24. After main bearing caps are removed, crankshaft is lifted from block. Lift hooks are covered with protective rubber hose to prevent crankshaft damage. *(Courtesy of Cummins Engine Company, Inc.)*

29. When they are thoroughly clean, remove the engine block and other parts from cleaning tank. With water, thoroughly rinse the cleaning fluid from all engine parts, giving particular attention to oil galleries, water jackets, and other passages.

30. Use compressed air to dry engine parts and remove water. Apply a light coat of engine oil to all machined surfaces to avoid rust.

PART 6   SELF-CHECK

1. In diagnosing engine problems, it must be determined whether the problem is a _____

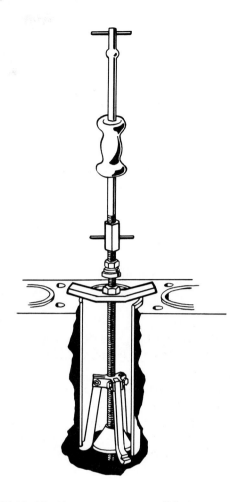

FIGURE 13-25. How to set up a slide hammer type of cylinder liner removing tool. *(Courtesy of Owatonna Tool Company)*

problem, a _____
problem, or both.

2. It is important that the engine disassembly be done _____
and in a proper _____

3. Before the pistons can be removed, the cylinder
_____
must be removed.

4. Main bearing caps and connecting rods are identified by _____

5. List three reasons why an engine could misfire.

6. List six reasons for low engine power.

7. Why should some engine parts not be put in a chemical cleaning solution?

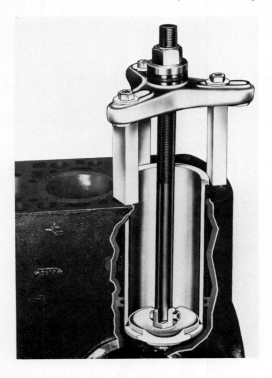

FIGURE 13-26. Screw type of cylinder liner puller setup. *(Courtesy of Owatonna Tool Company)*

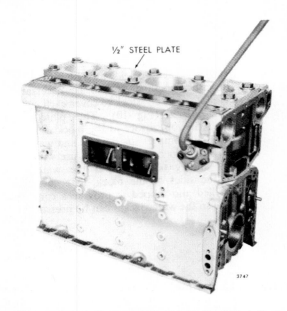

FIGURE 13-28. Pressure testing cylinder block for leaks requires sealing of all coolant passage openings. *(Courtesy of Detroit Diesel Allison)*

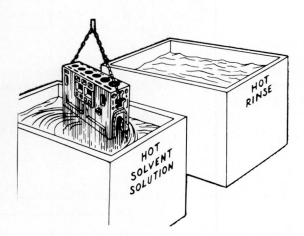

FIGURE 13-27. Proper cylinder block cleaning requires chemical hot tank cleaning and thorough rinsing. *(Courtesy of Deere and Company)*

Chapter 14

Cylinder Block Service

Performance Objectives

After thorough study of this chapter, sufficient practice on the appropriate training models and with the appropriate shop manuals, tools and equipment, you should be able to do the following:

1. Complete the self-check questions with at least 80 percent accuracy.

2. Follow all the normal general precautions outlined in this chapter.

3. Clean, inspect, and accurately measure engine block components to determine their serviceability.

4. Recondition or replace engine block components as required.

5. Correctly assemble and adjust all engine block components to meet manufacturer's specifications.

PART 1 RECONDITIONING PRECAUTIONS

It must be remembered when reconditioning engine components such as the following—

- surface grinding the cylinder heads
- surface grinding the block deck
- resizing the big end of the connecting rod
- line boring the main bearing bores—that precise dimensions specified in the manufacturer's service manual must be maintained for these engine measurements:

- crankshaft center line to the block deck
- connecting rod big end center to small end center
- connecting rod big end center to the top of piston
- top of piston to valves in the head
- injector protrusion through the head
- cylinder liner protrusion above the block

The consequences of not adhering to these tolerances can result in component damage and uneven power output between cylinders. In some cases, exceeding tolerances of 0.002 inch (.05 mm) can result in problems of this kind.

There is also some inconsistency in terminology with regard to cylinder block components. For purposes of clarification in this text, the terms *cylinder liner* and *cylinder sleeve* will be used interchangeably. The terms *cylinder liner bore, cylinder sleeve bore,* and *cylinder bore* will also be used interchangeably to designate the bore into which the piston and rings are fitted. The terms *cylinder block bore* and *block bore* will be used to designate the bore into which the cylinder liner or sleeve is fitted.

The sequence for installing cylinder liners, pistons, connecting rods, and crankshafts varies, depending on the make and model of the engine being serviced. The manufacturer's recommendations should always be followed to avoid problems. In some cases the crankshaft is installed before the piston, rod, and cylinder liner assemblies are installed. In other engines this sequence is reversed.

This text will cover these procedures in the following order: cylinder block, cylinder liners, crankshafts, auxiliary shafts, piston, and rod assemblies. Service procedures for these components are similar for most makes even though the order in which they are done may vary. Follow the sequence recommended by the engine manufacturer.

PART 2 CYLINDER BLOCK INSPECTION

Visual Inspection

After thorough cleaning of the block as outlined earlier, the block should be inspected for cracks or other damage. All threaded holes, dowel pins, and machined surfaces should be in good condition. Minor nicks and burrs can be removed with a fine file or stone. Cylinder block bores, seal ring grooves, and counterbores should also be inspected as should the cam follower bores, water jackets, oil passages, and the main bearing caps.

Pressure Testing and Magnetic Crack Detection

Prepare the block for pressure testing by closing all the cooling system openings as outlined in the appropriate service manual. Fill the water jackets completely with a mixture of about 20 percent ethylene glycol antifreeze. The antifreeze will help penetrate any possible cracks. Apply 40 to 80 psi (276 to 552 kPa) for about two hours. Examine all areas of the block for possible leakage and cracks. A cracked block should be replaced or repaired.

The block can be further tested for cracks by the magnetic inspection method. The suspected area is dusted with a special powder and the magnetic tester is positioned over the area. All flaws will become visible because of metal powder particles adhering to the area. The magnetic probe should be repositioned over the suspected area 90° from the original position to ensure complete coverage of the area.

PART 3   MEASURING THE CYLINDER BLOCK

Measuring Cylinder Block Surface

Measure the top of the block for flatness with a straightedge and feeler gauge. The block deck surface should be true within 0.001 inch (0.025 mm) across and no more than 0.003 inch (0.075 mm) variation lengthwise. These figures are typical. If not within limits specified in the service manual, the block deck must be refinished. The block height (crankshaft center to top of deck) must be within limits after block resurfacing. (See Fig. 14-1 and 14-2.)

Measuring Main Bearing Bores

Make sure that main bearing caps and block mating surfaces are absolutely clean. With bearing inserts removed, install caps and tighten main bearing cap bolts (cleaned and lubricated threads and bolt heads) to specifications. Bolts should be tightened in three stages: one-third of the specified torque, then two-thirds, and finally full torque. Main bearing bores can then be checked for alignment and out-of-round. (See Fig. 14-3 to 14-7.)

FIGURE 14-1. Block deck warpage must be checked with straightedge and feeler gauge. *(Courtesy of Detroit Diesel Allison)*

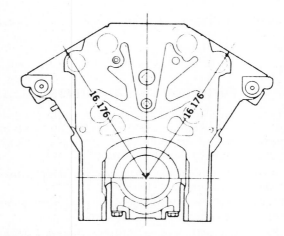

FIGURE 14-2. Cylinder block dimensions must be within limits specified in service manual after block is machined. *(Courtesy of Detroit Diesel Allison)*

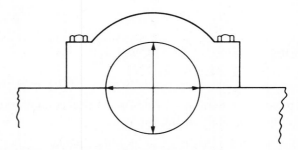

FIGURE 14-3. Main bearing bores can become out-of-round due to normal operation and incorrect tightening of bolts, can wear due to a bearing becoming loose and moving in the bore, and must be measured as indicated by arrows.

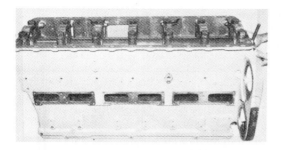

FIGURE 14-4. Measuring main bearing bore alignment with alignment bar. *(Courtesy of Cummins Engine Company, Inc.)*

Measure vertical and horizontal bore alignment with the straightedge held at the side of bores for horizontal alignment and at the top of the bores for vertical alignment. Use a feeler gauge of specified thickness to determine extent of misalignment. If bores are out of alignment more than the specified limit in the shop manual, the main bearing bores must be reconditioned. Maximum allowable misalignment is generally 0.001 inch (0.025 mm). Another method uses a main bearing bore alignment bar and feeler gauge.

Main bearing bores are reconditioned by grinding a specified amount from mating surfaces of main bearing caps on a specially designed grinder for the purpose. Usually no more than a few thousandths of an inch is removed. After grinding, the caps are again installed on the block and bolts tightened to specifications. The main bearing bores are then align bored. Block height from the crankshaft center line to the top of the block must be within limits after align boring, or the block should be replaced.

FIGURE 14-6. Align boring main bearing bores. *(Courtesy of Sunnen Products Company)*

FIGURE 14-5. Measuring main bearing bore out-of-round with dial gauge. *(Courtesy of Cummins Engine Company, Inc.)*

FIGURE 14-7. Align reaming main bearing bores. *(Courtesy of Cummins Engine Company Inc.)*

The main bearing bores should also be measured for roundness. This must be done with the caps in place as for checking alignment. A cylinder bore gauge type of dial indicator, an inside micrometer, or a telescopic gauge and outside micrometer are used for this purpose. Measurements are taken where specified in the shop manual, usually vertically and horizontally. The difference between the vertical and horizontal measurements is the amount of out-of-round. Maximum allowable out-of-round is usually 0.001 inch (0.025 mm). If out-of-round is excessive, main bearing bore must be reconditioned as described.

Oversize outside diameter main bearings are available for some engines. This allows align boring of the main bearing bores to specified oversize for which the bearings are available. This method ensures that block height from crankshaft to deck is maintained. New main bearing caps are available for some engines; however, these must be fitted and sized as outlined in the appropriate service manual.

Camshaft and Auxiliary Shaft Bores and Bearings

Remove the camshaft bearings, balance shaft bearings, or auxiliary shaft bearings with the special puller for this purpose. In some engines the bearings must be removed in a specific order or sequence; follow the sequence outlined in the service manual. (See Fig. 14–8 and 14–9.)

If any camshaft bearing bores or balance shaft bearing bores are slightly damaged or scored, they should be align bored to the specific oversize for which oversize diameter bearings are available. New bearings should be carefully installed, with the oil holes and grooves properly aligned as specified in the shop manual. Again, some engines require that the bearings be installed in the order specified in the shop manual. The bearings should also be installed to the correct depth in their bores for oil holes to be indexed properly and aligned and for shaft bearing journals to be supported fully.

Undersize inside diameter camshaft and balance shaft bearings are also available for some engines. Undersize bearings allow the use of camshafts and balance shafts with refinished journals.

Extreme care must be used when installing bearings to avoid bearing collapse, burring of bearing edges, or bearing distortion.

Camshaft and balance shaft bearing clearances are generally approximately 0.003 inch (0.075 mm) to 0.006 inch (0.150 mm).

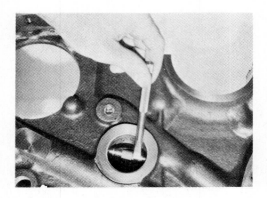

FIGURE 14–8. Measuring camshaft bearing bore diameter with an inside micrometer. *(Courtesy of Cummins Engine Company Inc.)*

Cylinder Block Bores

If cylinder block bores are eroded or damaged to the point where sealing of coolant is affected, the block must be reconditioned or replaced. In those engines with wet-type cylinder liners that have the seal ring grooves in the block, the block must be replaced if the sealing areas are damaged. On engines that have the seal ring grooves in the outer circumference of the cylinder liner, it may be possible to machine the lower block bore and install a sleeve to restore the lower bore sealing surface. Oversize outside diameter cylinder lines are available for some engines. (See Fig. 14–10 to 14–14.)

Engines that have dry cylinder liners may also be affected by block bore deterioration. If necessary, these blocks can be honed or rebored to specified oversize for which oversize outside diameter dry cylinder liners are available.

Counterbores

The counterbore in the cylinder block (Fig. 14–15) should be smooth and clean and free from erosion. If the block deck surface has been surface ground or if the counterbores are eroded or damaged, they must be reconditioned. Counterbores must be within specified limits for depth to provide specified cylinder liner protrusion above the block deck. This is necessary to ensure that all cylinders are properly sealed when the cylinder head and gaskets are installed. Counterbore depth variation limits between cylinders must also be met for the same reason. Maximum variation between cylinders should generally not exceed 0.001 inch (0.025 mm).

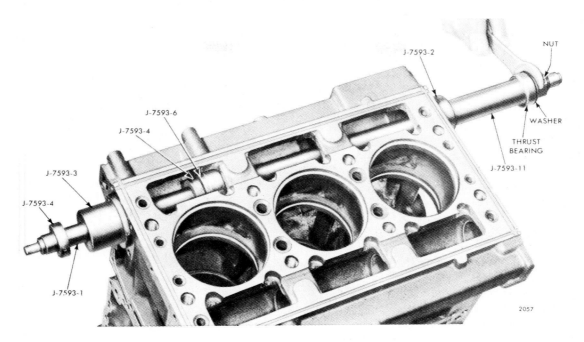

FIGURE 14-9. Removing or installing camshaft-balance shaft bearings. *(Courtesy of Detroit Diesel Allison)*

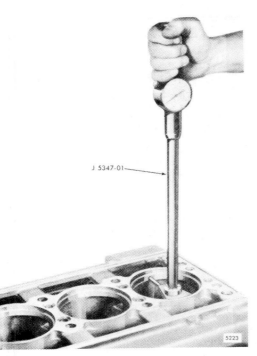

FIGURE 14-10. Measuring cylinder block bore diameter with a cylinder gauge. *(Courtesy of Detroit Diesel Allison)*

Counterbores can be reconditioned by machining them to a greater depth and installing shims to restore proper counterbore depth. The engine manufacturer's recommendations and specifications must be followed for proper counterbore reconditioning. Consult the appropriate service manual for any particular make or model.

PART 4   CYLINDER LINER INSPECTION AND HONING

Visual Inspection

Cylinder liners should be inspected for cracks, scoring, corrosion, flange irregularities, seal ring groove erosion, and in the case of dry liners, liner-to-block contact abnormalities. The outer circumference of dry cylinder liners must show evidence of good metal-to-metal contact to ensure good heat transfer to the water jackets. Liner-to-block bore clearance must be determined by measuring liner outside diameter and block bore diameter. The difference in these two measurements is the liner clearance. If this clearance is excessive, oversize outside diameter liners must be used. Follow the engine manufacturer's specifications. (See Fig. 14-16 and 14-17.)

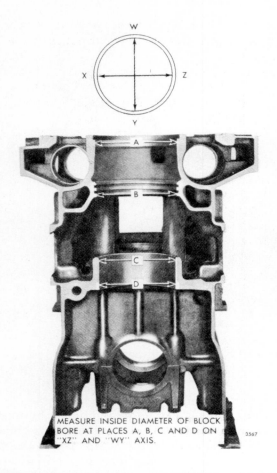

MEASURE INSIDE DIAMETER OF BLOCK BORE AT PLACES A, B, C AND D ON "XZ" AND "WY" AXIS.

FIGURE 14-11. Points where block bores should be measured on typical Detroit Diesel engine. *(Courtesy of Detroit Diesel Allison)*

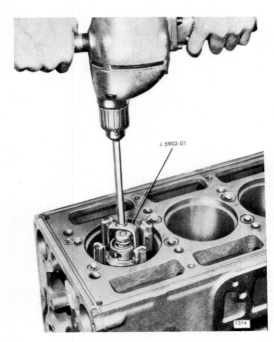

FIGURE 14-13. Honing cylinder block bore with rigid hone and electric drill. *(Courtesy of Detroit Diesel Allison)*

FIGURE 14-12. Measuring block bore in dry cylinder liner block. *(Courtesy of Detroit Diesel Allison)*

FIGURE 14-14. Reboring cylinder block bore to accommodate oversize sleeve or repair sleeve. *(Courtesy of Cummins Engine Company, Inc.)*

FIGURE 14-15. Measuring cylinder liner counterbore depth in engine block. *(Courtesy of Detroit Diesel Allison)*

Cylinder liners that show cracks, deep scoring or excessive corrosion, or flange irregularities must be replaced.

Cylinder liners that show signs of cavitation on the front but are otherwise serviceable may be repositioned by a quarter turn on the block. This increases the remaining life of the liner.

Liners must also be measured for inside diameter (cylinder bore size) to determine whether bore size is standard or oversize. Cylinder liner taper, out-of-round, and waviness must be measured to determine serviceability of the liner. Usually no more than a maximum of 0.001 inch (0.025 mm) of taper, out-of-round, or waviness is allowable. Liners with excessive taper, out-of-round, or waviness must be replaced or rebored if oversize pistons are available for the engine.

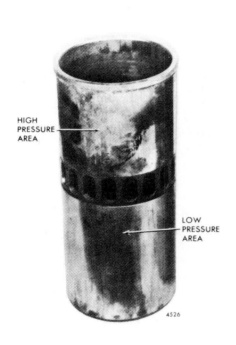

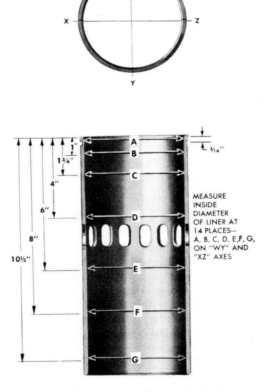

FIGURE 14-16. When the cylinder liner is removed from the cylinder block, it must be thoroughly cleaned and then checked and measured for cracks, scoring, poor contact on outer surface, flange irregularities, inside diameter, outside diameter, out-of-round, and taper. Typical for Detroit Diesel engine. *(Courtesy of Detroit Diesel Allison)*

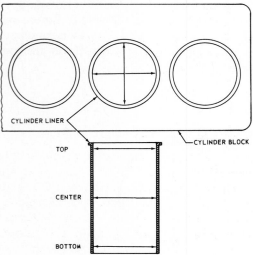

FIGURE 14-17. Points at which cylinder liner should be measured to determine taper and out-of-round. *(Courtesy of Mack Trucks Inc.)*

Cylinder Liner Taper

The difference in bore diameter at the top of the ring travel compared to bore diameter below ring travel is the amount of taper or wear. If excessive, cylinder liners should be rebored and fitted to new pistons or replaced. Some manufacturers allow a maximum of 0.001 inch (0.025 mm) before service is required.

Cylinder Liner Out-of-Round

Cylinder liners should be measured parallel to the piston pin and at right angles to the piston pin to determine cylinder out-of-round. This should be done at the top of ring travel, in the center of the cylinder, and at the bottom of the ring travel. If out-of-round exceeds the manufacturer's wear limits (approximately 0.001 inch (0.025 mm), the cylinder should be rebored and fitted to oversized piston or replaced.

Cylinder Liner Waviness

Cylinder waviness can be detected by carefully moving the cylinder gauge the full length of ring travel in the cylinder and observing the dial. If waviness is excessive, cylinders should be rebored and fitted to oversized piston or replaced.

Deglazing Cylinder Liners

If cylinder liner wear is minimal and if the instructions supplied by the piston ring manufacturer require it, the cylinder liners should be deglazed. This breaks up the glassy-smooth cylinder wall surface. The honed surface promotes good cylinder and ring lubrication during engine break-in and ensures good ring seating. A 60° (included angle) crosshatch pattern of a 25- to 30-microinch finish is desirable. Under no circumstances should the cylinder liner ever be clamped in a vise for honing or for any other reason. The liner should be placed in a block bore and clamped into place for honing or reboring with a suitable clamp that will not extend into the hone travel area.

A rigid deglazing hone with stones of 180 to 200 grit is used for this purpose. The hone is driven at about 300 rpm with a slow-speed electric drill while being moved up and down in the cylinder. If the angle of the crosshatch pattern is not steep enough, the hone is not being stroked fast enough or the drill speed is too fast. No excess material should be removed. Use solvent during honing to flush stones and keep them cutting.

Honing Cylinder Liners to Remove Taper

If taper is not excessive, cylinder liners may be honed with a rigid hone to restore parallelism. The piston-to-cylinder clearance limits specified by the manufacturer must be followed. (See Fig. 14-18 to 14-20.)

This type of cylinder liner service is usually done to extend the service life of an engine without the cost of purchasing new pistons and cylinder liners. The service life of this type of repair cannot be expected to be as long as that of an engine that has been rebored and fitted with new pistons. This type of repair is usually limited to cylinder taper of approximately 0.001 inch (0.025 millimeter) or less. This procedure also corrects out-of-round or wavy cylinder liner conditions if not excessive.

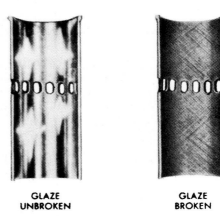

GLAZE
UNBROKEN

GLAZE
BROKEN

FIGURE 14-20. Cylinder liner (left) before honing and (right) after honing, showing 60° cross-hatch pattern produced by hone. *(Courtesy of Detroit Diesel Allison)*

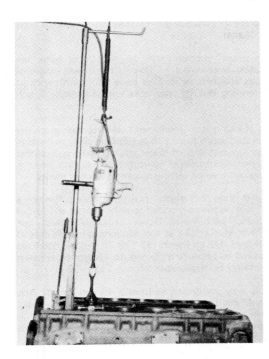

FIGURE 14-18. Setup required for honing cylinder liner in block. Liner must be secured to block to prevent liner rotation during honing. *(Courtesy of Cummins Engine Company Inc.)*

Reboring Cylinder Liners

Cylinder liners with excessive taper, out-of-round, or waviness may be rebored and fitted to oversized pistons. Maximum oversize is limited by liner wall thickness and is usually stated in the manufacturer's shop manual. Cylinder liners that are badly scored or otherwise damaged but not cracked may also be rebored if oversize pistons are available.

Cylinders are rebored or rough honed to within approximately 0.003 inch (0.08 millimeter) of finished size. They are then individually finish-honed to fit oversized pistons.

Core Hole Plugs and Water Hole Repairs

New core hole plugs should be installed with the proper tool. Plugs should be coated with sealer and then installed proper side out and to proper depth. Threaded plugs that were removed should also be sealed with sealer and installed. (See Fig. 14-21 and 14-22.)

Water holes in the top of the block may have become eroded sufficiently to affect sealing. If this is the case, they must be cut with a special cutting tool to receive a repair sleeve, which is available for the purpose. The sleeve is a press fit and should be coated with a sealer and installed with a bushing driver and hammer. Make sure that the sleeve bottoms in its bore. The sleeve will protrude slightly above the block after installation. Use a flat file to file down the sleeve level with the block. Make sure that the top of the block is not damaged during this operation.

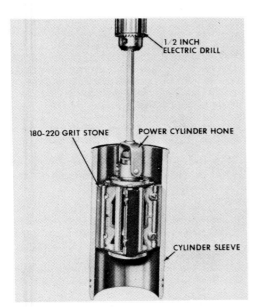

1/2 INCH ELECTRIC DRILL

180-220 GRIT STONE   POWER CYLINDER HONE

CYLINDER SLEEVE

FIGURE 14-19. Cutaway view of type of hone used in Figure 14-18. *(Courtesy of J. I. Case, a Tenneco company)*

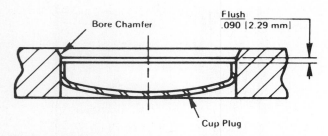

FIGURE 14-21. Cup type of core hole plug properly installed in cylinder block. *(Courtesy of Cummins Engine Company Inc.)*

Make sure that all threaded holes in the block are clean and free from all dirt and liquids. Oil or water in a blind hole can cause a block to crack when the bolt is tightened, due to extreme hydraulic pressures.

PART 5   INSTALLING CYLINDER LINERS

The cylinder block should be clean as outlined previously. (See Fig. 14-23 and 14-24.)

1. Wash the cylinder liners in solvent and blow dry with compressed air. Thoroughly lubricate the liners with the recommended lubricating oil. Wipe the cylinder liners with clean white wipers repeatedly until the wipers do not pick up any discoloration from the cylinder liners.

2. Install the liners in the block without the seals.

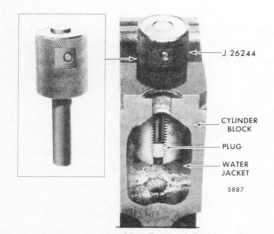

FIGURE 14-22. Threaded plug properly installed to correct depth in cylinder bolt hole prevents coolant leakage. *(Courtesy of Detroit Diesel Allison)*

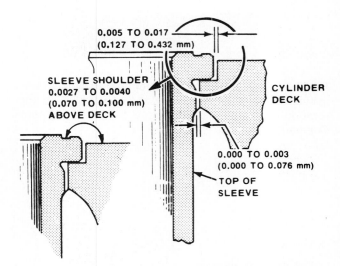

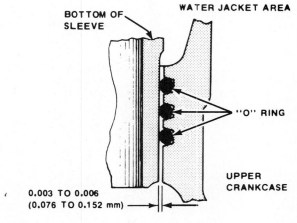

FIGURE 14-23. Wet type of cylinder sleeve installation details (typical), showing critical measurements and seal location. *(Courtesy of Mack Trucks Inc.)*

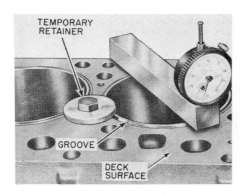

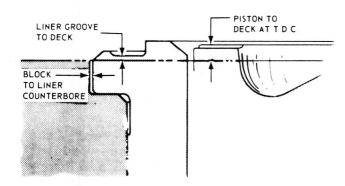

FIGURE 14-24. Measuring cylinder liner protrusion above block deck (top). Piston-to-liner-to-block surface relationship must be properly maintained (bottom). *(Courtesy of Mack Trucks Inc.)*

3. Install the tool for checking liner projection, including the liner holding tool.

4. Check the liner projection at four locations around the liner. The difference between the four measurements must not be more than 0.001 inch (0.025 mm).

5. The liner projection must be within manufacturer's specifications. The measurement between liners next to each other must not be more than 0.001 inch (0.025 mm) difference.

Turning the liner in the bore can make a difference in the liner projection if projection is not correct.

6. If the liner projection is not as specified by manufacturer, check the thickness of the liner flange and the depth of the bore in the block. The thickness of the liner flange must be as specified by the engine manufacturer.

If the liner counterbore in the block is worn and the measurement is not correct, the liner counterbore can be corrected with a cylinder block counter-

boring tool. Follow the tool manufacturer's directions.

7. Put a mark on the liner and block so that the liner can be installed in the same position from which it was removed.

8. Remove the liner and install new seals. Make sure that seals are not twisted when in place. Put recommended lubricant on the seals and install the liners in the cylinder block. If required, the liner should be seated by the method specified by the manufacturer's service manual.

9. To ensure that there is no out-of-round or waviness, the liner should be checked with a cylinder gauge as outlined earlier. If a problem exists, the liner should be removed and the necessary correction made.

Note: In some cases, the piston, rod, and liner are installed as an assembly. When this is the case, the piston and rod must be first properly serviced and installed in the liner. Refer to piston and rod service for this procedure.

PART 6   CRANKSHAFT SERVICE

Inspection And Cleaning

The crankshaft should be thoroughly inspected for damage to any machined surfaces. Main bearing journals, thrust surfaces, seal journal, and connecting rod bearing journals should appear smooth and polished without any nicks, burrs, grooves, ridges, scores, or cracks that could cause damage to bearings or shaft life. The keyway and other machined surfaces should also be in good condition. If the crankshaft passes a thorough visual inspection, it should be cleaned and measured.

Cleaning the crankshaft includes cleaning the oil passages with a rifle-type brush if necessary and washing in solvent. Care must be taken during cleaning and measuring not to damage any machined surfaces. Blow out all oil passages and blow dry the crankshaft with compressed air.

Measuring the Crankshaft

The crankshaft should be measured to determine main and rod bearing journal size, taper, out-of-round, and run-out. These measurements must be compared to the manufacturer's specifications and wear limits. If wear limits are not exceeded, the crankshaft may be used again. Minor imperfections may be touched up by polishing with crocus cloth. (See Fig. 14-25 to 14-33.)

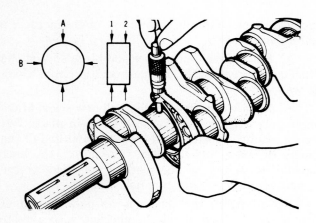

A VS B = VERTICAL TAPER
C VS D = HORIZONTAL TAPER
A VS C AND B VS D = OUT OF ROUND
CHECK FOR OUT-OF-ROUND AT EACH END OF JOURNAL

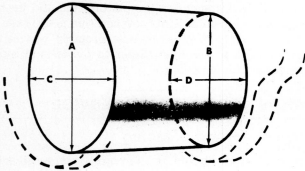

FIGURE 14-25. Measuring crankshaft bearing journal size, taper, out-of-round, and wear. *(Courtesy of Federal Mogul [top] and Ford Motor Company of Canada Ltd. [bottom])*

FIGURE 14-26. Measuring crankshaft main bearing journal run-out (crankshaft alignment) on V blocks with dial indicator. *(Courtesy of Cummins Engine Company Inc.)*

FIGURE 14-27. Measuring crankshaft thrust surface wear with inside micrometer. *(Courtesy of Cummins Engine Company Inc.)*

Maximum wear limits are usually no more than 0.001 inch (0.025 mm) for journal taper and out-of-round. Maximum crankshaft run-out should generally not exceed 0.002 inch (0.050 mm).

If the crankshaft does not meet specified wear limits, it must be replaced. Rebuilt crankshafts are normally used for replacement.

Crankshafts are reground to an undersize in engine rebuilding shops to provide good bearing journal surfaces. Crankshafts are straightened if required.

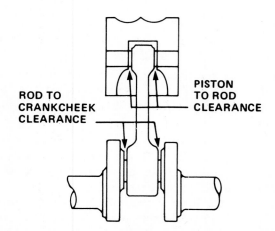

ROD TO CRANKCHEEK CLEARANCE

PISTON TO ROD CLEARANCE

IF ROD TO CRANKSHAFT CLEARANCE EXCEEDS PISTON TO ROD CLEARANCE ON EITHER SIDE, THE ROD MAY BOSS IN PISTON RATHER THAN ON CRANKSHAFT AS IT SHOULD. THIS MAY CAUSE COCKING OF PISTON AND EXCESSIVE OIL CONSUMPTION.

FIGURE 14-28. Piston-to-connecting-rod-to-crankshaft relationship must be maintained during engine overhaul. *(Courtesy of Cummins Engine Company, Inc.)*

FIGURE 14-29. Regrinding crankshaft journals. *(Courtesy of Van Norman Equipment)*

FIGURE 14-32. Installing crankshaft gear. *(Courtesy of Detroit Diesel Allison)*

FIGURE 14-30. Polishing crankshaft journals. *(Courtesy of Van Norman Equipment)*

Severely damaged journals are rebuilt by an automatic welding process, machined and hardened as required.

During the grinding and polishing process, no more material is removed than is needed to produce a good journal. Journals are finished to normal undersizes of 0.010, 0.020, or 0.030 inch (0.25, 0.51, or 0.76 millimeter). All rod journals are reground to the same undersize. The main journals are also ground to the same undersize but may differ from the rod journal undersize.

Main bearings and bearing bores in the block and caps must be absolutely clean. Place correctly sized bearings in both block and main bearing caps. To do this, place the bearing tang in the notch and

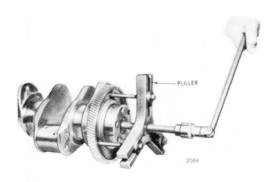

FIGURE 14-31. Removing crankshaft gear. *(Courtesy of Detroit Diesel Allison)*

FIGURE 14-33. Splitting crankshaft seal wear ring for replacement. *(Courtesy of Mack Trucks Inc.)*

then press the other side of the bearing down into place. If upper and lower bearing halves are not identical, make sure that the upper half with the oil hole is placed in the engine block. In some cases the only difference between the upper and lower bearing halves is that the lower half is several thousandths of an inch thicker. This may be indicated on the back of the bearing. If not, it must be measured to ensure correct installation. The flanged thrust bearing should be installed in the bore designed for this bearing. If the engine is equipped with washer-type thrust bearings, the bearings should be installed in the appropriate location as specified in the service manual. Oversize thrust bearings are available to correct excessive crankshaft end play.

PART 7   MEASURING BEARING CLEARANCE

After all bearings are in place, carefully install the crankshaft, but avoid crankshaft rotation to prevent scoring the bearings. Bearings and shaft must be free of oil during the procedure for checking bearing clearances. With the crankshaft in place (no rear main seal at this point), place a piece of Plastigage® across each bearing journal. (If bearing clearances are being checked with the engine in the vehicle or equipment, Plastigage® is placed across the bearings in bearing caps, and the crankshaft is raised against the upper bearing during this procedure.) (See Fig. 14–34 to 14–37.)

Install bearing caps and tighten bolts to specified torque (follow the manufacturer's torque pro-

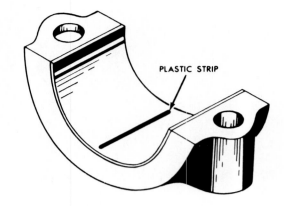

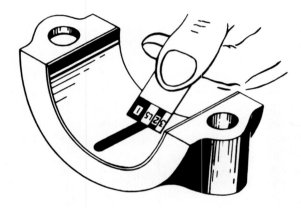

FIGURE 14–35. Using plastic strip to measure bearing clearance. *(Courtesy of Detroit Diesel Allison)*

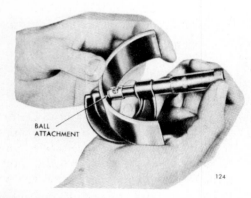

FIGURE 14–34. Measuring bearing wear with outside micrometer and special ball attachment. *(Courtesy of Detroit Diesel Allison)*

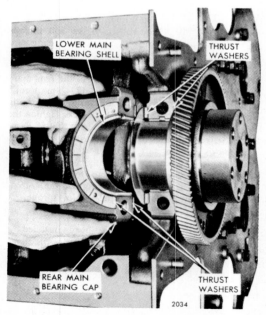

FIGURE 14–36. Installing lower main bearing and thrust washers. *(Courtesy of Detroit Diesel Allison)*

FIGURE 14-37. Measuring crankshaft end play with dial indicator. *(Courtesy of Detroit Diesel Allison)*

cedures and sequence). Some service manuals do not require following a specific order in which bolts are tightened; others may require tightening the center main bearing cap bolts first, then alternately tightening the remaining capscrews by working toward each end. Others may require starting at one end of the engine and tightening the capscrews in sequence toward the other end. Some manufacturers require a preliminary torque tightening procedure to seat all bearing inserts properly, then loosening all capscrews again before tightening bolts to specified torque. Specified torque may be achieved by using a torque wrench to tighten bolts to final torque, or bolts may have to be tightened to a preliminary torque and then tightened an additional amount in degrees of bolt turn as specified by the appropriate service manual. This is known as the torque-turn method. (Since procedures and specifications for tightening main bearing bolts vary considerably from one engine manufacturer to another, it is important that the procedures in the appropriate manufacturer's service manual be followed.)

Remove bearing caps and check clearance by measuring the width of Plastigage® with the scale provided on the envelope. Compare with the manufacturer's specifications. Clearance should be within limits.

Remove Plastigage® carefully without scratching bearings or journals. Install main bearing oil seals. This may require removing crankshaft on some models.

Thoroughly lubricate all bearings and seals with lubricating oil. Install crankshaft, main bearing caps, and bolts.

Tighten main bearing cap bolts or nuts in the sequence and method specified in the appropriate service manual.

Before tightening the thrust main bearing cap, pry the crankshaft back and forth to align thrust surfaces and then tighten as specified in the service manual.

At this point, crankshaft end play should be checked. There should be sufficient end play to avoid bearing and crankshaft damage; refer to specifications and correct if necessary. Install the flywheel housing.

The flywheel or converter flex plate should then be installed if in good condition. The starter ring gear and clutch friction surface should be in good condition and flywheel run-out within specifications. The drive bolt holes in the converter drive plate (flex plate) and starter ring gear should be in good condition and within run-out limits. All bolts should be tightened to specified torque. If the flywheel or flex plate is not in good condition, it should be replaced or reconditioned.

The crankshaft gear should be replaced if teeth show signs of excessive wear.

PART 8   CONNECTING ROD, PISTON, AND RING SERVICE

Disassembly

The piston rings are normally discarded and new rings installed when the engine is overhauled. The piston should never be clamped in a vise; however, the assembly may be held in a vise with soft jaws by clamping the connecting rod. Then the piston rings can be removed with the ring expander.

Next the connecting rod should be disconnected from the piston. This procedure will vary, depending on piston and rod design. In any case it requires removing the piston pin. In the full floating pin design, the pin retainers must be removed first, then the pin can be pushed out or tapped out with a soft drift and a hammer, while holding the assembly by hand. Heating the piston in hot water will expand an aluminum piston enough to make pushing the pin out by thumb push possible. The press fit pin must be removed with special equipment required to support the piston properly without damaging it, while pressing against the pin with either a hydraulic press or a threaded type of pin remover. The press fit pin may require several tons of applied force to remove the pin. This means that the piston must be properly supported to prevent piston damage during this procedure. In other designs the pin attaching bolts in the connecting rod must be removed to disconnect the pin and rod from the piston.

With the rod removed from the piston and the cap removed from the rod, the rod should be thoroughly washed in solvent and blown dry for inspection.

Rod Inspection

The connecting rods, caps, and bolts should be thoroughly inspected visually for any cracks, wear, or damage in the following areas: yoke and cap mating and bearing contact surfaces, bolt holes, bolt head seating area, and pin mounting area. If there is no visual evidence of cracks, wear, or damage, the rods and caps should be inspected further for cracks by the magnetic particle crack detection method (Magnaflux). (See Fig. 14–38 to 14–41.)

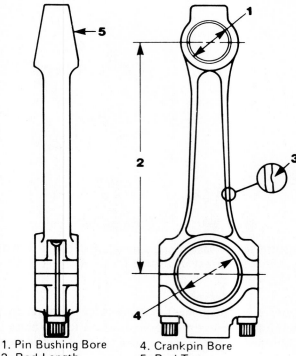

1. Pin Bushing Bore   4. Crankpin Bore
2. Rod Length   5. Rod Taper
3. Defect in Rod

FIGURE 14-39. Critical connecting rod inspection and measuring points. *(Courtesy of Cummins Engine Company, Inc.)*

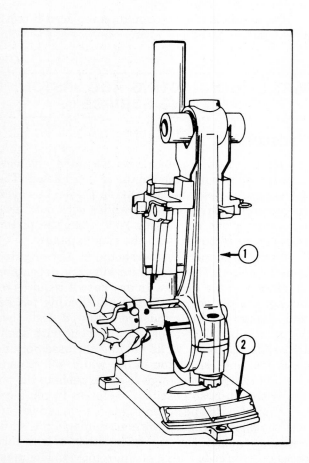

FIGURE 14-38. Measuring connecting rod alignment (bend and twist). *(Courtesy of Allis-Chalmers)*

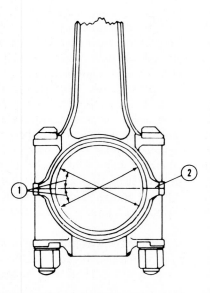

FIGURE 14-40. Connecting rod big end out-of-round measuring points. *(Courtesy of Cummins Engine Company, Inc.)*

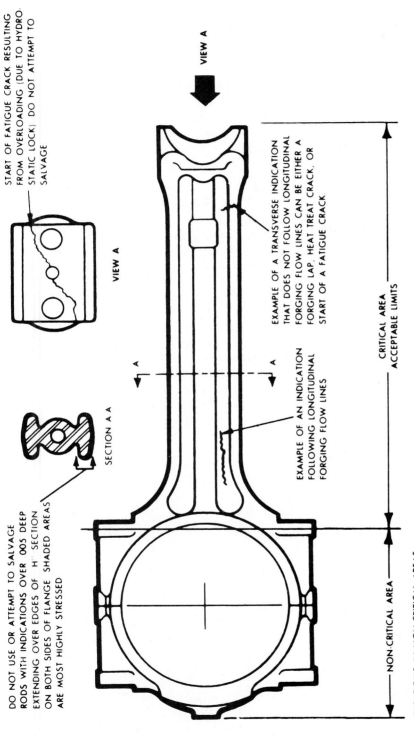

START OF FATIGUE CRACK RESULTING FROM OVERLOADING (DUE TO HYDROSTATIC LOCK) DO NOT ATTEMPT TO SALVAGE

VIEW A

EXAMPLE OF A TRANSVERSE INDICATION THAT DOES NOT FOLLOW LONGITUDINAL FORGING FLOW LINES CAN BE EITHER A FORGING LAP, HEAT TREAT CRACK, OR START OF A FATIGUE CRACK

CRITICAL AREA ACCEPTABLE LIMITS

VIEW A

SECTION A A

EXAMPLE OF AN INDICATION FOLLOWING LONGITUDINAL FORGING FLOW LINES

NON-CRITICAL AREA

DO NOT USE OR ATTEMPT TO SALVAGE RODS WITH INDICATIONS OVER .005 DEEP EXTENDING OVER EDGES OF "I" SECTION ON BOTH SIDES OF FLANGE SHADED AREAS ARE MOST HIGHLY STRESSED

INDICATIONS IN NON-CRITICAL AREAS ARE ACCEPTABLE UNLESS THEY CAN BE OBSERVED AS OBVIOUS CRACKS WITHOUT MAGNETIC INSPECTION.

LONGITUDINAL INDICATIONS FOLLOWING FORGED FLOW LINES ARE USUALLY SEAMS AND ARE NOT CONSIDERED HARMFUL IF LESS THAN 1/32 DEEP. DEPTH CAN BE DETERMINED BY GRINDING A SMALL AREA NEAR THE CENTER OF THE INDICATION.

TRANSVERSE INDICATIONS (ACROSS FLOW LINES), HAVING A MAXIMUM LENGTH OF 1/2, WHICH CAN BE REMOVED BY GRINDING NO DEEPER THAN 1/64 ARE ACCEPTABLE AFTER THEIR COMPLETE REMOVAL AN EXCEPTION TO THIS IS A ROD HAVING AN INDICATION WHICH EXTENDS OVER THE EDGE OF "I" SECTION AND IS PRESENT ON BOTH SIDES OF THE FLANGE. IN THIS CASE, MAXIMUM ALLOWABLE DEPTH IS .005 (SEE SECTION A-A)

GRINDING NOTES CARE SHOULD BE TAKEN IN GRINDING OUT INDICATIONS TO ASSURE PROPER BLENDING OF GROUND AREA INTO UNGROUND SURFACE SO AS TO FORM A SMOOTH CONTOUR

POOR PRACTICE   GOOD PRACTICE

FIGURE 14-41. Magnetic particle checking information for connecting rods. *(Courtesy of Detroit Diesel Allison)*

Rod Measuring

If the connecting rods pass the visual and magnetic inspection, the caps should be installed.

With the bearing removed from the connecting rod and the rod cap mating surfaces clean, the rod is clamped in a special holding fixture. The rod cap is then installed and the nuts tightened to specifications. The rod cap's numbered side must be on the numbered side of the rod. The special holding fixture prevents the connecting rod from being twisted while the nuts are tightened or removed.

The big end of the connecting rod can then be measured for any out-of-round condition, as illustrated. Maximum allowable out-of-round is usually 0.001 inch (0.03 millimeter). If out-of-round limits are exceeded, the connecting rod can be resized (usually a machine shop procedure). Several thousandths of an inch of material is ground from each side of the rod cap. The cap is installed and tightened to specifications. The big end is then honed to original size and roundness. This procedure is limited by the fact that the specified connecting rod length must be maintained.

The connecting rod must also be checked for twist or bend. Connecting rod alignment can be checked with the piston attached or removed on most alignment tools.

Twist or bend in the rod can be corrected by hand with the big end of the rod clamped in a vise with soft jaws. A long tapered rod specially made for this purpose is inserted in the hollow piston pin, and the rod is bent or twisted back to proper alignment and rechecked. The solid type of piston pin requires another attachment for straightening the rod.

Bushing Replacement

The small end or piston end of the connecting rod may require bushing replacement and piston pin replacement to restore pin-to-rod clearance and piston-to-pin clearance to meet specifications. (See Fig. 14-42 to 14-44.)

On connecting rods with the circular bushing, the bushing's inside diameter should be measured to determine wear. Maximum allowable wear is usually 0.0005 inch (0.0127 mm). If wear is excessive, the bushing should be replaced. This requires pressing or driving the bushing out of the rod with the proper size driver and installing a new bushing. The oil hole in the bushing must be aligned with the oil hole in the rod. In some cases, oil holes may have to be drilled in the bushing for piston cooling.

The new bushing may require expanding with a special tool to ensure that the bushing is in full con-

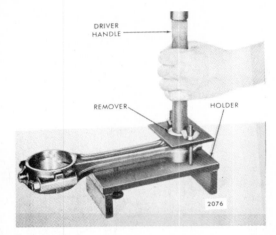

FIGURE 14-42. Removing or installing connecting rod small end bushing. *(Courtesy of Detroit Diesel Allison)*

tact with the bushing bore in the rod. In some cases the bushings are chamfered to remove any burrs, with a special reamer for the purpose. The terms used for bushing expanding include broaching or burnishing. Manufacturers may use any of these terms to describe this procedure.

On engines equipped with semicircular pin bushings, the bushing thickness is measured with an outside micrometer. If bushing wear is excessive, the bushing is replaced.

Piston pins should be measured for wear with an outside micrometer. Measurements should be made at both ends and in the middle, where wear

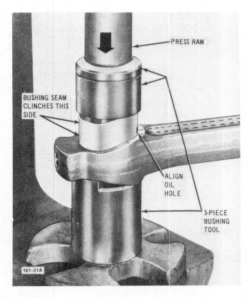

FIGURE 14-43. Installing connecting rod bushing in tapered rod. *(Courtesy of Mack Trucks Inc.)*

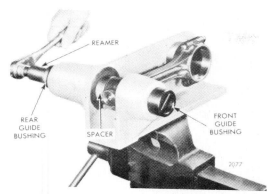

FIGURE 14-44. Reaming the connecting rod bushing in a special fixture is required to maintain accurate connecting rod length dimensions. *(Courtesy of Detroit Diesel Allison)*

normally occurs. Two measurements 90° apart should be made at each of the three wear areas. If wear is excessive, the pins should be replaced. Maximum allowable wear is usually 0.0005 inch (0.0127 mm). Follow the manufacturer's specifications.

Piston Service

Cleaning and Inspection

A good visual inspection of the pistons will determine whether they warrant cleaning and further service. With piston rings removed, ring grooves can be checked for excessive damage. Piston skirts should not have any evidence of major scoring or cracks. Piston pin retaining ring grooves should be inspected to make sure that they are in good condition. (See Fig. 14-45 to 14-56.)

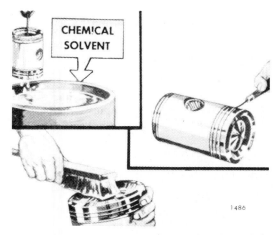

FIGURE 14-45. Cleaning pistons includes chemical cleaning, ring groove cleaning, and crown cleaning with a brush. *(Courtesy of Detroit Diesel Allison)*

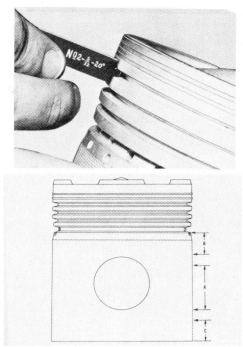

FIGURE 14-46. Checking piston groove wear (top). Piston diameter measuring points shown at bottom. *(Courtesy of Cummins Engine Company, Inc.)*

The piston head can be cleaned by using a scraper to remove most of the carbon. Aluminum pistons are easily damaged. Only careful cleaning will avoid piston damage. All cleaning tools must be used in a manner that will avoid scoring, gouging, or removal of metal.

A special ring groove cleaner is used to remove all the carbon from the ring grooves. Oil drain-back holes or slots in the oil ring groove must also be cleaned. These holes should not be enlarged during the cleaning process. The piston head and ring area may then be cleaned on the wire wheel of a bench grinder. Ring grooves and lands should not be rounded off during this process. Only light pressure of the piston against the wire wheel is used. The remainder of the piston can be cleaned with a soft, wire-bristle hand brush in solvent.

Piston Measuring

The piston should be measured at points specified by the manufacturer's shop manual and the measurements compared to specifications.

Areas to be measured include piston diameters, piston crown, ring grooves, pin bore diameters, and piston height.

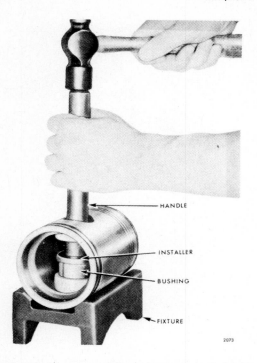

FIGURE 14-47. Installing piston pin bushing with proper equipment. *(Courtesy of Detroit Diesel Allison)*

Many manufacturers recommend replacing pistons that do not meet specifications. In some cases, ring grooves may be reconditioned by machining and installing a spacer above the ring to restore ring groove width. Ring groove wear should be measured with the appropriate gauge as recommended by the

FIGURE 14-49. Some engines require sealed piston pin retainers. A vacuum gauge is used to check for leakage. *(Courtesy of Detroit Diesel Allison)*

manufacturer. Ring side clearance should be measured with new rings and a feeler gauge. Follow the manufacturer's specifications. Clearance specifications vary considerably, depending on engine size and bore diameter.

Piston-to-bore clearance must also be measured and should be within specifications. Several methods are employed to determine clearance. One method is to measure piston diameter at its largest point and measure cylinder liner bore diameter. The difference between these two measurements is the

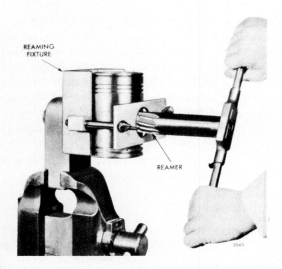

FIGURE 14-48. Reaming piston pin bushings in special piston holding fixture to prevent piston damage. *(Courtesy of Detroit Diesel Allison)*

FIGURE 14-50. Measuring piston to cylinder liner clearance with long feeler gauge and spring scale. *(Courtesy of Detroit Diesel Allison)*

FIGURE 14-51. Correct piston ring side clearance is a must and should be measured with a feeler gauge and new piston ring. *(Courtesy of Detroit Diesel Allison)*

piston-to-liner clearance. Another method is to use a long flat feeler gauge strip of specified thickness, attached at one end to a spring scale. The feeler gauge must be free of kinks or nicks. The feeler gauge strip is inserted full-length in the cylinder on the thrust side. The piston is then inserted in the same cylinder to specified depth in an inverted position. The number of pounds of pull required to remove the feeler gauge and the feeler gauge thickness determine piston-to-cylinder clearance. For example, a 4.820-inch (122.428-mm) diameter cylinder using a 0.004-inch (0.1016-mm) feeler gauge that requires 6 pounds (2.7216 kg) of pull to remove, in-

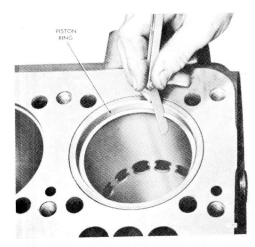

FIGURE 14-52. Measuring ring end gap in new cylinder liner. Ring must be positioned at bottom of ring travel in used liner for this measurement. *(Courtesy of Detroit Diesel Allison)*

FIGURE 14-53. Installing piston rings with ring expander. Rings should not be overexpanded. *(Courtesy of Detroit Diesel Allison)*

dicates a clearance of .005 inch (0.127 mm). Some service manuals require that this measurement be made at four evenly spaced locations around the cylinder.

The piston ring gap should be measured by pushing the ring into the lower cylinder with an inverted piston to position the ring squarely in the cylinder. If the cylinders have not been rebored and if there is any cylinder taper, the ring should be pushed to near the bottom of the cylinder to measure ring gap. Insufficient ring gap may cause ring ends to butt from expansion owing to heat and may cause severe ring and cylinder liner damage.

If ring gap does not meet the manufacturer's specifications, another ring set of the same size

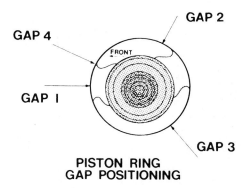

FIGURE 14-54. Typical piston ring gap positioning. *(Courtesy of Mack Trucks Inc.)*

OPERATION 1   OPERATION 2

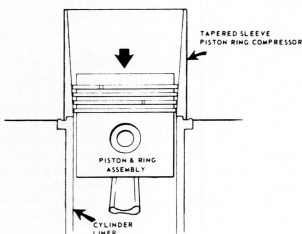

TAPERED SLEEVE
PISTON RING COMPRESSOR

PISTON & RING
ASSEMBLY

CYLINDER
LINER

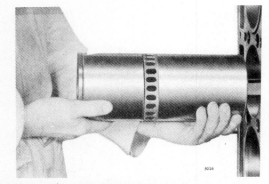

FIGURE 14-55. Installing piston in tapered ring compressor (top) and into cylinder liner (middle). Assembly is then installed in block (bottom). *(Courtesy of Detroit Diesel Allison)*

ROD TURNED TO AVOID HITTING SPRAY NOZZLE

FIGURE 14-56. Typical Mack truck piston-and-rod installation procedure. *(Courtesy of Mack Trucks Inc.)*

should be tried until the correct gap is achieved. In some cases careful filing of ring ends (when allowed by the manufacturer) is done to achieve proper ring gap.

The piston and connecting rod should then be assembled. The correct piston-to-rod relationship must be maintained during this procedure. Follow the manufacturer's recommended procedures.

Piston rings should be installed according to the ring manufacturer's instructions. Such items as placing the correct ring in its proper ring groove, right side up, and recommended ring gap positioning around the piston must be followed to ensure proper ring operation. With the three-piece type of oil ring, the oil ring spacer is installed first; then the top and bottom steel rails are spiraled into place. Oil rings of this type do not require special tools for installation. Compression rings must be installed with a special ring expander to avoid distortion and breakage. Be careful not to expand the rings any more than necessary for installation. Compression rings must not be twisted. Be careful not to score the pistons during ring installation. Never modify piston rings, spacer, or expanders in any way.

With the rod cap removed, the rings installed, and new bearing inserts properly installed (take note of oil hole alignment), the assembly is ready for installation.

Thoroughly lubricate the rings, piston, and pin by dunking the piston in a can of engine lubricating oil. (Leave the bearing and journal dry for checking bearing clearance with Plastigage®.) Install the ring compressor (notched side toward rod) just down far enough to cover all the rings, leaving enough piston uncovered to start it into the cylinder. Tighten the compressor carefully and completely to compress all rings fully without changing the location of the ring gaps.

If the tapered sleeve type of ring compressor is used, position the sleeve over the cylinder with the small diameter next to the cylinder. Position ring gaps around piston as required and carefully push piston into sleeve compressor and into cylinder. Do not allow any space between compressor and cylinder during this operation since this would allow a piston ring to expand, preventing the piston from entering the cylinder.

Cover the connecting rod bolts (to prevent crankshaft damage) and install the assembly into the cylinder block with the front of the piston toward the front of the engine. The crankpin for this cylinder should be at BDC for this operation. Carefully guide the connecting rod past the crankcase-mounted piston-cooling nozzles and onto the crankpin while you bump the piston into position with a wooden hammer handle. A special tool is recommended for this purpose by some service manuals.

The tool hooks over the bottom of the connecting rod, and the assembly can be pulled into the cylinder in this manner.

Check the bearing clearance with Plastigage® in the same manner as the main bearings. If clearance is correct, carefully remove Plastigage®, thoroughly lubricate bearing journal with engine lubricating oil, and reassemble the cap with new connecting rod bolt nuts tightened to specifications. New connecting rod nuts should be used since they are usually of the self-locking type, and old nuts may become loose during engine operation. Clamp the cylinder liner into place to prevent the liner from being pushed out when the crankshaft is turned.

Repeat this procedure for all piston-and-rod assemblies.

Measure connecting rod side clearance and compare to specifications. If excessive, connecting rods may have to be replaced.

It is good practice to turn the engine each time a piston-and-rod assembly is installed. If excessive effort is required to turn the crankshaft, the reason for this should be determined and corrected. This may require the removal and inspection of the last piston-and-rod assembly installed.

PART 9   CAMSHAFT, BALANCE SHAFT, AND DRIVE SERVICE

The camshaft should be thoroughly cleaned by washing in solvent and blowing dry with compressed air. Camshafts with lubrication passages may require removal of end plugs in order to clean the passages properly with a soft bristle brush and solvent. (See Fig. 14-57 to 14-63.)

The camshaft should be inspected after cleaning for excessive wear, pitting or chipped lobes. If the camshaft passes a thorough visual inspection, bearing journals should be measured for wear, out-of-round condition, and taper. Camshaft lobes should be measured for wear and lift. The entire camshaft should be inspected for cracks by the magnetic crack detection method.

If camshaft wear exceeds manufacturer's specified limits, the camshaft should be replaced. If crack detection reveals cracks that exceed allowable limits and locations, the camshaft should be replaced. If the engine is equipped with mushroom-type lifters, they must be installed before camshaft installation.

Cam lobes should be coated with a good high-pressure lubricant such as Lubriplate® before in-

FIGURE 14-57. Measuring cam lobe wear. *(Courtesy of Detroit Diesel Allison)*

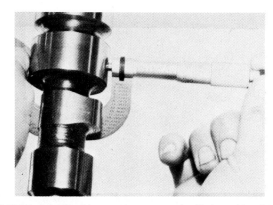

FIGURE 14-58. Measuring camshaft bearing journal wear with outside micrometer. *(Courtesy of Cummins Engine Company Inc.)*

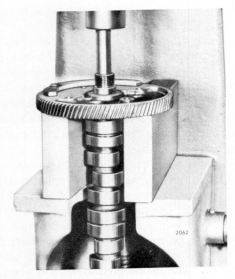

FIGURE 14-59. Removing camshaft gear in arbor press. *(Courtesy of Detroit Diesel Allison)*

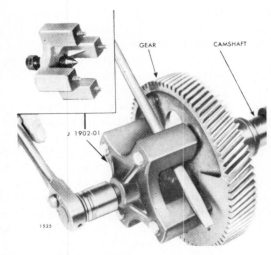

FIGURE 14-61. Removing camshaft gear with a special puller. *(Courtesy of Detroit Diesel Allison)*

stallation. Bearing journals and bearings should be lubricated with lubricating oil.

Camshafts that are held in place by a thrust plate or retainer must be checked for proper end play. If end play is incorrect, the thrust plate and gear may have to be replaced. The cam bearing bore plug should be coated with a good sealer and installed to the correct depth to avoid contact with the end of the camshaft.

Damage to bearings during camshaft installation should be avoided by properly supporting the

camshaft. Sharp edges of cam lobes may damage the bearings. A long bolt threaded into the drive end of the camshaft will allow the weight of the camshaft to be supported properly. The camshaft should be checked to see that it rotates freely without any binding after installation. Camshaft bearing clearance should be within limits specified in the appropriate service manual, generally in the range of from 0.003 inch (0.0762 mm) to 0.009 inch (0.2286 mm).

The camshaft drive gear should be installed with the necessary thrust plate or washers. In some cases the gear may have to be heated in oil to the specified temperature as outlined in the service manual to facilitate installation. Install the retaining bolt or nut and washer and tighten to specifications.

Install the camshaft and drive gears. Observe timing marks on all gears and install according to procedures and specifications outlined in the appropriate service manual. Be sure that gear backlash meets specifications.

Install balance shafts, auxiliary shafts, gears, and weights (depending on make and model) with all gears properly timed.

Install gear cover(s) and gasket and tighten all bolts in proper sequence and to proper torque.

General procedures are similar for overhead cam engines; however, the camshafts are installed after cylinder head installation. Service the oil pump, pickup tube and oil screen as outlined in Chapter 11. Install the pump and oil pan, following the proper tightening sequence and bolt torque as specified in the service manual.

FIGURE 14-60. Installing camshaft gear. *(Courtesy of Detroit Diesel Allison)*

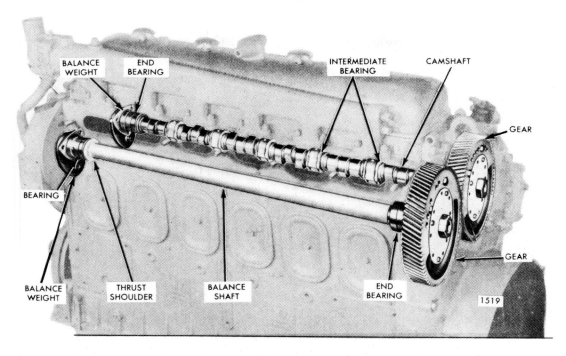

FIGURE 14-62. Camshaft and balance shaft positioned in block. *(Courtesy of Detroit Diesel Allison)*

FIGURE 14-63. Proper bolt-tightening sequence must be followed when tightening flywheel housing bolts. Follow specifications in the service manual. *(Courtesy of Detroit Diesel Allison)*

PART 10   SELF-CHECK

1. Cylinder liners are measured to determine (a) _____, (b) _____, and (c) _____.

2. Cylinder liner reconditioning requires that the cylinder wall have a _____pattern of _____inch finish.

3. Cylinder liners with excessive taper are _____.

4. What is the purpose of align boring main bearing bores?

5. How is deck warpage measured?

6. Camshaft bearings are always the same size on any engine. True or false?

7. A camshaft should be replaced if _____ or _____is excessive.

8. The crankshaft should be measured to determine main and rod bearing journal (a) _____, (b) _____, (c) _____, and (d) _____.

9. The inside diameter of undersized main bearings is (larger) (smaller) than standard main bearings.

10. The most convenient way to measure bearing clearance is to use _____.

11. The connecting rod should be checked for _____and _____.

12. A piston ring from a worn ring groove must be replaced. True or false?

Chapter 15

Cylinder Head and Valve Service

FIGURE 15-1. Using spring compressor to remove exhaust valves in Detroit Diesel engine. *(Courtesy of Detroit Diesel Allison)*

Performance Objectives

After thorough study of this chapter, sufficient practice on the appropriate training components and with the appropriate shop manuals, tools and equipment, you should be able to do the following:

1. Complete the self-check questions with at least 80 percent accuracy.

2. Follow the general precautions outlined in this chapter.

3. Clean, inspect, and accurately measure all cylinder head and valve components to determine their serviceability.

4. Recondition and replace all cylinder head and valve components as needed to meet manufacturer's specifications.

5. Correctly assemble and adjust all cylinder head and valve components to meet manufacturer's specifications.

PART 1   DISASSEMBLY AND CLEANING

Cylinder head and valve train service is often performed on engines without engine removal if lower engine service is not required. Procedures for cylinder head removal, service, and installation are similar whether the engine has been removed or not. (See Fig. 15-1 to 15-3.)

Once the cylinder heads have been removed, carefully remove the injectors from the heads. If no obvious damage is apparent, the cylinder head assembly can be cleaned in the degreaser (hot tank), as the engine block was in Chapter 14.

Preliminary cleaning can be done without the valves being removed. This protects the valve seats from damage when removing carbon deposits from the combustion chamber. Scrapers and a rotary wire brush are used for this purpose. Any deposits in the ports should also be removed.

Cylinder head disassembly includes removal of rocker arms and pivots, valve bridges or crossheads, valve springs, valves, and locks.

All parts should be kept in order for correct reassembly. A C-clamp type or lever type of valve-spring compressor is used to compress the valve springs far enough to allow the locks to be removed.

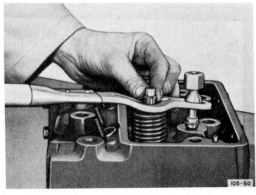

FIGURE 15-2. Removing valve locks with spring compressed. *(Courtesy of Mack Trucks Inc.)*

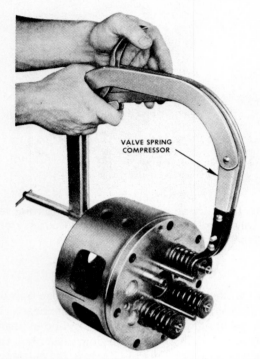

FIGURE 15-3. Using C-clamp type of spring compressor on pot type of cylinder head. *(Courtesy of Detroit Diesel Allison)*

Sometimes the locks and retainer are stuck to the valve stem, and the spring cannot be compressed with the spring compressor. In this case the spring retainer should be tapped with a soft-faced hammer while the spring compressor is applying some pressure to the spring. Tapping the retainer with the hammer will allow the locks and retainer to pop loose. *Caution:* Always keep compressed springs under full control.

If the valve stem tips are mushroomed from rocker arm action, they should be dressed before valve removal to avoid damage to the valve guides. A hand file can be used for this purpose.

Care must be exercised when using the valve-spring compressor to avoid slipping off the compressed spring. The compressed spring releases a powerful punch if it snaps out of place and can cause serious personal injury.

After disassembly, the cleaning process can be completed. Any remaining deposits that were not accessible while the valves were in place must be removed. The cylinder head should then be pressure tested or tested for cracks by the magnetic detection method. Follow the service manual procedures. Mark any points where leakage occurs for further inspection.

Valve guides should be cleaned with a carbon scraper type of valve guide cleaner, after which they should be cleaned with a rifle type of brush.

PART 2   INSPECTION AND MEASURING

Cylinder head warpage should be checked with a straightedge and feeler gauge; if excessive, it should be corrected. A maximum generally acceptable is 0.004 inch (0.101 mm) across and 0.003 inch (0.0762 mm) longitudinally for every two cylinders of cylinder head length, i.e., 0.006 inch (0.1524 mm) on a four-cylinder head. (See Fig. 15-4 and 15-5.)

The valves should be cleaned on the wire wheel of the bench grinder (use face mask). The entire valve should be cleaned this way; however, damage to the valve face should be avoided. Any valves that are obviously damaged, burned, or cracked or have insufficient margin should be replaced and do not require cleaning.

Valve stems should be measured for wear with an outside micrometer. Valves with stem wear in excess of 0.001 inch (0.3 millimeter) are usually replaced.

Caution: Do not damage sodium-filled valve stems. Sodium-filled valve stems are usually identifiable by their larger-diameter stems. The sodium in the valve stem explodes when exposed to atmosphere.

FIGURE 15-4. Pressure-testing cylinder head. *(Courtesy of Detroit Diesel Allison)*

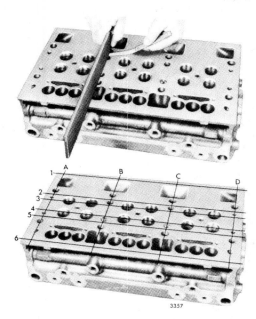

FIGURE 15-5. Checking cyliner head flatness (warpage) with straightedge and feeler gauge. *(Courtesy of Detroit Diesel Allison)*

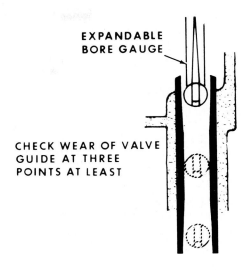

CHECK WEAR OF VALVE GUIDE AT THREE POINTS AT LEAST

EXPANDABLE BORE GAUGE

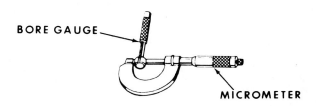

BORE GAUGE — MICROMETER

FIGURE 15-7. Measuring valve guide wear with small hole gauge and outside micrometer. Measurements are taken at three levels as indicated. *(Courtesy of J. I. Case, a Tenneco company)*

PART 3   VALVE GUIDE SERVICE

Valve stem-to-guide clearance can be measured in several different ways. Manufacturers differ in the method that they recommend. The specifications given in the manufacturer's shop manual apply only when the method that they recommend is followed. (See Fig. 15-6 to 15-8.)

A method that can be used on any make uses a dial indicator mounted on the cylinder head in such a way that valve stem side-to-side movement in the guide can be measured. The valve is inserted in the guide and the dial indicator mounted so that the plunger contacts the valve stem as close to the guide

as possible. With the valve off its seat about ¼ inch, side-to-side movement of the valve stem is measured. If this measurement is excessive (over 0.005 inch or 0.13 millimeter), the valve guide should be reconditioned, or replaced if not integral.

VALVE GUIDE CLEANERS
Expanding

VALVE GUIDE BRUSHES

FIGURE 15-6. Valve guide cleaners: scraper (top), brush (bottom).

FIGURE 15-8. Removing or installing valve guide. *(Courtesy of Mack Trucks Inc.)*

Another method that can be applied to all makes and models is to use an inside micrometer to measure the inside diameter of the valve guide. The valve stem diameter measurement is subtracted from the valve guide diameter measurement to determine valve stem-to-guide clearance.

When valve guide reconditioning or replacement is required, this must always be done before attempting to recondition the valve seats. This is necessary to maintain the correct relationship between the valve guides and seats.

Several methods of restoring correct valve stem-to-guide clearance are employed. On engines with replaceable guides, guides are removed with a special press or driver, and new guides installed to the correct depth. Valves with standard-diameter stems can then be used.

On cylinder heads with integral guides, the guides can be reamed to an oversize for which oversized stem valves are available. Another method is to machine the guides to allow new valve guide inserts of standard diameter to be installed.

A method known as *knurling* is also used to recondition valve guides. The appropriate knurling tool is inserted in the guide and turned. This causes the tool to spiral its way through the guide like the threads of a bolt. The knurling tool does not remove any metal but rather displaces the metal to reduce valve guide diameter. After knurling, the guides are reamed to restore the guide to original diameter and provide proper valve stem-to-guide clearance. If proper valve stem-to-guide clearance is not restored, the valves will not seat properly, and oil will get past the guides into the combustion chamber and exhaust system.

Valve Bridge or Crosshead and Guide

Inspect crossheads or valve bridges for wear, cracks, or damage. The magnetic crack detection method should be used. The guide bore in the crosshead or bridge should be measured for wear with a small bore gauge or inside micrometer. If wear exceeds limits or if cracks are evident, the bridge or crosshead should be replaced. Inspect the adjusting screw threads and nut. Threads should be in good condition. The rocker lever contact area should not show signs of excessive wear. (See Fig. 15.9 and 15-10.)

The bridge or crosshead guide should be measured for wear with an outside micrometer. If wear exceeds maximum allowable limits specified in service manual, replace the guide. The guide should also be checked for straightness; if not at right angle to the head surface, it should be replaced.

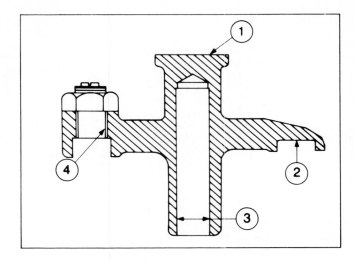

FIGURE 15-9. Valve crosshead bridge inspection and measuring points. *(Courtesy of Cummins Engine Company Inc.)*

PART 4   VALVE SEAT SERVICE

Seat reconditioning includes refacing the seat and correcting seat width in a manner that will provide proper seat-to-valve face contact as well as correct seat width. (See Fig. 15-11 to 15-16.)

Special equipment is required to correct valve seats. A high-speed driver (8000 to 12,000 rpm) is used to drive a grinding stone mounted on a pilot inserted tightly in the valve guide. Roughing stones, finishing stones, and special stones for induction-

FIGURE 15-10. Measuring the bridge guide for wear. *(Courtesy of Cummins Engine Company Inc.)*

FIGURE 15-11. Removing a valve seat insert. *(Courtesy of Detroit Diesel Allison)*

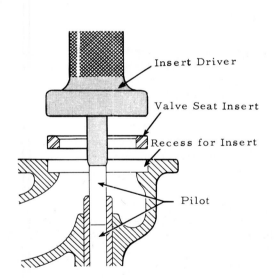

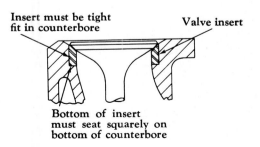

FIGURE 15-12. Installing valve seat insert (top). Properly installed seat (bottom). *(Courtesy of International Harvester Company)*

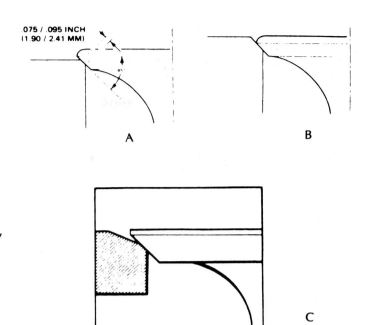

FIGURE 15-13. (A) Proper valve seat-to-face contact. (B) A seat that is too wide and requires overcutting to correct. (C) Corrected valve seat. (D) Method of valve seat narrowing. *(Courtesy of Sioux Tools Inc. [A, B, and C] and Ford Motor Company of Canada Ltd. [D])*

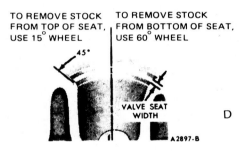

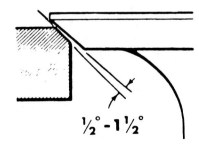

FIGURE 15-14. Grinding the valve at an angle of 1/2° to 1-1/2° less than the seat produces an interference angle as shown. This provides a high-pressure seal on the combustion chamber side of the valve and seat. *(Courtesy of Sioux Tools Inc.)*

FIGURE 15–15. Dressing the seat-grinding stone before grinding the valve seat. *(Courtesy of Detroit Diesel Allison)*

hardened or stellite seats are available. When seats require only a little grinding, the finishing stone only is used. Seats that need more grinding may require the use of a roughing stone first and then a finishing stone to complete the job. Seat grinding stones are available in a number of diameters and cutting angles to suit the various valve seat diameters and angles.

Replacement seats that are too badly damaged for correction by grinding can be replaced. Most in-

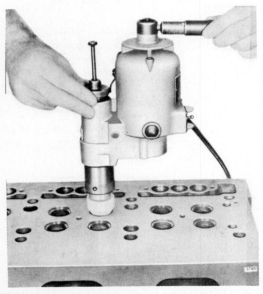

FIGURE 15–16. Eccentric type of valve seat grinder grinds only one area of seat at a time while moving progressively around entire seat during grinding process. *(Courtesy of Detroit Diesel Allison)*

tegral seat heads can be machined and a new valve seat insert installed. Replacement valve seat inserts are slightly larger in diameter than the machined opening (counter-bore) in the cylinder head. This provides an interference fit that assures good heat transfer and prevents the seat from becoming loose. This procedure usually requires cooling the seat insert sufficiently to cause it to shrink enough for easier installation. This type of work is normally done by specialty shops such as engine rebuilders or cylinder head rebuilders. After seat insert installation, seats must be ground to restore concentricity, squareness, correct seat width, and proper seat-to-valve face contact. Oversize valve seat inserts are available for some engines. This allows machining the insert bore that is damaged to specified oversize and then installing the appropriate oversize valve seat insert.

Seat Grinding

To grind the seats, select the proper diameter pilot and install it snugly in the valve guide. Select a seat grinding stone of the correct angle (same angle as the seat), usually, 60°, 45°, or 30°. The stone should be slightly larger in diameter than the valve head. Dress the stone in the stone dressing fixture frequently during the seat grinding procedure to ensure a good seat finish.

It is a good idea to wipe the seats clean by using a piece of fine emery cloth between the stone and the seat and giving it a good hard rub. This avoids contaminating the seat grinding stone with any oil or carbon residue that may be on the valve seat. Seat grinding stones should be handled in a manner that will keep them clean. Stones will soak up oil like a blotter. This causes them to become glazed and ineffective for seat grinding. Remove only as much material from the seat as required to provide a good finish of sufficient width all the way around the seat. Avoid any side pressure during the grinding process.

Grind the seat with short bursts only, checking frequently to inspect progress. Pressure of the stone against the seat must be precisely controlled to avoid chatter and to provide a good seat finish. Excessive pressure or chatter can destroy a valve seat very quickly. A drop of oil on the pilot and the star or hex drive can reduce the tendency to chatter. Avoid getting any oil on the seat or stone. *Note:* If the valve seat is too wide after grinding, it must be narrowed to specifications, usually 1/16 to 3/32 inch (1.6 mm to 2.3 mm), with the exhaust seat being the wider for better heat dissipation.

The objective in narrowing the valve seats is twofold: (1) the seat should be the correct width and (2) it should contact the center of the valve face. To determine whether overcutting or undercutting is required to narrow the valve seat, a new or reconditioned valve must be used as described below.

Overcutting is done in the same manner as seat grinding, except that a 15° stone is used for the purpose. This narrows the seat from the combustion chamber side and lowers the point of contact on the valve face (farther from the margin).

Undercutting is done similarly with a 60° stone, which narrows the seat from the port side.

To determine where the seat contacts the valve face, use the new or reconditioned valve. Mark the valve face with a series of pencil marks across the face of the valve all around the valve. Insert the valve in the guide, press down on the valve, and turn the valve one-quarter turn and back. Remove the valve and check the pencil marks. The pencil marks will be wiped out at the point where the seat contacts the valve. (Because of the interference angle, only the edge of the seat on the combustion chamber side will wipe out the pencil marks.) This should be about one-third of the way down on the face of the valve away from the margin, to center the seat on the valve face. Turning the valve only one-quarter turn while in contact with the seat provides the means for checking whether the seat is concentric. If the pencil marks are wiped out at one point all the way around the valve, seat concentricity is within limits. If the guide, seat, and valve all have been properly reconditioned, they will be concentric (centered in relation to each other). Concentricity is required to provide a good seal between the valve and seat. Some engines require valve head protusion limits above the cylinder head surface to be observed. If the valve protrudes too much, the valve or seat may be ground to lower the valve in the head.

PART 5   VALVE RECONDITIONING

After the valves have been thoroughly cleaned and inspected, those that passed the inspection should be reconditioned. Some valve-refacing equipment requires that valve reconditioning procedures follow a specific sequence. Valve refacers that support the valve tip in a coned shaft require the valve tip to be dressed and chamfered before it is refaced. If this is not done, the valve face will be ground off center. The equipment manufacturer's instructions for pro-

cedures and sequence should be followed. (See Fig. 15–17 to 15–27.)

To recondition a valve, the tip should be dressed and chamfered and the valve refaced. Remove only enough material to produce the desired results. If too much material is removed from the valve tip, there may be interference between the rocker arm and spring retainer or valve rotator. Follow the manufacturer's specifications for the allowable limits.

If the valve is bent or distorted, this will be easily noticed as the valve rotates in the machine. Valves damaged in this way should be replaced. Valves with an insufficient (1/32 inch or 0.8 millimeter minimum) margin after refacing should also be replaced.

Valve springs should be checked for acid etching, squareness, and pressures. Springs that do not meet the manufacturer's specifications should be replaced. Valve-spring installed height should be measured and corrected with shims if required (some models only).

PART 6   INJECTOR SLEEVES OR TUBES

Any indication of leakage around the injector tubes or sleeves while the cylinder head was being pressure tested requires removal of the sleeve for further inspection. A damaged sleeve should be replaced. New sleeves should be installed and checked for proper seating as outlined in the appropriate service manual. Procedures vary considerably. (See Fig. 15–28.)

Install new injector or injectors that have been properly serviced. (See appropriate section in this text and in service manual for injector service.) Make sure that injector protrusion (the amount that the injector protrudes below head surface when the injector is properly installed in a cylinder head) is within recommended limits specified in service manual. *Caution:* Fuel injectors are easily damaged. If injectors protrude below head surface, make sure that cylinder heads are handled in a manner that will avoid injector damage.

PART 7   ASSEMBLY AND INSTALLATION

All cylinder head parts, valves, springs and the like, should be absolutely clean for assembly. Valve and

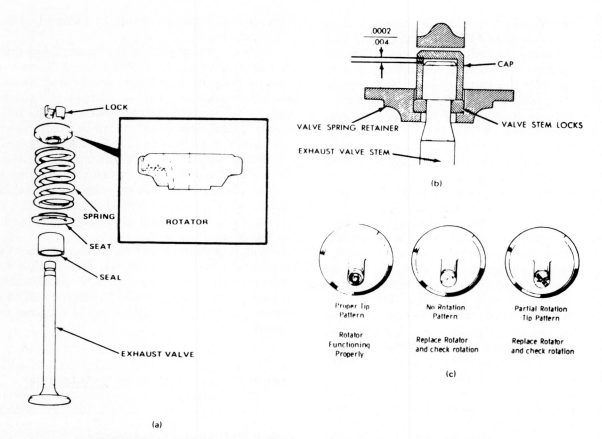

FIGURE 15-17. (A) Positive type of valve rotator. Valve rotation helps the valve and seat to maintain a positive seal. (B) Free type of valve rotator. (C) Wear patterns on the valve tip are good indicators of rotator malfunction. *(Courtesy of General Motors Company of Canada Ltd.)*

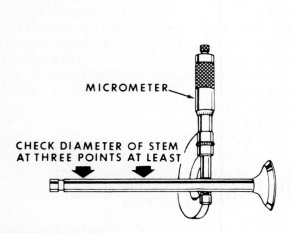

FIGURE 15-18. Measuring valve stem wear. *(Courtesy of J. I. Case, a Tenneco company)*

FIGURE 15-19. Dressing a worn valve stem tip. *(Courtesy of Sioux Tools Inc.)*

FIGURE 15-20. Chamfering a valve stem tip. *(Courtesy of Sioux Tools Inc.)*

FIGURE 15-22. Checking valve head to cylinder head surface relationship. If valve protrusion is excessive, seat or valve or both must be reground. *(Courtesy of Mack Trucks Inc.)*

seats should be lubricated with engine oil during assembly. Make sure that intake and exhaust valve springs are properly installed. Umbrella-type valve stem seals are installed on the valve stem before the spring is installed. Stem seals of the O-ring type that make the spring retainer act as an oil shedder are installed after the spring is compressed. The O-ring is installed in the groove closest to the head; then the valve locks are installed. (See Fig. 15-29 to 15-31.)

Install valve rotators above or below the spring as specified by the manufacturer. Valve-spring shims are installed between the spring and the head. Variable-rate springs are usually installed with the closest spaced coils toward the cylinder head.

On engines with overhead camshafts, the valve-operating mechanism and camshaft may require assembly before cylinder head installation and may require special tools. Follow the sequence specified by the manufacturer.

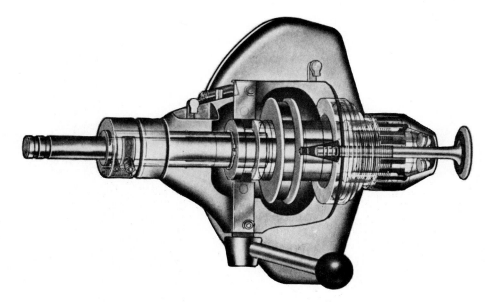

FIGURE 15-21. Valve properly positioned in valve refacer chuck. Chuck jaws must clamp valve on unworn machined surface of stem. Adjust floating shaft to support valve stem tip properly. *(Courtesy of Sioux Tools Inc.)*

FIGURE 15-23. Testing valve-spring pressures at specified height. *(Courtesy of Detroit Diesel Allison)*

FIGURE 15-25. Measuring valve-spring installed height. Material removed from valve face and seat can alter spring installed height and cause insufficient spring pressure. To correct, install appropriate shims between spring and head. Shims should not be used to correct springs that have failed the spring pressure test, Figure 15-23. *(Courtesy of Ford Motor Company of Canada Ltd.)*

Gasket surfaces, bolts, and bolt holes should be absolutely clean. Coat the new gasket and cylinder bolt threads with recommended sealer or lubricant (if specified by the manufacturer), and position the head gasket, seal rings, and fire rings on the engine block. Make sure that all gaskets and seal rings are properly installed and gaskets are right side up and correct end forward.

Carefully lower the cylinder head into place without bumping or shifting the gaskets. A slight nick in the gasket can cause gasket failure.

Install cylinder head bolts, making sure bolts are in the correct location if they are not of equal length. Tighten all bolts in the proper sequence in three steps and to specified torque to ensure good sealing.

If specified by the manufacturer's shop manual, cylinder head bolts should be retorqued to specifications after the engine has been run the specified distance or number of hours.

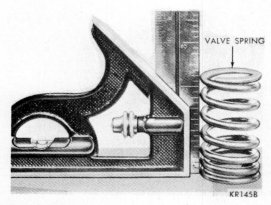

FIGURE 15-24. Checking valve springs for distortion. Spring should be checked at two locations 90° from each other. Maximum allowable distortion is usually no more than 1/16 inch. *(Courtesy of Chrysler Corporation)*

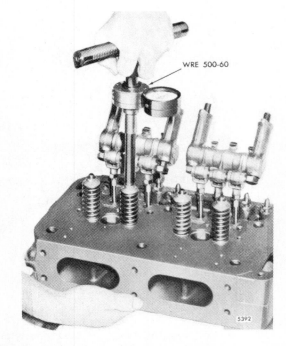

FIGURE 15-26. Checking pressure required to open exhaust valve. *(Courtesy of Detroit Diesel Allison)*

FIGURE 15-27. Checking for valve leakage with vacuum pump tester. *(Courtesy of Cummins Engine Company Inc.)*

PART 8 VALVE- AND INJECTOR- OPERATING MECHANISM SERVICE

Wash all parts in the appropriate cleaning fluid (fuel oil, lubricating oil, or solvent) as recommended in the service manual. Blow dry with compressed air. Inspect all parts visually for abnormal wear pat-

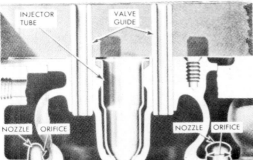

FIGURE 15-28. Water nozzles and injector tubes must be properly positioned and sealed according to the appropriate service manual for the engine. *(Courtesy of Detroit Diesel Allison)*

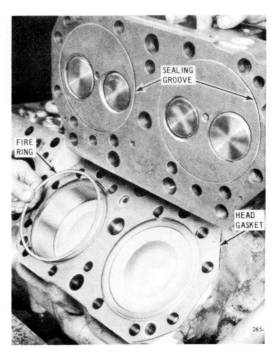

FIGURE 15-29. Installing fire ring gasket. *(Courtesy of Mack Trucks Inc.)*

terns. Wear patterns should be smooth and polished. There should be no evidence of chipping, cracks, corrosion, or pitting. Parts that do not pass the visual inspection should be replaced. (See Fig. 15-32 to 15-43.)

Other parts should be measured for wear. Refer to illustrations for typical examples of required

FIGURE 15-30. All gaskets and seal rings must be properly positioned and must not be disturbed during cylinder head installation. *(Courtesy of Detroit Diesel Allison)*

CYLINDER HEAD

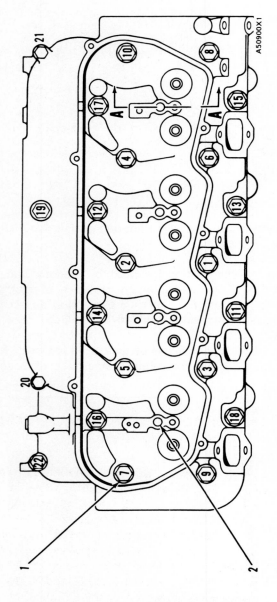

A50900X1

(1) Put 4S9416 Anti-Seize Compound on bolt threads and tighten bolts according to the following HEAD BOLT CHART:

HEAD BOLT CHART

Tightening Procedure	*EARLIER BOLTS (with six dash marks)	*LATER BOLTS (with seven dash marks)
Step 1. Tighten bolts 1 thru 18 in number sequence to:	60 ± 10 lb. ft. (80 ± 14 N·m)	60 ± 10 lb. ft. (80 ± 14 N·m)
Step 2. Tighten bolts 1 thru 18 in number sequence to:	95 ± 5 lb. ft. (130 ± 7 N·m)	110 ± 5 lb. ft. (150 ± 7 N·m)
Step 3. Again tighten bolts 1 thru 18 in number sequence to:	95 ± 5 lb. ft. (130 ± 7 N·m)	110 ± 5 lb. ft. (150 ± 7 N·m)
*See BOLT HEAD IDENTIFICATION pictures for EARLIER and LATER identification.		
Torque for head bolts 19 thru 22 (tighten in number sequence to)	32 ± 5 lb. ft. (43 ± 7 N·m)	

DASH MARKS

A50901X1

EARLIER (six dash marks)

LATER (seven dash marks)

BOLT HEAD IDENTIFICATION

FIGURE 15-31. Typical cylinder head bolt tightening sequence and specifications. Follow the sequence and specifications given in the appropriate service manual. *(Courtesy of Caterpillar Tractor Company)*

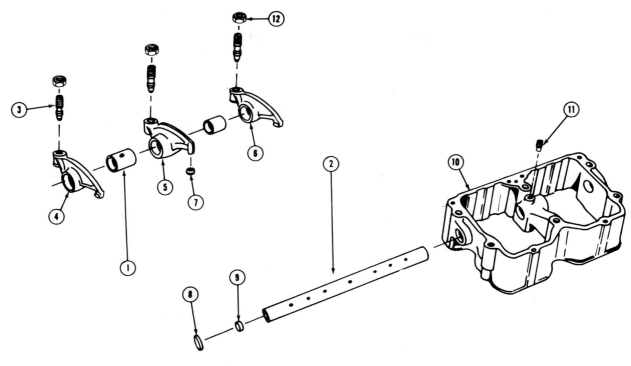

1. Bushing
2. Shaft
3. Adjusting Screw
4. Intake Lever
5. Injector Lever
6. Exhaust Lever
7. Injector Lever Socket
8. Housing Plug
9. Shaft Plug
10. Rocker Lever Housing
11. Shaft Setscrew
12. Adjusting Screw Locknut

FIGURE 15-32. Rocker shaft and housing assembly parts relationship. *(Courtesy of Cummins Engine Company Inc.)*

measurements for various types of cam followers and cam follower mounting parts. Parts that do not meet specifications must be replaced. Install cam followers thoroughly lubricated with engine lubricating oil. The procedure and sequence to follow will vary, depending on engine make and model. Follow service manual procedures for assembly. Install the push rods or tubes (and springs if so equipped), rocker arms, and shafts. All parts should be lubricated with engine lubricating oil at all wear points. Make sure that all valve adjustments and injector-operating adjustments are backed off before attempting to turn the crankshaft. This is necessary to avoid damage to valves and injectors. Assemble and install the compression release mechanism if so equipped. Install the injector control rack and linkage, if so equipped, making sure that all timing marks are properly aligned according to service manual specifications.

PART 9   PROCEDURE FOR ADJUSTING THE VALVE TRAIN

The objective in adjusting the valve train is to provide sufficient valve lash (clearance) to allow for any expansion of parts due to heat and still ensure that valves will be fully seated when closed. At the same time there must not be excessive lash, which would retard valve timing and cause rapid wear of valve train parts. (See Fig. 15-44 and 15-45.)

The static valve adjustment is made with the engine cold. A cold engine is an engine that has reached a stabilized temperature to within 10°F (5.56°C) of normal room temperature, 70°F (21.11°C). Further valve and injector-operating mechanism adjustment may be required on some engines after the engine has reached operating temperature. Due to extremely close piston-to-valve tol-

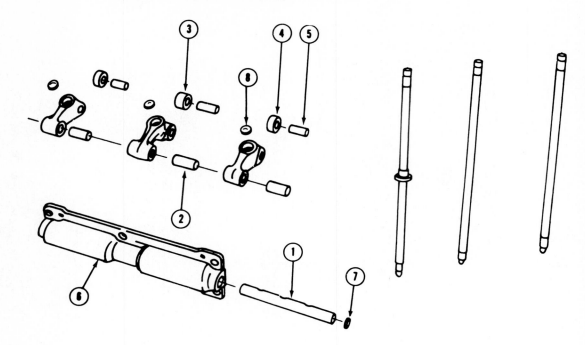

1. Shaft
2. Bushing
3. Injector Cam Roller
4. Valve
5. Valve Cam Roller Pins
6. Housing
7. Shaft Plug
8. Insert

FIGURE 15-33. Cam follower and housing assembly and push tubes. *(Courtesy of Cummins Engine Company Inc.)*

erance, some engine valve trains must not be adjusted with the engine running. Inserting the feeler gauge could cause the pistons to strike the valves, causing major damage.

The static valve adjustment is required after an engine overhaul to ensure proper engine starting and prevent damage to the valves from pistons hitting the valves. The valve bridge or crosshead adjustment must be made first, then the valve lash adjustment. On some engines the valve lash adjustment is made on one cylinder while the injector adjustment is made on the other.

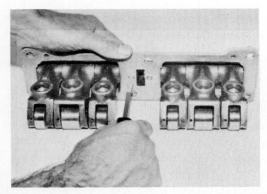

FIGURE 15-34. Removing cam follower shaft retaining screws. *(Courtesy of Cummins Engine Company Inc.)*

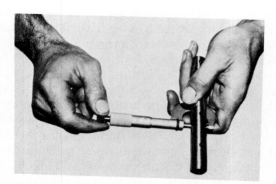

FIGURE 15-35. Measuring cam follower shaft wear. *(Courtesy of Cummins Engine Company Inc.)*

FIGURE 15-36. Measuring cam follower bushing diameter. *(Courtesy of Cummins Engine Company Inc.)*

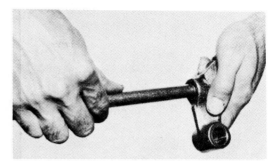

FIGURE 15-39. Checking push tube and push tube seat contact pattern using Prussian blue. There must be a minimum of 80 percent surface contact. *(Courtesy of Cummins Engine Company Inc.)*

Valve adjustment is sometimes done as a routine service procedure when there is no engine overhaul involved. This is often done with the engine running at a slow idle speed. When this is done, provision must be made to prevent oil from squirting and spraying over other engine parts.

FIGURE 15-40. Measuring cam follower roller inside diameter for wear. *(Courtesy of Cummins Engine Company Inc.)*

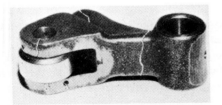

FIGURE 15-37. Cracked cam follower must be replaced. *(Courtesy of Cummins Engine Company Inc.)*

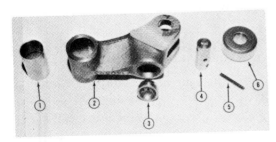

FIGURE 15-38. Disassembled cam follower. *(Courtesy of Cummins Engine Company Inc.)*

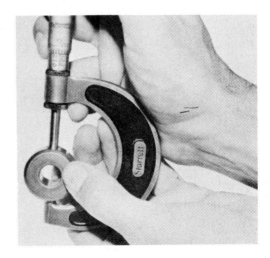

FIGURE 15-41. Measuring cam follower roller outside diameter for wear. *(Courtesy of Cummins Engine Company Inc.)*

FIGURE 15-42. Installing the cam follower roller pin using a feeler gauge to prevent collapse of the cam follower yoke. *(Courtesy of Cummins Engine Company Inc.)*

FIGURE 15-43. Measuring rocker shaft bushing wear. If wear is excessive, bushing must be replaced if rocker arm is acceptable otherwise. *(Courtesy of Cummins Engine Company Inc.)*

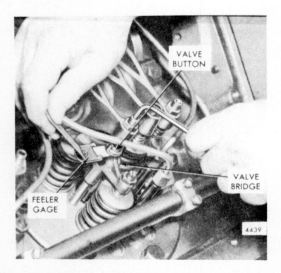

FIGURE 15-44. Adjusting the valve crosshead on a Detroit Diesel engine. *(Courtesy of Detroit Diesel Allison)*

FIGURE 15-45. Adjusting valve lash on a Caterpillar engine. (1) Adjusting screw, (2) lock nut, (3) feeler gauge. *(Courtesy of Caterpillar Tractor Company)*

Adjustment should only be made within the limits and sequence prescribed in the manufacturer's shop manual.

The valve train usually has an adjusting screw and lock nut provided at the push rod or the rocker arm. Clearance or lash is measured between the valve stem and rocker arm or crosshead or bridge. With the lock nut loosened off several turns, turning the adjusting screw will increase or decrease the amount of lash, depending on which direction the screw is being turned. Valves should be adjusted to specifications, then the lock nut should be tightened. Make sure that the adjusting screw does not turn while tightening the lock nut. Recheck the clearance with a feeler gauge and correct the adjustment if needed.

The proper sequence must be followed when adjusting valves in order to ensure that adjustment is being made while the cam follower is on the base circle or heel of the cam lobe.

On engines equipped with camshaft-operated injectors, adjust the injectors according to the recommended setting and sequence prescribed in the appropriate service manual.

PART 10   SELF-CHECK

1. What are the generally accepted maximum limits allowed for cylinder head warpage?

2. To measure valve stem-to-guide clearance, a ___ is used.

3. Name three methods used to correct excessive valve stem-to-guide clearance.

4. To recondition the valve guide, the seat must be reconditioned first. True or false?

5. A properly reconditioned valve seat must meet at least three conditions. These are (a) _____, (b) _____, and (c) _____.

6. A refaced valve must have at least _____ inch of margin to be serviceable.

7. Valve springs must be tested to assure (a) _____, (b) _____, and (c) _____.

8. Gaskets should be examined and, if in reasonable condition, used again. True or false?

9. A static valve adjustment is required after an engine overhaul to assure _____ and prevent _____.

Chapter 16

Induction and Exhaust Systems

Performance Objectives

After adequate study of this chapter and sufficient practical experience on appropriate training models and with proper tools, equipment, and shop manuals, you should be able to do the following:

1. Complete the self-check questions with at least 80 percent accuracy.

2. Describe the purpose, construction, and operation of the induction and exhaust systems.

3. Diagnose induction and exhaust system problems according to manufacturer's diagnostic procedures.

4. Recondition and replace faulty components as required to restore system efficiency.

5. Correctly adjust system components according to manufacturer's specifications.

6. Test the systems to determine the success of the service performed.

PART 1   INTRODUCTION

The induction system must provide the engine with an adequate supply of clean air at the proper temperature for good combustion (and for scavenging cylinders on two-stroke-cycle engines) for all operating speeds, loads, and operating conditions. Up to 1500 cubic feet of air per minute or more may be required, depending on engine size and load.

On a naturally aspirated four-stroke-cycle engine, the system includes the air cleaner, a precleaner (if used), the intake manifold, and the connecting tubing and pipes. On the two-stroke-cycle, the system also includes a blower for scavenging air and for combustion.

On a turbocharged engine, additional air is supplied by means of a turbocharger, which is exhaust gas-driven. On a supercharged engine a mechanically driven blower is used to provide additional air supply.

An air shut-off valve may be included to allow engine intake air to be shut off completely for emergency engine shutdown.

An intercooler or aftercooler may also be included in the induction system. Since cooler air is more dense, a greater amount of air is in fact supplied if the air is cooled. The intercooler is mounted to cool the intake air after it leaves the discharge side of the turbocharger and before it enters the engine (before it enters the blower on two-stroke-cycle diesels). The aftercooler is mounted in the two-stroke-cycle diesel engine block so that it will cool intake air after it leaves the blower and before it enters the cylinder ports.

PART 2   AIR CLEANERS AND INTAKE MANIFOLDS

The air cleaner is designed to remove moisture, dirt, dust, chaff, and the like from the air before it reaches the engine. It must do this over a reasonable time period before servicing is required. It also silences intake air noise.

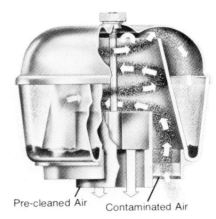

Pre-cleaned Air   Contaminated Air

FIGURE 16-1. Centrifugal action caused by air flow direction separates dirt from intake air. *(Courtesy of Donaldson Company Inc.)*

If dirt is allowed to enter the engine cylinders, the abrasive effects would result in rapid cylinder and piston ring wear. If the air cleaner is not serviced at appropriate intervals for the conditions in which it must operate, it will become restricted and prevent an adequate air supply for complete combustion from reaching the cylinders. Incomplete combustion results in carbon deposits on valves, rings, and pistons, which in turn results in engine wear and oil consumption problems.

Air cleaner types include (1) precleaners, (2) dry types of various designs, and (3) oil bath types. Air cleaner capacity (cubic feet per minute or liters per minute) may have to be up to twice as great on a turbocharged engine as compared to the same engine naturally aspirated.

Precleaners

Precleaners are mounted on the intake tube of the air cleaner. The simplest precleaner consists of a screened hood at the top of the air cleaner inlet. Other precleaners include a spirally vaned drum, which causes incoming air to spin and force dirt, which is heavier, to the outside. There the dirt falls into a dust cup or passes through a scavenging line connected to the exhaust pipe, where it is ejected into the atmosphere by exhaust gases. (See Fig. 16-1 and 16-3.)

When dust in the transparent dust cup reaches the level indicated by a line on the cup, it is removed, emptied, and reinstalled.

Dry-Type Air Cleaners

Dry-type air cleaners may have one or more replaceable filter elements and may include primary and secondary or safety filtering elements. Most dry-type air cleaners also include a vaned type of precleaning device. The vanes may be part of the filter element or they may be part of the air cleaner housing. (See Fig. 16-4 to 16-7.)

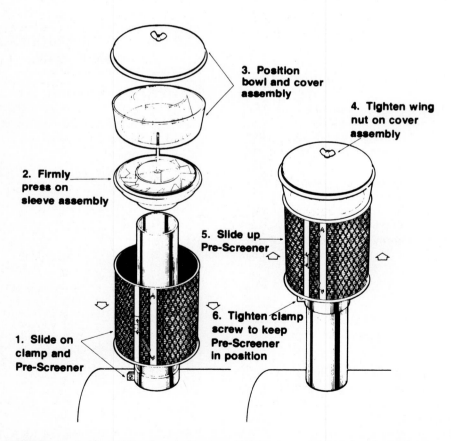

FIGURE 16-2. Assembly and installation procedure for typical precleaner. *(Courtesy of Donaldson Company Inc.)*

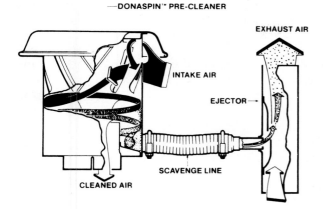

FIGURE 16-3. Precleaner with dirt ejector and scavenge line connected to exhaust system. *(Courtesy of Donaldson Company Inc.)*

As air enters the air cleaner, it passes over the vanes, which impart a swirling action to the air. The swirling action causes the heavier dust and dirt to be thrown outward centrifugally against the air cleaner housing where it goes to the dust collecter cup or bin. A one-way rubber discharge valve ejects dust and water from the dust cup into the atmosphere directly or through a scavenge line connected to the engine's exhaust. A one-way check valve in the exhaust scavenge line prevents engine exhaust gases from entering the air cleaner.

Air is further cleaned as it passes through the filter element or elements. Air cleaners that use primary and secondary filter elements have the advantage of protecting the engine in cases of primary filter element damage since air passes through the primary element first and then through the secondary or safety element.

Filtered air then passes through the air cleaner outlet and to the engine.

An air cleaner restriction indicator that reflects intake air vacuum may be mounted near the outlet side of the air cleaner. The restriction indicator may be of the self-contained type, which shows a red flag or card when restriction reaches filter replacement levels, or it may be of the vacuum-actuated light type. Another type is a gauge that continuously reads restriction in inches of water (H_2O) vacuum when the engine is in operation. The restriction indicator may be remote-mounted where it may be easily observed by the vehicle or engine operator.

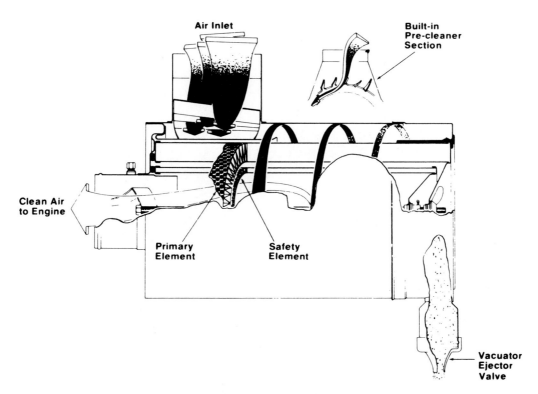

FIGURE 16-4. Heavy-duty Cycloflo3® type of dry air cleaner with safety element dust ejector showing air flow. *(Courtesy of Donaldson Company Inc.)*

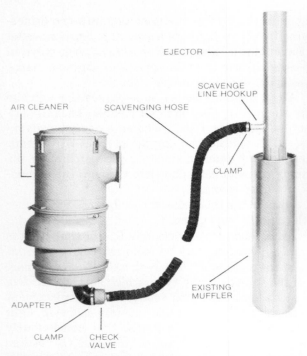

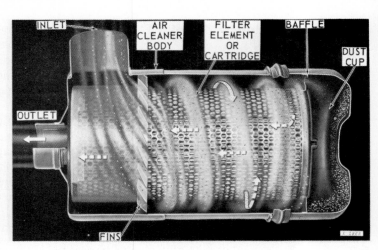

FIGURE 16-6. Air flow diagram of dry element type of air cleaner. Fins cause circular flow and centrifugal separation of dirt from air. *(Courtesy of Deere and Company)*

FIGURE 16-5. Air cleaner dirt ejector connected to exhaust system. Check valve prevents exhaust gases from entering air cleaner. *(Courtesy of Donaldson Company Inc.)*

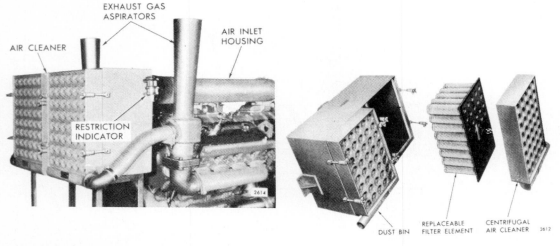

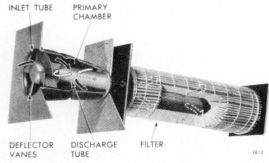

FIGURE 16-7. Typical large dry-type air cleaner mounting. (A) Parts identification. (B) Air flow through one section of cleaner. (C). *(Courtesy of Detroit Diesel Allison)*

Oil Bath Air Cleaner

In the oil bath air cleaner shown, air is drawn through the inlet and down through the center tube. At the bottom of the tube, the direction of air flow is reversed and oil is picked up from the oil reservoir cup. The oil-laden air is carried up into the separator screen where the oil, which contains the dirt particles, is separated from the air by collecting on the separator screen. (See Fig. 16–8.)

A low-pressure area is created toward the center of the air cleaner as the air passes a cylindrical opening formed by the outer perimeter of the central tube and the inner diameter of the separator screen. This low pressure is caused by the difference in air current velocity across the opening.

The low-pressure area, plus the effect of gravity and the inverted cone shape of the separator screen, causes the oil and dirt mixture to drain to the center of the cleaner cup. This oil is again picked up by the incoming air, causing a looping cycle of the oil; however, as the oil is carried toward another cycle, some of the oil will overflow the edge of the cup, carrying

the dirt with it. The dirt will be deposited in the outer area surrounding the cup. Oil will then flow back into the cup through a small hole located in the side of the cup. Above the separator screen, the cleaner is filled with a wire screen element, which will remove any oil that passes through the separator screen. This oil will also drain to the center and back into the pan. The clean air then leaves the cleaner through a tube at the side and enters the intake system.

Intake Manifold

The intake manifold is of cast-iron or cast-aluminum alloy construction. It is designed to direct air from the air cleaner or turbocharger to each cylinder intake port on four-stroke-cycle engines. On the two-stroke-cycle engine, it directs air from the air cleaner or turbocharger to the blower inlet opening.

The intake manifold is equipped with an emergency air shutdown valve on some engines. Closing the air shutdown valve prevents air from reaching

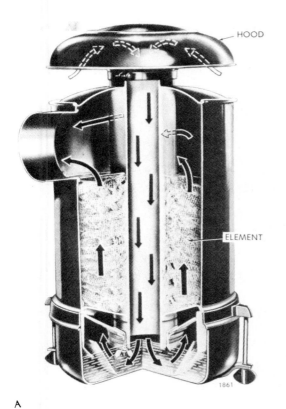

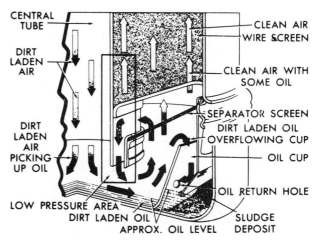

A   B

FIGURE 16–8. Cutaway views of oil bath type of air cleaner (A) and cleaner operation (B). *(Courtesy of Detroit Diesel Allison)*

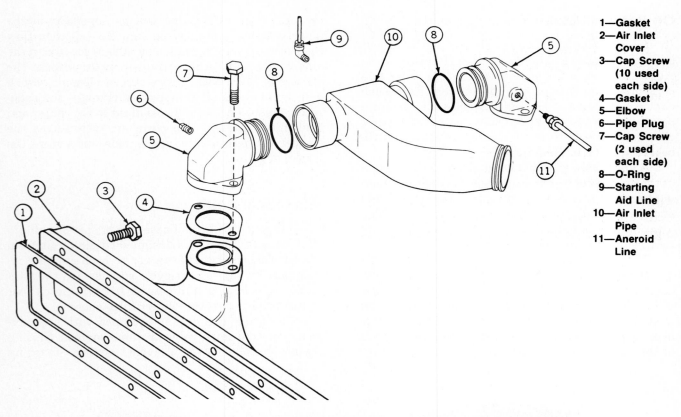

FIGURE 16-9. Air inlet system for V-type engine. *(Courtesy of Deere and Company)*

1—Gasket
2—Air Inlet Cover
3—Cap Screw (10 used each side)
4—Gasket
5—Elbow
6—Pipe Plug
7—Cap Screw (2 used each side)
8—O-Ring
9—Starting Aid Line
10—Air Inlet Pipe
11—Aneroid Line

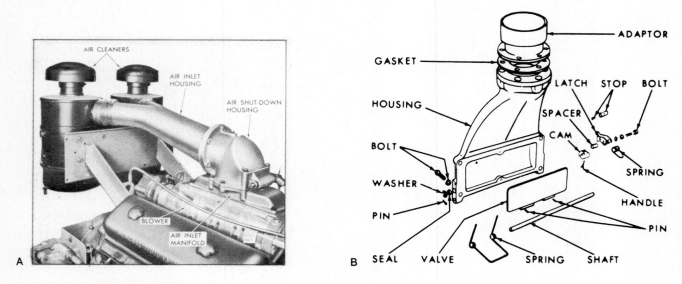

FIGURE 16-10. Air shutdown housing and intake manifold on two-stroke-cycle diesel engine (A). Exploded view of air shutdown housing and valve (B). *(Courtesy of Detroit Diesel Allison)*

the cylinders, thereby stopping combustion in the cylinders, which stops the engine.

Inside diameters of manifold air passages must be of sufficient diameter to provide adequate air for combustion for all engine-operating speeds and loads. (See Fig. 16–9 to 16–10.)

PART 3   TURBOCHARGERS AND SUPERCHARGERS

Turbochargers and superchargers are designed to increase the amount of air delivered to each of the engine's cylinders. In general superchargers are me-

chanically driven from the engine crankshaft while turbochargers are driven by waste exhaust gases from the engine's exhaust system. Turbochargers are the most common, although Detroit Diesel engines use a mechanically driven blower as well as a turbocharger. The blower is required to provide startup air and scavenging air on the two-stroke diesel.

Turbochargers

(The following description is typical of Schwitzer; others are similar.) The turbocharger compresses intake air to a density up to four times that of atmospheric pressure. This greater amount of dense

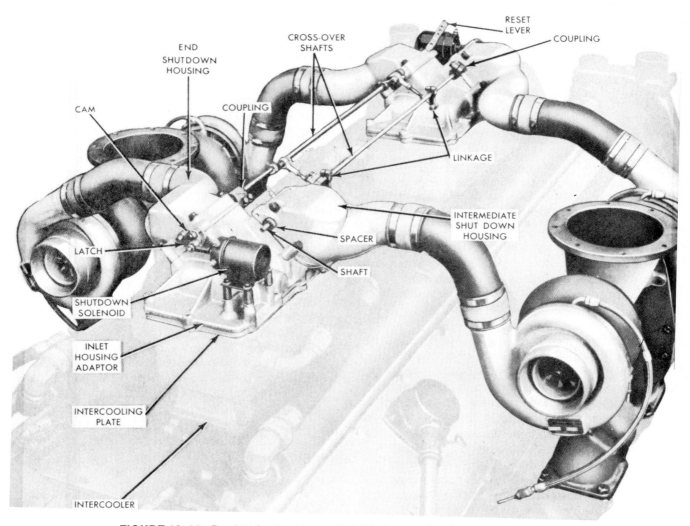

FIGURE 16–11. Dual turbocharger and air shutdown housing arrangement on larger Detroit Diesel engine. Note solenoids. *(Courtesy of Detroit Diesel Allison)*

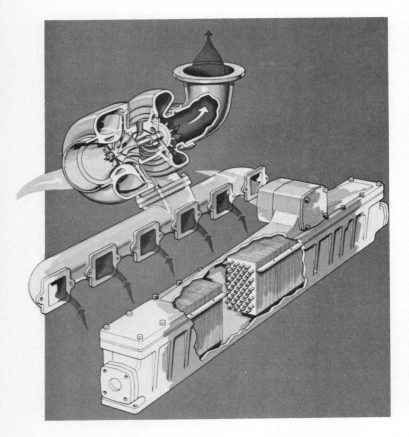

FIGURE 16-12. Turbocharger (top), exhaust manifold (center), and intake manifold and aftercooler (bottom) relationship on four-stroke-cycle engine. *(Courtesy of Cummins Engine Company, Inc.)*

air allows more fuel to be burned, thereby doubling the engine's power output. The turbocharger also reduces exhaust emissions and exhaust noise. It compensates for the less dense air encountered in higher altitude operation. (See Figures 16–11 to 16–17.)

A naturally aspirated or nonsupercharged engine has a limited supply of air for combustion. The air has only atmospheric pressure pushing it into the cylinders. A supercharger will provide pressurized air, which allows for more fuel and air to be packed into a cylinder for each firing. This provides more power and much better combustion efficiency. Thus, more power is realized from a given engine size, fuel economies improve, and emissions are reduced.

A centrifugal compressor pulls air through a rotating wheel at its center, accelerating the air to a high velocity, which flows radially outward through a shell-shaped housing. The air velocity is slowed after leaving the wheel, which converts velocity energy into pressure. This type of compressor is a high-

speed device. Current turbochargers run 80,000 to 130,000 revolutions per minute. A normal engine may only run 3000 to 4000 rpm; previous gear- or belt-driven compressors required expensive and power-robbing setups to drive at efficient speeds. Of the fuel energy available for an engine, 25 percent is used in cooling, 25 percent is converted for work at the flywheel, and 50 percent is wasted in the exhaust. A turbocharger uses some of this waste energy to drive its compressor.

The wheel of centrifugal compressor is attached to a shaft, which has a turbine wheel on the other end. This can be likened to a water wheel mounted in a flowing river with the force of the water turning the wheel, which in turn powers machinery. Substitute flowing exhaust gases for the river and you have the drive for the compressor wheel. The turbine wheel is enclosed by a shell-shaped housing much like the compressor section, but the flow is the reverse, or rapidly inward. Exhaust gases enter tangentially and flow toward the rotating wheel at the center. After flowing through the wheel, the gases exit at the center and continue through the exhaust pipe to the atmosphere. This turbine section works best at high speeds, which also makes it a good match to the compressor section. The common shaft runs in the sleeve bearings between the wheels. These bearings are carried in a bearing housing and comprise the bearing section. The compressor cover is attached to one end of the bearing housing, and the turbine housing is attached to the other end. The bearings are lubricated by engine oil. Only the wheels, shaft, and bearings rotate. The turbocharger unit can double on engine's horsepower, can fit in a 1-foot-cube space, and weighs approximately 45 pounds for a Schwitzer medium-size unit. The Schwitzer range at present will flow up to 1750 cubic feet of air per minute at pressure ratios up to four times atmospheric.

An engine may be equipped with one or two turbochargers. In the dual turbocharger arrangement each unit provides boosted air to one-half of the engine's cylinders. When connected in a series, boosted air from the primary turbocharger is fed to the secondary turbocharger, where it is boosted to even higher pressure before going to the aftercooler and engine.

The Turbine Section

Exhaust gas enters the turbine housing at the turbine inlet flange. The gas flows through the outer shell-shaped circular passage into the center located wheel. While in the rotating wheel, a turn is made

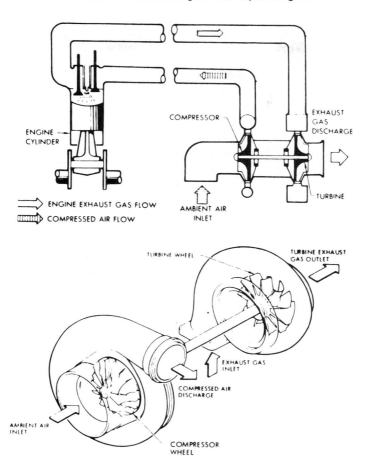

FIGURE 16-13. Intake air and exhaust gas flow diagram for turbocharger-equipped four-stroke-cycle engine. *(Courtesy of Airesearch Industrial Division of Garrett Corporation)*

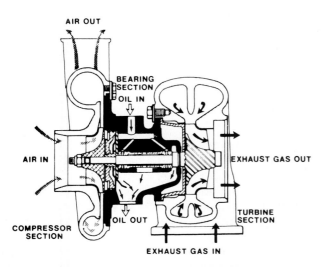

FIGURE 16-14. Cutaway view of turbocharger, showing air flow, exhaust flow, and oil flow. *(Courtesy of Schwitzer, Wallace Murray Corporation)*

and the gases flow out of the turbine housing. The turbine inlet flange is normally a four-bolt flange for direct mounting on the engine exhaust manifold. The gas flows from the flange connection through the turbine inlet passages to the area where the gas is free to enter the wheel. This area is known as the throat. The throat area wall separating the internal gas passage to the wheel and the turbine inlet passage is known as the tongue. The gas passage size constantly decreases from the throat area to the tongue area. The throat area is from the tongue outwards, and the tongue area is from the tongue to the wheel diameter. The throat area in square inches signifies the size of the housing.

The gas passage wall nearest the bearing housing is the base wall or X wall, while the opposite wall is the sidewall. The base wall is the back of the housing, while the sidewall is the front. The outer wall is the containment wall, which is usually thicker since

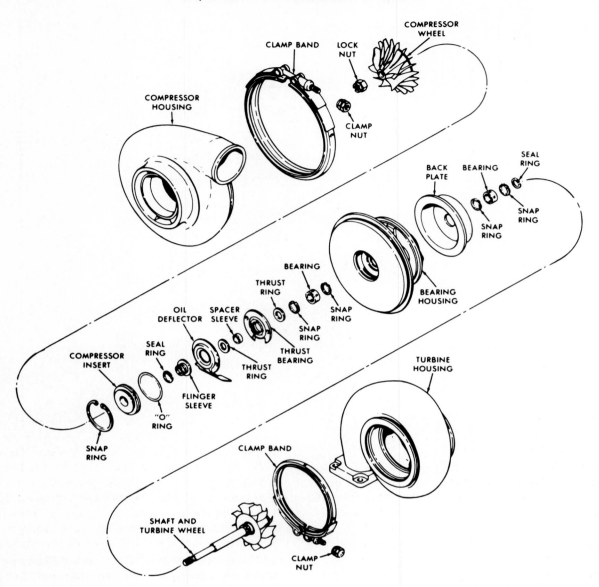

FIGURE 16-15. Exploded view and parts identification of typical turbocharger. *(Courtesy of Detroit Diesel Allison)*

it must contain the wheel fragments in the event of wheel failure.

For safety reasons each Schwitzer unit must "contain" its internal parts in case of failure. The passage width at the wheel outside diameter where the gases feed is the wheel gap. The radius at the gas flow turn in both the housing and the wheel is the contour radius. The diameter leading to the exhaust outlet on the wheel and housing is the exducer diameter. The exhaust outlet piping connection can be external, flanged for bolts or V-clamp bands, or internal, with connection by a slip joint in which a metal tube is formed with a bead for sealing.

The basic gas passage configuration in the turbine housing is undivided or single-flow. This means that all engine cylinders feed gas into a single passage for flow to the wheel for 360° in the housing. To gain efficiency, engine manifolds are split into varying patterns dividing flow to the turbine housing. The housings also have the gas passage split by a divider to provide separation of the pulsations of individual cylinder groupings. The

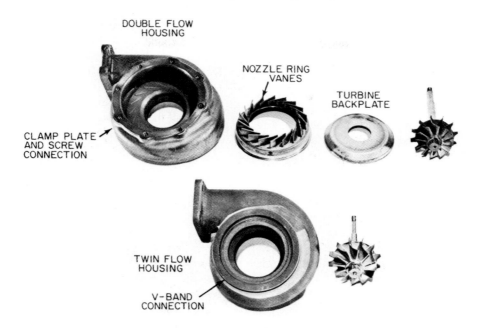

FIGURE 16–16. Double-flow and twin-flow turbine housing designs. *(Courtesy of Schwitzer, Wallace Murray Corporation)*

pulse acts like an aimed jet stream, which increases the power output of the wheel. There are two types of divided housings: double flow and twin flow. In the double flow, the throats are separated, and each passage feeds for only one-half of the turbine wheel circumference. In the twin flow, each passage starts feeding at a common throat and feeds with a divided passage for 360°. Some housings for the V-type engines have the turbine inlet flange separated into two flanges, each receiving flow from one bank of cylinders. These types have either double- or twin-flow feeds in the gas passages.

Some turbochargers have turbine housings with gas fed into the wheel directed by a nozzle ring. This consists of a series of curved metal extensions called vanes on a flange; these form nozzle passages leading from the housing gas passages to the wheel. The nozzle ring is sandwiched between the bearing housing and the turbine housing. These types are very efficient but have a narrow speed range and normally are used with a vaned collector-type compressor cover.

The turbine housing is attached to the bearing housing by a V-band clamp or clamp plates and screws. This hardware is stainless steel because of the 800°F to 1600°F (426.6°C to 871.11°C) exhaust temperatures. The turbine housing is cast from ductile iron or nickel-alloyed ductile irons.

Turbine wheels are bladed in radial fashion from the center of the wheel hub. The wheel hub extends as a wall to block the passage between the blades on the bearing housing side. This wall is the back-wall of the turbine wheel; the exhaust gas flows between the blades with one side of the passage formed by this wall and the other formed by the turbine housing. The wheel is made from a nickel-base alloy.

The turbine wheel is butt-welded to a shaft to form the shaft-and-turbine wheel assembly. Adjacent to the turbine wheel backwall is a shouldered shaft section, grooved to carry a piston ring followed by the area where the bearings run on the journal. A step-down in shaft diameter follows, and the shaft end is threaded. In assembly of the turbocharger, a sleeve or series of sleeves and washers and the compressor wheel are locked together in a column of parts by a nut. One of the sleeves carries a piston ring groove. All the parts are locked against the shoulder formed by the shaft step-down, thus making both wheels and the shaft a common rotating unit. All conventional shaft bearings rotate in operation. These bearings are free-floating, having an oil film on both the inside diameter and outside diameter. The pumping action of the oil flow and the shaft rotation causes the bearings to rotate at approximately one-third of the shaft speed. The piston rings do not rotate. The shaft and sleeve rotate around them as they are held stationary in the bearing housing bores.

The Compressor Section

Air enters the compressor cover at the center from an engine air cleaner. Piping connections will either be a hose connection or an O-ring connection. The hose connection is an external rubber hose retained by a clamp. The O-ring connection is an internal inlet connection on the cover in which a metal tube with an O-ring seal between the tube and cover is inserted. The passageway leading to the wheel and the wheel diameter in this passage area are called inducer diameters. The air makes a 90° turn going through the wheel. The radiused part of the cover and wheel is called the contour radius. Once through the wheel, the air is in a parallel-walled space called the diffuser. One wall is formed in the cover while the other is formed in the bearing housing. The air passes through the diffuser and makes another turn into the shell-shaped circular passage from which the air discharges through the compressor outlet. This outlet has piping connections for carrying the pressurized air to the engine. As with the inlet, this connection can be external by hose or internal by O-ring and tube.

The shell-shaped circular outer cover passages are of two types: collector and volute. The collector is made so that if you cut it open radially at any place leading to the discharge, the passage shape and size are identical. The volute, on the other hand, has a constantly expanding passage from starting point to discharge. The type of cover governs how wide and how efficient a range of air can be pumped at any speed level of the wheel. The volute type has a wider range and usually does not have vanes in the diffuser space. The volute-type diffusion, or the process of converting velocity to pressure, occurs through the diffuser and volute passage to the discharge. The collector's outer cover passages are like a storage vessel for the pressurized air, the majority of the diffusion is done in the diffuser, which normally carries vanes. These vanes are curved metal extensions of the cover, which closes off the diffuser space at intervals to form air passages, steering the air into the collector passage. These types can be very efficient over a limited speed range. Compressor covers and wheels are made from aluminum. The compressor cover is attached to the bearing housing by either a V-band clamp or by clamp plates and screws. The joint between the housing and cover may be sealed with an O-ring, depending on internal pressure or field usage.

Compressor wheel blades are either straight radial from the center of the wheel's hub or backward-leaning, in which the blading curves backward away from the directon of rotation. The wheel hub extends as a wall to block the passage between the blades on the bearing housing side. This wall is the backwall of the wheel, and the air flows between the blades with one side of the passage formed by this wall and the other formed by the cover.

Bearing Section

The bearing section is between the compressor section or "cold end" and the turbine section and "hot end." The bearing housing has flanges on both ends for attachment of the compressor cover and turbine housing. The hot end of the bearing housing usually has an "air space" for insulation, which is formed by a separate cup called the turbine back plate. This plate is sandwiched between the bearing housing and turbine housing at assembly and forms the wall next to the backwall of the turbine wheel. A piston ring bore is the bearing housing entrance for the shaft on the turbine side. An oil drain slot separates the piston ring bore from the bearing bore. This slot is a circular passage, opening into the oil drain cavity at the bottom of the housing. Oil enters from the engine oil supply under pressure at the oil inlet pad at the top of the bearing housing. This is usually a two-bolt flange pad from which a multiplicity of holes carries oil to the bearing at the bearing bore. Once the oil has left the bearings, it returns to the oil drain cavity by gravity flow to the oil drain pad at the bottom of the housing. This pad is usually a two-bolt flange pad to which piping is attached for returning the oil to the engine sump.

This flow of oil not only provides bearing lubrication but also removes considerable heat that comes from the adjacent hot turbine section.

On the compressor end of the bearing bore is a relatively large open pocket, which breaks into the oil drain cavity. This is known as the thrust bearing pocket. Assembled into this pocket will be the thrust bearing, oil deflector, and a cap called the insert, to close the bearing housing from the compressor section. The insert has a piston ring bore through which the rotating components exit the bearing housing. The insert also holds the other components of the thrust bearing pocket in place. The bearing housing is gray iron, the insert steel, and the turbine backplate is ductile iron.

Each turbocharger has a *journal* bearing system and a *thrust* bearing system. Between sizes of turbochargers, there are system variations.

The thrust bearing is stationary in all models. The wheels will run in balance under some conditions, but pressures, flows, and speed variations cause the shaft to move back and forth in the bearing system. The thrust system varies somewhat between large and small turbochargers, as will be pointed out in further discussion, but the principle will be the same. Two parallel thrust rings with a spacer sleeve between are mounted on the shaft, one on each side of the stationary thrust bearing. Oil films between the rings and bearing accept axial thrust loads in either direction. This is the system used in large turbochargers with two shaft bearings. On small units the shaft bearing is one piece. The bearing is contained between the shaft shoulder carrying the turbine end ring groove and the stationary thrust bearing, thus allowing the ends of the shaft bearing to accept thrust loads toward the compressor. A single thrust ring is mounted on the shaft to accept thrust loads toward the turbine end. All oil flowing through the thrust system is channeled by the oil deflector into the oil drain cavity on the compressor side and exits the shaft bearings at the oil drain slot on the turbine end.

The journal or shaft bearing system utilizes simple sleeve bearings of which the large turbochargers use two, due to the long span between lands. The small turbochargers use one bearing, which acts like two bearings connected by a relieved sleeve. The two bearings are restrained from axial movement by retaining rings on either side, allowing enough clearance for rotation. The single bearing is restrained between the shaft and thrust bearing as mentioned. Either system can be separately fed; each bearing or land has its individual oil supply delivered at midpoint. Oil then flows from the center of the bearing outward toward the oil drain slot and oil deflector and also flows toward the area between the bearings. Thus, "separately" fed bearing systems always have an oil drain between the bearings from the bearing bore to the oil drain cavity. Two bearing systems can also be end-fed where the oil supply comes into the bearing bore between the bearings. The oil flows from the inside out and does not require a bearing bore drain. A single bearing system fed "inside out" has a series of holes in the center of the bearing to allow oil entering the bearing bore to reach the inside diameter. This is called a center-fed system.

All systems have advantages and disadvantages; the operating characteristics of the engine determine which system to use in the application.

Gas is sealed from entering the bearing housing on both the compressor and turbine ends by stationary piston rings. These rings are not oil seals. On the turbine end, oil is kept away from the ring by the shaft's flinging action, directing the oil into a slot that opens into the oil drain cavity of the housing. On the compressor end, oil leaving the thrust bearing is channeled by a fixed oil deflector into the oil drain cavity. Any leakage past the deflector is picked up by the rotating flinger sleeve and thrown out into the drain cavity before reaching the ring seal.

These piston ring seals are very effective in operation with differential pressure between the housings and even low vacuum such as an air cleaner restriction on a diesel application. However, the oil must be kept away from the ring as any film, foam, or splash entering the seal area will leak out.

PART 4   BLOWER (TWO-STROKE-CYCLE ENGINE)

The blower (Fig. 16-18) supplies the fresh air needed for combustion and scavenging. Its operation is similar to that of a gear-type oil pump. Two hollow three-lobe rotors revolve with very close clearances in a housing bolted to the cylinder block. To provide continuous and uniform displacement of air, the rotor lobes are made with a helical (spiral) form.

Two timing gears, located on the drive end of the rotor shafts, space the rotor lobes with a close tolerance; therefore, as the lobes of the upper and lower rotors do not touch at any time, no lubrication is required.

Oil seals located in the blower end plates prevent air leakage and also keep the oil used for lubricating the timing gears and rotor shaft bearings from entering the rotor compartment.

Each rotor is supported in the end plates of the blower housing by a roller bearing at the front end and a two-row preloaded radial and thrust ball bearing at the gear end.

The blower rotor is driven by the blower drive shaft, which is coupled to the rotor timing gear by means of a flexible drive hub.

The ratio between the blower speed and the engine speed, and the number of teeth in the blower drive gears and reduction gears is dependent on engine type and size and whether naturally aspirated or turbocharged.

The blower rotors are timed by the two rotor gears at the rear end of the rotor shafts. This timing

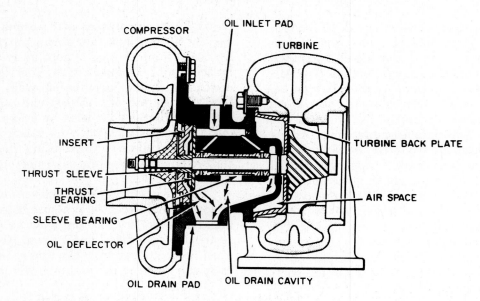

FIGURE 16-17. Oil inlet and outlet, showing turbine bearing lubrication.
(Courtesy of Schwitzer, Wallace Murray Corporation)

must be correct; otherwise the required clearance between the rotor lobes will not be maintained.

Normal gear wear causes a decrease in the rotor-to-rotor clearance between the edges of the rotor lobes. Clearance between the opposite sides of the rotor lobes is increased correspondingly.

Worn blower components can result in blower noise and increased friction. Rotor-to-rotor contact abrasion can cause abrasives to be ingested into the engine, which in turn would cause piston, piston ring, and cylinder liner damage. Normal preventive maintenance and inspection procedures should prevent this kind of damage from occurring, since deterioration would be detected in time.

PART 5   INTERCOOLERS AND AFTERCOOLERS

The intake air temperature increases considerably as a result of turbocharging. Booster air temperature may be in the 300°F (148.8°C) range or higher. Cooling the air makes the air more dense, which results in more air going into the cylinders. (See Fig. 16-19 to 16-23.)

The terms *intercooler* and *aftercooler* are often used interchangeably and refer to the same type of unit.

Cooling is accomplished either by using engine liquid coolant or air as the cooling medium. In the liquid type, engine coolant passes through a series of tubes (small radiator) back to the engine-cooling system. Air from the turbocharger flows between the tubes (usually in the opposite direction of liquid flow) and then to the cylinders.

In the air-to-air intercooler arrangement, a tip turbine fan driven by air from the turbocharger forces ambient air taken from the air cleaner over the intercooler core. Turbocharged air forced through the intercooler core is cooled in this way before it reaches the intake manifold.

PART 6   EXHAUST SYSTEM

The exhaust system is designed to collect the exhaust gases from the engine cylinders, direct them to the muffler where exhaust noise is reduced, and discharge them into the atmosphere. In addition, exhaust gases may be used to drive a turbocharger for improved air induction for combustion. The exhaust may also be used to eject dirt and dust from the air cleaner or precleaner into the atmosphere. Exhaust gas-driven turbine cargo unloaders for certain materials are used on some trucks. On some older model highway vehicles an operator-controlled valve in the exhaust system is used for braking purposes. (See Fig. 16-24 to 16-26.)

Exhaust system components include the exhaust manifolds, exhaust pipe, muffler, muffler ex-

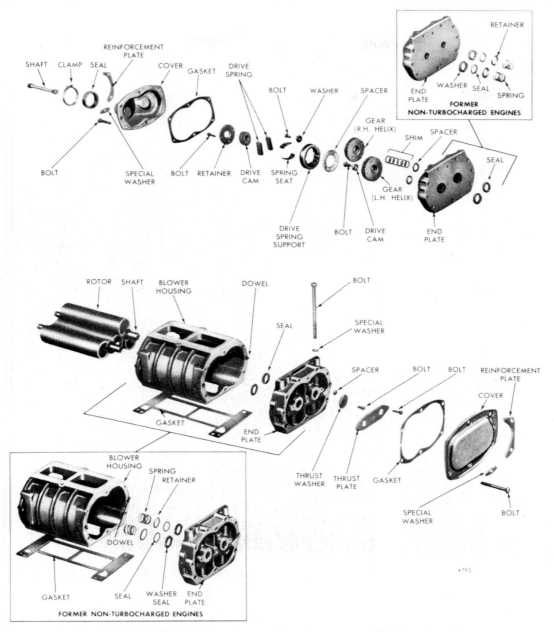

FIGURE 16–18. Typical Detroit Diesel blower parts identification. *(Courtesy of Detroit Diesel Allison)*

tension, and the connecting gasket and clamps. Exhaust pipes may include rigid steel tubing and flexible tubing. A hinged exhaust pipe cap is used on many vertical installations to prevent rain from entering the system. The cap closes automatically when the engine is shut off. A heat shield is used on many applications to prevent injury and component damage.

Exhaust system components must be of sufficient capacity to remove effectively exhaust gases produced by the engine at all operating speeds and loads.

Any restriction in the exhaust system through external damage or internal deterioration will affect the engine's performance.

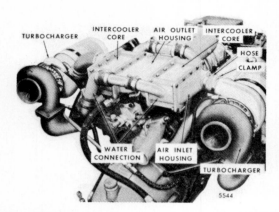

FIGURE 16-19. Typical Detroit Diesel V engine turbocharger and intercooler mounting. *(Courtesy of Detroit Diesel Allison)*

PART 7   AIR POLLUTION

Ingredients necessary to form photochemical smog are hydrocarbons (HC) and nitrogen oxides (NO$_x$) in the presence of continued sunlight. Both of these gases exist in many forms and come from a variety of sources. One major source of NO$_x$ and hydrocarbons is automobiles and other highway and rail transportation equipment.

The requirements for smog formation are sunshine and relatively still air. When the concentration of HC in the atmosphere becomes sufficiently high and NO$_x$ is present in the correct ratio, the action of sunshine causes them to react chemically, forming photochemical smog.

With a thermal inversion (where warmer air above prevents upward movement of cooler air near the ground), smog can accumulate under this lid all the way down to the ground within a few hours. The first effect of a smog buildup is reduced visibility and the blotting out of scenery in the distance; then, as the buildup of smog approaches ground level, its irritating effects on eyes, nose, and throat are sensed.

Vehicle emissions come from three different sources: exhaust, crankcase, and fuel evaporation. Numerous changes have been made to engines to meet emission standards. Emission-control requirements for the state of California vary from those of other states or provinces owing to its unique climate and location.

Control of exhaust emissions (hydrocarbons, carbon monoxide, and oxides of nitrogen) is accomplished by a combination of engine and fuel system modifications.

Three of the major engine emissions are described here.

Hydrocarbons (HC). Hydrocarbon emissions are the result of incomplete combustion. They are fuel left unburned, or partially burned, after combustion is completed.

Carbon Monoxide (CO). Carbon monoxide is the result of incomplete combustion of the fuel mixture due to an insufficient amount of oxygen in the air-fuel mixture.

Oxides of Nitrogen (NO$_x$). Nitrogen oxides are found when combustion temperatures reach high levels. The chemical nature of nitrogen and oxygen requires very high temperatures in order to combine both elements in any form. The x in NO$_x$ means that an oxide of nitrogen is formed when one molecule of nitrogen combines with any number of molecules of oxygen.

Devices used to control these emissions are described in the chapters on engines and fuel systems to which they apply.

Emission control methods include such devices as exhaust gas recirculation, positive crankcase ventilation, more precise control of injection timing and quantity of fuel injected, higher injection pressures for more complete combustion, more precise control of combustion temperatures, and the like.

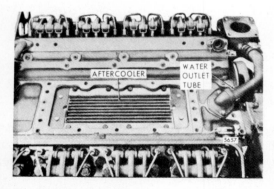

FIGURE 16-20. Detroit Diesel aftercooler cools intake air coming out of blower. Aftercooler is mounted beneath blower on V engine. *(Courtesy of Detroit Diesel Allison)*

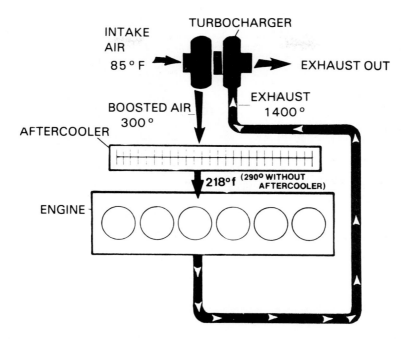

FIGURE 16-21. Typical intake air, boosted air, after-cooled air, and exhaust temperatures (85°F = 29.44°C, 300°F = 148.88°C, 218°F = 103.33°C, 1400°F = 760°C). *(Courtesy of Cummins Engine Company Inc.)*

PART 8   ENGINE SPEED AND INJECTION TIMING

The rated top speed of an engine is an important factor affecting exhaust emissions. Engines designed to run at higher speeds have less time for combustion to take place than lower speed engines. Consequently injection must begin earlier in order for the combustion process, cylinder pressure, piston position, and crank angle to be at the most advantageous state and position to provide as much push on the piston as possible and still not provide excessive emissions. The duration and rate of fuel injection are precisely controlled as well to aid in achieving the best results in engine performance, economy, and low emissions. This process involves injector design and injection pressures. Injector design will determine the spray pattern and the degree of fuel atomization achieved. Injection pressures generally have become higher in order to improve further the atomization of fuel as well as a thorough mixing of fuel particles and air.

Without a fuel delivery delay device, a greater amount of fuel is injected into the cylinders immediately upon acceleration of engine speed. On a turbocharged engine the amount of air delivered to the cylinders does not increase proportionately. Since the turbocharger is exhaust gas-driven, it takes a little more time for the air supply system to "catch up" to the fuel supplied to the cylinders. This results in excessive exhaust smoke and emissions. A fuel delivery delay device is used in the fuel system to overcome this problem. This device is sensitive to intake manifold (boost) air pressure to meter the amount of fuel delivered to the cylinders in direct proportion to the amount of air delivered to the cylinders. This device also provides for altitude compensation and air density.

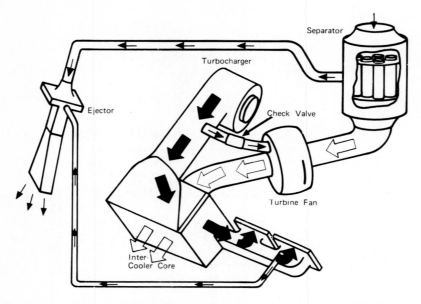

FIGURE 16-22. Typical flow diagram of air-to-air intercooler using tip turbine fan. Also shown is the dirt ejector from the air cleaner (separator). *(Courtesy of Mack Trucks Inc.)*

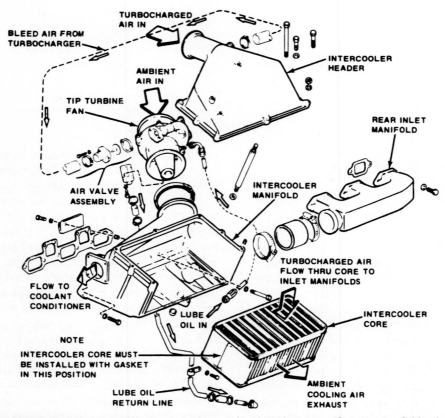

FIGURE 16-23. Air-to-air intercooler and related parts. *(Courtesy of Mack Trucks Inc.)*

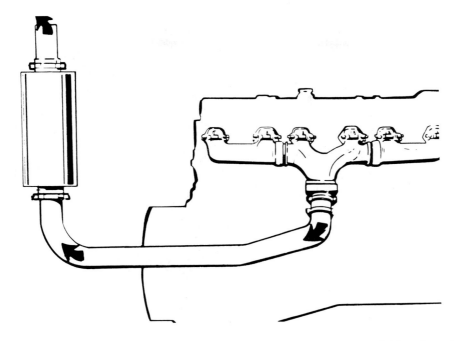

FIGURE 16-24. Typical exhaust system showing exhaust manifold, exhaust pipe, muffler, and extension. *(Courtesy of Cummins Engine Company Inc.)*

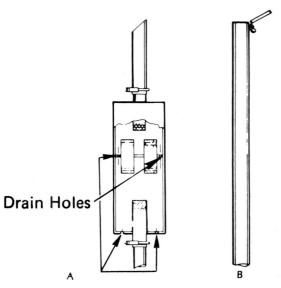

Drain Holes

FIGURE 16-25. (A) Moisture drain holes in muffler exhaust system prevent moisture from entering engine. Hinged rain cap (B) prevents moisture from entering exhaust system. Cap opens only when the engine is running and closes automatically when the engine is shut off. *(Courtesy of Cummins Engine Company Inc.)*

PART 9   SELF-CHECK

1. List the major components of the induction system of (a) a naturally aspirated engine and (b) a turbocharged engine.
2. What is the purpose of the blower on the two-stroke-cycle engine?
3. What functions does the air cleaner perform on the diesel engine?
4. List three results of operating an engine with an excessively restricted air cleaner.
5. Describe how the precleaner separates dirt from intake air.
6. What is the purpose of the air shut-down valve?
7. What effects does the supercharger or turbocharger have on air intake?
8. How is the turbocharger driven and at what speeds does it operate?
9. On a turbocharger, the compressor drives the turbine. True or false?

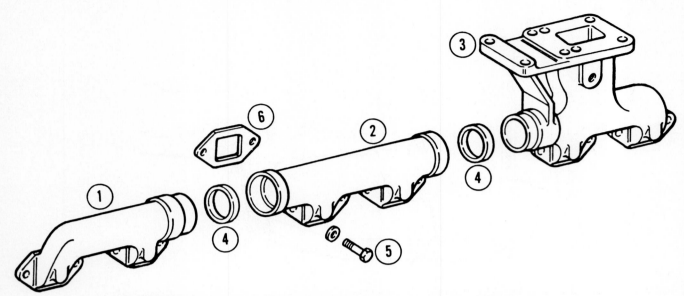

FIGURE 16-26. Exhaust manifold exploded view: (1) Rear section. (2) Center section. (3) Front section. (4) Seals between sections. (5) Attaching cap-screws. (6) Gaskets. *(Courtesy of Allis Chalmers)*

10. Explain the difference between series and parallel dual turbocharger installations.

11. How are the turbocharger shaft and bearings lubricated?

12. What precautions regarding turbocharger lubrication must be observed when starting or stopping the engine?

13. How is the two-stroke-cycle engine blower driven?

14. What is the importance of blower rotor timing?

15. Why are intercoolers or aftercoolers needed?

16. Explain why some engines do not need a muffler.

Chapter 17

Induction and Exhaust System Service

Performance Objectives

After thorough study of this chapter and sufficient practical work on the appropriate components, and with the necessary shop manuals, tools, and equipment, you should be able to do the following:

1. Complete the self-check with at least 80% accuracy.
2. Follow the accepted general precautions as outlined in this chapter.
3. Diagnose the basic induction system and exhaust system problems according to the diagnostic charts provided.
4. Safely remove, disassemble, clean, inspect, and accurately measure all induction system and exhaust system components.
5. Recondition all components as required.
6. Correctly assemble, install, and adjust all system components according to specifications provided.
7. Perform necessary tests and inspection to determine the success of the service performed.

It is critical to efficient engine operation that the air intake system and the exhaust system function in a manner that will not limit the engine's ability to take air in or to expel exhaust gases. Any restriction or leakage in either system will affect this ability and result in serious performance problems.

PART 1   GENERAL PRECAUTIONS

When servicing the induction and exhaust systems, the general precautions given at the beginning of Chapter 6 should be observed. In addition the following precautions should also be followed:

• Be careful not to damage parts by mishandling, improper storage, or improper use of tools and equipment. Parts such as air cleaner filter elements and turbocharger parts are easily damaged. Use the proper tools in the recommended way and in the sequence given in the appropriate service manual.

• Be aware of the danger of hot engine components, particularly cooling systems and exhaust systems.

• Be careful of rotating parts when checking engine components with the engine running. The engine fan, belts, and the turbocharger rotating assembly are examples. Do not get clothing, hands, electrical cords, hoses, or wiping rags near rotating parts.

• Be aware of the danger of an opened induction system's ability to inhale dirt, rags, and paper while the engine is running. Make sure that there is no possibility of this occurring before the engine is started. Any ingested material would seriously damage the blower, turbocharger, or engine.

• Always be sure that you know exactly how to stop an engine quickly in case of emergency.

PART 2   DIAGNOSTIC CHART

INDUCTION AND EXHAUST SYSTEMS

TROUBLE AND SYMPTOMS	PROBABLE CAUSES
Engine lacking power	1,4,5,6,7,8,9,10,11,18,20,21,22,25,26,27,28,29,30,A
Black smoke	1,4,5,6,7,8,9,10,11,18,20,21,22,25,26,27,28,29,30,A
Blue smoke	1,4,8,9,19,21,22,32,33,34,36,A
Excessive oil consumption	2,8,17,19,20,33,34,36,A
Excessive oil, turbine end	2,7,8,16,17,19,20,22,32,33,34,36
Excessive oil, compressor end	1,2,4,5,6,8,9,16,19,20,21,33,36,A
Insufficient lubrication	15,16,22,23,24,31,36
Oil in exhaust manifold	2,7,19,20,22,28,29,30,33,34
Damaged compressor wheel	3,6,8,20,21,23,24,36
Damaged turbine wheel	7,8,18,20,21,22,34,36
Drag or bind in rotating assembly	3,6,7,8,13,14,15,16,20,21,22,31,34,36
Worn bearings, journals, bearing bores	6,7,8,13,14,15,16,20,23,24,31,35,36
Noisy	1,3,4,5,6,7,8,9,10,11,18,20,21,22,A
Sludged or coked center housing	2,15,17

KEY TO PROBABLE CAUSE CODE NOS.

CODE NO.	PROBABLE CAUSE	CODE NO.	PROBABLE CAUSE
1	Dirty air cleaner element	22	Excessive carbon buildup behind turbine wheel
2	Plugged crankcase breathers	23	Too fast acceleration at initial start (oil lag)
3	Air cleaner element missing, leaking, not sealing correctly Loose connections to turbocharger		
4	Collapsed or restricted air tube before turbocharger	24	Too little warm-up time
5	Restricted-damaged crossover pipe turbocharger to inlet manifold	25	Fuel pump malfunction
6	Foreign object between air cleaner and turbocharger	26	Worn or damaged injectors
7	Foreign object in exhaust system (if from engine, check engine)	27	Valve timing
8	Turbocharger flanges, clamps, or bolts loose	28	Burned valves
9	Inlet manifold cracked; gaskets loose or missing; connections loose	29	Worn piston rings
10	Exhaust manifold cracked, burned; gaskets loose, blown, or missing	30	Burned pistons
11	Restricted exhaust system	31	Leaking oil feed line
12	Oil lag (oil delay to turbocharger at start-up)	32	Excessive engine pre-oil
13	Insufficient lubrication	33	Excessive engine idle
14	Lubricating oil contaminated with dirt or other material	34	Coked or sludged center housing
15	Improper type lubricating oil used	35	Oil pump malfunction
16	Restricted oil feed line	36	Oil filter plugged
17	Restricted oil drain line	A.	Oil bath air cleaner
18	Turbine housing damaged or restricted	1.	Air inlet screen restricted
19	Turbocharger seal leakage	2.	Oil pull-over
20	Worn journal bearings	3.	Dirty air cleaner
21	Excessive dirt buildup in compressor housing	4.	Oil viscosity low
		5.	Oil viscosity high

(Courtesy of Airesearch Industrial Division of The Garrett Corporation)

PART 3   DRY-TYPE AIR CLEANER SERVICE

No matter what system or method is used for determining air cleaner element service intervals, it should be geared around restriction figures. (See Fig. 17-1.)

The air cleaner elements should be serviced only when restriction reaches the maximum allowable limit, normally set by the engine manufacturers.

Restriction is the resistance to air flow through the air cleaner system into the engine.

Restrictions are best recorded by a water manometer, a dial indicator calibrated in inches of water, or an air cleaner service indicator.

Normally, restrictions are measured at high-idle, no-load on naturally aspirated or supercharged diesel engines and at full-load, wide-open throttle on turbocharged diesel engines.

Restrictions are measured in the air cleaner outlet tap (if provided), at a tap in the air transfer tube, or within the engine intake manifold.

Since some users do not have a water manometer or dial gauge available, the use of permanently mounted service indicators should be considered. The indicator can be mounted on the air cleaner or remote-mounted in an area where the operator can monitor the condition of the element constantly.

The element in the air cleaner should only be serviced when the maximum allowable restriction, established by the engine manufacturer, has been reached. The element should not be serviced on the basis of visual observation because this will lead to overservice.

The excess handling that is a result of overservice can cause

1. element damage,
2. improper installation of element,
3. contamination from ambient dust, and
4. increased service cost, time, and material.

As an air cleaner element becomes loaded with dust, the vacuum on the engine side of the air cleaner (at the cleaner outlet) increases. This vacuum is generally measured as "restriction in inches of water."

The engine manufacturer often places a recommended limit on the amount of restriction that the engine will stand without loss in performance before the element must be cleaned or replaced.

Mechanical gauges, warning devices, indicators, and water manometers are available to tell the

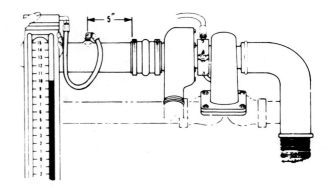

FIGURE 17-1. Measuring air inlet restriction with a water manometer. *(Courtesy of Cummins Engine Company Inc.)*

operator when the air cleaner restriction reaches this recommended limit. These gauges and devices are generally reliable, but the water manometer is the most accurate and dependable.

To use the manometer, hold it vertically and fill both legs approximately half full with water. One of the upper ends is connected to the restriction tap on the outlet side of the air cleaner by means of a flexible hose. The other end is left open to atmosphere.

Maximum restriction in the air cleaner occurs at maximum air flow. On a naturally aspirated or supercharged (not turbocharged) diesel, the maximum air flow occurs at maximum (high-idle) speed without regard for engine power. On a turbocharged diesel engine, the maximum air flow occurs only at maximum engine power.

With the manometer held vertically and the engine drawing maximum air, the difference in the height of the water columns in the two legs, measured in inches, is the air cleaner restriction. Restricton indicators are generally marked with the restriction at which the red signal flag "locks up."

Most engine manufacturers suggest a maximum restriction of 20 inches to 30 inches for diesels. Exceeding these maximums affects engine performance.

PART 4   CLEANING THE DRY ELEMENT

After removing the element or elements, they should be inspected for damage. Damaged elements should be replaced. If the element is not damaged, it should be cleaned as follows (Fig 17-2 and 17-3):

FIGURE 17-2. Changing dry-type air filter element. *(Courtesy of Cummins Engine Company Inc.)*

FIGURE 17-3. Cleaning a dry air filter element by washing in water and special cleaner. Never use water higher than 30 psi pressure for rinsing to prevent rupture of element. *(Courtesy of Deere and Company)*

Remove loose dirt from element with compressed air, or waterhose. *Caution:* Compressed air, 100 psi maximum with nozzles at least 1 inch away from element; water hose, 40 psi maximum without nozzle.

Mix 1 ounce (2 tablespoons) of D–1400 (Donaldson or other recommended solution) per 2 gallons of water, warm or cold, soft or hard (1 cup per 16 gallons). A stronger solution costs more and is not needed. Soak the element in solution for 15 minutes. Do not soak more than 24 hours.

Swish the element around in the solution to help remove dirt or use a cleaning tank with the proper solution and recirculating pump.

Rinse the element from "clean" side to "dirty" side with a gentle stream of water (less than 40 psi) to remove all suds and dirt.

Dry the element before reuse. Warm air (less than 160°F) must be circulated. Do not use a light bulb to dry the element.

Inspect for holes and tears by looking through the element toward a bright light. Check for damaged gaskets or dented metal parts. Do not reuse damaged elements.

Protect the element from dust and damage during drying and storage.

Install the elements.

Check to make sure that all gaskets are sealing. Examine for dust trails, which indicate leaks. Check to be certain that the wing nut is tight.

Examine the clean air transfer tubing for cracks, loose clamps, or loose flange joints.

Check air compressor connections (if used) to be certain that these are leak-free.

Check ether fittings (if used) to be certain that

no contaminants are entering through these connections.

PART 5   OIL BATH AIR CLEANER SERVICE

1. Remove the oil sump from the cleaner by loosening the retaining band (or wing nuts). (See Fig. 17–4.) Empty the sump and wash it with fuel oil to remove all of the sediment.

2. Remove the detachable screen by loosening the wing nuts and rotating the screen a one-quarter turn.

One of the most important steps in properly cleaning the tray-type oil bath air cleaner is the step that is most overlooked. Unless the filter tray is thoroughly cleaned, satisfactory performance of any engine cannot be realized. The presence of fibrous material found in the air is often underestimated and is the main cause of the malfunctioning of heavy-duty air cleaners. This material comes from plants and trees during their budding season and later from airborne seed from the same sources. Solid black areas in the mesh are accumulations of this fibrous material. When a tray is plugged in this manner, washing in a solvent or similar washing solution will not clean the tray satisfactorily. It must also be blown out with high-velocity compressed air or steam to remove the material that accumulates between the layers of screening. When a clean tray is held up to the light, an even pattern of light should be visible. Some trays have equally spaced holes in the retaining baffle. Check to make sure that they are clean and open. Dark spots in the mesh indicate the close overlapping of the mesh and emphasize the need for using compressed air or steam. It is suggested that users of heavy-duty air cleaners have a spare tray on hand to replace the tray that requires cleaning. Having an extra tray available makes for better service and the dirty tray can be cleaned thoroughly as recommended.

3. Remove the hood and clean it by brushing or by blowing out with compressed air. Push a lint-free cloth through the center tube to remove dirt or oil from the walls.

4. The fixed element should be serviced as operating conditions warrant. Remove the entire cleaner from the engine, soak the unit in fuel oil to loosen the dirt, then flush with clean fuel oil, and allow to drain thoroughly.

5. Clean and check all gaskets and sealing surfaces to ensure airtight seals.

6. Refill the oil cup *only* to the oil level marked. Use oil of the same grade as used in the engine crankcase. Do not overfill.

7. Install the removable screen in the housing and reinstall the housing.

8. Install the oil cup and the hood.

9. Check all of the joints and tubes and make sure that they are airtight.

All oil bath air cleaners should be serviced as operating conditions warrant. At no time should more than ½ inch of "sludge" be allowed to form in the oil cup or the area used for sludge deposit, nor should the oil cup be filled above the oil level mark.

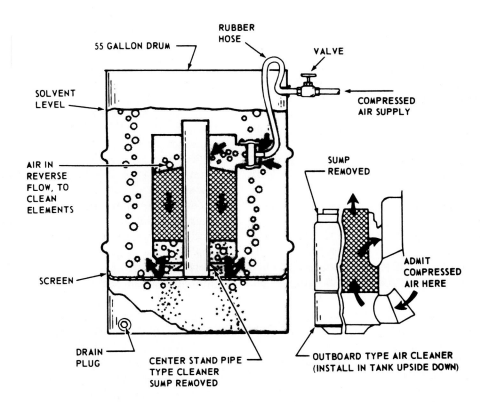

FIGURE 17-4. Cleaning oil bath type of air cleaner element in special cleaning fluid with compressed air supply. *(Courtesy of Mack Trucks Inc.)*

FIGURE 17-5. Measuring turbine shaft end play (axial play). Excessive end play indicates worn thrust surfaces requiring parts replacement. *(Courtesy of Detroit Diesel Allison)*

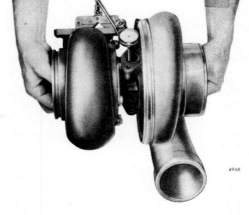

FIGURE 17-6. Measuring turbine shaft radial movement or clearance. Excessive radial play indicates bearing, shaft, and housing wear, requiring overhaul. *(Courtesy of Detroit Diesel Allison)*

PART 6   TURBOCHARGER SERVICE

Most turbocharger failures are caused by one of the three basic reasons: lack of lubricant, ingestion of foreign objects, and contamination of lubricant. Many turbochargers are removed needlessly because the following procedures are not followed. The purpose of system troubleshooting is to identify the reason for failure so that repair can be made before installing a new unit. (See Fig. 17-5 to 17-7.)

Common symptoms that may indicate possible turbocharger trouble are (1) the engine lacking power, (2) black smoke, (3) blue smoke and excessive engine oil consumption, and (4) noisy operation of the turbocharger.

We'll now examine each of these trouble symptoms in detail, starting with *the engine lacking power or black smoke*. Both of these complaints can result from insufficient air reaching the engine and can be caused by restrictions to the air intake or air leaks in the exhaust or induction systems. The first step in troubleshooting any turbocharger trouble is to start the engine and listen to the sound the turbo system makes. As you become more familiar with this characteristic sound, you will be able to identify an air leak between the compressor outlet and engine or an exhaust leak between engine and turbo, by a higher pitched sound. If the turbo sound cycles or changes in intensity, a plugged air cleaner, loose

material in the compressor inlet ducts, or dirt buildup on the compressor wheel and housing is the likely cause.

After listening, check the air cleaner for a dirty element. If in doubt, measure for restrictions per the engine manufacturer's shop manual. Next, with the engine stopped, remove the ducting from air cleaner to turbo and look for dirt buildup or foreign object damage. Then check for loose clamps on compressor outlet connections and check the engine intake system for loose bolts, leaking gaskets, etc. Then disconnect the exhaust pipe and look for restrictions or loose material. Examine the engine exhaust system for cracks, loose nuts, or blown gaskets. Then rotate the turbo shaft assembly. Does it rotate freely? Are there signs of rubbing or wheel-impact damage? Axial shaft play is end-to-end movement and radial shaft play is side-to-side. There is normally side-to-side play; however, if this play is sufficient to permit either of the wheels to touch the housing when the shaft is rotated by hand, then there is excessive wear. If none of these symptoms is present, the low-power complaint is not being caused by the turbocharger. Consult the engine manufacturer's troubleshooting procedures for the basic engine.

Next, we'll consider *blue smoke*, which is an indication of oil consumption and can be caused by either turbo seal leakage or other internal engine problems. First, check the air cleaner for restrictions per the engine manufacturer's shop manual.

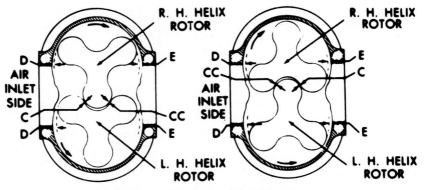

VIEWS FROM GEAR END OF BLOWER

ENGINE	BLOWER	A	B	C	CC	D	E
8 CYL.	TURBO	.007"	.014"	.010"	.004" TO .008"	.015"	.004"
8 CYL.	PUMPER	.012"	.019"	.010"	.004" TO .008"	.030"	.004"
8 CYL.	SMALL ROTOR	.007"	.014"	.010"	.004" TO .008"	.015"	.004"

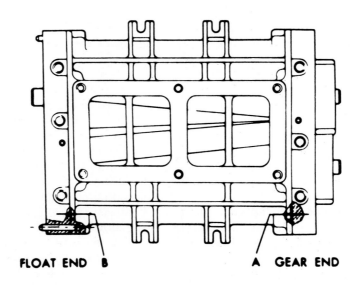

FLOAT END   B   A   GEAR END

NOTE: Time rotors to dimensions on chart for clearance between trailing side of R.H. Helix Rotor and leading side of L.H. Helix Rotor (cc) from both inlet and outlet side of blower.

FIGURE 17-10. Typical Detroit Diesel blower measurement specifications. For specific measurements refer to the appropriate service manual. *(Courtesy of Detroit Diesel Allison)*

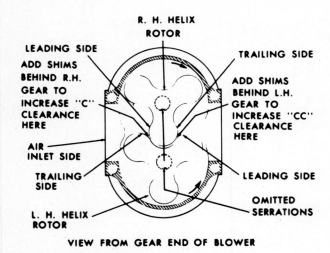

FIGURE 17-11. Proper location of shims determine rotor lobe clearances (also known as *rotor timing*). *(Courtesy of Detroit Diesel Allison)*

Inspect the blower rotor lobes, especially the sealing ribs, for burrs and scoring. If the rotors are slightly scored or burred, they may be cleaned up with emery cloth.

Examine the rotor shaft serrations for wear, burrs, or peening. Also inspect the bearing contact surfaces of the shafts for wear and scoring.

Inspect the inside surface of the blower housing for burrs and scoring. If the inside surface of the housing is slightly scored or burred, it may be cleaned up with emery cloth.

Check the finished ends of the blower housing for flatness and burrs. The end plates must set flat against the blower housing.

The finished inside face of each end plate must be smooth and flat. If the finished face is slightly scored or burred, it may be cleaned up with emery cloth.

Examine the serrations in the blower rotor gears for wear and peening; also check the teeth for wear, chipping, or damage. If the gears are worn to the point where the backlash between the gear teeth exceeds 0.004 inch or damaged sufficiently to require replacement, both gears must be replaced as a set.

Check the blower drive shaft serrations for wear or peening. Replace the shaft if it is bent.

Inspect the blower-drive coupling springs (pack) and the cam for wear.

Replace all worn or excessively damaged blower parts.

Clean the oil strainer in the vertical oil passage at the bottom of each blower end plate and blow out all oil passages with compressed air.

Assembly and Installation

Install new oil seals and sleeves. Install bearings, maintaining the same component relationship as originally, if parts are used again. Lubricate the seals and bearings with engine oil.

Install the blower rotors into the blower housing and align as specified in the service manual by positioning the housing over the rotors with the rear of the housing facing up. One end of the blower housing is marked rear on the outside face of the housing. Install the rear end plate. Make sure that bearings and seals are lubricated with engine oil. If necessary, tap the end plate lightly with a soft-face hammer. Install bearing retainers and tighten bolts to specifications. Install ball bearings on rotor shafts and front end plate. Make sure that bearings are lubricated with engine oil. Install bearing retainers and tighten bolts to specified torque.

Check rotor-to-end plate clearance with a feeler gauge. Check rotor-to-housing clearance. Make sure that clearances are within specifications.

Install the blower rotor gears as specified in the shop manual. Install and align gears and shims as outlined in the manual. Tighten all bolts to specified torque. Check backlash between rotor gears. Replace gears if backlash is over specifications.

Time the blower rotors to provide the specified clearance between rotor lobes by adding or removing shims between the gears and the bearings until specified clearances are obtained. Clearance should be measured from both the inlet and outlet sides of the blower.

Attach the remaining accessories to the blower and install the blower, using new gaskets and appropriate gasket sealer (sparingly) to keep the gasket in place during blower positioning. Tighten all bolts as specified. Attach all remaining accessories, blower drive, governor, alternator, air compressor, and the like as well as fuel lines and control linkages as required.

Connect all coolant lines and fill the engine with coolant as outlined in the engine section of this text. Attach intake and exhaust connections.

Adjust all linkages as outlined in the appropriate service manual. Recheck the entire installation before starting the engine.

PART 9   EXHAUST SYSTEM SERVICE

The entire exhaust system, manifolds, pipes, turbocharger connections, and mufflers should be inspected for leakage or external damage that could cause exhaust restriction. Any damaged component should be repaired or replaced. The system should also be checked for backpressure with a mercury or water manometer or an appropriately calibrated pressure gauge. This test reveals internal restriction that may otherwise not be evident. (See Fig. 17-12.)

If exhaust backpressure exceeds service manual specifications, early engine failure and poor performance may be expected. If no test fitting is provided, proceed as follows:

Select a point of measurement as close as possible to the manifold or turbocharger outlet flange in an area of uniform flow such as a straight section of pipe at least one-pipe diameter from any changes in flow area or flow direction.

At the point selected, weld a ⅛-inch (3.18 mm) pipe coupling to the exhaust tubing. Drill through tubing with a ⅛-inch (3.18 mm) drill.

Remove all burrs. Mount 90° Weatherhead fitting (4) to coupling. Then use 3 feet (0.9 m) of ⅛-inch (3.18 mm) inside diameter copper tubing (3) plus 10 feet (3 m) of 3/16-inch (4.76 mm) inside diameter soft rubber hose (2) to manometer. The manometer may be mercury-filled or water-filled.

Note: It is important that the line-to-manometer be as specified to minimize variation in reading

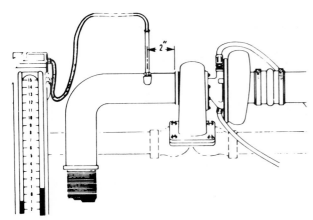

FIGURE 17-12. Measuring exhaust backpressure with manometer. *(Courtesy of Cummins Engine Company Inc.)*

due to a standing wave phenomenon that occurs in the manometer line. A change in the length or material of this line can significantly change the reading obtained.

Start the engine and operate until oil temperature reaches 140°F (60°C).

Take backpressure readings when the engine is developing its maximum horsepower at maximum engine speed.

Add the reading of mercury in both columns for the final figure.

Example: If mercury is 1 inch (25.4 mm) high in the left column and 1 inch (25.4 mm) low in the right column, there is 2 inches (50.8 mm) of pressure. If the mercury is 1 inch (25.4 mm) high in the right column and 1 inch (25.4 mm) low in the left column, there are 2 inches (50.8 mm) of vacuum. Maximum permissible backpressure is 3 inches (76.2 mm) Hg or 40.7 inch (1 m) of water.

If exhaust restriction is excessive, the faulty component should be identified and replaced.

PART 10   ANALYZING EXHAUST SMOKE

Diesel engine exhaust smoke is analyzed in terms of opacity (denseness). Engine exhaust smoke is analyzed when the engine coolant has reached the specified operating temperature, usually about 160°F (70°C). The exhaust smoke is then analyzed under specified operating modes such as acceleration and lugging modes. The exhaust smoke emitted is then compared to a smoke density chart held at arms length against exhaust smoke density or opacity. During the acceleration mode, for example, a 40 percent opacity may be the maximum acceptable whereas during the lugging mode a 20 percent maximum opacity may be allowed. See the density chart illustrated in Fig. 17-13.

PART 11   EMISSION STANDARDS

Emission standards are established by the federal Environmental Protection Agency (EPA) and the California Air Resources Board (CARB). The following emission standards are provided by the California Air Resources Board. It should be noted that the standards for the state of California are more stringent than those for other areas.

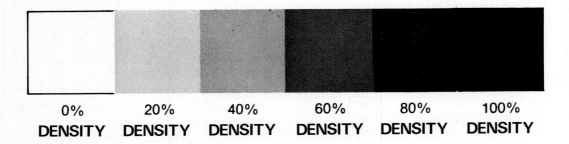

| 0%
DENSITY | 20%
DENSITY | 40%
DENSITY | 60%
DENSITY | 80%
DENSITY | 100%
DENSITY |

FIGURE 17-13. Exhaust smoke density is compared to the density chart to analyze exhaust smoke. Smoke opacity (density) must meet local emission standards.

New Vehicle Standards Summary

The following California standards, up through 1972, and federal standards, up through 1973, apply only to gasoline-powered medium-duty vehicles(1). California standards for 1973 and later year and federal standards for 1974 and later apply to both gasoline and diesel-powered medium-duty vehicles.

MEDIUM-DUTY VEHICLES—EXHAUST EMISSIONS STANDARDS

| YEAR | STANDARD | HYDROCARBONS (5) | | CARBON
MONOXIDE
CO | OXIDES OF
NITROGEN
NOx |
		NON- METHANE	TOTAL HC		
1969-77	Calif.	SEE HEAVY-DUTY STANDARDS			
1970-78	federal	SEE HEAVY-DUTY STANDARDS			
		50,000 MILE EXHAUST EMISSIONS STANDARDS CVS-75 Test Procedure (GRAMS PER MILE)			
1978-79	Calif.	no std.	(0.9)	17	2.3
1979	federal	See light-duty truck standards			
1980	Calif.	0.9	(0.9)	17	2.3
	federal	no std.	(1.7)	18	2.3
1981-82	Calif. (2)	0.39	(0.41)	9.0	1.0 (8)
	Calif. (3)	0.50	(0.50)	9.0	1.5
	Calif. (4)	0.60	(0.60)	9.0	2.0
	federal	no std.	(1.7)	18.0	2.3
1983 & Sub- sequent	Calif. (2)	0.39	(0.41)	9.0	0.4 (9)
	Calif. (2)	0.39	(0.41)	9.0	1.0 (10)
	Calif. (3)	0.50	(0.50)	9.0	1.0
	federal	no std.	(1.7)	18.0	2.3

CALIFORNIA 100,000 MILE EXHAUST EMISSIONS STANDARDS CVS-75 TEST PROCEDURE

YEAR	STANDARD	HYDROCARBONS (5)		CARBON MONOXIDE CO	OXIDES OF NITROGEN NOx
		NON-METHANE	TOTAL HC		
1981-82	(2) Option 1 (6) (7)	0.39	(0.41)	9.0	1.5
	(2) Option 2 (6) (7)	0.46	no std.	10.6	1.5
	(3)	0.50	(0.50)	9.0	2.0
	(4)	0.60	(0.60)	9.0	2.3
1983	(2) Option 1 (6)	0.39	(0.41)	9.0	1.5
	(2) Option 2 (6)	0.46	no std.	10.6	1.5
	(3)	0.50	(0.50)	9.0	2.0
	(4)	0.60	(0.60)	9.0	2.0
1984 & subsequent	(2) Option 1 (6)	0.39	(0.41)	9.0	1.0
	(2) Option 2 (6)	0.46	no std.	10.6	1.0
	(3)	0.50	(0.50)	9.0	1.5
	(4)	0.60	(0.60)	9.0	2.0

(1) Medium-duty vehicles as defined in Title 13 of the California Administrative Code means heavy-duty vehicles having a manufacturer's gross weight rating of 8,500 pounds or less (Manufacturers may elect to certify medium-duty vehicles up to 10,000 pounds (GVW)).

(2) 0-3999 pounds equivalent inertia weight (curb weight plus 300 pounds).

(3) 4000-5999 pounds equivalent inertia weight.

(4) 6000-8500 (or 10,000) pounds equivalent inertia weight.

(5) Hydrocarbon standards in parentheses indicate total hydrocarbon. When applicable, manufacturers may elect to certify vehicles to either the non-methane or total hydrocarbon standards.

(6) 100,000 mile options: Option 1 standards refer to the projected 50,000 mile emissions for hydrocarbons and carbon monoxide, while Option 2 standards refer to the projected 100,000 mile emissions for these pollutants. NOx emissions standards for both options refer to the 100,000 mile projected emissions.

(7) For 1981 model year vehicles with evaporative emissions values below 1.0 g/test, an adjustment to the exhaust hydrocarbon emissions standard may be granted by the Executive Officer (100,000 mile option only).

(8) 1.5 g/mi NOx for small manufacturers subject to "in lieu" standards pursuant to Section 1960.3 of Title 13, California Administrative Code. Production medium-duty vehicles must meet a 1.0 g/mi cumulative corporate average NOx standard based upon a full year's production.

(9) 1.0 g/mi NOx standard for small manufacturers subject to "in lieu" standards pursuant to Section 1960.4 of Title 13, California Administrative Code. Production medium-duty vehicles must meet a 0.7 g/mi cumulative coporate average NOx standard for the full model year's production.
0.7 g/mi NOx standard for small manufacturers for the 1984 and 1985 model years. Production medium-duty vehicles must meet a 0.7 g/mi cumulative corporate average NOx standard for each production quarter.
For the 1986 and subsequent model years, a small manufacturer must meet a 0.4 g/mi NOx standard for the prototype vehicle certification and for production vehicles per engine family on a quarterly basis.

(10) A manufacturer may choose to certify to this optional set of standards provided that the conditions set forth in Section 1960.15, Title 13, California Administrative Code are met. These conditions include a 7 year/75,000 mile recall provision for selected emissions control parts.

Additional Requirements:

Beginning with the 1981 model year, the maximum projected emissions of oxides of nitrogen measured on the federal Highway Fuel Economy Test shall be no greater than 2.0 times the applicable medium-duty vehicle oxides of nitrogen standards shown above.

Effective 1978, evaporative emissions standards are 6.0 grams per SHED test for the 1978-79 model years and 2.0 grams per SHED for 1980 and later model years.

SHED Test (Sealed Housing Evaporative Determination)—A method for measuring evaporative emissions from motor vehicles.

Previous evaporative emissions standards of 6.0 grams per test for 1970-71 and 2.0 grams per test for 1972 and later were based on the "carbon trap procedure" which became obsolete when the SHED test procedure was adopted for 1978 and subsequent model years.

Diesel-powered medium-duty vehicles are subject to the following 50,000 mile particulate exhaust emission standards: 0.6 g/mi for the 1982 through 1984 model years, 0.4 g/mi for the 1985 model year, 0.2 g/mi for the 1986 through 1988 model years, and 0.08 g/mi for 1989 and subsequent model years.

g/mi   –grams per mile.

CVS-75   –constant volume sample test which includes cold and hot starts.

The following is a summary of heavy-duty engine and vehicle (2) standards adopted by both the California Air Resources Board and federal Environ- mental Protection Agency, based on the 9-mode test procedures for gasoline engines and 13-mode test procedures for diesel engines. Also see footnote (5).

HEAVY-DUTY ENGINES AND VEHICLES – EXHAUST EMISSIONS STANDARDS
(DIESEL (1) AND GASOLINE)

YEAR	STANDARD	HYDRO-CARBONS	CARBON MONOXIDE	OXIDES OF NITROGEN	HYDRO-CARBONS & OXIDES OF NITROGEN
1969–71 (3)	Calif.	275 ppm	1.5%	no std.	no std.
1970–73 (4)	federal	275 ppm	1.5%	no std.	no std.
1972	Calif.	180 ppm	1.0%	no std.	no std.
(GRAMS PER BRAKE-HORSEPOWER-HOUR)					
1973–74	Calif.	no std.	40	no std.	16
1973–78	federal	no std.	40	no std.	16
1975–76	Calif.	no std.	30	no std.	10
1977–78	Calif. or	no std.	25	no std.	5
	Calif.	1.0	25	7.5	no std.
1979 (6)	Calif. or	1.5	25	7.5	no std.
	Calif. or	no std.	25	no std.	5
	Calif.	1.0	25	7.5	no std.
1979–1983	federal or	1.5	25	no std.	10
	federal	no std.	25	no std.	5
1980–83	Calif. or	1.0	25	no std.	6.0
	Calif.	no std.	25	no std.	5
1984 & later	Calif. or	0.5	25	no std.	4.5
	Calif. (5)	1.3	15.5	5.1	no std.
	federal (5)	1.3	15.5	10.7	no std.

(1) The above standards apply to diesel engines and vehicles sold in California on or after January 1, 1973, and na- tionwide on or after January 1, 1974.
(2) These standards apply to motor vehicles having a manufacturer's GVW rating of over 6000 pounds, excluding passenger cars and 1978 and later medium-duty vehicles.
(3) Applies to vehicles manufactured on or after January 1, 1969.
(4) Applies to vehicles manufactured on or after January 1, 1970.
(5) These standards are based upon the new transient cycle test procedures.
(6) For 1979 only, manufacturer using heated flame ionization detection (HFID) method of measuring hydrocarbons must meet the 1.5 g/BHP-hr standard; whereas, manufacturers using non-dispersive infrared (NDIR) method of measuring hydrocarbons must meet the 1.0 g/BHP-hr standard. Both standards are equivalent in stringency. Manufacturers may use either HFID or NDIR in meeting the combined hydrocarbon and oxides of nitrogen stan- dard of 5 g/BHP-hr. After 1979, manufacturers are required to sue HFID.
Additional Requirements:
Effective 1978, evaporative emissions standards are 6.0 grams per SHED test for the 1978–79 model years and 2.0 grams per SHED test for 1980 and later model years.
SHED Test (Sealed Housing Evaporative Determination) – A method for measuring evaporative emissions from mo- tor vehicles.
Previous evaporative emission standards of 6.0 grams per test for 1970–1971 and 2.0 grams per test for 1972 and later were based on the "carbon trap procedure" whice became obsolete when the SHED test procedure was adopted for 1978 and subsequent model years.
g/BHP-hr –grams per brake-horsepower-hour.
ppm   –parts per million.

PART 12   SELF-CHECK

1. How is engine performance affected by a restricted exhaust?

2. List six causes of black exhaust smoke.

3. List six causes of blue exhaust smoke.

4. List six causes of abnormal turbocharger noise.

5. What possible problems could occur from over-servicing the air cleaner?

6. What tool is used to measure air cleaner restriction?

7. A noisy blower in a two-cycle engine at idle indicates what kind of problem?

8. To inspect two-cycle engine blower rotors on the engine, remove the _____ and the _____.

9. Why must blower rotors be timed?

10. What approximate exhaust smoke density is allowed (a) during acceleration and (b) during the lugging mode?

Chapter 18

Final Assembly, Start-up, and Engine Break-in

Performance Objectives

After thorough study of this chapter and with sufficient practice on the appropriate training models and with the proper tools, equipment, and service manuals, you should be able to do the following:

1. Complete the self-check questions with at least 80 percent accuracy.

2. Install all engine parts required to complete the assembly of the engine.

3. Install the engine in the equipment from which it was removed.

4. Perform an engine break in using any of the methods described in this chapter.

5. Perform all work to meet manufacturer's specifications.

PART 1 FINAL ASSEMBLY AND INSTALLATION

The sequence for installing the remaining components will vary, depending on engine design and auxiliary equipment. These components include valve covers, fuel lines, fuel filters, fuel supply and injection pumps (make sure that fuel injection pump timing specifications are followed as recommended in the appropriate service manual), manifolds, water pump, fan, water filter, oil filters, oil coolers, start-ing motor, generator or alternator, compressor, drive belts (adjusted to proper tension), air intake and exhaust connections, turbocharger, governor, control linkages, and electrical connections before or after engine installation, depending on accessibility.

The engine should be installed, using the same lift equipment used for engine removal. All engine mounts and alignment devices should be properly aligned and tightened to specifications. Make sure that torque converter drive bolts, clutch assembly, and transmission or bell housing are properly aligned and tightened. Install the radiator and shroud, radiator shutters and controls, and all remaining electrical connections, control linkages, hydraulic connections, and the like.

Close all coolant drain plugs and valves. Open specified bleed valves in the cooling system and fill the system with coolant. Make sure that all air is bled from the system, then close the bleed or vent valves and top up coolant to specified level. Check for leaks and correct, if required.

Service or replace the air cleaner assembly as necessary. Make sure that all air system connections are airtight.

Fill the crankcase with the specified amount and type of lubricating oil. Pressurize and prime the lubrication system, using a pressure type of supply tank connected to the main oil gallery plug specified in the service manual. The turbocharger lubrication line may require priming separately; follow the recommended procedure in the service manual. Keep the system pressurized until the lubrication system is primed, then shut off the pressure line valve, and disconnect from engine. Replace the plug in the oil gallery. Check and correct the oil level in the crankcase. *Caution:* Do not overfill.

PART 2 START-UP PROCEDURE

Precautions

It must be stressed that the operator must be observant at all times, so that any malfunction that may develop will be detected. The instruments must be monitored constantly and all readings recorded.

If at any time during a run-in the engine develops any one of the following abnormal running characteristics, the engine must immediately be shut down and the cause investigated and corrected before continuing the run-in procedure:

1. any unusual noises such as knocks, scraping, etc.

2. the oil pressure taking a drastic drop. As a general rule normal engine oil pressure will be between 45 to 75 psi (310 to 517 kPa) depending on engine type, speed and oil viscosity.

3. the oil temperature reaching and exceeding 240°F (115.55°C)

4. the coolant temperature exceeding 200°F (93.33°C)

5. oil or coolant leaks

6. exhaust temperature exceeding the recommended maximum limits

In general the following procedures apply for break-in. Refer to the appropriate service manual for specific procedures and specifications. Bleed all air from the fuel system at locations specified by the manufacturer's service manual.

PART 3   ENGINE DYNAMOMETER BREAK-IN PROCEDURE

Note: This procedure may also be used for power-take-off dynamometers.

1. Crank the engine with the "stop" engaged until oil pressure is observed on the main gallery pressure gauge. Release the stop and, when engine starts, run at 800–1000 rpm, no-load for 3 to 5 minutes. Check for oil pressure, leaks, knocks, etc. Stop the engine and check oil and coolant levels.

2. Start the engine, set at 1600 rpm, and set the dynamometer at one-fourth to one-half load for the engine being tested. Operate for 15 minutes or until the thermostat opens and the oil temperature reaches 140°F (60°C).

3. Set throttle at full-load and adjust the dynamometer to obtain governed speed. Run-in engine at this speed/load for 1/2 hour.

4. With the throttle at rated speed, adjust the dynamometer load to obtain the speed and torque specified by the engine manufacturer. Run at this speed/load for time specified in manual.

5. Remove dynamometer load and check high idle and low idle of engine.

6. Allow the engine to cool down for 5 minutes at fast idle.

7. Shut down the engine and tighten hose clamps, manifold fasteners, etc.; retorque cylinder heads; and adjust valves according to the engine manufacturer's specifications.

PART 4   CHASSIS DYNAMOMETER BREAK-IN PROCEDURE

Note: Be sure that the dynamometer is in calibration before starting the test.

1. Crank the engine with the stop engaged until the oil pressure shows on the gauge. Release the stop, and, when engine starts, run at 800 to 1000 rpm, no-load for 3 to 5 minutes. Check for oil pressure, leaks, knocks, etc. Stop the engine and check oil and coolant levels.

2. Operate the chassis on the dynamometer in gear range and engine speed specified by the vehicle manufacturer or until the thermostat opens or the oil temperature reaches 140°F (60°C).

3. Operate in direct gear at or near full-load governed speed for 1/2 hour. *Note:* Depending on roller size, tire condition, size, etc., it may be necessary to make the 1/2 hour run in intervals with a cooling-off period in between.

4. Run at governed speed, full-load for 3 minutes. Repeat every decreased 200 rpm down through the speed range of the engine.

5. Shift into neutral and check high idle and low idle of the engine.

6. Allow the engine to cool down for 5 minutes at fast idle.

7. Shut down the engine; tighten hose clamps, manifold fasteners, etc.; retorque cylinder heads; and adjust valves according to the engine manufacturer's specifications.

PART 5   ROAD BREAK-IN PROCEDURE

1. Crank the engine with the stop engaged until oil pressure is observed on the oil pressure gauge. Release stop and, when the engine starts, run at 800 to 1000 rpm, no-load for 3 to 5 minutes. Check for oil pressure, leaks, knocks, etc. Stop the engine and check oil and coolant levels.

2. With a loaded truck body or trailer, operate through the gear ranges for 1/2 hour. Do not exceed specified rpm except as necessary to make gear changes.

3. On a suitable highway, operate at or near governed speed for ½ to 1 hour.

4. Check the engine high idle and low idle.

5. Allow the engine to cool down at fast idle for 5 minutes.

6. Shut down the engine and tighten loose clamps, manifold fasteners, etc.; retorque the cylinder heads; and adjust the valves according to the engine manufacturer's specifications.

PART 6   SELF-CHECK

1. List all the checks that must be made during engine start-up procedures after an engine overhaul.

2. Describe the engine dynamometer method of engine break-in.

3. What items must be checked and tightened after the break-in procedure?

Chapter 19

Diesel Fuels and Fuel Supply Systems

Performance Objectives

After thorough study of this chapter, the appropriate service manuals and training models, you should be able to do the following:

1. Complete the self-check questions with at least 80 percent accuracy.

2. List four types of diesel fuel injection systems.

3. Define the following terms as they relate to diesel fuels:

    (a) cetane number

    (b) volatility

    (c) viscosity

    (d) pour point

    (e) cloud point

    (f) heat value

    (g) flash point

    (h) sulphur content

    (i) diesel fuel classification

    (j) ignition delay period

4. State the causes of white, black, and blue exhaust smoke.

5. Store and handle fuel in a clean and safe manner.

6. Describe the diesel engine fuel supply system.

7. Describe the basic operation of the following supply pump types: diaphragm, gear, vane, plunger, and priming pumps.

8. State the purpose of the fuel filtering system.

9. Define primary, secondary, final, series, and parallel fuel filtering.

PART 1 INTRODUCTION

The diesel fuel system must provide the right amount of clean fuel to each cylinder at the correct time and must atomize the fuel adequately for good combustion. This must be achieved without entry of air into the fuel system. Injection of fuel into the cylinder must take place for a controlled period of time and must prevent leakage of fuel from the injector during non-injection time. Combustion must be controlled to limit exhaust emissions within pollution standards. (See Fig. 19–1 to 19–4.)

There are four basic types of diesel fuel injection systems to consider. These are (1) the multiple pump system, (2) the pressure time system, (3) the unit injector system, and (4) the distributor pump system.

The multiple pump system is used on many four-stroke-cycle engines such as Mack, Perkins, J.I. Case, Ford, Deere & Company, Fiat-Allis, International Harvester, Caterpillar, etc.

The pressure time system is used by the Cummins Engine Company, Inc. on its engines, which are used in many different manufacturers' products.

The unit injector system is used by Detroit Diesel Allison; for example, on its engines, both two-stroke and four-stroke, which are also used in many different manufacturers' products.

The distributor pump system is used on various makes of four-stroke engines.

PART 2 DIESEL FUELS

Conventional diesel fuels are distillates with a boiling range of about $+300°F$ to $+700°F$ ($149°C$ to $371°C$), obtained by the distillation of crude oil. They are predominantly straight-run fractions that contain the greatest amount of normal paraffins and naphthenes, and the least amount of isoparaffins and aromatics. Normal paraffins and naphthenes have superior diesel ignition qualities, but they have the disadvantages of higher pour points than isoparaffins and aromatics. (See Fig. 19–5.)

Diesel Fuel Quality

Cetane Number

The cetane number is a measure of the autoignition quality of a diesel fuel. The shorter the interval between the time when the fuel is injected and the time when it begins to burn, called the ignition delay period, the higher the cetane number. It is a measure of the ease with which the fuel can be ignited and is most significant in low-temperature starting,

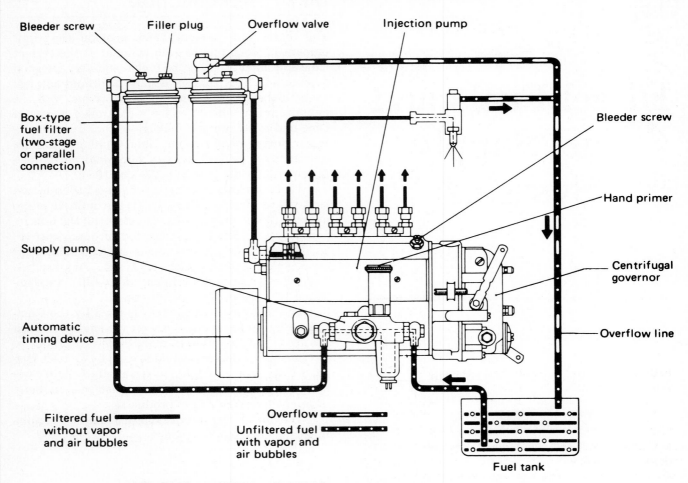

FIGURE 19-1. Typical Bosch in-line pump type of fuel system. *(Courtesy of Robert Bosch Canada Ltd.)*

warm-up and smooth even combustion. (See Fig. 19–6.)

Some hydrocarbons ignite more readily than others and are desirable because of this short ignition delay. The preferred hydrocarbons in order of their decreasing cetane number are normal paraffins, olefins, naphthenes, isoparaffins, and aromatics. This is the reverse order of their antiknock quality. The cetane number is measured in a single-cylinder test engine with a variable compression ratio. The reference fuels used are mixtures of cetane, which has a very short ignition delay, and alpha-methyl naphthalene, which has a long ignition delay. The percentage of cetane in the reference fuel, which gives the same ignition delay as the test fuel, is defined as the cetane number of the test fuel.

Formulas for estimating the cetane number have been developed, and these correlate fairly closely to the CFR engine test. However, such factors as the

aniline point, the API gravity, and the 50 percent point are needed.

Diesel engines whose rated speeds are below 500 rpm are classed as slow-speed engines, from 500 to 1200 rpm as medium-speed, and over 1200 rpm as high-speed. Cetane numbers of fuels readily available range from 40 to 55 with values of 40 to 50 most common. These cetane values are satisfactory for medium- and high-speed engines, while low-speed engines may use fuels in the 25 to 35 cetane number range.

Addition of certain compounds such as ethyl nitrate, acetone peroxide, and amyl nitrate will improve cetane number. Amyl nitrate is available commercially for this purpose.

Volatility

The distillation characteristics of the fuel describe its volatility. A properly designed fuel has the op-

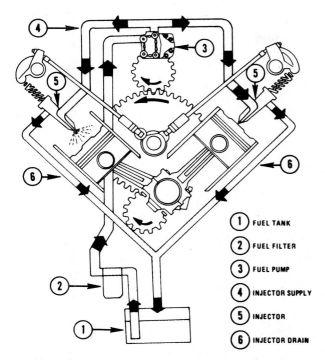

FIGURE 19-2. Typical Cummins pressure time (PT) fuel system. *(Courtesy of Cummins Engine Company Inc.)*

1. FUEL TANK
2. FUEL FILTER
3. FUEL PUMP
4. INJECTOR SUPPLY
5. INJECTOR
6. INJECTOR DRAIN

timum proportion of low-boiling components for easy cold starting and fast warm-up and heavier components, which provide power and fuel economy when the engine reaches operating temperature. Either too high or too low volatility may promote smoking, carbon deposits, and oil dilution due to the effect on fuel injection and vaporization in the combustion chamber. The 10, 50, and 90 percent points and the final boiling point are the principal volatility controls. Diesel engines in automotive, agricultural, and construction service use fuels with a final boiling point approaching 700°F (371°C). Urban buses generally use a fuel with a lower final boiling point to minimize exhaust smoke and odor. Detroit Diesel specifies a 550°F (288°C) maximum end point for its engines in city bus service.

Viscosity

The viscosity of the fuel affects atomization and fuel delivery rate. The viscosity of diesel fuel is normally specified at 100°F (38°C). Fuels for medium-speed and high-speed engines generally lie in the range from 1.4 to 4.3 centistokes viscosity at 100°F (38°C). The lubricating properties of some low-viscosity, low-pour winter fuels can be improved by the addition of 1 percent crankcase oils or lubricity additives. This is important where injection pumps and injectors depend on fuel oil for lubrication and the fuel oil viscosity is below 1.3 centistokes at 100°F (38°C).

Fuels with viscosities over 6 centistokes at 100°F (38°C) are limited to use in slow-speed engines and may require preheating for injection.

Pour Points and Cloud Points

Before the pour point of a fuel is attained, the fuel will become cloudy due to the formation of wax crystals. This usually occurs some 10°F to 15°F (5°C to 8°C) above the pour point and is referred to as the cloud point. Wax crystals may begin to plug fuel filters when the fuel temperature drops to the cloud point. How critical this is in winter operation depends on the design of the fuel system with regard to fuel line bore, freedom from bends, size and location of filters, and degree of warm fuel recirculation as well as the amount and kind of wax crystals.

Additives known as *flow improvers* are being used successfully to improve the fuel fluidity at low temperatures. Flow improvers modify wax crystal growth so that the wax, which forms at low temperatures, will pass through the fine (typically 10 to 20 microns) fuel filter screens. The addition of typically 0.1 percent flow improver can result in satisfactory fuel flow at 15°F (−9°C) colder temperatures than possible with untreated fuel.

Heat Value

The heat value of fuel is a general indication of how much power the fuel will provide when burned. This is obtained by burning the oil in a special device known as a calorimeter. With such equipment, a measured quantity of fuel is burned and the amount of heat is carefully measured in BTUs per pound of fuel. In the metric system the heat unit is known as a joule. To convert the BTU into joules, multiply by 1054.8. (See Fig. 19-7.)

Flash Points

The flash point is determined by heating the fuel in a small enclosed chamber until the vapors ignite when a small flame is passed over the surface of the liquid. The temperature of the fuel at this point is the flash point. The flash point of a diesel fuel has no relation to its performance in an engine nor to its auto ignition qualities. It does provide a useful check on suspected contaminants such as gasoline, since as little as 0.5 percent of gasoline present can lower the flash point of the fuel very markedly.

Shipping, storage, and handling regulations are predicated on minimum flash point categories.

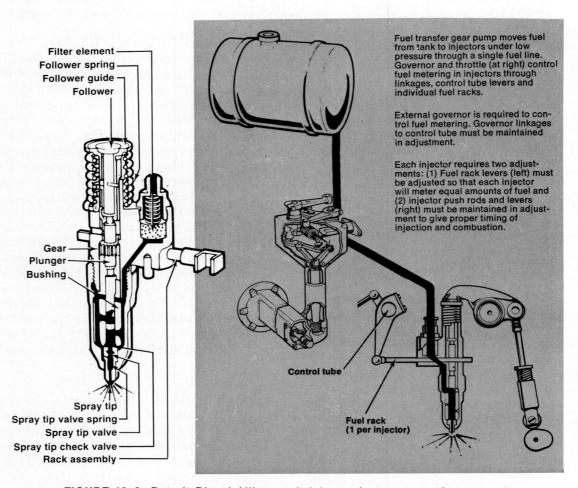

Filter element
Follower spring
Follower guide
Follower

Gear
Plunger
Bushing

Spray tip
Spray tip valve spring
Spray tip valve
Spray tip check valve
Rack assembly

Fuel transfer gear pump moves fuel from tank to injectors under low pressure through a single fuel line. Governor and throttle (at right) control fuel metering in injectors through linkages, control tube levers and individual fuel racks.

External governor is required to control fuel metering. Governor linkages to control tube must be maintained in adjustment.

Each injector requires two adjustments: (1) Fuel rack levers (left) must be adjusted so that each injector will meter equal amounts of fuel and (2) injector push rods and levers (right) must be maintained in adjustment to give proper timing of injection and combustion.

Control tube

Fuel rack
(1 per injector)

FIGURE 19–3. Detroit Diesel Allison unit injector fuel system. *(Courtesy of Detroit Diesel Allison)*

Sulfur Content

Sulfur in diesel fuel can cause combustion chamber deposits, exhaust system corrosion, and wear on pistons, rings and cylinders, particularly at low water-jacket temperatures. Sulfur tolerance by an engine is dependent on the type of engine and the type of service.

A fuel sulfur content above 0.4 percent is generally considered as medium or high whereas fuel sulfur content below 0.4 percent is considered low. Summer grades of commercially available diesel fuel are commonly in the 0.2 to 0.5 percent sulfur range.

Winter grades often have less than 0.2 percent sulfur. Some slow- or medium-speed engines in stationary service are designed to operate on heavy fuels, which have sulfur contents up to 1.25 percent or even higher.

Diesel-engine crankcase oils are formulated to combat various levels of fuel sulfur content. It is important to use the engine builders' recommended crankcase oil quality and oil change intervals, which often relate to the sulfur level of the fuel as well as to other service conditions.

Most diesel fuels are low in sulfur content; by using the engine builders' recommended crankcase

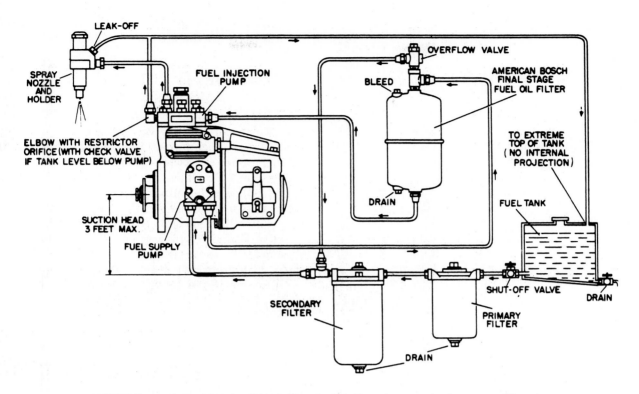

FIGURE 19-4. Typical American Bosch distributor pump fuel system. *(Courtesy of American Bosch, subsidiary of United Technologies Corporation)*

Typical Inspections
Esso Diesel Fuel

	Summer	Winter
Gravity °API	35.9	38.4
Flash °F.	145	137
cSt at 100°F.	3.0	2.4
Cloud °F.	+30	0
Pour °F.	+20	−15
% S	0.3	0.2
Cetane index	49	45
Distillation		
10%	405	365
50%	530	482
90%	633	580
FBP	690	650

Particle settling rate

Product	Gravity °API	Kinematic Viscosity, cs at 100°F. (38°C.)	Time in Hours for a 5-Micron Particle To Settle One Foot
Gasoline	65	0.55	0.7
Kerosene	42	1.3	3
Diesel Fuel	35	2.0-3.5	8 +

FIGURE 19-5. *(Courtesy of Imperial Oil Ltd.)*

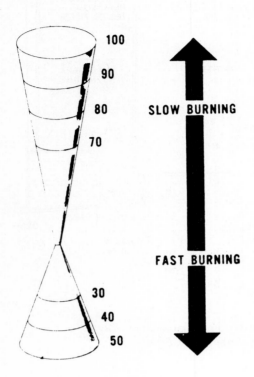

FIGURE 19-6. The octane rating of gasoline and cetane rating of diesel fuel. Note that the higher the octane rating number of the gasoline, the slower it burns; the higher the cetane number of diesel fuel, the faster it burns. *(Courtesy of Ford Motor Company of Canada Ltd.)*

1D Diesel	137,000
2D Diesel	141,800
Gasoline	125,000
Butane	103,000
Propane	93,000

FIGURE 19-7. Heat value per gallon in BTUs.

oil quality, and oil change intervals, there should be no concern with the effects of fuel sulfur.

PART 3   DIESEL FUEL CLASSIFICATION

The American Society for Testing and Materials (ASTM) has set minimum quality standards for diesel fuel grades as a guide for engine operators. These are grade 1-D, which specifies winter fuels, and grade 2-D, which defines summer fuel. These definitions are very broad and the diesel fuels marketed generally will meet one of these definitions.

The Canadian government 3-GP-6d diesel fuel specification recognizes five categories of diesel fuels, with more restrictive standards than ASTM.

The specifications for fuels required by major diesel engine manufacturers are summarized in Figure 19-8.

It is not practical to supply diesel fuel that meets all details of the specifications shown in Figures 19-5 and 19-8 without carrying special grades for certain builder requirements. However, quality requirements such as sulphur, ash, water, sediments, and corrosion ratings are met by diesel fuel that is blended on a seasonal and geographical basis to satisfy anticipated temperature conditions.

The Exxon diesel fuel typical inspections represent fuels supplied in summer across the country and in winter in the more moderate zones. In northern areas the pour, cloud, and low-temperature flow properties are adjusted to the requirements of winter conditions.

Other Grades of Diesel Fuel

In addition to the grades of diesel fuel listed in the accompanying table, there are special fuels for railroad locomotives, buses, the military, etc.

PART 4   DIESEL FUEL OPERATING PROBLEMS

Why Do Diesels Smoke?

White smoke is caused by tiny droplets of unburned fuel. It is usually caused by low engine temperatures and disappears when the engine warms up.

Black smoke is caused by a mechanical defect such as a faulty injector or by engine overload and/or overfuel. Blue/gray smoke is the result of burning lube oil. It indicates a mechanical defect.

Summary of Principal Fuel Specifications of Diesel Engine Makers*

Maker	Cetane No. (Diesel Index)	Sulfur, Mass %	Distillation	Flash Point	Pour Point (Cloud Point)	Viscosity
ALLIS-CHALMERS	40 min.	0.5 max.	90%, 640°F. max.	100°F. or legal	10°F. below ambient	1.4-5.8 cSt at 100°F.
J. I. CASE	50 min.	0.10 max.	10%, 348-400°F. 50%, 445 max. 95%, 465 min. End pt., 550 max.	120°F. min.	−25°F.	1.4 cSt at 100°F. min.
CATERPILLAR PREFERRED	40 35 on PC. engines	0.4 with normal oil change 1.0 max. outside U.S.A.		100°F. or legal	10°F. below fuel temp. (Cloud ambient)	20 cSt max. in pumps and injections not above
CUMMINS REGULAR	40 min.	1 max.	90% 675°F. 100% 725°F.	125°F. or legal	(Below lowest temp. expected)	1.4-5.8 cSt at 100°F.
EMERGENCY FUELS	35 min. above 32°F. 40 min. below 32°F.	2 max.	90% 740°F.	Legal or higher	(Highest temp. fuel-handling equipment will allow)	1.2-13.1 cSt at 100°F.
JOHN DEERE	50 45	0.3 max. 0.5 max.	% Max. 10 50 90 FBP 420 470 550 575°F. 480 540 620 700°F.	110°F. 140°F.	−25(−20) 15(0)	29-33 SSU at 100°F. 33-40 SSU at 100°F.
DETROIT DIESEL ALLISON DIV., GENERAL MOTORS CITY BUSES OTHER APPLNS.	45 min. 45 min. 40 min.	0.30 max. 0.50 max. 0.50 max.	FBP 550°F. max. FBP 675°F. max. FBP 675°F. max.			
MERCEDES-BENZ	45 min.	0.5 max.	Vol. % at 300°C. 60 min. Vol. % at 360°C. 90 min. End pt. 385°C. max.	55°C. min. (Abel-Pensky)		2-8 cSt at 20°C.
MASSEY-FERGUSON	45 min.					
PERKINS	45 min.					
WHITE MOTOR— HERCULES DIV.						
WHITE SUPERIOR	35-50	1.0 max.	10% 475°F. max. 90% 675°F. max.	150°F. min. or legal	20°F. below lowest expected temp.	35-50 SSU at 100°F.
WORTHINGTON	42-47 (above 500 rpm) 37-47 (300-500 rpm)	0.60 max.		150°F. min. or legal	0°F. max.	30-33 SSU at 100°F. min. 37-40 SSU at 100°F. max.

NOTE* Refer to the specifications of individual equipment supplier for detailed requirements not necessarily noted in the summary.

FIGURE 19-8. *(Courtesy of Imperial Oil Ltd.)*

Gravity	Ash, Mass %	Water and/or Sediment, Vol. %	Carbon Residue, Conradson, on 10% btms. Mass %	Cu Corrosion, 3 h at 100°C.	Other
30-40° API	0.02 max.	W&S, 0.10 max.	0.35 max.	No. 3 max.	
42° API min.		W&S, 0.05 max.	0.03 max.	2a max.	Saybolt Color 12-21 Doctor Test—sweet
		BS&W 0.1 max.		No. 3 max.	ASTM D975 and D396 Grades 1, 2, 1-D, 2-D
30-42° API at 60°F. (0.815-0.875 Sp.gr.)	0.02 max.	W&S 0.1 max. (Mass %)	0.25 max. (Ramsbottom)	No. 2 max. after 3 h at 122°F.	
15-57° API at 60°F.	0.05 max.	W&S 0.5 max. (Mass %)	5.0 max. (Ramsbottom or Conradson)	No. 2 after 1 h at 212°F.	
	0.01	0.01	.35	2	ASTM D975 Grade 1-D
	0.01	0.03	.35	3	ASTM D975 Grade 2-D
					ASTM D975 Grade 1-D ASTM D975 Grade Winter 2-D ASTM D975 Grade Summer 2-D
Gravity at 15°C. 0.820-0.855 g/ml	0.02 max.	Sed. 0.01 max. (mass %) Water 0.05 max.	0.05 max.	Not more than slight varnish after 3 h at 50°C.	Calorific value: 10 200 kcal/kg min. Filterability (Hagemann-Hammeric) Summer: 0°C. min. Winter: −15°C. min. Neut. No.: 0.30 mg KOH/g max. Zn corrosion: 4 mg wt loss max. Vanadium: 1 mg/kg max.
					ASTM D975-66T Grades 1-D, 2-D VV-F-800, Grades DF-A, DF-1, DF-2
					ASTM D975-66T Grades 1-D, 2-D VV-F-800, Grades DF-A, DF-1, DF-2
					ASTM D975 Grade 2-D
30° API min.	0.02 max.	BS&W 0.1 max.	0.5 max.		
33-37° API	0.01 max.	W&S 0.20 max.	0.15 max.		High heating value 19 500 Btu/lb min.

FIGURE 19-8. (Continued)

Diesel engines producing smoke also tend to produce objectionable odors.

Dirt and Water (Ice)

Diesel fuel line filters and injector filters are in the 10-2 micron range. Fuel tankcaps, hoses, and nozzles must be kept clean, and water must be minimal to prevent plugging of these filters with dirt or emulsions.

Ice and water are usually formed from condensation or from dissolved water that forms cloud ice at prolonged low temperatures and may cause filter plugging. Small quantities of alcohol can be used to alleviate fuel line or filter ice-plugging.

Diesel Fuel Handling

One of the most critical parts of the diesel engine is the fuel injection system. Many of the components of this system are highly finished and operate with clearances as small as 0.0001 inch (.00254 mm). Since any foreign material that finds its way into the system can damage these parts and seriously impair engine performance, clean fuel is essential to the proper operation of the engine.

Producers use reasonable care to ensure cleanliness in the production, storage, and shipment of diesel fuel. However, the consumer's storage facilities often contain small quantities of fine rust particles and condensed water from humid air. These contaminants become well dispersed in the diesel fuel during the filling operation.

The best way to protect a diesel engine from such contaminants is simply to allow them to settle out in the storage tank. The time required for this settling will vary with the size of tank used. The table (Fig. 19–5) indicates how fast a very small particle (5 microns in size) will settle in diesel and other fuels. It is evident from these settling times that diesel fuel should be allowed to stand for a day or at least overnight before it is removed from storage vessels. A properly designed storage tank is essential if clean fuel is to be delivered to the engine. The impurities that settle out must remain separated from the fuel. To facilitate this, the following practices should be observed:

1. The tank should be sloped slightly, and a drain positioned at the low end. Through this drain, settled impurities can be removed easily.

2. The fuel suction outlet should be located well above any possible level of settled impurities so that only clean fuel is pumped.

3. A filter should be provided in the suction line to remove any impurities that have not had time to settle.

4. The storage tank itself and any interior tank coating should be insoluble in the fuel and nonreactive with the fuel.

5. The tank should be as large as possible, consistent with the user's normal fuel requirements since the use of a number of small containers increases the chance of contamination.

6. Proprietary additives should not be blended into the diesel fuel without first consulting the fuel supplier.

Clean fuel in the storage tank does not ensure clean fuel in the engine. Care must be taken to minimize contamination when the engine fuel tanks are filled. Also, engine tanks should be filled at the end of each day to reduce the amount of water that will condense in the tanks overnight. The engine tank should be provided with a drain, and the engine tank discharge line should have a filter. Both the drain and the filter should be serviced regularly to eliminate water and dirt that have accumulated and thereby ensure that the filter will not become "plugged."

PART 5   FUEL SUPPLY SYSTEM

The fuel supply system includes a fuel tank, supply pump, and fuel filters. The fuel supply system is similar in most diesel fuel systems.

Fuel Tanks

The fuel supply tank is constructed of steel or aluminum and may be round or rectangular in shape. Many units have more than one fuel tank. In this case the operator is able to select which tank is to supply the fuel by a fuel control valve. The fuel tank is provided with a vented filler cap, which allows the tank to "breathe" as the fuel expands and contracts due to temperature change and as fuel is consumed.

Also provided is a fuel supply line fitting to which the fuel line is connected. A second fuel line fitting may also be provided for fuel systems that use an injector "leak off" fuel return line. The fuel tank is usually equipped with a fuel shut-off valve at the fuel outlet fitting. Some fuel tanks have a water trap that is equipped with a drain tap. Water from condensation in the fuel tank collects in the water trap where it is periodically drained.

It is of the utmost importance that diesel fuel be kept absolutely clean. This means that the entry of dirt, moisture, and other foreign material must be kept out of the system when handling fuel or fuel system components.

The fill caps and fill nozzles should be cleaned before removal and refueling.

Fuel Supply Pumps

A fuel supply (or transfer) pump is needed in the fuel system to provide an adequate supply of fuel at specified pressure from the fuel tank to the fuel injection pump or the unit injectors via the necessary fuel filters. Several types of fuel supply pumps are being used. Camshaft-operated diaphragm pumps, gear pumps, plunger pumps, and hand-priming pumps are used, depending on the type of fuel system.

Diaphragm Pump

The mechanically operated fuel pump consists of a lever-operated spring-loaded diaphragm, an inlet valve, and an outlet valve. The lever is operated from an eccentric on the engine camshaft. A push rod may be used between the eccentric and the lever. The lever pivots on a pin to provide rocker arm action. The other end of the lever is connected to the diaphragm link with a sliding or slotted connection. This allows the diaphragm to be pulled by the lever. (See Fig. 19-9 and 19-10.)

When the diaphragm is lifted, it creates a low-pressure area in the pump. Atmospheric pressure on the fuel in the tank causes fuel to be forced into the low-pressure area in the pump past the one-way inlet valve. Spring pressure keeps the outlet valve closed during the pump intake stroke.

When the lever moves down, the diaphragm spring forces the diaphragm down and pressurizes the fuel. The pressure closes the inlet valve and opens the outlet valve.

Spring pressure forces the diaphragm down to supply the system with sufficient fuel for all operating conditions.

Fuel pressure is dependent on diaphragm spring pressure.

Some diaphragm pumps are equipped with a lever that is connected to the diaphragm so that the pump can be operated manually. This allows the operator to prime the fuel system manually and purge

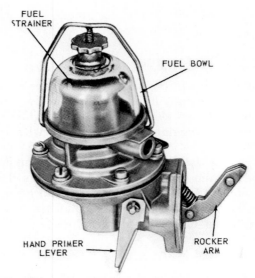

FIGURE 19-9. Diaphragm type of fuel transfer pump with hand-priming lever. *(Courtesy of Deere and Company)*

all air from the system before starting the engine after fuel system service.

Plunger Pump

The plunger type of supply pump is operated from a cam on the camshaft of the fuel injection pump. As the injection pump camshaft rotates, the plunger follows the camshaft profile by moving up and down

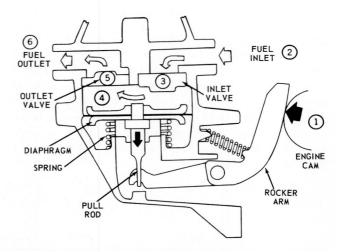

FIGURE 19-10. Typical diaphragm pump operation. *(Courtesy of Deere and Company)*

in its bore. The plunger spring keeps the plunger roller against the cam and controls output pressure. (See Fig. 19–11.)

As the plunger moves down, a low-pressure area is created in the pump body fuel chamber. This causes the outlet valve to close and the inlet valve to open. Atmospheric pressure in the fuel tank forces fuel to flow in past the inlet valve and fills the chamber above the plunger. At the same time fuel in the chamber below the plunger is forced out to the filter and then to the injection pump. As the injection pump cam continues to turn, the plunger moves up, creating pressure on the fuel above the plunger. This closes the inlet valve and opens the outlet valve, forcing fuel through the outlet valve to the chamber below the plunger. During this portion of plunger travel, fuel is simply transferred from above the plunger to the area below the plunger. This cycle of events is then repeated as the injector pump camshaft continues to rotate.

Gear Pump

The gear type of fuel supply pump is driven from the fuel injection pump camshaft or from the air blow drive on two-stroke-cycle engines. (See Fig. 19–12.)

Fuel enters the pump body on the inlet side of the gears when the rotation of the gears creates low pressure in this area. Atmospheric pressure in the fuel tank forces fuel out of the tank through the line to the fuel supply pump. Fuel trapped between the teeth of each gear is carried around to the outlet side

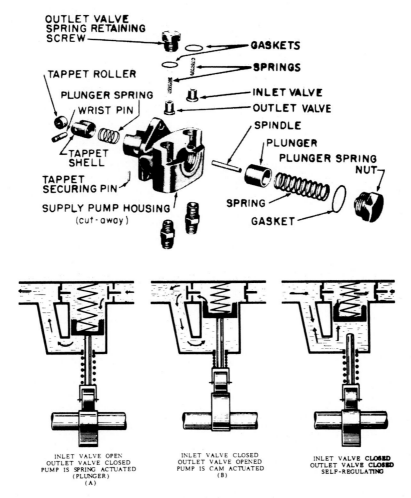

FIGURE 19–11. Plunger type of fuel transfer pump: (A) exploded view. (B) Operation. *(Courtesy American Bosch, subsidiary of United Technologies Corporation)*

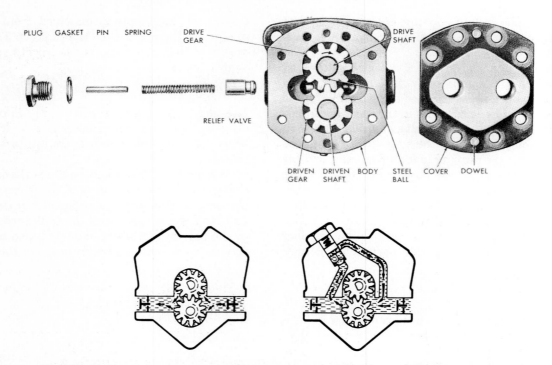

PLUG   GASKET   PIN   SPRING   DRIVE GEAR   DRIVE SHAFT

RELIEF VALVE

DRIVEN GEAR   DRIVEN SHAFT   BODY   STEEL BALL   COVER   DOWEL

FIGURE 19–12. Gear type of supply (transfer pump). *(Courtesy of Detroit Diesel Allison [A], American Bosch, subsidiary of United Technologies Corporation [B])*

of the housing. As the teeth of the two gears mesh, fuel is forced out from between the gear teeth and out of the fuel outlet to the filter.

A passage from the outlet is connected to the pressure-regulating valve. A spring-loaded valve is forced back against spring pressure as pressure builds up. When output pressure increases sufficiently, the relief valve is forced back far enough to open a connecting passage to the inlet side of the pump. Spring pressure on the pressure-regulating valve thereby controls pump output pressure.

Vane Pump

The vane-type fuel supply pump consists of a stationary liner or race, a set of spring-loaded vanes or blades, and a slotted drive rotor. The vanes fit into the slots of the rotor and are pushed away from center into contact with the outer race or liner by the springs. Since the liner is offset or eccentric in relation to the rotor shaft, rotating the rotor causes the vanes to move in and out of their slots. This movement changes the volume between the vanes. (See Fig. 19–13.)

Turning the rotor causes an increase in volume between the vanes positioned over the fuel inlet passage. Increasing the volume creates a low-pressure area between the vanes, allowing atmospheric pressure in the fuel tank to force fuel into the pump low-pressure area. As the rotor continues to turn, the vanes pass the inlet passage. Continued rotation forces the vanes back into their respective slots in the rotor. This pressurizes the fuel trapped between the vanes. At this point, rotor rotation has positioned this area over the fuel outlet passage and fuel is forced out of the pump.

This action takes place in turn between each pair of vanes as the rotor is forced to turn. The outlet passage connects to a pressure regulator valve, which regulates pump output pressure in the same manner as the gear pump pressure regulator. Output pressure is controlled by regulator valve-spring pressure.

The vane pump is usually driven by the fuel injection pump and may be part of the injection pump itself.

Priming Pumps

Some fuel systems are equipped with a hand-operated priming pump. This is a plunger type of pump designed to prime the system manually with fuel to

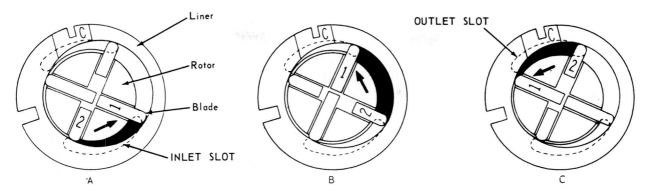

FIGURE 19-13. Vane type of fuel transfer pump, showing pump operation. *(Courtesy of Deere and Company)*

eliminate air from the system before starting the engine after any fuel system has admitted air. (See Fig. 19-14.)

When the plunger is lifted, it creates a low-pressure area in the plunger barrel. Atmospheric pressure acting on the fuel in the fuel tank forces fuel into the plunger barrel by opening the check valve on the inlet side. The outlet check valve is closed by this action. As the plunger is forced down in the barrel, the pressure closes the inlet valve and opens the outlet valve, permitting pressurized fuel to flow to the fuel supply line of the injection system. This pumping action is repeated until high pressure on the plunger is felt, indicating that the system is fully primed and all air is removed.

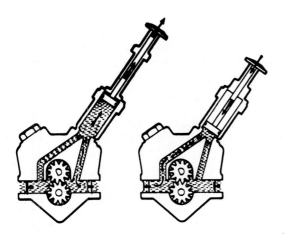

FIGURE 19-14. Hand-priming pump attachment for gear type of fuel transfer pump. At left, plunger is raised to fill cylinder. Right shows plunger pushed down, discharging fuel past the outlet check valve. *(Courtesy of American Bosch, subsidiary of United Technologies Corporation)*

Most diaphragm-type supply pumps have a hand-operated lever provided to operate the diaphragm as a priming pump.

PART 6   FUEL FILTERS

Effective filtering of the fuel oil is most important for long troublefree fuel injection system operation. Dirt is the worst enemy of the fuel injection equipment. Even the most microscopic particles of abrasive foreign material can destroy the finely machined and lapped parts of the injection system. Clearances between moving parts of the injection system can be as low as from 0.000007 to 0.000008 inch (0.0001778 mm to 0.0002032 mm). Dirt of such dimensions that will affect such close tolerances is often suspended in the atmosphere. As the fuel tank "breathes" this air as a result of fuel level fluctuations, this type of dirt enters the fuel system even though very careful handling of the fuel may prevent entry of dirt through other means. (See Fig. 19-15 to 19-17.)

For these reasons a very efficient filtering system that will remove these impurities and still allow free fuel flow is necessary. Several types of filters are used for this purpose: the strainer type, the surface type, and the depth type.

The porosity of the filtering material will determine the size of impurities that the filter will remove. This is stated in microns. One micron is 0.000039 inch in size or 0.000001 m. The clearance between an injector plunger and barrel may be as little as 2 microns. If the plunger is held in the hand and warmed, it will expand enough so that it cannot be inserted into the barrel until the temperature of the two units is equalized.

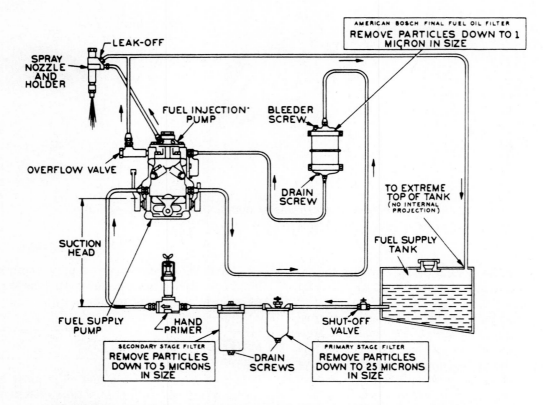

LEAK-OFF

SPRAY NOZZLE AND HOLDER

FUEL INJECTION PUMP

BLEEDER SCREW

AMERICAN BOSCH FINAL FUEL OIL FILTER
REMOVE PARTICLES DOWN TO 1 MICRON IN SIZE

OVERFLOW VALVE

DRAIN SCREW

TO EXTREME TOP OF TANK
(NO INTERNAL PROJECTION)

FUEL SUPPLY TANK

SUCTION HEAD

FUEL SUPPLY PUMP

HAND PRIMER

SHUT-OFF VALVE

SECONDARY STAGE FILTER
REMOVE PARTICLES DOWN TO 5 MICRONS IN SIZE

DRAIN SCREWS

PRIMARY STAGE FILTER
REMOVE PARTICLES DOWN TO 25 MICRONS IN SIZE

FIGURE 19-15. Three-stage fuel-filtering system showing primary, secondary, and final filters. *(Courtesy of American Bosch, subsidiary of United Technologies Corporation)*

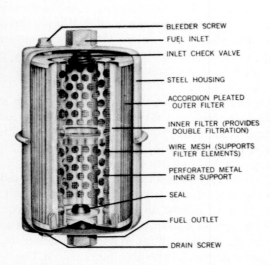

BLEEDER SCREW
FUEL INLET
INLET CHECK VALVE
STEEL HOUSING
ACCORDION PLEATED OUTER FILTER
INNER FILTER (PROVIDES DOUBLE FILTRATION)
WIRE MESH (SUPPORTS FILTER ELEMENTS)
PERFORATED METAL INNER SUPPORT
SEAL
FUEL OUTLET
DRAIN SCREW

FIGURE 19-16. Cross-sectional view of final-stage filter, showing construction. *(Courtesy of American Bosch, subsidiary of United Technologies Corporation)*

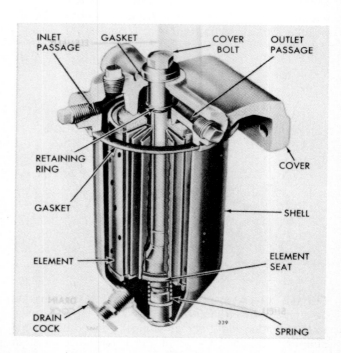

INLET PASSAGE
GASKET
COVER BOLT
OUTLET PASSAGE
RETAINING RING
GASKET
COVER
SHELL
ELEMENT
ELEMENT SEAT
DRAIN COCK
SPRING

329

FIGURE 19-17. Detroit Diesel 92 engine fuel filter. *(Courtesy of Detroit Diesel Allison)*

The strainer in the fuel system may be simply a finely woven screen or it may be a series of stacked disc-type screens. This screens out the larger impurities in the fuel.

The surface type of filter material is usually of the paper element accordion bellows design. Fuel oil passing through the paper element leaves impurities deposited on the element surface.

The depth type of filter is of the cotton fiber-filled canister element type or the "bathroom tissue" type of rolled paper element. Impurities are collected at various depths in this type of filter as fuel oil passes through.

Filters are further classified as primary and secondary or final filters when they are connected in series. Series-connected filters cause fuel to pass through the primary filter first, then to the secondary and final filter. All the fuel must first pass through the primary filter before it can go to the secondary filter. In some cases, fuel filters are arranged in parallel. Fuel is able to flow through both filters at the same time. If one filter becomes plugged, the other could still function. Filters may be of the spin-on/throw-away type or of the replaceable element canister type.

PART 7   SELF-CHECK

1. List four types of fuel injection systems.
2. Define the following diesel fuel terms:

    (a) cetane number

    (b) volatility

    (c) viscosity

    (d) pour point and cloud point

    (e) heat value

    (f) flash point

    (g) sulfur content

    (h) diesel fuel classification

    (i) ignition delay period

3. State the causes of white, blue, and black exhaust smoke.
4. List the major components of a diesel fuel supply system as found on a diesel engine.
5. Describe the basic operation of the following fuel supply or transfer pumps: diaphragm, gear, vane, plunger, and priming pumps.

Chapter 20

Diesel Fuel Injection

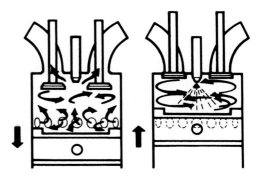

FIGURE 20-1. Action in open combustion chamber, direct injection four-stroke-cycle engine.

Performance Objectives

After thorough study of this chapter, the appropriate training models and service manuals, you should be able to do the following:

1. List four requirements that a diesel fuel injection system must meet.

2. Define the ignition delay period.

3. List five combustion chamber designs and briefly describe each one.

4. Describe the operation of the hole type and the pintle type injection nozzles.

5. List the causes and effects of:

    (a) nozzle opening pressure too high

    (b) nozzle opening pressure too low

    (c) incorrect nozzle spray pattern

    (d) nozzle dribble or leakage

6. Remove, clean, test, adjust, and install fuel injection nozzles according to manufacturer's specifications.

7. Complete the self-check questions with at least 80 percent accuracy.

PART 1　FUEL INJECTION PRINCIPLES

Every diesel fuel injection system must have the ability to inject the fuel into the cylinders at sufficiently high pressures to assure good atomization and vaporization of the fuel in the cylinders. It must accurately meter the quantity of fuel injected to correspond with the engine's speed and load. It must do this at exactly the correct time in relation to piston position and crankshaft angle. It must inject this fuel over an accurately determined period of time to provide the correct rate of pressure rise in the cylinder and to maintain as much as possible a pressure level in the cylinder to provide the needed power and torque output. This requires a correct beginning, duration, and ending of injection related to engine speed.

Engine efficiency and the combustion process are most favorable when the amount of exhaust smoke is minimal. If combustion is incomplete or too much fuel is injected, the engine "smokes." This may result in overheating, with subsequent engine damage, such as piston and cylinder damage and oil dilution.

The preparation of the fuel in the cylinder for proper combustion takes a certain amount of time. The period from the time injection begins until the time when combustion begins is known as the ignition delay period. The injected amount of fuel

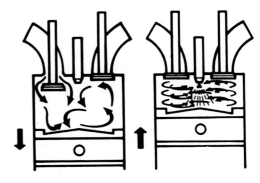

FIGURE 20-2. Direct injection, open combustion chamber action in cylinder of two-stroke-cycle engine.

317

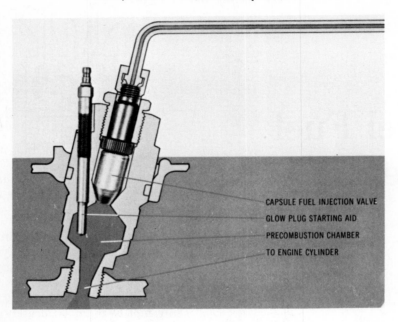

FIGURE 20-3. Cross-sectional view of Caterpillar injection valve and precombustion chamber. *(Courtesy of Caterpillar Tractor Company)*

should be small during this period, which is about 0.001 second. The objective is to obtain (as much as possible) constant pressure in the cylinder by means of a controlled rate of injection and combustion. This is not possible in high-speed diesel engines but can more nearly be achieved in large slow-speed engines. In direct injection diesel engines the injected amount of fuel should be generally constant per degree of crank angle for this reason.

Different manufacturers achieve these results in different ways. Some use an injection pump with injectors connected to each pumping element by thick-walled fuel injection lines of exactly equal inside diameter and length. In this system a separate pumping element, cam-operated, is used for each cylinder. The pumping elements may be contained in a single-pump housing or in separate pump housings, one for each cylinder. The Bosch, Caterpillar Scroll and CAV are examples of this system.

Other systems combine the high-pressure pumping element and the injector, one for each cylinder. One system, the Detroit Diesel unit injector system, uses a similar pumping and metering element as the Bosch and CAV plunger and barrel systems. Another, the Cummins pressure time system, uses a plunger and cup in each injector as the pumping element.

Other systems used include a plunger and barrel with a metering sleeve as the pumping and metering element. In the Caterpillar sleeve-metering

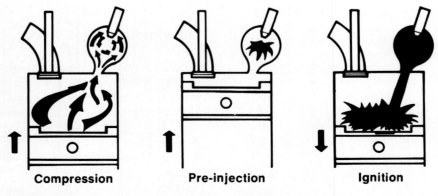

FIGURE 20-4. Prechamber indirect injection action.

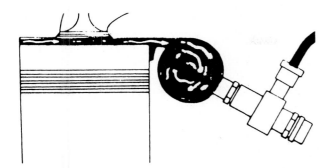

FIGURE 20-5. Swirl chamber action with indirect injection.

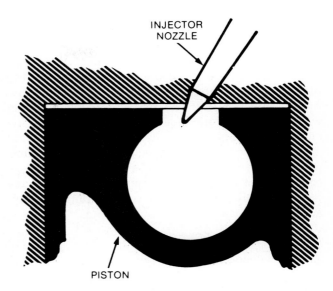

FIGURE 20-7. M type of combustion chamber.

system, each cylinder is provided with a pumping element with all the elements contained in one pump housing. In the American Bosch distributor system only one pumping element and metering sleeve are provided, with high-pressure fuel delivered to each cylinder by a distributing head and a rotating plunger.

PART 2 COMBUSTION CHAMBER DESIGN

The principle of fuel injection is to inject a metered quantity of fuel into each cylinder at exactly the correct time in a properly atomized and definite spray pattern for the type of combustion chamber used, in such a way as to provide the most efficient means of combustion possible. Regardless of cylinder pressures, which can be quite high, fuel injection pressure must be considerably higher to inject fuel properly into the pressurized cylinder. Fuel injection pressures may be as high as 20,000 pounds per square inch (137,900 kPa) in some systems.

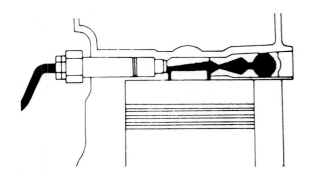

FIGURE 20-6. Power cell (energy cell) type of combustion chamber and injection.

To achieve this purpose, different manufacturers have produced several different types of fuel systems and combustion chamber designs.

Combustion chamber designs include the open, precombustion, turbulence, energy cell, and M types.

In order to burn all the fuel injected into the cylinder, there must be adequate atomization and vaporization of the fuel. There must also be enough air around each fuel particle to allow complete combustion.

The injectors spray fuel into the cylinders in a very fine mist or spray under very high pressures (up to 20,000 pounds per square inch or 137,900 kPa). This provides the necessary vaporization; turbulence is provided by the combustion chamber design.

Open Combustion Chamber

In the open combustion chamber design, fuel is injected directly into the combustion chamber. The cylinder head and valves provide a flat cover or top for the chamber. The design of the piston compresses the air in the cylinder. Various piston head designs are used to accomplish this action.

Piston designs for the open combustion chamber include a flanged-domed design, a dished design with a raised center area, and other irregular shaped designs. Advantages claimed for the open combustion direct injection design include good fuel economy and simplicity of design.

Precombustion Chamber

The precombustion chamber design consists of two interconnected chambers. The smaller chamber is located in the cylinder head or block close to the top of the cylinder. The fuel injector is mounted so that fuel is injected into the precombustion chamber where combustion begins. The precombustion chamber is connected to the area above the piston and, as fuel injection and combustion continue, pressure is forced through the connecting passage to the top of the piston. The precombustion chamber design allows the use of a wider range of fuels and provides very smooth combustion.

Turbulence Chamber

The turbulence chamber resembles the precombustion chamber in that a separate turbulence chamber is connected to the area above the piston. There is very little room above the piston when it is at TDC. As the piston moves up on the compression stroke, compressed air is forced into the turbulence chamber. The turbulent air promotes good mixing of fuel and air and provides good combustion.

Energy Cell

The energy cell is a combination of the precombustion chamber and the turbulence chamber. It is also known as the Lanova combustion chamber. Combustion takes place mostly in the main figure-eight chamber. This design depends on a great deal of turbulence to provide the necessary mixing of air and fuel and the distribution of the air-fuel mixture in the cylinder. Most of the combustion chamber is in the direct path of the intake and exhaust valves. Turbulence in this design is dependent on thermal action rather than piston speed as in the case of the open chamber.

M-Type Chamber

The M type of combustion chamber consists of a spherical chamber in the head of the piston. This chamber has a small opening at the top. The fuel injector is positioned so that fuel will be injected into this chamber. This type of chamber has the advantage of eliminating the well-known diesel knock. It is also capable of using a variety of different fuels such as diesel fuels, kerosene, gasoline, etc.

Special Intake Valve

Another method used to create a swirling action of incoming air is a specially designed intake valve. This valve has a vane type of protrusion on the bottom of its head deflecting incoming air into a swirling action when the valve is open. This helps promote turbulence. One disadvantage of this design is that the valve is not allowed to turn in the valve guide and therefore the even wear distribution of rotating valves is lost.

Regardless of the type of combustion chamber used on any given engine, it should be remembered that the combustion chamber and injector are matched and work as a team. The combination of combustion chamber design (shape), engine compression ratio, fuel injection spray pattern, and fuel injection pressure is very carefully selected by the engine and fuel system manufacturers for good combustion, power, fuel economy, and low exhaust emissions.

PART 3   COMBUSTION KNOCK

How smoothly a diesel engine operates depends to a large extent on the rate of pressure rise in the engine's cylinders during combustion, more specifically the rate of pressure increase in pounds per square inch (kilopascals) per degree of crankshaft rotation. The higher the rate of pressure increase, the rougher the engine. Engine roughness increases proportionally with an increasing rate of pressure rise.

The ignition delay period is a factor in the degree of roughness or combustion knock of an engine. The longer the delay period, the greater the amount of fuel that is injected into the cylinder before ignition occurs. The larger the amount of fuel injected before ignition begins, the greater the engine knock. When ignition finally occurs after the delay period, the large quantity of fuel in the cylinder burns very rapidly, resulting in extremely high cylinder pressures. The sudden violent combustion and consequent rapid rise in cylinder pressure causes the vibrations and noise known as combustion knock or diesel knock.

Factors that decrease the ignition delay period are:

1. Higher cetane number of diesel fuel

2. Increased charge air pressure in the cylinder at the time of fuel injection

3. Increased turbulence of air in the combustion chamber

4. Increased temperature of charge air in the cylinder at the time of fuel injection

5. Reduced size of fuel droplets injected into the cylinder.

PART 4   INJECTION NOZZLES

The fuel injection nozzle directs the metered quantity of fuel received from the injection pump into the combustion chamber of the engine. Fuel is discharged in a finely atomized pattern in such a manner that it is thoroughly mixed with compressed air in the combustion chamber.

These nozzles are held by a cap nut in a device known as a nozzle holder, which holds the nozzle in position in the cylinder head of the engine. In addition, the holder conveys the fuel from the injection tubing to the nozzle. The nozzle holder also contains the mechanism for controlling the nozzle opening pressure and leakoff ducting. A copper gasket is used between the injector and the cylinder head to seal compression and promote maximum heat transfer.

For maximum power and economy, it is essential that the fuel be consumed completely without excessive smoke or hydrocarbons, carbon monoxide or oxides of nitrogen. This is accomplished by dispersing the fuel in a mist of infinitesimally fine droplets. In that way the surface area of each droplet is greatly reduced and maximum combustion can result.

To obtain these characteristics, the design details of the nozzle are dependent largely on features of the engine, such as its speed, configuration of combustion chamber, size of engine, compression ratio, and intended use.

The following description of injection nozzle design and operation is typical of most diesel injection systems using an engine-driven injection pump with external high-pressure fuel injection lines used to deliver fuel to the injectors. This includes such systems as United Technologies Diesel Systems (formerly American Bosch), Robert Bosch, CAV, Diesel Systems Group, Stanadyne Inc. (formerly Roosa Master), and Caterpillar. The Cummins, PT, and Detroit Diesel unit injectors are described in the chapters dealing with those fuel injection systems.

Nozzle Operation and Design*

Nozzles

The Bosch injection nozzle consists of a nozzle body and needle valve. Nozzle body and valve are made of high-grade steel. The valve is a lapped fit in its body. These two parts are considered as a unit for replacement purposes.

*This section courtesy of Robert Bosch Canada Ltd.

Operation

The nozzle is operated by fuel pressure. The pressure generated by the injection pump acts on the exposed annular area of the nozzle valve and lifts it from its seat as soon as the force of the pressure spring in the nozzle holder is exceeded. The fuel is then injected through the nozzle orifices into the combustion chamber. (See Fig. 20-8.)

During injection, the fuel takes the following path: fuel injection tubing to the fuel passage of the nozzle holder (annular groove) inlet passages, pressure chamber, and spray orifice(s) of the nozzle to the combustion chamber. Fuel leaking past the valve stem is returned to the fuel tank via the leak-off connection on the nozzle holder and a return line.

After injection of the fuel delivered by the injection pump, the pressure spring again forces the valve back on its seat via the spindle and the valve stem. The injector is then closed for the next pressure stroke. No fuel dribble should occur, as might happen if the nozzle showed excessive wear.

Designs and Application of Injection Nozzle

There are two main types:

hole-type nozzles—for open chamber engines (also known as direct injection engines)

pintle nozzles—for precombustion chamber, turbulence chamber, and air-cell engines

However, numerous deviating forms exist within the two main types, based on the diversity of engines.

Hole-Type Nozzles

The valve of the hole-type nozzle (Fig. 20-10) has a cone at its end, which serves as seat. There are single-hole and multihole nozzles. Single-hole nozzles have only one orifice, which may be drilled centrally or laterally. In the case of multihole nozzles, the orifices form an angle—the spray angle (up to 180°). To obtain optimum fuel distribution in the combustion chamber, up to 12 orifices (usually symmetrical) are provided. Orifice diameter and orifice length affect shape and penetration of the spray. Conventional nozzle designs are available with spray orifice diameters starting at 0.2 mm and increasing in steps of 0.02 mm. The different sizes of hole-type nozzles are distinguished by the letters *S, T, U, V,* and *W.*

Hole-type nozzles are used on engines with direct injection. The nozzle opening pressure usually amounts to between 2100 to 3600 psi (14,479 kPa to 24,822 kPa).

INJECTOR

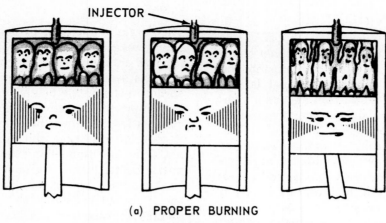

(a) PROPER BURNING

(Fuel Charge Ignites Early and Burns Evenly
To Overcome Knocking)

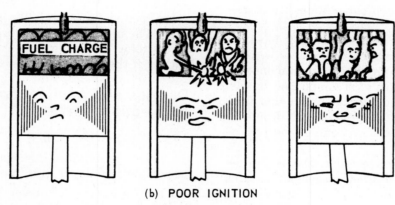

(b) POOR IGNITION

(Ignition Of Fuel Charge Is Delayed,
Followed By A Small Explosion)

FIGURE 20-8. Knock in diesel engines. In diesel engines, knock is due to the
fuel igniting too slowly. It should start to burn almost as soon as it is injected.
(A). If there is much delay, a fuel buildup results, which burns with explosive
force (B) and causes knocking.

Cooled Hole-Type Nozzles

In engines operating on heavy fuels, the temperature on the nozzles may become so high that they require cooling. (See Fig. 20-11.)

The nozzle body of these nozzles is equipped with a passage for the fuel inlet while the two others serve for coolant inlet and outlet. The lower end of the body has a double thread, which is closed to the outside by a cooling jacket. The coolant is discharged from the feed duct of the nozzle holder into the inlet passage of the nozzle, then through one thread of the double thread into the annulus; from there it is forced through the outlet passage into the outlet duct of the nozzle holder.

Oil as well as oil-water emulsions that do not attack steel can serve as coolants.

Pintle Nozzles

The valve of the pintle nozzle has a specially shaped pintle at its end, which projects into the spray hole of the nozzle body with a slight clearance. The spray pattern can be changed as needed by different dimensions and profile of the pintle. Furthermore, the pintle keeps the spray orifice free from carbon deposits. (See Fig. 20-12 to 20-14.)

Pintle nozzles are used in precombustion chamber and turbulence chamber engines. In these engines, the fuel is prepared mainly by the air turbulence supported by a suitable shape of the

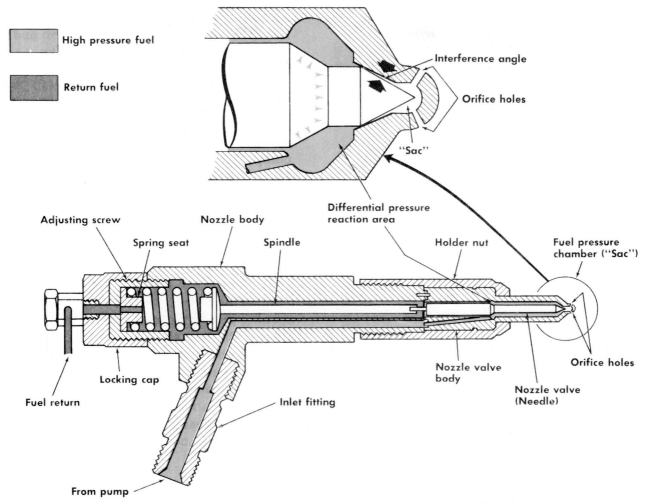

High pressure fuel

Return fuel

Interference angle

Orifice holes

"Sac"

Differential pressure
reaction area

Fuel pressure
chamber ("Sac")

Adjusting screw

Spring seat

Nozzle body

Spindle

Holder nut

Locking cap

Fuel return

Inlet fitting

From pump

Nozzle valve
body

Nozzle valve
(Needle)

Orifice holes

FIGURE 20-9. Typical nozzle operation. *(Courtesy of Deere and Company)*

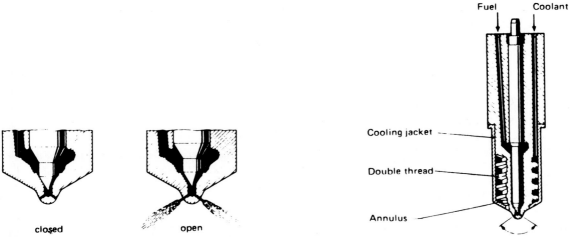

closed

open

FIGURE 20-10. Typical spray pattern for Robert Bosch hole-type nozzle as used in open combustion chamber direct injection design. *(Courtesy of Robert Bosch Canada Ltd.)*

Fuel   Coolant

Cooling jacket

Double thread

Annulus

FIGURE 20-11. Cooled hole-type nozzle as used on engines burning heavier fuel, which causes injectors to be subjected to more heat. *(Courtesy of Robert Bosch Canada Ltd.)*

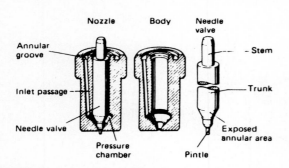

FIGURE 20-12. Injector nozzle components, pintle-type nozzle. *(Courtesy of Robert Bosch Canada Ltd.)*

FIGURE 20-13. Inward-opening pintle nozzle with a narrow spray pattern for precombustion chamber engines. *(Courtesy of Robert Bosch Canada Ltd.)*

FIGURE 20-14. Inward-opening nozzle used with precombustion chamber-type engines. Cone-shaped pintle tip provides slightly wider spray pattern than Figure 20-13. *(Courtesy of Robert Bosch Canada Ltd.)*

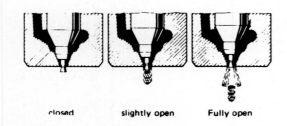

FIGURE 20-15. Inward-opening throttling-type pintle nozzle providing only a small amount of fuel at the start of injection, allowing progressive opening of the nozzle (throttling effect). *(Courtesy of Robert Bosch Canada Ltd.)*

spray. For pintle nozzles, the nozzle opening pressure is usually between 1100 psi and 1800 psi (7584 kPa to 12,411 kPa).

Throttling Pintle Nozzles

The throttling pintle nozzle is a pintle nozzle with special pintle dimensions. (See Fig. 20-15.)

A pilot injection is obtained by the shape of the pintle tip. The valve frees only a very narrow annular orifice during opening, allowing only little fuel to pass (throttling effect). With further opening (produced by a pressure increase), the annular orifice area is increased and the main portion is injected only near the end of the valve lift. Normally, combustion and engine operation become smoother with the throttling pintle nozzle because the pressure in the combustion chamber increases slowly. The desired throttling effect is obtained by the pintle configuration together with the characteristics of the pressure spring (in the nozzle holder) and the clearance in the throttling gap. (See Fig. 20-16 to 20-22.)

PART 5   INJECTOR PROBLEMS

Effects of Faulty Injector Operation

1. Opening pressure too high: tends to decrease the amount of fuel injected; tends to retard injection timing; loss of power and efficiency.

2. Opening pressure too low: decreases fuel atomization; may increase amount of fuel injected; can cause nozzle dribble, exhaust smoke, and loss of efficiency.

3. Incorrect spray pattern: exhaust smoke; carbon deposits; loss of power.

4. Leaky nozzle, nozzle dribble: exhaust smoke; detonation; loss of power; crankcase oil dilution; carbon deposits.

Causes of Faulty Injector Operation

1. Opening pressure too high: incorrect pressure setting; nozzle valve sticking; plugged nozzle spray holes.

2. Opening pressure too low: incorrect pressure setting; faulty pressure spring; nozzle valve sticking.

3. Incorrect spray pattern: nozzle valve eroded; plugged nozzle spray holes; nozzle valve sticking; dirt in nozzle.

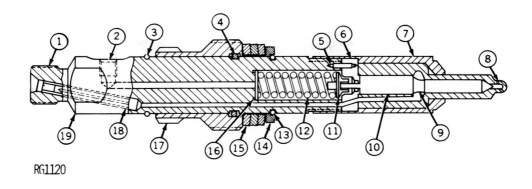

RG1120

1—Fuel Inlet
2—Leak-Off Connection
3—Snap Ring
4—O-Ring
5—Dowel Pin
6—Intermediate Plate

7—Nozzle Retaining Nut
8—Nozzle Orifice
9—Annulus
10—Nozzle Valve
11—Spring Seat
12—Spring

13—Snap Ring
14—Retainer
15—Conical Washers
16—Shims
17—Gland Nut
18—Edge-Type Filter
19—Holder

FIGURE 20-16. Cross-sectional view of Bosch KDEL-21mm injection nozzle. *(Courtesy of Deere and Company)*

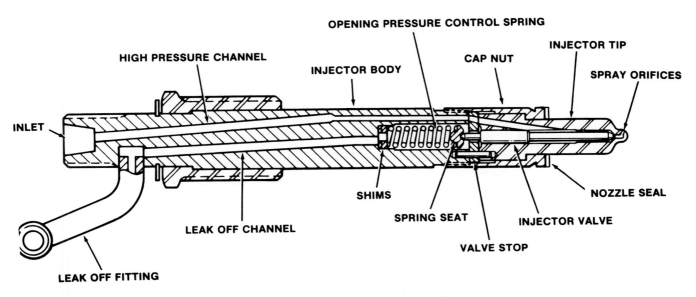

FIGURE 20-17. Cross section of shim adjustable nozzle. *(Courtesy of J. I. Case, a Tenneco company)*

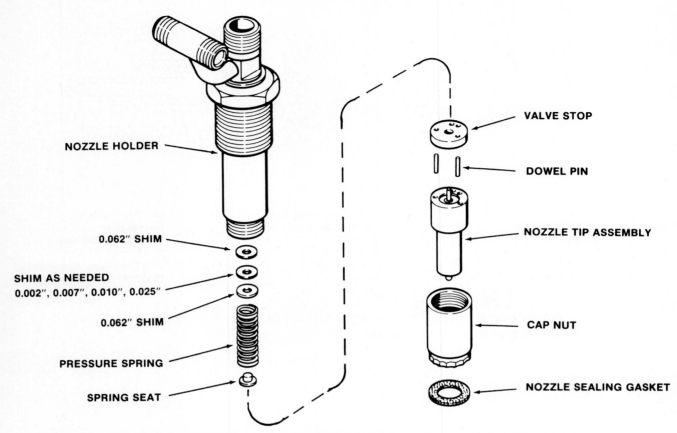

FIGURE 20-18. Exploded view of threaded injection nozzle holder and nozzle. *(Courtesy of J. I. Case, a Tenneco company)*

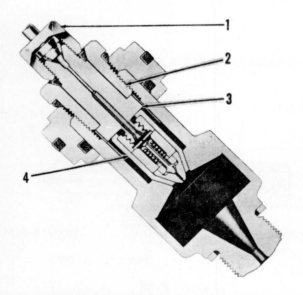

FUEL INJECTION VALVE CROSS-SECTION
1–Fuel line assembly. 2–Nut. 3–Body. 4–Nozzle assembly.

FIGURE 20-19. Caterpillar precombustion chamber and injection valve. *(Courtesy of Caterpillar Tractor Company)*

4. Leaky nozzle, nozzle dribble: nozzle valve or seat damaged; dirt in nozzle; broken pressure spring or adjusting screw; nozzle valve sticking.

PART 6   INJECTOR SERVICE

Injector service varies, depending on the particular make and model. (See Fig. 20–23 to 20–35.) All removal, disassembly, inspection, overhaul, reassembly, testing, adjusting, and installation procedures for any given make and model should be followed as recommended in the appropriate manufacturer's shop service manual.

It is critical to good engine performance that all injectors for any engine be of the same type and size. Never mix injectors or mated and lapped injector parts.

Some injectors are not repairable and must be replaced if defective. Others can be overhauled and adjusted. Detroit Diesel injector and Cummins injector service require calibration for metered fuel delivery.

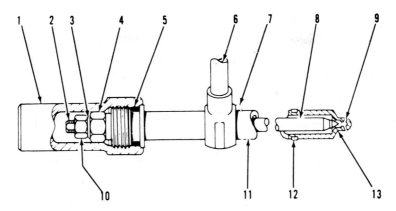

FIGURE 20-20. Fuel injection nozzle—Caterpillar pencil type used in open combustion chamber, direct injection engines. 1. Cap. 2. Lift adjustment screw. 3. Pressure adjustment screw. 4. Lock nut for pressure adjustment screw. 5. O-ring seal. 6. Fuel inlet. 7. Compression seal. 8. Valve. 9. Orifices (four). 10. Lock nut for lift adjustment screw. 11. Nozzle body. 12. Carbon dam. 13. Nozzle tip. *(Courtesy of Caterpillar Tractor Company)*

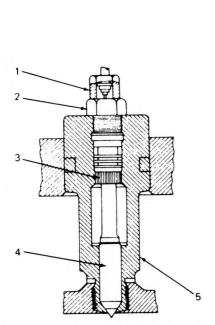

FIGURE 20-21. Direct injection nozzle component identification: (1) Fuel line nut. (2) Nozzle nut. (3) Nozzle body. (4) Nozzle valve. (5) Adapter used with direct injection engine nozzle. *(Courtesy of Caterpillar Tractor Company)*

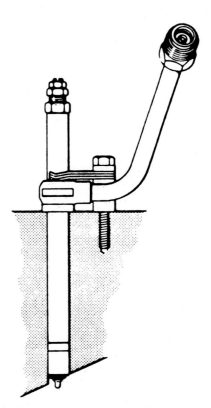

FIGURE 20-22. Pencil-type injector and injector mounting clamp. *(Courtesy of Roosa Master Stanadyne Inc./ Hartford Division)*

SPECIAL TOOLS

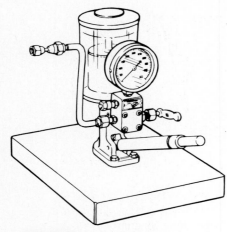

CAS-10091
DIESEL FUEL INJECTION
NOZZLE TESTER (BENCH MODEL)

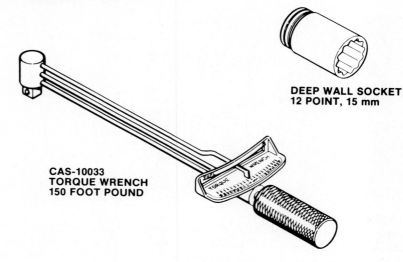

CAS-10033
TORQUE WRENCH
150 FOOT POUND

DEEP WALL SOCKET
12 POINT, 15 mm

BOX WRENCH
12 POINT, 15 mm

CROWSFOOT WRENCH
SNAP-ON TOOL AN-8506-8

MAGNIFYING GLASS
66-0135

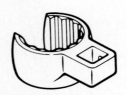

CROWSFOOT WRENCH
SNAP-ON TOOL AN-8508-11
SNAP-ON TOOL AN-8508-14

ORIFICE
CLEANING WIRES
66-0036

65-0476
ADAPTER ELBOW

CAS-1411
INJECTOR BORE
CLEANING TOOL

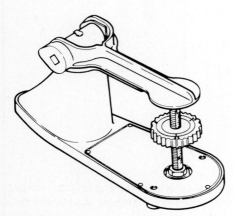

CAS-1357
COMPRESSION SPRING TESTER

65-0878
TEST PUMP ADAPTER

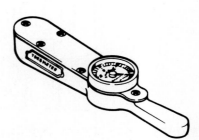

CAS-10032
TORQUE WRENCH
200 INCH POUND

FIGURE 20-23. Typical nozzle service tool kit. *(Courtesy of J. I. Case, a Tenneco company)*

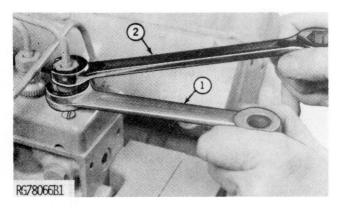

FIGURE 20-24. Always use two wrenches when loosening or tightening fuel lines: (1) to hold the fitting adapter and prevent it from turning and (2) to turn the fitting nut. *(Courtesy of Deere and Company)*

While many service shops are equipped to provide injector rebuilding, other shops rely on replacing injectors with rebuilt or new injectors.

The following service procedures apply in general to injectors. In addition the appropriate service manual must be used for all special procedures and all service data.

General Procedure for Injector Service

1. Clean the area around the injectors before removal in order to prevent dirt and other foreign

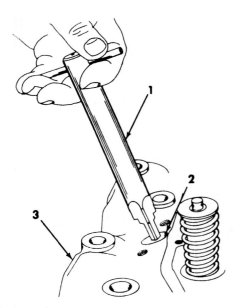

FIGURE 20-25. Cleaning the nozzle bore. (1) Bore cleaning tool. (2) Nozzle bore. (3) Cylinder head. *(Courtesy of Allis-Chalmers)*

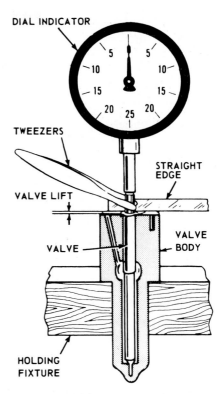

FIGURE 20-26. Using dial indicator to measure nozzle valve lift. *(Courtesy of Mack Trucks Inc.)*

matter from entering the fuel system. Any high-pressure washing system is good for this job.

2. Disconnect all fuel lines from injectors. Cap all openings. Disconnect all injector operating mechanisms (Cummins and Detroit Diesel).

3. Remove the injector-retaining device. Remove the injector. Some injectors must be removed using a special puller. Follow service manual procedures.

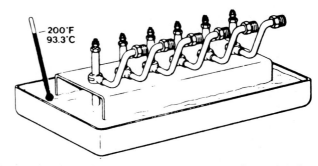

FIGURE 20-27. Controlled temperature of nozzles immersed in controlled temperature fluid is important for accurate nozzle testing. *(Courtesy of Diesel Systems Group—Stanadyne Inc.)*

FIGURE 20-28. Adjusting nozzle opening pressure. *(Courtesy of Diesel Systems Group—Stanadyne Inc.)*

4. If the problem diagnosis procedure has established that an engine compression test is required, this should be done when the injectors are removed. Record the results of the compression test. If an engine compression problem is indicated, repair the engine as necessary.

5. Disassemble and clean the injectors according to the appropriate manufacturer's service manual. Do not mix parts from one injector with those of another injector.

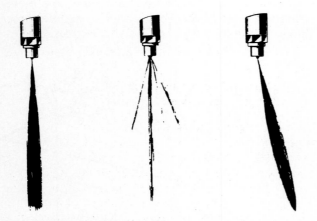

FIGURE 20-29. Injector spray pattern on left is the correct pattern for injector used with energy cell type of combustion chamber. Other two patterns are not acceptable. *(Courtesy of J. I. Case, a Tenneco company)*

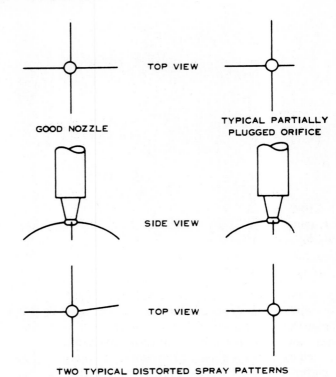

FIGURE 20-30. Nozzle spray pattern analysis for injectors used in direct injection/open-type combustion chambers. *(Courtesy of Caterpillar Tractor Company)*

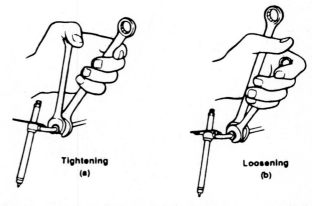

FIGURE 20-31. Always use two wrenches when tightening or loosening injection line fittings. *(Courtesy of Diesel Systems Group—Stanadyne Inc.)*

FIGURE 20-32. Using a soft brass bristle brush for cleaning nozzle tip prevents damage. *(Courtesy of Diesel Systems Group—Stanadyne Inc.)*

6. Inspect carefully all injector parts and compare wear and damage to service limits indicated in the appropriate manufacturer's service manual. Inspection may require the use of a magnifying glass due to the extremely close tolerances and finely honed surfaces of injector parts. Some injector parts are not available separately. Mating parts such as nozzles must be purchased as an assembly due to the very close tolerances of these parts.

7. Assemble the injector according to the procedures given in the appropriate service manual. *Never apply undue pressure or force.* All parts must be at normal room temperature and be properly lubricated with the specified fuel oil. If parts are not at equal temperatures, it may be impossible to assemble them.

8. Mount the injector on the specified test stand and, using the specified test fluid, which should be at the proper testing temperature, proceed with injector testing and adjustment. See Part 7.

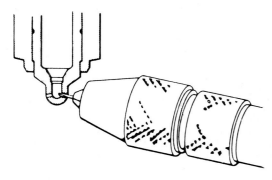

FIGURE 20-33. Cleaning injector nozzle orifices with special tool of specified diameter. *(Courtesy of Diesel Systems Group—Stanadyne Inc.)*

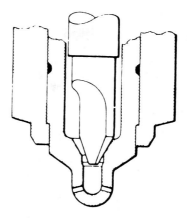

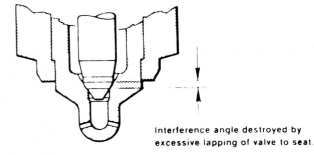

Interference angle destroyed by excessive lapping of valve to seat.

FIGURE 20-34. Cleaning nozzle seat with special tool (top). Effects of excessive lapping of nozzle valve to seat (bottom). *(Courtesy of Diesel Systems Group—Stanadyne, Inc.)*

9. Testing and adjusting must be done to achieve the following (depending on make and model of injector):

a. specified nozzle-opening pressure

b. correct spray pattern

c. proper chatter

d. no nozzle drip at specified fuel pressure

e. no excessive leak back internally under specified pressure

f. specified amount of fuel delivery (fuel metering—Detroit Diesel and Cummins)

10. Prepare cylinder head, injector openings, precombustion chamber, and air cells (as required) for injector installation. This includes making sure that there is no copper gasket left in the injector mounting hole; cleaning carbon from the injector opening, air cell, and precombustion chamber; removing of old O-rings; and installing of new O-rings and copper gaskets. Follow the appropriate shop manual to ensure a positive seal around the injector as well as specified injector protrusion into the cylinder head.

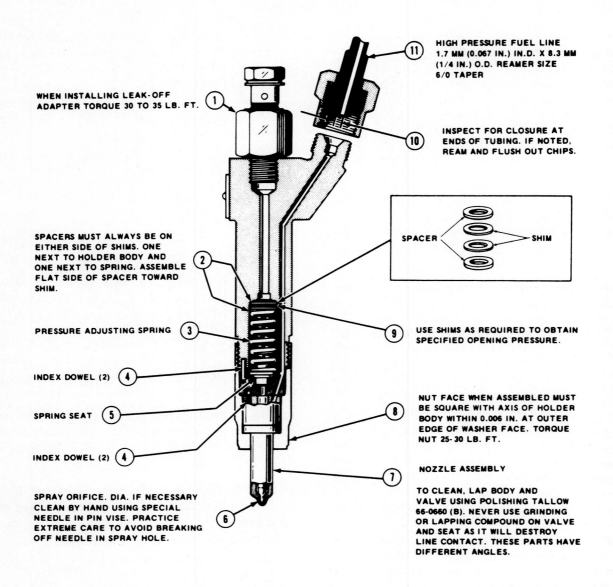

WHEN INSTALLING LEAK-OFF ADAPTER TORQUE 30 TO 35 LB. FT.

HIGH PRESSURE FUEL LINE 1.7 MM (0.067 IN.) IN.D. X 8.3 MM (1/4 IN.) O.D. REAMER SIZE 6/0 TAPER

INSPECT FOR CLOSURE AT ENDS OF TUBING. IF NOTED, REAM AND FLUSH OUT CHIPS.

SPACER   SHIM

SPACERS MUST ALWAYS BE ON EITHER SIDE OF SHIMS. ONE NEXT TO HOLDER BODY AND ONE NEXT TO SPRING. ASSEMBLE FLAT SIDE OF SPACER TOWARD SHIM.

PRESSURE ADJUSTING SPRING

USE SHIMS AS REQUIRED TO OBTAIN SPECIFIED OPENING PRESSURE.

INDEX DOWEL (2)

SPRING SEAT

NUT FACE WHEN ASSEMBLED MUST BE SQUARE WITH AXIS OF HOLDER BODY WITHIN 0.006 IN. AT OUTER EDGE OF WASHER FACE. TORQUE NUT 25-30 LB. FT.

INDEX DOWEL (2)

NOZZLE ASSEMBLY

SPRAY ORIFICE. DIA. IF NECESSARY CLEAN BY HAND USING SPECIAL NEEDLE IN PIN VISE. PRACTICE EXTREME CARE TO AVOID BREAKING OFF NEEDLE IN SPRAY HOLE.

TO CLEAN, LAP BODY AND VALVE USING POLISHING TALLOW 66-0660 (B). NEVER USE GRINDING OR LAPPING COMPOUND ON VALVE AND SEAT AS IT WILL DESTROY LINE CONTACT. THESE PARTS HAVE DIFFERENT ANGLES.

Engine Series	Nozzle Opening Pressure - PSI	Valve Lift-In	Injection Pump	Holes	Hole Diameter Inches (mm)
ENDT676	4200 to 4400	0.019	PES	5	0.0126 (0.320)
ENDT675C	3800 to 3950	0.014	PES	5	0.0126 (0.320)
ETAZ673	4200 to 4400	0.019	PES	5	0.0126 (0.320)
ET673*	3800 to 3950	0.014	PES	5	0.0126 (0.320)
ETAY676	4200 to 4400	0.019	PES	5	0.0126 (0.320)
ETY675	3800 to 3950	0.014	PES	5	0.0126 (0.320)
ETZ675	3800 to 3950	0.014	PES	5	0.0126 (0.320)
ETAY673	4200 to 4400	0.019	PES	5	0.0126 (0.320)
ETY673E	3800 to 3950	0.014	PES	5	0.0126 (0.320)

FIGURE 20-35. Bosch injector specifications for Mack truck engines. Note shim method of adjusting nozzle opening pressure. *(Courtesy of Mack Trucks Inc.)*

11. Position injectors into cylinder heads as required to achieve proper fuel line connections and injector-operating mechanism connections (time injectors as specified in Cummins and Detroit Diesel engines). Tighten the injector mounting device to specified torque, following the proper sequence and in proper increments as specified in the appropriate service manual.

12. Connect all fuel lines and tighten to specifications.

13. Bleed all air from fuel system (some models only).

14. Start the engine and inspect for fuel leaks.

PART 7   TESTING INJECTORS

The injectors should be removed periodically from the engine in accordance with the engine manufacturer's instructions for routine testing and examination. Disassembly, cleaning, and reconditioning are carried out according to the condition of the injectors.

Each injector should be checked on a lever-operated test pump, before and after any cleaning or dismantling work is carried out, to determine the general condition of the injector when the injection pressure is applied. Injection setting pressures, together with spray hole and needle lift details, are given in manufacturer's service manuals for all types of injectors.

Warning: When the injectors are being tested, care must always be exercised not to allow the fuel spray to come into contact with the hands; otherwise the spray will penetrate the skin.

Procedure

1. Expel the air from the test pump, which has been filled with test fluid. Mount the injector on the test pump and check the pressure at which the needle valve opens. Compare with the specified setting pressure.

2. Examine the spray when pumping about 60 strokes per minute. The correct spray should appear as a fine mist without distortion and without visible streaks of unvaporized fuel.

3. Check the nozzle tip, which should remain dry on completion of the fuel injection. Nozzle dribble indicates seat leakage.

4. Check for leakage at the nozzle nut. Leakage may result if the joint faces of nozzle and nozzle holder do not register correctly to form a fuel-tight joint, or if there is dirt between the faces.

If the injector operates satisfactorily on preliminary testing, it should not be dismantled. The nozzle should be cleaned with a brass wire brush to remove any carbon deposit. The exterior of the injector should then be cleaned with benzine and dried off with compressed air. After cleaning the exterior, the injector should be checked again for operation on the test pump.

If the injector is faulty on preliminary testing, it must be dismantled for cleaning the interior, examination, and further testing.

PART 8   SELF-CHECK

1. List four requirements that good diesel fuel must meet.

2. Define the ignition delay period.

3. What is diesel knock?

4. Name five combustion chamber designs and describe each one briefly.

5. Describe the hole type and pintle type nozzle operation.

6. List the causes and effects of:
   (a) nozzle opening pressure too high
   (b) nozzle opening pressure too low
   (c) incorrect nozzle spray pattern
   (d) nozzle dribble or leakage.

7. Name the tests required when checking nozzle operation on a pop tester.

Chapter 21

Governing Fuel Delivery

PART 1   INTRODUCTION

The delivery of fuel to the engine must be controlled to provide the following operating characteristics, depending on the type of service the engine is designed to perform:

• control of excess fuel for starting

• control of engine low idle speed

• limiting of maximum engine speed

• maintenance of any desired engine speed regardless of changes in load

• delay of fuel delivery during acceleration on turbocharged engines

• control of fuel delivery to match air density at different altitudes

• fuel shut-off to stop engine

• torque control

The following information on governor design and operation (except for Part 4) is provided courtesy of Robert Bosch. Other governors are covered in the appropriate chapters on fuel injection systems.

The governor is responsible for ensuring that the engine maintains a certain speed under various load conditions, that the engine speed does not exceed a certain level as a protection against self-destruction, and that the engine does not stop during pauses in loaded operation, i.e., during idling. The governor accomplishes all this by controlling the amount of fuel injected into the engine.

The diesel engine draws air in only during the intake stroke. During the compression stroke this air is heated to such a high temperature that the diesel fuel injected into the engine toward the end of the compression stroke ignites of its own accord. The fuel is metered by the fuel injection pump and is injected under high pressure through the injection nozzle into the combustion chamber.

Fuel injection must take place:

• in an accurately metered quantity corresponding to the engine load,

• at the correct instant in time,

• for a precisely determined period of time, and

• in a manner suited to the particular combustion process concerned.

Maintenance of these conditions is the function of the fuel injection pump and the governor. The quantity of fuel injected into the engine during each plunger lift is approximately proportional to the torque of the engine. This fuel delivery is adjusted by turning of the pump plungers, each of which has an inclined helix machined into it. As a plunger is turned, its effective stroke is varied. The plungers are turned by means of the control rod acting through either a set of gear teeth or some other transmission part. In a motor vehicle, the control rod is connected to the accelerator pedal through the governor and a linkage; when the accelerator pedal is pressed down, the pedal travel is converted to corresponding control-rod travel. Stationary engines can be operated with the governor control lever or by an electric speed-control device.

PART 2   NEED FOR GOVERNORS

In a diesel engine there is no fixed position of the control rod at which the engine will maintain its speed accurately without a governor. During idling, for example, the engine speed without a governor would either drop to zero or would increase continuously until the engine races and runs completely out of control. The latter possibility results from the fact that the diesel engine operates with an excess of air; consequently, effective throttling of the cylinder charge does not take place as the speed increases. (See Fig. 21-1 to 21-3.)

If a cold engine is started, for example, by the starting motor, and if it is permitted to continue idling with a corresponding amount of fuel injected, the inherent friction in the engine, as well as the transmission resistance of parts driven by the engine (such as the generator, air compressor, fuel injection pump, etc.), decreases after a certain length of time. As a result, if the position of the control rod were to remain unchanged without a governor, the

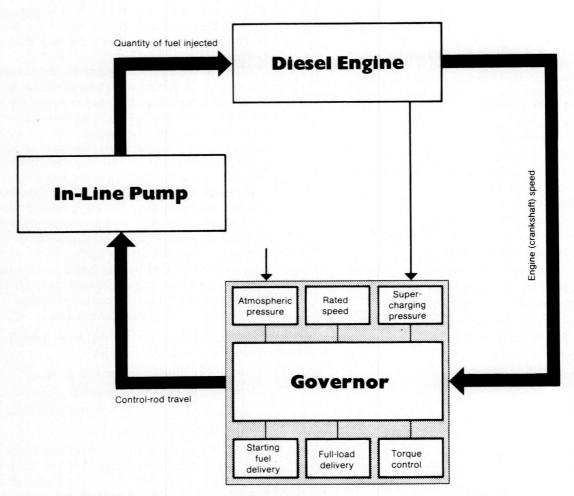

FIGURE 21-1. This diagram shows the factors that affect governor operation. *(Courtesy of Robert Bosch Canada Ltd.)*

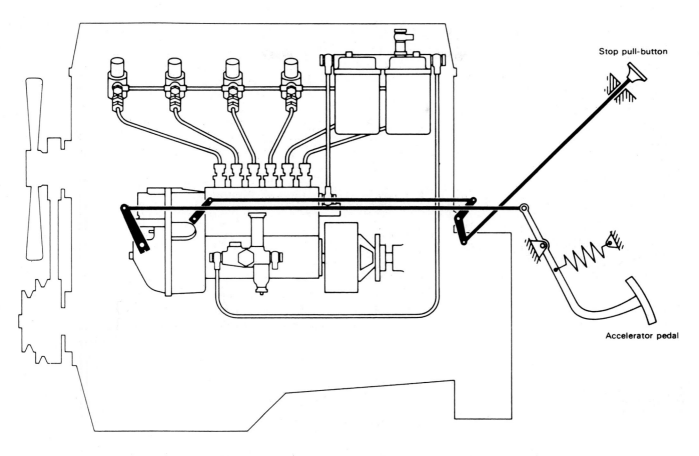

FIGURE 21-2. Operator controls and linkage to injection pump and governor. *(Courtesy of Robert Bosch Canada Ltd.)*

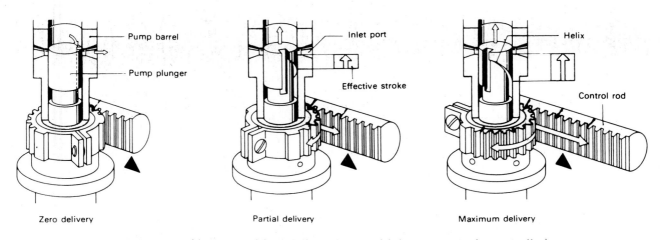

FIGURE 21-3. Metering of fuel delivery in a multiplunger pump is controlled by pump plunger position, which in turn is controlled by the control rod (rack). The control rod is connected to the governor, which moves the rack in response to engine speed and load as well as throttle position. *(Courtesy of Robert Bosch Canada Ltd.)*

engine speed would constantly increase and could rise to a level at which the engine would ultimately destroy itself.

Because of the reasons stated, a governor is required for operation of the injection pump.

The governor operates dependent either on the rotational speed of the engine (mechanical governor and hydraulic governor) or on the intake manifold pressure (pneumatic governor). In both cases, the governor varies the amount of fuel injected into the engine and thereby regulates the engine speed.

PART 3   FUNCTIONS OF THE GOVERNOR

The basic function of every governor is to limit the high-idle speed; i.e., it must ensure that the speed of the diesel engine does not exceed the maximum value specified by the manufacturer. Depending on the type of governor, further functions can be the maintenance of certain specified speeds, e.g., the idle speed or speeds within a particular rotational-speed range or the entire range between idle speed and high-idle (maximum) speed.

• maximum-speed governors. These governors are designed to limit the maximum speed only.

• minimum-maximum-speed governors. These governors control the idle speed as well as the maximum speed.

• variable-speed governors. These governors control the idle and maximum speeds as well as the speed range between them.

In addition to its basic function, the governor must also fulfill control functions, such as automatically providing or cutting off the starting fuel delivery (the increased fuel quantity that is required for starting) and varying the full-load delivery as a function of *speed* (torque control), *charge-air-pressure*, or *atmospheric pressure*.

In order to carry out these functions, supplementary equipment is required in some cases, which will be described later.

Maximum-Speed Regulation

When the load is removed from the engine, the maximum full-load speed may rise no higher than rated (high-idle or no-load speed) in accordance with the permissible speed droop. The governor accomplishes this by drawing back the control rod in the shut-off direction.

The range is designated as the maximum-speed regulation.

The greater the speed droop, the greater is the increase in speed from the maximum full-load speed to the maximum noted no-load (high-idle) speed.

Intermediate-Speed Regulation

If required by the intended application of the governor (for example, in vehicles with an auxiliary drive), the governor can also maintain constant (within certain limits) various speeds between the idle and maximum speeds.

Depending on the load, therefore, the speed would only fluctuate between full load and no load within the performance range of the engine.

Low-Idle-Speed Control

The speed of a diesel engine can also be regulated in the lowest speed range. If the control rod returns from the starting position after a cold diesel engine is started, the frictional resistance of the engine is still relatively high. The amount of fuel required to keep the engine in operation is therefore somewhat larger, and the speed is somewhat slower, than would normally correspond to the idle-speed adjustment point.

After the friction during the warm-up period has diminished, the speed increases; the control rod moves back when the idle speed for the warm engine is reached.

The various demands made on governors have led to the development of the following different types:

Maximum-Speed Governors

Maximum-speed governors are designed for diesel engines that drive machines or machine systems—for example, engine-generator sets—at a fixed rated speed. Here the governor must ensure only that the maximum speed is maintained; there is no idle control or control of a particular starting fuel delivery. If the engine speed rises above the rated speed, as a result of a decreasing load on the engine, the governor shifts the control rod in the shut-off direction; i.e., the control-rod travel becomes smaller, and the fuel delivery decreases. The high-idle speed is reached when the entire load is removed from the engine.

Minimum-Maximum-Speed Governors

With diesel engines used in trucks, speed control in the range between idle and the maximum speed is often not required. In this rotational-speed range the driver operates the control rod in the fuel injection pump directly by means of the accelerator pedal and thus sets the required torque. The governor ensures that the engine does not stall in the idle-speed range, and it also controls the maximum speed.

The cold engine is started with the starting fuel delivery. The driver has pressed the accelerator pedal all the way down. If the driver releases the accelerator pedal, the control rod returns to the idle position. During the warm-up period, the idle speed fluctuates along the idle-speed control curve and evens out.

When the engine has warmed up, the greatest starting fuel delivery is generally not required for a restart; many engines can even be started when the governor control lever is in the idle position.

If the driver presses the accelerator pedal all the way down with the engine operating, the control rod moves to the full-load delivery position. As a result, the engine speed increases, torque control of the fuel delivery starts, and the full-load delivery is slightly reduced.

The full-load delivery is injected into the engine with the accelerator pedal pressed all the way down until the maximum full-load speed is reached. Full-load speed regulation corresponding to the speed droop of the governor then starts. The engine speed rises somewhat, the control-rod travel is reduced, and, as a result, the fuel delivery is decreased. The high-idle speed is reached when the entire load is removed from the engine. During overrun (e.g., vehicle traveling downhill), the control-rod travel can become zero and the engine speed can increase somewhat further.

Variable-Speed Governors

Vehicles with auxiliary drives (for example, for cistern pumps or for extending fire-fighting ladders) and agricultural tractors, which must maintain a certain operating speed, as well as boats and stationary assemblies of equipment, are equipped with variable-speed governors.

These governors control not only the idle and maximum speeds but also speeds between them independent of the engine load. The desired speed is set with the control lever. The governor operates as follows: starting of the engine with the starting fuel delivery, variation of the full-load regulation along the full-load characteristic curve, and extending the torque-control range until the onset of speed regulation at the maximum full-load speed.

PART 4   GOVERNOR TYPES AND TERMINOLOGY

Governor Types

Four types of governors are used on diesel engines:

1. *Mechanical governors:* Control fuel delivery by mechanical means only. All operate with centrifugal flyweights and springs that act on the fuel metering system through mechanical linkage only.

2. *Hydraulic governors:* Control fuel delivery through hydraulic pressure. These also operate with flyweights and springs, however, the flyweights and springs operate a control valve which controls the amount of hydraulic pressure allowed to act on a power piston. The power piston, in turn, acts on the fuel metering system to control fuel delivery. Engine lubricating oil pressure or pressure from a governor oil pump provides hydraulic pressure.

3. *Pneumatic governors:* Used on some automotive engines equipped with an air throttle valve. These governors operate on pressure differences between atmosphere and engine vacuum.

4. *Isochronous governor:* Uses compressed air from the vehicle storage tank to actuate a power piston to control fuel delivery. The isochronous governor incorporates a design feature which provides zero speed droop, holding the engine at constant speed regardless of engine load changes within the load range of the engine.

A tailshaft or torque converter governor is used on some vehicle applications to ensure that constant output shaft speed is maintained. It consists of a mechanical shaft-speed-sensitive device mounted on the engine torque converter to monitor output shaft speed. It is mechanically connected to the engine governor in a manner that maintains constant output shaft speed regardless of torque load as long as maximum torque load is not exceeded.

Governor Terminology

Several terms that apply to governor operation should be defined in order to better understand governor action.

1. *Speed droop:* The difference in engine speed between no load and full load speed. Speed droop is expressed as a percentage of no-load speed and may be as high as 10 percent on some governors. Governors with speed droop are non-isochronous.

2. *Hunting:* An undesirable fluctuation of engine speed (alternately slowing down and speeding up). Hunting is the result of overcontrol by the governor, which requires adjustment or cleaning (since it may be sticking) to correct.

3. *Stability:* The ability of the governor to maintain the desired engine speed without hunting.

4. *Sensitivity:* The change in engine speed that occurs before the governor acts on the fuel control linkage.

5. *Promptness:* The time required for the governor to move the fuel control mechanism from a no-load position to a full-load position. Promptness is determined by governor design and its ability to overcome the resistance to movement of the control linkage.

PART 5   SPEED DROOP

Every engine has a torque characteristic curve corresponding to its maximum loading capacity. A certain maximum torque is associated with every speed. If the load on an engine is removed with no change in the position of the control lever, the engine speed may increase within the control range by only a certain permissible amount as determined by the engine manufacturer (for example, from any full-load speed to any idle speed). The increase in speed is proportional to the change in load; i.e., the greater the reduction in load, the greater the increase in speed. Conversely, of course, when the engine is idling and a load is applied, the speed will decrease somewhat; hence the designation of this characteristic as "speed droop."

The speed droop of the governor is generally related to the maximum full-load speed (rated speed). As the speed decreases, the speed droop increases and is at its greatest in the idle-speed range.

Generally, more stable behavior of the entire control circuit (governor, engine, and driven machine or vehicle) can be attained by a fairly large speed droop. On the other hand, the speed droop is limited by operating conditions, for example, to about 2–5 percent for generators, about 6–10 percent for vehicles, and about 10–15 percent for excavators with a storage flywheel.

With the speed set to a constant value, the actual speed varies within the speed-droop range as the load on the engine is changed (resulting, for example, from a change in the slope of the road).

Because of these changes in engine speed resulting from changes in the load, the speed droop was also known previously as cyclic irregularity.

PART 6   TORQUE CONTROL

Optimum exploitation of the engine torque can be achieved by means of torque control. Torque control is not an actual control process but is one of the regulation functions carried out by the governor. It is designed for the full-load delivery, i.e., the maximum amount of fuel delivered in the loadable range of the engine that can burn smokefree. (See Fig. 21-4 to 21-6.)

The fuel requirement of the non-turbocharged diesel engine generally decreases as the speed increases (lower relative rate of air flow, thermal limiting conditions, changed mixture formation). At the same time the amount of fuel delivered by the injection pump increases within a certain range as the speed rises, as long as the control rod remains in the same position, because of the throttling effect at the control port in the pump plunger-and-barrel assembly. If too much fuel is injected into the engine, smoke will develop as the engine overheats.

The amount of fuel injected into the engine must be matched to the actual fuel requirement. As the speed increases the fuel delivery decreases (positive torque control); as the speed drops, the fuel delivery increases.

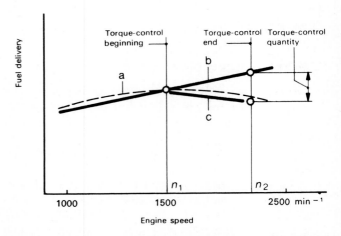

FIGURE 21-4. Fuel requirement and fuel-delivery characteristic with torque control: (a) fuel requirement of engine, (b) full-load delivery without torque control, and (c) full-load delivery with torque control. *(Courtesy of Robert Bosch Canada Ltd.)*

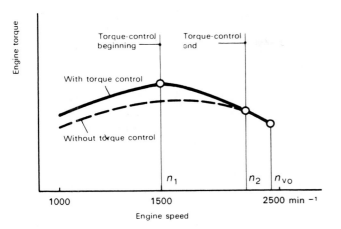

FIGURE 21-5. Torque characteristic of diesel engine, with and without torque control. *(Courtesy of Robert Bosch Canada Ltd.)*

Torque-control systems are arranged and designed according to the particular type of governor.

In engines with an exhaust turbocharger achieving a high measure of supercharging, the fuel requirement for full load in the lower speed range rises so sharply that the natural increase in fuel delivered by the fuel injection pump is no longer adequate. In such cases, torque control must be carried out as a function of speed or charge-air pressure; depending on the prevailing conditions, this can be accomplished with the governor, the manifold-pressure compensator, or both operating together.

This form of torque control is called negative. Negative torque control means an increase in the fuel delivery as the speed rises.

This type of torque control is clearly the opposite of the common positive torque control in which the quantity of fuel injected is reduced as the speed increases.

Designation	Type of governor	Operating principle	For pump size	Used for
RQ	Minimum-maximum-speed	Centrifugal	A, P	Road vehicles, locomotives, assemblies of equipment
RQU*	or maximum-speed only		ZW	
EP/RS	Minimum-maximum-speed	Centrifugal	A, P	Road vehicles
RQV	Variable-speed and combination	Centrifugal	A, P	Vehicles with auxiliary drive, trucks
RQUV*			ZW	Machine systems
RQV-K	Variable-speed with any type of torque control	Centrifugal	A, P	Road vehicles
EP/RSV	Variable-speed	Centrifugal	A, M, P	Tractor vehicles, machine systems, motor vehicles
EP/RSUV*			P, ZW	Machine systems
EP/M	Variable-speed	Pneumatic	A, M	Automobiles, tractor vehicles

* with step-up gear for slow-speed engines

FIGURE 21-6. Robert Bosch governor types and application. *(Courtesy of Robert Bosch Canada Ltd.)*

PART 7   MECHANICAL GOVERNORS

Mechanical governors, i.e., governors employing the principle of centrifugal force, are used most widely with larger diesel engines. (See Fig. 21-7 and 21-8.)

The Bosch mechanical governor is mounted on the fuel injection pump. The injection-pump control rod is connected with the governor linkage through a flexible joint, and the connection to the accelerator pedal is made through the governor control lever.

Two different designs are used in mechanical governors:

• RQ, RQV, with the governor springs built into the flyweights.

• RS, RSV, with the centrifugal force acting through a system of levers on the governor spring located outside the two flyweights.

In mechanical governor types RQ and RQV, each of the two flyweights acts directly on a spring set, which is designed specifically for a given nominal speed.

In mechanical governors of type RS/RSV the action of the two flyweights presses the sliding bolt against the tensioning lever, which is drawn in the opposite direction by the governor spring. When the speed is set by the control lever, the governor spring is tensioned by an amount corresponding to the desired speed.

In both types of metering unit, the governor springs are so selected that at the desired speed the centrifugal force and the spring force are in equilib-

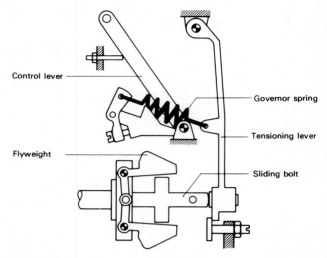

FIGURE 21-8. Metering unit for RS and RQV governor. *(Courtesy of Robert Bosch Canada Ltd.)*

rium. If this speed is exceeded, the increasing centrifugal force of the flyweights acts through a system of levers to move the control rod, and the fuel delivery system is decreased.

PART 8   MINIMUM-MAXIMUM-SPEED GOVERNOR RQ

Construction

The fuel-injection pump camshaft drives the governor hub through a vibration damper. The two flyweights with their bell cranks are supported in the governor hub, with one spring set built into each flyweight. By means of the bell cranks, the radial travel of the flyweights is converted to axial movements of the sliding bolt, which in turn transmits these movements to the slider. Movement of the slider is held in a straight line by the guide pin; the slider itself, operating through the fulcrum lever, forms the connection between the flyweight mechanism and the control rod. The lower end of the fulcrum lever is fastened to the slider. (See Fig. 21-9 and 21-10.)

In the fulcrum lever the movable guide block is guided radially by the linkage lever; this lever is connected with the control lever. The control lever is operated either manually or through a linkage system from the accelerator pedal.

When the control lever is moved, the guide block is shifted, and the control lever is tilted around the pivot at the slider; if the governor takes effect, the pivot for the fulcrum lever is at the guide block. The

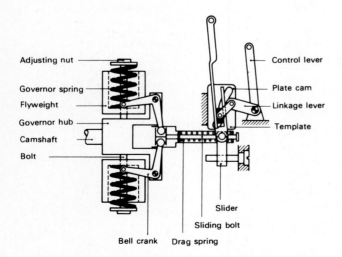

FIGURE 21-7. Metering unit for RQ and RQV governor. *(Courtesy of Robert Bosch Canada Ltd.)*

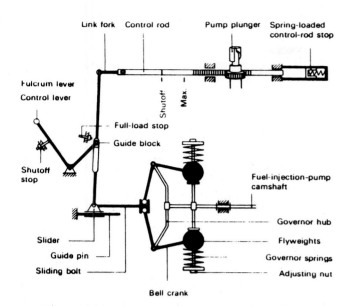

FIGURE 21–9. RQ minimum-maximum speed governor in full shut-off position. *(Courtesy of Robert Bosch Canada Ltd.)*

transmission ratio of the fulcrum lever changes. As a result, even in the idle range where the centrifugal forces are still low, there is more than sufficient adjustment force for the control rod.

The spring sets (governor springs) built into the flyweights consist generally of three cylindrical helical springs arranged as shown.

The outer spring is supported between the flyweight and the outer spring seat, while the two in-

ner springs are positioned between the outer and inner spring seats. During low-idle-speed control, only the outer spring (the idle-speed spring) takes effect. As the speed increases, the flyweights, after surpassing the idle-speed travel path at the inner spring seat, are pressed against the inner spring seat and remain in this position until maximum-speed regulation begins. During control of the maximum speed, all springs act together. The two inner springs are designated maximum-speed control springs.

Operating Characteristics

Engine Stopped. The control lever is positioned against the shut-off stop, the control rod is in the shut-off position, and the flyweights are positioned all the way in.

Start Position (control lever at full load). (Fig. 21–11) After the control rod overcomes the force of the return spring in the spring-loaded control-rod stop, it moves all the way to the starting-fuel-delivery position.

Idle Position. (Fig. 21–12) After the engine starts to operate and the control lever (accelerator pedal) is released, this lever returns to the idle position (a corresponding stop should be provided in the vehicle or on the engine). The control rod also returns to the idle position, which is determined by the now operating governor. The term *idle speed* of an engine is understood here to mean the lowest speed at which the engine will continue to operate

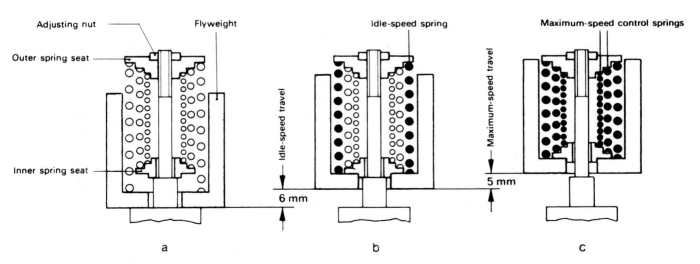

FIGURE 21–10. Flyweight travel and governor springs in the RQ governor. *(Courtesy of Robert Bosch Canada Ltd.)*

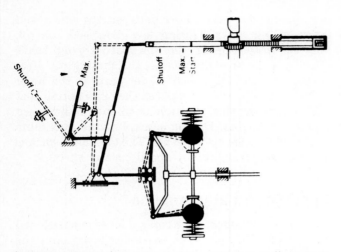

FIGURE 21-11. RQ governor in start (maximum fuel) position. *(Courtesy of Robert Bosch Canada Ltd.)*

reliably under no load; in this condition the engine is loaded only by its own internal friction and by the other equipment permanently connected to it such as the generator, fuel injection pump, fan, etc. In order to be able to overcome this idle load, the engine requires a certain amount of fuel and receives this fuel at a position of the control lever corresponding to the specified idle position.

Part-Load Position.(Fig. 21–13) When the engine is loaded (between no-load and full-load). As the driver presses the accelerator pedal down somewhat, the engine speeds up. As a result, the flyweights are forced outward. In other words, the governor initially tends to prevent this increase in engine speed. However, after the speed exceeds the

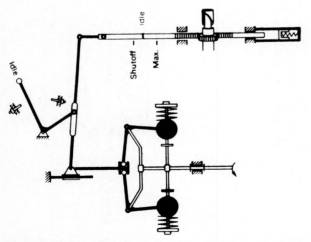

FIGURE 21-12. RQ governor in idle position. *(Courtesy of Robert Bosch Canada Ltd.)*

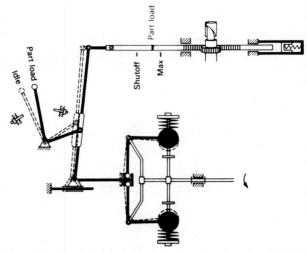

FIGURE 21-13. RQ governor in part-load position. Flyweight position remains unchanged (due to spring tension) until maximum engine speed is reached. *(Courtesy of Robert Bosch Canada Ltd.)*

idle speed by only a slight amount, the flyweights are brought up against the spring seats, which are loaded by the maximum-speed control springs. They remain in this position until the maximum speed is reached, because the maximum-speed control springs yield to the centrifugal force only when the engine tends to exceed its nominal speed. For this reason, the governor has no effect between the idle speed and the maximum speed. In this intermediate range the position of the control rod, and thus the torque developed by the engine, is influenced only by the driver.

Maximum Speed Regulation at Full Load.(Fig. 21–14) In the highest speed range of the engine (maximum-speed regulation), maximum-speed regulation begins when the engine exceeds the nominal speed, at full-load or part-load, depending on the position of the control lever. For this reason, as soon as maximum-speed regulation has started, the position of the control rod no longer depends solely on the driver but also on the governor. The maximum-speed regulation travel of the flyweights is 5 mm. This results in a control-rod travel of about 16 mm (with a lever ratio of 1:3.23), which is sufficient to regulate fuel delivery from maximum delivery to shut-off.

Torque-Control Mechanism in the RQ Governor

In the RQ governor the torque-control mechanism is built into the flyweights, between the inner spring seat and the maximum-speed control springs (Fig.

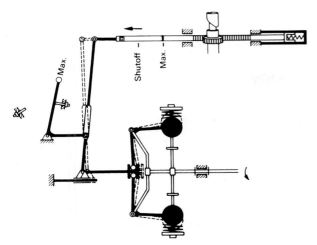

FIGURE 21–14. RQ governor maximum speed regulation at full load. Control rod begins to move to the shut-off position. *(Courtesy of Robert Bosch Canada Ltd.)*

21–15). The torque-control spring is installed in a spring retainer, on the outside of which the two maximum-speed control springs are supported. For operating purposes, therefore, the torque-control spring is connected in series with the maximum-speed control springs. The space between the inner spring seat and the spring retainer is the torque-control travel (0.3 to 1 mm). The width of this space can be adjusted with compensating shims.

The start of torque control depends on the fuel-requirement characteristic curve of the engine. At a point somewhat below the maximum speed, the torque-control spring is compressed to the extent that the inner spring seat and the spring retainer are pressed against each other. Without the torque-control spring, the governor has no effect between the idle speed and the maximum speed. However, since the torque-control springs yield, the fly-

weights can move outward by the distance of the torque-control travel in the range between the idle and maximum speeds, and therefore shift the control rod accordingly in the shut-off direction (positive torque control).

Minimum-Maximum-Speed Governor RQU

The RQU governor is suitable for controlling very low speeds. It is fitted with a transmission gear effecting a speed-increasing ratio (of about 3:1) between the fuel-injection-pump camshaft and the governor hub. Governor-type RQU was developed for PE. .Z fuel injection pumps, which are used for larger, usually slow-speed engines. The operation and operating characteristics of this governor are similar to those of the RQ governor.

The linkage lever in the RQU governor is constructed of two parts, as in the RQV governor, and is guided by a plate cam, also in the RQV governor. (See Fig. 21–16.)

Maximum-Speed Governors RQ and RQU

Construction of the maximum-speed governor differs from the construction of the minimum-maximum-speed governor primarily in that the idle-speed stage is not incorporated. In operation, the maximum-speed governor functions in a similar manner to the maximum-speed stage of the RQ or RQU minimum-maximum-speed governor; i.e., maximum-speed regulation starts if the engine exceeds the maximum rated speed.

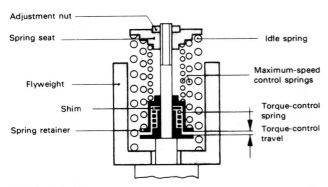

FIGURE 21–15. Torque control mechanism in the RQ governor. *(Courtesy of Robert Bosch Canada Ltd.)*

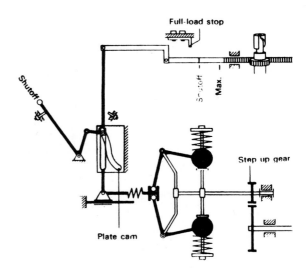

FIGURE 21–16. RQU type of governor (minimum-maximum speed) in the shut-off position. *(Courtesy of Robert Bosch Canada Ltd.)*

PART 9   VARIABLE-SPEED GOVERNOR RQV

Construction

The RQV governor is constructed similarly to the RQ governor, but it is not identical (Fig. 21-17 and 21-18). In the RQV, the governor springs are built into the flyweights, but the flyweights move continuously outward within the specified adjustment range as the speed increases. A certain speed at which speed regulation begins is associated with each position of the control lever. Movements of the control lever are transmitted through the two-part linkage lever and the guide block to the fulcrum lever, and thus to the control rod. The pivot point of the fulcrum lever can be shifted. In addition, the pivot point is guided in a plate cam fastened on the governor housing so that the transmission ratio of the fulcrum lever changes in the range of 1:1.7 to 1:5.9.

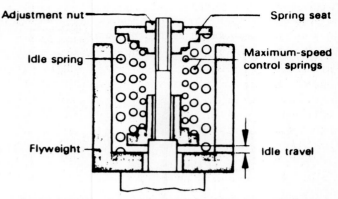

FIGURE 21-18. Flyweight and springs in RQV governor. *(Courtesy of Robert Bosch Canada Ltd.)*

The sliding bolt, being the connecting element between the flyweight assembly and the fulcrum lever, is spring-loaded for pressure and tension (drag spring).

As in the RQ governor, the spring sets built into the flyweights consist generally of three concentri-

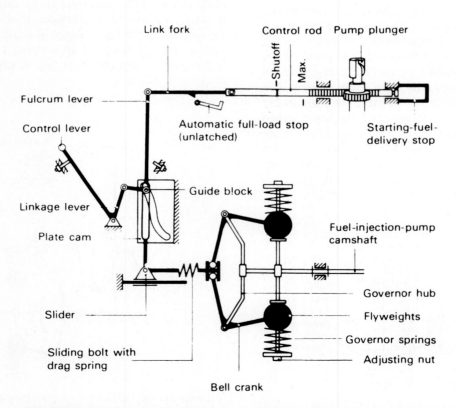

FIGURE 21-17. Schematic drawing of RQV variable-speed governor. *(Courtesy of Robert Bosch Canada Ltd.)*

cally arranged cylindrical helical springs. The outer spring serves for low-idle-speed control and is supported between the flyweight and the adjusting nut for the spring preload. After moving across the short idle-speed travel path, the flyweight is positioned against the spring seat; the inner springs, which are installed between the spring seat and the adjusting nut, also take effect. From this point on, all the springs act together to control the speeds set by the control lever.

Operating Characteristics

Engine Stopped. The control lever is positioned against the shut-off stop, and the control rod is in the shut-off position.

Starting Position. (Fig. 21–19) The control lever is positioned against the maximum-speed stop; the control rod moves to the starting-fuel-delivery stop.

Idle Position. (Fig. 21–20) After the engine has started to operate and the control lever (accelerator pedal) has been released, this lever returns to the idle position. The control rod also returns to the idle position, which is determined by the now operating governor.

Load on Engine. (Fig. 21–21 and 21–22) If the load on the engine at any speed set by the control lever (accelerator pedal) is increased or decreased, the variable-speed governor maintains the set speed by increasing or decreasing the amount of fuel delivered within the associated speed droop.

For example, the driver has moved the control lever from the idle position to a position intended to

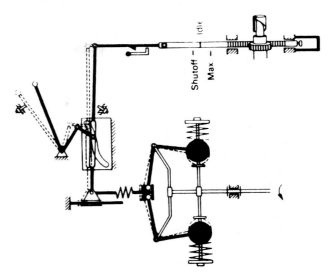

FIGURE 21–20. RQV governor in idle position. *(Courtesy of Robert Bosch Canada Ltd.)*

correspond to a desired vehicle speed. The movement of the control lever is transmitted through the linkage to the fulcrum lever. The transmission ratio of the fulcrum lever is variable and immediately above the idle-speed range becomes so large that even a relatively small part of the total control-lever or flyweight travel is adequate to shift the control rod to the set full-load stop. A control-rod stop (fixed or adjusted by hand; in no case spring-loaded) must therefore be available. Additional swiveling move-

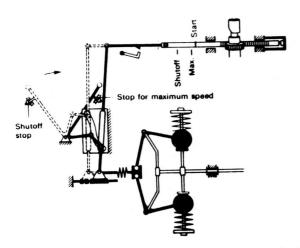

FIGURE 21–19. RQV governor in start (maximum fuel) position. *(Courtesy of Robert Bosch Canada Ltd.)*

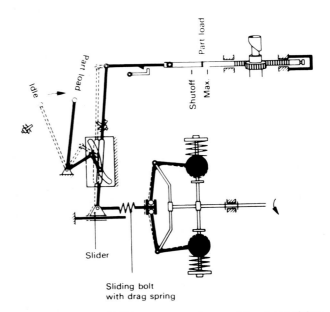

FIGURE 21–21. RQV governor in part-load position. *(Courtesy of Robert Bosch Canada Ltd.)*

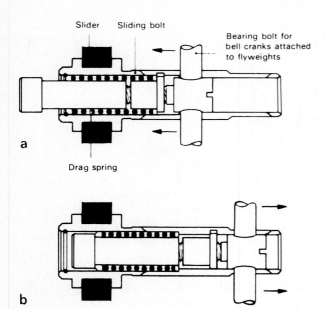

FIGURE 21-22. Sliding bolt with drag spring. (a) During acceleration or when the engine is overloaded, the flyweights have shifted the control rod to the full-load stop. The drag spring is tensioned. (b) Vehicle traveling downhill, engine driven by vehicle. The flyweights have shifted the control rod to the shutoff stop, the drag spring is tensioned and absorbs further movement of the flyweights.

ment of the control lever results in the drag spring being tensioned. The control rod remains temporarily in the maximum-fuel-delivery position, and this results in a rapid increase in engine speed. This in turn forces the flyweights outward, but the control rod remains in the maximum-fuel-delivery position until the tension on the drag spring is released. Then the flyweights start to act on the fulcrum lever, and the control rod is thereby shifted in the shut-off direction. As a result, the amount of fuel delivered becomes smaller and the engine speed is restricted. This engine speed limit corresponds to the position of the control lever and to the flyweight sotp.

During operation, therefore, one specific rotational-speed range is associated with every position of the control lever as long as the engine is not overloaded or driven by the vehicle when traveling downhill (overrun). If the engine loading becomes somewhat greater, for example when traveling uphill, the engine and governor speeds decrease. As a result, the flyweights move inward and shift the control rod to the maximum-delivery direction. This holds the engine speed constant at a level determined by the position of the control lever and by the

speed droop. However, if the uphill slope (loading) is so great that, even though the control rod is shifted all the way to the maximum-fuel-delivery stop, the speed nevertheless still decreases, the flyweights are brought further inward in accordance with this speed, and they shift the sliding bolt to the left.

The flyweights therefore tend to shift the control rod farther in the maximum-fuel-delivery direction. However, since the control rod is already positioned against the full-load stop and cannot move farther in the maximum-fuel-delivery direction, the drag spring is tensioned. This means that the engine is overloaded. In this case the driver must shift to a lower gear.

When traveling downhill the opposite situation prevails. The engine is driven by the vehicle and its speed increases. As a result, the flyweights are forced outward and the control rod is shifted in the shut-off direction until it reaches the stop. If the engine speed then increases still further, the drag spring is tensioned in the opposite direction (control rod in shut-off position).

Such behavior of the governor applies basically for all positions of the control lever should for any reason the engine loading or speed change so greatly that the control rod is brought up against one of its terminal stops, i.e., maximum fuel delivery or shut-off.

Torque Control

In the RQV governor, the torque-control mechanism is built into the control-rod stop. For additional information on this point see control-rod stops described later in this section.

Full-Load Speed Regulation

If the engine exceeds its maximum speed, full-load speed regulation starts. During this process the flyweights move outward and the control rod shifts in the shut-off direction. If the entire load on the engine is removed, the high-idle speed is attained. (See Fig. 21-23.)

Variable-Speed Governor RQUV

The RQUV governor is used for regulating very low rotational speeds, for example, the speeds at which marine engines operate. It is a variant of the RQUV governor and affects two different speed-increasing ratios (about 1:2.2 or 1:3.76) between the driving element (i.e., the fuel-injection-pump camshaft) and the governor hub. Similar to the lever ratio in the RQV

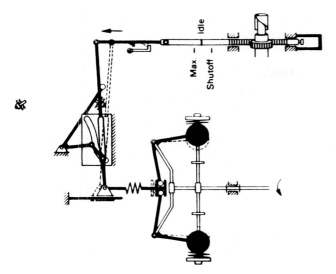

FIGURE 21-23. RQV governor maximum-speed regulation under full load. *(Courtesy of Robert Bosch Canada Ltd.)*

governor, the ratio of the fulcrum lever is variable here also (from 1:1.85 to 1:7). The RQUV governor can be used for PE. .P and PE. .ZW fuel-injection pumps. The operation and operating characteristics of this governor are similar to those of the RQV. (See Fig. 21-24.)

Variable-Speed Governor RQV-K

The RQV-K governor has basically the same flyweight assembly, with governor springs built into

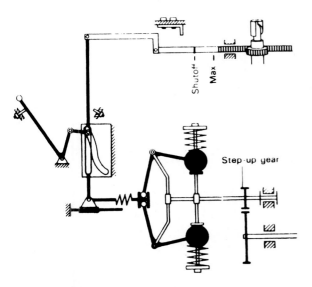

FIGURE 21-24. Schematic drawing of RQUV variable-speed governor. *(Courtesy of Robert Bosch Canada Ltd.)*

the flyweights, as the RQV governor. The essential point of difference lies in the type of torque control. While in all other governors, torque control is based for practical purposes on a certain reduction in the amount of fuel delivered at full load and when the speed increases, the full-load delivery with the RQV-K design can be increased somewhat as well as decreased (Fig. 21-25).

Torque control takes place in the RQV-K governor because the rocker on the upper end of the fulcrum lever runs along a rocker guide (i.e., a curved track) at the full-load stop, the guide being designed to reflect the varying fuel requirement of the engine. The strap forming the connection between the fulcrum lever and the control rod transmits this movement to the control rod. As a result, a full-load fuel delivery corresponding to the desired variations in torque is developed.

Depending on the shape of the curve, the fuel delivery can be increased as well as decreased.

The full-load stop can be shifted in a longitudinal direction by an adjusting screw in order to set the fuel delivery; by means of another adjusting screw, the inclination of the rocker guide—and thus the steepness of the torque-control characteristic—can also be varied.

Operating Characteristics

Engine Stopped. The control lever is positioned against the shut-off stop, and the control rod is in the shut-off position.

Starting the Engine. (Fig. 21-26) Move the governor control lever to the maximum-speed position. As this is done, the rocker swings under the full-load stop, and the control rod shifts to the starting-fuel-delivery position. A stop for the starting fuel delivery is located on the fuel injection pump.

When the starting motor has been switched on, the fuel injection pump delivers the quantity of fuel required for starting (starting fuel delivery) through the injection nozzle into the engine.

Idle. (Fig. 21-27) When the engine has started to operate independently, the control lever is brought back to the idle position. As this takes place, the spring-loaded rocker slides back under the full-load stop to the idle position. The engine then operates at its idle speed.

Full-Load Delivery at Low Speed. (Fig. 21-28) If the control lever is moved (e.g., from the idle position to the maximum-speed position), the guide block moves along the rocker guide in the plate cam and simultaneously downward in the fulcrum lever

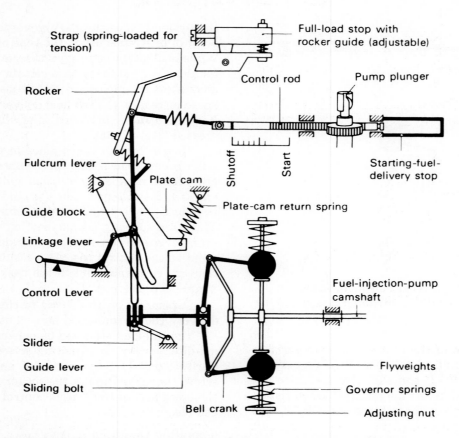

FIGURE 21-25. Schematic drawing of RQV-K variable-speed governor. *(Courtesy of Robert Bosch Canada Ltd.)*

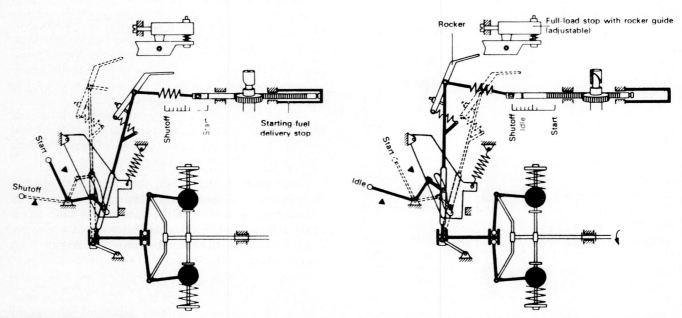

FIGURE 21-26. RQV-K governor in start (maximum fuel) position. *(Courtesy of Robert Bosch Canada Ltd.)*

FIGURE 21-27. RQV-K governor in idle position. *(Courtesy of Robert Bosch Canada Ltd.)*

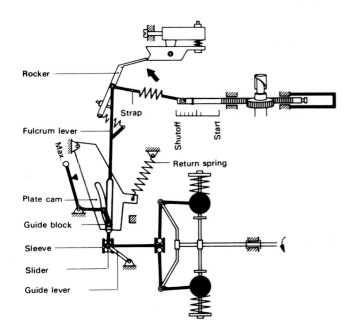

FIGURE 21-28. RQV-K governor at full-load delivery—low speed, start of negative torque control. *(Courtesy of Robert Bosch Canada Ltd.)*

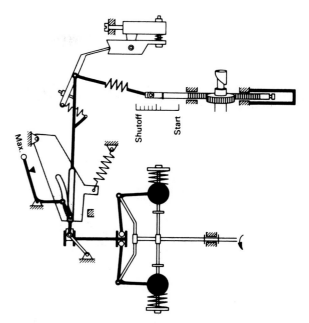

FIGURE 21-29. RQV-K governor at full-load delivery—medium speed, reversal of torque control. *(Courtesy of Robert Bosch Canada Ltd.)*

guide. As this movement takes place, the fulcrum lever swings to the right around the pivot point at the slider and, acting through the strap, shifts the control rod in the full-load direction. Fuel delivery increases and the engine speed rises.

The flyweights move outward and the sleeve is shifted somewhat to the right. As a result, a swiveling movement is developed; the guide lever and the fulcrum lever are raised so that the rocker slides along the rocker guide at the full-load stop.

As the slider moves downward in the fulcrum-lever guide track, the plate cam is raised from its stop on the housing against the force of the return spring.

Full-Load Delivery at Medium Speed. (Fig. 21-29) [with torque control]). If the speed continues to rise, the flyweights are forced farther outward and the rocker slides along the rocker guide at the full-load stop. Until the curve changes direction at B, torque control takes place in the sense of increasing the full-load fuel delivery as the speed rises (negative torque control); after the curve changes direction, torque control takes place in the sense of decreasing the full-load delivery (positive torque control, B-C on the governor characteristic curve.

End of Torque Control, Start of Speed Regulation. (Fig. 21-30) At the end of torque control, when

speed regulation starts, the plate cam is again positioned against the stop on the housing.

If the speed continues to increase, regulation of the high-idle speed starts. The flyweights move farther outward, and the sleeve moves correspondingly to the right. As a result, the fulcrum lever swings around the pivot point at the guide block with its upper part to the left. As these actions take place, the control rod moves in the shut-off direction (C-D on the governor characteristic curve).

During operation, one specific rotational-speed range is associated with every position of the control lever as long as the engine is not overloaded or driven by the vehicle when traveling downhill. If the engine loading becomes somewhat greater (for example, when traveling uphill), the engine and governor speeds decrease. As a result, the flyweights move inward and shift the control rod in the maximum-delivery direction. This holds the engine speed constant at a level determined by the position of the control lever (or accelerator pedal). However, if the engine loading (uphill slope) is so great that, even though the control rod is shifted all the way to the maximum-fuel-delivery stop, the speed nevertheless decreases, the flyweights are brought farther inward in accordance with this speed, and they shift the sleeve farther in the maximum-fuel-delivery direction.

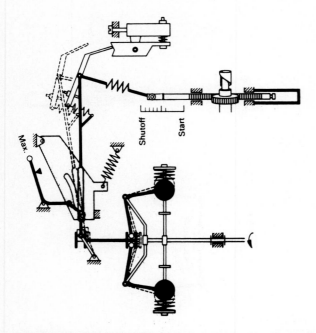

FIGURE 21–30. RQV-K governor at maximum full-load speed end of positive torque control (dotted pattern: speed regulation). *(Courtesy of Robert Bosch Canada Ltd.)*

However, since the control rod cannot move any farther in the maximum-fuel-delivery direction, the lower part of the fulcrum lever moves to the left against the resistance of the return spring for the plate cam and thereby raises the plate cam from its stop.

When traveling downhill, the opposite situation prevails: The engine is driven by the vehicle and its speed increases. As a result, the flyweights are forced outward and the control rod is shifted in the shut-off direction. If the engine speed then increases further (control rod in the shut-off position), the spring-loaded strap, which connects the fulcrum lever with the control rod, yields. If the driver slows the vehicle down somewhat by applying the brakes or shifts into a higher gear, the strap is again shortened to its normal length.

This behavior of the governor applies basically for all positions of the control lever should for any reason the engine loading or speed change so greatly that the control rod is brought up against one of its terminal stops, i.e., maximum-fuel-delivery or shut-off.

Variable-Speed Governor EP/RSV

Construction

Governor EP/RSV is constructed differently from the comparable type RQV (Fig. 21–31). It has only one governor spring, and this spring can swivel. When setting the speed at the control lever, the position and tension of this spring change so that the effective torque at the tensioning lever is in equilibrium with the torque developed by the flyweights at the desired speed. All adjustments of the control lever and the flyweight travel are transmitted through the governor linkage to the control rod.

The starting spring attached to the upper end of the fulcrum lever pulls the control rod to the starting position, automatically setting the starting fuel delivery.

A full-load stop and a torque-control mechanism are built into the governor. In order to stabilize the idle speed, an auxiliary idle-speed spring with an adjusting screw is built into the governor cover. The speed droop can be varied within certain limits by means of the governing spring adjusting screw. Lighter flyweights are required for higher speed ranges; when using springs that are under less tension, it is possible to set a smaller speed droop at lower speeds.

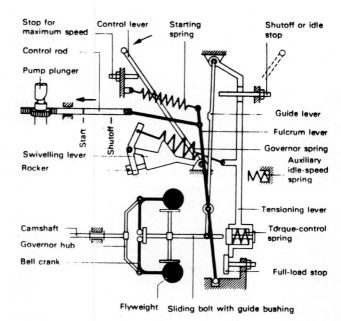

FIGURE 21–31. EP/RSV governor schematic. *(Courtesy of Robert Bosch Canada Ltd.)*

Operating Characteristics

Starting the Engine. (Fig. 21-32) When the engine is stopped, the control rod is in the starting-fuel-delivery position so that the engine can be started with the control lever in the idle position.

Idle Position. (Fig. 21-33) The control lever is positioned against the idle stop. As a result, the governor spring is almost completely relaxed and stands almost vertically. It has a very weak effect, so that the flyweights swing outward at even a low speed. The sliding bolt, and with it the guide lever, therefore moves to the right. In turn, the guide lever swings the fulcrum lever to the right, so that the control rod is moved in the shut-off direction to the idle position. The tensioning lever is positioned against the auxiliary idle-speed spring, which assists the low-idle-speed control.

Regulation of Low Speeds. (Fig. 21-34) Even a relatively small shift of the control lever from the idle position is sufficient to move the control rod from its initial position to its full-load position. The fuel injection pump delivers the full load fuel quantity into the engine cylinders, and the speed rises. As soon as the centrifugal force is greater than the tension of the governor spring corresponding to the position of the control lever, the flyweights swing outward and shift the guide bushing, sliding bolt,

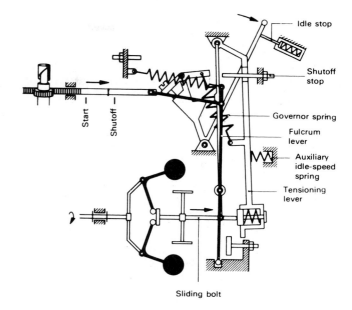

FIGURE 21-33. EP/RSV governor in idle position. *(Courtesy of Robert Bosch Canada Ltd.)*

fulcrum lever, and control rod back to a point of smaller fuel delivery. The speed of the engine does not increase further and is held constant by the governor as long as external conditions remain uniform.

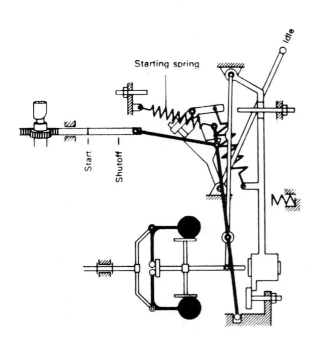

FIGURE 21-32. EP/RSV governor in start (maximum fuel) position. *(Courtesy of Robert Bosch Canada Ltd.)*

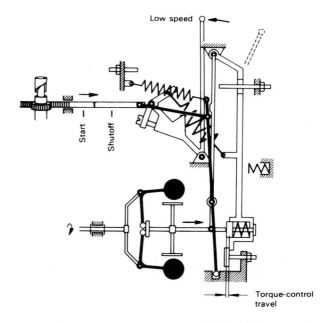

FIGURE 21-34. EP/RSV governor at full-load, low-speed, start of torque control position. *(Courtesy of Robert Bosch Canada Ltd.)*

Regulation at High-Idle Speed. (Figs. 21-35 and 21-36) If the control lever is moved to the maximum-speed stop, the governor operates in basically the same manner as described above. In this case, however, the swiveling lever tensions the governor spring completely.

The governor spring thus acts with a greater force, drawing the tensioning lever to the full-load stop and the control rod to maximum fuel delivery. The engine speed increases and the centrifugal force steadily rises.

In governors equipped with torque control, as soon as the tensioning lever is positioned against the full-load stop, the torque-control spring is steadily compressed as the speed increases. As a result, the guide lever, fulcrum lever, and control rod move in the shut-off direction accordingly and "torque-control" the fuel delivery; i.e., they reduce the delivery by an amount corresponding to the torque-control travel.

When the maximum full-load speed is reached, the centrifugal force overcomes the governor spring tension, and the tensioning lever is deflected to the right. The sliding bolt with the guide lever and the control rod, coupled through the fulcrum lever, move in the shut-off direction, until, under the new loading conditions, a correspondingly lower fuel delivery has been established.

If the entire load on the engine is removed, the high-idle speed is attained.

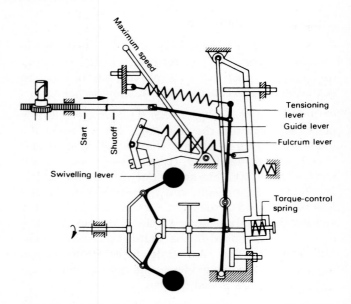

FIGURE 21-36. EP/RSV governor at no-load, regulated from full load. *(Courtesy of Robert Bosch Canada Ltd.)*

Stopping the Engine: With the Control Lever. (Fig. 21-37) Engines with governors that do not have a special stopping mechanism are stopped by moving the governor control lever to the shut-off position. As this is done, the lugs on the swiveling lever (inclined arrow) press on the guide lever. This lever swings to the right, taking the fulcrum lever, and thus the control rod as well, to the shut-off po-

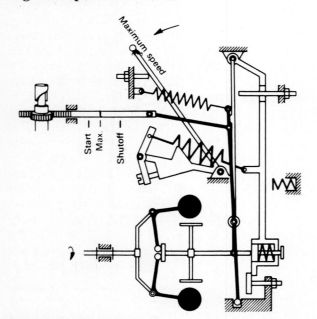

FIGURE 21-35. EP/RSV governor at full-load, maximum speed, end of torque control and start of full-load speed regulation. *(Courtesy of Robert Bosch Canada Ltd.)*

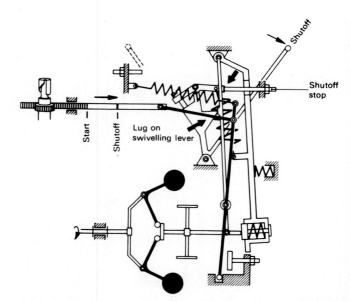

FIGURE 21-37. EP/RSV governor, stopping the engine with the governor control lever. *(Courtesy of Robert Bosch Canada Ltd.)*

sition with it. Since the tension exerted by the governor springs on the sliding bolt is released, the flyweights swing outward.

With the Stop Lever. (Fig. 21–38) In the case of governors fitted with a special shut-off mechanism, the control rod can be set to shut off if the stop lever is moved to the shut-off position.

When the stop lever is pressed to shut-off, the upper part of the fulcrum lever is swung to the right around the pivot point C in the guide lever. As a result, the control rod is drawn by the strap to shut-off. When the stop lever is released, a return spring not shown in the drawing brings it back into the initial position.

Variable-Speed Governor EP/RSUV

The RSUV governor is used to control very low speeds, for example, those at which low-speed marine engines operate (Fig. 21–39 and 21–40). In terms of its construction, it differs essentially from the EP/RSV governor in its transmission gear for speed-increasing ratio (step-up gear), which is installed between the driving element (i.e., the fuel-injection-pump camshaft), and the governor hub. The operation of this governor is basically the same as that of the EP/RSV. It is used with the fuel injection pumps of sizes P and Z.

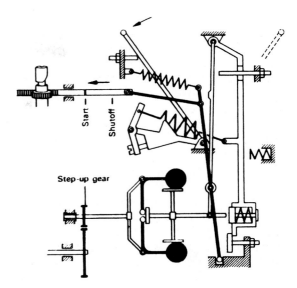

FIGURE 21–39. Schematic drawing of the variable-speed governor, EP/RSUV, maximum-speed position. *(Courtesy of Robert Bosch Canada Ltd.)*

PART 10   MINIMUM-MAXIMUM-SPEED GOVERNOR EP/RS

Construction

The EP/RS minimum-maximum-speed governor is a governor with only slight control-lever forces; it was developed from the EP/RSV variable-speed governor. The control lever that tensions the swiveling spring in the EP/RSV governor, thus serving to set the speed, is blocked in the current design of the EP/RS in the maximum-speed position by an adjustable stop on the governor cover. It is also possible to set an intermediate speed, for example, in the case of vehicles with an auxiliary drive. The stop lever used with the EP/RSV governor operates in the EP/RS as an accelerator pedal (lever) with a reversed actuating direction. (See Fig. 21–41.)

In addition to the torque-control spring, a supplementary idle-speed spring is built into the spring retainer for low-idle-speed control; this supplementary spring brings the control rod to the starting-fuel-delivery position, thus controlling the idle speed. The idle-speed stop screw and the auxiliary idle-speed spring in the ER/RSV governor have been eliminated.

Operating Characteristics

Start Position. (Fig. 21–42) Set the accelerator lever (pedal) to the full-load position. The idle-speed

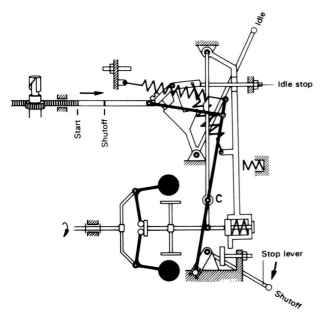

FIGURE 21–38. EP/RSV governor, stopping the engine with the shut-off mechanism. *(Courtesy of Robert Bosch Canada Ltd.)*

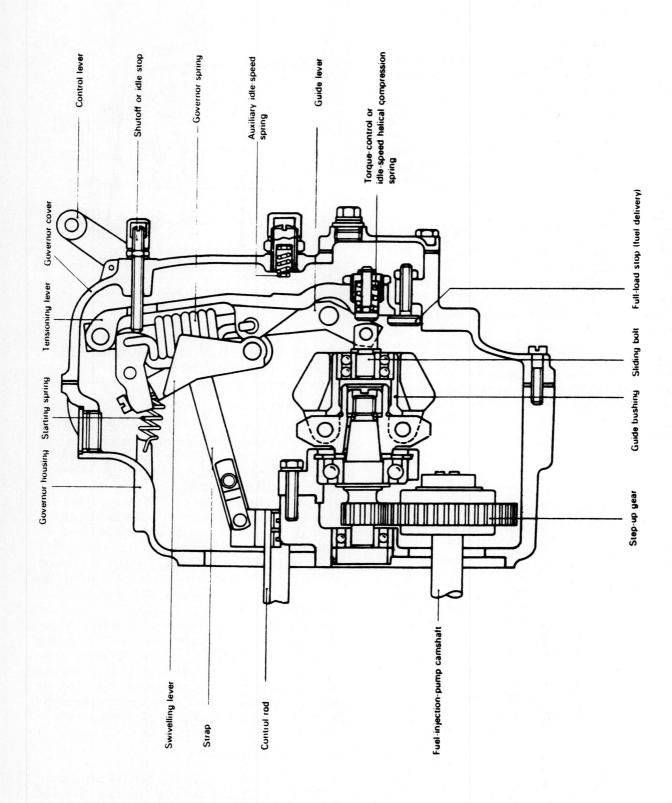

Control lever

Shutoff or idle stop

Governor spring

Auxiliary idle-speed spring

Guide lever

Torque-control or idle-speed helical compression spring

Governor cover

Tensioning lever

Full-load stop (fuel delivery)

Sliding bolt

Guide bushing

Governor housing   Starting spring

Swivelling lever

Strap

Control rod

Step-up gear

Fuel-injection-pump camshaft

FIGURE 21-40. EP/RSUV variable-speed governor, cutaway view. *(Courtesy of Robert Bosch Canada Ltd.)*

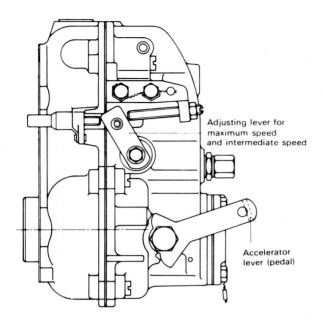

FIGURE 21-41. Minimum-maximum speed EP/RS governors, external view. *(Courtesy of Robert Bosch Canada Ltd.)*

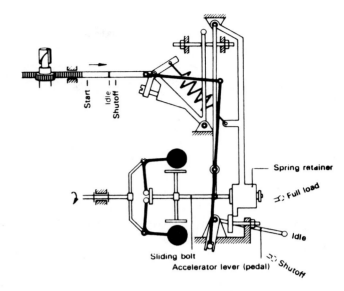

FIGURE 21-43. EP/RS governor in idle position. *(Courtesy of Robert Bosch Canada Ltd.)*

spring in the spring retainer shifts the control rod, acting through the sliding bolt, guide lever, fulcrum lever, and strap, to the start position.

Idle-Speed Position. (Fig. 21-43) After the engine has started to operate, the accelerator lever (pedal) is moved to the idle-speed position. The idle-speed spring built into the spring retainer presses against the sliding bolt through the thrust pin and controls the idle speed.

Loading the Engine (Between the Idle Speed and Maximum Speed). (Fig. 21-44) In this range the driver sets the fuel delivery with the accelerator pedal according to the torque required. If the driver wants to increase the driving speed or if the person

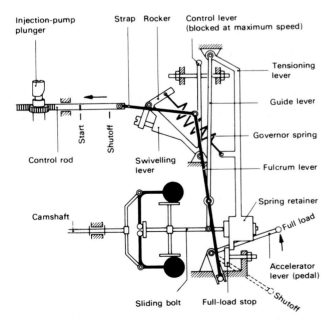

FIGURE 21-42. EP/RS governor in start position. *(Courtesy of Robert Bosch Canada Ltd.)*

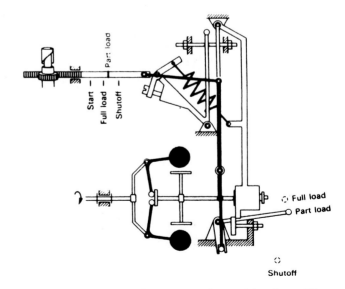

FIGURE 21-44. EP/RS governor in part-load position. *(Courtesy of Robert Bosch Canada Ltd.)*

must drive up a hill, the driver must accelerate; similarly one must ease off the accelerator pedal if a lower engine power is required. Torque control of the fuel delivery takes place because the centrifugal force acting on the sliding bolt exceeds the force of the torque-control spring built into the spring retainer. The torque-control spring yields in accordance with the centrifugal force acting through the sliding bolt, and the fulcrum lever shifts the control rod in the shut-off direction by the amount of the torque-control travel.

Maximum-Speed Regulation at Full Load. (Fig. 21–45) The accelerator lever (pedal) is in the full-load position, the maximum speed is reached. Now the centrifugal force exceeds the force of the governor spring. For this reason, the sliding bolt and the guide lever move to the right. The fulcrum lever pulls the control rod in the shut-off direction. When the entire load is removed from the engine, the engine reaches the high-idle speed.

Stopping the Engine. (Fig. 21–46) Move the accelerator lever (pedal) to the shut-off position. The fulcrum lever swings around the bearing point C and pulls the control rod to shut-off. The flyweights move inward.

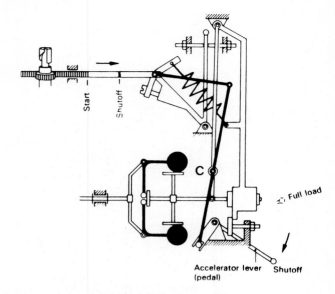

FIGURE 21-46. EP/RS governor in shut-off position. *(Courtesy of Robert Bosch Canada Ltd.)*

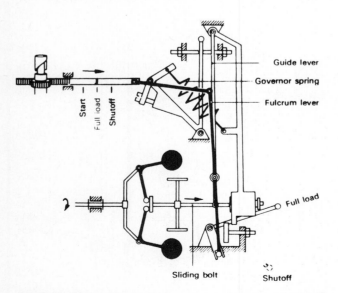

FIGURE 21-45. EP/RS governor start-of-speed regulation, at maximum speed and full load. *(Courtesy of Robert Bosch Canada Ltd.)*

PART 11   CONTROL-LEVER AND CONTROL-ROD STOPS FOR MECHANICAL GOVERNORS

Control-Lever Stops

On the governor cover are two stop screws, one for shut-off and one for the full-load delivery (maximum speed) (Fig. 21–47).

If desired, and depending on the type of governor (RQ or RQV), a stop can also be installed for the low-idle speed or for an intermediate speed (fuel delivery), which is lower than the full-load delivery.

Spring-Loaded Idle-Speed Stop

The spring-loaded idle-speed stop consists of a sleeve with an external thread and from which a bolt under spring tension projects. (See Fig. 21–48.)

At the idle-speed fuel delivery, the stop lever is positioned against the spring-loaded bolt. In order to stop the engine, the governor control lever must be moved into the shut-off position against the force of the helical compression spring until the engine has come to a stop.

Reduced-Delivery Stop

This stop serves as a fixed adjustment point for a fuel delivery lower than the full-load delivery or for an intermediate speed (depending on the type of governor). It is mounted on the governor cover and

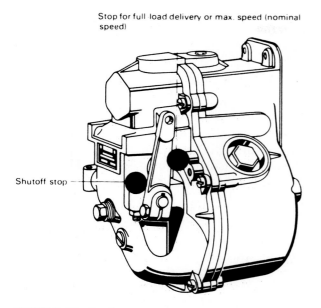

Stop for full load delivery or max. speed (nominal speed)

Shutoff stop

FIGURE 21-47. Shut-off and full-load stop screws location. *(Courtesy of Robert Bosch Canada Ltd.)*

acts together with a short lever fastened on the control-lever shaft in such a way that it can be adjusted. (See Fig. 21–49 and 21–50.)

The schematic cross-sectional drawing shows a spring-loaded bolt that can be shifted in the housing by means of a shaft with a furrow. In one end position, the lever strikes against the bolt and limits the travel of the control lever.

In the other end position, the bolt releases the lever, and the control lever can reach its end position.

Control-Rod Stops

In addition to the stops for shut-off and for full-load delivery or maximum speed (parts required in every governor to define the control-lever travel), a special stop is required for the control rod which limits the travel of this rod at full-load or starting fuel delivery. Depending on the particular purpose and use, there are various designs of control-rod stop: rigid and spring-loaded designs; stops for the full-load delivery, with mechanical or electromagnetic unlocking for the starting fuel delivery; and stops with a built-in torque-control mechanism. In addition, there

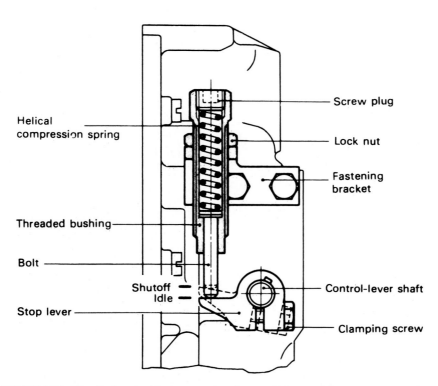

Helical compression spring

Threaded bushing

Bolt

Shutoff
Idle

Stop lever

Screw plug

Lock nut

Fastening bracket

Control-lever shaft

Clamping screw

FIGURE 21-48. Spring-loaded control rod stop, RQ and RQV governors. *(Courtesy of Robert Bosch Canada Ltd.)*

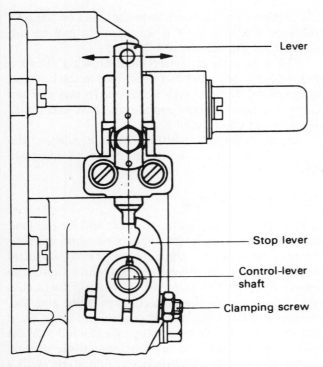

FIGURE 21–49. Reduced fuel delivery stop. *(Courtesy of Robert Bosch Canada Ltd.)*

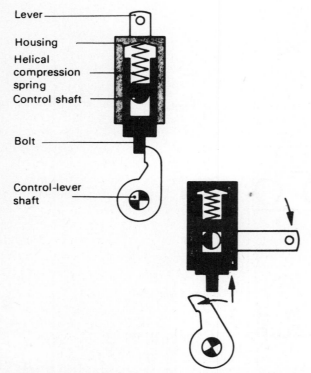

FIGURE 21–50. Reduced delivery stop cross-sectional view and operation. *(Courtesy of Robert Bosch Canada Ltd.)*

are full-load stops designed to carry out special compensation functions. Control-rod stops are produced for mounting on the fuel injection pump or on the governor.

Rigid Excess-Fuel Stop for Starting

The rigid stop for starting-fuel delivery is used mainly in RQ governors with a low-idle speed. When the engine is running, the excess fuel for starting is withdrawn through the governor and cannot have a damaging effect (such as would be caused by the development of smoke). (See Fig. 21–51.)

Spring-Loaded Control-Rod Stop for RQ Governors

If the accelerator pedal is pressed all the way down during the starting process, the stop bolt moves against the resistance of the spring to the set starting-fuel-delivery position. The spring built into the stop acts against the idle-speed spring and thus causes an early shift of the control rod back from the start position. This prevents a brief interim period of starting fuel delivery if the engine is accelerated rapidly from idling.(See Fig. 21–52.)

Automatic Full-Load Control-Rod Stop

When the engine is at rest, the flyweight governor springs, acting through the sliding bolt, press on the rocker-arm spring; as a result, the rocker-arm forces the stop strap with the full-load stop lug downward.

Therefore, when the engine is started, the control rod can be shifted to the start position when the accelerator pedal is pressed down (Figure 21–53.)

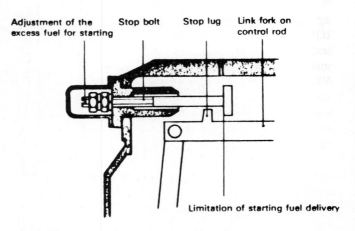

FIGURE 21–51. Rigid control rod stop for RQ governors to limit the starting fuel delivery. *(Courtesy of Robert Bosch Canada Ltd.)*

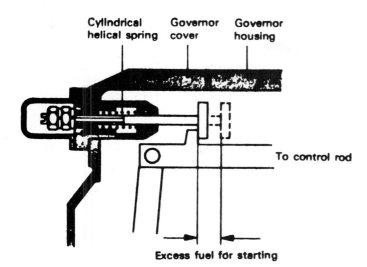

FIGURE 21-52. Spring-loaded control rod stop for RQ governors to limit the starting fuel delivery. *(Courtesy of Robert Bosch Canada Ltd.)*

After the engine has been started, the sliding bolt moves away from the rocker arm under the influence of centrifugal force. For the same reason, the control rod moves back from the starting-fuel-delivery position to a position of smaller delivery. As a result, the rocker-arm spring presses the rocker arm with its long lever arm upward, and the lug on the stop strap again limits the travel of the control rod at the stop piece on the link fork to the full-load delivery.

Control-Rod Stop with External Torque-Control Mechanism (For RQV Governors)

The control-rod stop with a draw lever for the starting fuel delivery and with torque control is used with RQV governors. When the control rod travel is limited to the full-load delivery, the adjusting screw is positioned against the edge of the locking bolt. When the draw lever is pulled, this locking bolt is turned 90°. As a result, the control-rod can shift toward the starting position by the dimension of the milled section on the locking bolt (release of the excess fuel for starting). (See Fig. 21-54.)

Torque control takes place as a result of the interaction between the drag spring in the governor and the torque-control spring; these two springs must be matched exactly to each other for this purpose.

If a higher speed is set at the control lever, the drag spring is tensioned for the duration of the acceleration; as a result, the torque-control spring is also compressed and sets a greater control-rod travel. As the speed increases, the flyweights move outward and the drag spring is relaxed. The force of the torque-control spring is then stronger, so that this spring shifts the control rod in the direction of smaller fuel delivery.

Adjustment of Torque Control

The start of torque control can be set by varying the initial tension of the torque-control spring. This initial tension of the torque-control spring is varied with the adjusting screw. The torque-control spring is screwed onto a thread bushing by one or more thread turns, and the torque-control characteristic can be varied by turning this bushing to change the number of threads holding the spring. The torque-control travel is set by using shims of different thicknesses.

Control-Rod Stop with Internal Torque-Control Mechanism (For RQV Governors)

In those cases where there is not enough room available for installation of the control-rod stop for RQV governors with external torque control, the control-rod stop with internal torque control can be used. The installed length of this control-rod stop is less than a quarter of the installed length of the stop with external torque control. (See Fig. 21-55 to 21-57.)

When the engine accelerates, the control rod is shifted in the maximum-fuel-delivery direction by the fulcrum lever acting through the strap. At the same time the drag spring in the sliding bolt is tensioned, the stop bolt is brought up against the full-load stop, and the torque-control spring is compressed through the rocker (start of torque control). When the flyweights move outward as the speed increases, the drag spring relaxes again. Starting at a certain speed, the force of the torque-control spring overcomes the force of the drag spring and the torque-control spring relaxes. As a result, the rocker turns and the stop bolt positioned at the full-load stop pulls the control rod in the shut-off direction until it reaches the other end of the torque-control path. The rocker is then positioned against its upper stop (end of torque control).

PART 12   MANIFOLD-PRESSURE COMPENSATOR (LDA)

In pressure-charged engines the full-load delivery is determined according to the charge-air pressure. In

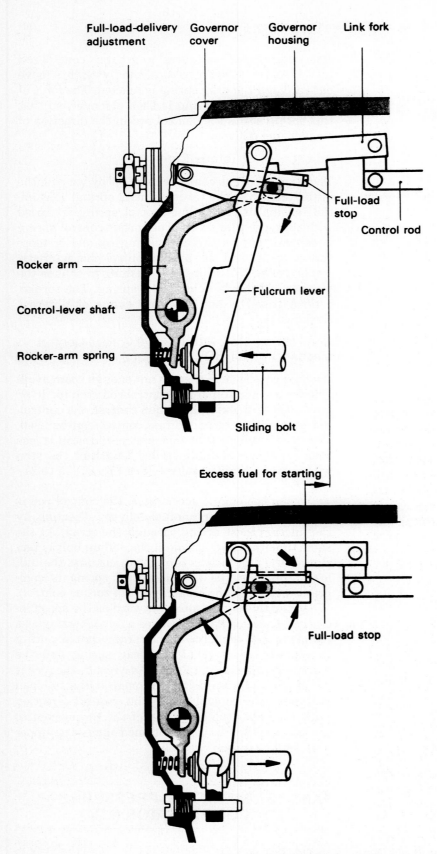

FIGURE 21-53. Automatic full-load control rod stop for RQ governors. Release of starting fuel delivery (top) and limitation to full-load fuel delivery (bottom). *(Courtesy of Robert Bosch Canada Ltd.)*

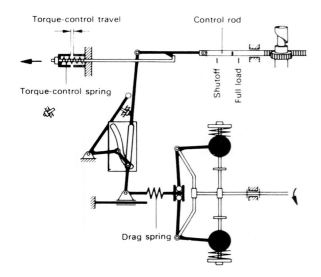

FIGURE 21–54. Schematic drawing of the control rod stop with torque control mechanism (RQV) and torque control springs stronger than the drag spring. *(Courtesy of Robert Bosch Canada Ltd.)*

the lower speed range, however, the charge-air pressure is lower; therefore, the weight of the air charge in the engine cylinders is also lower. For this reason, the full-load delivery must be matched in a corresponding ratio to the reduced weight of the air. This function is carried out by the manifold-pressure compensator (LDA), which reduces the full-load delivery in the lower speed range starting from a certain (selectable) charge-air pressure (Fig. 21–58 to 21–60). Various designs of the LDA are produced for mounting on fuel-injection pumps and on either the side or the top of governors. The following description refers to an LDA designed for mounting on the RSV governor.

The construction of all these special control-rod stops is basically identical. Between the housing bolted onto the top of the governor and a suitable cover is a diaphragm that is tensioned and air-tight. A connector fitting for the charge-air pressure is located in the cover. From below, a helical compression spring acts on the diaphragm; this spring is supported at its other end on a guide bushing attached to the housing by means of a thread. The initial tension of this spring can therefore be changed within certain limits.

A threaded pin is attached to the diaphragm through a plate washer and a guide washer; at the lower end of this pin, which projects out of the housing, a screw with a lock nut is attached. The head of this screw is set to a certain distance from the housing surface and transmits the movement of the

threaded pin through a bell crank to the control rod. This distance is preset, but after the LDA has been mounted corrections can be made with the headless setscrew.

If charge-air pressure is applied to the diaphragm, the threaded pin moves against the force of the helical compression spring, traveling the greatest distance at the full charge-air pressure. Movement of the threaded pin is transmitted through the bell crank, which is supported in the governor housing on an axle so that it can turn, to the strap attached to the fuel-injection-pump control rod. As the charge-air pressure decreases, the control rod is moved in the shut-off direction.

In order to permit the control rod to be moved into the starting-fuel-delivery position when the engine is started, the bell crank can be disengaged from the strap by lateral movement of the control shaft. This can be done manually either with a control cable or through a linkage system. Governor designs also exist, however, with electromagnetic activation of the control shaft; in these designs, the electromagnet takes effect only during starting.

PART 13   ALTITUDE-PRESSURE COMPENSATOR (ADA)

In countries or regions where road traffic is subject to extremely wide variations in altitude, the amount of fuel injected into the engine must be matched to the worsening air charge in the engine cylinders from a certain altitude upward. This function is carried out by the altitude-pressure compensator (ADA) (see Figures 21–61 and 21–62).

The ADA is used in conjunction with mechanical governors RQ or RQV and is mounted on the governor cover. Basically, the ADA consists of an aneroid capsule, installed vertically in a housing; the capsule can be set to a certain altitude by means of an adjusting screw and an opposing spring-loaded threaded bolt. Within the effective range of the aneroid capsule, the length of the cell increases as the air pressure decreases. The spring-loaded threaded bolt at the bottom of the aneroid capsule and the fork attached to the threaded bolt transmit the changes in length to the swivel-mounted cam plate. This cam plate acts on the bolt that is connected to the fuel-injection-pump control rod.

As the aneroid capsule expands, the cam plate swings downward. The bolt connected with the stop strap pulls the control rod in the shut-off direction, and the fuel delivery decreases; if the length of the

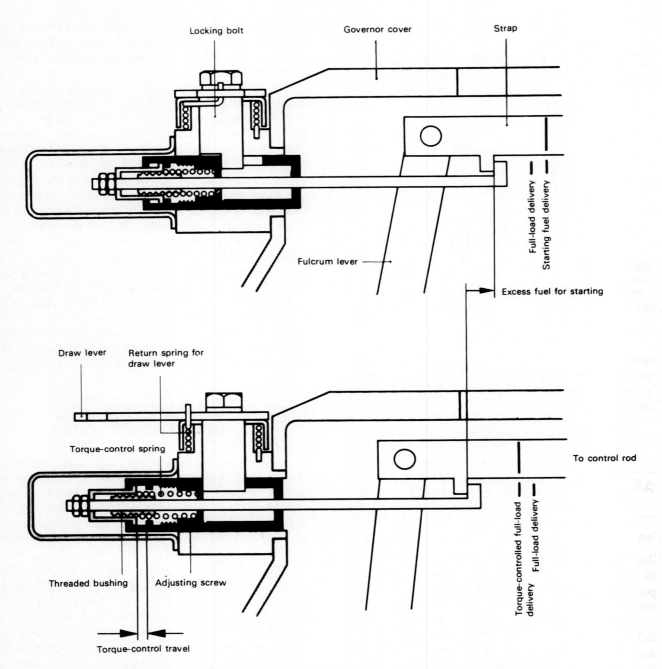

FIGURE 21-55. Control-rod stop for RQV governors with draw lever for starting fuel delivery and with torque-control mechanism. Upper drawing: starting-fuel-delivery position; lower drawing: full-load-delivery position with torque control.

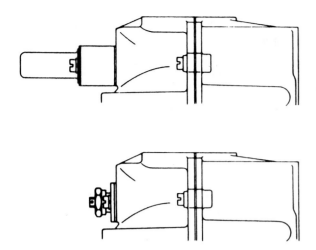

FIGURE 21-56. External torque control mechanism (above) and internal torque control mechanism (below) on RQV governor. *(Courtesy of Robert Bosch Canada Ltd.)*

aneroid capsule shortens, the amount of fuel delivered increases. In order to adjust the full-load delivery, the cam plate is adjustable in the horizontal plane by means of a screw.

PART 14   ELECTRIC SPEED-CONTROL DEVICE

Use

Electric speed-control devices (Fig. 21–63 to 21–65) are used for remote-controlled adjustment of the speeds of engines used with assemblies of equipment.

Various mounting parts are required to attach such a device to a fuel-injection-pump governor; the exact choice of mounting parts will depend on the type of governor concerned.

Design

The electric speed-control device is fitted with a base plate on which is mounted a 24-volt motor; this motor is connected through a clutch with a threaded spindle, which can move a threaded nut back and forth in guide rails, depending on the direction of motor rotation. The governor control lever is at-

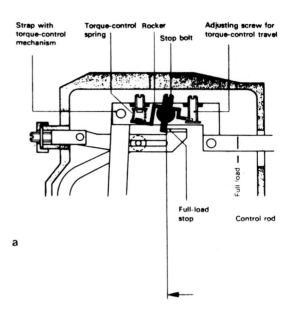

a

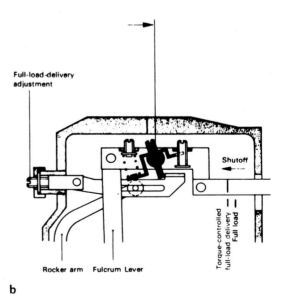

b

FIGURE 21-57. RQ governor with internal torque control mechanism: (a) start of torque control; (b) end of torque control. *(Courtesy of Robert Bosch Canada Ltd.)*

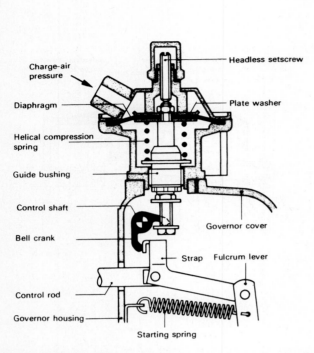

FIGURE 21-58. Cross-sectional drawing of manifold pressure compensators (LDA). *(Courtesy of Robert Bosch Canada Ltd.)*

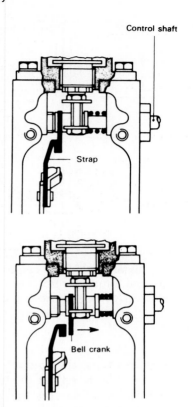

FIGURE 21-60. Manifold pressure compensator. Top: operating position. Bottom: start position. *(Courtesy of Robert Bosch Canada Ltd.)*

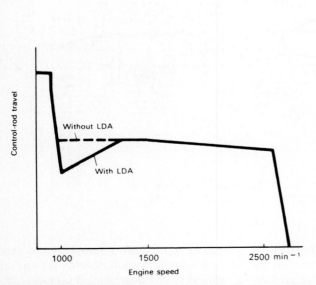

FIGURE 21-59. Effect of manifold pressure compensator (LDA) on control rod travel. *(Courtesy of Robert Bosch Canada Ltd.)*

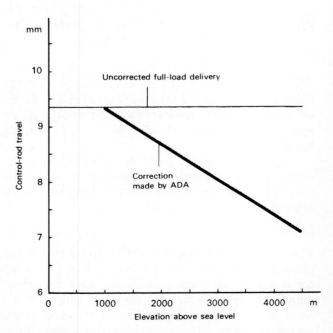

FIGURE 21-61. Example of correction of control rod travel by the altitude compensator (ADA). *(Courtesy of Robert Bosch Canada Ltd.)*

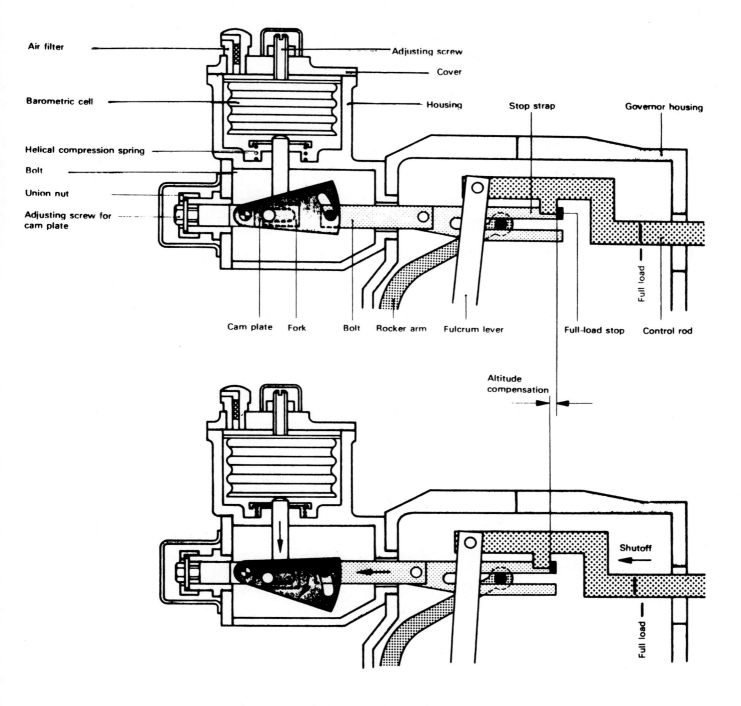

FIGURE 21-62. Altitude pressure compensator (ADA); upper—normal position; lower—operating position at low atmospheric pressure. *(Courtesy of Robert Bosch Canada Ltd.)*

tached to the threaded nut by a positive mechanical connection developed either by a torsion spring mounted on the governor-control-lever shaft (RQV and RQUV governors) or by a tension spring attached to the control lever. The spring forces the control lever in the full-load direction in the RQV and RQUV and in the shut-off direction in the RSV and RSUV governors. The motor and the adjustment spindle are joined by a releasable clutch fitted with overload protection. When the motor clutch has been released, the adjustable spindle can be turned by the handwheel.

The adjustment travel is restricted by electric limit switches in both directions.

In order to stop the diesel engine, the governor control lever is drawn to the shut-off position. The shut-off stop is adjusted in similar fashion to the full-load stop on the governor cover. Two models of the electric speed-control device are produced: one for left-hand mounting and one for right-hand mounting.

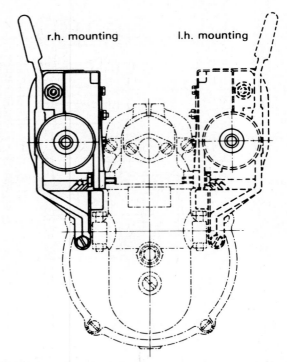

FIGURE 21-64. End view of speed control device for RQV governor. *(Courtesy of Robert Bosch Canada Ltd.)*

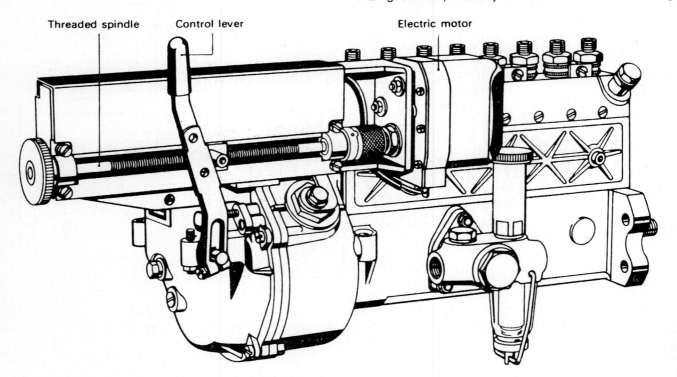

FIGURE 21-63. Electric speed control device, left-hand mounting. *(Courtesy of Robert Bosch Canada Ltd.)*

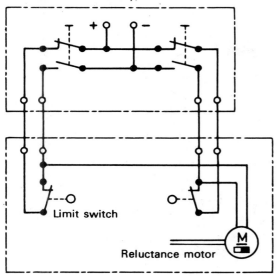

Push-button switch
(not included in delivery)

Limit switch

Reluctance motor

Electric speed-control
device

FIGURE 21-65. Wiring diagram of speed control device. *(Courtesy of Robert Bosch Canada Ltd.)*

PART 15   PNEUMATIC GOVERNOR

Variable-Speed Governor EP/M

Operating Principle. The pneumatic governor consists of two main parts:

• the venturi assembly fastened to the engine intake manifold on the inlet side

• the diaphragm block mounted on the fuel injection pump

Air drawn into the engine flows through the air filter and then through the venturi-shaped channel in the venturi assembly. At the narrowest point in this channel are a throttle valve and the connector fitting for the vacuum line leading to the diaphragm block. The throttle valve is connected with the accelerator pedal through the control lever and the linkage system.

Depending on the position of the throttle valve, the vacuum necessary for regulation is set (at full-load about 400 mm water column) at a lower, me-

dium, or higher speed. The diameter of the venturi channel must be so chosen that when the throttle valve is completely open the required nominal speed can still be reached without bother. A stop screw serves to set the nominal speed accurately (limitation of the throttle valve opening). Pneumatic governors are used mainly in automobiles and agricultural tractors. The purpose of the auxiliary venturi at the point where the vacuum is taken is to ensure that if the engine should start to operate in the reverse direction, it does not run out of control; i.e., it can be stopped.

A diaphragm, connected with the fuel-injection-pump control rod through a linkage system, divides the diaphragm block into two chambers:

• the vacuum chamber (connected by a hose or pipe with the venturi assembly) in which the governor spring tends to shift the diaphragm, and thus the control rod, in the maximum fuel direction

• the atmospheric chamber, which is connected with the outside air

When the engine is operating, the position of the diaphragm, and consequently the position of the

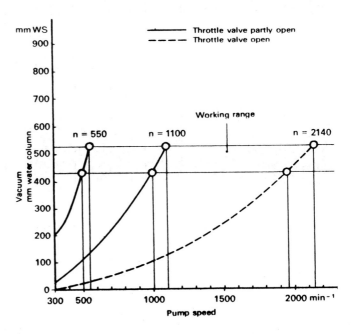

FIGURE 21-66. Vacuum in the vacuum chamber at various speeds and positions of the throttle valve of a diesel engine with a pneumatic governor. *(Courtesy of Robert Bosch Canada Ltd.)*

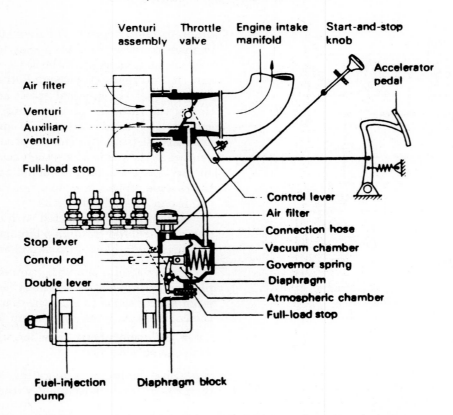

FIGURE 21-67. Schematic drawing of the pneumatic governor in the at-rest position. *(Courtesy of Robert Bosch Canada Ltd.)*

control rod, depends on the magnitude of the difference between the pressures prevailing on the two sides of the diaphragm and developed by the engine loading at a particular time. This is because the engine speed rises or falls respectively if the load on the engine is increased or decreased at a certain position of the throttle valve. This results in the differing pressures in the vacuum chamber. If the initial tension of the governor spring is greater than the vacuum acting on the diaphragm, the control rod is shifted in the direction of increased fuel delivery. If the vacuum increases, the diaphragm is moved by the effect of the atmospheric air pressure against the spring pressure, and the control rod is shifted in the shut-off direction.

Speed regulation starts when the engine reaches that speed at which the vacuum is able to overcome the pressure exerted by the governor spring, or vice versa. The pneumatic governor is effective from the idle speed to the maximum speed.

Operating Characteristics

Starting the Engine. (Fig. 21-68) A stop for excess-fuel delivery during starting is built into the diaphragm block; it can be adjusted by means of a thread.

In order to start the engine, the start-and-stop lever is moved in the start direction. As a result, a spring-loaded stop pin is pressed into the adjusting screw by the two-armed lever (double lever); the governor spring can shift the diaphragm, and thus the control rod, in the maximum fuel-delivery direction. More fuel is therefore fed to the engine during starting than at full load.

Idle-Speed Range. (Fig. 21-69) When the engine is idling, the control lever at the throttle valve is positioned against the adjustable idle stop, and the venturi channel is almost completely closed. Even at the idle speed, a vacuum is developed in the vacuum chamber, which is sufficient to draw the con-

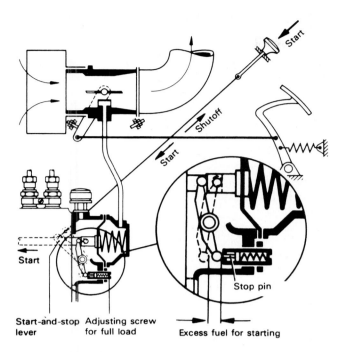

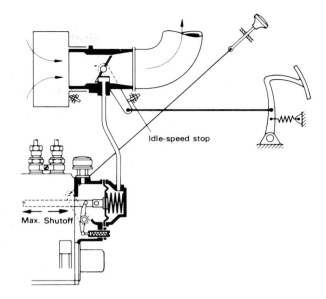

FIGURE 21-69. Idle position of pneumatic governor. *(Courtesy of Robert Bosch Canada Ltd.)*

FIGURE 21-68. Diaphragm housing in pneumatic governor with built-in control rod stops for full load and for excess fuel for starting. *(Courtesy of Robert Bosch Canada Ltd.)*

trol rod, against the pressure of the governor spring, into its idle position. If the load on the engine is reduced, the engine accelerates and the vacuum therefore increases. As a result, the diaphragm shifts the control rod still farther in the shut-off direction, and the engine runs more slowly again. If, on the other hand, the load on the engine is increased, the engine is slowed down, and the vacuum decreases; the governor spring therefore shifts the control rod in the maximum-fuel-delivery direction, and the engine runs faster again. In other words, the governor limits the idle speed in both an upward and downward direction; i.e., it regulates the idle speed.

Top-Speed Range (Full-Load Speed Regulation). *(Fig. 21-70)* If the driver wants to bring the engine up to the nominal speed (full output power), the accelerator pedal must be pressed all the way down. The control lever at the throttle valve is then brought up against its (adjustable) full-load stop, and the throttle valve itself is completely open. Initially, only a slight vacuum prevails in the vacuum chamber, and the vacuum required to regulate the

maximum speed is only developed at the nominal speed. When the nominal speed is exceeded, the control rod moves away from its full-load stop and is shifted in the shut-off direction until the amount of fuel delivered has become so small that the high-idle speed can no longer be exceeded.

Between the Idle Speed and the Maximum Speed. The pneumatic governor also maintains—just as a mechanical variable-speed governor does—every speed between the idle speed and the maximum speed constant within the limits of the speed droop. The farther the accelerator pedal (and thus the throttle valve) is shifted in the maximum-fuel-delivery direction, the higher the engine speed rises.

Stopping the Engine. (Fig. 21-71) The stop lever at the diaphragm block can be connected to a shut-off device, which is combined with the glow-plug-and-starter switch. When the engine is stopped, the double lever connected to the stop lever forces the control rod to shut-off. Delivery of fuel is thus cut off and the engine stops (for an exception to this see reverse operation of the engine, immediately following).

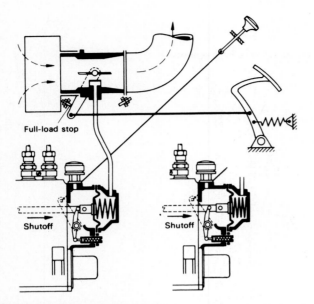

FIGURE 21-70. Full-load speed regulation. Left: nominal speed, full load. Right: high-idle speed. *(Courtesy of Robert Bosch Canada Ltd.)*

Reverse Operation of the Engine. (Fig. 21-72) In most cases, the engine manufacturer installs, as a standard feature, some method of protecting the engine against starting to operate in the reverse di-

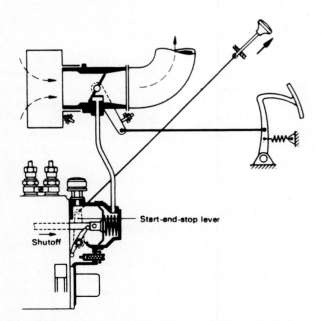

FIGURE 21-71. Pulling the stop button up stops the engine. *(Courtesy of Robert Bosch Canada Ltd.)*

rection. However, cases can arise where engines without this protection (especially prechamber engines) have warmed up and started to operate in the reverse direction, for example, as a result of a recoil during starting or rolling backward down a slope. In such a case the pneumatic governor does not operate correctly because, when an engine runs in the reverse direction, the intake line becomes the exhaust. Since the throttle valve is only slightly opened during idling, the exhaust gases would accumulate in the intake line. However, an overpressure in the intake line means that an overpressure also develops in the governor vacuum chamber. No suction forces then act on the diaphragm; instead, pressure forces act on it. Assisted by the governor spring, the control rod is therefore shifted very strongly in the maximum-fuel-delivery direction. The result of these actions is that the engine accelerates rapidly and tends to run out of control; under these conditions the engine cannot be stopped even with the stop draw knob because of the large counteracting forces.

In order to prevent the accumulation of exhaust gases in the intake line in such a case, an auxiliary venturi is built into the venturi assembly at the connection point where the vacuum is transmitted to the governor. This auxiliary venturi bypasses the throttle valve so that even when the throttle valve is closed the exhaust gases can escape during reverse operation of the engine. If, however, the throttle valve is opened briefly during reverse operation of the engine, the engine tends to run out of control even if the throttle valve is immediately closed again.

An engine that starts to operate in the reverse direction, recognizable by development of dense smoke from under the vehicle hood, must be stopped immediately; otherwise bearings can be damaged as a result of inadequate lubrication, and the air filter can burn up. The most effective and reliable method to stop the engine quickly in this case is to engage a gear (preferably third or fourth), press the brake pedal down as far as it will go, and then force the engine to come to a stop by releasing the clutch. Other possible methods of stopping the engine are to hold the exhaust pipe closed or to operate the shut-off device and at the same time press the accelerator pedal all the way down.

Speed Droop in the Pneumatic Governor. In the pneumatic governor speed droop is fairly uniform throughout the entire speed range. Speed droops

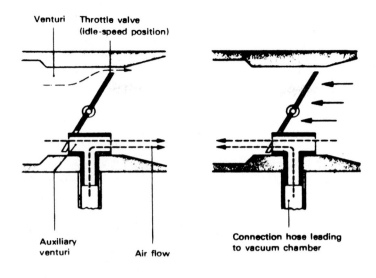

FIGURE 21-72. Air flow in the venturi during forward engine operation (left) and reverse engine operation (right). *(Courtesy of Robert Bosch Canada Ltd.)*

common in motor vehicle engines range from 6 to 12 percent.

Special Designs

Diaphragm Block with Auxiliary Idle-Speed Spring (Spring Capsule). (Fig. 21-73) The auxiliary idle-speed spring operates only when the engine is idling. This spring is much stiffer than the maximum-speed control spring and serves to stabilize the idle speed. Figures 21-74 and 21-75 show a diaphragm block in which the auxiliary spring is automatically switched into operation by a cam during idling and switched off at full load. As a result, reliable speed regulation is achieved, for example, during overrun.

Damping of Fluctuations. (Fig. 21-76 to 21-78) In engines with four or fewer cylinders, excessive fluctuations in the vacuum can occur in the diaphragm block, causing irregular operation of the engine. In order to keep these fluctuations within bounds, a throttle is installed in the connection screw for the vacuum line at the diaphragm block.

In order to prevent any small diaphragm fluctuations that may still occur from being transmitted to the control rod, the control rod and the diaphragm bolt are not connected rigidly together. The connector bolt joining the control rod and the diaphragm is therefore not fixed in a hole diaphragm bolt. The diaphragm bolt can thus move ax-

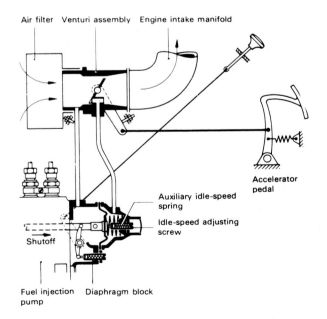

FIGURE 21-73. Schematic drawing of pneumatic governor; diaphragm block with auxiliary idle-speed spring in the idle-speed adjusting screw.

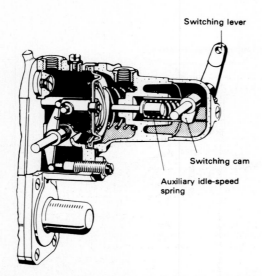

FIGURE 21-74. Diaphragm block EP MZ 60 A (with auxiliary spring), adjusting screw inside, adjustable from outside. *(Courtesy of Robert Bosch Canada Ltd.)*

ially back and forth a certain distance without taking the control rod with it. This measure results in smoother operation of the engine.

In order to dampen noises during idling (in governors with an auxiliary idle-speed spring), the diaphragm bolt is fitted with a rubber damper on the vacuum side.

Torque Control with the Pneumatic Governor

If the full engine power is required, the driver must press the accelerator pedal all the way down (Fig. 21-79). As this is done, the butterfly (throttle) valve

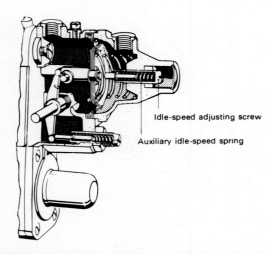

FIGURE 21-75. Diaphragm block EP MN 60 A with auxiliary idle-speed spring and switching cam. *(Courtesy of Robert Bosch Canada Ltd.)*

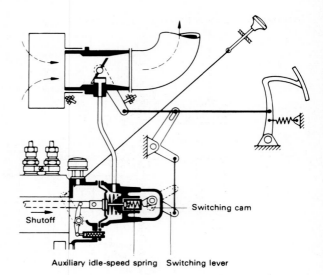

FIGURE 21-76. Schematic drawing of a pneumatic governor: diaphragm block with auxiliary idle-speed and switching cam. *(Courtesy of Robert Bosch Canada Ltd.)*

is fully opened and a slight overpressure develops in the vacuum chamber because of the relatively low speed. As a result, the governor spring presses the control rod against its full-load stop and the torque-control spring is compressed until the torque-control travel is zero; i.e., the control rod travels farther in the maximum-fuel-delivery direction by the amount of the torque-control travel. The engine

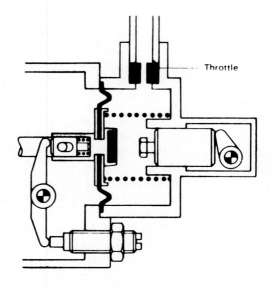

FIGURE 21-77. Throttling device dampens vacuum fluctuations on engines with four or fewer cylinders. *(Courtesy of Robert Bosch Canada Ltd.)*

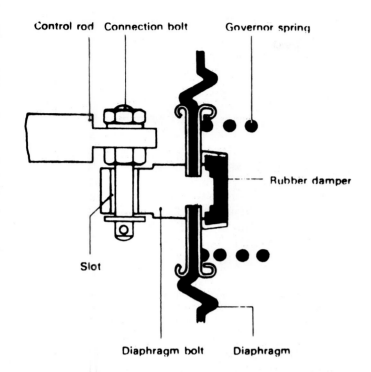

FIGURE 21-78. Coupling between control rod and diaphragm bolt. *(Courtesy of Robert Bosch Canada Ltd.)*

FIGURE 21-79. Installation of auxiliary idle-speed spring. *(Courtesy of Robert Bosch Canada Ltd.)*

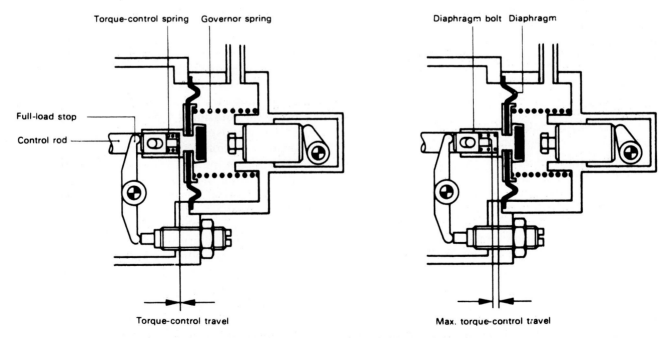

FIGURE 21-80. Pneumatic governor with torque control mechanism. *(Courtesy of Robert Bosch Canada Ltd.)*

reacts to the additional fuel by speeding up. The more the speed increases, the greater the vacuum becomes; as a result the pressure exerted by the governor spring on the torque-control spring decreases. The torque-control spring begins to relax and shifts the diaphragm with the control rod in the shut-off direction by the amount of the torque-control travel.

PART 16   SUMMARY

1. The governor for the diesel-fuel injection pump is as important as the fuel injection pump itself because without a governor, which controls the amount of fuel injected into the diesel engine, operation of the engine is not possible. A diesel engine without a governor would either come to a stop in the low rotational-speed range or accelerate to the point of self-destruction.

2. The Bosch governor is attached to the PE-type in-line fuel injection pumps and, depending on the design, is controlled either by centrifugal force or by the vacuum in the diesel-engine intake manifold.

3. Different types of governors have been developed corresponding to their various intended uses or functions:

a. Maximum-speed governors control only the maximum speed, i.e., the maximum full-load speed of the engine.

b. Minimum-maximum-speed governors control the idle speed and the maximum speed.

c. Variable-speed governors control all speeds from the idle speed to the maximum speed, depending on the setting of the control lever. These governors can be designed to be controlled either mechanically or pneumatically.

d. Combination governors are variants of the variable-speed governor. With this type of governor, a certain speed range between the idle and the maximum speeds remains uncontrolled; depending on the particular design of the governor, this uncontrolled range can begin immediately after the idle speed or it can begin at some intermediate speed and extend to the maximum speed. In the remainder of the rotational-speed range, the speed of the engine is set directly by the control lever.

4. The control process is basically the same in all types of governor. If the speed of the engine exceeds a speed set with the control lever within the control range, the governor acts to draw the fuel-injection-pump control rod to a position of smaller fuel delivery; if the engine speed falls below the set speed, the governor acts to shift the control rod to a position of greater fuel delivery.

5. Increases or decreases in the rotational speed, and the corresponding changes in the amount of fuel delivered, take place in accordance with the so-called speed droop of the governor. The speed droop indicates, as a percentage, the decrease in rotational speed of an engine when a load is suddenly applied to it, and is determined by the engine manufacturer. The change in fuel delivery is approximately proportional to the change in rotational speed.

6. In addition to its basic function(s), other control functions must be carried out by the governor, such as automatically releasing or cutting off the larger amount of fuel required for starting (the starting fuel delivery), as well as correcting the full-load delivery as a function of rotational speed (torque control), charge-air pressure, or atmospheric pressure. For these purposes, supplementary equipment, which can be mounted either inside or on the governor, is required.

7. Most governors are fitted with a torque-control mechanism, i.e., a mechanism to match the amount of fuel delivered by the fuel injection pump to the full-load amount of fuel burned in the engine with no development of smoke above a certain rotational-speed range. Torque control is required because the fuel requirement of the nonsupercharged diesel engine decreases as the rotational speed increases, but on the other hand, the amount of fuel delivered by the fuel injector pump increases within a certain range with the control rod in the same position as the rotational speed increases. The torque-control mechanism decreases the fuel delivery within a certain rotational-speed range as the speed increases. This type of torque control is designated *positive torque control* or torque control in the "sense of control." In engines fitted with exhaust gas turbochargers with a fairly high degree of supercharging, the fuel requirement of the engine at full load rises so sharply above a certain speed that the natural increase in fuel delivered by the fuel injection pump is no longer adequate. In such a case the fuel delivery must be increased correspondingly as the rotational speed increases ("negative torque control").

Both positive and negative torque control within one family of governor-operating characteristics can be achieved.

8. Suitable stops are provided on the governor or on the fuel-injection pump to set the full-load delivery, idle delivery, starting fuel delivery, full-load speed, or intermediate speeds. In addition, the following control-rod stops are produced for special compensation functions:

a. The manifold-pressure compensator. It reacts to the charge-air pressure of an exhaust turbosupercharger and is designed to match the full-load delivery continuously to the reduced charge-air pressure in the lower rotational-speed range.

b. The altitude-pressure compensator. It ensures that the full-load delivery of fuel injected into the engine is matched to the air density, which decreases as the altitude increases.

PART 17   SELF-CHECK

1. What fuel delivery control functions do governors perform?
2. Why is a governor needed on a diesel engine?
3. What is speed droop?
4. Speed droop is greatest at what engine speed range?
5. Define positive and negative torque control.
6. What is meant by the term isochronous?
7. How can governor hunting be corrected?
8. What is governor stability?
9. What is a tailshaft governor used for?
10. What does an altitude compensator do?

Chapter 22

Robert Bosch, American Bosch, and CAV Port and Helix Injection Pumps

Performance Objectives

After thorough study of this chapter, the appropriate training models and service manuals and with the necessary tools and equipment, you should be able to do the following with respect to diesel engines equipped with port and helix type fuel injection pumps:

1. Complete the self-check questions with at least 80 percent accuracy.

2. Describe the basic construction and operation of a port and helix fuel injection pump including the following:

    (a) method of fuel pumping and metering

    (b) delivery valve action

    (c) lubrication

    (d) injection timing control

3. Diagnose basic fuel injection system problems and determine the required correction.

4. Clean, inspect, test, measure, repair, replace, and adjust fuel injection system components as needed to meet manufacturer's specifications.

The most common diesel fuel injection pump worldwide is probably the type using the port and helix plunger and barrel method of pumping and metering fuel. The applications of this design include individual pumping elements mounted separately on the engine and the multiple plunger in line- and V-configuration injection pumps.

Although these pumps are produced by various manufacturers around the world, they differ only slightly in their operation and service.

Robert Bosch pumps of this type include the PE and PF series. United Technologies American Bosch pumps of this type include the in-line and V-type APE series, the flange-mounted single-element APF series, and the smaller PLB. The model 200 offers either mechanical or microprocessor control of fuel delivery. CAV pumps of this type include the Majormec, Minimec, and Micromec series.

With the exception of Parts 12, 14, and 15 the following is based on information provided courtesy of Robert Bosch. Most of the balance is based on information provided courtesy of United Technologies American Bosch.

PART 1 CONSTRUCTION AND OPERATION

Figure 22–1 shows a PE injection pump with mechanical governor (1), fuel supply pump (2) and timing device (3). The supply pump transfers the fuel from the fuel tank and delivers it through the fuel filter into the suction gallery of the injection pump. The pump plunger driven by the injection pump camshaft delivers the fuel through the delivery valve and the fuel-injection tubing to the injection nozzle. After completion of the plunger lift, the spring-loaded delivery valve closes the fuel-injection tubing and the plunger is returned to its starting position by the plunger return spring.

Pumping Element (Plunger and Barrel Assembly)

The number of pumping elements in the PE injection pump corresponds to the number of cylinders in the engine. (See Fig. 22–2 to 22–4.)

Each pumping element consists of a plunger and a barrel. The plunger is fitted into the barrel so perfectly (clearance of 1/10,000 inch or a few 1/1000 mm) that it seals even at very high pressure and low speeds. Therefore, only complete pumping elements can be replaced, certainly not just the plunger or the barrel only.

The plunger is provided with a vertical groove as well as a helical cut. The edge formed on the upper section of the plunger is known as the "helix."

The barrel is usually provided with two radial ports opposite to one another, through which the

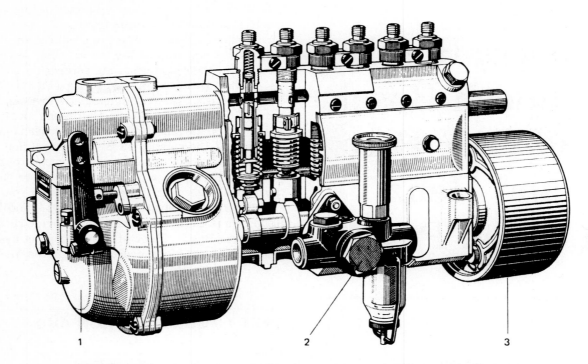

FIGURE 22-1. Robert Bosch model PE fuel injection pump: 1. Mechanical governor. 2. Fuel supply pump. 3. Timing device and drive coupling. *(Courtesy of Robert Bosch Canada Ltd.)*

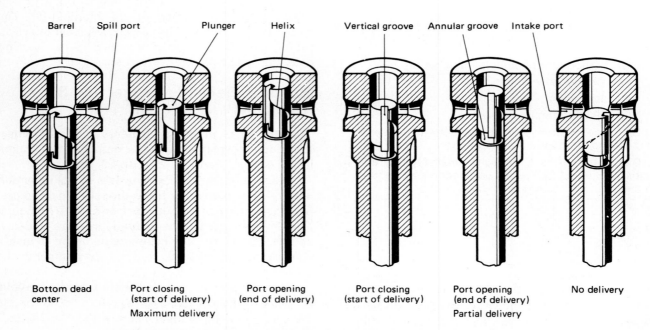

| Barrel | Spill port | Plunger | Helix | Vertical groove | Annular groove | Intake port |

| Bottom dead center | Port closing (start of delivery) | Port opening (end of delivery) | Port closing (start of delivery) | Port opening (end of delivery) | No delivery |
| | Maximum delivery | | | Partial delivery | |

FIGURE 22-2. Dual orifice (port) pumping element plunger and barrel. *(Courtesy of Robert Bosch Canada Ltd.)*

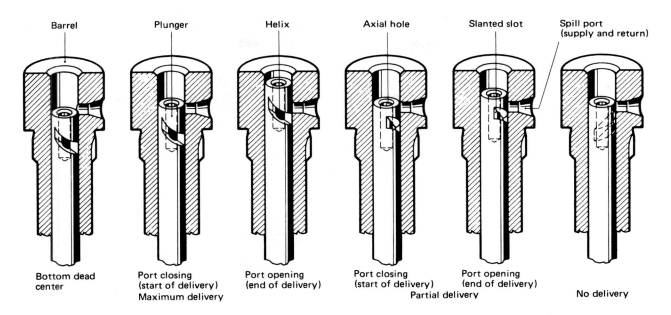

FIGURE 22-3. Single orifice (port) pumping element plunger and barrel. *(Courtesy of Robert Bosch Canada Ltd.)*

fuel flows from the suction gallery to the pressure chamber (intake and spill port) and through which it returns when the delivery ceases due to port opening.

However, there are also "single-port elements." Their barrels have only one lateral port (spill port). The plunger has an axial hole in place of the vertical groove and a slanted slot instead of a milled helix.

In the lowest plunger position, the pressure chamber above the plunger is filled with fuel which

has entered from the suction gallery through the lateral barrel ports. As the plunger moves upward, the barrel ports are closed and fuel is discharged through the delivery valve into the fuel-injection tubing. Fuel delivery ceases when the helix registers with the spill port because from this moment on, the pressure chamber of the barrel is connected to the suction gallery—through the vertical and annular groove or the axial hole and slanted slot. Thus, the fuel is returned to the suction gallery.

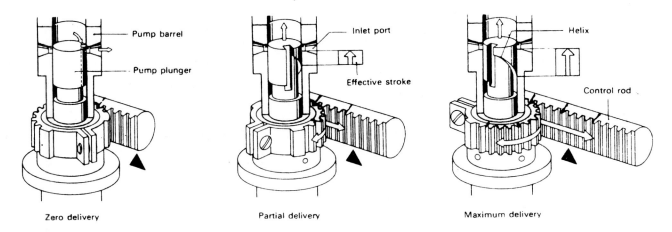

FIGURE 22-4. Metering of fuel delivery is controlled by pump plunger position, which in turn is controlled by the control rod (rack). The control rod is connected to the governor, which moves the rack in response to engine speed and load as well as throttle position. *(Courtesy of Robert Bosch Canada Ltd.)*

Since the end of delivery is reached when the plunger helix opens the spill port, the effective stroke can be varied by turning the pump plunger. If the plunger is turned until the vertical groove or the helix opens the spill port, the fuel in the pressure chamber is returned to the suction side during the plunger lift: no fuel is delivered.

In pump sizes PE (S) A, CW, Z and ZW, the control sleeve slipped over the pump plunger supports at its upper end a clamped-on gear segment (control sleeve gear); at the bottom it is provided with two vertical slots in which the plunger control arms slide. The teeth of the gear segment engage the teeth of the control rack. The pump plungers thus can be rotated by the control rack during operation so that the effective stroke and therefore the fuel delivery of the pump can be infinitely varied from zero to maximum delivery.

The marking STOP and the arrow on the control rack (only provided on PF pumps) indicate the direction in which the control rack must be moved to make the fuel delivery zero. The opposite end position of the control rack gives maximum fuel delivery which generally is limited by an adjustable stop.

The control rack of the pump must be connected to the governor so that no lateral or torsional forces are transmitted to the control rack. If the control rack moves with difficulty or jams, the governing is not stable; if the control rack is stuck in the full-load position, the engine may overspeed. If a linkage is necessary, the connection between control rack and governor is best made by a fork head.

For the pump sizes PE (S) M and P, the metering principle for fuel delivery is identical to that described above except that a link connected to the control rod is used for rotating the plunger in place of the rack-and-pinion.

Pumping Element with an Upper and a Lower Helix

There are pumping elements in which the plunger not only has a lower helix (for port opening) but also an upper one for control of the port closing. Rotating the plunger changes the start of delivery as a function of load. (See Fig. 22-5.)

These elements are used in the precombustion and turbulence chamber engines because noise reduction can be obtained when the speed-dependent variant in timing (by the timing device) is combined with a load-dependent variation in timing.

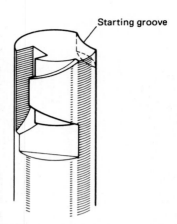

FIGURE 22-5. Pump plunger with upper and lower helixes as well as starting groove. *(Courtesy of Robert Bosch Canada Ltd.)*

Starting Groove

Some diesel engines start more easily when fuel injection timing is retarded during the starting process relative to normal operation.

For this purpose, the plunger is provided with a starting groove at the top which results in a port closing with a 5–10° retard. As soon as the engine reaches operating speed, the governor pulls the control rod to the normal operating position.

Delivery Valve (Retraction Type)

As soon as the helix of the plunger opens the spill port, the pressure in the pump barrel drops. The higher pressure in the line to the injection nozzles and the valve spring force the delivery valve onto its seat. It closes the high-pressure delivery line against the pump barrel until fuel delivery starts again in the next delivery stroke. (See Fig. 22-6.)

The delivery valve furthermore has the task of "relieving" the high-pressure delivery line. Relief of the high-pressure line is necessary to obtain rapid closing of the nozzle valve and to prevent the dribbling of fuel into the combustion chamber. The "retraction volume" depends on the length of the high-pressure line and the quantity of fuel delivered. The delivery valve is guided with its stem in the valve body. During the delivery process, it is lifted from its seat, so that the fuel can be discharged into the delivery valve holder through flutes terminating in an annulus. Above the annulus there is a short cylindrical shaft section (retraction piston); it fits tightly into the valve body and is followed by the conical valve.

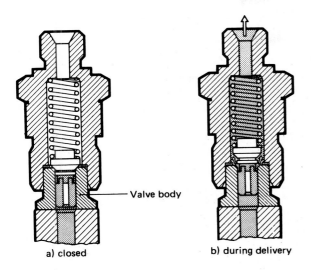

FIGURE 22-6. Delivery valve operation. *(Courtesy of Robert Bosch Canada Ltd.)*

At the end of delivery, the retraction piston first enters the bore in the valve body and closes the high-pressure delivery line from the pressure chamber. Only then will the conical valve seat. Thereby the volume available for fuel in the high-pressure line is increased by the volume of the retraction piston. The fuel in the high-pressure line can expand quickly and the nozzle valve closes instantaneously.

PART 2   CONTROL ROD STOPS AND SMOKE LIMITER

The control rod travels the greatest distance when the driver depresses the accelerator pedal fully.

Control rod travel is usually limited by an adjustable stop (full-load, fuel quantity). Various control rod stops exist; examples are explained more fully in Chapter 21.

Smoke Limiter

The smoke limiter (Fig. 22-7 and 22-8) is a full-load stop, controlled by the boost pressure, for turbocharged engines. Its task is to reduce the full-load delivery in the lower speed range starting at a given (optional) charge-air pressure. It has a cast housing with a diaphragm clamped airtight at the top by means of the cover. In the center, the cover has a threaded connection for the pressure line to the engine intake manifold. The diaphragm is forced against the cover by a coil spring. The latter rests with its bottom on the guide bushing; the initial tension of the coil spring can be changed by rotating the sleeve. The start of response of the smoke limiter can thus be adjusted to different charge-air pressure values. The guide bushing is locked by the indexing lockwasher.

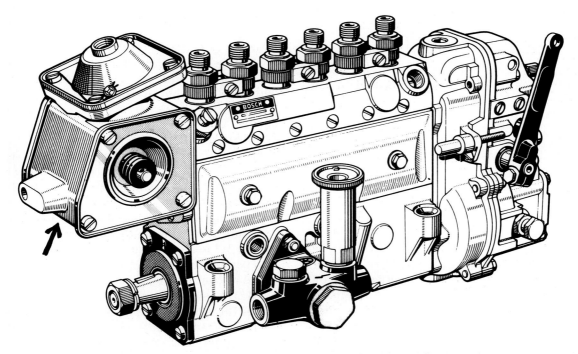

FIGURE 22-7. Location of smoke limiter on injection pump. *(Courtesy of Robert Bosch Canada Ltd.)*

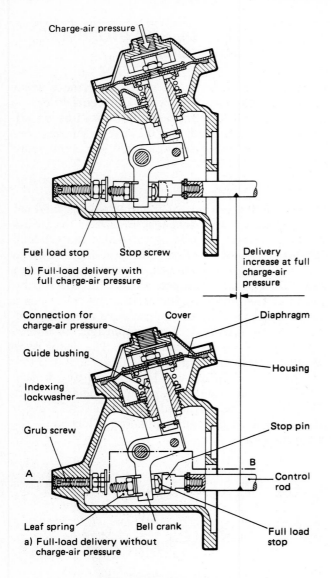

Longitudinal section through the smoke limiter

FIGURE 22–8. Smoke limiter. *(Courtesy of Robert Bosch Canada Ltd.)*

The diaphragm is articulated with the bell crank by a bolt. The fulcrum of the bell crank is the shaft which can be rotated and shifted in the housing.

The lower arm of the bell crank carries the stop screw secured by a lock. The lock simultaneously fastens the leaf spring on the bell crank. The full-load stop screw with lock is located on the housing. The stop pin is screwed to the end of the control rod.

When the bell crank is in the rest position, the stop screw on the bell crank serves as full-load stop. In this position, and after loosening the lock nut, the stop screw can be adjusted to the specified full-load delivery (without charging air).

When the diaphragm is pressurized by the charging air, the bell crank rotates until its stop screw abuts against the full-load stop fixed in the housing. In this position, the fuel delivery must correspond to the maximum charge-air value.

Upon completion of the adjustment, the full-load stop is secured from outside by inserting the grub screw, and the tap hole with the closing disc is closed in a tamper-proof way (by calking). In the case of unauthorized tampering, the full-load stop can then be adjusted only in the direction of "less fuel delivery."

Actuation of the Excess-Fuel Device for Starting

During the starting process, the control lever on the governor (or the accelerator pedal) must be placed in full-load position. In this position, the stop pin attached to the control rod rests on the "full-load stop without charge-air pressure."

The starting cable is now actuated, i.e., the shaft is pulled out in the longitudinal direction by about 10 mm by a Bowden cable or a solenoid. The bell crank and stop screw follow this movement and permit the control rod to travel the distance of the starting quantity.

After completion of the starting process, the shaft together with the bell crank is forced back into the initial position by spring tension.

Protection Against Actuation of the Excess-Fuel Device During Operation

The control rod can only reach the start position when the control lever on the governor or the accelerator pedal has been placed at full load. If the starting cable is actuated while the control rod is not at full-load position, the leaf spring is placed in the path of the control rod and the latter cannot move to the starting position.

If the driver attempts to clamp the starting cable in the excess-fuel position, the control rod can no longer reach the full-load or starting positions after deceleration or shifting into another gear, and the engine thus receives even less fuel than the full-load quantity.

PART 3   INSTALLATION AND DRIVE

PE injection pumps are mounted either on a base plate or in a segment-shaped pan. Pumps with front flange (PES..) are generally flanged to the gear case of the engine.

With engines which are operated for long periods of time on a steep gradient (for example, Caterpillar bulldozers), an injection pump with a lubrication system independent of the installation position should be used. Such pumps are connected to the lubrication system of the engine.

Injection pumps must be driven positively. Flexible couplings (steelflex or rubber couplings) are not acceptable because of the danger of delay in the beginning of delivery and the discharge rate curve. To equalize the small differences (lateral and angular misalignment) between engine drive shaft and pump camshaft, the use of a disc coupling is recommended; this can be supplied on special order.

The injection pump with a front flange is generally driven by a gear or a timing device with gear, made adjustable for accurate timing. See Part 10 of this chapter.

The drive speed in four-cycle engines corresponds to the camshaft speed and in two-cycle engines to the speed of the engine crankshaft.

Injection pumps with symmetrical cams can be used for clockwise as well as counterclockwise rotation, but it should be noted that the injection sequence differs depending on the direction of rotation. Injection sequence and firing order of the engine must coincide.

PART 4   INJECTION PUMP TIMING

For timing the pump to the engine, marks for the beginning of delivery are provided on the engine as well as on the injection pump. Usually, engine cylinder 1 (compression stroke) serves as the reference, although the manufacturer's instructions should be followed in each case.

On the diesel engine, the mark for the beginning of delivery is usually located on the flywheel (upper dead center mark and divisions with the mark for beginning of delivery), on the V-belt pulley, or on the vibration damper. In the injection pump, the delivery starts from pump cylinder 1, if the mark on the fixed coupling member or on the timing device coincides with the mark on the pump housing.

Normally, the first cylinder on the fan side of the engine is engine cylinder No. 1. The pump barrel next to the pump drive is pump barrel No. 1.

Before installation, the mark for the beginning of delivery on the injection pump must be made to coincide with the mark on the housing by turning the pump shaft in the direction of rotation.

PART 5   LUBRICATION

Today automatic lubrication is used almost exclusively for injection pumps and governors. Such pumps are connected to the lubrication system of the engine. Another method is the oil pump lubrication with a common oil supply, as well as oil sump lubrication with a separate oil sump for injection pump and governor, used for example, in the case of inclined installation. Pumps with oil sump lubrication are not maintenance-free.

Engine Lubrication

Filtered engine oil is supplied to the cam and spring chamber through a pressure line and an inlet port via the roller tappet clearance or through a special oil supply valve; it also reaches the governor through the camshaft bearing or through a passage. In the case of base mounting, lubricating oil returns to the engine by a return line; with flange mounted pumps, the oil returns through the camshaft bearing (without oil seal).

Pumps designed for connection to the engine lubrication system must be filled with the same lubricating oil as the engine before being put into operation.

Sump Lubrication

The lubricating oil (identical to that of the engine) must be poured in manually, through the cover, after removal of the venting cap or air filter. The oil level is controlled at the time of the engine oil change as specified by the engine manufacturer (after about 2750 miles, 4500 km, or 150 operating hours) by loosening the oil level screw (center of camshaft). Excess oil (increased by fuel leakage) is drained; oil shortage is made up.

Lubricating oil change is recommended in the case of removal of pump and governor or of a complete engine overhaul as specified by the manufacturer (after about 60,000 miles, 100,000 km, or 3000 operating hours).

PART 6   INJECTION PUMP TYPE PES..M..

The injection pump PES..M.. (for example, PES 4 M 50..), is the smallest of PE type in outside dimensions. The essential distinguishing feature compared to the conventional injection pumps PE..A or Z, however, is the type of delivery control. In pump types PES..M.. (Fig. 22–9), the pump plunger is not rotated by a gear segment, but via a lever rigidly connected to the control sleeve, the so-called control arm. The control rod consists of round steel flattened on one side. In the control rod are mounted clamps with a slot in which the control arm slides with its riveted bolt. Individual adjustment is made by shifting the clamp on the control rod.

Other distinguishing features are the following:

The pump plunger rests on the roller cam without adjusting screw; adjustment of port closing is made with tappet rollers of different diameters.

The governor housing for installation of a mechanical or a pneumatic governor is integral with the pump housing.

Lubricating oil is poured in through the threaded hole for the air filter on top of the governor housing (in place of a dip stick).

PART 7   INJECTION PUMP TYPES PE..P.. AND PES..P..

The injection pump PE..P.. and PES..P.. differs from the conventional PE and PES pumps externally as well as in design, although its mechanical operation is basically the same. (See Fig. 22–10.)

The main distinguishing features are the following:

Each pumping element is combined into a unit with delivery valve and delivery valve holder by a flanged bushing.

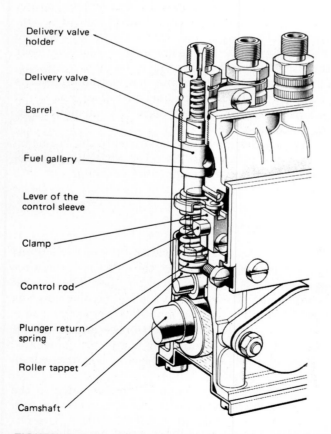

Delivery valve holder
Delivery valve
Barrel
Fuel gallery
Lever of the control sleeve
Clamp
Control rod
Plunger return spring
Roller tappet
Camshaft

FIGURE 22-9. Sectional diagram of injection pump PES 4 M. *(Courtesy of Robert Bosch Canada Ltd.)*

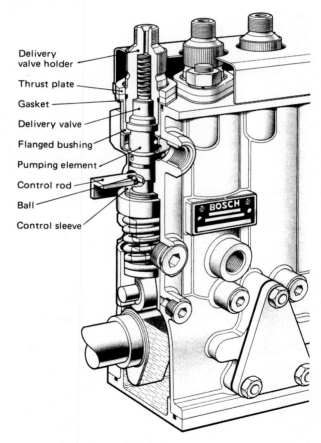

Delivery valve holder
Thrust plate
Gasket
Delivery valve
Flanged bushing
Pumping element
Control rod
Ball
Control sleeve

Sectional diagram of injection pump PE 6 P..

FIGURE 22-10. Sectional diagram of injection pump PE 6P. *(Courtesy of Robert Bosch Canada Ltd.)*

The basic adjustment of the fuel delivery is made by rotating the flanged bushing, and timing is adjusted by placing shims under the flanged bushing. The pump housing does not have an inspection cover for adjustment. It is of closed design.

Control of delivery is effected via the control rod consisting of an angle iron with slots which mate with balls mounted on the control sleeve.

The pump is connected to the lubrication system of the engine.

PART 8   INJECTION PUMP TYPE PF

Construction and Operation

The operation of the PF type unit pumps is fundamentally the same as that of the PE type pumps. (See Fig. 22–11.)

The PE injection pump design does not have a built-in camshaft like the PE design and the engine designer must therefore provide a separate drive for each pumping element. Normally, PF injection pumps are operated by an additional cam on the engine camshaft. PF injection pumps are mounted with a flange which may have varying forms. PF injection pumps usually are of the single plunger type; however, pumps in sizes K, A and B are also delivered in multiple plunger models.

In the PF pump, as in the PE pump, the port closing serves as the basis for timing the pump to the engine. In the PF pumps with a timing window, the plunger is in port closing position when the timing marks on the timing window and on the guide sleeve coincide.

In PF pumps without a timing window, the value "a" is shown on the mounting flange or on the name plate. It refers to the distance between the underside of guide sleeve or roller tappet and the flange contact surface with the pump installed and pump drive in B.D.C. position. An adjusting screw on the roller tappet or, in the PFR pump, shims placed under the pump flange serve for the exact adjustment.

Lubrication

Before installation, the guide sleeve in the pump must be lubricated. During operation, it receives oil for lubrication from the engine through the drive. Consequently, the pump needs to be lubricated only when it has been removed or disassembled.

PART 9   GOVERNORS

Depending on the intended application, engines are expected either to maintain a certain speed independent of the respective load or, when they operate within a certain speed range; not to substantially exceed this in either direction.

To maintain the desired speed under a variable load, the delivered quantities of fuel must be metered corresponding to the required torque. In diesel engines, this is done by changing the fuel quantity injected; the control rod of the injection pump must therefore be appropriately moved whereby the injected quantity is "controlled." In motor vehicles

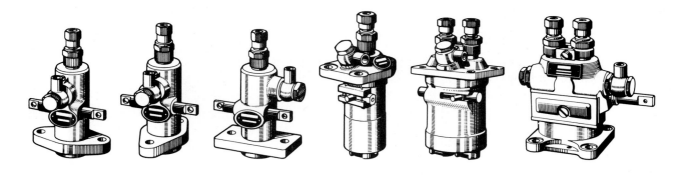

FIGURE 22–11. PF single- and dual-plunger pumps with different flange positions. *(Courtesy of Robert Bosch Canada Ltd.)*

this is done by the driver with the accelerator pedal between idling and maximum speed. However, the idling speed and maximum speed must be automatically limited so that the engine does not die during idle or exceed the maximum permissible speed. Consequently, the PE pumps have been provided with automatic governors which satisfy these requirements.

There are mechanical governors and pneumatic governors. The mechanical governor which is driven by the diesel engine, is a speed sensitive control, while the pneumatic governor is a control device responsive to the air flow (vacuum) in the intake manifold of the engine.

The movements of the flyweights of the mechanical governor or those of the diaphragm of the pneumatic governor are transmitted in a suitable manner to the control rod of the injection pump. The control rod thus is either moved into the STOP or MAXIMUM FUEL direction so that the engine speeds are automatically regulated within the desired ranges.

A distinction is made according to the intended application:

1. Maximum-minimum speed governors: for vehicle engines;

2. Constant speed governors which are operated only at constant speed (machine tools, generating sets);

3. Variable speed governors: for engines which must be controlled automatically over the entire speed range (marine engines, tractor engines, engines driving electrical generators and sometimes also locomotive engines).

See Chapter 21 for details on governor types and operation.

PART 10   TIMING DEVICES

Function

During the injection process, the nozzle is opened by a pressure wave which is propagated through the fuel-injection tubing at sound velocity. While the necessary (opening) time is constant independent of engine speed, the injection lag in crank degrees increases with speed. (See Fig. 22–12 to 22–15.)

Similar conditions apply to the ignition process. The delay period from the beginning of injection to the initiation of ignition is approximately constant in time and independent of engine speed. For this

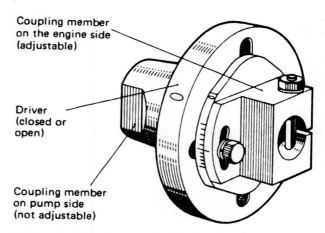

FIGURE 22-12. Adjustable drive coupling for a Bosch injection pump without an automatic timing device. *(Courtesy of Robert Bosch Canada Ltd.)*

reason the ignition lag in crank degrees also increases with speed.

Since the most favorable combustion and highest performance of a diesel engine are obtained only at a determined piston position referred to the degree of crank angle, it is recommended to advance the beginning of delivery of the injection pump with increasing speed. This is provided by an automatic timing device. During operation it rotates the pump shaft up to 8° with respect to the engine shaft, so that the beginning of injection is advanced by the same value. The timing device is generally recommended in engines with a wide speed range (motor vehicle engines) and longer fuel-injection tubings.

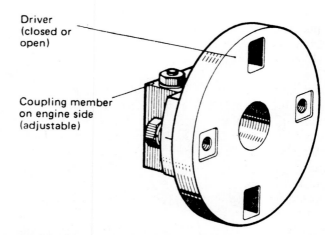

FIGURE 22-13. Adjustable drive coupling for a Bosch injection pump with an automatic timing device. *(Courtesy of Robert Bosch Canada Ltd.)*

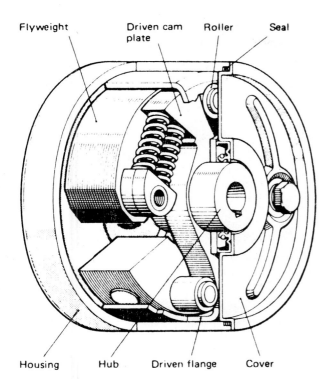

Flyweight   Driven cam   Roller   Seal
plate

Housing   Hub   Driven flange   Cover

FIGURE 22-14. Mechanical automatic injection pump timing device for Robert Bosch injection pump. *(Courtesy of Robert Bosch Canada Ltd.)*

There are two designs of automatic timing devices, however, their principle of operation is essentially the same. The four-spring design will be described below.

Principle of Operation

The automatic timing device operates responsive to speed. It uses centrifugal force for adjustment. Its drive is taken from the engine via the housing in which two flyweights are mounted on studs. Each flyweight has a roller serving as cam follower in contact with the profiled surface of the cam plate.

The driven flange is movably supported in the housing. The driven cam plate with hub is rigidly connected to the driven flange; a cap nut serves for mounting the hub (on the camshaft of the injection pump). The cam plate has four guide pins for the springs which abut against the cam plate and the spring counter-bearings on the flyweight anchor studs. The springs force the flyweights with their rollers against the profiled surfaces of the driven cam plate.

With increasing speed the flyweights are displaced outwardly by centrifugal force and their roll-

ers press on the profiled surfaces of the cam plate. In this process, they overcome the spring tension and rotate the driven flange together with the hub which is rigidly connected to the camshaft of the injection pump. The injection timing is thus advanced.

For lubrication and vibration damping, the timing device is filled with oil. It is closed with a cover fastened with screws to the flyweight anchor studs. The cover is sealed against the housing and the hub.

PART 11   FUEL SUPPLY PUMP

To have good charge of diesel fuel in the pressure chamber of the pump barrel during the intake stroke, the fuel must be delivered to the injection pump under pressure (about 1 bar—14.22 psi). Only in stationary engines can a gravity tank be installed high enough to provide a sufficient pressure head. In most other diesel vehicles, the fuel must be pumped to the injection pump. For this purpose, a fuel supply pump is directly mounted to the injection pump. The supply pump is driven by the camshaft of the injection pump by means of an eccentric placed between two cams. (See Fig. 22-16.)

The hand primer screwed into the housing of the supply pump serves to fill the low-pressure side of the injection system during start-up or for refilling the lines and venting after repair of the injection system, for example, change of a fuel filter element.

PART 12   ELECTRONICALLY-CONTROLLED INJECTION PUMP—UNITED TECHNOLOGIES MODEL 200

The United Technologies Model 200 fuel injection pump (Fig. 22-17 and 22-18) is available in a mechanically or electronically controlled version. It is a half-engine-speed in-line multi-plunger pump. Its application includes class 6 and 7 medium duty trucks, heavy duty farm tractors, and construction and industrial equipment. The model 200 represents state-of-the-art design engineering for in-line injection pumps. It has a new self-contained timing device capable of up to 20 engine degrees of timing advance. The standard system features microprocessor control of fuel quantity and timing. An optional model with mechanical control is also available.

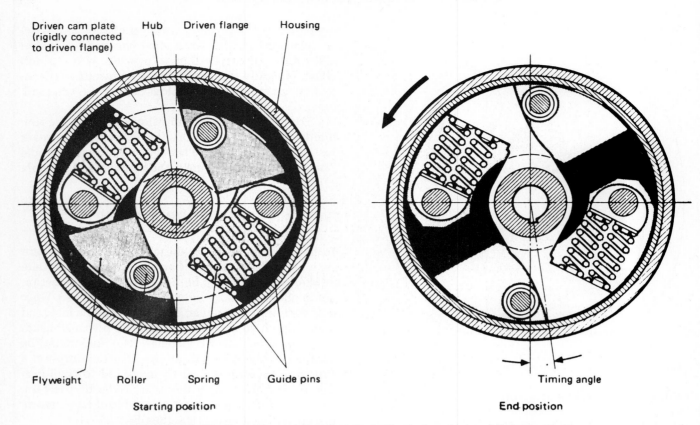

FIGURE 22-15. Operation of Bosch automatic timing device. Minimum (left) and maximum (right) advance positions shown. *(Courtesy of Robert Bosch Canada Ltd.)*

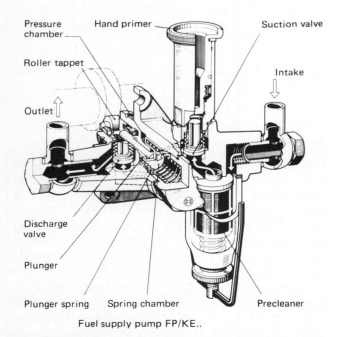

FIGURE 22-16. Plunger type fuel supply pump. *(Courtesy of Robert Bosch Canada Ltd.)*

Electronic controls and special features include the following:

Microprocessor control of timing advance and fuel quantity for maximum performance and economy.

Pump capability of 15,000 psi (1000 bar) injection pressure for optimum injection spray characteristics.

Helix type spill control fuel metering.

Reduced smoke through optional turbocharger control.

Self-diagnostic feature with provision for visual display on vehicle dashboard.

On-engine service diagnostic port.

The model 200 pump line includes 4, 6, and 8 cylinder versions for engines ranging in power from 150 to 400 bhp (112 to 300 kW) with a maximum power rating of 50 bhp (37 kW) per cylinder.

The helix type spill control design also provides the option of mechanically varying the injection

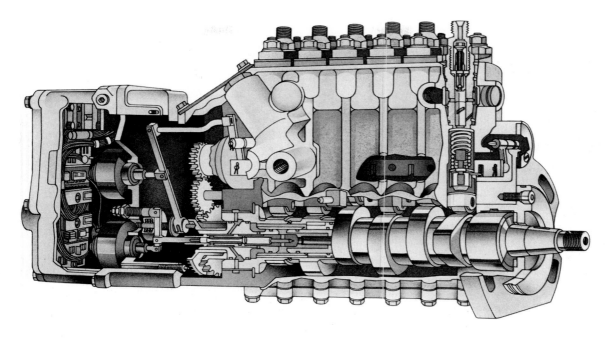

FIGURE 22-17. United Technologies Model 200 electronically controlled fuel injection pump. *(Courtesy of United Technologies—American Bosch)*

timing with respect to load through the upper helix design.

Delivery valves or snubbers are incorporated individually or in combination to control injection characteristics and ensure prevention of injection system cavitation. The self-priming supply pump is built in and can provide up to 50 psi (350 kPa) pressure. The pivoted fuel control rack produces minimum friction loading for improved governor control, even under extreme temperature conditions.

The self-contained timing device integral with the camshaft provides up to 20 engine degrees of timing advance either microprocessor or mechanically controlled.

The microprocessor control provides all the basic features of a diesel pump governor as well as the following:

- excess fuel for starting
- timing control for starting
- road speed control
- fuel scheduling
- torque limiting
- injection timing
- overspeed protection
- a self-health diagnostic output

- a "personality" module to ease pump-to-engine matching

Other features available for control of turbocharged engines requirements include:

- manifold pressure sensing
- ambient condition sensing
- independent control of power take-off speed and regulation

System Operation

Fuel is drawn from the fuel tank through the primary filter and into the supply pump. Pressurized fuel is pumped from the supply pump through the final stage filter and then to the injection pump fuel gallery and the overflow valve. Fuel from the overflow valve and nozzle leak-off drains back into the fuel tank. Pressurized fuel also flows through the injection barrel ports into the plunger bore of the injection barrel.

As the camshaft lifts, the tappet assemblies the plungers are pushed upward with the top of the plunger closing the ports. Continued upward movement of the plunger pressurizes the fuel in the plunger bore. This high-pressure fuel opens the delivery valve and flows through the injection tubing

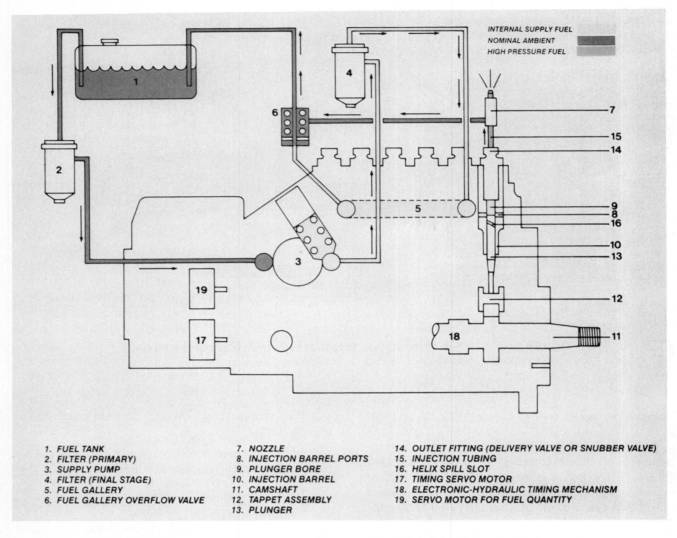

INTERNAL SUPPLY FUEL
NOMINAL AMBIENT
HIGH PRESSURE FUEL

1. FUEL TANK
2. FILTER (PRIMARY)
3. SUPPLY PUMP
4. FILTER (FINAL STAGE)
5. FUEL GALLERY
6. FUEL GALLERY OVERFLOW VALVE

7. NOZZLE
8. INJECTION BARREL PORTS
9. PLUNGER BORE
10. INJECTION BARREL
11. CAMSHAFT
12. TAPPET ASSEMBLY
13. PLUNGER

14. OUTLET FITTING (DELIVERY VALVE OR SNUBBER VALVE)
15. INJECTION TUBING
16. HELIX SPILL SLOT
17. TIMING SERVO MOTOR
18. ELECTRONIC-HYDRAULIC TIMING MECHANISM
19. SERVO MOTOR FOR FUEL QUANTITY

FIGURE 22-18. Model 200 pump schematic diagram. *(Courtesy of United Technologies—American Bosch)*

and the injector nozzle into the combustion chamber.

Additional upward movement of the plunger causes the helix spill slot to open the injection barrel ports and terminate fuel injection. The plunger lift dimension from port closing to port opening is a function of the rotary position of the plunger controlled by the fuel control rack.

An electronically controlled servo-motor adjusts the fuel quantity by moving the fuel control rack and linkage in the required direction. A second electronically controlled servo-motor adjusts the injection timing by precise positioning of the control element of the timing mechanism. Both of these functions are performed in response to signals from the microprocessor in response to the various input sensors and the vehicle operator through the throttle control.

PART 13   INJECTION EQUIPMENT FOR MULTI-FUEL ENGINES

General Information

Diesel engines which can be operated not only with diesel fuel but also with other fuels, such as gasoline, kerosene, etc., are known as multi-fuel diesel engines. (See Fig. 22-19.)

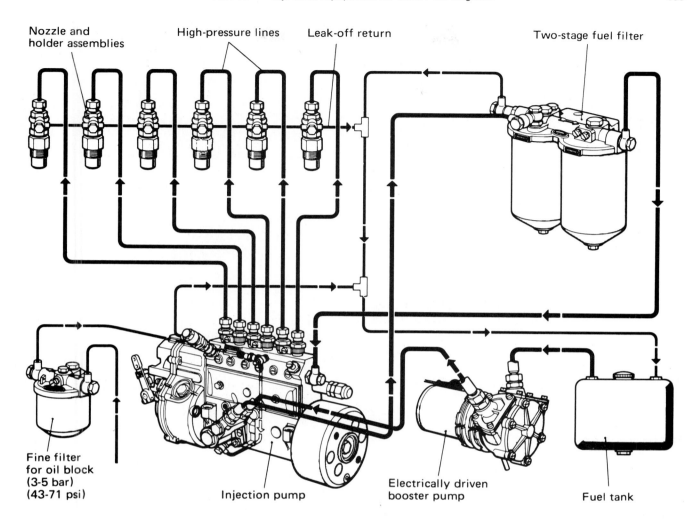

FIGURE 22-19. Injection system for dual or multifuel diesel engine. *(Courtesy of Robert Bosch Canada Ltd.)*

The injection equipment of the multi-fuel engines consists of the following parts as in the standard diesel engine:

Injection pump with governor

Fuel supply pump

Fuel filter

Nozzle and holder assemblies

Lines and accessories

The additional equipment consists of an electrically driven booster pump and a lubricating oil fine filter for the oil block. This equipment was developed from proven Bosch products for standard diesel engines. Particular attention had to be given to three characteristics of lower-boiling fuels distinguishing them from the diesel fuel: lower viscosity, lower boiling point, and lower specific weight.

Lower viscosity gives rise to the risk of greater leakage losses of fuel leaking past the plunger of the injection pump. Consequently, each pumping element was equipped with an oil block. Lubricating oil (sealing oil) for the oil block is obtained from the engine lubricating system and consequently must be thoroughly filtered.

The *lower boiling point* increases the danger of vapor formation within the injection pump. This was solved by increasing the pressure in the gallery of the injection pump to about 1.5 bar (21 psi). Furthermore, lowering of the fuel temperature also counteracts the vaporization tendency. Consequently, considerably more fuel is delivered to the injection pump than is injected into the engine. At the same time, any vapor bubbles are rapidly purged from the pump by these measures. The excess fuel delivered returns to the fuel tank.

Stopping the hot engine unavoidably produces vapor bubbles again because the injection pump is no longer flushed by the mechanically driven supply pump and consequently is no longer cooled; furthermore, the excess pressure in the gallery drops. When the engine is to be restarted, the vapor bubbles must be either condensed rapidly by overpressure or purged rapidly from the gallery. This is accomplished by an electrically driven, self-priming, booster pump, which is started before actuation of the starting motor and thus flushes the suction chamber with fuel.

In a conversion from diesel fuel to light fuel, the *lower specific weight* of the lower-boiling fuels would result in a lower fuel weight per delivery cycle. To compensate for this, the full-load stop is equipped with an adjustable stop which can be easily flipped up for operation with light fuel and thus guarantees a larger delivery quantity per delivery stroke.

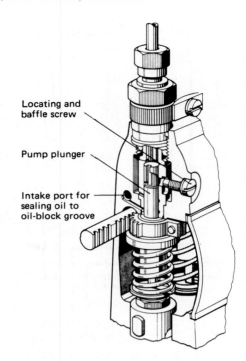

FIGURE 22–20. Sectional view of an injection pump for multifuel operation. *(Courtesy of Robert Bosch Canada Ltd.)*

Fuel Injection Pump

The injection pump for the multi-fuel engine differs from the injection pump of a standard diesel engine primarily by the oil block. The pump plunger is equipped with two annular grooves. The upper groove is connected to the fuel gallery of the pump through a port. The lower groove with an inlet port for pressurized oil serves as oil block. Oil from the lubricating circuit of the engine is forced into this groove under a pressure which in the range of normal operating speeds is higher than that in the fuel gallery of the injection pump. (See Fig. 22–20 and 22–21.)

At idling and low part-load speeds, the oil block pressure may drop below fuel gallery pressure. A check valve has therefore been mounted on the inlet of the sealing oil into the injection pump. It prevents transfer of fuel into the lubricating circuit of the engine via the sealing oil line.

Additional characteristics of the injection pump are as follows:

Locking the pump barrel with locating screws which simultaneously serve as baffle screws

slight excess pressure in the fuel gallery due to the overflow valve

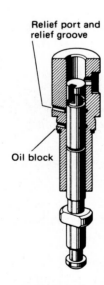

FIGURE 22–21. Sectional view of a pumping element for multifuel operation. *(Courtesy of Robert Bosch Canada Ltd.)*

double lobe cam for operating the supply pump check valve on the oil block

Fuel Supply Pump

The temperature of the fuel supply pump should be as low as possible during operation with gasoline and it is for this reason that the supply pump is mounted on the injection pump on a thick insulating pad for heat insulation. (See Fig. 22–22.)

Furthermore, like the injection pump, the supply pump is equipped with an oil block feature.

The tappet plunger combines roller tappet, pump plunger and suction valve. It has four holes on its circumference for fuel inlet and two vertical slots for guidance by the locating screw. The tappet spring on the inside of the plunger forces the latter against the cam on the injection pump. The spring rests on the locating screw. A suction valve is screwed into the end of the tappet plunger. It has a large cross-section and a plane sealing surface. Gas bubbles formed by fuel heating can escape into the high-pressure line at standstill.

The discharge valve, like the suction valve, has a large cross-section and a plane sealing surface. It is installed in the pump housing.

The tappet plunger is connected in parallel with a second piston—the compensating piston. The lat-

ter acts as an elastic intermediate component and serves to maintain the most constant pressure possible.

The compensating piston is forced against the plug screw by spring tension. It is equipped with a stud on the piston bottom to prevent obstruction of the drain hole.

Operation

At the end of the suction stroke the valve stem of the suction valve plate rests on the locating screw, so that the valve is not completely closed. Fuel in the suction line is still in motion from the suction stroke and consequently can flow into the pressure chamber.

In the discharge stroke the tappet plunger is pushed to the outside by the driving cam against the tension of the tappet spring. This closes first the suction valve and opens the discharge valve. Fuel on the pressure side flows through the fuel supply line of the fuel filter.

During the discharge stroke the compensating piston returns against the coil spring tension.

At the end of the discharge stroke, fuel stored above the compensating piston is forced by the compensating piston into the above mentioned fuel supply line of the filter or the injection pump. At T.D.C., the tappet plunger rests on the discharge valve body which is still lifted. The latter can thus return to its seat only after the start of the suction stroke. During the suction stroke, the force stored in the tappet spring serves as the propelling force; fuel flows into the pressure chamber during the suction stroke.

Oil-Block Filter

The lubricating oil used for the oil block must be finely filtered just like the fuel, so that the sensitive precision parts of the injection pump and nozzles will be kept free from contaminants that might cause damage or premature wear. The lubricating oil is therefore passed through a fine filter, the most important part of which—the filter element—is installed in the bowl in such a way that the lubricating oil can only reach the oil block after having passed through the element.

The oil pressure in the supply line should be about 3 to 5 bar (43 to 71 psi). If the pressure in the engine lubricating oil system is higher, a suitable throttle must be installed in the supply line to the filter or an appropriately adjusted overflow valve must be used. (See Fig. 22–23 to 22–25.)

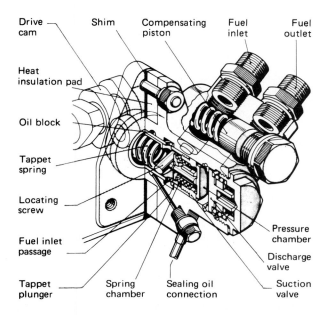

FIGURE 22-22. Fuel supply pump for multifuel engine. *(Courtesy of Robert Bosch Canada Ltd.)*

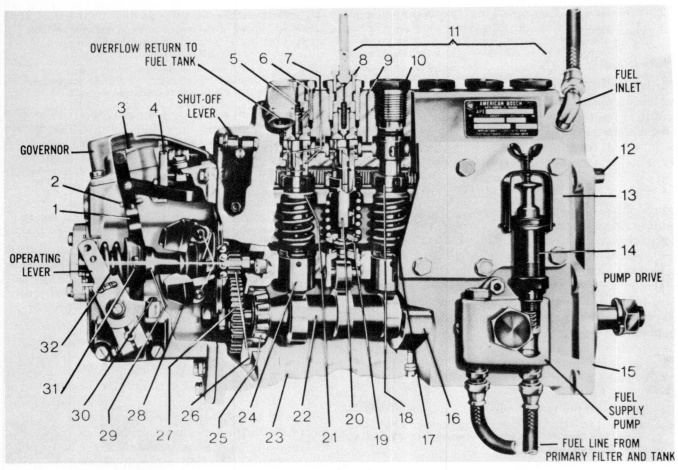

FIGURE 22-23. Sectional view of pump. *(Courtesy of United Technologies— American Bosch)*

PART 14 FUEL INJECTION SYSTEM SERVICE

General Precautions

Follow all the normal shop safety precautions. The following general precautions should also be observed during the fuel system service.

Servicing cab-over and cab-forward truck engines requires additional safety precautions as follows:

Before Tilting the Cab

• Check clearance above and in front of the truck cab.

• Keep tool chests and workbenches away from the front of the cab.

• Inspect the sleeper and cab interior for loose luggage, tools, and liquid containers that could fall forward.

• Check "buddy" or "jump" seats on the right side of the cab to be sure that they are secured in place.

While Tilting the Cab

• Be sure that the engine is not running.

• Check the position of the steering shaft U joint to prevent binding.

• Never work under a partially tilted cab, unless it is properly supported.

• When the cab is tilted past the over-center position, use cables or chains to prevent it from falling. Do not rely on the cab hydraulic lift mechanism to

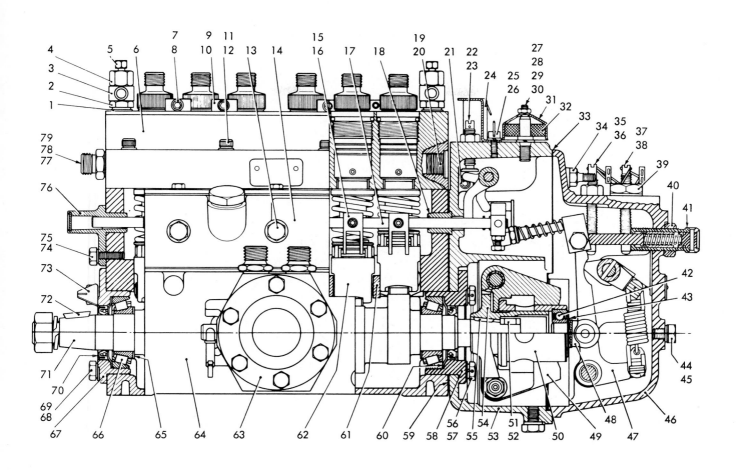

1	Joint Washer	28	Nut	54	Governor Hub
2	Air Vent Body	29	Spring Washer	55	Hub Woodruff Key
3	Air Vent Banjo	30	Spring Washer	56	Governor Housing Screw
4	Nut	31	Breather Cap	57	Spring Washer
5	Air Vent Valve	32	Filter Capsule	58	Bearing Housing
6	Pump Body	33	Jointing	59	Jointing
7	Clamp Screw	34	Pillar Stud	60	Washer
8	Spring Washer	35	Idling Speed Adjusting Screw	61	Tappet Locating T—Piece
9	Clamp	36	Nut	62	Tappet Assy.
10	Clamp	37	Max. Speed Adjusting Screw	63	Feed Pump
11	Pump Body Screw	38	Nut	64	Cambox
12	Sealing Washer	39	Oil Filler Plug	65	Shims
13	Cover Bolt	40	Locknut	66	Taper Roller Bearing
14	Inspection Cover	41	Damper Screw	67	Bearing Housing
15	Control Rod Fork	42	Ball Bearing	68	Screw
16	Control Rod Fork Screw	43	Shims	69	Spring Washer
17	Control Rod	44	Governor Oil Level Plug	70	Oil Seal
18	Control Rod Bush	45	Joint Washer	71	Camshaft
19	Gallery Plug	46	Governor Cover	72	Coupling Woodruff Key
20	Gallery Plug Joint Washer	47	Crank Lever	73	Timing Indicator
21	Trip Pin	48	Fork	74	Screw
22	Max. Fuel Stop Screw	49	Governor Flyweight	75	Spring Washer
23	Locknut	50	Sleeve	76	Control Rod Cover
24	Sealing Cover	51	Slotted Nut	77	Spill Gallery Union
25	Screw	52	Spring Washer	78	Inlet Gallery Union
26	Spring Washer	53	Governor Housing	79	Joint Washers
27	Air Breather Pillar				

FIGURE 22-24. Sectional view of pump. *(Courtesy of CAV Limited)*

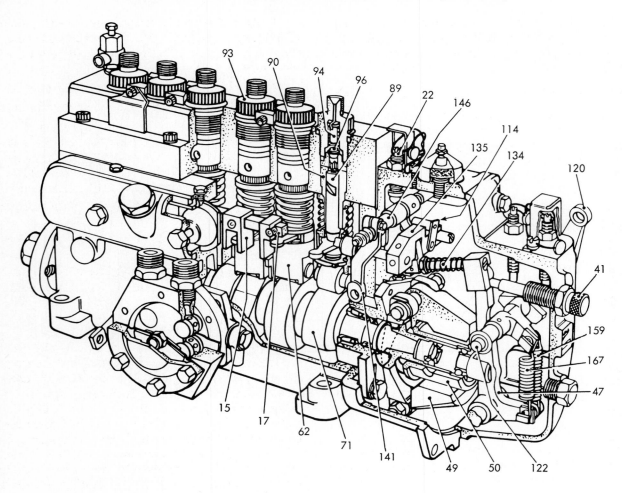

15	Control Fork	93	Delivery Valve Holder
17	Control Rod	94	Volume Reducer
22	Max. Fuel Stop Screw	96	Delivery Valve
41	Damper	114	Trip Lever
47	Crank Lever	120	Speed Control Lever
49	Governor Flyweight	122	Speed Lever Shaft
50	Governor Sleeve	134	Telescopic Link
62	Tappet Assy.	135	Bridge Link
71	Camshaft	141	Stop Control Lever
89	Plunger	146	Excess Fuel Device
90	Barrel	159	Governor Idling Spring
		167	Governor Main Spring

FIGURE 22-25. Cutaway section of typical 6-cylinder Majormec pump. *(Courtesy of CAV Limited)*

retain or break the fall of the cab in any position beyond the over-center point.

While the Cab is Tilted

• If it's necessary to open a door, take care to avoid damage to door hinges and/or window glass.

While Lowering the Cab

• Be sure that the engine is not running.

• Be sure that the cab lowers properly on mounting and locating pins. Twisted or misaligned cabs may miss locating pins, and the cab latch will not secure the cab in a locked position.

- Lock the cab-latching mechanism when the cab is all the way down. Failure to lock the cab will allow it to swing forward when the truck is stopped suddenly.

Industrial Attachments

- Be sure that all hydraulic, mechanical, or air-operated equipment such as loaders, buckets, and the like are lowered to their at-rest position on the floor or are adequately and safely supported by other means before attempting to do any fuel system service.

Other Precautions

- Be aware of the potential danger of disconnecting any pressurized systems such as hydraulics, fuel, air, or air conditioning, including refrigerant-handling precautions.
- Disconnect the battery ground cable at the battery to avoid electrical system damage.
- Be careful when handling hot cooling systems (the system is pressurized, and radiator cap removal can cause the system to boil and overflow, causing burns to the hands and face). Turn the cap to the first notch only to relieve system pressure (use a rag) and then remove the cap.
- Be careful of hot engine parts.
- Engine oil for pump lubrication may be hot enough to cause burns; handle with care to avoid injury.
- Use proper methods and equipment for handling parts to avoid personal injury and parts damage.
- Fuel system components have many precision machined surfaces; handle them with extreme care to avoid damage.
- Observe cleanliness of all parts and the entire area; this is essential during assembly. A very small piece of dirt can cause complete failure. Abrasives damage friction surfaces severely. Extremely close tolerances are destroyed by small particles of dirt or foreign material.
- Use only filtered compressed shop air for cleaning fuel system parts after washing.
- Use only filtered and temperature-controlled testing fluids for injector and injection pump service.
- Perform all injector and injection pump service in a controlled environment—a lab with controlled temperature and filtered air.

- Very high fuel injection pressures (15,000 psi, 103,425 kPa) will penetrate skin easily and cause serious damage and infection; avoid all contact with fuel under injection pressure.
- Avoid inhaling fuel and test fluid vapors.
- Have all test equipment accurately calibrated for test results to be valid.
- Be absolutely sure that the correct specifications are being used for any given make or model of fuel system and fuel system components. Refer to the appropriate service manual for specifications.
- Be sure to follow the specifications in the appropriate shop service manual for installation, alignment, and timing of injectors and injection pumps.
- Always keep all fuel line and fitting openings properly capped when disconnected to prevent entry of dirt or moisture.

Preventive Maintenance

The cost of operating a diesel engine can be greatly reduced by following a preventive maintenance program of careful inspection and checking. Many troubles are eliminated and anticipated before they can become costly problems. This applies to any diesel engine, regardless of where it may be operating—a truck, tractor, large earth moving equipment, electric power generating equipment, a pleasure yacht, ocean-going ships, transcontinental bus, off-highway equipment, oil well drilling equipment, regardless of size and horsepower.

General Cleanliness

This is basic in all diesel work and is most often overlooked. Remember that any particle of dirt that gets into the injection pump shortens its life. Before opening or loosening any part, such as a fuel line, filter or supply tank, the exterior of the immediate area should be cleaned. In severe conditions, after wiping away the accumulated dirt, brush the surface with appropriate solvent and blow dry with compressed air.

Fuel

Fuel must be clean and of the type recommended by the manufacturer. Make sure, when fuel is pumped into the supply tank, every precaution is taken so no dirt enters the supply tank. Using inferior or "discount" fuels is false economy, as they result in early failure of parts and increased maintenance costs. The pour point of the fuel being used should be at least $10°F (-12°C)$ below the lowest operating ambient temperature.

TROUBLESHOOTING

PROBLEM

CAUSE	A. Fuel not reaching pump.	B. Fuel delivered from supply pump but not to nozzles.	C. Fuel reaching nozzles but engine won't start.	D. Engine starts hard.	E. Engine starts and stops.	F. Erratic engine operation – surge misfiring, poor governor regulation.	G. Engine idles imperfectly.	H. Engine does not develop full power or speed.	I. Engine smokes black.	J. Engine smokes blue.	REMEDY
No fuel.	2										Fill fuel tank and bleed system
Foreign material in fuel tank.	3										Clean fuel tank.
Fuel cap vent holes clogged.	4										Clean vent holes or replace cap.
Rack binding.		4									* Replace or repair pump.
Broken camshaft.		6									* Replace or repair pump.
Fuel lines clogged or restricted.	7	2		4	1	2	8	4			Blow out all fuel lines with filtered air. Replace if damaged. Remove and inspect all flexible lines.
Air leaks on suction side of system.	8			5	6	8	2	5			Pressurize system with air to locate leaks. Repair as needed.
Supply pump malfunction.	6			11	7	18	4	15			Replace supply pump.
Plungers sticking.		5		18	9	19	14	19			* Disassemble and inspect for burrs, corrosion or varnishes.
Delivery valves dirty or sticky.		7									* Inspect and clean delivery valves and holders.
Governor spring worn or broken.						16	10	8			* Remove and replace.
Governor linkage broken.							12	9			* Remove, replace and readjust.
Tank valve closed.	1										Open valve.

Numbers in "Problem" check chart indicate the numerical order in which to check possible "Causes" of problem. Start with No. 1, then 2, etc.

* Refer to Disassembly and Reassembly of Pump and Governor.

(Courtesy of International Harvester Co.)

Numbers in "Problem" check chart indicate the numerical order in which to check possible "Causes" of problem. Start with No. 1, then 2, etc.

CAUSE	A. Fuel not reaching pump.	B. Fuel delivered from supply pump but not to nozzles.	C. Fuel reaching nozzles but engine won't start.	D. Engine starts hard.	E. Engine starts and stops.	F. Erratic engine operation — surge misfiring, poor governor regulation.	G. Engine idles imperfectly.	H. Engine does not develop full power or speed.	I. Engine smokes black.	J. Engine smokes blue.	REMEDY
Cam roller sticking.		9		8							* Remove, check for size and burrs and reassemble.
Cranking speed too low.			1	1							Charge or replace batteries.
Lube oil too heavy at low temperature.			16	7							Refer to Operator's Manual.
Nozzles faulty or sticking.	3		13	22		9	7	11	5		Replace or correct nozzles.
Intake air temperature low.			4	2							Provide starting aids. Refer to Operator's Manual.
Engine compression poor.			15	9				24	9		Correct compression. Refer to Engine Manual.
Pump timed incorrectly to engine.			2	3		4	3	7	4	3	Correct timing.
Excessive fuel leakage past plungers (worn or badly scored).			11	19			13	20			* Replace.
Supply pump faulty, pressure too low.				12				16			* Remove and inspect parts.
Filters clogged.	5			6	5	3		6			Remove and replace clogged elements.
Cam rollers worn.			9	17				18	10		* Remove and replace.
Automatic advance faulty or not operating.			12	21		13		14	7	4	Replace.
Governor linkage out of adjustment.		1		14		17	6	12			Adjust or replace linkage.
Governor not operating; parts or linkage worn, sticking or binding, or incorrectly assembled.		8		15		15	5	10			* Disassemble, inspect parts, replace if necessary and reassemble.

* Refer to Disassembly and Reassembly of Pump and Governor.

(Courtesy of International Harvester Co.)

Numbers in "Problem" check chart indicate the numerical order in which to check possible "Causes" of problem. Start with No. 1, then 2, etc.

CAUSE	A. Fuel not reaching pump.	B. Fuel delivered from supply pump but not to nozzles.	C. Fuel reaching nozzles but engine won't start.	D. Engine starts hard.	E. Engine starts and stops.	F. Erratic engine operation — surge misfiring, poor governor regulation.	G. Engine idles imperfectly.	H. Engine does not develop full power or speed.	I. Engine smokes black.	J. Engine smokes blue.	REMEDY
Maximum fuel setting at low limit or too low.		10	16					17			Reset to pump specifications.
Engine valves faulty or out of adjustment.			23		10	11		6			Correct valves or valve adjustments as in engine manual.
Water in fuel.		5	8	2	5	1	23				Drain fuel system and pump housing, provide new fuel, prime system.
Improper rack adjustment.						7					Adjust rack.
Air intake restricted.					3			2			Check. Refer to Engine Manual.
Delivery valve spring broken.						12					* Replace.
Low octane fuel.		14	10			6	9	21	8		Provide fuel per engine specifications.
Injection pipes leaking or connected to wrong engine cylinders.		6				1		13			Relocate pipes for correct engine firing sequence.
Nozzle return lines clogged.						11					Remove lines, blow-out, inspect and reassemble.
Loose nut on end of pump camshaft.						14					Tighten to specified torque.
Shut-off device interfering with governor linkage.		7	13					2			Check and adjust governor linkage dimension.
Governor high-idle adjustment incorrect.								3			Adjust to pump specifications.
Aneroid stop plate incorrectly adjusted.		8	20						12		Adjust to specification.

* Refer to Disassembly and Reassembly of Pump and Governor.

(Courtesy of International Harvester Co.)

Numbers in "Problem" check chart indicate the numerical order in which to check possible "Causes" of problem. Start with No. 1, then 2, etc. CAUSE	A. Fuel not reaching pump.	B. Fuel delivered from supply pump but not to nozzles.	C. Fuel reaching nozzles but engine won't start.	D. Engine starts hard.	E. Engine starts and stops.	F. Erratic engine operation – surge misfiring, poor governor regulation.	G. Engine idles imperfectly.	H. Engine does not develop full power or speed.	I. Engine smokes black.	J. Engine smokes blue.	REMEDY
Throttle arm travel not sufficient.			3					1			Check installation and adjust throttle linkage.
Maximum fuel setting too high.									11		Reset pump.
Engine overheating.					4			3			Correct as in Engine Manual.
Exceeding rated load.								1			Reduce load on engine.
Engine cold.										1	Check thermostat, warm to operating temperature. Refer to Engine Manual.
Lube oil pumping past valve guides or piston rings in engine.										5	Correct as in Engine Manual.
Excess lube oil in engine air cleaner.										2	Remove, check and reduce oil quantity to specified level. Refer to Engine Manual.
Aneroid incorrectly adjusted.								22	13		Readjust.

(Courtesy of International Harvester Co.)

Pump Lubricating Oil

As the injection pump is lubricated by the engine oil, it is important that engine oil filters and engine oil be changed regularly to ensure that clean oil reaches the pump. The oil used in the engine should be the type recommended by the engine manufacturer.

Fuel Tank

Because moisture will condense and accumulate in the fuel supply tank, it is important that the tank be drained of such water and any sediment. In areas of high humidity, such draining must be done more frequently. In general, draining the fuel tank once each month is satisfactory.

Supply Hoses, Suction Hoses and Fittings

Frequency of inspection will vary with the conditions under which the equipment is operated. Under severe operating conditions, such as heavy earth moving equipment, once a week is desirable. Inspect for dented, crimped or collapsed lines, which can restrict the flow of fuel. Check for abraded or cracked hoses and damaged or cracked fittings, which can cause suction air leaks. Make certain that all hose connections and fittings are tight.

Primary Filter(s)

The oil filter or diesel filters should be replaced in accordance with the engine manufacturer's recommendations. Remember, it costs far less to change a filter than it does to change an injection pump. Only filters approved by the engine manufacturer should be used.

Secondary Filter

Follow the engine manufacturer's instructions regarding the servicing of this filter.

Hand Priming Pump

It is necessary to bleed air from the low pressure side of a system whenever a filter or filter element has been replaced or a line disconnected. The plunger retaining strap of the hand primer must always be kept in place.

High Pressure Lines

High pressure fuel lines are made of high tensile strength steel, seamless and cold drawn. On each installation, the fuel lines are of equal length and inside diameter. Any variation will result in unequal quantities of fuel being delivered to the individual cylinders of the engine. Therefore, when replacement is necessary, the new lines must be of identical specifications in order to avoid impairment of injection characteristics.

Both ends of each fuel line must be inspected for possible crimping of the bore before an injection pump or nozzle holder is re-installed. The inside diameter of the fuel line should be checked with a drill of the proper size as indicated by the manufacturer's specifications. Re-drill or ream if necessary and carefully flush the entire line.

The tubings must be securely fastened to prevent rubbing against each other or parts of the engine. Suitable clamps are provided to prevent such vibration and motion of the fuel lines.

Injection Pump Timing

Good preventive maintenance includes frequent inspection of the pump timing to be sure it is correctly timed to the engine in accordance with the manufacturer's specifications.

Injection Pump Mounting

It is important to make certain that injection pump mounting bolts are correctly torqued to the manufacturer's specification. Loose bolts result in vibration and consequent leakage of fuel line connections and wear of pump parts.

Throttle and Stop Linkage

Improperly adjusted throttle and stop linkage frequently results in complaints of low power. The throttle linkage must be adjusted so that the injection pump operating lever moves from the idle stop to the full load stop (maximum speed stop). "Breakaway" throttle levers deflect slightly at each end of their travel.

Worn, bent and otherwise damaged throttle linkage and worn pivot points and joints must be replaced.

Stop linkage must also be checked periodically for proper operation. The stop linkage must be adjusted in accordance with the manufacturer's specifications. When making the adjustment or inspection, be sure the stop mechanism does not interfere with normal operation of the injection pump and governor.

Injection Nozzles

To insure maximum performance and economy of operation, the nozzle and nozzle holder assemblies should be removed from the engine on a scheduled basis and checked for proper opening pressure, chatter and spray pattern. If nozzle does not operate properly, it must be cleaned and repaired or replaced. If necessary, reset nozzle opening pressures. Specialized equipment is required for making such checks. See Chapter 20 for injector service.

Air Induction System

The ratio of the amount of air to the amount of fuel is seriously affected when the air intake is restricted. This, in turn, results in smoking exhaust, loss of power and reduced fuel MPG operation. To overcome this condition, it is essential that the air cleaner be kept clean. The frequency of inspecting and cleaning the air cleaner is dependent largely on the conditions under which the diesel equipment operates. The manufacturer's instructions cover normal conditions. For dusty conditions, inspect and clean the air cleaner more frequently.

Overflow Valves

The overflow valves, including orifice types, should be cleaned periodically in Varsol or an equivalent solvent. At the same time, the various parts should be inspected carefully for excessive wear and scoring. Orifice valves are flow checked in accordance to special procedures and replaced if defective. Special attention should be given to the valve and valve seat. If nicked or scored, they should be replaced. Rough idle, surge or loss of power will result if there is valve leakage, if stuck open or if there is a broken valve spring.

Pump-Cleaning and Flushing

Cleaning Procedure

1. Thoroughly wash external surfaces of the injection pump with cleaning fluid. Make certain that all foreign matter is removed.

2. Remove inspection window cover or timing window cover. If applicable, remove or open drain plugs or fittings.

3. Remove governor top cover.

4. Flush lube oil compartments in pump and governor with approved cleaning fluid, such as Varsol, to remove all traces of dirt, grease, carbon and other foreign matter.

Note: After removing inspection covers, check components of governor for damage or binding of parts. Make certain fulcrum lever operates freely. Also check pump components for faulty or damaged parts, including broken spring, broken or worn control unit or sleeves, binding control rod or rack, bro-

ken plunger(s) or camshaft, cracked or damaged housing, etc.

5. Re-install covers.

Flushing Procedure

1. Mount pump on approved test stand.

Note: It may be necessary to replace the original pump drive hub and/or coupling with a coupling and/or hub that will fit the particular test stand being used.

2. Lubricate pump as follows: APE type—if pump is normally lubricated by engine oil, add approximately one pint (0.473 liter) of SAE 30 lube oil to camshaft compartment to lubricate cam lobes, tappet rollers, etc.; if not engine oil lubricated, add sufficient SAE 30 oil to camshaft compartment to maintain oil plug level. PSB, PSJ & PSM type—use a pressurized lube oil system (system and ducted adapters and fixtures available from Bacharach Instrument Co., Pittsburgh, Pa.) filled with SAE 30 oil, or equivalent, to properly lubricate the pump. The lube oil pressure must be 20 psi (1.38 Bars) minimum or, if pump includes lube oil pressure adjusting devices, a pressure of 35–40 psi (2.41–2.76 Bars) is required.

Note: Do not splash-lubricate PSB, PSJ or PSM type pumps, as this provides insufficient lubrication to certain components.

Important: Do not use calibrating oil or fuel oil as a lubricant.

Caution: Do not allow lube oil to mix with or contaminate calibrating or fuel oil.

3. Connect a set of high pressure tubings to the pump outlets and direct the opposite ends of the tubings into a pail, thereby bypassing the nozzle and holder assemblies and test stand tank.

4. To flush pump sump, proceed as follows: APE type—operate at about 600 RPM with the control rack in the mid-position for approximately five minutes. PSB, PSJ and PSM types—operate at a speed of about 1000 RPM with the operating lever in the full load position for approximately three minutes.

This procedure removes any fuel oil and foreign material from the pump sump and eliminates the possibility of contaminating the test oil or damaging the nozzles during the calibration check.

5. Stop test stand and remove the high pressure tubing.

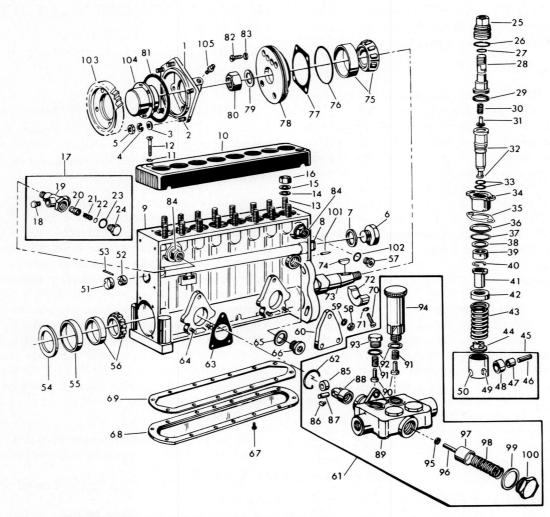

FIGURE 22-26. Exploded view of Robert Bosch inline fuel injection pump. *(Courtesy of International Harvester Company)*

PART 15   INJECTION PUMP OVERHAUL, PHASING, AND CALIBRATION GUIDELINES

Disassembly

For periodic disassembly and inspection, the normal procedure is to remove the inspection cover, pump body, delivery valves, pumping elements, tappets, drive coupling, governor cover, governor, and camshaft. The extent of disassembly thereafter depends on pump condition. (See Fig. 22-26 to 22-37.)

Prior to disassembly, the exterior of the pump should be cleaned with cleaning solvent and any lu-

bricant drained. During disassembly it is critical that all parts pertaining to each pumping element be kept together and separate from other pumping element parts. Separate containers or a tray with separate compartments are recommended. Do not mix parts among pumping elements. These parts include the barrel and plunger, plunger spring, delivery valve, guide and holder, tappet, phasing spacers or adjusters, and the like.

The barrel and plunger of each element are selectively fitted or mated and must be kept together at all times during pump service. The delivery valves and guides are also selectively fitted. Phasing adjustment spacers are of selective thickness for each element.

1. Adapter	36. Spacer	71. Washer
2. Stud	37. "O" ring	72. Center support
3. Washer	38. Washer	73. Camshaft
4. Washer	39. Cap	74. Key
5. Nut	40. Snap ring	75. Bearing
6. Plug	41. Sleeve	76. "O" ring
7. Gasket	42. Upper spring seat	77. Shim
8. Rack	43. Spring	78. Housing
9. Housing	44. Lower spring seat	79. Washer
10. Cover	45. Tappet assembly	80. Nut
11. Washer	46. Pin	81. "O" ring
12. Screw	47. Bushing	82. Screw
13. Stud	48. Roller	83. Lock washer
14. Washer	49. Block	84. Fitting
15. Washer	50. Tappet	85. Roller
16. Nut	51. Retainer	86. Pin
17. Check valve assembly	52. Rack guide	87. Lock pin
18. Plug	53. Rack stop pin	88. Tappet
19. Body	54. Shim	89. Housing
20. Piston	55. Rear bearing spacer	90. Valve
21. Spring	56. Bearing	91. Spring
22. Washer	57. Plug	92. Gasket
23. Gasket	58. Nut	93. Plug
24. Plug	59. Washer	94. Hand pump
25. Cap	60. Cover	95. "O" ring
26. "O" ring	61. Fuel supply pump	96. Spindle
27. "O" ring	62. Snap ring	97. Plunger
28. Holder	63. Gasket	98. Spring
29. Gasket	64. Stud	99. Gasket
30. Spring	65. Gasket	100. Plug
31. Delivery valve	66. Plug	101. Pointer
32. Plunger and barrel	67. Screw	102. Gasket
33. "O" ring	68. Cover	103. Gear
34. Sleeve	69. Gasket	104. Adjuster assembly
35. Shim	70. Screw	105. Plug

FIGURE 22-26. *(Cont'd.)*

Cleaning and Inspection

Clean all parts in cleaning solvent and blow dry with compressed air. Examine all parts visually for signs of wear or damage and replace any that are excessively worn or damaged.

The upper parts of the barrel and plunger usually show the most wear. Fine scoring marks are normal after even a short period of operation, and the element should not be replaced because of them. If the scoring is deeper and the edge of the spill groove is damaged, the element should be replaced.

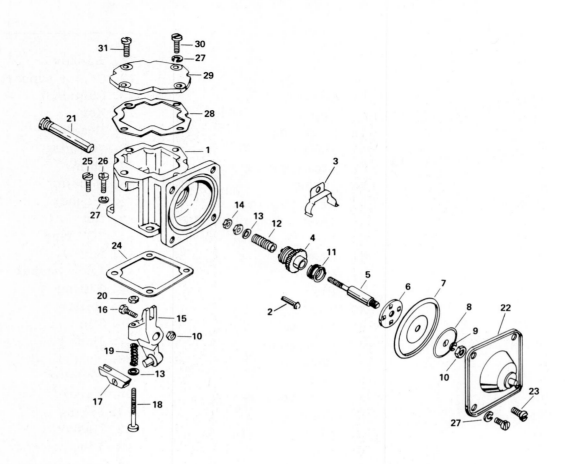

1. Housing	17. Stop plate
2. Pin	18. Screw
3. Clip	19. Spring
4. Guide bushing	20. Nut
5. Diaphragm bolt	21. Sliding link bolt
6. Spring upper seat	22. Diaphragm cover
7. Diaphragm	23. Screw
8. Diaphragm washer	24. Gasket
9. Lock washer	25. Screw
10. Nut	26. Screw
11. Diaphragm return spring	27. Washer
12. Anti-lash spring	28. Gasket
13. Washer	29. Aneroid top cover
14. Lock nut	30. Screw
15. Sliding link	31. Screw
16. Screw	

FIGURE 22-27. Exploded view of aneroid for Robert Bosch inline injection pump. *(Courtesy of International Harvester Company)*

FIGURE 22-28. Removing delivery valve clamps (7, 8, 9 & 10) and pump body screws. *(Courtesy of CAV Ltd.)*

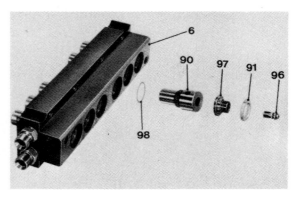

FIGURE 22-31. Using a soft hammer, the plungers (90) are removed by applying a sharp tap to the bottom of each plunger. *(Courtesy of CAV Ltd.)*

FIGURE 22-29. Pump body removed from cam box. Parts from each pumping element are kept separate during entire operation. *(Courtesy of CAV Ltd.)*

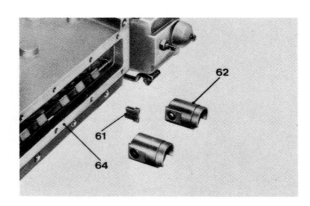

FIGURE 22-32. Removing tappet assemblies from cam box. *(Courtesy of CAV Ltd.)*

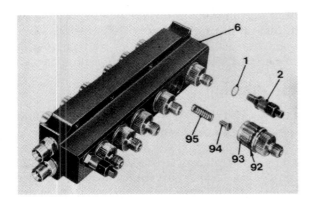

FIGURE 22-30. Removing the air vent (2) and delivery valve holder (93) and valves (94, 95) from pump body. *(Courtesy of CAV Ltd.)*

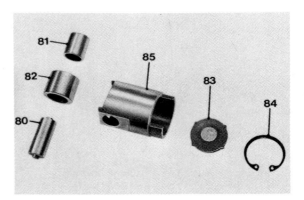

FIGURE 22-33. Exploded view of tappet assembly; bushing—81, roller—82, pin—80, body—85, spacer—83, retainer—84. *(Courtesy of CAV Ltd.)*

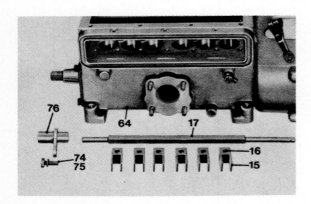

FIGURE 22-34. Control rod components removed from cam box. *(Courtesy of CAV Ltd.)*

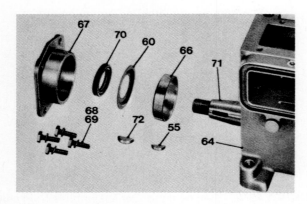

FIGURE 22-35. Bearing housing (67), oil seal (70), and bearing outer race (66) removed. *(Courtesy of CAV Ltd.)*

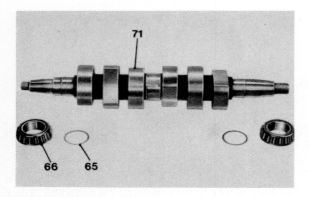

FIGURE 22-36. Camshaft and bearings removed. Shims (65) are used to establish correct camshaft end play. Note center bearing journal. *(Courtesy of CAV Ltd.)*

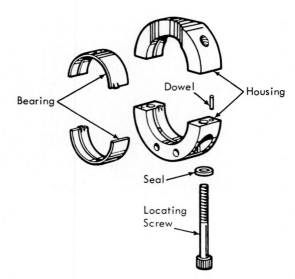

FIGURE 22-37. Details of camshaft split center bearing. *(Courtesy of CAV Ltd.)*

A complete set of elements should be installed to ensure equal fuel delivery to all cylinders.

Inspect the delivery valves for pitting or damage. Replace if needed. Inspect the camshaft and camshaft bearings. Excessive wear of the cams will affect injection characteristics. Replace a worn camshaft and worn or damaged bearings. Inspect the governor weights, pins, ramp, rollers, collar, and spring for wear or damage. Examine all bushings for wear or damage. Replace any parts that are worn or damaged.

Reassembly

Assemble the pump in the reverse order. Replace all gaskets, O rings, seals, lock tab washers, lock nuts, and sealing washers. Cleanliness is extremely important during assembly. Lubricate all moving parts with fuel oil during assembly. This includes plungers and barrels, delivery valves and guides, camshafts and bearings, and governor components.

Pump Phasing

Phasing is the procedure for checking and adjusting the phase angles between successive injections. For example, on a six-cylinder four-cycle engine there are 360 degrees of pump camshaft rotation for 720 degrees of engine crankshaft rotation. Since all six cylinders fire in 360 degrees of pump camshaft rotation, the phase angle is 360 ÷ 6, or 60 degrees between the start of injection of all six elements. (See Fig. 22-38 to 22-41.)

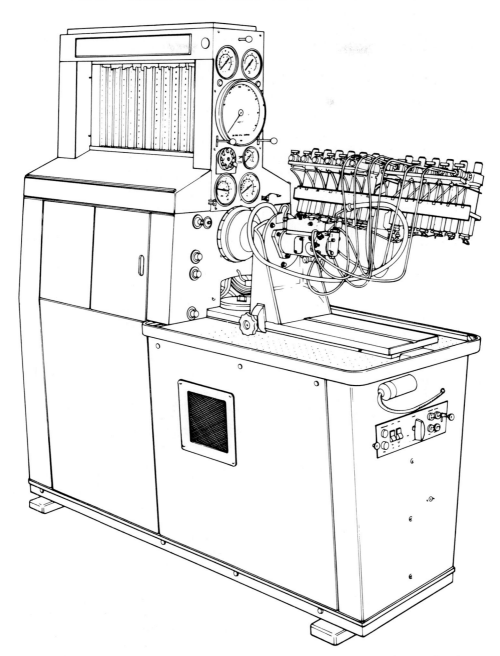

FIGURE 22-38. Typical injection pump test stand with test injectors in place and a distributor, pump for a six-cylinder engine mounted on stand. *(Courtesy of CAV Ltd.)*

Phasing is done manually or on a special test machine that uses a round degree plate graduated in degrees to measure the angular motion of the pump camshaft. Adjustment is provided by shims or adjusters for each pumping element. Phasing is necessary if any of the pumping element parts have been replaced. Calibration data is provided in the pump manufacturer's service manual data sheet.

Very precise handling and observation of the degree plate and adjustments is required to achieve accurate phasing and balanced fuel injection timing for all cylinders.

FIGURE 22-39. Measuring head clearance with dial indicator. Head clearance must be checked and adjusted to prevent the plunger from striking the delivery valve body before pump timing can be done. Adjustment is by shim, spacer or tappet screw. *(Courtesy of CAV Ltd.)*

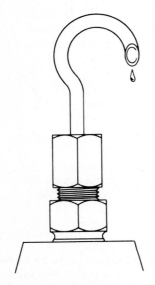

FIGURE 22-40. Spill pipe (swan neck tube) used for spill timing (phasing) the injection pump. Port opening (upper helix) and port closing (lower helix) are checked. *(Courtesy of CAV Ltd.)*

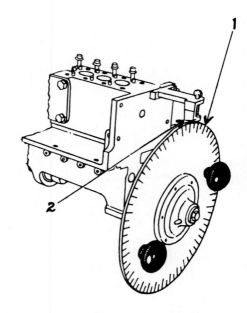

FIGURE 22-41. Injection pump ready for phasing. 1. Degree wheel. 2. Pointer or edge of housing.

Spill Timing Method of Pump Timing and Phasing

Spill timing is a manual method of pump timing and phasing. Depending on pump plunger design, spill timing is done for port closing or for port opening. Spill timing for port closing is done on pumps with a lower helix plunger design. This design provides for a variable ending and a constant beginning of fuel delivery. Spill timing for port opening is done on pumps with an upper helix plunger design, which provides for a variable beginning and a constant ending of fuel delivery.

In either case, the procedure requires that a swan-neck tube be installed in place of the number 1 delivery valve of the pump. This allows observation of the beginning or ending of fuel delivery during the procedure. The injection pump is mounted to a suitable base plate and the base plate is clamped securely in a vise. A degree wheel for the purpose is mounted to the drive end of the pump. The degree wheel provides the means to turn the pump camshaft and indicates exact camshaft position. A clean gravity type of fuel supply is required with a shutoff valve in the supply line. The fuel supply is connected to the pump fuel inlet fitting.

Spill Timing Pump with Lower Helix

1. Move the fuel control rod through its travel range and place it in the stop position. This leaves

the inlet port at the plunger open. Turn the camshaft in the proper drive direction to position No. 1 cam lobe at the BDC position. Open the fuel supply valve to allow fuel to flow until air-free fuel flows from the swan-neck tube. As solid fuel flows from the tube, slowly and carefully continue to turn the camshaft with the degree wheel. As fuel flow diminishes during port closing, continue to turn the camshaft very slowly until flow decreases to zero. Stop turning at that precise point. Repeat this procedure several times to ensure accuracy.

2. When the port closing position has been established, adjust the degree wheel to index zero at the pointer. Check the timing line at the drive coupling to see if it coincides with the mark on the pump housing. If not, place a new timing mark on the coupling to reflect the newly established timing of port closing.

To complete the pump phasing, repeat the procedure above for each pumping element in the order of the engine injection sequence. Port closing for each succeeding element in the injection sequence should be exactly 45 degrees later than the preceding element for an eight-cylinder engine. For a six-cylinder engine it is 60 degrees, and for a four-cylinder it is 90 degrees.

Adjustment is made by means of a tappet adjusting screw, shims, or spacers, depending on pump design. If the phase angle to the succeeding element is too great, the plunger is too low and must be raised. If the phase angle is too small, the plunger is too high and it must be lowered.

Spill Timing Pump with Upper Helix

The procedure for spill timing of port opening is very similar to that described above for port closing.

With the injection pump prepared as above for checking the number 1 pumping element, open the fuel supply valve and turn the pump camshaft very slowly in the proper drive direction until fuel flow from the swan-neck tube stops. Then very slowly continue to turn the camshaft until fuel just begins to flow from the tube. Repeat this procedure several times to establish accurately the exact point of port opening. When the point of port opening has been established, check the timing mark on the drive coupling as in step 2 above.

To complete the pump phasing, repeat the procedure for establishing port closing for each pumping element in the injection sequence to ensure that port closing for each succeeding element in the injection sequence will be exactly the number of degrees specified.

Calibration

Caution: Never drive a fuel injection pump in a direction other than that shown on the pump or serious damage will result.

Calibration is the procedure for adjusting the quantity of fuel delivery of each of the pumping elements so that each element will deliver precisely the same specified quantity of fuel. Adjustment is made by setting the control sleeve gear on each element or by setting the control forks on the control rod, depending on pump design.

Mount the pump on the test machine and be sure that there is sufficient clearance at the drive coupling. Connect the fuel supply line to the pump inlet and the delivery pipes to the delivery valve holders. Connect the lubrication line from the tester to the pump or fill pump sump with lubricating oil if pump is not engine-oil lubricated.

Set the specified (usually No. 6 or No. 1) element sleeve gear or control fork to the dimension specified in the data sheet. Select the appropriate pump drive direction (cw, left hand or ccw, right hand).

Open the fuel supply valve to pressure flow. Loosen the pump bleed screw and start the pump drive. Run the pump until all air has been expelled; then tighten the bleed screw.

Move the speed lever to the maximum speed position. Maximum fuel delivery from the specified element is adjusted first to specifications. Run the pump at the specified speed and note the delivery from the specified element in its test tube for the specified number of shots. Adjust the maximum fuel stop screw until the maximum delivery is as specified. Repeat several times to ensure accuracy.

Adjust the position of the other sleeve gears or forks so that the maximum delivery from them is exactly the same as that from the specified element. Check the operation of the excess fuel device and fuel stop control, and check fuel delivery at the excess fuel position and at idle as specified.

After calibration install all plugs, covers, and the like with new gaskets and sealing washers as needed. Install tamper-proof locks and seals as specified.

PART 16   SELF-CHECK

1. How is sealing achieved between the pump plunger and barrel?

2. During operation, injection stops when the helix registers with the _____.

3. How is the effective stroke of the plunger changed to control fuel delivery?

4. What is the purpose of the vertical groove in the plunger?

5. The lower helix in a plunger controls port (closing) or (opening).

6. The purpose of plunger designs using an upper and a lower helix is to vary port opening and port closing. True or False.

7. What is the purpose of the delivery valve?

8. Why is a smoke limiter used on diesel engines?

9. How is lubrication provided to the injection pump?

10. How are pumping plungers actuated on injection pumps that do not have a camshaft?

11. Why is it necessary to advance injection timing with increased engine speed?

12. Microprocessor control of fuel injection in the United Technologies model 200 pump includes control of seven pump functions. Name six of these.

13. Why should parts from pumping elements not be mixed?

14. What is injection pump phasing?

15. What is the phase angle for an injection pump for a four-cycle six-cylinder engine?

16. Define injection pump calibration.

Chapter 23

Detroit Diesel Unit Injector Fuel System

Performance Objectives

After thorough study of this chapter, the appropriate training models and service manuals, and with the necessary tools and equipment you should be able to do the following with respect to diesel engines equipped with the Detroit Diesel Unit Injector fuel system:

1. Complete the self-check questions with at least 80 percent accuracy.

2. Describe the basic construction and operation of Detroit Diesel fuel system components including the following:

    (a) method of fuel pumping and metering

    (b) injector lubrication

    (c) injection timing control

    (d) fuel control rack

    (e) governor types

3. Diagnose basic fuel injection system problems and determine the correction required.

4. Remove, clean, inspect, test, measure, repair, replace, and adjust fuel injection system components as needed to meet manufacturer's specifications.

Most of this chapter is based on information provided courtesy of General Motors Corporation, Detroit Diesel Allison Division.

The Detroit Diesel fuel injection system combines high pressure pumping, fuel metering, and injection in one assembly called a *unit injector*. Each engine cylinder is provided with one injector. The helix type plungers are controlled by the fuel control rack, which in turn is controlled by a foot or hand operated throttle (depending on engine application) and the governor. The unit injector plungers are actuated for pumping by a special lobe on the engine camshaft, a cam follower, push rod, and rocker arm. Fuel is injected into each cylinder once every camshaft (and crankshaft) revolution. (See Fig. 23-1.)

PART 1   UNIT INJECTOR OPERATION

The *Detroit Diesel unit injector* pump plunger and helix operate as follows (Fig. 23-2 to 23-8). The fuel is injected into the cylinder by means of a plunger, which is forced downward by a rocker arm. As the plunger moves downward, it forces some fuel to spray into the cylinder combustion chamber in a fine accurately timed spray. The plunger moves the same distance on each stroke, and a circumferential groove in the plunger determines the timing and amount of fuel injected. The upper edge of this groove is cut in the shape of a helix. The lower edge may be straight or a helix. When the plunger is rotated by the fuel control rack, it changes the position of the helix, and this changes the fuel output of the injector and the timing.

The plunger moves in a bushing inside the injector. There are two ports on opposite sides of the bushing, one higher than the other. Fuel is delivered to the injector under low pressure from a fuel delivery pump and may flow into or out of the bushing through either port. As long as either port is uncovered by the plunger, fuel is free to flow out of the bushing through either port; thus no pressure can be developed by the plunger. As the plunger descends at the start of an injection stroke, the lower end of the plunger shuts off the lower port hole. As the plunger continues to descend, it reaches a point where the helix of the upper land closes off the upper port. With both ports shut off by the plunger, no fuel may flow from the injector except to be forced out through the tip. Thus injection starts and continues under high pressure as the plunger moves downward until the lower land finally passes the lower port. At that point the groove is open to the lower port, and fuel may escape up through the drilled hole in the center of the plunger. At this

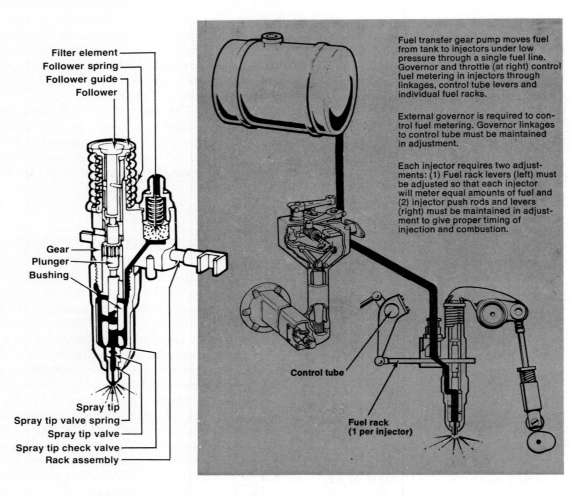

Fuel transfer gear pump moves fuel from tank to injectors under low pressure through a single fuel line. Governor and throttle (at right) control fuel metering in injectors through linkages, control tube levers and individual fuel racks.

External governor is required to control fuel metering. Governor linkages to control tube must be maintained in adjustment.

Each injector requires two adjustments: (1) Fuel rack levers (left) must be adjusted so that each injector will meter equal amounts of fuel and (2) injector push rods and levers (right) must be maintained in adjustment to give proper timing of injection and combustion.

Filter element
Follower spring
Follower guide
Follower

Gear
Plunger
Bushing

Spray tip
Spray tip valve spring
Spray tip valve
Spray tip check valve
Rack assembly

Control tube

Fuel rack
(1 per injector)

FIGURE 23-1. Detroit Diesel Allison unit injector fuel system. *(Courtesy of Detroit Diesel Allison)*

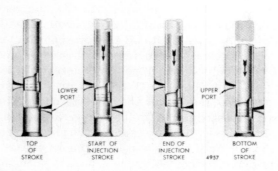

FIGURE 23-2. High-pressure pumping action of downward moving plunger in Detroit Diesel unit injector. *(Courtesy of Detroit Diesel Allison)*

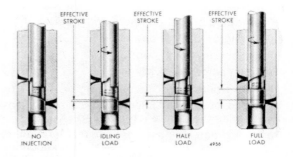

FIGURE 23-3. Metering of fuel is achieved by rotating the plunger. Plungers are controlled by fuel rack. Plunger helix and port relationship determines amount of fuel (metering) delivered by each plunger stroke. *(Courtesy of Detroit Diesel Allison)*

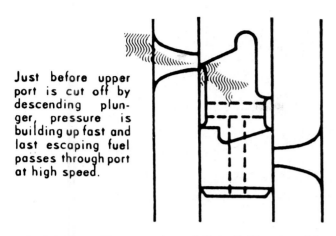

Just before upper port is cut off by descending plunger, pressure is building up fast and last escaping fuel passes through port at high speed.

FIGURE 23-4. Close-up view of Detroit Diesel unit injector plunger and barrel, showing T-hole (dotted lines) in plunger. Injection ends when downward moving plunger aligns the T-slot hole with the spill port on the right. *(Courtesy of Detroit Diesel Allison)*

point, of course, injection stops even though the plunger continues downward for a short distance before rising again to its starting position.

Thus, the start of injection depends on the point at which the upper land, with its helix edge, shuts off the upper port. The sooner this port is shut off, the sooner injection starts and the longer it takes; thus, more fuel is injected. If the plunger is turned

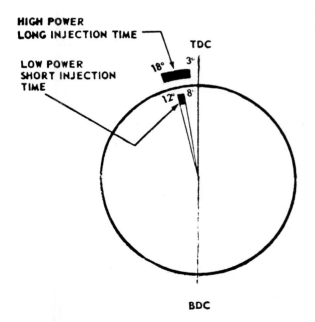

FIGURE 23-5. Typical injection timing diagram for variable beginning and variable ending helix Detroit Diesel unit injectors. *(Courtesy of Detroit Diesel Allison)*

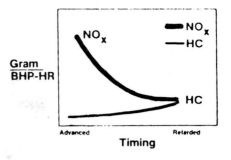

FIGURE 23-6. Effects of injection timing on exhaust emissions. *(Courtesy of Cummins Engine Company Inc.)*

so that the helix edge shuts off the port later, injection starts later, continues for a shorter length of time, and injects a smaller amount of fuel.

The full-fuel position occurs when the lower part of the upper helix is positioned at the upper port so that this port is shut off; injection begins almost immediately after the lower port is shut off. The no-fuel position occurs when the plunger is positioned so that the upper helix is at its extreme upper position adjacent to the upper port; by the time the upper port is closed, the lower port has already been opened.

Rotation of the plunger between its no-fuel and full-fuel positions is accomplished by using a rack to rotate a gear on the plunger. The gear has a flat on its internal diameter to match a flat surface on the upper part of the plunger. The plunger is free to slide up and down in the opening in the gear.

The angle at which the upper and lower helices are ground on GM diesel injectors is controlled accurately to afford precise control of fuel injection. The relationship of the upper and lower helices determines the amount of fuel injected for any given setting of the fuel control. If this engineered relationship is altered, the performance of the injector is changed. If such alteration is different from one plunger to another, engine operation at part-throttle settings will be rough and erratic.

As explained, the upper helix controls the start-of-injection timing. At low power, injection starts late; at high power, it starts early. In the no-fuel position, the upper helix cuts off the upper port after the lower port is exposed, and thus no fuel is injected.

The lower helix controls the end-of-injection timing. There are three types of lower-land machining. One is flat (actually not a helix). It obviously always opens the lower port at the same point in the stroke regardless of fuel control position and thus

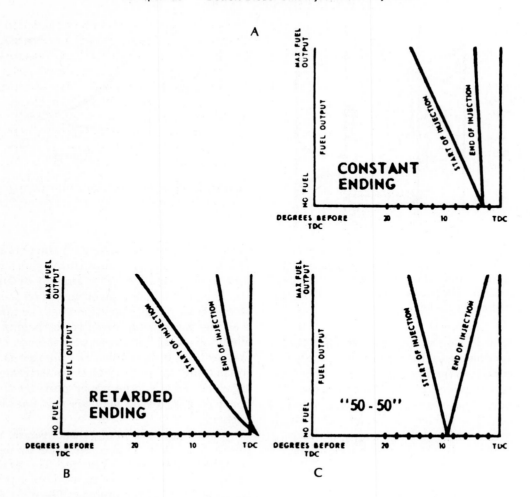

FIGURE 23-7. Detroit Diesel unit injector injection timing graphs for different designs of injector helixes. These graphs represent maximum speed and load (maximum fuel) at the top to the bottom. A represents a helix design with a variable beginning and a constant ending. This provides advanced injection timing and longer injection duration with increased speed and load. B represents a helix design with a variable beginning and a variable ending. This provides advanced beginning and ending of injection as well as increased injection duration with speed and load increase. Both beginning and ending of injection are retarded as engine speed and load decrease. C represents a helix design with a variable beginning and a variable ending. This provides advanced beginning and a retarded ending of injection as well as increased duration as engine speed and load increase. Beginning of injection is retarded, and the ending of injection is advanced as engine speed and load decrease.

always stops injection at the same time. This is known as the *constant ending* type of injector.

In the second type of injector, known as the *retarded* type, the lower helix is a right-hand spiral (as in the upper helix) but has a much shallower angle than the upper helix. Thus, as the upper helix advances the start-of-injection timing for greater fuel

delivery, the lower helix also advances the shut-off timing but to a lesser extent.

In the third type of injector, known as the *50-50* type, the lower helix is a left-hand spiral (opposite to the upper helix) so that increased fuel delivery at the higher fuel settings is a combination of an earlier starting point and a later stopping point.

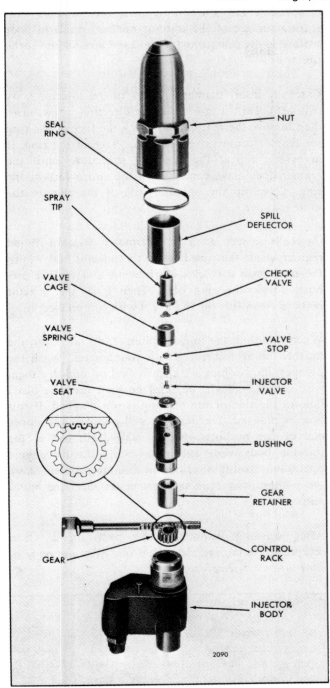

FIGURE 23-8. Injector rack, gear, spray tip, and valve assembly details and relative location of parts. *(Courtesy of Detroit Diesel Allison)*

In a diesel engine cylinder, combustion starts a short time after injection starts; if all engine power strokes are to be spaced evenly to provide smooth running, all injectors must begin injection at exactly the same point in relation to piston travel. If there is a variation from one cylinder to the next in the start or ending of injection, or in the amount of fuel injected, the engine will run roughly.

Fuel economy and good engine performance also depend on careful control of injection timing. Fuel should be sprayed in at an even rate so that it can mix thoroughly with the air and burn completely. Of course, the faster the engine runs, the faster the fuel must be injected. But, at a given engine speed, the best combustion efficiency is achieved by always injecting the fuel at a rate at which it can mix completely and uniformly with the compressed air in the cylinder.

PART 2   PLUNGER, BARREL, AND HELIX CONSTRUCTION

The plunger and bushing assembly in the GM diesel injector is an outstanding example of precision manufacturing. The mating surfaces of plunger and bushing are so smooth, uniform, and accurately fitted that you can hold the plunger, with the bushing assembled to it, spin the bushing—and it will continue to spin seemingly endlessly, just like a fine ball bearing that is almost frictionless. Yet you can take this same assembly, place it in a test fixture, and test it for leakage with high-pressure air; the leakage may be almost immeasurable because the surfaces fit together so closely and uniformly.

The bushing (barrel) is bored and reamed to an accurate inside diameter, and the two ports are accurately located, drilled, and countersunk. The port is through-hardened. Then the sealing surface at the lower end is ground flat. The bushing is case-hardened to improve its strength, wear resistance, and corrosion resistance. Then a locating pin is inserted, the mating surface at the lower end is lapped, and the inside diameter is honed and lapped to make it smooth, round, and straight.

Machine operations on the plunger start with an axial-drilled hole in each end. Then the circumferential groove with the helix edge is cut on an accurate cam-controlled milling machine, and a cross-hole is drilled to intersect the groove and the axial hole in the lower end. Two beveled undercuts are machined, one in the middle of the plunger and one at the upper end under the head, and the head is beveled. The plunger is through-hardened and then ground to approximately final diameter, with the "gear flat" on the upper section being ground to an accurate form. Then the plunger is lapped to the finished diameter.

The internal diameter of the bushings (barrels) is measured with an air gauge, and the bushings are segregated into groups to the nearest 12-millionths of an inch (0.0003048 mm). The plungers are measured and bushings are mated very closely to maintain a specified clearance that never exceeds 60 millionths of an inch (0.001524 mm). Each assembly is checked for leakage with compressed air before being sent on to final inspection.

PART 3   DETROIT DIESEL INJECTION TIMING

Detroit Diesel unit injector basic timing is achieved by proper alignment of the timing marks on the control racks and gears in relation to the piston and crankshaft position. Variation of injection timing in relation to engine speed is achieved through plunger helix design and position as shown in Figure 23–7 and described in Part 1 of this Chapter.

Precise adjustment of the injector-operating mechanism is also required for accurate and even injection timing for all cylinders.

PART 4   DETROIT DIESEL UNIT INJECTORS (CROWN VALVE AND NEEDLE VALVE TYPES)

The fuel injector is a lightweight compact unit, which enables quick, easy starting directly on diesel fuel and permits the use of a simple open-type combustion chamber. The simplicity of design and operation provides for simplified controls and easy adjustment. No high-pressure fuel lines or complicated air-fuel mixing or vaporizing devices are required. (See Fig. 23–9 to 23–13.)

The fuel injector performs four functions:

1. creates the high fuel pressure required for efficient injection

2. meters and injects the exact amount of fuel required to handle the load

3. atomizes the fuel for mixing with the air in the combustion chamber

4. permits continuous fuel flow

Combustion required for satisfactory engine operation is obtained by injecting, under pressure, a small quantity of accurately metered and finely atomized fuel oil into the cylinder.

Metering of the fuel is accomplished by an upper and lower helix machined in the lower end of the injection plunger.

The continuous fuel flow through the injector serves, in addition to preventing air pockets in the fuel system, as a coolant for those injector parts subjected to high combustion temperatures.

To vary the power output of the engine, injectors having different fuel output capacities are used. The fuel output of the various injectors is governed by the helix angle of the plunger and the type of spray tip used.

Since the helix angle on the plunger determines the output and operating characteristics of a particular type of injector, it is imperative that the correct injectors are used for each engine application. If injectors of different types are mixed, erratic operation will result and may cause serious damage to the engine or to the equipment that it powers.

Note: Do not intermix injectors with other types of injectors in an engine.

Each fuel injector has a circular disc pressed into a recess at the front side of the injector body for identification purposes. The identification tag indicates the nominal output of the injector in cubic millimeters.

Each injector control rack is actuated by a lever on the injector control tube, which, in turn, is connected to the governor by means of a fuel rod. These levers can be adjusted independently on the control tube, thus permitting a uniform setting of all injector racks.

The fuel injector combines in a single unit all of the parts necessary to provide complete and independent fuel injection at each cylinder.

Operation

Fuel, under pressure, enters the injector at the inlet side through a filter cap and filter. From the filter, the fuel passes through a drilled passage into the supply chamber, that area between the plunger bushing and the spill deflector, in addition to that area under the injector plunger within the bushing. The plunger operates up and down in the bushing, the bore of which is open to the fuel supply in the annular chamber by two funnel-shaped ports in the plunger bushing.

The motion of the injector rocker arm is transmitted to the plunger by the follower, which bears against the follower spring. In addition to the reciprocating motion, the plunger can be rotated, during operation, around its axis by the gear, which meshes with the control rack. For metering the fuel, an up-

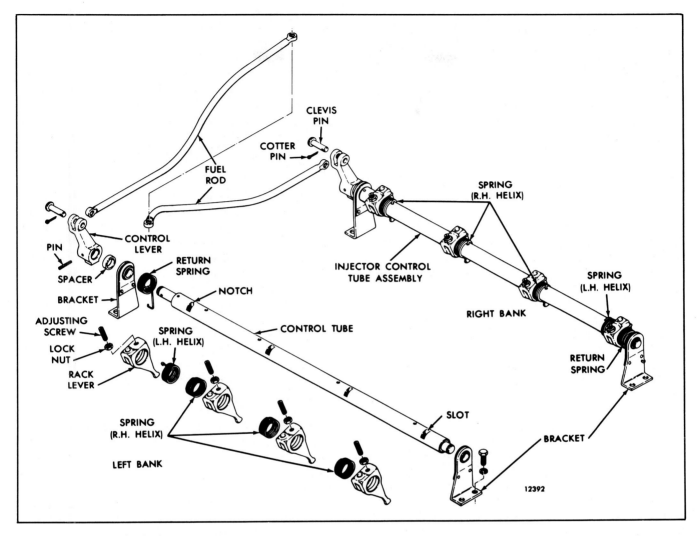

FIGURE 23-9. Injector control tube assembly, spring-loaded type. *(Courtesy of Detroit Diesel Allison)*

per helix and a lower helix are machined in the lower part of the plunger. The relation of the helices to the two ports changes with the rotation of the plunger.

As the plunger moves downward, under pressure of the injector rocker arm, a portion of that fuel trapped under the plunger is displaced into the supply chamber through the lower port until the port is closed off by the lower end of the plunger. A portion of the fuel trapped below the plunger is then forced up through a central passage in the plunger into the fuel metering recess and into the supply chamber through the upper port until that port is closed off by the upper ports, both closed off. The remaining fuel under the plunger is subjected to increased pressure by the continued downward movement of the plunger.

Needle Valve Type Only

When sufficient pressure is built up, it opens the flat, nonreturn check valve. The fuel in the check valve cage, spring cage, tip passages, and tip fuel cavity is compressed until the pressure force acting upward on the needle valve is sufficient to open the valve against the downward force of the valve spring. As soon as the needle valve lifts off of its seat, the fuel is forced through the small orifices in the spray tip and atomized into the combustion chamber.

The fuel injector outlet opening, through which the excess fuel oil returns to the fuel return manifold and then back to the fuel tank, is directly adjacent to the inlet opening.

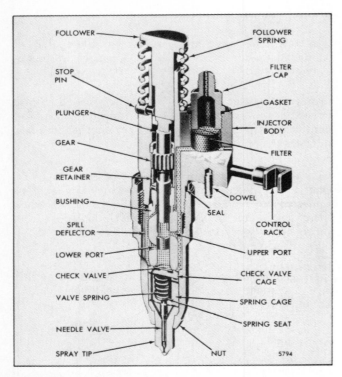

FIGURE 23-10. Needle valve fuel injector. *(Courtesy of Detroit Diesel Allison)*

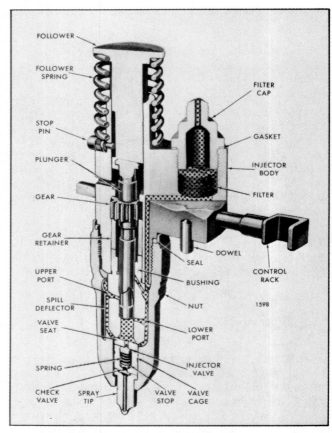

FIGURE 23-11. Crown valve fuel injector. *(Courtesy of Detroit Diesel Allison)*

Changing the position of the helices, by rotating the plunger, retards or advances the closing of the ports and the beginning and ending of the injector period. At the same time, it increases or decreases the amount of fuel injected into the cylinder. With the control rack pulled out all the way (no injection), the upper port is not closed by the helix until after the lower port is uncovered. Consequently, with the rack in this position, all of the fuel is forced back into the supply chamber, and no injection of fuel takes place. With the control rack pushed all the way in (fuel injection), the upper port is closed shortly after the lower port has been covered, thus producing a maximum effective stroke and maximum injection. From this no-injection position to full-injection position (full rack movement), the contour of the upper helix advances the closing of the ports and the beginning of injection.

When the lower land of the plunger uncovers the lower port in the bushing, the fuel pressure below the plunger is relieved and the valve spring closes the needle valve, ending injection.

A pressure relief passage has been provided in the spring cage to permit bleed-off of fuel leaking past the needle pilot in the tip assembly.

A check valve, directly below the bushing, prevents leakage from the combustion chamber into the fuel injector in case the valve is accidentally held open by a small particle of dirt. The injector plunger is then returned to its original position by the injector follower spring.

Crown Valve Type Only

When sufficient pressure is built up, the injector valve is lifted off its seat, and the fuel is forced through small orifices in the spray tip and atomized into the combustion chamber.

A check valve, mounted in the spray tip, prevents air leakage from the combustion chamber into the fuel injector if the valve is accidentally held open by a small particle of dirt. The injector plunger is then returned to its original position by the injector follower spring.

Crown Valve and Needle Valve

On the return upward movement of the plunger, the high-pressure cylinder within the bushing is again

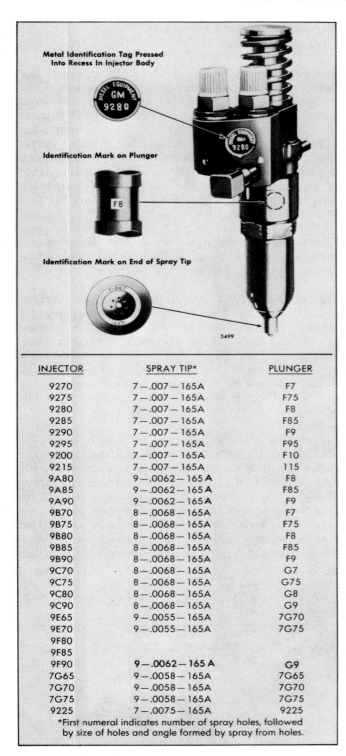

INJECTOR	SPRAY TIP*	PLUNGER
9270	7 — .007 — 165A	F7
9275	7 — .007 — 165A	F75
9280	7 — .007 — 165A	F8
9285	7 — .007 — 165A	F85
9290	7 — .007 — 165A	F9
9295	7 — .007 — 165A	F95
9200	7 — .007 — 165A	F10
9215	7 — .007 — 165A	115
9A80	9 — .0062 — 165 A	F8
9A85	9 — .0062 — 165 A	F85
9A90	9 — .0062 — 165 A	F9
9B70	8 — .0068 — 165A	F7
9B75	8 — .0068 — 165A	F75
9B80	8 — .0068 — 165A	F8
9B85	8 — .0068 — 165A	F85
9B90	8 — .0068 — 165A	F9
9C70	8 — .0068 — 165A	G7
9C75	8 — .0068 — 165A	G75
9C80	8 — .0068 — 165A	G8
9C90	8 — .0068 — 165A	G9
9E65	9 — .0055 — 165A	7G70
9E70	9 — .0055 — 165A	7G75
9F80		
9F85		
9F90	9 — .0062 — 165 A	G9
7G65	9 — .0058 — 165A	7G65
7G70	9 — .0058 — 165A	7G70
7G75	9 — .0058 — 165A	7G75
9225	7 — .0075 — 165A	9225

*First numeral indicates number of spray holes, followed by size of holes and angle formed by spray from holes.

FIGURE 23-12. Injector identification, typical. *(Courtesy of Detroit Diesel Allison)*

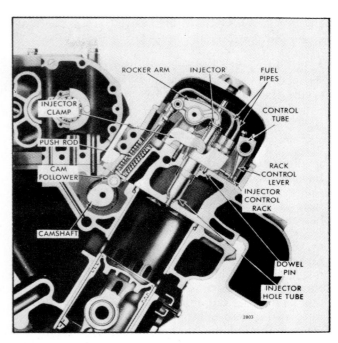

FIGURE 23-13. Unit injector mounting. *(Courtesy of Detroit Diesel Allison)*

filled with fuel oil through the ports. The constant circulation of fresh cool fuel through the injector renews the fuel supply chamber, helps cool the injector, and also effectively removes all traces of air, which might otherwise accumulate in the system and interfere with accurate metering of the fuel.

PART 5   GOVERNORS AND CONTROL DEVICES

Horsepower requirements on an engine may vary due to fluctuating loads; therefore, some method must be provided to control the amount of fuel required to hold the engine speed reasonably constant during load fluctuations. To accomplish this control, a governor is introduced in the linkage between the throttle control and the fuel injectors. The governor is mounted on the front end of the blower and is driven by one of the blower rotors. (See Fig. 23-14 to 23-18.) The following types of mechanical governors are used:

1. Limiting Speed Mechanical Governor
2. Variable Speed Mechanical Governor

Engines requiring a minimum and maximum speed control, together with manually controlled inter-

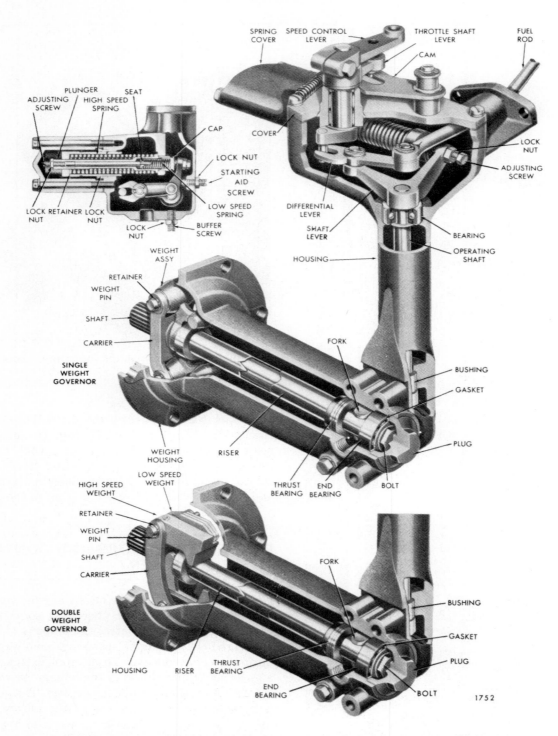

FIGURE 23-14. Detroit Diesel limiting-speed mechanical governor. *(Courtesy of Detroit Diesel Allison)*

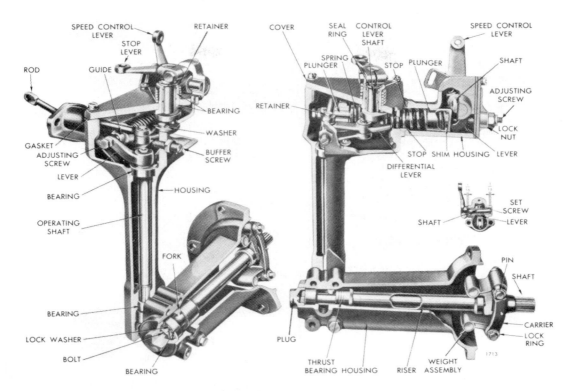

FIGURE 23–15. Detroit Diesel variable-speed mechanical governor. *(Courtesy of Detroit Diesel Allison)*

mediate speeds, are equipped with a limiting speed mechanical governor.

Engines subjected to varying load conditions that require an automatic fuel compensation to maintain a near constant engine speed, which may be changed manually by the operator, are equipped with a variable speed mechanical governor.

Each type of governor has an identification plate located on the control housing, containing the governor assembly number, type, idle speed range and drive ratio. The maximum engine speed, not shown on the identification plate, is stamped on the option plate attached to one of the valve rocker covers.

Fast Idle Cylinder

The limiting speed governor equipped with a fast idle air cylinder is used on vehicle engines where the engine powers both the vehicle and auxiliary equipment.

The fast idle system consists of a fast idle air cylinder installed in place of the buffer screw and a throttle locking air cylinder mounted on a bracket fastened to the governor cover. An engine shutdown air cylinder, if used, is also mounted on the governor cover.

The fast idle air cylinder and the throttle locking air cylinder are actuated at the same time by air from a common air line. The engine shutdown air cylinder is connected to a separate air line.

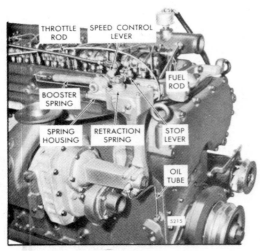

FIGURE 23–16. Variable-speed governor mounted on inline engine. *(Courtesy of Detroit Diesel Allison)*

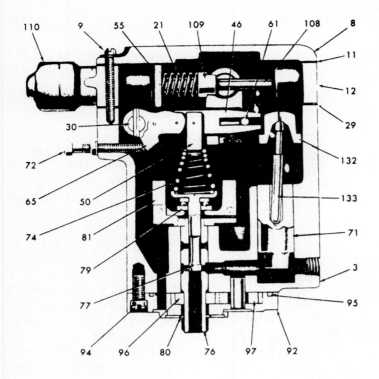

110   9   55   21   109   46   61   108   8

11
12
29

30

72

132

65

133

74   50

71

81

3

79

95

77

94   96   80   76   97   92

1. Housing—governor
8. Cover—governor
9. Screw—cover
11. Gasket—cover
12. Subcap
21. Spring—fuel rod
29. Gasket—housing to subcap
30. Shaft—speed adjusting
46. Lever—floating
50. Fork—spring
55. Pine—speed adjusting lever stop
61. Bracket—droop adjusting
65. Lever—speed adjusting
71. Piston—power
72. Screw—maximum speed adjusting
74. Spring—speeder
76. Ball head assembly
77. Plunger—pilot valve
79. Bearing—plunger
80. Lock ring
81. Flyweight
92. Base—governor
94. Screw—base to housing
95. Ring—housing to base seal
96. Gear—oil pump drive
97. Gear—oil pump driven
108. Rod—spring guide
109. Seat—guide rod spring
110. Cover—spring pad
132. Lever—terminal
133. Pin—terminal lever to piston
(Courtesy of Detroit Diesel Allison)

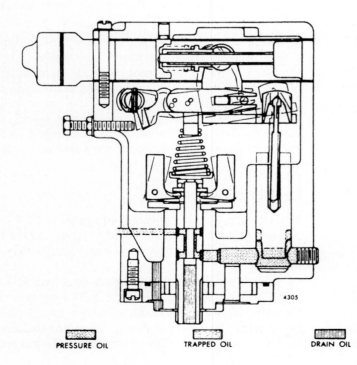

4305

PRESSURE OIL   TRAPPED OIL   DRAIN OIL

FIGURE 23–17. (Top) SG variable-speed hydraulic governor component identification. (Bottom) position of governor as load increases and engine speed tends to decrease.

426

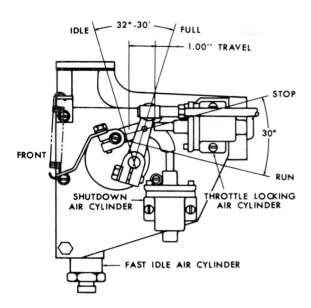

FIGURE 23-18. Governor with fast-idle cylinder. *(Courtesy of Detroit Diesel Allison)*

The air supply for the fast idle air cylinder is usually controlled by an air valve actuated by an electric solenoid. The fast idle system should be installed so that it will function only when the parking brake system is in operation to make it tamper-proof.

The vehicle accelerator-to-governor throttle linkage is connected to a yield link so the operator cannot overcome the force of the air cylinder holding the speed control lever in the idle position while the engine is operating at the single fixed high idle speed.

Operation

During highway operation, the governor functions as a limiting speed governor.

For operation of auxiliary equipment, the vehicle is stopped and the parking brake set. Then, with the engine running, the low speed switch is placed in the ON position. When the fast idle air cylinder is actuated, the force of the dual idle spring is added to the force of the governor low-speed spring, thus increasing the engine idle speed.

The governor now functions as a constant speed governor at the high idle speed setting, maintaining a near constant engine speed regardless of the load within the capacity of the engine. The fast idle system provides a single fixed high idle speed that is not adjustable, except by disassembling the fast idle air cylinder and changing the dual idle spring. As with all mechanical governors, when load is applied, the engine speed will be determined by the governor droop.

Fuel Modulator

The fuel modulator, used on certain turbocharged aftercooled engines, maintains the proper fuel to air ratio in the lower speed ranges where the mechanical governor would normally act to provide maximum injector output. It operates in such a manner that, although the engine throttle may be moved into the full-speed position, the injector racks cannot advance to the full-fuel position until the turbine speed is sufficient to provide proper combustion. (See Fig. 23-19.)

The fuel modulator will reduce exhaust smoke and also will help to improve fuel economy. The modulator mechanism is installed on the left bank between the No. 1 and No. 2 cylinders.

A fuel modulator consists of a cast housing containing a cylinder, piston, cam and spring mounted on the cylinder head. A lever and roller which controls the injector rack is connected to the injector control tube. Tubes run from the air box to the housing to supply pressure to actuate the piston.

The modulator tells the fuel system how much fuel the engine can efficiently use based on air box pressure. Increased air box pressure forces the piston and cam out of the cylinder bore allowing the rack to move toward full fuel.

Whenever the fuel injector rack control levers are adjusted, the fuel (air box) modulator lever and roller assembly must first be positioned free of cam contact. This is done by loosening the clamp screw.

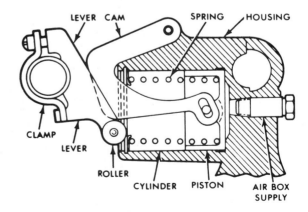

MODULATOR MECHANISM

FIGURE 23-19. Typical fuel modulator. *(Courtesy of Detroit Diesel Allison)*

Throttle Delay Mechanism

The throttle delay mechanism is used to retard full-fuel injection when the engine is accelerated. This reduces exhaust smoke and also helps to improve fuel economy. (See Fig. 23–20.)

The throttle delay mechanism is installed between the No. 1 and No. 2 cylinders on the right-bank cylinder head. It consists of a special rocker arm shaft bracket (which incorporates the throttle delay cylinder), a piston, throttle delay lever, connecting link, orifice plug, ball check valve and U-bolt.

A yield link replaces the standard operating lever connecting link in the governor.

Oil is supplied to a reservoir above the throttle delay cylinder through an oil supply fitting in the drilled oil passage in the rocker arm shaft bracket. As the injector racks are moved toward the no-fuel position, free movement of the throttle delay piston is assured by air drawn into the cylinder through the ball check valve. Further movement of the piston uncovers an opening which permits oil from the reservoir to enter the cylinder and displace the air. When the engine is accelerated, movement of the injector racks toward the full-fuel position is momentarily retarded while the piston expels the oil from the cylinder through an orifice. To permit full accelerator travel, regardless of the retarded injector rack position, a spring loaded yield link replaces the standard operating lever connecting link in the governor.

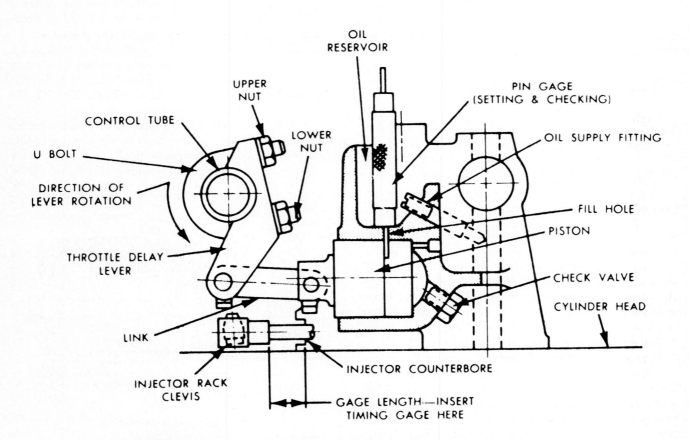

FIGURE 23-20. Throttle delay cylinder components and adjustment. *(Courtesy of Detroit Diesel Allison)*

PART 6   FUEL SYSTEM DIAGNOSIS AND SERVICE

Chart 2

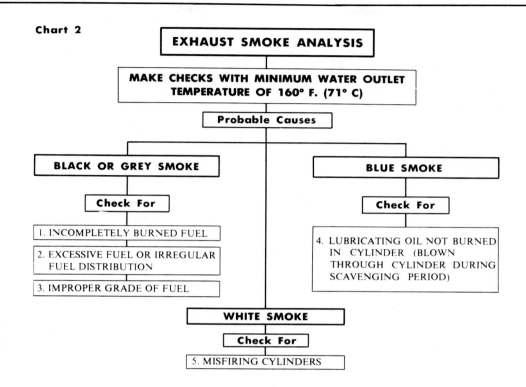

SUGGESTED REMEDY

1. High exhaust back pressure or a restricted air inlet causes insufficient air for combustion and will result in incompletely burned fuel.

High exhaust back pressure is caused by faulty exhaust piping or muffler obstruction and is measured at the exhaust manifold outlet with a manometer. Replace faulty parts.

Restricted air inlet to the engine cylinders is caused by clogged cylinder liner ports, air cleaner, or blower air inlet screen. Clean these items. Check the emergency stop to make sure that it is completely open and readjust it if necessary.

2. If the engine is equipped with a throttle delay, check for the proper setting, leaky check valve, and restricted filling of the piston cavity with oil from the reservoir.

If the engine is equipped with a fuel modulator, check the cam to determine if it is stuck in the full fuel position. Verify tightness of the roller lever clamp on the control tube. Determine correctness of the installed fuel modulator piston spring and check if the spring has taken a permanent "set" or if the spring rate is too low.

The above affects only excessive acceleration smoke, but does not affect smoke at constant speed.

Check for improperly timed injectors and improperly positioned injector rack control levers. Time the fuel injectors and perform the appropriate governor tune-up.

Replace faulty injectors if this condition still persists after timing the injectors and performing the engine tune-up.

Avoid lugging the engine as this will cause incomplete combustion.

3. Check for use of an improper grade of fuel.

4. Check for internal lubricating oil leaks.

5. Check for faulty injectors and replace as necessary. Check for low compression.

The use of low cetane fuel will cause this condition. Refer to *Fuel Specifications*.

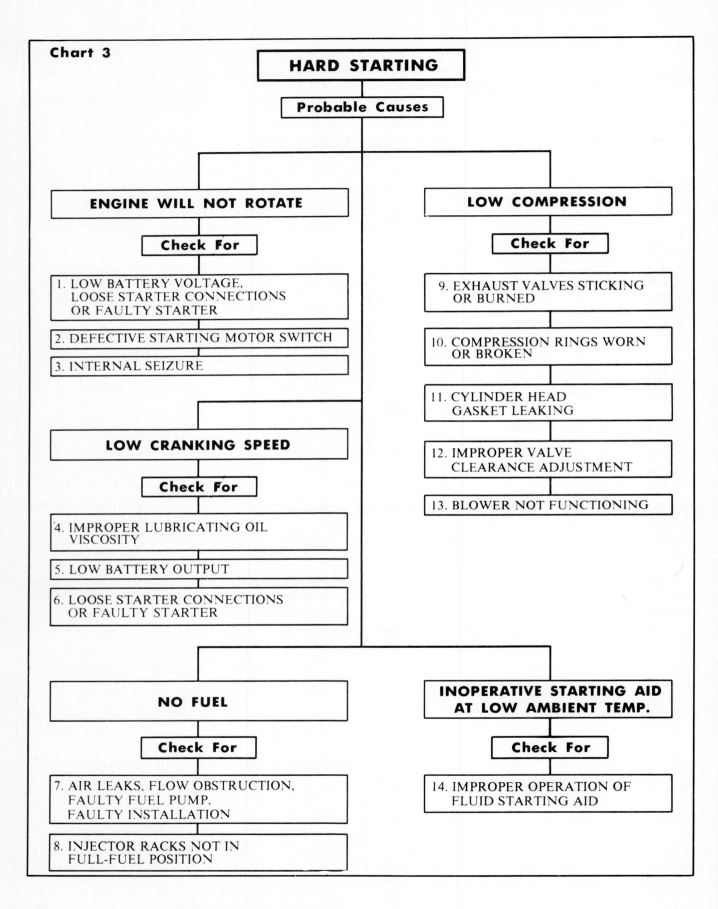

Chart 3

HARD STARTING

Probable Causes

ENGINE WILL NOT ROTATE

Check For

1. LOW BATTERY VOLTAGE, LOOSE STARTER CONNECTIONS OR FAULTY STARTER

2. DEFECTIVE STARTING MOTOR SWITCH

3. INTERNAL SEIZURE

LOW CRANKING SPEED

Check For

4. IMPROPER LUBRICATING OIL VISCOSITY

5. LOW BATTERY OUTPUT

6. LOOSE STARTER CONNECTIONS OR FAULTY STARTER

LOW COMPRESSION

Check For

9. EXHAUST VALVES STICKING OR BURNED

10. COMPRESSION RINGS WORN OR BROKEN

11. CYLINDER HEAD GASKET LEAKING

12. IMPROPER VALVE CLEARANCE ADJUSTMENT

13. BLOWER NOT FUNCTIONING

NO FUEL

Check For

7. AIR LEAKS, FLOW OBSTRUCTION, FAULTY FUEL PUMP, FAULTY INSTALLATION

8. INJECTOR RACKS NOT IN FULL-FUEL POSITION

INOPERATIVE STARTING AID AT LOW AMBIENT TEMP.

Check For

14. IMPROPER OPERATION OF FLUID STARTING AID

Chart 3

HARD STARTING

SUGGESTED REMEDY

1. Refer to Items 2, 3 and 5 and perform the operations listed.

2. Replace the starting motor switch.

3. Hand crank the engine at least one complete revolution. If the engine cannot be rotated a complete revolution, internal damage is indicated and the engine must be disassembled to ascertain the extent of damage and the cause.

4. Refer to *Lubrication Specifications*

5. Recharge the battery if a light load test indicates low or no voltage. Replace the battery if it is damaged or will not hold a charge.

Replace terminals that are damaged or corroded.

At low ambient temperatures, use of a starting aid will keep the battery fully charged by reducing the cranking time.

6. Tighten the starter connections. Inspect the starter commutator and brushes for wear. Replace the brushes if badly worn and overhaul the starting motor if the commutator is damaged.

7. To check for air leaks, flow obstruction, faulty fuel pump or faulty installation, consult the *No Fuel or Insufficient Fuel* chart.

8. Check for bind in the governor-to-injector linkage. Readjust the governor and injector controls if necessary.

9. Remove the cylinder head and recondition the exhaust valves.

10. Remove the air box covers and inspect the compression rings through the ports in the cylinder liners. Overhaul the cylinder assemblies if the rings are badly worn or broken.

11. To check for compression gasket leakage, remove the coolant filler cap and operate the engine. A steady flow of gases from the coolant filler indicates either a cylinder head gasket is damaged or the cylinder head is cracked. Remove the cylinder head and replace the gaskets or cylinder head.

12. Adjust the exhaust valve clearance.

13. Remove the flywheel housing cover at the blower drive support. Then remove the snap ring and withdraw the blower drive shaft from the blower. Inspect the blower drive shaft and drive coupling. Replace the damaged parts. Bar the engine over. If the blower does not rotate, remove the air inlet adaptor and visually inspect the blower rotors and end plates. **If visual distress is noted, remove the blower.**

14. Operate the starting aid according to the instructions under *Cold Weather Starting Aids* in the Service Manual.

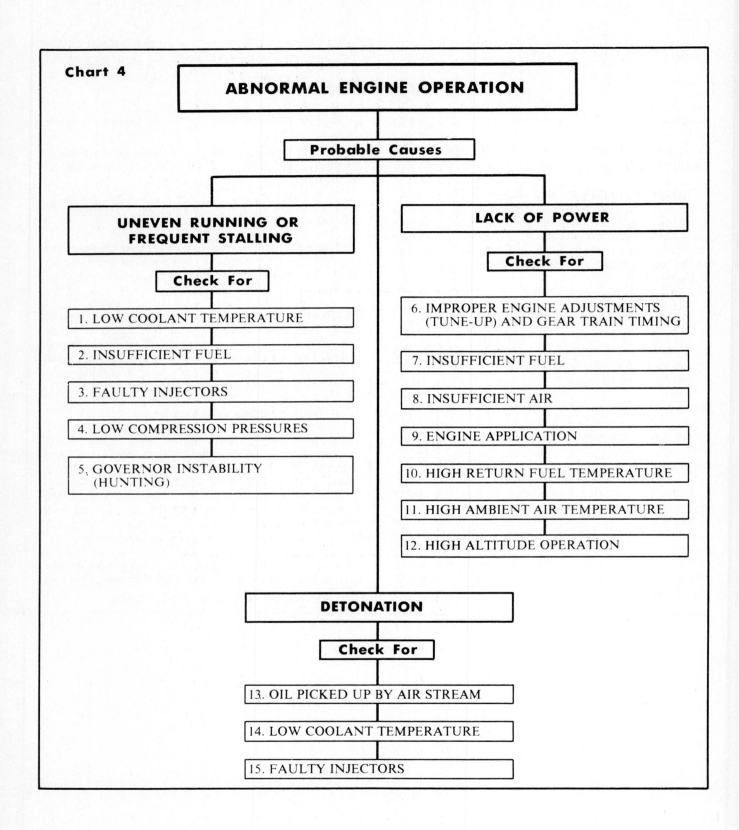

Chart 4

ABNORMAL ENGINE OPERATION

Probable Causes

UNEVEN RUNNING OR FREQUENT STALLING

Check For

1. LOW COOLANT TEMPERATURE

2. INSUFFICIENT FUEL

3. FAULTY INJECTORS

4. LOW COMPRESSION PRESSURES

5. GOVERNOR INSTABILITY (HUNTING)

LACK OF POWER

Check For

6. IMPROPER ENGINE ADJUSTMENTS (TUNE-UP) AND GEAR TRAIN TIMING

7. INSUFFICIENT FUEL

8. INSUFFICIENT AIR

9. ENGINE APPLICATION

10. HIGH RETURN FUEL TEMPERATURE

11. HIGH AMBIENT AIR TEMPERATURE

12. HIGH ALTITUDE OPERATION

DETONATION

Check For

13. OIL PICKED UP BY AIR STREAM

14. LOW COOLANT TEMPERATURE

15. FAULTY INJECTORS

1. Check the engine coolant temperature gage and if the temperature does not reach 160-185 °F (71-85 °C) while the engine is operating, consult the *Coolant Temperature* chart.

2. Check engine fuel spill back and if the return is less than specified, consult the *No Fuel or Insufficient Fuel* chart.

3. Check the injector timing and the position of the injector racks. If the engine was not tuned correctly, perform an engine tune-up. Erratic engine operation may also be caused by leaking injector spray tips. Replace the faulty injectors.

4. Check the compression pressures within the cylinders and consult the *Hard Starting* chart if compression pressures are low.

5. Erratic engine operation may be caused by governor-to-injector operating linkage bind or by faulty engine tune-up. Perform the appropriate engine tune-up procedure as outlined for the particular governor used.

6. If the engine is equipped with a throttle delay, check for the proper setting, binding or burrs on the piston or bracket, and a plugged discharge orifice.

If equipped with a fuel modulator, determine if there is any interference with the roller assembly or roller contact with the cam at wide open throttle (WOT) position. Check for burrs and binding on the piston and bracket bore. Determine correctness of the installed fuel modulator spring and check if the spring has taken a permanent "set," or if the spring rate is too high.

Perform an engine tune-up if performance is not satisfactory.

Check the engine gear train timing. An improperly timed gear train will result in a loss of power due to the valves and injectors being actuated at the wrong time in the engine's operating cycle.

7. Perform a *Fuel Flow Test* and, if less than the specified fuel is returning to the fuel tank, consult the *No Fuel or Insufficient Fuel* chart.

8. Check for damaged or dirty air cleaners and clean, repair or replace damaged parts.

Remove the air box covers and inspect the cylinder liner ports. Clean the ports if they are over 50% plugged.

Check for blower air intake obstruction or high exhaust back pressure. Clean, repair or replace faulty parts.

Check the compression pressures.

9. Incorrect operation of the engine may result in excessive loads on the engine. Operate the engine according to the approved procedures.

10. Refer to Item 13 on Chart 4.

11. Check the ambient air temperature. A power decrease of .15 to .50 horsepower per cylinder, depending upon injector size, for each 10 °F (6 °C) temperature rise above 90 °F (32 °C) will occur. Relocate the engine air intake to provide a cooler source of air.

12. Engines lose horsepower with increase in altitude. The percentage of power loss is governed by the altitude at which the engine is operating.

13. Fill oil bath air cleaners to the proper level with the same grade and viscosity lubricating oil that is used in the engine.

Clean the air box drain tubes and check valve (if used) to prevent accumulation that may be picked up by the air stream and enter the engine cylinders. Inspect the check valve as follows:

1. Disconnect the drain tube between the check valve and the air box drain tube nut at the air box cover.

2. Run the engine and note the air flow through the valve at idle engine speed.

3. If the check valve is operating properly, there will be no air flow at engine speeds above idle.

Inspect the blower oil seals by removing the air inlet housing and watching through the blower inlet for oil radiating away from the blower rotor shaft oil seals while the engine is running. If oil is passing through the seals, overhaul the blower.

Check for a defective blower-to-block gasket. Replace the gasket, if necessary.

14. Refer to Item 1 of this chart.

15. Check injector timing and the position of each injector rack. Perform an engine tune-up, if necessary. If the engine is correctly tuned, the erratic operation may be caused by an injector check valve leaking, spray tip holes enlarged or a broken spray tip. Replace faulty injectors.

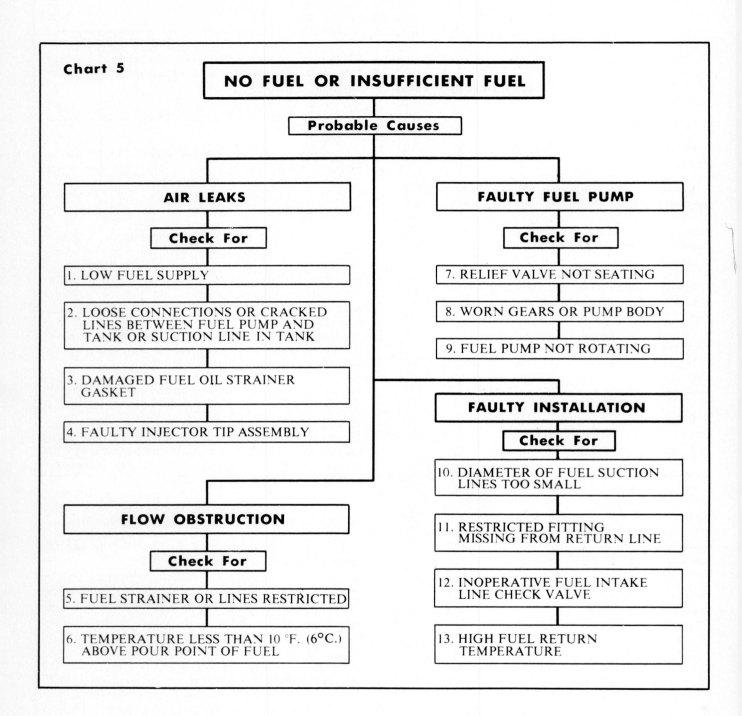

Chart 5

NO FUEL OR INSUFFICIENT FUEL

Probable Causes

AIR LEAKS

Check For

1. LOW FUEL SUPPLY

2. LOOSE CONNECTIONS OR CRACKED LINES BETWEEN FUEL PUMP AND TANK OR SUCTION LINE IN TANK

3. DAMAGED FUEL OIL STRAINER GASKET

4. FAULTY INJECTOR TIP ASSEMBLY

FLOW OBSTRUCTION

Check For

5. FUEL STRAINER OR LINES RESTRICTED

6. TEMPERATURE LESS THAN 10 °F. (6°C.) ABOVE POUR POINT OF FUEL

FAULTY FUEL PUMP

Check For

7. RELIEF VALVE NOT SEATING

8. WORN GEARS OR PUMP BODY

9. FUEL PUMP NOT ROTATING

FAULTY INSTALLATION

Check For

10. DIAMETER OF FUEL SUCTION LINES TOO SMALL

11. RESTRICTED FITTING MISSING FROM RETURN LINE

12. INOPERATIVE FUEL INTAKE LINE CHECK VALVE

13. HIGH FUEL RETURN TEMPERATURE

Chart 5

NO FUEL OR INSUFFICIENT FUEL

SUGGESTED REMEDY

1. The fuel tank should be filled above the level of the fuel suction tube.

2. Perform a *Fuel Flow Test* and, if air is present, tighten loose connections and replace cracked lines.

3. Perform a *Fuel Flow Test* and, if air is present, replace the fuel strainer gasket when changing the strainer element.

4. Perform a *Fuel Flow Test* and, if air is present with all fuel lines and connections assembled correctly, check for and replace faulty injectors.

5. Perform a *Fuel Flow Test* and replace the fuel strainer and filter elements and the fuel lines, if necessary,

6. Consult the *Fuel Specifications* for the recommended grade of fuel.

7. Perform a *Fuel Flow Test* and, if inadequate, clean and inspect the valve seat assembly.

8. Replace the gear and shaft assembly or the pump body.

9. Check the condition of the fuel pump drive and blower drive and replace defective parts.

10. Replace with larger tank-to-engine fuel lines.

11. Install a restricted fitting in the return line.

12. Make sure that the check valve is installed in the line correctly; the arrow should be on top of the valve assembly or pointing upward. Reposition the valve if necessary. If the valve is inoperative, replace it with a new valve assembly.

13. Check the engine fuel spill-back temperature. The return fuel temperature must be less than $150°F$ ($66°C$) or a loss in horsepower will occur. This condition may be corrected by installing larger fuel lines or relocating the fuel tank to a cooler position.

PART 7   UNIT INJECTOR SERVICE

Injector Removal

1. Clean and remove the valve rocker cover (See Fig. 23-21).

2. Remove the fuel pipes from both the injector and the fuel connectors.

   Note: Immediately after removal of the fuel pipes from an injector, cover the filter caps with shipping caps to prevent dirt from entering the injector. Also protect the fuel pipes and fuel connectors from entry of dirt or foreign material.

3. Crank the engine to bring the outer ends of the push rods of the injector and valve rocker arms in line horizontally.

4. Remove the two rocker shaft bracket bolts and swing the rocker arms away from the injector and valves.

5. Remove the injector clamp bolt, special washer and clamp.

6. Loosen the inner and outer adjusting screws (certain engines have only one adjusting screw and lock nut) on the injector rack control lever and slide the lever away from the injector.

7. Lift the injector from its seat in the cylinder head.

8. Cover the injector hole in the cylinder head to keep foreign material out.

9. Clean the exterior of the injector with clean fuel oil and dry it with compressed air.

Control Rack and Plunger Movement Test

Place the injector in the injector fixture and rack freeness tester. Place the handle on top of the injector follower. (See Fig. 23-22.)

If necessary, adjust the contact screw in the handle to ensure the contact screw is at the center of the follower when the follower spring is compressed.

With the injector control rack held in the no-fuel position, push the handle down and depress the follower to the bottom of its stroke. Then very slowly release the pressure on the handle while moving the control rack up and down as shown until the follower reaches the top of its travel. If the rack does not fall freely, loosen the injector nut, turn the tip, then retighten the nut. Loosen and retighten the nut a couple of times if necessary. Generally this will free the rack. Then, if the rack isn't free, change the injector nut. In some cases it may be necessary to disassemble the injector to eliminate the cause of the misaligned parts.

Installing Fuel Injector in Tester

1. Select the proper clamping head. Position it on the clamping post and tighten the thumb screw into the lower detent position.

2. Connect the test oil delivery piping into the clamping head.

3. Connect the test oil clear discharge tubing onto the pipe on the clamping head.

4. Locate the adaptor plate on top of the support bracket by positioning the 3/8″ diameter hole at the far right of the adaptor plate onto the 3/8″ diameter dowel pin. This allows the adaptor plate to swing out for mounting the fuel injector.

FIGURE 23-21. Removing Detroit Diesel unit injector. Note rocker arms removed to gain access to injector. Injector retaining clamp, fuel line, and control rack have also been disconnected. *(Courtesy of Detroit Diesel Allison)*

5. Mount the injector through the large hole and insert the injector pin in the proper locating pin hole.

6. Swing the mounted injector and adaptor plate inward until they contact the stop pin at the rear of the support bracket.

Injector Valve Opening and Spray Pattern Test

This test determines spray pattern uniformity and the relative pressure at which the injector valve opens and fuel injection begins. (See Fig. 23-24.)

1. Clamp the injector properly and purge the air from the system.

Move lever 4 down and operate pump lever 1 to produce a test oil flow through the injector. When air bubbles no longer pass through the clear discharge tubing, the system is free of air and is now ready for testing.

2. Move lever 4 *down.*

3. Position the injector rack in the full-fuel position.

4. Place pump lever 1 in the vertical position.

5. Move lever 3 to the forward detent position.

6. The injector follower should be depressed rapidly (at 40 to 80 strokes per minute) to simulate operation in the engine. Observe the spray pattern to see that all spray orifices are open and dispersing the test oil evenly. The beginning and ending of injection should be sharp and the test oil should be finely atomized with no drops of test oil forming on the end of the tip.

FIGURE 23-22. Checking the control rack and plunger to ensure that they move freely. *(Courtesy of Detroit Diesel Allison)*

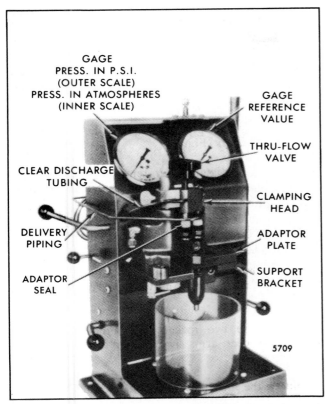

FIGURE 23-23. Unit injector installed in tester for valve opening and spray pattern test. *(Courtesy of Detroit Diesel Allison)*

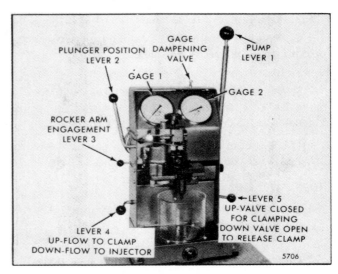

FIGURE 23-24. Identification of injector tester operating levers. *(Courtesy of Detroit Diesel Allison)*

The highest pressure reference number shown on gauge 2 will be reached just before injection ends. Use the reference values in the service manual to determine the relative acceptability of the injector.

Spray Tip Test

1. Move lever 4 *down* and operate pump lever 1 rapidly with smooth even strokes (40 strokes per minute) simulating the action of the tip functioning in the engine.

2. Note the pressure at which the needle valve opens on gauge 1. The valve should open at specified pressure. The opening and closing action should be sharp and produce a normal, finely atomized spray pattern.

If the valve opening pressure is below specifications and/or atomization is poor, the cause is usually a weak valve spring or a poor needle valve seat.

If the valve opening pressure is within specifications proceed to check for spray tip leakage as follows:

a. Actuate pump lever 1 several times and hold the pressure at 1500 psi (10 335 kPa) for 15 seconds.

b. Inspect the spray tip for leakage. There should be no fuel droplets, although a slight wetting at the spray tip is permissible.

Injector High Pressure Test

This test checks for leaks at the filter cap gaskets, body plugs and nut seal ring.

1. Clamp the injector properly and purge the air from the system.

2. Close the Thru-Flow valve, but do not over-tighten.

> *Note:* Make sure lever 4 is in the *down* position before operating pump lever 1.

3. Operate pump lever 1 to build up to 1600 to 2000 psi (11 024 to 13 780 pKa) on gauge 1. Check for leakage at the injector filter cap gaskets, body plugs and injector nut seal ring.

Injector Pressure Holding Test

This test determines if the body-to-bushing mating surfaces in the injector are sealing properly and indicates proper plunger-to-bushing fit.

1. Clamp the injector properly and purge the air from the system.

2. Close the Thru-Flow valve, but do not over-tighten.

3. Move lever 2 to the rear, horizontal position.

4. Operate pump lever 1 until gauge 1 reads approximately 700 psi (4 823 kPa).

5. Move lever 4 to the *up* position.

6. Time the pressure drop between 450 to 250 psi (3 100 to 1 723 kPa). If the pressure drop occurs in less than 15 seconds, leakage is excessive.

If the fuel injector passes all of the above tests, proceed with the *Fuel Output Test.*

Injector Fuel Output Test

To check the fuel output, operate the injector in calibrator as follows (Fig. 23–25):

Note: Place the cam shift index wheel and fuel flow lever in their respective positions. Turn on the test fuel oil heater switch and preheat the test oil to 95–105°F (35–40°C).

1. Place the proper injector adaptor between the tie rods and engage it with the fuel block locating pin. Then slide the adaptor forward and up against the fuel block face.

2. Place the injector seat into the permanent seat (cradle handle in vertical position). Clamp the injector into position by operating the air valve.

Note: Make sure the counter on the calibrator is preset at 1000 strokes. If for any reason this setting has been altered, reset the counter to 1000 strokes by twisting the cover release button to the left and hold the reset lever in the full up position while setting the numbered wheels. Close the cover. Refer to the calibrator instruction booklet for further information.

3. Pull the injector rack out to the no-fuel position.

4. Turn on the main power control circuit switch. Then start the calibrator by turning on the motor starter switch.

Note: The low oil pressure warning buzzer will sound briefly until the lubricating oil reaches the proper pressure.

5. After the calibrator has started, set the injector rack into the full-fuel position. Allow the injector to operate for approximately 30 seconds to purge the air that may be in the system.

6. After the air is purged, press the fuel flow start button (red). This will start the flow of fuel into

the vial. The fuel flow to the vial will automatically stop after 1000 strokes.

7. Shut the calibrator off (the calibrator will stop in less time at full-fuel).

8. Observe the vial reading and refer to specifications to determine whether the injector fuel output falls within the specified limits.

Note. The calibrator may be used to check and select a set of injectors which will inject the same amount of fuel in each cylinder at a given throttle setting, thus resulting in a smooth running, well balanced engine.

An injector that passes all of the above tests may be put back into service. However, an injector which fails to pass one or more of the tests must be rebuilt and checked on the calibrator.

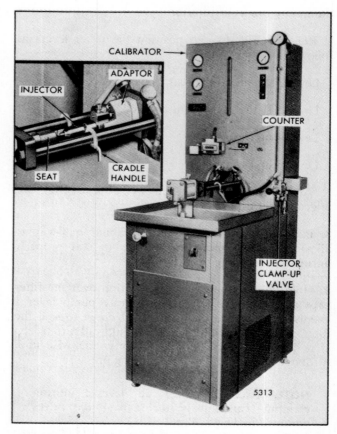

FIGURE 23–25. Injector in calibrator for fuel output test. *(Courtesy of Detroit Diesel Allison)*

Disassemble Injector

If required, disassemble an injector as follows (Fig. 23-26 to 23-28):

1. Support the injector upright in injector holding fixture and remove the filter caps, gaskets and filters.

Note: Whenever a fuel injector is disassembled, discard the filters and gaskets and replace with new filters and gaskets. *In the offset injector, a filter is used in the inlet side only. No filter is required in the outlet side.*

2. Compress the follower spring. Then raise the spring above the stop pin with a screw driver and withdraw the pin. Allow the spring to rise gradually.

3. Remove the plunger follower, plunger and spring as an assembly.

4. Invert the fixture and, using special socket loosen the nut on the injector body.

5. Lift the injector nut straight up, being careful not to dislodge the spray tip and valve parts. Remove the spray tip and valve parts from the bushing and place them in a clean receptacle until ready for assembly.

FIGURE 23-27. Removing injector nut with special tool. *(Courtesy of Detroit Diesel Allison)*

When an injector has been in use for some time, the spray tip, even though clean on the outside, may not be pushed readily from the nut with the fingers. In this event, support the nut on a wood block and drive the tip down through the nut, using tool J 1291-02.

FIGURE 23-26. Removing or installing filter cap. *(Courtesy of Detroit Diesel Allison)*

FIGURE 23-28. Removing or installing plunger follower, plunger, and spring. *(Courtesy of Detroit Diesel Allison)*

6. Remove the spill deflector. Then lift the bushing straight out of the injector body.

7. Remove the injector body from the holding fixture. Turn the body upside down and catch the gear retainer and gear in your hand as they fall out of the body.

8. Withdraw the injector control rack from the injector body. Also remove the seal ring from the body.

Clean Injector Parts

Since most injector difficulties are the result of dirt particles, it is essential that a clean area be provided on which to place the injector parts after cleaning and inspection.

Wash all of the parts with clean fuel oil or a suitable cleaning solvent and dry them with clean, filtered compressed air. *Do not use waste or rags for cleaning purposes.* Clean out all of the passages, drilled holes and slots in all of the injector parts.

Carbon on the inside of the spray tip may be loosened for easy removal by soaking for approximately 15 minutes in a suitable solution prior to the external cleaning and buffing operation. Methyl Ethyl Ketone J 8257 solution is recommended for this purpose.

Clean the spray tip with tool J 9464-01.

Note: Care must be exercised when inserting the carbon remover in the spray tip to avoid contacting the needle valve seat in the tip.

Wash the tip in fuel oil and dry it with compressed air. Clean the spray tip orifices with pin vise J 4298-1 and the proper size spray tip cleaning wire. Use wire J 21460 to clean .0055" diameter holes and wire J 21461 to clean .006" diameter holes.

Before using the wire, hone the end until it is smooth and free of burrs and taper the end a distance of 1/16" with stone J 8170. Allow the wire to extend 1/8" from tool J 4298-1.

The exterior surface of an injector spray tip may be cleaned by using a brass wire buffing wheel. To obtain a good polishing effect and longer brush life, the buffing wheel should be installed on a motor that turns the wheel at approximately 3000 rpm. A convenient method of holding the spray tip while cleaning and polishing is to place the tip over the drill end of the spray tip cleaner tool J 1243 and hold the body of the tip against the buffing wheel. In this way, the spray tip is rotated while being buffed.

Note: Do not buff excessively. *Do not use a steel wire buffing wheel or the spray tip holes may be distorted.*

When the body of the spray tip is clean, lightly buff the tip end in the same manner. This cleans the spray tip orifice area and will not plug the orifices.

Wash the spray tip in clean fuel oil and dry it with compressed air.

Clean and brush all of the passages in the injector body, using fuel hole cleaning brush and rack hole cleaning brush. Blow out the passages and dry them with compressed air.

Carefully insert reamer in the injector body. Turn it in a clockwise direction a few turns, then remove the reamer and check the face of the ring for reamer contact over the entire face of the ring. If necessary, repeat the reaming procedure until the reamer does make contact with the entire face of the ring. Clean up the opposite side of the ring in the same manner.

Carefully insert a .375" diameter straight fluted reamer inside the ring bore in the injector body. Turn the reamer in a clockwise direction and remove any burrs inside the ring bore. Then wash the injector body in clean fuel oil and dry it with compressed air.

Remove the carbon deposits from the lower inside diameter taper of the injector nut with carbon remover. Use care to minimize removing metal or setting up burrs on the spray tip seat. Remove only enough metal to produce a clean uniform seat to prevent leakage between the tip and the nut. Carefully insert carbon remover in the injector nut. Turn it clockwise to remove the carbon deposits on the flat spray tip seat.

Wash the injector nut in clean fuel oil and dry it with compressed air. Carbon deposits on the spray tip seating surfaces of the injector nut will result in poor sealing and consequent fuel leakage around the spray tip.

When handling the injector plunger, do not touch the finished plunger surfaces with your fingers. Wash the plunger and bushing with clean fuel oil and dry them with compressed air. Be sure the high pressure bleed hole in the side of the bushing is not plugged. If this hole is plugged, fuel leakage will occur at the upper end of the bushing where it will drain out of the injector body vent and rack holes, during engine operation, causing a serious oil dilution problem. *Keep the plunger and bushing together as they are mated parts.*

After washing, submerge the parts in a clean receptacle containing clean fuel oil. *Keep the parts of each injector assembly together.* Do not mix parts of one injector with those of another.

Inspect Injector Parts

Inspect the teeth on the control rack and the control rack gear for excessive wear or damage. Also check for excessive wear in the bore of the gear and inspect the gear retainer. Replace damaged or worn parts. (See Fig. 23-29 to 23-32.)

Inspect the injector follower and pin for wear.

Inspect both ends of the spill deflector for sharp edges or burrs which could create burrs on the injector body or injector nut and cause particles of metal to be introduced into the spray tip and valve parts. Remove burrs with a 500 grit stone.

Inspect the follower spring for visual defects. Then check the spring with spring tester.

Check the seal ring area on the injector body for burrs or scratches. Also check the surface which contacts the injector bushing for scratches, scuff marks or other damage. If necessary, lap this surface. A faulty sealing surface at this point will result in high fuel consumption and contamination of the lubricating oil. Replace any loose injector body plugs or a loose dowel pin. Install the proper number tag on a service replacement injector body.

Inspect the injector plunger and bushing for scoring, erosion, chipping or wear. Check for sharp edges on that portion of the plunger which rides in the gear. Remove any sharp edges with a 500 grit stone. Wash the plunger after stoning it. Injector Bushing Inspectalite can be used to check the port holes in the inner diameter of the bushing for cracks or chipping. Slip the plunger into the bushing and check for free movement. *Replace the plunger and bushing as an assembly if any of the above damage is noted, since they are mated parts.* Use new mated factory parts to ensure the best performance from the injector.

Injector plungers cannot be reworked to change the output. Grinding will destroy the hardened case at the helix and result in chipping and seizure or scoring of the plunger.

Examine the spray tip seating surface of the injector nut and spray tip for nicks, burrs, erosion or brinelling. Reseat the surface or replace the nut or tip if it is severely damaged.

The injector valve spring plays an important part in establishing the valve opening pressure of the injector assembly. Replace a worn or broken spring.

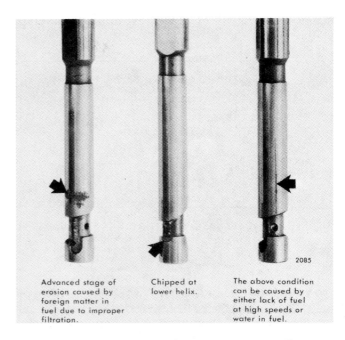

Advanced stage of erosion caused by foreign matter in fuel due to improper filtration.

Chipped at lower helix.

The above condition can be caused by either lack of fuel at high speeds or water in fuel.

FIGURE 23-29. Unusable injector plungers. *(Courtesy of Detroit Diesel Allison)*

Inspect the sealing surfaces of the injector parts. Examine the sealing surfaces with a magnifying glass for even the slightest imperfections will prevent the injector from operating properly. Check for burrs, nicks, erosion, cracks, chipping and excessive wear. Also check for enlarged orifices in the spray tip.

Replace damaged or excessively worn parts. Check the minimum thickness of the lapped parts as noted in the chart in the service manual.

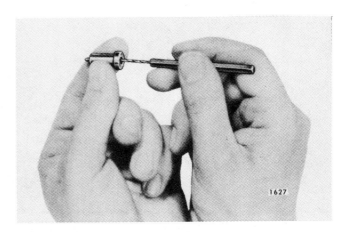

FIGURE 23-30. Cleaning injector spray tip. *(Courtesy of Detroit Diesel Allison)*

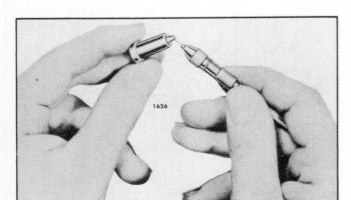

FIGURE 23-31. Cleaning spray tip orifices. *(Courtesy of Detroit Diesel Allison)*

Examine the seating area of the needle valve for wear or damage. Also examine the needle quill and its contact point with the valve spring seat. Replace damaged or excessively worn parts.

Examine the needle valve seat area in the spray tip for foreign material. The smallest particle of such material can prevent the needle valve from seating properly. Polish the seat area with polishing stick J 22964. Coat only the tapered end of the stick with polishing compound J 23038 and insert it directly into the center of the spray tip until it bottoms. Rotate the stick 6 to 12 times, applying a light pressure with the thumb and forefinger.

Note: Be sure that no compound is accidentally placed on the lapped surfaces located higher up in the spray tip. The slightest lapping action on these surfaces can alter the near-perfect fit between the needle valve and tip.

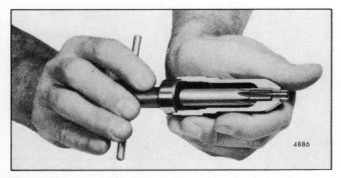

FIGURE 23-32. Cleaning injector nut spray tip seat. *(Courtesy of Detroit Diesel Allison)*

Before reinstalling used injector parts, lap all of the sealing surfaces. It is also good practice to lightly lap the sealing surfaces of new injector parts which may become burred or nicked during handling.

Note: The sealing surface of current spray tips is precision lapped by a new process which leaves the surface with a dull satin-like finish; the lapped surface on former spray tips was bright and shiny. It is not recommended to lap the surface of the *new* current spray tip.

Assemble Injector

Use an extremely clean bench to work on and to place the parts when assembling an injector. Also be sure all of the injector parts, both new and used, are clean. (See Fig. 23-33 to 23-35.)

Assemble Injector Filters

Always use new filters and gaskets when reassembling an injector.

1. Insert a new filter, dimple end down, slotted end up, in each of the fuel cavities in the top of the injector body.

Note: Install a new filter in the inlet side (located over the injector rack) in a fuel injector with an offset body. No filter is required in the outlet side of the offset body injector.

2. Place a new gasket on each filter cap. Lubricate the threads and install the filter caps. Tighten the filter caps to specified torque.

3. Purge the filters after installation by directing compressed air or fuel through the filter caps.

4. Install clean shipping caps on the filter caps to prevent dirt from entering the injector.

Assemble Rack and Gears

Note the drill spot marks on the control rack and gear. Then proceed as follows:

1. Hold the injector body, bottom end up, and slide the rack through the hole in the body. Look into the body bore and move the rack until you can see the drill marks. Hold the rack in this position.

2. Place the gear in the injector body so that the marked tooth is engaged between the two marked teeth on the rack.

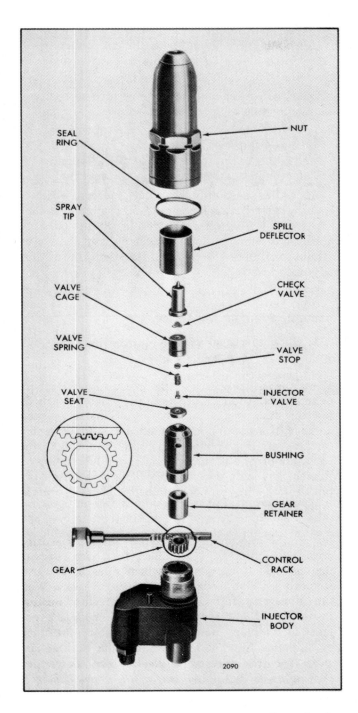

FIGURE 23-33. Injector rack, gear, spray tip and valve assembly details, and relative location of parts. *(Courtesy of Detroit Diesel Allison)*

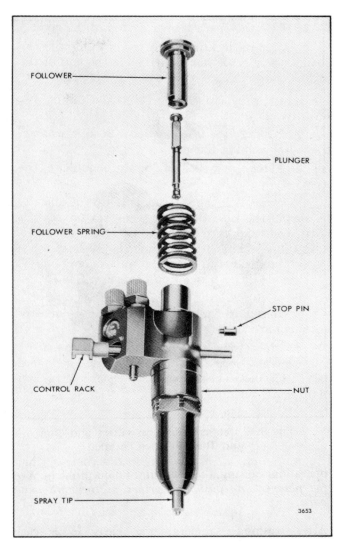

FIGURE 23-34. Injector plunger, follower, and relative location of parts. *(Courtesy of Detroit Diesel Allison)*

FIGURE 23-35. Tightening injector nut with torque wrench. *(Courtesy of Detroit Diesel Allison)*

3. Place the gear retainer on top of the gear.

4. Align the locating pin in the bushing with the slot in the injector body, then slide the end of the bushing into place.

Assemble Spray Tip, Spring Cage and Check Valve Assemblies

1. Support the injector body, bottom end up, in injector holding fixture.

2. Place a new seal ring on the shoulder of the body.

Note: Wet the seal ring with test oil and install the ring all the way down past the threads and onto the shoulder of the injector body. This will prevent the seal from catching in the threads and becoming shredded. Use a seal ring protector if available.

3. Install the spill deflector over the barrel of the bushing.

4. Place the check valve (without the .010″ hole) centrally on the top of the bushing. Then place the check valve cage over the check valve and against the bushing.

Note: The former and new check valve and check valve cage are not separately interchangeable in a former injector.

5. Insert the spring seat in the valve spring, then insert the assembly into the spring cage, spring seat first.

Note: Install a new spring seat in a former injector if a new design spray tip assembly is used.

6. Place the spring cage, spring seat and valve spring assembly (valve spring down) on top of the check valve cage.

Note: When installing a new spray tip assembly in a former injector, a new valve spring seat must also be installed. The current needle valve has a shorter quill.

7. Insert the needle valve, tapered end down, inside of the spray tip. Then place the spray tip and needle valve on top of the spring cage with the quill end of the needle valve in the hole in the spring cage.

8. Lubricate the threads in the injector nut and carefully thread the nut on the injector body by hand. Rotate the spray tip between your thumb and first finger while threading the nut on the injector body. Tighten the nut as tight as possible by hand. At this point there should be sufficient force on the spray tip to make it impossible to turn with your fingers.

9. Use socket and a torque wrench to tighten the injector nut to specified torque.

Note: Do not exceed the specified torque. Otherwise, the nut may be stretched and result in improper sealing of the lapped surfaces in a subsequent injector overhaul.

10. After assembling a fuel injector, always check the area between the nut and the body. If the seal is still visible after the nut is assembled, try another nut which may allow assembly on the body without extruding the seal and forcing it out of the body-nut crevice.

Assemble Plunger and Follower

1. Slide the head of the plunger into the follower.

2. Invert the injector in the assembly fixture (filter cap end up) and push the rack all the way in. Then place the follower spring on the injector body.

3. Place the stop pin on the injector body so that the follower spring rests on the narrow flange of the stop pin. Then align the slot in the follower with the stop pin hole in the injector body. Next align the flat side of the plunger with the slot in the follower. Then insert the free end of the plunger in the injector body. Press down on the follower and at the same time press the stop pin into position. When in place, the spring will hold the stop pin in position.

Check Spray Tip Concentricity

To assure correct alignment, check the concentricity of the spray tip as follows (Fig. 23-36):

1. Place the injector in the concentricity gauge and adjust the dial indicator to zero.

2. Rotate the injector 360° and note the total runout as indicated on the dial.

3. If the total runout exceeds .008″, remove the injector from the gauge. Loosen the injector nut, center the spray tip and tighten the nut to specified torque. Recheck the spray tip concentricity. If, after several attempts, the spray tip cannot be positioned satisfactorily, replace the injector nut.

Test Reconditioned Injector

Before placing a reconditioned injector in service, perform all of the tests (except the visual inspection of the plunger) previously outlined under *Test Injector*.

The injector is satisfactory if it passes these tests. Failure to pass any one of the tests indicates that defective or dirty parts have been assembled. In this case, disassemble, clean, inspect, reassemble and test the injector again.

Install Injector

Before installing an injector in an engine, remove the carbon deposits from the beveled seat of the injector tube in the cylinder head. This will assure correct alignment of the injector and prevent any undue stresses from being exerted against the spray tip.

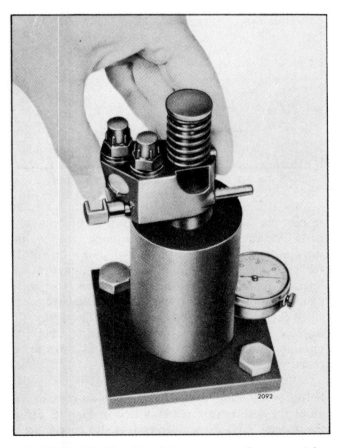

FIGURE 23-36. Checking injector spray tip concentricity. *(Courtesy of Detroit Diesel Allison)*

Use injector tube bevel reamer to clean the carbon from the injector tube. Exercise care to remove ONLY the carbon so that the proper clearance between the injector body and the cylinder head is maintained. Pack the flutes of the reamer with grease to retain the carbon removed from the tube.

Be sure the fuel injector is filled with fuel oil. If necessary, add clean fuel oil at the inlet filter cap until it runs out of the outlet filter cap.

Install the injector in the engine as follows:

1. Insert the injector into the injector tube with the dowel pin in the injector body registering with the locating hole in the cylinder head.

2. Slide the injector rack control lever over so that it registers with the injector rack.

3. Install the injector clamp, special washer (with curved side toward injector clamp) and bolt. Tighten the bolt to specified torque. Make sure that the clamp does not interfere with the injector follower spring or the exhaust valve springs.

> *Note:* Check the injector control rack for free movement. Excess torque can cause the control rack to stick or bind.

4. Move the rocker arm assembly into position and secure the rocker arm brackets to the cylinder head by tightening the bolts to the torque specified.

> *Note:* On four valve cylinder heads, there is a possibility of damaging the exhaust valves if the exhaust valve bridge is not resting on the ends of the exhaust valves when tightening the rocker shaft bracket bolts. Therefore, note the position

> of the exhaust valve bridge before, during and after tightening the rocker shaft bolts.

5. Remove the shipping caps. Then install the fuel pipes and connect them to the injector and the fuel connectors. Use socket to tighten the connections to specified torque.

> *Important:* Do not bend the fuel pipes and do not exceed the specified torque. Excessive tightening will twist or fracture the flared end of the fuel line and result in leaks. Lubricating oil diluted by fuel oil can cause serious damage to the engine bearings.

6. Perform an engine tune up as outlined in the next section Part 8.

PART 8   ENGINE TUNE-UP

To tune-up an engine completely, perform all of the adjustments, in the applicable tune-up sequence given below after the engine has reached normal operating temperature. Since the adjustments are normally made while the engine is stopped, it may be necessary to run the engine between adjustments to maintain normal operating temperature.

Note: Before starting an engine after an engine speed control adjustment or after removal of the engine governor cover, the serviceman must determine that the injector racks move to the no-fuel position when the governor stop lever is placed in the stop position. Engine overspeed will result if the injector racks cannot be positioned at no-fuel with the governor stop lever.

Caution: An overspeeding engine can result in engine damage which could cause personal injury.

Tune-Up Sequence for Mechanical Governor

1. Adjust the exhaust valve clearance.
2. Time the fuel injectors.
3. Adjust the governor gap.
4. Position the injector rack control levers.
5. Adjust the maximum no-load speed.
6. Adjust the idle speed.
7. Adjust the Belleville spring for "TT" horsepower.
8. Adjust the buffer screw.
9. Adjust the throttle booster spring (variable speed governor only).
10. Adjust the supplementary governing device, if used.

Tune-Up Sequence for Hydraulic Governor

1. Adjust the exhaust valve clearance.
2. Time the fuel injectors.
3. Adjust the governor linkage.
4. Position the injector rack control levers.
5. Adjust the load limit screw.
6. Compensation adjustment (PSG governors only).

7. Adjust the speed droop.
8. Adjust the maximum no-load speed.

PART 9   INJECTOR TIMING

To time an injector properly, the injector follower must be adjusted to a definite height in relation to the injector body. (See Fig. 23-37.)

All of the injectors can be timed in firing order sequence during one full revolution of the crankshaft. Refer to the *General Specifications* at the front of the service manual for the engine firing order.

Time Fuel Injector

After the exhaust valve clearance has been adjusted time the fuel injectors as follows:

1. Place the governor speed control lever in the *idle* speed position. If a stop lever is provided, secure it in the *stop* position.

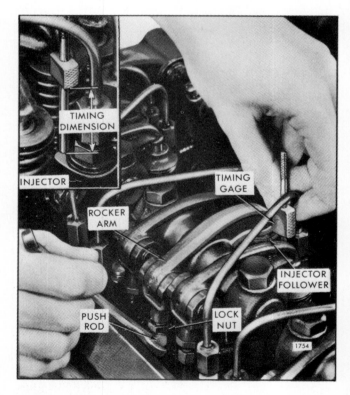

FIGURE 23-37. Checking and adjusting injector timing. *(Courtesy of Detroit Diesel Allison)*

2. Rotate the crankshaft, manually or with the starting motor, until the exhaust valves are fully depressed on the particular cylinder to be timed.

Note: If a wrench is used on the crankshaft bolt at the front of the engine, do not turn the crankshaft in a left-hand direction of rotation or the bolt may be loosened.

3. Place the small end of the injector timing gauge in the hole provided in the top of the injector body with the flat of the gauge toward the injector follower. Refer to the service manual for the correct timing gauge.

4. Loosen the injector rocker arm push rod lock nut.

5. Turn the push rod and adjust the injector rocker arm until the extended part of the gauge will just pass over the top of the injector follower.

6. Hold the push rod and tighten the lock nut. Check the adjustment and, if necessary, readjust the push rod.

7. Time the remaining injectors in the same manner.

A hole located in the injector body, on the side opposite the identification tag, may be used to visually determine whether or not the injector rack and gear are correctly timed. When the rack is all the way in (full-fuel position), the flat side of the plunger will be visible in the hole, indicating that the injector is "in time." If the flat side of the plunger does not come into full view and appears in the "advanced" or "retarded" position, disassemble the injector and correct the rack-to-gear timing.

PART 10   ADJUST INJECTOR RACK CONTROL LEVERS

The position of the injector rack control levers must be correctly set in relation to the governor. (See Fig. 23-38 to 23-40.) Their positions determine the amount of fuel injected into each cylinder and ensure equal distribution of the load. Properly positioned injector rack control levers with the engine at full load will result in the following:

1. Speed control lever at the maximum speed position.

2. Governor low speed gap closed.

3. High speed spring plunger on the seat in the governor control housing.

4. Injector fuel control racks in the full-fuel position.

Procedure (Typical for V71 Engine)

1. Disconnect any linkage attached to the governor speed control lever.

2. Turn the idle speed adjusting screw until there is no tension in the idle spring.

Note: A false fuel rack setting may result if the idle speed adjusting screw is not backed out as noted above or removed.

Injector rack must be adjusted so the effort to move the throttle from the idle speed position to the maximum speed position is uniform. A sudden increase in effort usually near the full-fuel position can result from:

(a) Injector racks adjusted too tight causing the yield link to separate.

(b) Binding of the fuel rods on the cylinder head.

(c) Failure to back out the idle screw.

3. Remove the clevis pin from the fuel rod at the right bank injector control tube lever.

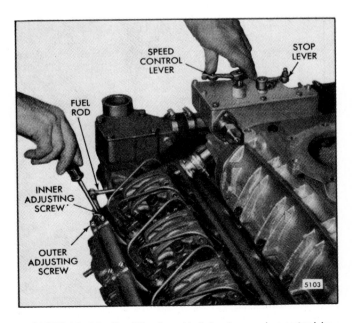

FIGURE 23-38. Positioning 1L injector rack control lever (two-screw assembly). *(Courtesy of Detroit Diesel Allison)*

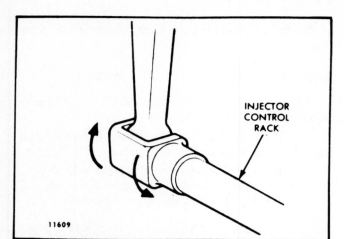

FIGURE 23-39. Checking rotating movement of injector control rack. *(Courtesy of Detroit Diesel Allison)*

4. Loosen all of the injector rack control lever adjusting screws and lock nuts on both cylinder heads enough to determine the freeness of the control lever and injector rack. Be sure all of the injector rack control levers are free on the injector control tubes.

5. Move the speed control lever to the maximum speed position with light finger presure. Turn down the adjusting screw of the No. 1L injector rack control tube lever until the injector rack clevis is bottomed against the injector body. The injector

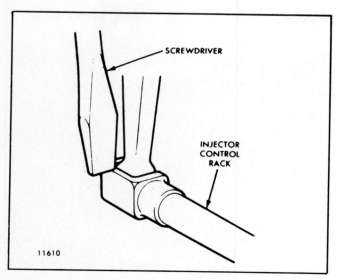

FIGURE 23-40. Checking injector control rack spring. *(Courtesy of Detroit Diesel Allison)*

rack should be easily rotated but not moved in and out. This will place the No. 1L injector rack in the full-fuel position.

The preceding steps should result in placing the governor linkage and control tube assembly in the same position they will attain while the engine is operating at full load.

6. To verify proper injector rack adjustment, hold the speed control lever in the maximum speed position using a screwdriver or finger tip and note that the injector rack clevis rotates freely but the rack will not move inboard or outboard.

If the rack and lever moves inboard or outboard, it is too loose. The injector rack is too tight if the rack clevis springs back after being depressed with a screwdriver. Again, verify the injector rack adjustment.

7. Remove the clevis pin from the fuel rod at the left bank injector control tube lever.

8. Insert the clevis pin in the fuel rod at the right cylinder bank injector control tube lever and position the No. 1L injector rack control lever as previously outlined in Procedures 5 and 6 for the No. 1L injector rack control lever.

Insert the clevis pin in the fuel rod at the left bank injector control tube lever.

9. Verify that the adjustment for the No. 1L and 1R injector racks are equal. Move the speed control lever to the maximum speed position. Rotate the clevis pins at the injector control tube levers and note a *slight* drag or resistance. With the fingertips the pin should move freely back and forth with no fuel rod deflection. This slight drag should be equal for both pins. If the drag is not equal turn either the 1R or 1L rack adjusting screw until both 1L and 1R pins are the same. Move the speed control lever back to the idle position and then back to the full-fuel position and note that the fuel rods do not deflect. If they do deflect the rack adjustment for either bank is too tight and *must* be readjusted as outlined previously.

Once the No. 1L and No. 1R injector rack control levers are adjusted, do not try to alter their settings. All adjustments are made on the remaining control racks.

10. To adjust the remaining injector rack control levers, remove the clevis pins from the fuel rod at the injector control tube levers. Hold the left bank

injector control racks in the full-fuel position by means of the lever on the end of the control tube and proceed as follows:

a. Tighten the adjusting screw of the No. 2L injector rack control lever until the injector rack clevis is observed to roll up or an increase in effort to turn the screwdriver is noted. Now you can have "kickup" which you should not have had before. Securely lock the adjusting screw locknut.

b. Verify the injector rack adjustment on No. 1L. If No. 1L does not "spring" back upward, turn the No. 2L adjusting screw counterclockwise slightly until the No. 1L injector rack returns to its full-fuel position and secure the adjusting screw locknut. Verify proper injector rack adjustment for both No. 1L and No. 2L injectors. Turn clockwise or counterclockwise the No. 2L injector rack adjusting screw until both No. 1L and No. 2L injector racks are in the full-fuel position when the locknut is securely tightened.

c. Adjust the remaining injectors using the procedures outlined in Step "B," always verifying proper injector rack adjustment.

11. When all of the injector rack control levers are adjusted, recheck their settings. With the control tube lever in the full-fuel position, check each injector control rack as outlined in Step 10. All of the injector control racks must have the same spring condition with the control tube levers in the full-fuel position as described in Item 10, a and b.

12. Insert the clevis pins in the fuel rods at the injector control tube lever and secure with a cotter pin. Recheck the rack settings as described in Item 5.

13. Turn the idle speed adjusting screw in until it projects approximately 3/16″ from the locknut to permit starting of the engine.

PART 11   ADJUST GOVERNOR

Adjust Maximum No-Load Engine Speed

If the governor has been reconditioned or replaced and to ensure the engine speed will not exceed the recommended no-load speed as given on the engine option plate, set the maximum no-load speed as follows (Fig. 23–41):

Note: Be sure the buffer screw projects approximately 5/8″ from the locknut to prevent interference while adjusting the maximum no-load speed.

1. Loosen the spring retainer locknut and back off the high speed spring retainer approximately five turns.

2. With the engine running at operating temperature and no-load on the engine, place the speed control lever in the maximum speed position. Turn the high speed spring retainer until the engine is operating at the recommended no-load speed.

3. Hold the high speed spring retainer and tighten the locknut.

Adjust Idle Speed

With the maximum no-load speed properly adjusted, adjust the idle speed as follows (Fig. 23–42):

1. With the engine running, at normal operating temperature and with the buffer screw backed out to avoid contact with the differential lever, turn

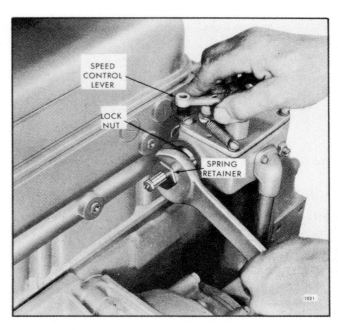

FIGURE 23-41. Adjusting maximum no-load speed on inline engine. *(Courtesy of Detroit Diesel Allison)*

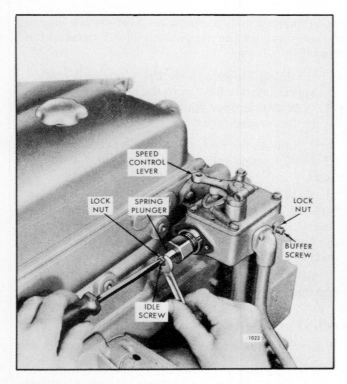

FIGURE 23-42. Adjusting engine idle speed on inline engine. *(Courtesy of Detroit Diesel Allison)*

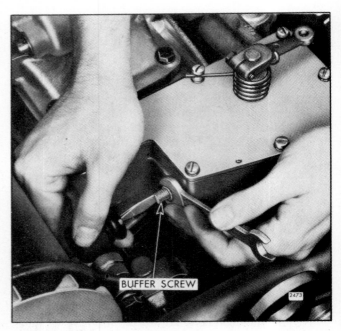

FIGURE 23-43. Adjusting buffer screw. *(Courtesy of Detroit Diesel Allison)*

the idle speed adjusting screw until the engine operates at approximately 15 rpm below the recommended idle speed.

Note: It may be necessary to use the buffer screw to eliminate engine roll. Back out the buffer screw, after the idle speed is established, to the previous setting (5/8″).

2. Hold the idle screw and tighten the locknut.

3. Install the high speed spring retainer cover and tighten the two bolts.

Adjust Buffer Screw

With the idle speed properly set, adjust the buffer screw as follows (Fig. 23-43):

1. With the engine running at normal operating temperature, turn the buffer screw in so it contacts the differential lever as lightly as possible and still eliminates engine roll.

Note: Do not increase the engine idle speed more than 15 rpm with the buffer screw.

2. Recheck the maximum no-load speed. If it has increased more than 25 rpm, back off the buffer screw until the increase is less than 25 rpm.

3. Hold the buffer screw and tighten the locknut.

PART 12   ADJUST THROTTLE DELAY

Adjustment (Current Throttle Delay)

Whenever the injector rack control levers are adjusted, disconnect the throttle delay mechanism by loosening the U-bolt which clamps the lever to the injector control tube. After the injector rack control levers have been positioned, the throttle delay mechanism must be readjusted. With the engine stopped, proceed as follows (Fig. 23-44):

1. Insert the timing gauge between the injector body and the shoulder on the injector rack clevis of the injector nearest the throttle delay cylinder. Refer to service manual for the proper gauge to set a Certified Vehicle engine.

2. Hold the governor throttle in the maximum speed position. This should cause the injector rack to move toward the full-fuel position and against the gauge.

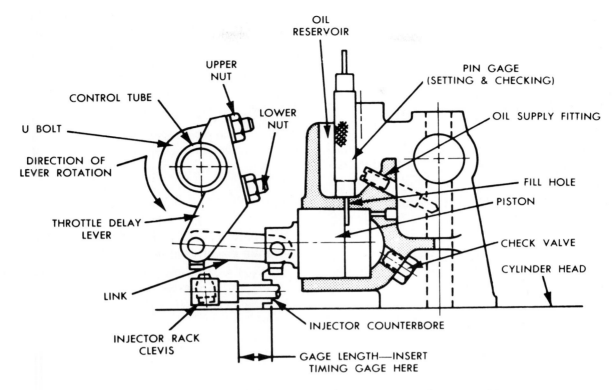

FIGURE 23-44. Adjusting throttle delay cylinder (current). *(Courtesy of Detroit Diesel Allison)*

3. Insert the pin gauge larger diameter end in the cylinder fill hole.

4. Rotate the throttle delay lever until further movement is limited by the piston contacting the pin gauge.

5. Tighten the U-bolt nuts while exerting a slight amount of torque on the lever; in the direction of rotation. Be careful not to bend the gauge or damage the piston by using excessive force.

6. Check the setting as follows:

a. Remove the pin gauge.

b. Attempt to reinstall the pin gauge. It should not be possible to reinsert the gauge without moving the injector racks toward the no-fuel position.

c. Reverse the pin gauge smaller diameter end and insert it in the cylinder fill hole. It should enter the cylinder without resistance.

Note: If the larger diameter end of the pin gauge enters the fill hole (Step 6b), increase the torque on the upper U-bolt nut. If the smaller diameter will not enter the fill

hole (Step 6c) without resistance, increase the torque on the lower U-bolt nut.

7. Release the governor throttle and remove the timing gauge and pin gauge.

8. Move the injector control tube assembly from the no-fuel to the full-fuel position to make sure there is no bind.

Adjustment (Former Throttle Delay)

Whenever the injector rack control levers are adjusted, disconnect the throttle delay mechanism by loosening the U-bolt which clamps the lever to the injector control tube. After the injector rack control levers have been positioned, the throttle delay mechanism must be readjusted. With the engine stopped, proceed as follows:

1. Insert gauge J 23190 (.454″ setting) between the injector body and the shoulder on the injector rack. Then exert a light pressure on the injector control tube in the direction of full fuel.

2. Align the throttle delay piston so it is flush with the edge of the throttle delay cylinder.

3. Tighten the U-bolt on the injector control tube and remove the gauge.

4. Move the injector rack from the no-fuel to the full-fuel position to make sure it does not bind.

PART 13   ADJUST FUEL MODULATOR

After completing the injector rack control lever and governor adjustment, adjust the fuel modulator, with the engine stopped, as follows (Fig. 23–45):

1. Insert gauge J 23190 (.454″ setting) between the injector body and the shoulder on the injector rack.

2. Position the governor speed control lever in the maximum speed position and the governor run stop lever in the RUN position.

3. Rotate the air box modulator lever assembly and clamp on the injector control tube until the lever roller contacts the modulator cam with sufficient force to take up the pin clearance.

4. Check to make sure only the roller contacts the cam and not the lever stamping. Tighten the lever and clamp screw.

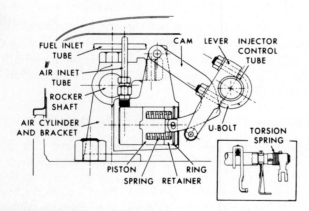

FIGURE 23–45. Fuel modulator assembly. *(Courtesy of Detroit Diesel Allison)*

5. Remove the gauge from between the injector body and the shoulder on the injector rack.

PART 14   SELF-CHECK

1. What functions does the unit injector perform?

2. The amount of fuel injected and the change in injection timing are determined by what?

3. What type of plunger is used in the unit injector?

4. How is the beginning and the end of injection controlled by the plunger?

5. At low power, injection starts (early/late) and at high power, it starts (early/late).

6. What is the difference between a constant-ending and a retarded-ending injector?

7. How is the governor driven on a V71 engine?

8. The idle adjusting screw acts on the _____ spring of the governor.

9. Maximum no-load speed is adjusted by turning the _____ on the governor.

10. What is the purpose of a fast idle cylinder?

11. The fuel modulator is a fuel delivery delay device used on turbocharged engines. True or False.

12. The throttle delay mechanism performs what function?

13. List three reasons for black exhaust smoke.

14. List five reasons for low engine power.

15. Give two reasons for detonation in an engine.

16. List the six tests required for unit injectors.

17. Why must parts from one injector not be mixed with those from another?

18. A timing gauge is required to set injector timing. True or False.

19. The rack control lever adjustment must be made before injector timing is done. True or False.

20. The fuel modulator adjustment must be made after the governor adjustments. True or False.

Chapter 24

Cummins PT Fuel Injection System

Performance Objectives

After thorough study of this chapter, the appropriate training models and service manuals, and with the necessary tools and equipment you should be able to do the following with respect to diesel engines equipped with the Cummins PT fuel injection system:

1. Complete the self-check questions with at least 80 percent accuracy.
2. Describe the basic construction and operation of a Cummins PT fuel injection system including the following:
    (a) method of fuel pumping and metering in the injector
    (b) fuel pump operation
    (c) injector and pump lubrication
    (d) injection timing control
    (e) governor types
3. Diagnose basic fuel injection system problems and determine the correction required.
4. Remove, clean, inspect, measure, repair, replace, and adjust fuel injection system components as needed to meet manufacturer's specifications.

This chapter is based on information provided courtesy of the Cummins Engine Co. Inc.

PART 1   INTRODUCTION

The Cummins PT fuel injection system is used exclusively on Cummins engines. A low-pressure positive-displacement pump supplies fuel to the injectors at a pressure that varies with engine speed. The pump also contains the governor assembly, which controls fuel pressure, engine idle speed, maximum speed, and torque. A single low-pressure fuel line provides fuel for all injectors. Cummins diesel engines are four-cycle engines and the injectors are operated from special lobes on the engine camshaft through cam followers, push rods, and rocker arms. The camshaft turns at one-half crankshaft speed. The injectors meter the fuel and inject it into the cylinder at high pressure. (See Fig. 24–1.)

Fuel Pump

The fuel pump is coupled to the compressor or fuel pump drive which is driven from the engine gear train. The fuel pump main shaft turns at engine crankshaft speed and in turn drives gear pump, governor and tachometer shaft.

The fuel pump assembly is made up of three main units (Fig. 24–2 to 24–4):

1. Gear pump, which draws fuel from supply tank through a fitting on top of fuel pump housing or an inlet fitting in gear pump and forces it through pump filter screen to governor.

2. Governor, which controls flow of fuel from gear pump, as well as maximum and idle engine speeds.

3. Throttle, which provides a manual control of fuel flow to injectors under all conditions in operating range.

Gear Pump and Pulsation Damper

Gear pump and pulsation damper are located at rear of fuel pump (rear being end farthest from drive coupling).

Gear pump is driven by pump main shaft and contains a single set of gears to pick up and deliver fuel throughout fuel system. A pulsation damper mounted to gear pump contains a steel diaphragm which absorbs pulsations and smoothes fuel flow through fuel system. From gear pump, fuel flows through filter screen and to governor assembly.

Throttle

Throttle provides a means for operator to manually control engine speed above idle as required by varying operating conditions of speed and load.

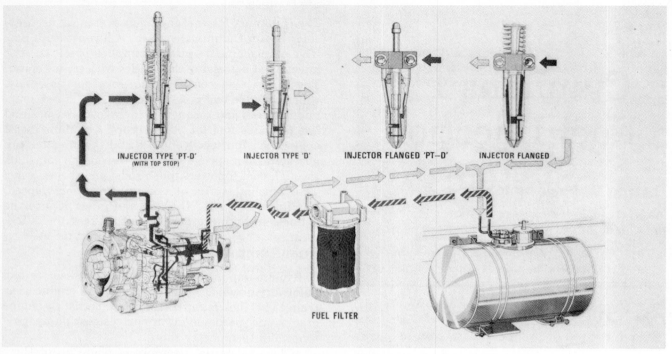

FIGURE 24-1. Fuel flow in Cummins PT fuel system. Note different injector types. *(Courtesy of Cummins Engine Company)*

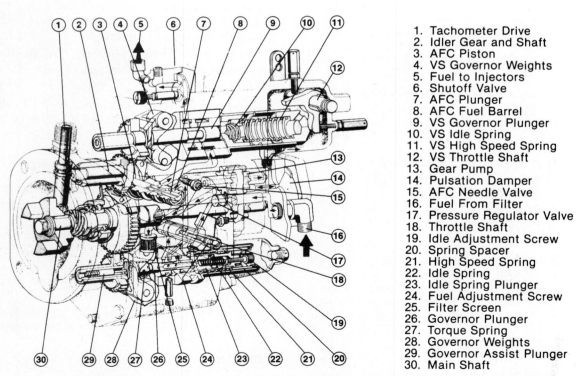

1. Tachometer Drive
2. Idler Gear and Shaft
3. AFC Piston
4. VS Governor Weights
5. Fuel to Injectors
6. Shutoff Valve
7. AFC Plunger
8. AFC Fuel Barrel
9. VS Governor Plunger
10. VS Idle Spring
11. VS High Speed Spring
12. VS Throttle Shaft
13. Gear Pump
14. Pulsation Damper
15. AFC Needle Valve
16. Fuel From Filter
17. Pressure Regulator Valve
18. Throttle Shaft
19. Idle Adjustment Screw
20. Spring Spacer
21. High Speed Spring
22. Idle Spring
23. Idle Spring Plunger
24. Fuel Adjustment Screw
25. Filter Screen
26. Governor Plunger
27. Torque Spring
28. Governor Weights
29. Governor Assist Plunger
30. Main Shaft

FIGURE 24-2. Cummins PTG fuel pump with AFC (Air Fuel Control) and variable speed governor. Fuel flow is indicated by arrows. *(Courtesy of Cummins Engine Company)*

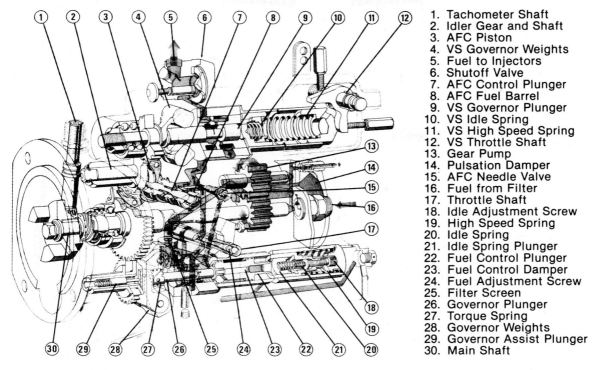

1. Tachometer Shaft
2. Idler Gear and Shaft
3. AFC Piston
4. VS Governor Weights
5. Fuel to Injectors
6. Shutoff Valve
7. AFC Control Plunger
8. AFC Fuel Barrel
9. VS Governor Plunger
10. VS Idle Spring
11. VS High Speed Spring
12. VS Throttle Shaft
13. Gear Pump
14. Pulsation Damper
15. AFC Needle Valve
16. Fuel from Filter
17. Throttle Shaft
18. Idle Adjustment Screw
19. High Speed Spring
20. Idle Spring
21. Idle Spring Plunger
22. Fuel Control Plunger
23. Fuel Control Damper
24. Fuel Adjustment Screw
25. Filter Screen
26. Governor Plunger
27. Torque Spring
28. Governor Weights
29. Governor Assist Plunger
30. Main Shaft

FIGURE 24–3. Cummins PTH fuel pump with AFC and VSG. Fuel flow indicated by arrows. *(Courtesy of Cummins Engine Company)*

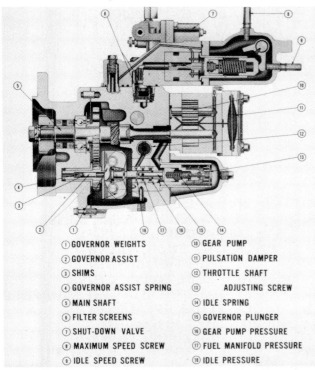

① GOVERNOR WEIGHTS
② GOVERNOR ASSIST
③ SHIMS
④ GOVERNOR ASSIST SPRING
⑤ MAIN SHAFT
⑥ FILTER SCREENS
⑦ SHUT-DOWN VALVE
⑧ MAXIMUM SPEED SCREW
⑨ IDLE SPEED SCREW
⑩ GEAR PUMP
⑪ PULSATION DAMPER
⑫ THROTTLE SHAFT
⑬ ADJUSTING SCREW
⑭ IDLE SPRING
⑮ GOVERNOR PLUNGER
⑯ GEAR PUMP PRESSURE
⑰ FUEL MANIFOLD PRESSURE
⑱ IDLE PRESSURE

FIGURE 24–4. Cummins PTG fuel pump cross section with parts identified. *(Courtesy of Cummins Engine Company)*

In PT (type G) fuel pump, fuel flows through governor to throttle shaft. At idle speed, fuel flows through idle port in governor barrel, around throttle shaft sleeve. Above idle speed, fuel flows through main governor barrel port to throttling hole in throttle shaft and onward to injectors.

Shut-Down Valve

Either a manual or an electric shut-down valve is used on Cummins fuel pumps.

With a manual valve, the control lever must be fully clockwise or open to permit fuel flow through the valve.

With the electric valve, the manual control knob must be fully counterclockwise to permit the solenoid to open the valve when the "switch key" is turned on. For emergency operation in case of electrical failure, turn manual knob clockwise to permit fuel to flow through the valve.

PART 2   FUEL PUMP AND GOVERNOR

The PT-G fuel pump has a central main housing containing the governor in the lower half. All of the subassemblies are bolted to the main housing. The front cover contains the drive parts for both the gear pump and the governor. (See Fig. 24–5 to 24–9.)

The gear pump, shut-down valve, tachometer drive, and filter are attached as subassemblies to the main housing. The throttle shaft is assembled in a bore at right angles to the drive shaft in the main housing.

The governor consists of a set of flyweights driven in direct proportion to engine speed; a plunger, which rotates with the governor weights and reciprocates axially in a fixed sleeve; and a governor spring pack acting in opposition to the governor weights at the opposite end of the governor plunger.

The basic functions of the fuel pump are:

1. to provide the fuel pressure to the injectors as required to produce the desired full-load torque curve;

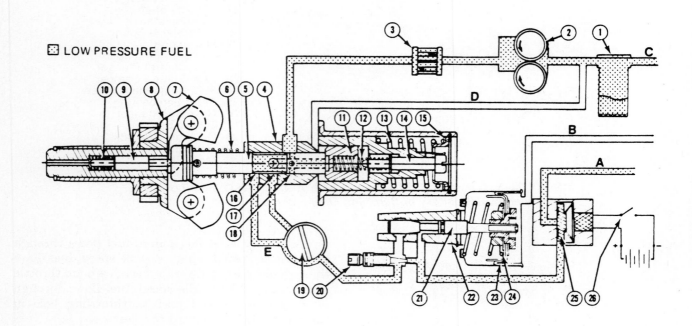

1	PRIMARY FUEL FILTER	17	MAIN GOVERNOR PORT
2	GEAR PUMP	18	GOVERNOR DUMP PORTS
3	FILTER SCREEN	19	THROTTLE
4	GOVERNOR SLEEVE	20	AFC NEEDLE VALVE
5	GOVERNOR PLUNGER	21	AFC CONTROL PLUNGER
6	TORQUE CONTROL SPRING	22	AFC BARREL
7	GOVERNOR WEIGHTS	23	DIAPHRAGM (BELLOWS)
8	GOVERNOR WEIGHT CARRIER	24	AFC SPRING
9	WEIGHT ASSIST PLUNGER	25	SOLENOID VALVE
10	WEIGHT ASSIST SPRING	26	IGNITION SWITCH
11	IDLE SPRING PLUNGER	A	FUEL TO INJECTORS
12	IDLE SPEED SPRING	B	AIR FROM INTAKE MANIFOLD
13	MAXIMUM SPEED GOVERNOR SPRING	C	FUEL FROM TANK
14	IDLE SPEED ADJUSTING SCREW	D	BY-PASSED FUEL
15	MAXIMUM SPEED GOVERNOR SHIMS	E	IDLE FUEL PASSAGE
16	IDLE SPEED GOVERNOR PORT		

FIGURE 24–5. Cummins PTG-AFC fuel pump components. *(Courtesy of Cummins Engine Company Inc.)*

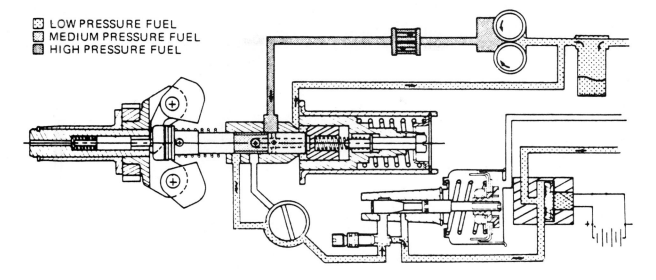

FIGURE 24-6. PTG fuel pump starting and idle conditions. Note fuel pressures and flow. Position of governor plunger meters fuel at this point in response to engine speed. Air fuel control plunger meters fuel in response to changes in intake manifold air pressure. *(Courtesy of Cummins Engine Company Inc.)*

2. to provide a means for part throttle operation;

3. to limit the maximum speed of the engine with stability and consistent regulation;

4. to control the idle speed with closed throttle and varying load.

Pressure Control

Figure 24-5 shows a simplified version of the pressure-regulating system in the fuel pump with which the governor weight controls the maximum fuel pressure as a function of engine speed.

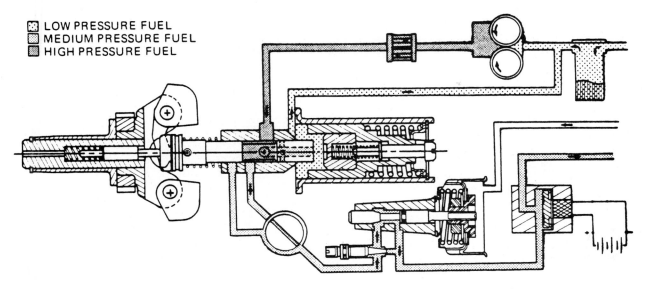

FIGURE 24-7. Fuel flow in PTG pump during normal driving conditions. *(Courtesy of Cummins Engine Company Inc.)*

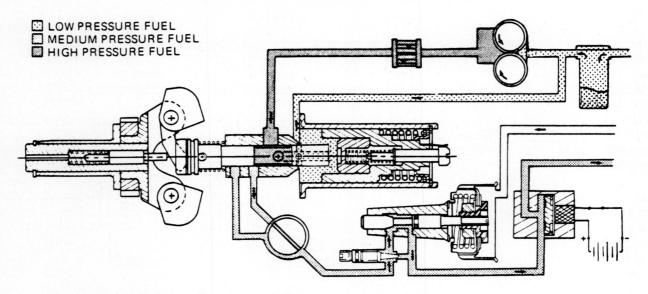

FIGURE 24-8. Fuel flow during beginning of high-speed governing. *(Courtesy of Cummins Engine Company Inc.)*

The basic parts of the flow-regulating system are:

1. a source of flow and pressure, which in the PT-G fuel pump is a positive displacement gear pump;

2. a governor and plunger with porting and a relieved groove on the plunger that enables the fuel to flow through the governor when the port and grooves are appropriately indexed;

3. a throttle valve between the governor and the fuel pump discharge to reduce the pressure in the fuel manifold and therefore provide part-load control;

4. the fuel conduit to the injection; and

5. a fuel bypass circuit from the governor plunger fuel groove through an axial drilling out of the end of the plunger and back to the inlet of the fuel supply pump.

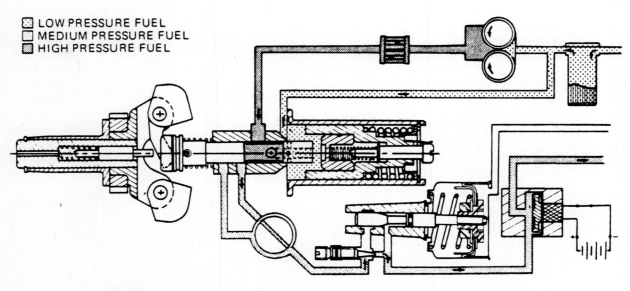

FIGURE 24-9. Fuel flow at complete high-speed governing. Note change in fuel flow at AFC. *(Courtesy of Cummins Engine Company Inc.)*

The pressure delivered by the fuel pump is determined by the forces acting to close the gap between the ends of the plunger and the pressure control button. This pressure within the fuel pump is independent of throttle position but is directly related to engine speed. All fuel that is not delivered to the injectors is discharged out of the gap formed by the end of the governor plunger and the adjacent pressure control button. Some fuel is bypassed under all operating conditions.

The pressure force that is acting at the bypass gap is opposed by the centrifugal force of the governor weights applied axially to the driven end of the plunger.

The fuel pressure required to separate the pressure control button from the governor plunger is determined by the cross-sectional area of the recess in the pressure control button. The equilibrium condition is reached when the forces generated by the fuel pressure in the recess area in the pressure button are equal to the governor weight force on the governor plunger.

At a given engine speed, the pressure delivered by the pump is a function of the governor weight force. Since the governor weight varies directly as the square of the engine speed, the pressure delivered by the pump also varies in this manner. This characteristic will produce approximately constant engine torque. This flat torque curve provides a basis for obtaining almost *any desired torque characteristic by means of minor adjustments*, which is another important and unique feature of the PT-G fuel pump.

Total Flow Control

The supply pump for the PT fuel system is designed to deliver several times more fuel than required to operate the engine at maximum power. The excess fuel is bypassed and returned to the pump inlet. Hence, several different engine models can use the same size supply pump.

Different engine models such as six- or eight-cylinder engines of the same basic design may require the same fuel pressure to the injector but different volumes of fuel flow. For this variation, no change in pump parts is required since the bypassed fuel quantity simply changes to accommodate the different flows. It also follows that variations in pump capacity caused by tolerances and pump wear have no effect on the pressure delivered by the fuel pump. *The ability of the fuel pump to accommodate the changes in supply pump delivery throughout the life of the fuel pump is an important and unusual feature, providing unchanged engine characteristics through the life of the engine.*

In case a change of injector supply pressure is desired, this change can be done by replacing the control button with one that has a different size recess.

Engine operation at part load is accomplished simply by throttling the fuel pressure between the fuel pump delivery and the injector. The throttle consists of a shaft with a cross-drilled fuel passage, which is indexed with holes in the bore of the pump housing. It should be noted that the pressure within the fuel pump is unaffected by throttle position; the throttle affects only the downstream pressure.

PART 3   GOVERNOR OPERATION

Idling and High-Speed Mechanical Governor: This mechanical governor, often called "automotive governor," is actuated by a system of springs and weights, and has two functions. First, the governor maintains sufficient fuel for idling with throttle control in idle position; second, it cuts off fuel to injectors above maximum rated rpm. The idle springs in the governor spring pack position the governor plunger so the idle fuel port is opened enough to permit passage of fuel to maintain engine idle speed.

During operation between idle and maximum speeds, fuel flows through governor to injectors in accordance with engine requirements as controlled by throttle and limited by size of idle spring plunger counterbore on PT (type G) fuel pumps. When the engine reaches governed speed, governor weights move the governor plunger, and fuel passages to throttle and injectors are shut off. At the same time another passage opens and dumps the fuel into main pump body. In this manner engine speed is controlled and limited by governor regardless of throttle position. Fuel leaving the governor flows through throttle, shutdown valve, inlet supply lines into internal drillings in cylinder heads and on to injectors.

PT (Type G) Variable-Speed Governors

There are two mechanical variable speed governors used with PT (type G) fuel pump. The "Mechanical Variable Speed (MVS)" governor which is mounted directly on top of, or remotely near, the fuel pump; and "Special Variable Speed (SVS)" governor which is a special spring-pack assembly at lower rear of fuel pump.

These governors may be added to existing engines as required; such additions to existing engines will require recalibration of fuel pumps. For all applications in which operator does not control engine speed by maintaining constant touch with throttle, it is recommended MVS governors be used.

Mechanical Variable Speed (MVS) Governor

This governor supplements the "standard or automotive governor" to meet the requirements of machinery on which engine must operate at a constant speed, but where extremely close regulation is not necessary.

Adjustment for different rpm can be made by means of a lever control or adjusting screw. At full-rated speed, this governor has a speed droop between full-load and no-load of approximately eight percent.

As a variable-speed governor, this unit is suited to varying speed requirements of cranes, shovels, etc., in which same engine is used for propelling unit and driving a pump or other fixed-speed machine.

As a constant-speed governor, this unit provides control for pumps, nonparalleled generators and other applications where close regulation (variation between no-load and full-load speeds) is not required.

The (MVS) governor assembly mounts on top of the fuel pump, and fuel solenoid is mounted to the governor housing. The governor also may be remote mounted.

Fuel from fuel pump body enters variable-speed governor housing and flows to governor barrel and plunger. Fuel flows past plunger to shut-down valve and on into injector according to governor lever position, as determined by operator.

The variable-speed governor cannot produce engine speeds in excess of automotive governor setting. The governor can produce idle speeds below automotive pump idle speed setting, but should not be adjusted below this speed setting when operating as a combination automotive and variable-speed governor.

Special Variable Speed (SVS) Governor

The SVS governor provides many of the same operational features of MVS governor, but is limited in application. An overspeed stop should be used with SVS governors in unattended applications; in attended installations, a positive shutdown throttle arrangement should be used if no other overspeed stop is used.

Marine applications require automotive throttle of fuel pump to be locked open during operation and engine speed control is maintained through SVS governor lever.

Power take-off applications use SVS governor lever to change governed speed of engine from full-rated speed to an intermediate power take-off speed. During operation as an automotive unit, SVS governor is in high-speed position. See operation instructions for further information.

Hydraulic governor applications, not having variable-speed setting provisions, use the SVS governor to bring engine speed down from rated speed for warm-up at or slightly above 1000 rpm.

Hydraulic Governor

Hydraulic governors are used on stationary power applications where it is desirable to maintain a constant speed with varying loads.

The Woodward SG Hydraulic Governor uses lubricating oil, under pressure, as an energy medium.

The governor acts through oil pressure to increase fuel delivery. An opposing spring in governor control linkage acts to decrease fuel delivery.

In order that its operation may be stable, speed droop is introduced into governing system. Speed droop means the characteristic of decreasing speed with increasing load. The desired magnitude of this speed droop varies with engine applications and may easily be adjusted to cover a range of approximately one-half of one percent to seven percent.

Assume a certain amount of load is applied to the engine. The speed will drop, flyballs will be forced inward and will lower pilot valve plunger. This will admit oil pressure underneath power piston, which will rise. The movement of power piston is transmitted to terminal shaft by terminal lever. Rotation of terminal shaft through linkage to the fuel pump causes fuel setting of engine to be increased.

Air/Fuel Control (AFC)

The Air/Fuel Control is an acceleration exhaust smoke control device built internally in the fuel pump. It restricts fuel proportionally to engine air intake manifold pressure during engine acceleration.

During acceleration, turbocharger speed (intake manifold air pressure) lags behind the almost instantaneous fuel delivering capability of the fuel pump and injectors, thus supplying an overrich fuel to air mixture which causes excessive exhaust smoke.

The Air/Fuel Control provides a more completely combustible fuel to air mixture by continuously monitoring turbocharger air pressure and proportionally responding to load or acceleration changes.

PART 4   INJECTORS

PT (Type D) Injectors

The injector provides a means of introducing fuel into each combustion chamber. It combines the acts of metering, timing and injection. Principles of operation are the same for inline and V-engines but injector size and internal design differ slightly.

Fuel supply and drain flow are accomplished through internal drillings in the cylinder heads. A radial groove around each injector mates with the drilled passages in the cylinder head and admits fuel through an adjustable (adjustable by burnishing to size at test stand) orifice plug in the injector body. A fine mesh screen at each inlet groove provides final fuel filtration.

The fuel grooves around the injectors are separated by "O" rings which seal against the cylinder head injector bore. This forms a leak-proof passage between the injectors and the cylinder head injector bore surface.

Fuel flows from a connection atop the fuel pump shut-down valve through a supply line into the lower drilled passage in the cylinder head. A second drilling in the head is aligned with the upper injector radial groove to drain away excess fuel. A fuel drain allows return of the unused fuel to the fuel tank.

The injector contains a ball check valve. As the injector plunger moves downward to cover the feed opening, an impulse pressure wave seats the ball and at the same time traps a positive amount of fuel in the injector cup for injection. As the continuing downward plunger movement injects fuel into the combustion chamber, it also uncovers the drain opening and the ball rises from its seat. This allows free flow through the injector and out the drain for cooling purposes and purging gases from the cup.

PT (Type D) Injector Top-Stop

The "Top-Stop" injector functions like the standard PT (type D) injector except the upward travel of the injector plunger is limited by an adjustable stop. The stop is set before the injector is installed in the engine.

When the injector is installed and properly adjusted in the engine, plunger spring load is carried against the stop which allows engine lubricating oil to better lubricate the sockets, reducing wear in the injection train. Consequently, the injector train remains in adjustment longer.

PTE Injector

In the Cummins PT fuel system fuel enters the injectors through an adjustable orifice at low pressure as determined by the fuel pump throttle and/or governor. The metering orifice controls the quantity of fuel that enters the injector nozzle at a pressure determined by the fuel pump and the time interval during which the metering orifice in the nozzle is uncovered by the injector plunger. The detachable plunger tip valve is held in contact with the injector plunger by the outer tip valve spring.

When the plunger moves down fuel entry through the metering orifice in the nozzle is cut off. The plunger continues down forcing fuel out of the nozzle, at high pressure, through the injector spray holes as a fine spray. As the tip valve approaches its seat the area between the tip valve and the seat becomes smaller than the total spray hole area of the injector and the tip valve snaps down on its seat to provide a fast positive end of injection. The continued downward movement of the plunger causes a hydraulic force on top of the tip valve to assist the inner spring to keep the tip valve seated preventing secondary injection.

With the tip valve seated the lower groove in the plunger uncovers the spill port in the nozzle to relieve the high fuel pressures as the plunger overtravels to the end of its stroke. The fuel spill is accomplished through the drilling in the center of the tip valve and cross holes in the plunger connecting to the lower groove.

After the initial downward movement of the plunger during the injection cycle the upper groove in the plunger connects the scavenge port to the drain passage and fuel begins to flow freely through the return line to the fuel tank.

After injection the tip valve remains seated until the next metering cycle and the fuel is flowing freely through the drain ports as in the PT (type D) injectors. The timing of metering and injection is also determined by the engine camshaft for optimum operation at all engine speeds.

PART 5   PRESSURE TIME (PT) PUMPING AND METERING

Figure 24–10 shows the cycle of events in the injector. The plunger is retracted during about one-half of each engine cycle. The fuel, from the "common rail" or fuel passage, enters into the injector through an orifice plug. This adjustable orifice provides a means for calibrating the injector for a desired identical fuel delivery. The PT-D injector contains a ball valve. When the plunger is retracted and the metering or feed orifice is open, the fuel pressure opens the check ball valve and the fuel enters into the cup.

The amount of fuel entering into the cup is—as stated before—a function of the fuel pressure and the time that the metering orifice is open. As the injector plunger moves downward, a pressure wave closes the check ball valve; shortly after the metering edge of the plunger closes the metering orifice, the crossover groove on the plunger opens the crossholes, and the fuel circulation or purging begins.

The metering portion of the cycle is followed by the injection in which the plunger moves further down and the cam-operated positive displacement plunger then generates whatever injection pressure is required to discharge the fuel through the spray holes into the combustion chamber. During injec-

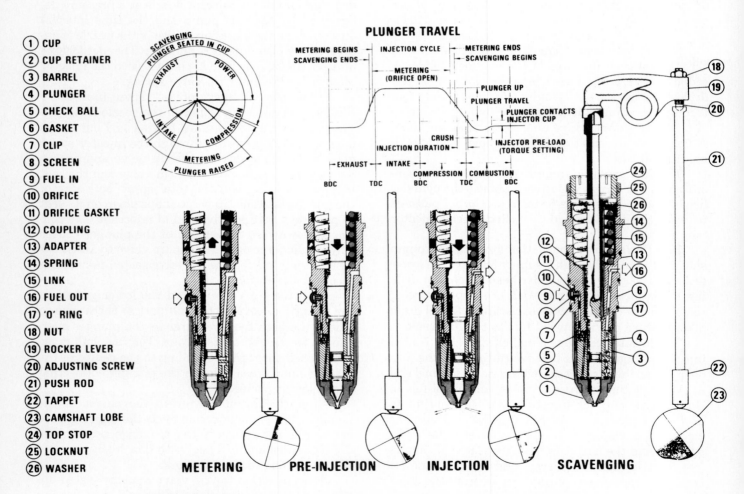

① CUP
② CUP RETAINER
③ BARREL
④ PLUNGER
⑤ CHECK BALL
⑥ GASKET
⑦ CLIP
⑧ SCREEN
⑨ FUEL IN
⑩ ORIFICE
⑪ ORIFICE GASKET
⑫ COUPLING
⑬ ADAPTER
⑭ SPRING
⑮ LINK
⑯ FUEL OUT
⑰ 'O' RING
⑱ NUT
⑲ ROCKER LEVER
⑳ ADJUSTING SCREW
㉑ PUSH ROD
㉒ TAPPET
㉓ CAMSHAFT LOBE
㉔ TOP STOP
㉕ LOCKNUT
㉖ WASHER

METERING   PRE-INJECTION   INJECTION   SCAVENGING

FIGURE 24–10. In the Cummins PT system the high-pressure pumping is done in the injector while metering is a function of fuel pressure available at the injector and the duration or "time" of metering. Study these diagrams to determine how the Cummins system works. *(Courtesy of Cummins Engine Company Inc.)*

tion and after completion of injection, the purging cycle continues, until the retracted plunger once again closes the crossover holes. It is important to note that there is no purging of fuel taking place during the metering cycle. The fuel circulation within the injector starts only after metering of the fuel is completed, thereby eliminating the effect of drain flow fluctuation or drain restriction during metering. This is the so-called closed drain metering cycle.

The PT-D injector cup is made of two pieces: the cup itself and the cup retainer. There is ample clearance between the cup and the retainer to allow the cup to center on the injector plunger. Because of this self-centering feature, the seating of the plunger in the cup is assured, preventing leakage between the cup and the plunger.

The PT-D injector plunger has a relieved diameter, which operates in the cup with a small clearance and the inside diameter of the plunger remains engaged in the cup so the metered fuel enters the volume between the cup and the metering edge on the plunger. There is always a fuel film in the annulus below the metering chamber, which works as a *liquid seal*, sealing off the metering portion of the injector, to prevent carboning.

While the metering orifice is open-metered, fuel is delivered into the injector cup, but this fuel charge never completely fills the cavity. The remaining portion is filled with compressed and heated air, which enters the cup through the spray holes from the engine cylinder. Thus a mixture of heated air and fuel is prepared in the cup before injection.

Because of the vaporous mixture within the cup, actual injection does not commence when the camshaft starts the downward movement of the injector plunger; injection cannot begin until the pressure in the cup equals and then exceeds that in the combustion chamber. At the beginning of the plunger movement, a mixture of air and fuel is being compressed so that the pressure rise in the cup follows the law of compressible fluids, producing a relatively slow discharge from the cup. Soon, however, the downward movement of the plunger compresses the fuel to a substantially "solid mass" and the rate of pressure rise in the cup increases sharply to inject the main charge. With all other things equal, the rate of the injector plunger travel during the injection of the main charge determines the degree of atomization, penetration, and distribution of the fuel emerging into combustion chamber. Since the small pilot charge is displaced and arrives in the combustion chamber slightly ahead of the main charge, it

passes through the usual ignition delay but has already started the combustion when the main charge arrives. Thus, the condition for minimum ignition delay and fast smooth combustion of the main charge has been created by the initial or pilot injection characteristic of the injector.

It should also be noted that the beginning of the injection is a variable depending on the quantity of fuel in the injector cup cavity. At light load, the charge is relatively small and the injector plunger reaches the solid fuel level relatively late, i.e., near the top-dead-center of the piston travel, as desired. As the load increases, the injected quantity increases and the solid fuel level is met earlier, before the top-dead-center. The pilot injection, during the compressible phase, always precedes the main charge injection. Thus, the *actual injection timing varies automatically* in the proper direction to suit varying engine load conditions; this is an inherent characteristic of the injector. At constant injected fuel quantity, however, the beginning of the injection remains the same regardless of engine speed. This characteristic is a significant advantage over other conventional fuel systems where variable timing devices are needed to compensate for the pressure wave propagation or injection lag due to high-pressure fuel lines between the fuel pump and injectors.

As with most injector systems, if good combustion is to be obtained, the rate of fuel injection must be maintained right up to the end of the injection process. This is accomplished in the design of the injector cam. The injection cam incorporates a "nose," which theoretically projects the travel of the injector plunger beyond the cup seat by the amount below the dwell level. This nose therefore compensates for the elasticity of the injection mechanism during actual injection and ensures holding the plunger tight against its seat while relatively high pressures exist in the engine cylinder. This "nose" does not increase the maximum load on the injector mechanism but prevents the maximum load, or the injection pressure, from decreasing near the end of injection.

The fuel quantity delivered through the metering orifice is regulated by changing the fuel pressure at the entrance of the metering orifice in the injector.

The second metering variable is time. Since the injector plunger is controlled by the camshaft, the metering time increment per cycle decreases as engine speed increases. For example, doubling the engine speed reduces the absolute metering time per cycle by one-half. For constant pressure, the fuel per cycle decreases with increasing engine speed but the total engine fuel rate remains unchanged.

In combining both variables of metering pressure and metering time, it will be evident that constant engine torque throughout the engine speed range requires the fuel metering pressure to vary as the square of the engine speed. *The important feature of the PT-G fuel pump is its inherent ability to produce a fuel-metering pressure that varies as the square of the engine speed.* This is achieved by taking unique advantage of the fact that a rotating mechanical governor flyweight also has a square-law characteristic.

PART 6   CUMMINS INJECTION TIMING

Cummins injection timing is controlled by the amount of fuel metered into the injector cup. With less fuel in the cup (low engine speed) injection begins later in relation to piston travel than it does when more fuel is metered into the injector cup (higher engine speeds). In other words, less fuel means less injection timing advance, which means lower engine speeds; more fuel means more injection timing advance which means higher engine speeds. Precise adjustment of the injector-operating mechanism is also required to obtain accurate and even injection timing between all cylinders.

Start upstroke (fuel circulates)

Fuel at low pressure enters the injector at (A) and flows through the inlet orifice (B), internal drillings, around the annular-groove in the injector cup and up passage (D) to return to the fuel tank. The amount of fuel flowing through the injector is determined by the fuel pressure before the inlet orifice (B). Fuel pressure in turn is determined by engine speed, governor and throttle.

Upstroke complete (fuel enters injector cup)

As the injector plunger moves upward, metering orifice (C) is uncovered and fuel enters the injector cup. The amount is determined by the fuel pressure. Passage (D) is blocked, momentarily stopping circulation of fuel and isolating the metering orifice from pressure pulsations.

Downstroke (fuel injection)

As the plunger moves down and closes the metering orifice, fuel entry into the cup is cut off. As the plunger continues down, it forces fuel out of the cup through tiny holes at high pressure as a fine spray. This assures complete combustion of fuel in the cylinder. When fuel passage (D) is uncovered by the plunger undercut, fuel again begins to flow through return passage (E) to the fuel tank.

Downstroke complete (fuel circulates)

After injection, the plunger remains seated until the next metering and injection cycle. Although no fuel is reaching the injector cup, it does flow freely through the injector and is returned to the fuel tank through passage (E). This provides cooling of the injector and also warms the fuel in the tank.

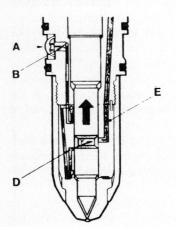 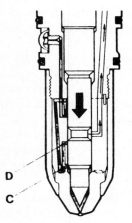 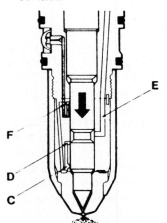 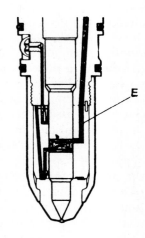

FIGURE 24–11. Cummins injector operation. *(Courtesy of Cummins Engine Company Inc.)*

Cummins also uses automatic injection-timing advance mechanisms to improve timing control. The system, introduced in 1980, requires the engine to be equipped with an air compressor. This system is known as *mechanical variable timing*. Another more recent system, known as *hydraulic variable timing*, does not require air pressure to operate. It uses engine lubricating oil instead.

Mechanical Variable Timing

Mechanical variable timing varies the injection timing by changing the location of the injector cam followers relative to the camshaft. This is accomplished by mounting an eccentric rigidly to the cam follower shaft. The shaft changes the location of the injector cam followers on the camshaft when the cam follower shaft is rotated. To rotate the cam follower shaft, an air cylinder with a rack-and-pinion gear arrangement is used. (See Fig. 24–12 to 24–17.)

The MVT actuator is an air-operated cylinder. The cylinder shaft has a rack of gear teeth, which mesh with the pinion gear mounted to the cam follower shaft. Raising and lowering the cylinder piston cause the pinion to rotate. The cylinder piston is held in the down position by a large spring. When air pressure is applied to the cylinder, the piston will rise against the spring pressure. An oversize first tooth on the rack provides positive alignment with the pinion.

Mounted to the cam follower shaft are eccentrics. These eccentrics and the cam followers are held to the shaft with a set screw. Rotation of the cam follower shaft provides the in-and-out movement of the followers.

Air flow to and from the cylinder is controlled by a three-way solenoid control valve mounted to the actuator. The solenoid valve is activated by an electrical pressure switch located in the discharge side of the fuel pump pressure line. The air supply to the actuator is plumbed from the dry storage tank of the truck.

When the engine is first started, there usually is no truck system air pressure. The spring will hold the piston down in the cylinder, and the engine starts with retarded injection timing. When air pressure builds up in the truck system, this pressure will enter the cylinder and raise the piston. This will cause the injection timing to advance.

Air will be directed to the actuator as long as the fuel pressure is below the preset pressure point or when the engine is from 0 to 25 percent of load. With these conditions the electrical pressure switch in the fuel pump will remain activated, supplying current to the three-way solenoid valve. The solenoid valve remains activated, allowing air pressure to enter the actuator.

When the driver puts the truck under load, fuel pressure will increase. When fuel pressure reaches the preset point or approximately 25 percent of load, the electrical pressure switch will deactivate. The three-way solenoid valve will shift, closing off the port that supplies air to the actuator and allows trapped air in the cylinder to be exhausted.

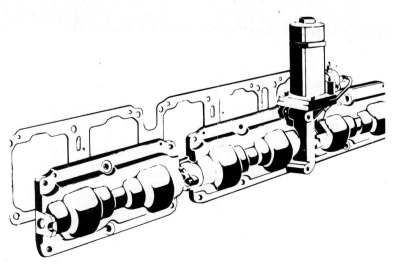

FIGURE 24-12. External view of Cummins mechanical variable timing (MVT) device. *(Courtesy of Cummins Engine Company Inc.)*

...exhaust port of the solenoid valve ...internal spring in the actuator will force ...piston down, discharging the trapped air. So that the injection timing does not go from advance to retard too rapidly, an orifice is located in the discharging line. The air is gradually discharged from the cylinder. Thus the change in timing is very gradual.

All three cam follower housings and the actuator are splined together on a shaft. The whole assembly weighs around 80 pounds and is mounted to the engine completely assembled. A one-piece cam follower gasket is used. It is the same thickness all the way across.

The camshaft also has been modified. The camshaft injector lobes have been redesigned. To eliminate power increases, a slow retraction cam is used. This camshaft will yield no power advantage of advanced engine timing. It limits the amount of fuel that can enter into the injector cup. This is accomplished by decreasing the metering time in the advance mode.

Hydraulic Variable Timing

The Hydraulic Variable Timing (HVT), a step-timing system, is relatively simple. It is made up of a special PT (type D) Top-Stop injector with a hydraulic tappet between the injector lever and the plunger. (See Figures 24–18 to 24–20.)

At any engine speed the system allows the engine to operate at *advanced* injection timing under

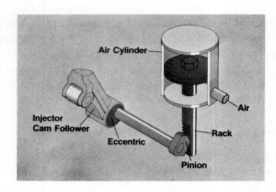

FIGURE 24-14. To rotate the follower shaft, an air cylinder with a rack-and-pinion gear arrangement is used, operated by air from the air brake system. *(Courtesy of Cummins Engine Company Inc.)*

light-load conditions, and at *retarded* timing during *high-load* conditions.

To meet strict California emission laws Mechanical Variable Timing (MVT) was introduced in the NH/NT automotive California engines in 1980. This system was too complex and expensive to be incorporated in the V engines and the K-6 engine.

Meanwhile, Cummins' engineers had designed and tested HVT which, while quite different from MVT, accomplishes the same results.

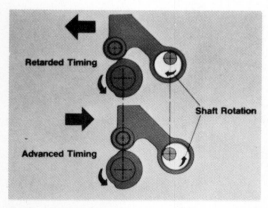

FIGURE 24-13. This is how mechanical variable timing comes into use. Depending on the engine load, it varies the injection timing by changing the location of the injector cam followers relative to the camshaft. *(Courtesy of Cummins Engine Company Inc.)*

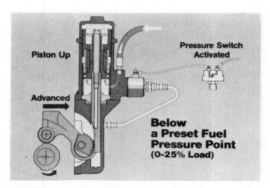

FIGURE 24-15. Air will be directed to the actuator as long as the fuel pressure is below the preset pressure point or when the engine is from 0 to 25 percent of the load. With these conditions the electrical pressure switch in the fuel pump will remain activated, supplying current to the three-way solenoid valve. The solenoid valve remains activated, allowing air pressure to enter the actuator, advancing injection timing. *(Courtesy of Cummins Engine Company Inc.)*

Advantages and Improvements

The HVT

- improves cold-weather idling characteristics
- reduces white smoke in cold climates
- reduces injector carboning

HVT Oil Flow (Advanced Timing)

1. When engine fuel pressure (5) is less than 32 psi, the fuel pressure switch (6) is *closed;* the HVT oil control valve (3) is *open* (Fig. 24-20).

2. Engine oil, under pressure, flows from the oil filter head (2) to the open oil control valve (3).

3. The pressure relief valve (4) returns a small amount of oil to the oil pan.

4. Oil flows from the control valve to the oil manifold (7).

5. From the manifold oil flows through the oil transfer connection (9) and into the HVT tappet assembly (8).

6. The tappet fills with oil, causing injection timing to be advanced.

7. At the end of the injection cycle, the increased oil pressure in the tappet moves the load-

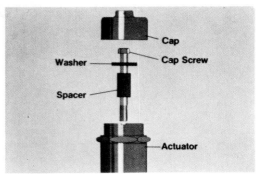

FIGURE 24-17. Injection timing can be adjusted in two ways. A long and a short spacer are available. The longer spacer provides more range between minimum and maximum injection timing advance compared to the shorter spacer. If base timing is incorrect, it can be adjusted by turning the cap, then locking this adjustment with the lock nut. *(Courtesy of Cummins Engine Company Inc.)*

cell check ball from its seat. The oil then drains from the tappet through the drain holes in the injector adapter.

HVT Oil Flow (Retarded Timing)

1. When engine fuel pressure (5) is more than 32 psi, the fuel pressure switch (6) is *open;* the oil control valve (3) is *closed*. (See Figure 24-20.)

2. Engine oil, under pressure, flows from the oil filter head (2) to the closed oil control valve (3).

3. The pressure relief valve (4) prevents oil pressure in the oil manifold from exceeding 6 psi. This oil pressure must be 10 psi or more before it can flow into the tappet. Injection timing is retarded when there is no oil in the tappet.

PART 7   FUEL SYSTEM DIAGNOSIS AND SERVICE

The following diagnostic charts and remedies on the Cummins PTG-AFC air fuel control fuel system are provided courtesy of the Cummins Engine Company, Inc. Other Cummins fuel system charts and remedies are similar. For appropriate overhaul and calibration procedures and service data, refer to the appropriate manufacturer's shop service manual.

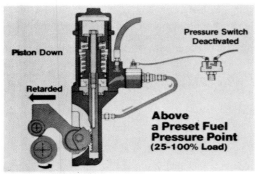

FIGURE 24-16. When the driver puts the truck under load, fuel pressure will increase. When fuel pressure reaches the preset pressure point or approximately 25 percent of load, the electrical pressure switch will deactivate. The three-way solenoid valve will shift, closing off the port that supplies air to the actuator and allows trapped air in the cylinder to be exhausted, retarding injection timing. *(Courtesy of Cummins Engine Company)*

AFC TROUBLE-SHOOTING

COMPLAINT	POSSIBLE CAUSE	CHECKING PROCEDURE	CORRECTION
Fuel leakage	AFC barrel out of location	Check AFC barrel retaining ring. Heavy leakage.	Install retaining ring; shim if necessary; make sure locks in position.
	Damaged seal on AFC plunger	Check glyd ring and O-ring for small leakage; check barrel and plunger finish.	Carefully install new seals, barrel, and plunger, using service tools.
	Scratched AFC barrel plunger bore	Check for small leakage from vent screw.	Replace barrel, plunger, and seals; make sure parts are clean.
	Dirt entering from loose vent screw media	Check if media loose in screw.	Install new vent screw, barrel, plunger, and seals. Shield vent if in dirt stream.

Note: Check glyd ring for wear, scratches, or misinstallation by visual inspection preferably with magnifying glass, a good light, and fuel wiped off with a clean cloth.

DRIVEABILITY: DRIVEABILITY IS CLASSIFIED INTO TWO GENERAL AREAS, EACH WITH SEPARATE COMPLAINTS.
1. Starting and/or sluggish acceleration. 2. Low power and/or no acceleration.

COMPLAINT	POSSIBLE CAUSE	CHECKING PROCEDURE	CORRECTION
Startability and/or sluggish acceleration			
Hard starting or engine not starting	No-air needle valve shut-off or set too lean	Apply 25 psi air pressure to AFC unit.	If engine starts, reset "no-air" needle valve.
		Caution: Excessive pressure will rupture bellows.	If engine does not start, AFC is not at fault.
	AFC plunger too short	Length should be 3.340/3.360 inches.	Replace plunger.
	AFC cover center die diameter too deep	Check if face of cover to bottom of die hole 0.190/0.200 inch deep.	Replace cover.
	Barrel meter hole too deep	Check distance to hole, 1.615/1.635 inches.	Replace barrel.
No-air valve not set to limits	Set before AFC plunger adjustment not after	Reset.	Make AFC adjustment, then no-air valve.
	Pressure too high	Reset.	Loosen jam nut; turn valve (cw) until pressure correct; tighten jam nut.
	Pressure too low	Reset.	Loosen jam nut; turn valve (ccw) until pressure correct; tighten jam nut.

Note: After any adjustment to pump, readjust AFC and no-air valve settings.

COMPLAINT	POSSIBLE CAUSE	CHECKING PROCEDURE	CORRECTION
AFC pressure not set to limits	No-air needle valve not bottomed during adjustment	Reset. Check thread depth so valve seats.	Loosen jam nut on valve and bottom valve against seat; check for no flow before adjusting AFC. Bottom tap threads 5/16–24 UNF.
	Pressure set too high	Reset.	Loosen jam nut; turn AFC plunger (ccw) until pressure is correct; tighten jam nut.
	Plunger sticking	Differential pressure should not exceed 15 psi.	See section on AFC doesn't repeat.
Scored plunger and barrel	Vent screw filter media loose	Check if media loose in screw.	Install new vent screw, barrel, and plunger seals.
AFC not repeating within recheck limits; AFC recheck +3 psi; no-air recheck +2 psi.	High plunger-to-barrel seal movement resistance	Set AFC to limits, increase air to 25 psi and decrease to setting, record value. Must not exceed 15 psi difference in values.	Work AFC plunger in and out several times. Check barrel surface finish and replace glyd ring.
Sluggish acceleration	Plugged vent capscrew	Blow through freely.	Clean or replace each time removed, and at maintenance check.
	Loose AFC barrel	Add parts noted, as necessary.	Correct installation or add 139585 spring to bottom of barrel and add S-16240 snap ring.

COMPLAINT	POSSIBLE CAUSE	CHECKING PROCEDURE	CORRECTION
Low power	Air pressure required to actuate AFC higher than specifications; bellows	Apply 25 psi air pressure to AFC unit; large air leakage will be noted at vent capscrew.	Replace bellows. Bellows action forces air out. A small air leak is expected.
	Wrong AFC spring or setting	Check AFC spring setting and/or color code.	Adjust AFC setting or change spring.
	AFC plunger stuck in starting position	Apply 25 psi pressure to AFC unit and check for rise in fuel pressure as throttle is advanced.	Remove and check plunger and barrel and clean. Check seals.
	Ruptured bellows.	Check for large air leak.	Install new bellows. Watch; small air leak may be bellows action.
	AFC restricting early; power falling off rapid near torque peak	Check AFC setting.	Reset. AFC control should occur at least 100 rpm below torque peak.
	Plugged filter vent capscrew	Blow through freely.	Clean or replace each time removed.
Bellows leak	Improper use of AFC service tool or overtightened cover	Check for frayed edges, elongated holes, or stretched holes. Torque capscrew.	Index tool sockets over nuts and tighten to 30/55 in-lb. only.
	Bellows upside down; not aligned with bolt pattern.	Visually check.	Change or replace.
	Bubbles, cracks, separated bellows	Visually check.	Replace bellows.
Fall off in power	AFC barrel movement	See sluggish acceleration section.	
Engine not accelerating properly.	Air line to AFC leaking; plugs not connected	Check for loose line, connections, or plugged line.	Tighten connection lines or replace line.
	Ruptured bellows	Apply 25 psi air pressure to AFC air leakage at vent screw.	Replace bellows.
	Plugged vent capscrew.	Remove screw and check AFC operation.	Clean or replace.
	No-air needle valve set low	Reset.	Reset.
	Sticking AFC plunger	Remove cover and check plunger movement. Also see AFC doesn't repeat section.	Remove and check plunger and barrel and clean. Check seals.
	Wrong AFC setting, delay too much	Check plunger setting.	Reset to specifications.
Air leakage at AFC vent screw	Bellows damaged	Large air leakage from vent screw.	Replace bellows.
	AFC seal gasket (cork) not compressed, damaged, or missing	Small air leakage from vent screw.	Tighten nut on piston assembly to 30/40 in-lb. Center bolt in piston so gasket seats. Replace gasket.
	Telfon tape seal on AFC plunger threads	Remove plunger nut and check for tape or sealant.	Retape plunger thread.
Excessive smoke	Fuel pump flow exceeding calibration specifications	Connect ST-435 pressure gauge to pump shutdown valve. Apply 25 psi air pressure to AFC unit. Take a snap pressure reading and check pressure against calibration specifications.	If fuel pressure is within specification, fuel rate is satisfactory. If above specification, adjust fuel pump pressure.
Also check: injectors, turbocharger, engine timing.	Wrong AFC spring, plunger setting, or no-air setting	Check plunger and no-air settings on test stand.	Change spring; adjust plunger or no-air needle valve.

| Low engine fuel pressure | AFC not wide open | With gauge at shut-off valve, idle engine, then snap throttle wide open. Compare psi reached to no-air snap rail check in calibration data. | Reset AFC. May use for other pump problems as trouble-shooting aid on engine. |
| Fuel leakage, no no-air needle valve | No-air needle valve O-ring | Check O-ring for damage and valve bore for burrs. | Replace O-ring. Remove burrs as necessary using appropriate tools. |

Notes:
1. When checking AFC barrel inside diameter, a polished appearance is normal, but deep scratches will cause leakage even with a new seal; replace barrel.
2. AFC plungers will show a polished appearance. Light scratches are normal, but severe scratches require plunger replacement.
3. When any fuel pump component is replaced, special attention should be given to cleanliness and lubrication. With the AFC section be sure to use glyd ring installation, forming tools, and torque wrenches and make sure the hysteresis check described under AFC doesn't repeat within recheck limits section.

Complaint and Corrections—PT (Type G)

Charted on the following pages are the complaints, with the items to check for correction of the complaints if the fuel pump test stand has been properly maintained, leaving no test stand error. Each check is numbered, so that you may go immediately to the tabulated description of the causes and corrective action as necessary.

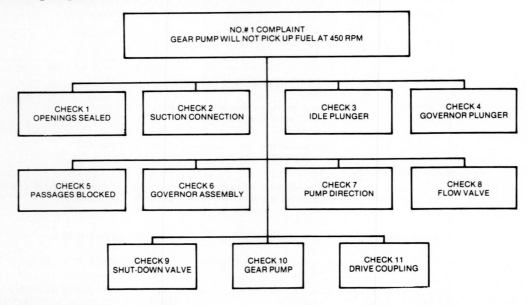

CAUSE	CORRECTION	CAUSE	CORRECTION
Check 1: Openings not sealed correctly	Seal all openings and use new gaskets where necessary.	Check 5: Blocked fuel passages	Clean fuel passages so that they all are open.
Check 2: Suction connection not tight or is damaged	Tighten suction connection or replace if mutilated.	Check 6: Faulty governor assembly	Check governor assembly for proper assembly.
Check 3: Idle plunger dirty	Check face of idle plunger (pressure control button) for any foreign material.	Check 7: Pump turning wrong direction	Check pump for right- or left-hand rotation and set test stand accordingly.
Idle plunger worn	Change idle plunger to give a square fit with governor plunger (use the same idle plunger code number).	Check 8: Flow valve not open	Open test stand flow valve to allow fuel to enter gear pump.
Check 4: Governor plunger dirty	Clean idle plunger mating face of foreign material.	Check 9: Shutdown valve not open	Open shutdown valve on top of fuel pump.
Governor plunger worn	Change governor plunger to give a square fit with idle plunger.	Check 10: Gear pump worn	Replace gear pump if it will not deliver required flow.
		Check 11: Drive coupling not in mesh	Mesh fuel pump and test stand drive couplings.

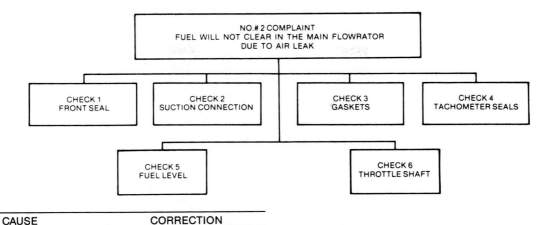

NO.# 2 COMPLAINT
FUEL WILL NOT CLEAR IN THE MAIN FLOWRATOR
DUE TO AIR LEAK

CHECK 1
FRONT SEAL

CHECK 2
SUCTION CONNECTION

CHECK 3
GASKETS

CHECK 4
TACHOMETER SEALS

CHECK 5
FUEL LEVEL

CHECK 6
THROTTLE SHAFT

CAUSE	CORRECTION
Check 1: Front seal leakage. This can be determined by covering "weep hole" in front cover with Lubriplate®, which stops air entrainment.	Remove fuel pump from test stand, then remove front cover and install new seals in cover.
Check 2: Suction connection not tight or is damaged. This can be determined by pouring lube oil over suction connection.	Tighten suction connection or replace if mutilated.
Check 3: Main housing or spring pack housing gasket leaking air.	Replace gaskets as required.

CAUSE	CORRECTION
Check 4: Tachometer drive seals leaking. Check by pouring lube oil over tachometer drive housing.	Remove the fuel pump from the test stand and disassemble the pump enough to replace tachometer drive oil seal in main housing.
Check 5: Fuel level in test stand reservoir low.	Fill fuel reservoir with Cummins test oil.
Check 6: Throttle shaft O-rings or housing leakage, determined by pouring fuel oil over the housing.	Replace O-ring on the throttle shaft or replace the housing if leaking.

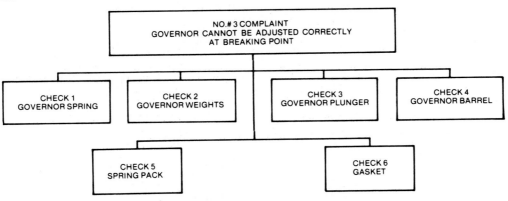

NO.# 3 COMPLAINT
GOVERNOR CANNOT BE ADJUSTED CORRECTLY
AT BREAKING POINT

CHECK 1
GOVERNOR SPRING

CHECK 2
GOVERNOR WEIGHTS

CHECK 3
GOVERNOR PLUNGER

CHECK 4
GOVERNOR BARREL

CHECK 5
SPRING PACK

CHECK 6
GASKET

CAUSE	CORRECTION
Check 1: Governor spring incorrect due either to wear or incorrect governor spring	Replace spring if worn beyond limits.
Check 2: Loose or broken weights; Broken weld, weight pins, or carrier Governor weights incorrect for that specific pump	Replace with new parts as necessary. Governor weights, of the correct weight, should be installed (heavy or shaved).
Check 3: Governor plunger improper fit in governor barrel	Refit the governor plunger to the barrel. This usually requires a plunger one or two classes larger than previously used and must be lapped to fit with no. 80 fine grit lapping compound. Remove all lapping compound after use.

CAUSE	CORRECTION
Sheared governor plunger drive tangs	Replace drive tangs on plunger assembly.
Check 4: Governor barrel not located in housing correctly, preventing fuel passages from lining up	Line up the fuel passages so as not to restrict fuel flow. This may be done by heating the housing in an oven at 300°F (149°C) and removing the barrel and then reinstalling in the housing.
Governor barrel not pinned into position	Make sure that fuel passages are lined up and install the pin into the governor barrel.
Check 5: Spring pack lock ring out of position	The lock ring must be in the groove to adjust the governor correctly.
Check 6: Gasket leakage between the fuel pump housing and gear pump	The gasket should be replaced or relocated. Correct gasket must be used.

471

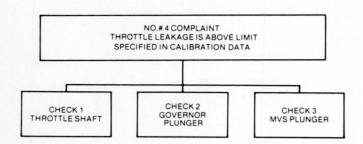

CAUSE	CORRECTION
Check 1: Throttle shaft scored or incorrect fit in throttle sleeve	Install the next size larger throttle shaft, if necessary. Fit to bore must be free without sticking tendency when rotating or moving in or out of bore by hand. If oversize shaft does not correct leakage, send the housing to rebuild center.
Check 2: Governor plunger incorrect fit in governor barrel	Install the next size larger plunger. Fit to bore must be free without sticking tendency when rotating or moving in or out of the bore by hand.
Check 3: Leakage past MVS plunger if MVS is used	Install the next size larger plunger or remove shims between the snap ring and governor housing.

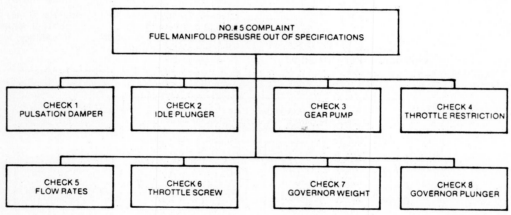

CAUSE	CORRECTION
Check 1: Low-fuel manifold pressure	Replace fractured pulsation damper diaphragm.
Check 2: Fuel manifold pressure too high or too low because of incorrect idle plunger (button) or surface finish	Replace idle plunger (button) with correct plunger if incorrect plunger was used. Polish surface of plunger if rough burrs or chipped areas are found on surface of plunger.
Check 3: Gear pump fails to obtain delivery and pressure	Replace gear pump.
Check 4: Wrong throttle restriction	Set throttle restriction to correct values.
Check 5: Test stand set at wrong flow rate	Set test stand at flow rate indicated in calibration data pertinent to fuel pump being calibrated.
Check 6: Throttle screw out of adjustment	Adjust throttle screw.
Check 7: Governor weight carrier assembly incorrect or faulty	Replace with correct new governor weight carrier assembly.
Check 8: Scored governor plunger	Replace with new governor plunger and lap to fit.

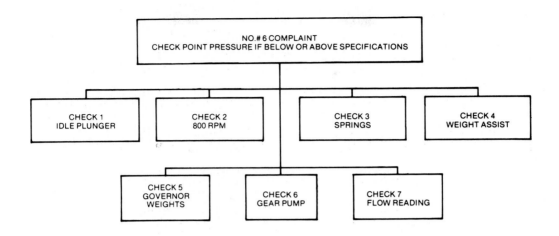

CAUSE	CORRECTION
Check 1: Idle plunger or governor plunger rough or has voids	Polish surface with oil stone or replace if necessary.
Check 2: Check lowest rpm checkpoint	Check lowest checkpoint under complaint no. 7 to be sure that it is within specifications before proceeding.
Check 3: Incorrect torque or governor spring	Remove front cover and check for proper torque or governor spring and free length of spring. Shim torque spring.
Check 4: Incorrect weight assist protrusion	Make correct weight assist setting by proper shimming or replace front cover assembly.
Check 5: Incorrect weights or worn weight carrier assembly	Replace with correct new weight carrier assembly.
Check 6: Gear pump delivery low	Check gear pump delivery.
Check 7: Flow reading incorrect	Adjust flow-meter valve.

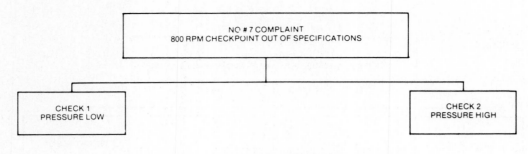

CAUSE	CORRECTION
Check 1: Checkpoint pressure too low	If weight assist protrusion is within specifications, one or more shims may be added to assembly.
Check 2: Checkpoint pressure too high	Remove weight assist shims to decrease pressure. If no shims can be removed, install new weight assist assembly or front cover.

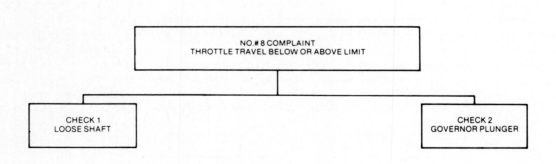

CAUSE	CORRECTION
Check 1: Throttle shaft too loose	Remove and check fit in throttle sleeve.
Check 2: Pressure drop across throttle shaft over 20 percent	Check fit of plunger and install correct size plunger to give correct pressure drop for degree of throttle travel.
	Note: Be sure to re-mark throttle or governor sleeve if a different size shaft of plunger has been installed.

PT (type G) FUEL SYSTEM TROUBLESHOOTING

POSSIBLE CAUSE	Acceleration Slow	Air Leaks	Carboned Valves, Injector Cups	Deceleration Slow	Failure To Pick Up Fuel	Fuel Consumption Excessive	Fuel Pressure High	Fuel Pressure Low	Governed Speed High	Governed Speed Low	High Speed Surge MVS	Idle Speed Too High	Idle Surge MVS	Idle Undershoot MVS	Low Power	Rough Operation	Smoke Black, Low Speed	Throttle Leakage Excessive	Wear Rate High
Air Signal Attenuator Filter Plugged	●										●		●		●	●			
Air Leaks	●	●											●		●	●	●	●	●
Aneroid Not Opening	●			●	●			●		●					●		●	●	
Aneroid Stuck Open	●				●			●					●						
Aneroid Valve Stuck Open	●				●			●					●						
Cooling Line By-Passing					●														
Cranking Speeds Slow					●														
Filter Suction Restricted	●				●			●							●				
Fuel Dirty	●		●												●		●		
Fuel With Water	●				●														●
Fuel, Wrong Type	●							●									●		●
Gear Pump Worn	●			●				●							●				
Governor Plunger Chamfer, Inadequate											●								
Governor Plunger Scored					●						●		●						●
Governor Plunger, Wrong/Worn/Sticking	●									●	●		●	●	●				●
Governor Plunger, Worn/Scored	●			●							●		●		●				●
Governor Spring Shims Low									●										
Governor Spring Shims High									●										
Governor Weights Incorrect (Heavy)	●								●	●	●			●					
Governor Weight, Pin Wear									●				●				●		
High Speed Spring Shimming Wrong	●								●										
Idle Plunger (Button) Wrong	●			●								●							
Idle Plunger Spring Weak												●	●						
Idle Spring Wrong												●	●	●					
Injector Adjustment Loose	●					●									●	●	●		●
Injector Cup Cracked, Wrong, Damaged			●			●										●	●		
Injector Flow High	●					●											●		
Incorrect Injector	●			●		●									●		●		
Injector Orifice Size Wrong				●		●		●							●		●		
Injector Plunger Worn																	●		
Pressure Valve Failure	●	●																	
Reversed Rotation, Drive Failure					●														
Screw Adjust, Incorrect							●	●	●	●	●	●		●		●		●	
Shutoff Valve Restriction								●				●	●						
Speed Settings, Unmatched auto/MVS or VS									●	●	●			●					
Spring Fatigue									●					●					
Throttle Leakage Excessive					●			●				●	●					●	
Throttle Linkage	●			●				●		●					●		●		
Throttle Shaft Restricted								●					●		●	●			
Throttle Shims Excessive				●				●										●	
Throttle Shims Insufficient	●							●		●									
Torque Spring Wrong	●				●			●		●		●	●		●				
Weight Assist Setting High									●		●		●				●		
Weight Assist Set Wrong								●	●	●	●		●	●			●		
Torque Limiting Valve Coil Not Energized	●									●	●					●			
Torque Limiting System (Air Valve) Switch Closed									●				●						

PART 8   INJECTOR SERVICE

Locating a Faulty Injector

A faulty or misfiring injector can be located as follows.

1. Run the engine at the speed where the problem occurs.

2. Hold one injector at a time down on its seat by pushing down on the rocker arm at the injector end.

3. Listen for a change in the sound of the engine. A good injector held down will noticeably change the sound of the engine. A bad injector will have little or no effect on the engine when held down.

Injector Removal

1. Loosen the injector by adjusting locknut on the rocker arm and backing out the adjusting screw.

2. Push the rocker arm down against the injector and remove the push rod.

3. Remove the injector link and hold-down bolts.

4. Use the appropriate injector puller and remove the injector from the head.

Testing Injectors

Cummins injectors are tested on a special test stand designed for this purpose. Procedures vary somewhat among different makes of testers, therefore instructions given for any particular tester must be

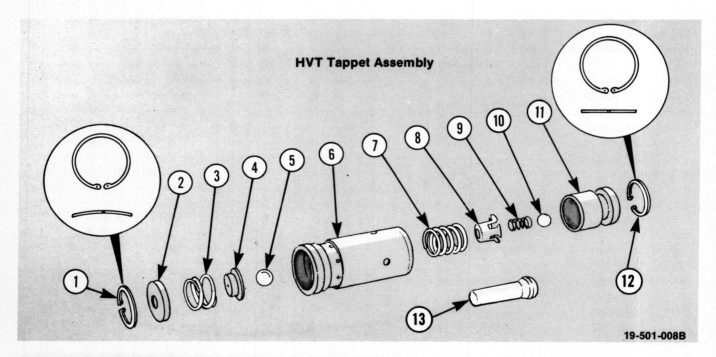

Ref. No.	Description	Ref. No.	Description	Ref. No.	Description
(1)	Lower Retaining Ring (Bowed)	(6)	Sleeve	(10)	Inlet Check Ball
		(7)	Plunger Return Spring	(11)	Plunger
(2)	Socket			(12)	Upper Retaining Ring (Flat)
(3)	Load Cell Spring	(8)	Spring Retainer		
(4)	Ball Guide	(9)	Inlet Check Ball Spring	(13)	Upper Injector Link

FIGURE 24–18. Hydraulic valve tappet for Cummins Hydraulic variable timing system. *(Courtesy of Cummins Engine Company)*

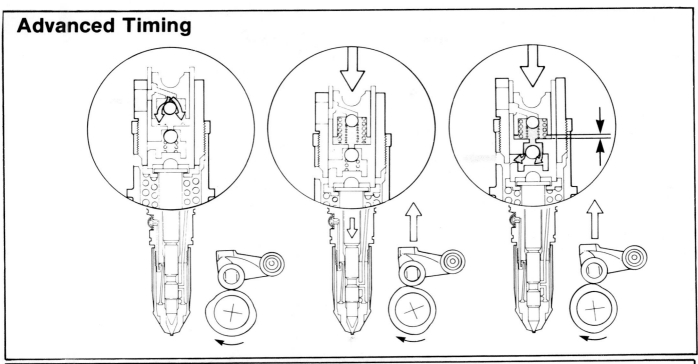

Advanced Timing

Retarded Timing

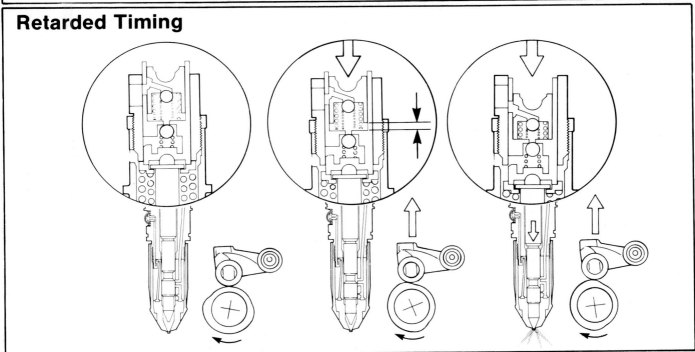

FIGURE 24-19. Oil flow—advanced and retarded timing. *(Courtesy of Cummins Engine Company Inc.)*

Advanced Timing

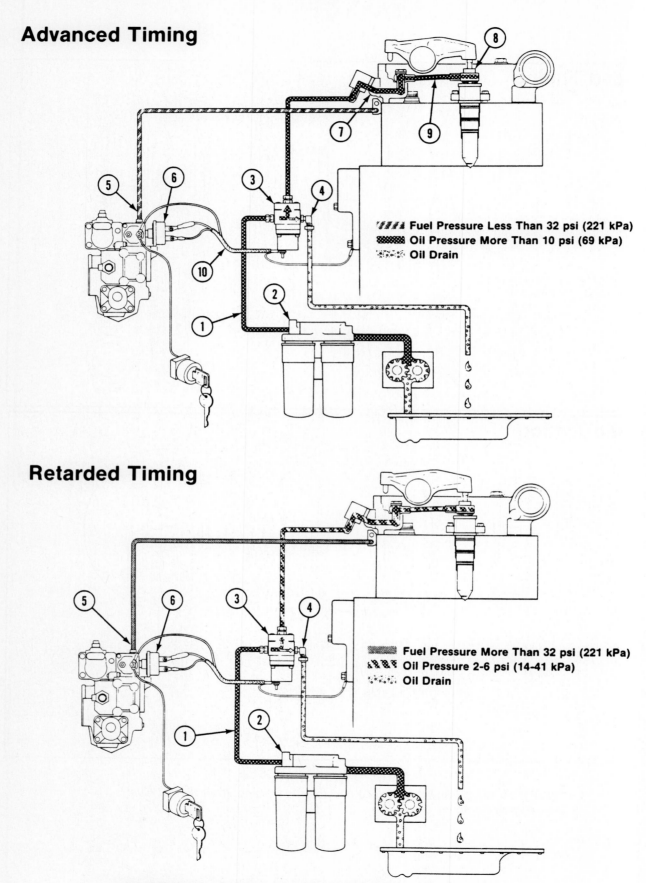

Fuel Pressure Less Than 32 psi (221 kPa)
Oil Pressure More Than 10 psi (69 kPa)
Oil Drain

Retarded Timing

Fuel Pressure More Than 32 psi (221 kPa)
Oil Pressure 2-6 psi (14-41 kPa)
Oil Drain

FIGURE 24-20. *(Courtesy of Cummins Engine Company Inc.)*

FIGURE 24-21. Removing or installing injector. *(Courtesy of Cummins Engine Company Inc.)*

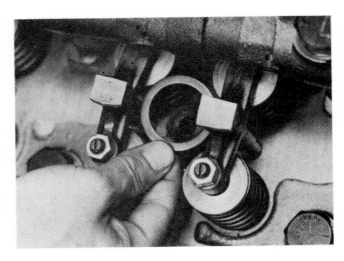

FIGURE 24-22. Installing injector retaining ring. *(Courtesy of Cummins Engine Company Inc.)*

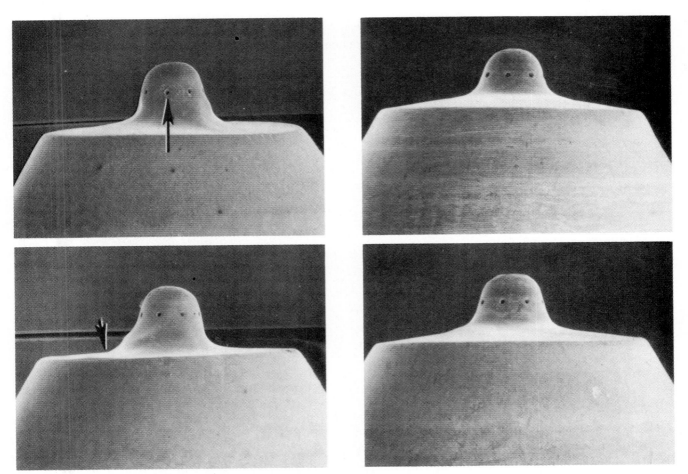

FIGURE 24-23. Typical injector cup and tip problems. *(Courtesy of Cummins Engine Company Inc.)*

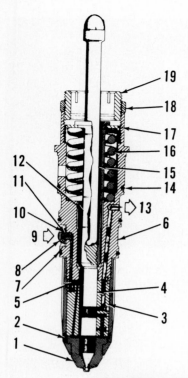

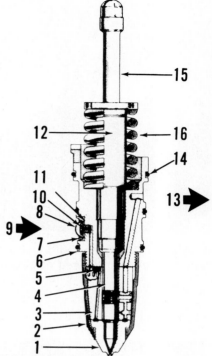

1. Cup
2. Cup Retainer
3. Barrel
4. Plunger
5. Check Ball
6. Adapter
7. Screen Clip
8. Fuel Screen
9. Fuel In
10. Orifice Plug
11. Orifice Gasket
12. Coupling
13. Fuel Out
14. O-ring
15. Link
16. Spring
17. Retainer
18. Locknut
19. Top Stop Screw

FIGURE 24-24. Cummins injector components. *(Courtesy of Cummins Engine Company Inc.)*

followed for the tests to be valid. Follow the equipment manufacturer's instructions for testing injectors.

The following tests are performed on Cummins injectors (See Fig. 24-25 to 24-28):

1. Check for sticking plunger.
2. Check spray pattern.
3. Check injector leakage at
   (a) plunger to barrel
   (b) plunger to cup
   (c) check ball
4. Calibration and flow testing.

Injectors that do not pass these tests must be disassembled, cleaned, inspected, repaired, tested, and calibrated according to Cummins service manual procedures and specifications.

Injector Installation

1. Install new O rings on the injector.
2. Lubricate the O rings with clean oil.
3. Install and align the injector in the cylinder head.

FIGURE 24-25. Checking for sticking injector plunger. *(Courtesy of Cummins Engine Company Inc.)*

FIGURE 24-26. Testing injector spray pattern. *(Courtesy of Cummins Engine Company Inc.)*

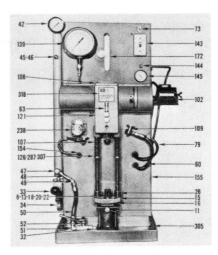

FIGURE 24-28. Injector flow tester-calibrator. *(Courtesy of Cummins Engine Company Inc.)*

PART 9   ADJUSTING INJECTORS

1. Turn the crankshaft in the direction of rotation until No. 1–6 VS mark on the vibration damper aligns with the pointer on the front cover. In this position, both the intake and exhaust valves must be closed for cylinder No. 1. If not, turn the crankshaft one revolution, and align the mark with the pointer. Begin the adjustment on cylinder No. 1 (both intake and exhaust valves closed) and follow the firing order. (See Fig. 24–29 to 24–31.)

4. Use injector removing–installing tool to seat injector in sleeve.

5. Install retaining ring, clamp, and capscrew. Tighten capscrew to specified torque.

6. Install the link and push rod.

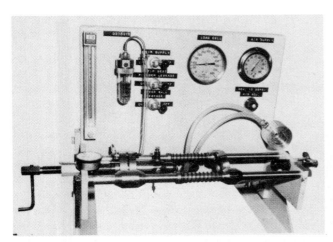

FIGURE 24-27. Injector leakage tester. *(Courtesy of Cummins Engine Company Inc.)*

FIGURE 24-29. Injection timing fixture mounted in place. *(Courtesy of Cummins Engine Company Inc.)*

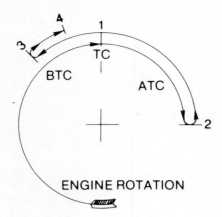

ENGINE ROTATION

FIGURE 24-30. Engine injection timing procedure. *(Courtesy of Cummins Engine Company Inc.)*

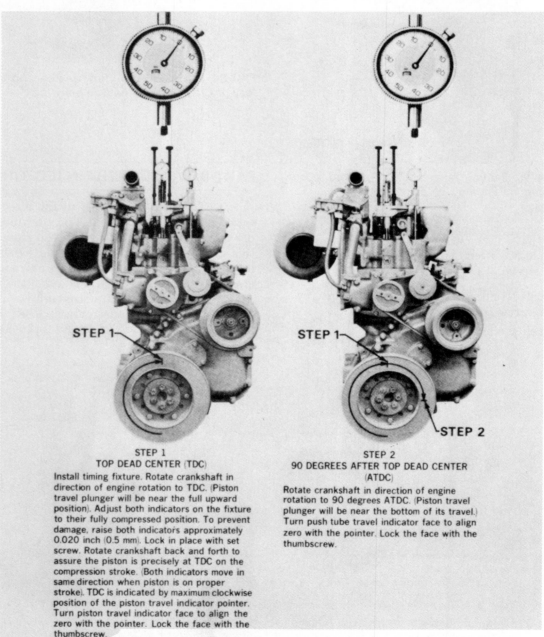

STEP 1

STEP 1

STEP 1

STEP 2

STEP 1
TOP DEAD CENTER (TDC)

Install timing fixture. Rotate crankshaft in direction of engine rotation to TDC. (Piston travel plunger will be near the full upward position). Adjust both indicators on the fixture to their fully compressed position. To prevent damage, raise both indicators approximately 0.020 inch (0.5 mm). Lock in place with set screw. Rotate crankshaft back and forth to assure the piston is precisely at TDC on the compression stroke. (Both indicators move in same direction when piston is on proper stroke). TDC is indicated by maximum clockwise position of the piston travel indicator pointer. Turn piston travel indicator face to align the zero with the pointer. Lock the face with the thumbscrew.

STEP 2
90 DEGREES AFTER TOP DEAD CENTER
(ATDC)

Rotate crankshaft in direction of engine rotation to 90 degrees ATDC. (Piston travel plunger will be near the bottom of its travel.) Turn push tube travel indicator face to align zero with the pointer. Lock the face with the thumbscrew.

FIGURE 24-31. Cummins injector timing. *(Courtesy of Cummins Engine Company Inc.)*

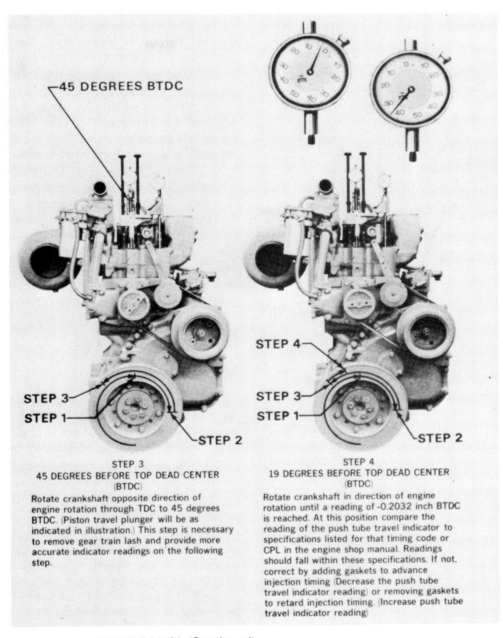

STEP 3
45 DEGREES BEFORE TOP DEAD CENTER
(BTDC)
Rotate crankshaft opposite direction of
engine rotation through TDC to 45 degrees
BTDC. (Piston travel plunger will be as
indicated in illustration.) This step is necessary
to remove gear train lash and provide more
accurate indicator readings on the following
step.

STEP 4
19 DEGREES BEFORE TOP DEAD CENTER
(BTDC)
Rotate crankshaft in direction of engine
rotation until a reading of -0.2032 inch BTDC
is reached. At this position compare the
reading of the push tube travel indicator to
specifications listed for that timing code or
CPL in the engine shop manual. Readings
should fall within these specifications. If not,
correct by adding gaskets to advance
injection timing (Decrease the push tube
travel indicator reading) or removing gaskets
to retard injection timing. (Increase push tube
travel indicator reading)

FIGURE 24-31. (Continued)

Note: The valve set marks on the damper are sometimes hard to see. Make sure
the correct mark is at the pointer before adjusting the valves at that cylinder.

2. Set up the ST-1170 Dial Indicator and support with the indicator extension on top of the injector plunger at the No. 1 cylinder.

3. Make sure the indicator extension is tight
in the indicator stem and is not touching the rocker
lever.

4. Turn the injector lever adjusting screw
clockwise until the injector plunger hits the cup.

5. Turn the adjusting screw counterclockwise
1/2 turn. Turn the adjusting screw clockwise again
until the injector plunger touches the injector cup.
Set the dial on the indicator to zero.

6. Turn the adjusting screw counterclockwise
until the dial indicator shows a total reading as indicated in service manual. Use the correct value according to the rocker lever ratio.

Custom Rated NHC-250: Three horsepower ratings of 250 hp @ 2100 rpm, 225 hp @ 2100 rpm, and 225 hp @ 1950 rpm are available to meet specific load and road conditions. Change your power rating by a simple button change in the PT fuel pump. 1950 rpm setting also requires a governor adjustment.

225 H.P. @ 1950 R.P.M.	225 H.P. @ 2100 R.P.M.	250 H.P. @ 2100 R.P.M.

FIGURE 24-32. Method of changing power ratings of Cummins engine. *(Courtesy of Cummins Engine Company Inc.)*

7. Tighten the adjusting screw locknuts to specified torque.

8. Actuate the injector plunger several times as a check of the adjustment.

Note: The same engine position used for setting injectors is used to adjust the intake and exhaust valves on the same cylinder.

9. Adjust the valve clearance by putting the correct feeler gauge between the rocker lever and crosshead contact pads. Turn the adjusting screw down until the rocker lever touches the feeler gauge. See the service manual for valve clearance values.

Caution: Make sure the valve tappet rollers are against the lobe on the camshaft before adjusting the valves.

10. Continue through the firing order until all the injectors and valves have been adjusted.

PART 10   FUEL PUMP SERVICE

Checking and Adjusting the Fuel Pump on the Engine

Before making the fuel system checks or adjustments on the engine, make sure of the following:

1. The engine must be at the operating temperature. The fuel temperature is not above 110°F [43°C].

2. The engine parts are the same as those in the Control Parts List and in good condition. The timing, valves and injectors are correctly adjusted.

3. The instruments (gauges and tachometers) must have high accuracy.

4. The control linkage of the vehicle throttle is adjusted for full throttle travel. When released, the throttle is stopped by the throttle adjustment screw (throttle leakage adjustment screw).

Note: The control linkage of the vehicle throttle must have a maximum throttle stop. When the fuel pump is in the full throttle position there must not be any pressure on the throttle shaft.

5. When the fuel pump is correctly calibrated, very little adjustment is required after the installation on the engine. A small adjustment of the idle setting and the fuel rail pressure is acceptable.

Governor Settings

Idle Speed

1. After the fuel pump installation, the engine must be run to get all of the air out of the fuel system. The engine must be up to the operating temperature (165°F [74°C] oil temperature or higher).

Note: The idle speed adjustment must never be made on a cold engine.

2. Remove the pipe plug from the spring pack cover.

3. The idle adjustment screw is held in position by a spring clip. Turn the screw in to increase or out to decrease the speed. Use Tool No. ST-984 o 3375981 to adjust the idle speed while the engine is running. This tool will not let the spring pack cover leak when the idle is adjusted.

4. Replace the pipe plug when the idle speed is correct.

Note: Beginning in 1978, the engine speed is on the engine dataplate. It is under the exhaust emission control information on the dataplate.

5. On the VS, SVS and MVS and the PT (type H) the maximum and idle adjustment screws are on the governor cover.

a. To adjust the idle, loosen the rear idle adjustment screw locknut.

b. Turn the adjustment screw in or out to get the speed requirement. Do not set the SVS governor idle in a power takeoff application, at less than 1100 rpm.

c. Tighten the adjustment screw locknut immediately after the adjustment to stop the air from getting in the fuel pump.

6. The MVS governor fuel pumps with a stub shaft are set as follows. Adjust the idle on the automotive governor. The MVS governor idle screw will be adjusted 10 to 12 psi [69 to 83 kPa] above the specifications.

a. If a surge occurs at idle, it is acceptable to turn the standard idle screw all the way in. Set the VS or MVS governor idle screw to specifications.

b. If the surge does not stop on the MVS, replace the Part No. 153240 Idle Spring with Part No. 201116. Use a Part No. 154461 Series Plunger.

c. If the surge occurs on a VS pump, change the VS plunger. If the upper governor plunger is Part No. 213610, replace it with a plunger Part No. 212350. This plunger has a 10 degree chamfer.

7. Some problems with high engine vibrations have occurred at engine idle speeds in vehicles that also have power takeoffs. This is found in cement mixture applications. Adjust the idle for a smooth engine operation.

8. The idle speeds in the service manual are to be used as a reference point. Small changes can be made from these speeds. Large changes in the idle speed can cause other engine or application problems.

9. Problems like difficult gear changes will be found with high idle speeds. Moving a load from a stop can be a problem if the idle speeds are adjusted too low.

High Speed

1. The tachometer and fuel pressure gauge must be of high accuracy. The engine fuel system must have all of the air removed. The engine must be at the operating temperature.

2. The correct method of checking the governor setting is to "load" the engine on an engine or chassis dynamometer.

3. The maximum engine speed is adjusted by adding or removing the shims under the high speed

governor spring. Normally, this adjustment is made on the fuel pump test stand as the fuel pump is calibrated. It usually does not need to be changed on the engine.

Fuel Pump Removal, Installation, and Testing

The fuel pump should be removed only after it has definitely been established through testing that it is at fault. Fuel pump removal and installation is a simple procedure since no pump timing is required.

To remove the fuel pump, first clean pump and surrounding area thoroughly. Disconnect the fuel lines and cap all openings to prevent the entry of dirt. Disconnect the throttle linkage. Remove the pump mounting bolts and then the pump. To install the pump, simply follow the above procedures in reverse order, making sure that all connections and mounting bolts are tightened to specifications and that there are no fuel leaks.

A faulty fuel pump must be disassembled, cleaned, inspected, repaired, and reassembled. After assembly the pump is installed on a test stand for testing and calibration. The procedures and specifications of the test equipment manufacturer and the appropriate Cummins service manual must be applied during pump overhaul, testing, and calibration.

The power rating of the engine can be changed by changing to a different fuel control button. (See Fig. 24-32.)

PART 11   SELF-CHECK

1. The Cummins PT fuel injection system is used only on Cummins engines. True or False.

2. What three main units make up the Cummins PT pump?

3. What are the four functions of the PT pump?

4. Name the three types of governors used on PT pumps.

5. The air fuel control device restricts fuel delivery during engine acceleration. True or False.

6. Two factors control the amount of fuel injected by the PT system. What are they?

7. Actual injection does not start with the beginning of the downward movement of the injector plunger. Why not?

8. The Cummins mechanical variable timing system changes injection timing by using air pressure to change the position of the _____ relative to the _____.

9. List five causes of low fuel pressure in the PT fuel system.

10. List three reasons for excessive throttle leakage in the PTG fuel pump.

11. What four tests must be performed on Cummins injectors?

12. Idle and high speed governor adjustments must be made with the engine at operating temperature. True or False.

Chapter 25

Caterpillar Fuel Injection Systems

Performance Objectives

After thorough study of this chapter, the appropriate training models and service manuals, and with the necessary tools and equipment, you should be able to do the following with respect to diesel engines equipped with Caterpillar fuel injection systems:

1. Complete the self-check questions with at least 80 percent accuracy.

2. Describe the basic construction and operation of Caterpillar fuel injection systems, including the compact housing, sleeve metering, and unit injection systems and including the following for each system:

    (a) method of fuel pumping and metering

    (b) delivery valve action

    (c) lubrication

    (d) injection timing control

    (e) governor types

3. Diagnose basic fuel injection system problems and determine the needed correction.

4. Remove, clean, inspect, test, measure, repair, replace, and adjust fuel injection system components as needed to meet manufacturer's specifications.

Caterpillar diesel engines are equipped with one of the following four fuel injection systems depending on engine model and size.

This chapter is based on information provided through the courtesy of the Caterpillar Tractor Company.

1. Flanged body injection pump
2. Compact housing injection pump
3. Sleeve metering injection pump
4. Unit injector system

Each of these systems is described separately in this chapter.

PART 1 FLANGED BODY FUEL INJECTION PUMP

The injection pump plungers (3) and the lifters (8) are lifted by cams on the injection pump camshaft (9) and always make a full stroke (Figure 25-1). Each pump (1) measures the amount of fuel to be injected into its respective cylinder and delivers it to the fuel injection valve. The amount of fuel pumped per stroke can be varied by turning the plunger in the barrel. The plunger is turned by the governor action through the rack (6) which meshes with the gear segment (4) on the bottom of the pump plunger.

Figure 25-2 illustrates the functioning of an injection pump as the plunger makes a stroke.

In A the plunger is down and the inlet port (2) is uncovered. Fuel flows into the space above the plunger through the slot and into the recess around the plunger.

In B the plunger has started up and the port is covered. The fuel is trapped and will be forced through the check valve, fuel line and injection valve as the plunger moves upward.

In C the plunger has risen until the port is uncovered by the recess in the plunger. The fuel can now escape back through the port into the fuel manifold and injection will cease.

It will be noted that the recess in the pump plunger forms a helix around the upper end of the plunger. Figure 25-3 illustrates how rotating the pump plunger affects the quantity of fuel injected.

In D the plunger has been rotated into the shutoff position. The slot connecting the top of the plunger with the recess is in line with the port; therefore, no fuel can be trapped and injected.

In E the plunger has been rotated into the idling position. The narrow part of the plunger formed by the helix will cover the port for only a short part of the stroke. This permits only a small amount of the fuel to be injected per stroke.

In F the plunger has been rotated into the full load position. The wide part of the plunger formed by the helix covers the port for a longer part of the stroke. This permits a larger amount of fuel to be injected per stroke.

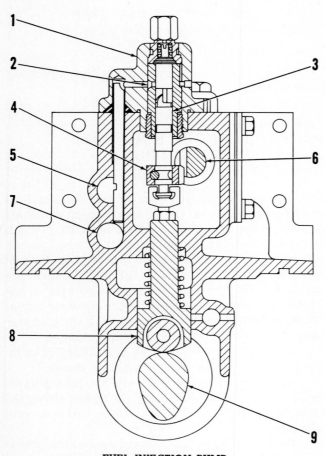

FUEL INJECTION PUMP

1–Pump. 2–Inlet port. 3–Pump plunger. 4–Gear segment.
5–Bleeder manifold. 6–Rack. 7–Fuel passage. 8–Lifter.
9–Camshaft.

FIGURE 25-1. Cross section of flanged body injection pump. *(Courtesy of Caterpillar Tractor Company)*

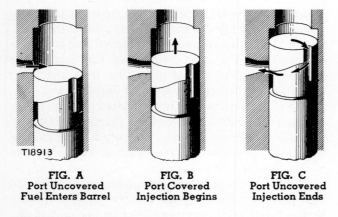

FIG. A
Port Uncovered
Fuel Enters Barrel

FIG. B
Port Covered
Injection Begins

FIG. C
Port Uncovered
Injection Ends

FIGURE 25-2. *(Courtesy of Caterpillar Tractor Company)*

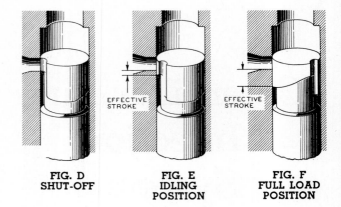

FIG. D
SHUT-OFF

FIG. E
IDLING
POSITION

FIG. F
FULL LOAD
POSITION

FIGURE 25-3. *(Courtesy of Caterpillar Tractor Company)*

PART 2   COMPACT HOUSING (SCROLL) INJECTION PUMP

In this fuel injection pump design there is one injection pump for each cylinder. The injection pumps are in the pump housing on the top front of the engine. The injection valves are in the precombustion chambers or adapters (for engines with direct injection) under the valve covers. (See Fig. 25-4 and 25-5.)

The transfer pump brings fuel from the fuel tanks through the primary filter and sends it through the priming pump, main filter and to the manifold of the injection pump housing. The fuel in the manifold of the injection pump housing goes to the injection pumps. The injection pumps are in time with the engine and send fuel to the injection valves under high pressure.

Some of the fuel in the manifold is constantly sent back through the return line to the fuel tank to remove air from the system. On the outlet elbow of the injection pump there is a damper to reduce shock loads, and a restriction orifice to keep fuel pressure high and to control the amount of fuel that goes back to the fuel tank.

The fuel priming pump is used to fill the system with fuel and to remove air from the fuel filter, fuel lines and components.

The transfer pump has a bypass valve and a check valve. The bypass valve (lower side) controls the maximum pressure of the fuel. The extra fuel goes to the inlet of the pump. The check valve allows the fuel from the tank to go around the transfer pump gears when the priming pump is used.

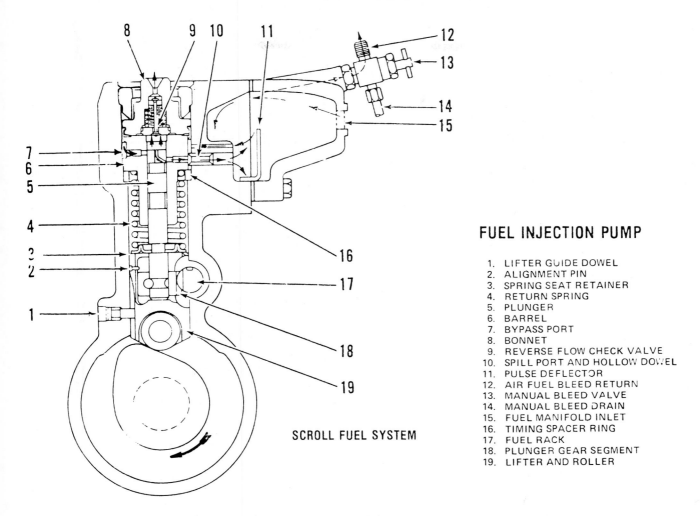

FUEL INJECTION PUMP

1. LIFTER GUIDE DOWEL
2. ALIGNMENT PIN
3. SPRING SEAT RETAINER
4. RETURN SPRING
5. PLUNGER
6. BARREL
7. BYPASS PORT
8. BONNET
9. REVERSE FLOW CHECK VALVE
10. SPILL PORT AND HOLLOW DOWEL
11. PULSE DEFLECTOR
12. AIR FUEL BLEED RETURN
13. MANUAL BLEED VALVE
14. MANUAL BLEED DRAIN
15. FUEL MANIFOLD INLET
16. TIMING SPACER RING
17. FUEL RACK
18. PLUNGER GEAR SEGMENT
19. LIFTER AND ROLLER

SCROLL FUEL SYSTEM

FIGURE 25-4. Cross-sectional view and parts identification of Caterpillar Scroll (helix) type injection pump. *(Courtesy of Caterpillar Tractor Company)*

Fuel Injection Pump Operation

The rotation of the cams on the camshaft causes the lifters and pump plungers to move up and down. The stroke of each pump plunger is always the same. The force of springs holds the lifters against the cams of the camshaft.

The pump housing is in line or "V" shape (similar to the engine cylinder block), with four pumps on each side.

The pump plunger is designed with a helix or scroll type of relief passage, similar to the flanged body pumps, to control fuel metering.

When the pump plunger is down, fuel from the fuel manifold goes through inlet passage and fills the chamber above pump plunger. As the plunger moves up it closes the inlet passage.

The pressure of the fuel in the chamber above the plunger increases as the plunger rises until it is high enough to cause the check valve to open. Fuel under high pressure flows out of the check valve through the fuel line to the injection valve until the inlet passage opens into the pressure relief passage in the plunger. The pressure in the chamber decreases and the check valve closes.

The longer the inlet passage is closed the larger the amount of fuel which will be forced through the check valve. The period for which the inlet passage is closed is controlled by the pressure relief passage. The design of the passage makes it possible to change the inlet passage closed time by rotation of the plunger. When the governor moves the fuel racks they move the gears that are fastened to the plungers. This causes a rotation of the plungers.

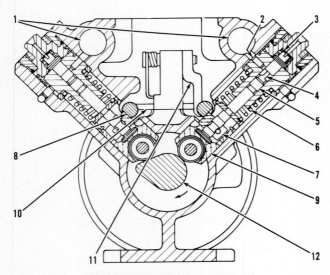

**CROSS SECTION OF THE FUEL
INJECTION PUMP HOUSING**

1. Fuel manifold.   2. Inlet passage.   3. Check valve.   4.
Pressure relief passage.   5. Pump plunger.   6. Spring.   7.
Gear.   8. Fuel rack (left).   9. Lifter.   10. Link.   11. Lever.
12. Camshaft.

FIGURE 25-5. V-type fuel injection pump. *(Courtesy of
Caterpillar Tractor Company)*

The governor is connected to the left rack. The
spring load on the lever removes the play between
the racks and link. The fuel racks are connected by
linkage. They move in opposite directions (when one
rack moves in, the other rack moves out).

Fuel Injection Valves

Fuel, under high pressure from the injection pumps,
is sent through the injection valves. The injection
valves change the fuel to the correct fuel spray pattern for good combustion in the cylinders. See Chapter 20 for more details.

The fuel injection valves are installed in the precombustion chambers in engines equipped with precombustion chambers. An adapter takes the place
of the precombustion chamber in engines equipped
with direct injection. The precombustion chambers
or adapters are installed in the cylinder heads.

PART 3   GOVERNOR OPERATION

Hydra-Mechanical Governor

The earlier and later governors operate the same.
The earlier governors have two levers and the later

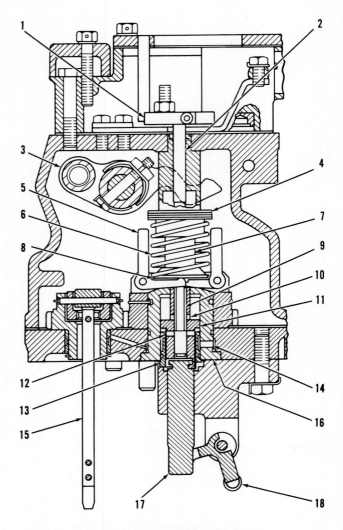

HYDRA-MECHANICAL GOVERNOR (LATER)

1. Collar.   2. Bolt.   3. Lever assembly.   4. Upper spring
seat.   5. Weights.   6. Governor spring.   7. Lower spring
seat.   8. Thrust bearing.   9. Valve.   10. Upper oil passage
in piston.   11. Piston.   12. Lower oil passage in piston.   13.
Sleeve.   14. Oil passage in cylinder.   15. Drive assembly.
16. Cylinder.   17. Pin.   18. Lever.

FIGURE 25-6. Cross section of Hydra-Mechanical governor. *(Courtesy of Caterpillar Tractor Company)*

governors have a one piece lever. The shut-off solenoid and air-fuel ratio control have been moved from
the injection pump housing to the governor housing
on the later governors. (See Fig. 25-6 and 25-7.)

The accelerator pedal, or governor control, is
connected to the control lever on the engine governor. The governor controls the amount of fuel needed
to keep the desired engine rpm.

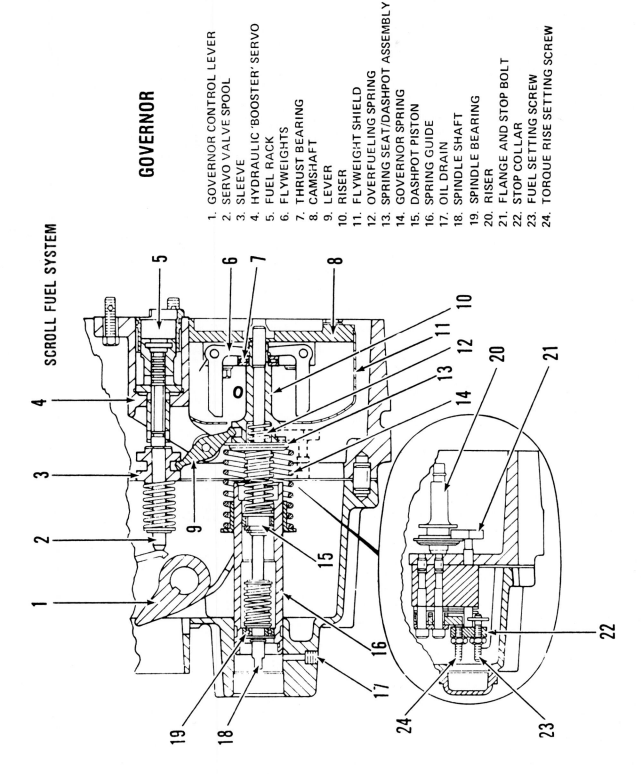

SCROLL FUEL SYSTEM

GOVERNOR

1. GOVERNOR CONTROL LEVER
2. SERVO VALVE SPOOL
3. SLEEVE
4. HYDRAULIC 'BOOSTER' SERVO
5. FUEL RACK
6. FLYWEIGHTS
7. THRUST BEARING
8. CAMSHAFT
9. LEVER
10. RISER
11. FLYWEIGHT SHIELD
12. OVERFUELING SPRING
13. SPRING SEAT/DASHPOT ASSEMBLY
14. GOVERNOR SPRING
15. DASHPOT PISTON
16. SPRING GUIDE
17. OIL DRAIN
18. SPINDLE SHAFT
19. SPINDLE BEARING
20. RISER
21. FLANGE AND STOP BOLT
22. STOP COLLAR
23. FUEL SETTING SCREW
24. TORQUE RISE SETTING SCREW

FIGURE 25-7. Caterpillar Scroll (compact housing) pump governor components. *(Courtesy of Caterpillar Tractor Company)*

The governor has governor weights driven by the engine through the drive assembly. The governor has a governor spring, valve and piston. The valve and piston are connected to the fuel rack through a pin and levers. The pressure oil for the governor comes from the governor oil pump, on top of the injection pump housing. The oil used is from the engine lubrication system. Pressure oil goes through passage and around the sleeve. The accelerator pedal, or governor control, controls only the compression of governor spring. Compression of the spring always pushes down to give more fuel to the engine. The centrifugal force (rotation) of governor weights always pulls to get a reduction of fuel to the engine. When these two forces are in balance, the engine runs at the desired rpm (governed rpm).

When the engine load increases, the engine rpm decreases and the rotation of governor weights will get slower. (The governor weights will move toward each other.) The governor spring moves valve down. This lets the oil flow from the lower passage around the valve and through the upper passage to fill the chamber behind the piston. This pressure oil pushes the piston and pin down to give more fuel to the engine. (The upper end of the valve stops the oil flow through the top of the piston, around the valve.) Engine rpm goes up until the rotation of the governor weights is fast enough to be in balance with the force of the governor spring.

When there is a reduction in engine load, there will be an increase in engine rpm and the rotation of governor weights will get faster. This will move the valve up. This stops oil flow from the lower passage and oil pressure above the piston goes out through the top, around valve. Now, the pressure between the sleeve and piston pushes the piston and pin up. This causes a reduction in the amount of fuel to the engine. Engine rpm goes down until the centrifugal force (rotation) of the governor weights is in balance with the force of the governor spring. When these two forces are in balance, the engine will run at the desired rpm (governed rpm).

To stop an engine equipped with a shutoff solenoid, turn the switch to the "OFF" position. This will cause the shutoff solenoid to move a lever that moves the fuel rack to the fuel closed position. With no fuel to the engine cylinders, the engine will stop.

Oil from the governor pump gives lubrication to the governor weight support, gear, thrust bearing, and drive gear bearing. The other parts of the governor get lubrication from "splash-lubrication" (oil thrown by other parts). Oil from the governor runs down into the housing for the fuel injection pumps.

PART 4   MECHANICAL AUTOMATIC TIMING ADVANCE

The automatic timing advance unit is installed on the front of the camshaft for the fuel injection pump and is gear driven through the timing gears. The drive gear for the fuel injection pump is connected to the injection pump camshaft through a system of weights, springs, slides, and a flange. Each of the two slides is held on the gear by a pin. The two weights can move in guides inside the flange and over the slides, but the notch for the slide in each weight is at an angle with the guides for the weight in the flange. As centrifugal force (rotation) moves the weights away from the center, against the springs, the guides in the flange and the slides on the gear make the flange turn a little in relation to the gear. Since the flange is connected to the camshaft for the fuel injection pump, the fuel injection timing is also changed. No adjustment can be made to these automatic timing advance units. (See Fig. 25–8.)

PART 5   HYDRAULIC VARIABLE TIMING UNIT

This unit is designed to provide automatic advance of injection timing in relation to the increase in engine speed. (See Fig. 25–9.)

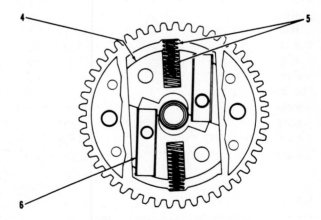

AUTOMATIC TIMING ADVANCE UNIT
4. Weights.   5. Springs.   6. Slides.

FIGURE 25–8. Caterpillar automatic mechanical injection pump advance unit. This unit is part of the engine camshaft gear, which drives the injection pump gear. *(Courtesy of Caterpillar Tractor Company)*

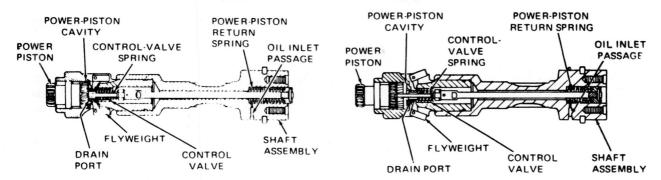

FIGURE 25-9. Caterpillar automatic hydraulic-mechanical timing advance mechanism. Centrifugal flyweights operate control valve, which controls hydraulic pressure applied to power piston. Helically splined power piston and pump drive advances injection timing as piston moves out. Engine lubricating oil provides hydraulic force. *(Courtesy of Caterpillar Tractor Company)*

Two flyweights act on a sliding control valve in opposition to spring pressure. This valve controls hydraulic pressure acting on one side of a power piston, while spring pressure acts on the other side. The power piston is splined to the pump camshaft with straight splines that allow axial movement of the power piston. At the other end, the power piston is splined to the drive shaft through helical or angled splines. The helical splines provide the advance and retard of the injection pump camshaft.

Control oil pressure is supplied from the engine lubrication system. As engine speed increases, a point is reached when the flyweights acting on the control valve shut off the oil drain back through the unit. This causes a pressure rise to act on the power piston, moving it axially on its bore. This causes the drive shaft to advance the injection pump through the helical splines which, in turn, advances injection timing. As engine speed decreases and the control valve once again opens the oil drain-back passage, the hydraulic pressure acting on the power piston decreases and spring pressure returns the power piston and pump camshaft to the retarded position. The degree of advance provided is a result of the balance of forces acting on the power piston.

PART 6   FUEL RATIO CONTROL

Operation

The fuel ratio control coordinates the movement of the fuel rack with the amount of air available in the inlet manifold, keeping exhaust smoke to a minimum. (See Fig. 25-10.)

Later controls have a manually operated override lever to allow unrestricted rack movement during cold starts. After the engine starts, the override automatically resets in the RUN position.

A collar (6) mechanically connects with the fuel rack. The head of bolt assembly (7) latches through a slot in collar (6). An air line joins the chamber

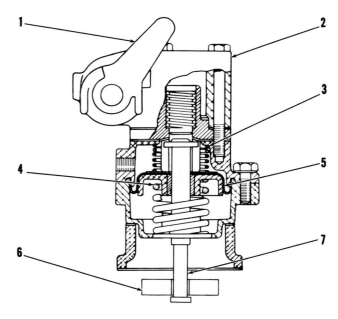

FIGURE 25-10. Fuel ratio control cross section. 1. Lever. 2. Housing. 3. Spring. 4. Spring. 5. Diaphragm. 6. Collar. 7. Bolt assembly. *(Courtesy of Caterpillar Tractor Company)*

above diaphragm (5) with the air in the engine inlet manifold.

When the operator moves the governor control to increase engine rpm, the governor spring moves collar (6), contacting the head of bolt assembly (7). The bolt assembly restricts the movement of the collar until spring (3) and the turbocharger boost of air pressure in the cover forces diaphragm, (5), spring (4) and bolt assembly (7) to relieve the bolt head restriction to the collar. This will allow the fuel rack to move toward the increased fuel position as turbocharger air pressure increases with the increase in engine rpm. The characteristics of springs (3) and (4) are such that a balance is maintained between increase of fuel, increase of load, increase of engine rpm and increase of air for combustion in the engine. The system stays in balance to assure minimum smoke and give maximum engine acceleration or load acceptance.

PART 7   SLEEVE METERING FUEL INJECTION PUMP

The sleeve metering fuel system is a pressure type fuel system. The name for the fuel system is derived from the method used to control the amount of fuel sent to the cylinders. This fuel system has an injection pump for each cylinder of the engine. It also has a fuel transfer pump on the front of the injection pump housing. The governor is on the rear of the injection pump housing.

The drive gear for the fuel transfer pump is on the front of the camshaft for the injection pumps. The carrier for the governor weights is bolted to the rear of the camshaft for the injection pumps. The injection pump housing has a bearing on each end to support the camshaft. The camshaft for the sleeve metering fuel system is driven by the timing gears at the front of the engine.

The injection pumps, lifters and rollers, and the camshaft are all inside the pump housing. The pump housing and the governor housing are full of fuel at transfer pump pressure (fuel system pressure). **Diesel fuel is the only lubrication for the moving parts in the transfer pump, injection pump housing, and the governor. The injection pump housing must be full of fuel before turning the camshaft.**

Operation of Fuel Injection Pumps

The main components of a fuel injection pump in the sleeve metering fuel system are barrel (A), plunger

(B), and sleeve (D). Plunger (B) moves up and down inside the barrel (A) and sleeve (D). Barrel (A) is stationary while sleeve (D) is moved up and down on plunger (B) to make a change in the amount of fuel for injection. (See Fig. 25–11 and 25–12.)

When the engine is running, fuel under pressure from the fuel transfer pump goes in the center of plunger (B) through fuel inlet (C) during the down stroke of plunger (B). Fuel can not go through fuel outlet (E) at this time because it is stopped by sleeve (D), (see position 1).

Fuel injection starts (see position 2) when plunger (B) is lifted up in barrel (A) enough to close fuel inlet (C). There is an increase in fuel pressure above plunger (B), when the plunger is lifted by camshaft (4). The fuel above plunger (B) is injected into the engine cylinder.

Injection will stop (see position 3) when fuel outlet (E) is lifted above the top edge of sleeve (D) by camshaft (4). This movement lets the fuel that is

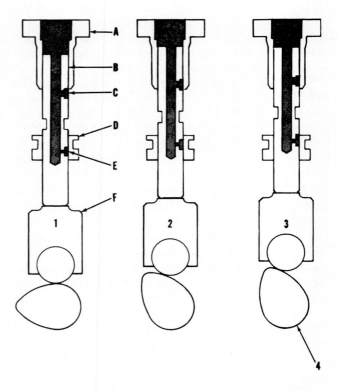

FIGURE 25–11. Fuel injection sequence. 1,2,3. Injection stroke (positions) of a fuel injection pump. 4. Injection pump camshaft. A. Barrel. B. Plunger. C. Fuel inlet. D. Sleeve. E. Fuel outlet. F. Lifter. *(Courtesy of Caterpillar Tractor Company)*

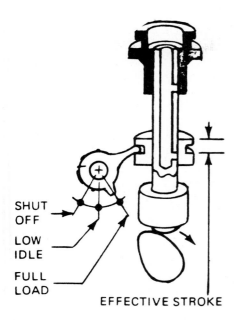

SHUT
OFF →

LOW
IDLE →

FULL
LOAD →

EFFECTIVE STROKE

FIGURE 25-12. Effective stroke of plunger is determined by vertical positioning of the sleeve. The lower the sleeve, the less fuel will be injected since the spill port is uncovered sooner. *(Courtesy of Caterpillar Tractor Company)*

above, and in, plunger (B) go through fuel outlet (E) and return to the fuel injection pump housing.

When the sleeve (D) is raised on plunger (B), fuel outlet (E) is covered for a longer time, causing more fuel to be injected in the engine cylinders. If sleeve (D) is low on plunger (B), fuel outlet (E) is covered for a shorter time, causing less fuel to be injected.

PART 8   GOVERNOR OPERATION

This fuel system has governor weights, a thrust collar and two governor springs. One governor spring is for high idle and the other governor spring is for low idle. Rotation of the shaft for governor control, compression of the governor springs, movement of connecting linkage in the governor and injection pump housing controls the amount of fuel sent to the engine cylinders.

As shown in Figure 25-13, fuel under pressure moves from fuel transfer pump (11) and fills the housing for the fuel injection pumps (14). Pressure of the fuel in the housing (14) is controlled by a bypass valve. Pressure of the fuel at FULL LOAD is

30 ± 5 psi (205 ± 35 kPa). If the pressure gets too high, the bypass valve will open to let some of the fuel return to the inlet of fuel transfer pump.

Lever (15) for the governor is connected by linkage and governor springs (25) to the sleeve control shafts (19). Any movement of lever (22) will cause a change in the position of sleeve control shafts (19).

When lever (15) is moved to give more fuel to the engine, lever (22) will put governor springs (25) in compression and move thrust collar (28) forward. As thrust collar (28) moves forward, the connecting linkage will cause sleeve control shafts (19) to turn. With this movement of the sleeve control shafts, levers (31) will lift sleeves to make an increase in the amount of fuel sent to the engine cylinders.

When starting the engine, the force of over fueling spring (27) is enough to push thrust collar (28) to the full fuel position. This lets the engine have the maximum amount of fuel for injection when starting. At approximately 400 rpm, governor weights (30) make enough force to push spring (27) together. Thrust collar (28) and spring seat (26) come into contact. From this time on, the governor works to control the speed of the engine.

When governor springs (25) are put in compression, the spring seat at the front of the governor springs will make contact with load stop lever (29). Rotation of the load stop lever moves load stop pin (17) up until the load stop pin comes in contact with the stop bar or stop screw. This stops the movement of thrust collar (28), the connecting levers, and sleeve control shafts (19). At this position, the maximum amount of fuel per stroke is being injected by each injection pump.

The carrier for governor weights (30) is held on the rear of camshaft (13) by bolts. When engine rpm goes up, injection pump camshaft (13) turns faster. Any change of camshaft rpm will change the rpm and position of governor weights (30). Any change of governor weight position will cause thrust collar (28) to move. As governor weights (30) turn faster, thrust collar (28) is pushed toward governor springs (25). When the force of governor springs (25) is balanced by the centrifugal force of the governor weights, sleeves of the injection pumps are held at a specific position to send a specific amount of fuel to the engine cylinders.

The parts of the dashpot work together to make the rpm of the engine steady. The dashpot works as piston (23) moves in the cylinder which is filled with fuel. The movement of piston (23) in the cylinder either pulls fuel into the cylinder or pushes it out. In either direction the flow of fuel is through orifice

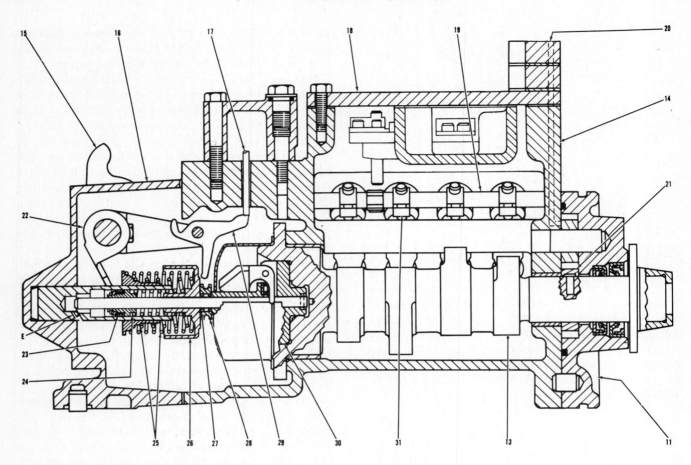

11. Fuel transfer pump. **13.** Camshaft. **14.** Housing for fuel injection pumps. **15.** Lever. **16.** Governor housing. **17.** Load stop pin. **18.** Cover. **19.** Sleeve control shafts (two). **20.** Inside fuel passage. **21.** Drive gear for fuel transfer pump. **22.** Lever on governor shaft. **23.** Piston for dashpot governor. **24.** Spring for dashpot governor. **25.** Governor springs (inner spring is for low idle: outer spring is for high idle). **26.** Spring seat. **27.** Over fueling spring. **28.** Thrust collar. **29.** Load stop lever. **30.** Carrier and governor weights. **31.** Sleeve levers. **E.** Orifice for dashpot.

FIGURE 25–13. Cross section of fuel system with dashpot governor. *(Courtesy of Caterpillar Tractor Company)*

(E). The restriction to the flow of fuel by orifice (E) gives the governor its function.

When the load on the engine decreases, the engine starts to run faster and governor weights (30) put force against springs (25). This added force puts more compression on springs (25) and starts to put spring (24) in compression. Spring (24) is in compression because the fuel in the cylinder behind piston (23) can only go out through orifice (E). The rate of flow through orifice (E) controls how fast piston (23) moves. As the fuel is pushed out of the cylinder by piston (23), the compression of spring (24) becomes gradually less. When springs (25) and spring (24) are both in compression, their forces work together against the force of weights (30). This gives the effect of having a governor spring with a high spring rate. A governor spring with a high spring rate keeps the engine speed from having oscillations during load changes.

When the load on the engine increases, the engine starts to run slower. Governor weights (30) put less force against springs (25). Springs (25) start to push seat (26) to give engine more fuel. Seat (26) is connected to piston (23) by spring (24). When seat (26) starts to move, the action puts spring (24) in tension. As piston (23) starts to move, a vacuum is made inside the cylinder. The vacuum will pull fuel into the cylinder through orifice (E). The rate of fuel flow through orifice (E) again controls how fast piston (23) moves. During this condition, spring (24) is

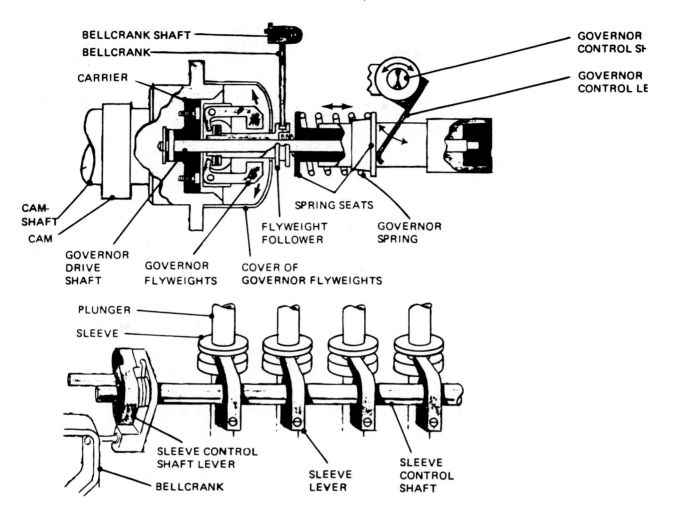

FIGURE 25-14. Caterpillar mechanical governor (top) and sleeve control shaft. The mechanical governor is connected to the camshaft. When the governor control lever is moved clockwise, the increased spring force pushes the flyweight follower to the left. *(Courtesy of Caterpillar Tractor Company)*

pulling against springs (25). This makes the movement of seat (26) and springs (25) more gradual. This again gives the effect of a governor spring with a high spring rate.

When the governor control lever is turned toward the FUEL-OFF position with the engine running, there is a reduction of force on governor springs (25). The movement of the linkage in the governor will cause sleeve control shafts (19) to move sleeves down, and less fuel will be injected in the engine cylinders.

To stop the engine, turn the ignition switch to the "OFF" position. This will cause the shut-off solenoid to move linkage in the fuel pump housing.

Movement of the linkage will cause sleeve levers (31) to move sleeves down, and no fuel is sent to the engine cylinders. With no fuel going to the engine cylinders, the engine will stop.

Constant Bleed Valve

The constant bleed valve lets approximately 9 gallons of fuel per hour go back to fuel tank. This fuel goes back through the return line for the constant bleed valve. This flow of fuel removes air from the housing and also helps to cool the fuel injection pump. A check valve restricts the flow of fuel until pressure in the housing is at 8 ± 3 psi (55 ± 20 kPa).

PART 9 HYDRAULIC AIR-FUEL RATIO CONTROL

With the engine stopped, the valve is in the fully extended position. The movement of fuel rack linkage is not limited by the valve. (See Fig. 25-15.)

When the engine is started oil flows through the oil inlet into the pressure oil chamber. From the chamber the oil flows through the large oil passages inside the valve, and out through the small oil passages to the oil outlet.

A hose assembly connects the inlet air chamber to the inlet air system. As the inlet air pressure increases it causes the diaphragm assembly to move down. Valve (2), that is part of the diaphram assembly, closes the large and small oil passages. When these passages are closed oil pressure increases in the chamber. This increase in oil pressure moves the valve up. The control is now ready for operation.

When the governor control is moved to increase fuel to the engine, the valve limits the movement of the fuel rack linkage in the "Fuel On" direction. The oil in the chamber acts as a restriction to the movement of the valve until inlet air pressure increases.

As the inlet air pressure increases the valve moves down and lets oil from the chamber drain through large oil passages and out through oil drains. This lets the valve move down so that the fuel rack linkage can move gradually to increase fuel to the engine. The control is designed not to let the fuel increase until the air pressure in the inlet manifold is high enough for complete combustion. It prevents large amounts of exhaust smoke caused by an air-fuel mixture with too much fuel.

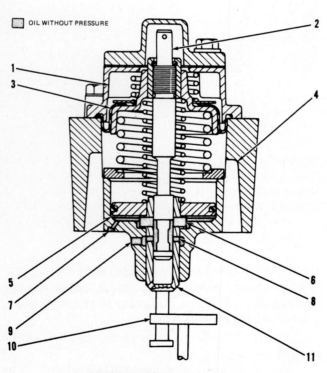

AIR-FUEL RATIO CONTROL
(Engine Stopped)

1. Inlet air chamber. 2. Valve. 3. Diaphragm assembly. 4. Oil drains. 5. Pressure oil chamber. 6. Large oil passages. 7. Oil inlet. 8. Small oil passages. 9. Oil outlet. 10. Fuel rack linkage. 11. Valve.

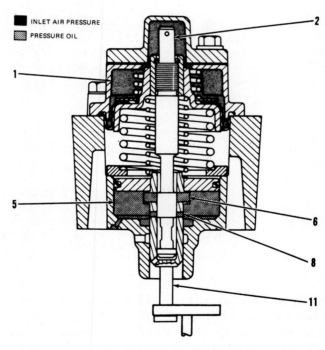

AIR-FUEL RATIO CONTROL
(Ready for operation)

1. Inlet air chamber. 2. Valve. 5. Pressure oil chamber. 6. Large oil passages. 8. Small oil passages. 11. Valve.

FIGURE 25-15. *(Courtesy of Caterpillar Tractor Company)*

RIGHT SIDE OF ENGINE

FIGURE 25-16. External fuel system components of unit injection system. 1. Fuel manifold (right hand). 2. Pressure regulating valve. 4. Priming pump. 5. Fuel filters. *(Courtesy of Caterpillar Tractor Company)*

The control movements take a very short time. No change in engine acceleration (rate at which speed increases) can be felt.

PART 10   CATERPILLAR UNIT INJECTOR FUEL SYSTEM

General Description

As shown in Figure 25-17, the fuel transfer pump (6) is located on the right front side of the engine. The lower shaft of oil pump drives the gear type transfer pump. Fuel from the supply tank goes in the transfer pump at opening (7).

The transfer pump has a check valve and a bypass valve. The check valve prevents fuel flow back through the transfer pump when priming pump (4) is used. The bypass valve is located behind plug (10). The bypass valve limits the maximum pressure of the fuel. It will open the outlet side of the pump to the pump inlet if the fuel pressure goes up to 520 kPa (75 psi). This will help prevent damage, to fuel system components, caused by too much pressure.

The transfer pump pushes fuel through fuel filters (5) to fuel manifolds (1). The fuel manifolds have two sections. The fuel flows through the top section of the manifold to tubes connected to the right side of each cylinder head. There are fuel strainers in the fittings where fuel goes in each head. A drilled hole in the cylinder head takes fuel to a circular chamber around the injector. The chamber is made by O-rings on the outside diameter of the injector and the injector bore in the head.

Only part of the fuel in the chamber is used for injection. Approximately 21 liter/min (5.5 U.S. gpm) flows through the chamber to a drilled hole in the left side of the cylinder head. This passage is connected by a tube to the bottom section of the fuel manifold. A constant flow of fuel around the injectors helps to cool them.

The fuel flows back through the bottom section of each fuel manifold to pressure regulating valve (2), on the front of the right fuel manifold. The fuel flows through this valve then back to the tank.

Pressure regulating valve (2) is a ball check type valve between the bottom section of the fuel manifolds and the line that returns fuel to tank. The valve keeps the pressure of the fuel at a constant 240 kPa (35 psi). The valve has a high resistance to the flow of fuel up to 240 kPa (35 psi), but little resistance to air. Air can be easily removed from the system by the priming pump.

The adapter that holds pressure regulating valve (2) has a small orifice that connects the inlet and outlet passages. This orifice acts as a syphon break when filters are changed. Normally it will not be necessary to use priming pump (4) to force air from the system after the filters are changed. The priming pump must be used when the lines are dry. For example: after an overhaul or other major fuel system work.

Fuel Injection Control Linkage

A fuel injector (1) is located in each cylinder head (Fig. 25-18). The position of rack (5) controls the amount of fuel injected into the cylinder. Pull the rack out of the injector for more fuel, push it in for less fuel.

Rack position is changed by bellcrank (3). The bellcrank is moved by push rod (4). The push rods have an adjustment screw on the top. The adjustment screw is used to synchronize the injectors. The push rods are spring loaded. If the rack of one injector sticks (will not move) it will still be possible for the governor to control the other racks, so the engine can be shut down. Each push rod on the right

FUEL SYSTEM

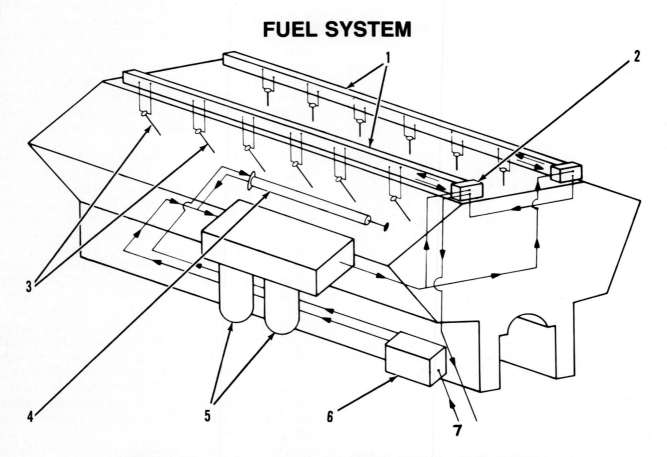

FIGURE 25-17. Fuel flow schematic. 1. Fuel manifolds. 2. Pressure regulating valve. 3. Injectors. 4. Priming pump. 5. Fuel filters. 6. Fuel transfer pump. *(Courtesy of Caterpillar Tractor Company)*

side of the engine is connected by a lever (6) to the torsion shaft (8).

When the rotation of governor shaft (7) is clockwise as seen from in front of the engine, the action of the governor linkage moves torsion shaft (8) counterclockwise. That is, in the fuel-on direction.

Right torsion shaft (8) and left torsion shaft (2) are connected by cross shaft (9). The linkage between the injectors on the left side of the engine and torsion shaft (2) is similar to the linkage on the right side.

Should the linkage become disconnected from the governor, the weight of the control linkage will move the rack pin to fuel shutoff, and the engine will stop.

Fuel Injector

The injector is held in position by clamp (3) (Fig. 25-19). Fuel is injected when rocker arm (2) pushes the top of the injector down. The movement of the rocker arm is controlled by the camshaft through follower (7) and push rod (4). The amount of fuel injected is controlled by rack (5). Movement of the rack causes rotation of a gear fastened to plunger (6). Rotation of the plunger changes the effective stroke (that part of the stroke during which fuel is actually injected) of the plunger.

Injection timing is a product of two factors: the angular location of camshaft (8) and the location of plunger (6). The angular location of the camshaft is

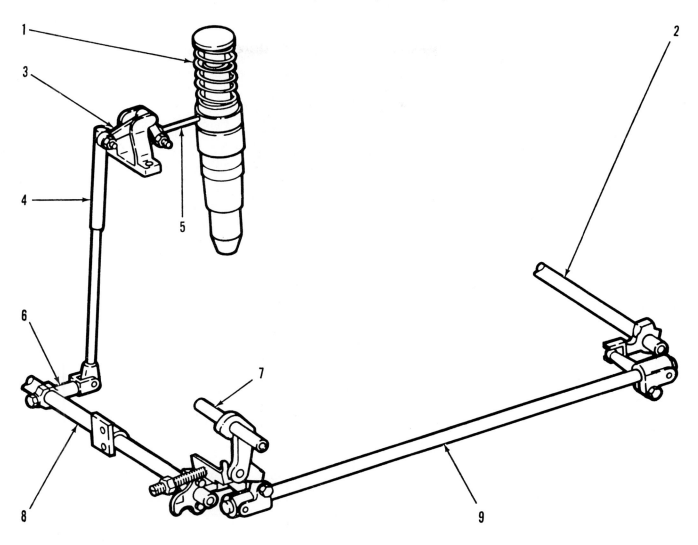

FIGURE 25-18. Fuel injector control linkage. 1. Injector. 2. Torsion shaft (left side). 3. Bellcrank. 4. Push rod. 5. Rack. 6. Lever. 7. Governor shaft. 8. Torsion shaft (right side). 9. Cross shaft. *(Courtesy of Caterpillar Tractor Company)*

controlled by the camshaft drive gears at the rear of the engine. The location of the plunger can be adjusted with screw (1).

Governor

The UG8 Dial Governor (Fig. 25-20) is a mechanical-hydraulic governor. A hydraulic activated power piston is used to turn the output terminal shaft of the governor. The terminal shaft is connected by linkage to the fuel torsion shafts. The rotation of the fuel torsion shafts controls rack movement at each injection pump. Make reference to FUEL INJECTOR CONTROL LINKAGE. The governor oil pump and ballhead are driven by a bevel gear set in the governor drive housing. The bevel gear set is driven by the front gear train.

The oil pump gives pressure oil to operate the power piston. The drive gear of the oil pump has a bushing in which the pilot valve plunger moves up and down. The driven gear of the oil pump is also the drive for the ballhead.

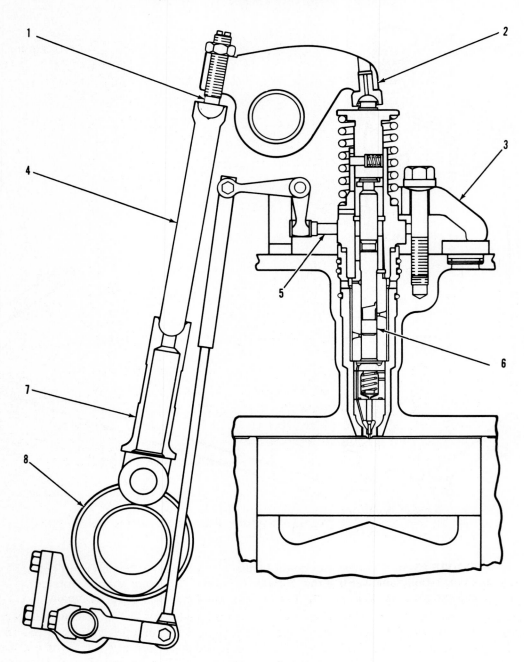

FIGURE 25-19. Fuel injector operation. 1. Screw. 2. Rocker arm. 3. Clamp. 4. Push rod. 5. Rack. 6. Plunger. 7. Follower. 8. Camshaft. *(Courtesy of Caterpillar Tractor Company)*

An accumulator is used to keep a constant oil pressure of approximately 830 kPa (120 psi) to the top of the power piston and to the pilot valve.

The power piston is connected by a lever to the output terminal shaft. There is oil pressure on both the top and bottom of the power piston. The bottom of the piston has a larger area than the top.

Less oil pressure is required on the bottom than on the top to keep the piston stationary. When the oil pressure is the same on the top and bottom of

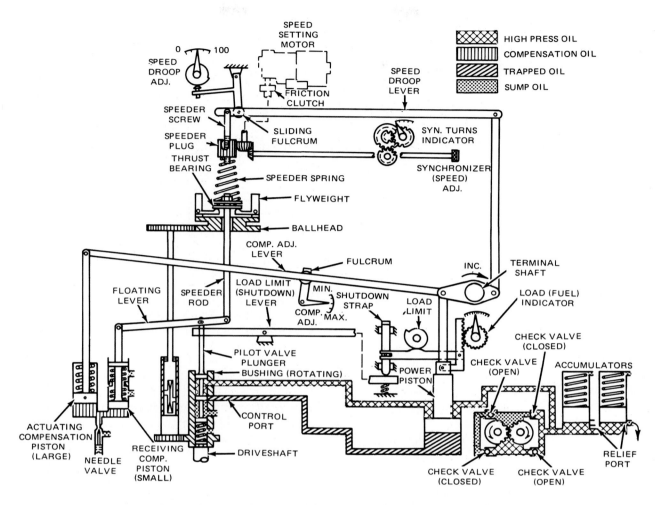

FIGURE 25-20. Model UG8 Dial governor schematic. *(Courtesy of Caterpillar Tractor Company)*

the piston, the piston will move up and cause the output terminal shaft to turn in the increase fuel direction. When oil pressure on the bottom of the piston is directed to the sump, the piston will move down and cause the output terminal shaft to turn in the decrease fuel direction. Oil to or from the bottom of the power piston is controlled by the pilot valve.

The pilot valve has a pilot valve plunger and a bushing. The bushing is turned by the governor drive shaft. The rotation of the bushing helps reduce friction between the bushing and the plunger. The pilot valve plunger has a land that controls the oil flow through the ports in the bushing. When the pilot valve plunger is moved down, high pressure oil goes to the bottom of the power piston and the power piston will move up. When the pilot valve plunger is moved up, the oil on the bottom of the

power piston is released to the sump and the power piston moves down. When the pilot valve plunger is in the center (balance) position, the oil port to the bottom of the power piston is closed and the power piston will not move. The pilot valve plunger is moved by the ballhead assembly.

The ballhead assembly has a ballhead, flyweights, speeder spring, thrust bearing, speeder plug, and speeder rod. The ballhead assembly is driven by a gear and shaft from the driven gear of the oil pump. The speeder rod is fastened to the thrust bearing which is on the toes of the flyweights. The speeder rod is connected to the pilot valve plunger with a lever. The speeder spring is held in position on the thrust bearing by the speeder plug.

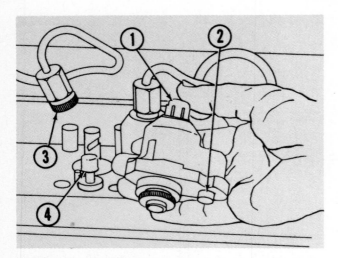

FIGURE 25-21. Removing flanged body injection pump. 1. Cover. 2. Plug. 3. Fuel line plug. 4. Ferrule cap. *(Courtesy of Caterpillar Tractor Company)*

As the ballhead turns, the flyweights move out due to centrifugal force. This will make the flyweight toes move up and cause compression of the speeder spring. When the force of the speeder spring and the force of the flyweights are equal the engine speed is constant. The speeder plug can be moved up or down manually to change the compression of the speeder spring and will change the speed of the engine.

The compensation system gives stability to engine speed changes. The compensation system has a needle valve and two pistons—an actuating piston and a receiving piston. The actuating piston is connected to the output terminal shaft by the compensation adjusting lever. A fulcrum that is adjustable is on the lever. When the position of the fulcrum is changed, the amount of movement possible of the actuating piston is changed.

The receiving piston is connected to the pilot valve plunger and the speeder rod by a lever.

The needle valve makes a restriction to oil flow between the oil sump and the two pistons.

When the actuating piston moves down, the piston forces the oil under the receiving piston and moves it up. When the receiving piston moves up it raises the pilot valve plunger to stop the flow of oil to the bottom of the power piston.

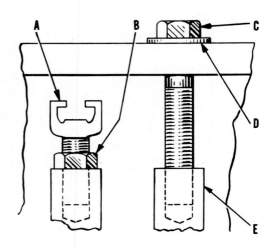

FIGURE 25-23. Removing lifter yokes and raising lifters. A. Lifter yoke. B. Lock nut. C. Bolt. D. Washer. E. Threaded end of lifter. *(Courtesy of Caterpillar Tractor Company)*

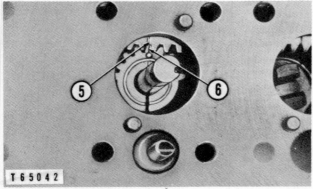

FIGURE 25-22. Alignment of timing marks of flanged body pump. 5. Mark on rack. 6. Mark on gear. *(Courtesy of Caterpillar Tractor Company)*

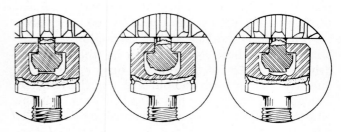

FIGURE 25-24. Wear between yoke and plunger of flanged body pump. *(Courtesy of Caterpillar Tractor Company)*

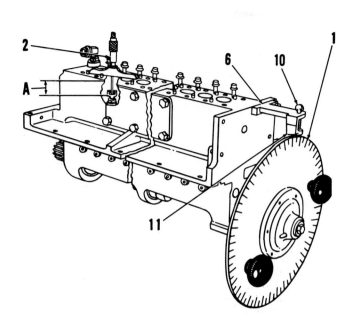

FIGURE 25-25. Fuel pump lifter setting off engine. 1. 1F8747 timing plate. 2. 6F6922 micrometer depth gauge. 6. 8M9015 adapter. 10. Lockscrew. 11. Edge of boss. A-Dimension to be measured. *(Courtesy of Caterpillar Tractor Company)*

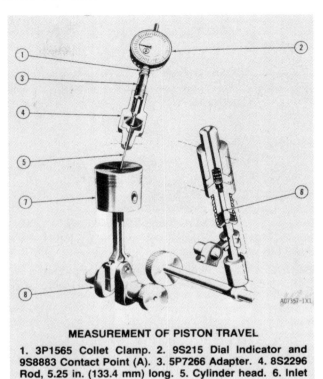

MEASUREMENT OF PISTON TRAVEL

1. 3P1565 Collet Clamp. 2. 9S215 Dial Indicator and 9S8883 Contact Point (A). 3. 5P7266 Adapter. 4. 8S2296 Rod, 5.25 in. (133.4 mm) long. 5. Cylinder head. 6. Inlet port. 7. Piston.

FIGURE 25-27. Determining piston TDC position for spill timing fuel injection pump. *(Courtesy of Caterpillar Tractor Company)*

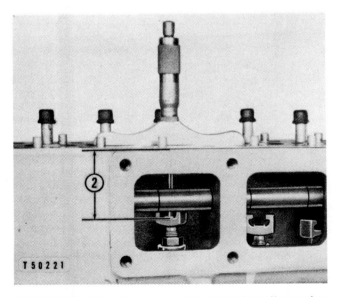

FIGURE 25-26. Measuring lifter setting dimension. (Typical Illustration.) 2. Distance to be measured. *(Courtesy of Caterpillar Tractor Company)*

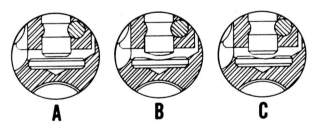

WEAR BETWEEN LIFTER WASHER AND PLUNGER

Fig. A shows the contact surfaces of a new pump plunger and a new lifter washer. In Fig. B the pump plunger and lifter washer have worn a large amount. Fig. C shows how the flat end of a new plunger makes bad contact with a worn lifter washer, causing rapid wear to both parts.

FIGURE 25-28. Compact housing injection pump lifter and plunger wear. *(Courtesy of Caterpillar Tractor Company)*

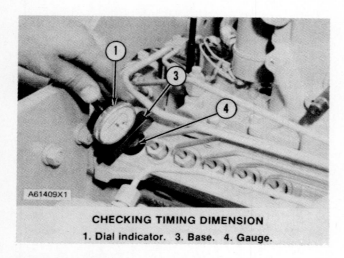

CHECKING TIMING DIMENSION
1. Dial indicator. 3. Base. 4. Gauge.

FIGURE 25-29. Checking injection pump timing with dial indicator. *(Courtesy of Caterpillar Tractor Company)*

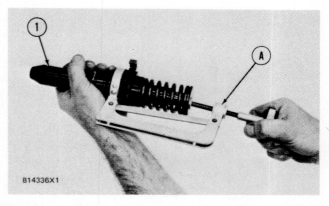

FIGURE 25-31. Tool A is used to compress injector spring to check that rack moves freely at all spring heights. *(Courtesy of Caterpillar Tractor Company)*

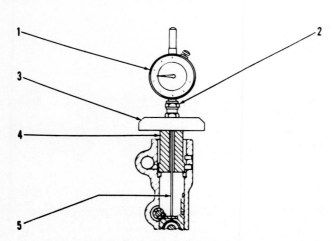

CHECKING TIMING DIMENSION

1. 8S3158 Dial Indicator. 2. 3P1565 Collet Clamp. 3. 5P4156 Base. 4. 5P4158 Gauge — 2 in. (50.8 mm) long. 5. 5P4163 Contact Point, 4.75 in. (120.7 mm) long.

FIGURE 25-30. Cross section of timing dimension check. *(Courtesy of Caterpillar Tractor Company)*

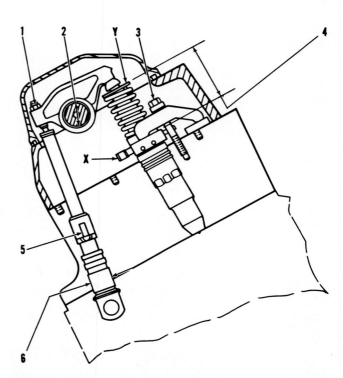

FIGURE 25-32. Service check points on unit injector system: 1. Rocker arm adjustment and lock nut torque. 2. Rocker arm bearing to shaft clearance. 3. Injector clamp bolt torque. 4. Injector timing dimension set with timing gauge. 5. Guide spring must be replaced when injectors are serviced. Springs must not be used again. 6. Valve lifter diameter. Y. Do not tap or hit at this point to install injector. X. Injector rack must move freely after servicing. *(Courtesy of Caterpillar Tractor Company)*

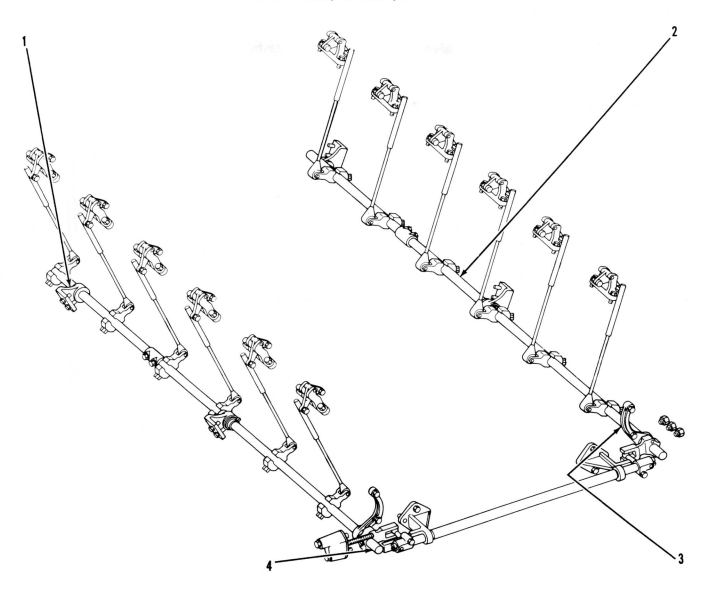

FIGURE 25-33. Points to check on control linkage: 1. Diameter of control shaft bearing bores. 2. Diameter of control shafts at wear points. 3. Diameter of bearing bores in support assemblies. 4. Clearance between side shafts and cross shafts. *(Courtesy of Caterpillar Tractor Company)*

PART 11   CATERPILLAR FUEL SYSTEMS SERVICE

The following diagnosis charts and corrective procedures are provided courtesy of the Caterpillar Tractor Company.

This troubleshooting guide will be an aid to the serviceman in finding the cause of fuel system problems.

Be sure to get a good description of the problem. The operator and/or the person who owns the truck can most likely give you a good description of the problem. What that person tells you about the problem can save you time and make the repair job faster and easier.

The probable causes of the problems are given in the charts in the order they most commonly take place. Check the probable causes in the order that they are given.

When troubleshooting, use the section on recommended procedure, which follows each chart, to check and make the necessary corrections for each probable cause.

Refer to the appropriate Caterpillar service manual for repair procedures, adjustments, and specifications for the particular engine and fuel system being serviced.

Refer to Chapter 20 for injection nozzle service.

Misfiring and Running Rough

Recommended Procedure

1. Air in Fuel System. With air in the fuel system, the engine will normally be difficult to start, run rough, and release a large amount of white smoke. If the engine will not start, loosen a fuel injection line nut at the valve cover base. With the governor lever in the shutoff position, operate the fuel priming pump until the flow of fuel from the loosened fuel injection line is free of air. Tighten the fuel line nut. Fasten the priming pump and start the engine. If the engine still does not run smooth or releases a large amount of white smoke, loosen the fuel line nuts one at a time at the valve cover base until the fuel that comes out is free of air. Tighten the fuel line nuts. If the air can not be removed in this way, put 5 psi (35 kPa) of air pressure to the fuel tank.

Caution: Do not use more than 8 psi (55 kPa) of air pressure in the fuel tank or damage to the tank may result.

Check for leaks at the connections between the fuel tank and the fuel transfer pump. If leaks are found, tighten the connections or replace the lines. If there are no visible leaks, remove the fuel supply line from the tank and connect it to an outside fuel supply. If this corrects the problem, the suction line (standpipe) inside the fuel tank has a leak.

2. Valve Adjustment Not Correct. Check and make necessary adjustments as per Testing and Adjusting section of this Service Manual. Intake valve clearance is .015 in. (0.38 mm) and exhaust valve clearance is .030 in. (0.76 mm). Also check for a bent or broken push rod.

3. Fuel Injection Timing Not Correct. Check and make necessary adjustments.

4. Automatic Timing Advance Does Not Operate Correctly. Check with engine warm. Use the 1P3500 Timing Light Group. If the timing light is not available, make rapid "acceleration" (increase in speed) from low idle to high idle. Engine must have smooth acceleration. A timing advance that does not operate correctly can cause delays of the engine acceleration at some rpm before high idle, or possibly cause the engine to run rough and have exhaust noise (backfire) during acceleration. This condition is difficult to find if engine acceleration is slow or at a constant engine rpm.

5. Bad Fuel Nozzle(s). Find a bad nozzle by running engine at the rpm where it runs rough. Loosen the fuel line nut at the cylinder head enough to stop fuel supply to that cylinder. Each cylinder must be checked this way. If a cylinder is found where loosening of the nut makes no difference in the rough running, test the nozzle for that cylinder.

6. Valve Leakage: Wear or Damage to Pistons and/or Piston Rings; Wear or Damage to Cylinder Walls. The cylinder head will have to be removed to make a visual inspection of these inside problems.

7. Cylinder Head Gasket Leakage. Leakage at the gasket of the cylinder head can show as an outside leak or can cause loss of coolant through the radiator overflow. Remove the radiator filler cap and, with the engine running, check for air bubbles in coolant caused by exhaust gases.

8. Engine Camshaft Timing Not Correct. Engine camshaft timing can be checked by finding the top center position for number one cylinder.

9. Fuel Leakage from Fuel Injection Line Nut. Tighten nut to 30 ± 5 lb. ft. (40 ± 7 N·m). Again check for leakage. Be sure to check fuel injection lines inside the valve cover base.

10. Fuel Has a High "Cloud Point." The fuel "cloud point" is the temperature at which wax begins to form in the fuel. If the atmospheric temperature is lower than the "cloud point" of the fuel, wax will form and plug the filter. Change the filter and drain the tank and the complete fuel system. The replacement fuel must be of a better grade with a lower "cloud point."

Too Much Exhaust Smoke

Black or Gray

Engine Runs Smoothly

Recommended Procedure

1. Engine Used in a Lug Condition. "Lugging" (when the truck is used in a gear too high for engine rpm to go up as accelerator pedal is pushed farther down, or when the truck is used in a gear

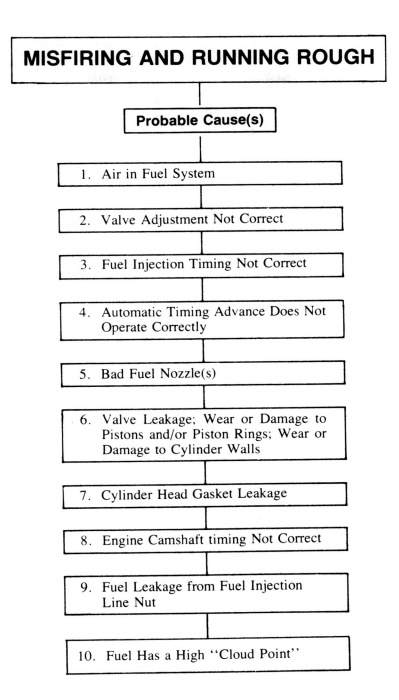

MISFIRING AND RUNNING ROUGH

Probable Cause(s)

1. Air in Fuel System

2. Valve Adjustment Not Correct

3. Fuel Injection Timing Not Correct

4. Automatic Timing Advance Does Not Operate Correctly

5. Bad Fuel Nozzle(s)

6. Valve Leakage; Wear or Damage to Pistons and/or Piston Rings; Wear or Damage to Cylinder Walls

7. Cylinder Head Gasket Leakage

8. Engine Camshaft timing Not Correct

9. Fuel Leakage from Fuel Injection Line Nut

10. Fuel Has a High "Cloud Point"

where engine rpm goes down with accelerator pedal at maximum travel) the engine causes a reduction in the intake of air with full fuel delivery to the cylinders. Because there is not enough air to burn all the fuel, the fuel that is not used comes out the exhaust as black smoke. To prevent lugging the engine, use a gear where engine can have "acceleration" (increase in speed) under load.

2. Dirty Air Cleaner. If the air cleaner has a restriction indicator, see if the red piston is in view. If there is no restriction indicator, restriction can be checked with a water manometer or a vacuum gauge (which measures in inches of water). Make a connection to the piping between the air cleaner and the inlet of the turbocharger. Check with the engine running at full load rpm. If a gauge is not available,

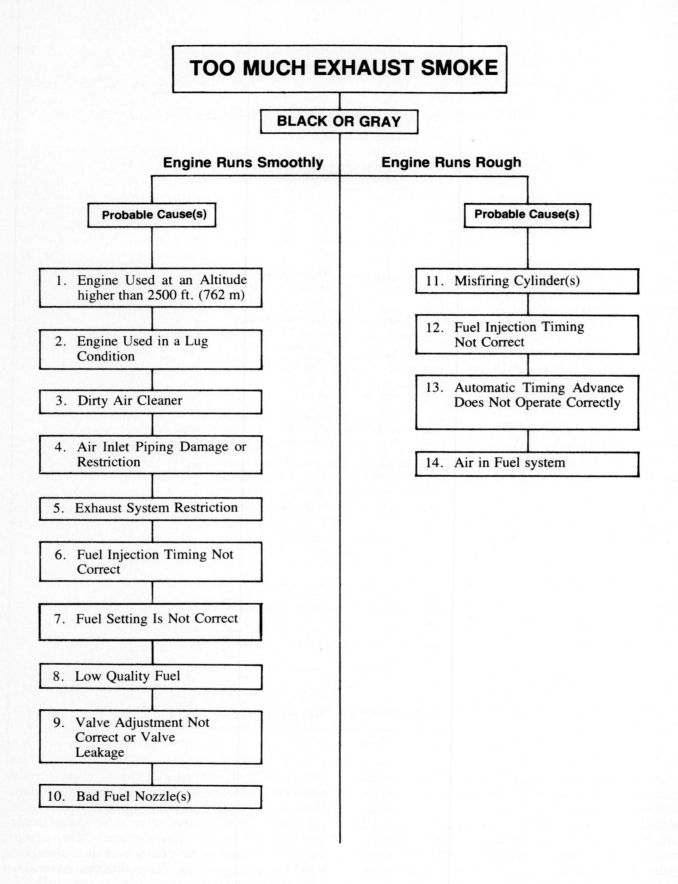

TOO MUCH EXHAUST SMOKE

BLACK OR GRAY

Engine Runs Smoothly

Probable Cause(s)

1. Engine Used at an Altitude higher than 2500 ft. (762 m)

2. Engine Used in a Lug Condition

3. Dirty Air Cleaner

4. Air Inlet Piping Damage or Restriction

5. Exhaust System Restriction

6. Fuel Injection Timing Not Correct

7. Fuel Setting Is Not Correct

8. Low Quality Fuel

9. Valve Adjustment Not Correct or Valve Leakage

10. Bad Fuel Nozzle(s)

Engine Runs Rough

Probable Cause(s)

11. Misfiring Cylinder(s)

12. Fuel Injection Timing Not Correct

13. Automatic Timing Advance Does Not Operate Correctly

14. Air in Fuel system

visually check the air cleaner element for dirt. If the element is dirty, clean the element or install a new element.

3. Air Inlet Piping Damage or Restriction. Make a visual inspection of the air inlet system and check for damage to piping, rags in the inlet piping, or damage to the rain cap or the cap pushed too far on the inlet pipe. If no damage is seen, check inlet restriction with a clean air cleaner element.

4. Exhaust System Restriction. Make a visual inspection of the exhaust system. Check for damage to piping or for a bad muffler. If no damage is found, you can check the system by checking the back pressure from the exhaust (pressure difference measurement between exhaust outlet and atmosphere). You can also check by removing the exhaust pipes from the exhaust manifolds. With the exhaust pipes removed, start and load the engine on a chassis dynamometer to see if the problem is corrected.

5. Fuel Injection Timing Not Correct. Check and make necessary adjustments.

6. Fuel Setting Is Not Correct. Check and make necessary adjustments.

7. Low Quality Fuel. Test the engine with fuel according to recommendations by the Caterpillar Tractor Co.

8. Valve Adjustment Not Correct or Valve Leakage. Check and make necessary adjustments. Valve leakage normally causes the engine to "misfire" (injection not regular) and run rough.

9. Bad Fuel Nozzle(s). Bad fuel nozzles will normally cause the engine to "misfire" (injection not regular) and run rough, but can cause too much smoke with engine still running smooth. Remove the fuel nozzles and test.

Engine Runs Rough

10. Misfiring Cylinder(s). See Misfiring and Running Rough.

11. Fuel Injection Timing Not Correct. Check and make necessary adjustments.

12. Automatic Timing Advance Does Not Operate Correctly. Check with engine warm. Use the 1P3500 Timing Light Group. If the timing light is not available, make rapid "acceleration" (increase in speed) from low idle to high idle. Engine must have smooth acceleration. A timing advance that does not operate correctly can cause delays of the engine acceleration at some rpm before high idle, or possibly cause the engine to run rough and have exhaust noise (backfire) during acceleration. This condition is difficult to find if engine acceleration is slow or at a constant engine rpm.

13. Air in Fuel System. With air in the fuel system, the engine will normally be difficult to start, run rough, and release a large amount of white smoke. If the engine will not start, loosen a fuel injection line nut at the valve cover base. With the governor lever in the shutoff position, operate the fuel priming pump until the flow of fuel from the loosened fuel injection line is free of air. Tighten the fuel line nut. Fasten the priming pump and start the engine. If the engine still does not run smooth or releases a large amount of white smoke, loosen the fuel line nuts one at a time at the valve cover base until the fuel that comes out is free of air. Tighten the fuel line nuts. If the air can not be removed in this way, put 5 psi (35 kPa) of air pressure to the fuel tank.

Caution. Do not use more than 8 psi (55 kPa) of air pressure in the fuel tank or damage to the tank may result.

Check for leakage at the connections between the fuel tank and the fuel transfer pump. If leaks are found, tighten the connections or replace the lines. If there are no visible leaks, remove the fuel supply line from the tank and connect it to an outside fuel supply. If this corrects the problem, the suction line (standpipe) inside the fuel tank has a leak.

Too Much Exhaust Smoke

White Smoke

Recommended Procedure

1. Cold Outside Temperature. When the air outside is cold, the cylinder temperature is cooler. Not all the fuel will burn in the cylinders. The fuel which does not burn comes out the exhaust as white smoke. White smoke is normal in cold temperatures until the engine operates long enough to become warm. There will be less white smoke if No. 1 diesel fuel is used.

2. Long Idle Periods. When an engine runs at idle speed for a long period of time, the cylinders cool and all of the fuel does not burn. Do not idle an engine for a long period of time. Stop an engine when it is not in use. If long idle periods are necessary, use No. 1 diesel fuel.

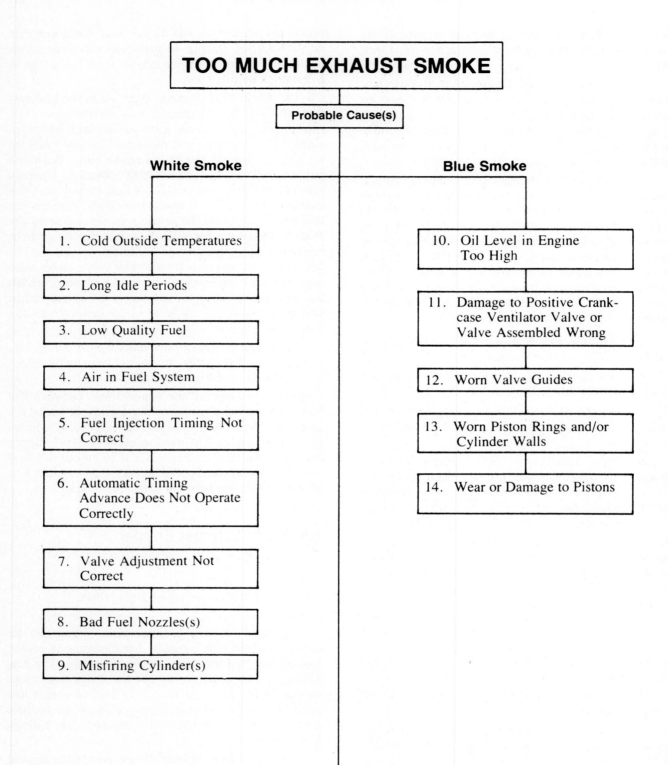

TOO MUCH EXHAUST SMOKE

Probable Cause(s)

White Smoke

1. Cold Outside Temperatures

2. Long Idle Periods

3. Low Quality Fuel

4. Air in Fuel System

5. Fuel Injection Timing Not Correct

6. Automatic Timing Advance Does Not Operate Correctly

7. Valve Adjustment Not Correct

8. Bad Fuel Nozzles(s)

9. Misfiring Cylinder(s)

Blue Smoke

10. Oil Level in Engine Too High

11. Damage to Positive Crankcase Ventilator Valve or Valve Assembled Wrong

12. Worn Valve Guides

13. Worn Piston Rings and/or Cylinder Walls

14. Wear or Damage to Pistons

3. Low Quality Fuel. Test the engine using fuel according to recommendations by Caterpillar Tractor Co.

4. Air in Fuel System. With air in the fuel system, the engine will normally be difficult to start, run rough, and release a large amount of white smoke. If the engine will not start, loosen a fuel injection line nut at the valve cover base. With the governor lever in the shutoff position, operate the fuel priming pump until the flow of fuel from the loosened fuel injection line is free of air. Tighten the fuel line nut. Fasten the priming pump and start the engine. If the engine still does not run smooth or releases a large amount of white smoke, loosen the fuel line nuts one at a time at the valve cover base until the fuel that comes out is free of air. Tighten the fuel line nuts. If the air can not be removed in this way, put 5 psi (35 kPa) of air pressure to the fuel tank.

Caution: Do not use more than 8 psi (55 kPa) of air pressure in the fuel tank or damage to the tank may result.

Check for leakage at the connections between the fuel tank and the fuel transfer pump. If leaks are found, tighten the connections or replace the lines. If there are no visible leaks, remove the fuel supply line from the tank and connect it to an outside fuel supply. If this corrects the problem, the suction line (standpipe) inside the fuel tank has a leak.

5. Fuel Injection Timing Not Correct. Check and make necessary adjustments.

6. Automatic Timing Advance Does Not Operate Correctly. Check with engine warm. Use the 1P3500 Timing Light Group. If the timing light is not available, make rapid "acceleration" (increase in speed) from low idle to high idle. Engine must have smooth acceleration. A timing advance that does not operate correctly can cause delays of the engine acceleration at some rpm before high idle, or possibly cause the engine to run rough and have exhaust noise (backfire) during acceleration. This condition is difficult to find if engine acceleration is slow or at a constant engine rpm.

7. Valve Adjustment Not Correct. Check and make necessary adjustments.

8. Bad Fuel Nozzle(s). Bad fuel nozzles will normally cause the engine to "misfire" (injection not regular) and run rough, but can cause too much smoke and the engine still be running smooth. Remove the fuel nozzles and test.

9. Misfiring Cylinder(s). See Misfiring and Running Rough.

Blue Smoke

10. Engine Oil Level Too High. Do not put too much oil in the crankcase. If the oil level in the crankcase goes up as the engine is used, check for fuel in the crankcase.

11. Worn Valve Guides. See the Specifications for the maximum permissible wear of the valve guides.

12. Worn Piston Rings and/or Cylinder Walls. Worn piston rings and/or cylinder walls can be the cause of blue smoke and can cause a loss of compression. Make a visual inspection of the cylinder walls and piston rings. If necessary, measure the cylinder walls and piston rings. High wear at low mileage is normally caused by dirt coming into the engine with the inlet air.

13. Wear or Damage to Pistons. Check the piston ring grooves for wear. Pistons which have worn grooves and pistons with damage or defects can cause blue smoke and too much oil consumption. Make sure the oil return holes under the oil ring are open.

Difficult Starting

Engine Crankshaft Turns Freely

Exhaust Smoke Can Be Seen While Starting

Recommended Procedure

1. Cold Outside Temperatures. It can be necessary to use starting aids, or to heat engine oil or coolant at temperatures below 10°F (−12°C) for D.I. engines and 50°F (10°C) for P.C. engines.

2. Air in Fuel System. With air in the fuel system, the engine will normally be difficult to start, run rough, and release a large amount of white smoke. If the engine will not start, loosen a fuel injection line nut at the valve cover base. With the governor lever in the shutoff position, operate the fuel priming pump until the flow of fuel from the loosened fuel injection line is free of air. Tighten the fuel line nut. Fasten the priming pump and start the engine. If the engine still does not run smooth or releases a large amount of white smoke, loosen the fuel line nuts one at a time at the valve cover base until the fuel that comes out is free of air. Tighten the fuel line nuts. If the air can not be re-

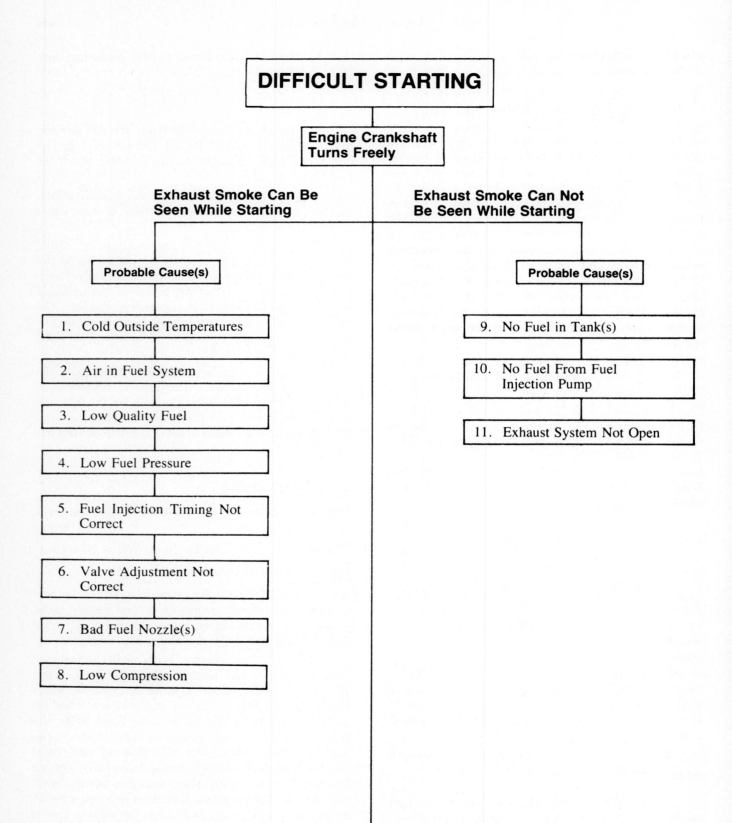

DIFFICULT STARTING

**Engine Crankshaft
Turns Freely**

**Exhaust Smoke Can Be
Seen While Starting**

Probable Cause(s)

1. Cold Outside Temperatures

2. Air in Fuel System

3. Low Quality Fuel

4. Low Fuel Pressure

5. Fuel Injection Timing Not Correct

6. Valve Adjustment Not Correct

7. Bad Fuel Nozzle(s)

8. Low Compression

**Exhaust Smoke Can Not
Be Seen While Starting**

Probable Cause(s)

9. No Fuel in Tank(s)

10. No Fuel From Fuel Injection Pump

11. Exhaust System Not Open

moved in this way, put 5 psi (35 kPa) of air pressure to the fuel tank.

Caution: Do not use more than 8 psi (55 kPa) of air pressure in the fuel tank or damage to the tank may result.

Check for leakage at the connections between the fuel tank and the fuel transfer pump. If leaks are found, tighten the connections or replace the lines. If there are no visible leaks, remove the fuel supply line from the tank and connect it to an outside fuel supply. If this corrects the problem, the suction line (standpipe) inside the fuel tank has a leak.

3. Low Quality Fuel. Remove a small amount of fuel from the tank and check for water in the fuel. If there is water in the fuel, remove fuel from the tank until it is free of water and fill with a good quality fuel. Change the fuel filter and "prime" (remove the air and/or low quality fuel from the fuel system) the fuel system with the fuel priming pump. If there is no water in the fuel, prime and start the engine by using an outside source of fuel. If engine starts correctly using different fuel, remove all fuel from the tank and fill with good quality fuel. Prime the fuel system if necessary.

4. Low Fuel Pressure. Change the fuel filter. If the pressure is still low, check the bypass valve in the fuel transfer pump. Debris in the system can make the valve become stationary in the open position.

5. Fuel Injection Timing Not Correct. Check and make necessary adjustments.

6. Valve Adjustment Not Correct. Check and make necessary adjustments.

7. Bad Fuel Nozzle(s). Remove the fuel nozzles and test.

8. Low Compression. See Misfiring and Running Rough.

Exhaust Smoke Can Not Be Seen While Starting

9. No Fuel in Tank(s). Check fuel level visually (do not use the fuel gauge only). Be sure tank selection valve is open to the tank with fuel in it. Be sure valve in fuel line between the tanks is open.

10. No Fuel from Fuel Injection Pump. Loosen a fuel injection line nut at the camshaft base. With ignition switch in the ON position and accelerator in the FUEL ON position, turn the engine with the starter to be sure there is no fuel from the fuel in-jection pump. To find the cause for no fuel, follow Steps (a) through (d) until the problem is corrected.

a. Prime the fuel system as shown in Procedure No. 2.

b. Check shut-off solenoid by turning ignition switch to ON position. You must hear a sound when the plunger opens. If no sound is heard, make sure there is battery voltage at the solenoid. If the solenoid does not work, install a new solenoid.

c. If you are not using a good quality of fuel at temperatures below 10°F (−12°C), it is possible that the fuel in the system can "wax" (not have correct flow characteristics) and cause a restriction in the fuel system. Install a new fuel filter. It may be necessary to drain the complete fuel system and replace with a No. 1 grade of fuel.

d. Check for fuel supply line restriction by removing the fuel supply line for the fuel filter base. Put 5 psi (35 kPa) of air pressure to the fuel tank.

Caution: Do not use more than 8 psi (55 kPa) of air pressure in the fuel tank or damage to the tank may result.

If there is no fuel, or only a weak flow of fuel from the fuel supply line, there is a restriction in the fuel supply line and/or the fuel tank.

11. Exhaust System Not Open. Loosen the exhaust pipe from the exhaust manifold. If engine will now start, check the exhaust system for damage and/or restrictions.

PART 12 SELF-CHECK

1. What type of pumping and metering element is used in the flanged body pump?

2. How is the amount of fuel injection controlled by the flanged body pump?

3. What is the major difference between the flanged body pump and the compact housing pump?

4. The hydra-mechanical governor uses engine lubricating oil to operate a piston. True or False.

5. In the hydra-mechanical governor what controls the compression of the governor spring?

6. The centrifugal weights in the governor always try to reduce fuel delivery to the engine. True or False.

7. The mechanical automatic timing advance is connected to the ———————.

8. What is the purpose of the fuel ratio control?

9. How is fuel metering accomplished in the sleeve metering injection pump?

10. Caterpillar unit injector pumping action is the result of engine camshaft action. True or False.

11. How is fuel metering achieved in the unit injector?

12. List five causes of engine misfiring.

13. List five causes of engine difficult starting.

Chapter 26
United Technologies (American Bosch) and Robert Bosch Distributor Injection Pumps

Performance Objectives

After thorough study of this chapter, the appropriate training models and service manuals, and with the necessary tools and equipment, you should be able to do the following with respect to diesel engines equipped with American Bosch and Robert Bosch distributor type injection pumps:

1. Complete the self-check questions with at least 80 percent accuracy.

2. Describe the basic construction and operation of the American Bosch and Robert Bosch distributor injection pumps including the following:

    (a) method of fuel pumping and metering

    (b) delivery valve action

    (c) lubrication

    (d) injection timing control

    (e) governor types

3. Diagnose basic fuel injection system problems and determine the correction needed.

4. Remove, clean, inspect, test, measure, repair, replace fuel injection system components as required to meet manufacturer's specifications.

This chapter is based on information courtesy of United Technologies American Bosch, Robert Bosch Canada Ltd., and International Harvester Company.

PART 1 UNITED TECHNOLOGIES (AMERICAN BOSCH) DISTRIBUTOR PUMPS

American Bosch single plunger distributor pumps of the PS type include the PSB, PSJ, PSU and Model 100 pumps. These pumps are compact units designed particularly for applications with relatively broad speed requirements such as military vehicles, medium and heavy duty trucks, farm equipment and tractors. In addition to small size, other design features are ease of servicing, low cost and simplified calibration. All pumps, with the exception of the PSU, are designed for flange mounting to the engine; the PSU is mounted within the engine block. (See Fig. 26-1 to 26-3.)

The prominent feature of the Bosch PS series is the single precision plunger, which is actuated by the pump camshaft. It reciprocates for pumping action and rotates continuously for distribution of fuel via discharge outlets to individual cylinders of the engine. The precise rotating and reciprocating action of the single plunger distributes a uniform volume of fuel per cylinder for smooth power and fuel economy. There is no need to make fuel delivery adjustments for any one cylinder.

The plunger is lapped and fitted into the head and metering sleeve, thereby making these three components an inseparable assembly. The lower extension of the plunger is specially machined for positioning with the plunger face gear by a drive that fixes the angular position of the plunger in relation to the face gear.

The plunger makes one complete revolution for every two revolutions of the pump camshaft, which is operating at engine crankshaft speed. The plunger rotates continuously while moving vertically through the pumping cycle; therefore, on an eight-cylinder engine, the four-lobe cam actuates the plunger eight times for every two revolutions of the pump camshaft. On a six-cylinder engine, the three-lobe cam actuates the plunger six times for every two revolutions of the pump camshaft.

All PS series pumps are engine-oil lubricated. Internal ducting allows an adequate amount of lubricating oil to be supplied under pressure to the tappet assembly, camshaft bushings, thrust washer faces, and governor shaft. All other parts are splash-lubricated. The internal timing device (intra-vance) is actuated by lubricating oil pressure as is the excess-fuel-starting device. Lubricating oil is returned to the crankcase through the pump housing and the external timing device, when used.

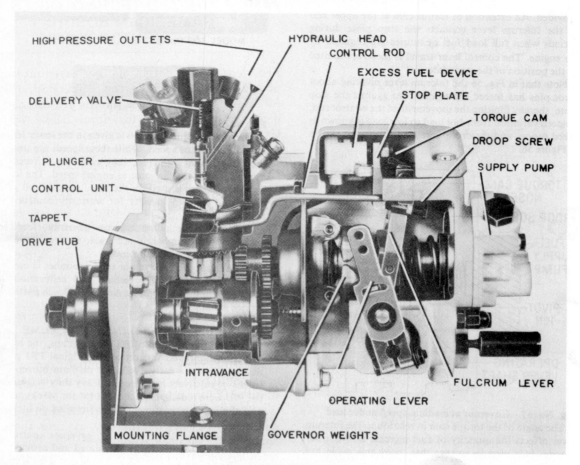

FIGURE 26-1. Partial cutaway view of United Technologies Model 100 distributor pump. *(Courtesy of American Bosch, subsidiary of United Technologies Corporation)*

Fuel Flow

Fuel enters the pump from the supply system through the pump housing inlet connection. This action fills the sump area and the head cavity between the top of the plunger and the bottom of the delivery valve when the plunger is at the bottom of its stroke. (See Fig. 26-4 and 26-5.)

As the continuously rotating plunger moves upward under cam action, it closes two horizontal galleries, which contain the inlet ports. This traps the fuel and builds up pressure until the spring-loaded delivery valve is caused to open. Further movement of the plunger forces the fuel through the delivery valve and is carried through the intersecting duct to the annulus and distributing slot in the plunger. The vertical distributing slot on the plunger connects to the outlet duct, which is then registered as the plunger rotates.

The rotary and vertical movement of the plunger are so placed in relation to the outlet ports that the vertical distributing slot overlaps only one outlet duct during the effective portion of each stroke.

After sufficient vertical movement of the plunger, its metering port passes the edge of the metering sleeve, and the fuel under pressure is forced down the center of the plunger and into the sump surrounding the metering sleeve. This completes a pumping cycle for one cylinder.

Fuel Metering

The quantity of fuel delivered per stroke is controlled by the position of the metering sleeve in relation to the fixed port closing position. That is the point at which the top of the plunger covers the top

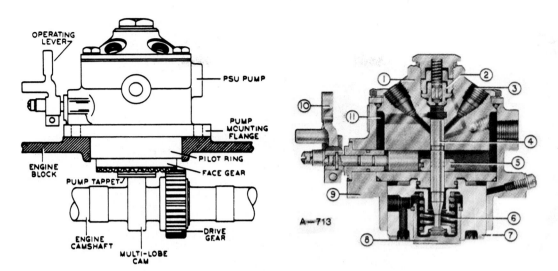

FIGURE 26-2. American Bosch PSU model injection pump is engine camshaft-operated.

1. Nitrided steel hydraulic head
2. Delivery valve assembly
3. Fuel outlets
4. Lapped and fitted through-hardened steel plunger
5. Fuel metering control sleeve
6. Plunger tappet return spring
7. Face gear
8. Tappet assembly
9. Pump mounting flange
10. Pump operating lever
11. Fuel pump

(Courtesy of American Bosch, subsidiary of United Technologies Corporation)

of the horizontal galleries or fill ports, as they are called. (See Fig. 26-6.)

As the horizontal metering hole, during the plunger stroke, passes over the top edge of the metering sleeve, pumping pressure is relieved down through the center hole of the plunger and out into the sump surrounding the metering sleeve. Fuel delivery terminates despite the continued upward movement of the plunger.

When the metering sleeve is in its lowest position, the metering hole in the plunger is uncovered by the top edge of the sleeve before the upper end of the plunger can cover the horizontal galleries. In this condition, no pressure can be built up even after the galleries are covered. As a result, no fuel can be delivered. This is the shut-off position.

If the metering sleeve is moved to mid-position, the hole in the plunger is uncovered later in the stroke by the sleeve; hence, the effective stroke of the plunger is longer and fuel is delivered. If the metering sleeve position is raised further, the horizontal hole in the plunger remains covered by the sleeve until relatively late in the plunger stroke, thereby increasing the effective stroke of the plunger and fuel delivery.

To sum up, the upward movement of the metering sleeve increases, and its down movement decreases the quantity of fuel pumped per stroke.

Delivery Valve Operation

The delivery valve assembly is located directly above the plunger. Its function is to assist the injection of fuel by preventing irregular loss of fuel from the delivery to the supply side of the system between pumping strokes. (See Fig. 26-7.)

The delivery valve assembly consists of a valve with a conical seat and a mating valve body. Opening pressure is controlled by the delivery valve spring that sits on top of the valve.

Pressure is created after the plunger on the upward stroke closes the horizontal galleries (scallops), which form the inlet ports. When the pressure overcomes the force of the valve spring, the valve opens and fuel under pressure flows through the distributing passage into the injection tubing.

When the horizontal hole in the plunger passes the top edge of the metering sleeve, there is a sudden drop in fuel pressure below the delivery valve; the force of the valve spring, combined with the dif-

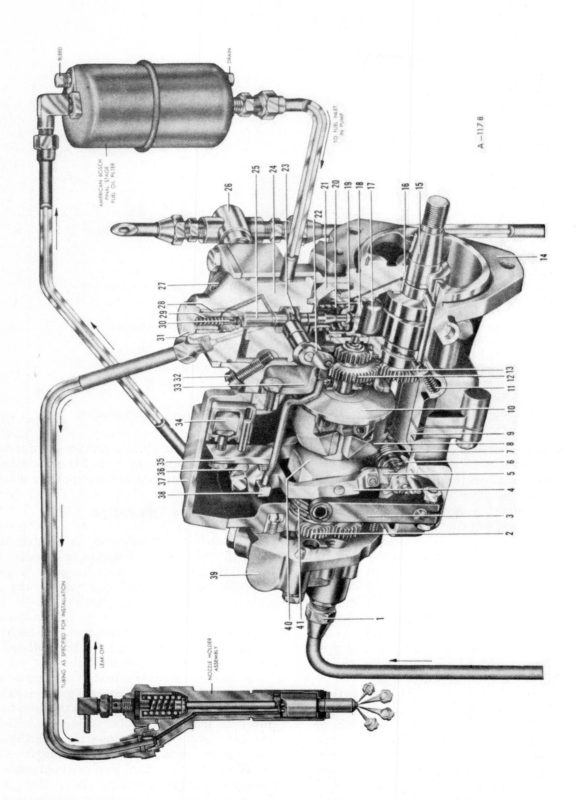

FIGURE 26-3. American Bosch single plunger, sleeve-metering distributor pump, and governor components. *(Courtesy of American Bosch, subsidiary of United Technologies Corporation)*

NOMENCLATURE

1. Fuel supply fitting	22. Control unit assembly
2. Supply pump drive gear	23. Fuel metering sleeve
3. Operating lever	24. Hydraulic head
4. Fulcrum lever with droop screw	25. Pump plunger
5. Fulcrum lever bracket	26. Overflow valve
6. Governor lever stop	27. Fuel discharge outlet
7. Operating lever spring	28. Delivery valve holder
8. Governor weights	29. Delivery valve spring
9. Governor housing	30. Delivery valve
10. Governor weight spider	31. Delivery valve cap screw
11. Lube oil inlet	32. Locating screw
12. Governor drive gear	33. Ball bearing support plate
13. Camshaft gear	34. Excess fuel starting device
14. Pump housing	35. Stop plate
15. Camshaft	36. Smoke limit cam
16. Camshaft bearing	37. Governor cover
17. Tappet roller	38. Control rod
18. Tappet guide	39. Fuel supply pump
19. Lower spring seat	40. Sliding sleeve
20. Plunger return spring	41. Outer governor spring
21. Face gear	

FIGURE 26-3 (Continued)

ferential in pressure, starts to return the valve to its seat.

As the valve starts down into its body, the lower edge of the retraction piston enters the valve bore and, at that moment, the flow of fuel to the bore in the head stops. Further downward movement of the piston increases the volume on the high-pressure side by the amount of piston movement (its displacement volume) and consequently reduces the residual pressure in the affected injection tubing, nozzle, and nozzle holder. This lowered pressure assists the rapid closing of the injection nozzle valve and diminishes the effect of hydraulic pressure waves that exist in the tubing between injections. As a result, the possibility of the nozzle opening is reduced before the next regular delivery cycle.

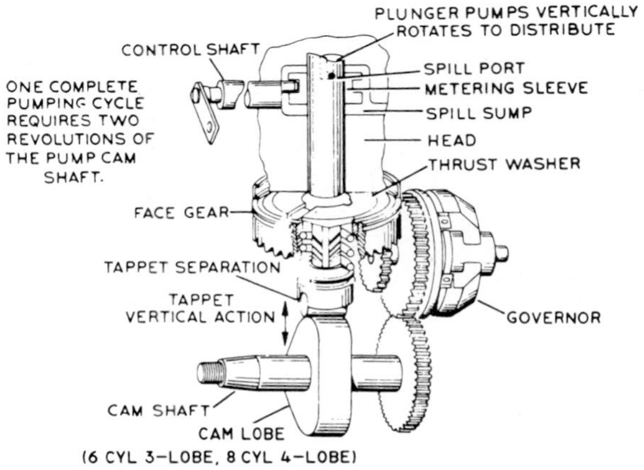

FIGURE 26-4. American Bosch distributor pump operation. Note sleeve metering and plunger rotation for distribution to each cylinder. (*Courtesy of American Bosch, subsidiary of United Technologies Corporation*)

PART 2   DISTRIBUTOR PUMP TIMING

An external or internal timing mechanism may be incorporated in PS series distributor-type fuel injection pumps. These timing mechanisms provide a means for automatically varying injection timing in relation to top-dead-center as engine speed increases to obtain optimum engine combustion efficiency. These devices are adaptable to various fuel injection pumps, provide a timing advance of up to 20°, and become operative at speeds from 300 to 3200 rpm, depending on application and the type used. (See Fig. 26-8.)

There are three types of the externally mounted timing mechanisms, which, except for some parts

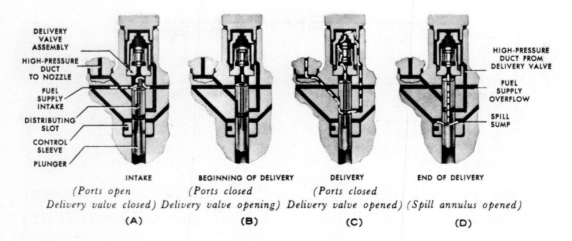

FIGURE 26-5. Details of Bosch distributor pumping action and high-pressure passage to cylinder. *(Courtesy of American Bosch, subsidiary of United Technologies Corporation)*

differences, are similar and function in the same manner.

In operation, centrifugal force, due to the rotation of the shaft and spindle, tends to open the weights and is resisted by the pressure of the springs. At a predetermined speed, the pressure of the springs and the centrifugal force are balanced. When that speed is exceeded, the movement of the weights causes the sliding gear to move longitudinally. This, in turn, causes the pump camshaft to rotate slightly. In so doing, the timing of the pump is advanced the desired amount.

Internal Advance Mechanism

An advance mechanism is built inside the cam to alter the relationship between the input shaft and the cam surface. The camshaft is supported between two bearings and the drive shaft is supported inside the camshaft. The camshaft gear is machined on the camshaft. The amount of cam advance is determined by the longitudinal position of the sliding double splined sleeve. This sleeve has helical splines on both the inner and outer surfaces to engage with corresponding helical splines on the drive shaft and

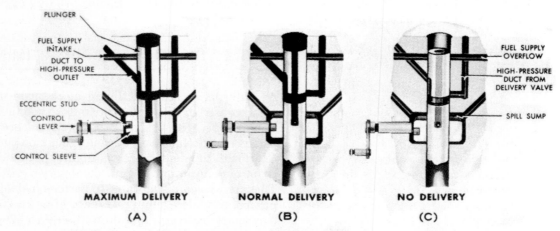

FIGURE 26-6. Position of sleeve (up or down) determines effective length of plunger stroke (metering) and therefore the amount of fuel delivered to the injector. *(Courtesy of American Bosch, subsidiary of United Technologies Corporation)*

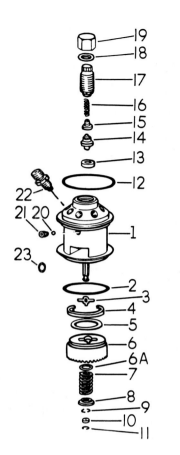

1. Hydraulic head	12. Gasket
2. Gasket	13. Spacer
3. Plunger guide	14. Delivery valve assembly
4. Gear retainer	15. Guide
5. Thrust washer	16. Spring
6. Face gear	17. Delivery valve holder
6A. Plunger spring washer	18. Gasket
7. Plunger spring	19. Cap nut
8. Lower spring seat	20. Sealing ball
9. Split ring	21. Set screw
10. Plunger button	22. Discharge fitting
11. Retaining ring	23. Gasket

FIGURE 26-7. Hydraulic head components. *(Courtesy of American Bosch, subsidiary of United Technologies Corporation)*

the camshaft. The sleeve position is determined by a hydraulic servo valve. (See Fig. 26-9 and 26-10.)

The control action is as follows, assuming that the entire mechanism is initially at equilibrium at some intermediate speed. If the speed increases, the weights move outward and the hydraulic servo valve moves to the right. Lube oil which is supplied through the right bearing flows through the valve to the working chamber and forces the sleeve to the left. The sleeve is sealed on both inner and outer surfaces by piston rings. As the sleeve moves to the left, the inner follow-up rod also moves to the left, increasing the spring force. Movement of the sleeve and follow-up rod continues until the spring force overcomes the higher centrifugal force. The movement of the sleeve causes the cam to advance with respect to the drive shaft by the helix angle of the splines and the spring rate. The spring rate and preload are tailored to the engine. The advance curve is dependent on speed only and is not affected by fuel delivery, lube oil temperature, viscosity, etc.

When the speed decreases, the opposite occurs. The weights move inward causing the hydraulic servo valve to move to the left. Lube oil is released from the working chamber through the left hand valve annulus and out through the drain hole in the shaft. As the oil leaves, the sleeve and the follow-up rod move to the right to reduce the spring force. This action continues until the force is reduced and the weights again resume their equilibrium position.

PART 3   GOVERNOR OPERATION

The variable governor portion of single plunger distributor fuel injection pumps is a mechanical, centrifugal device that is assembled as a unit in the governor housing. The primary purpose of the governor is to serve as a means of presetting and maintaining within close limits the diesel engine over a nominal speed range, irrespective of engine load. In addition, the governor controls engine idling speed

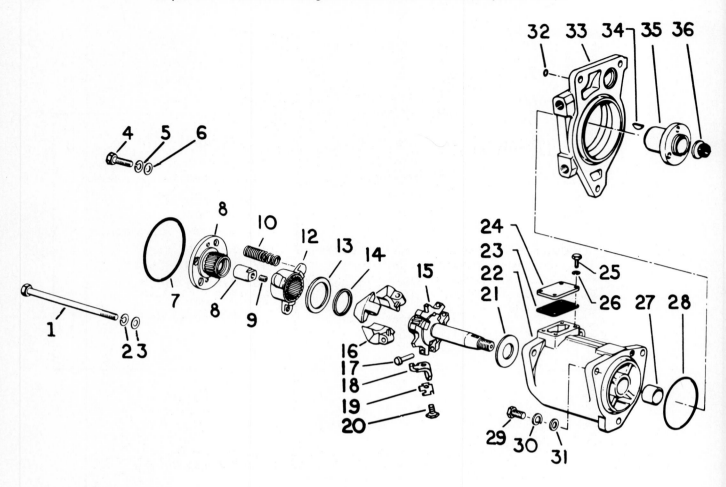

FIGURE 26-8. Exploded view of external timing device. *(Courtesy of American Bosch, subsidiary of United Technologies Corporation)*

1. Mounting Screw	10. Spring	19. Tab Washer	28. Gasket
2. Lockwasher	11. Set Screw	20. Screw	29. Screw
3. Washer	12. Sliding Gear	21. Thrust Plate	30. Lockwasher
4. Mounting Screw	13. Spacer	22. Housing	31. Washer
5. Lockwasher	14. Spacer	23. Gasket	32. Lube Duct Gasket
6. Washer	15. Spider	24. Timing Window	33. Adapter Flange
7. Gasket	16. Weight	25. Screw	34. Key
8. Hub	17. Weight Pin	26. Lockwasher	35. Drive Hub
9. Camshaft Nut	18. Pin Retainer	27. Bushing	36. Locking Nut

to prevent stalling and maximum speed to prevent overspeeding. (See Fig. 26–11 to 26–13.)

The governor is driven at engine speed by a gear mounted on the governor drive shaft that engages a gear mounted on the pump camshaft. The governor drive shaft assembly consists of the weights, pivot pins, spider, shaft and friction clutch.

Friction clutch slippage adjustment is accomplished by adding or removing spacers from between the leaf springs and the spider.

The design of the friction clutch permits the weight and spider assembly to momentarily slip on its hub whenever sudden speed or load changes occur. This action dampens torsional vibrations in the

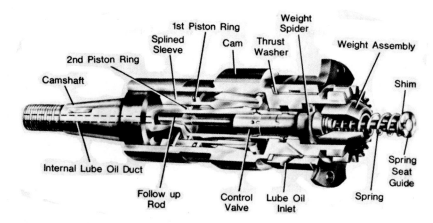

FIGURE 26-9. American Bosch distributor pump automatic injection timing device known as *Intravance.* This device provides up to 20 degrees of injection timing advance. Device is effective from 300 to 3200 rpm, depending on engine application. *(Courtesy of American Bosch, subsidiary of United Technologies Corporation)*

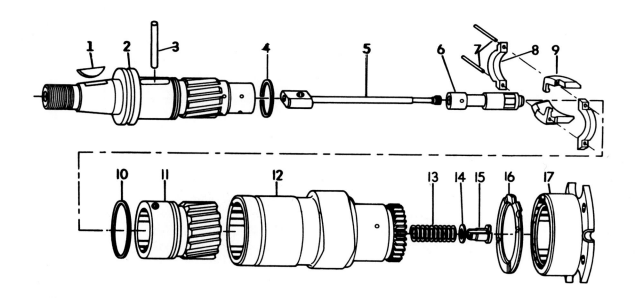

1. Woodruff key
2. Drive shaft
3. Cross pin
4. Seal ring
5. Follow-up rod
6. Control valve
7. Weight pins
8. Spiders
9. Weights
10. Seal ring
11. Splined sleeve
12. Cam
13. Spring
14. Shim
15. Nut
16. Thrust washer
17. Bearing plate assembly

FIGURE 26-10. Exploded view of internal advance mechanism. *(Courtesy of American Bosch, subsidiary of United Technologies Corporation)*

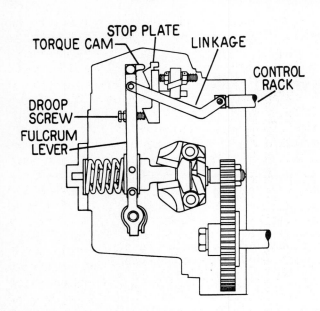

FIGURE 26-11. Governor at medium speed under load. *(Courtesy of American Bosch, subsidiary of United Technologies Corporation)*

pump camshaft and drive, and protects the pump and governor components. It also helps to minimize wear.

As the weights of the governor revolve, the centrifugal force tends to move them outward. The movement is opposed by the governor springs acting through the sliding sleeve assembly. If the speed is decreased, the spring force will exceed the centrifugal force of the weights and move the sliding sleeve toward the injection pump. Thus, any change in speed changes position of the sliding sleeve assembly.

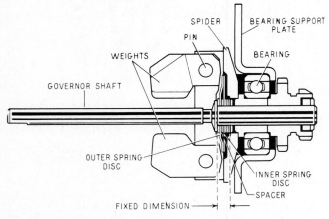

FIGURE 26-12. Governor friction clutch. *(Courtesy of American Bosch, subsidiary of United Technologies Corporation)*

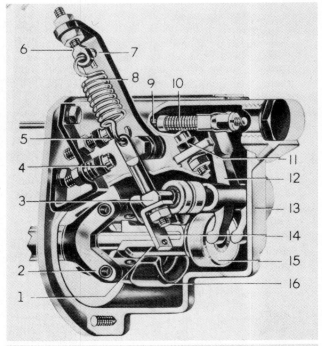

1. Sliding Sleeve	9. Control Rod
2. Spider Assembly	10. Control Rod Spring
3. Fulcrum Lever Shaft	11. Full Load Screw
4. Shutoff Screw	12. Fulcrum Lever
5. Idle Screw	13. Governor Housing
6. Spring Adjusting Screw	14. Shutoff Screw Assy.
7. Operating Lever	15. Thrust Bearing
8. Extension Spring	16. Flyweight

FIGURE 26-13. Governor with external spring used in applications where close regulation of torque is not required. *(Courtesy of American Bosch, subsidiary of United Technologies Corporation)*

The fulcrum lever bracket is attached to the fulcrum lever by means of a pin. The fulcrum lever bracket fits over the operating lever shaft but is not directly connected to it. The torsion spring hub is firmly secured to the operating shaft and connected through the torsion spring to the fulcrum lever bracket.

The ends of the spring straddle the tongue of the spring hub and fulcrum lever bracket with the result that the spring tension tends to keep these two parts in line with each other. It is normal for them to become separated during operation of the governor at full load position.

When the engine returns to normal speed as determined by foot pedal position, the torsion spring will bring the tongue of the spring hub and the fulcrum lever bracket in line.

A movement of the operating lever will rotate the spring hub and the fulcrum lever bracket, thereby causing the fulcrum lever to turn about the pivot pin. This movement changes the position of the pump control unit because the upper end of the fulcrum lever is connected by linkage to the control unit lever. The fuel quantity will be increased or decreased, depending on the direction of operating lever movement.

When the engine is operating under load and the load is removed, the speed would increase, the weights would move outward forcing the sleeve away from the pump, and the fulcrum lever would pull the control rod to a position of less fuel delivery. This condition is known as "high idle."

An adjustable maximum fuel delivery stop plate is provided. An extension of torque cam at the upper end of the fulcrum lever contacts the stop plate during periods when full load fuel quantities are required by the engine. The control lever travel is therefore limited by the position of the stop plate.

A speed variation from full load to low load is known as speed droop, and a droop screw adjustment is provided.

The angle of the torque cam in relation to the fulcrum lever affects the quantity of fuel increase at the lower speeds. It is very important that, once this angle has been set by the factory, no changes should be made in the angle since damage to the engine might result due to excess fuel delivery.

With the governor at medium speed under full load the reduced speed has caused the weight to collapse and the sleeve has moved toward the pump by the spring forces. The fulcrum lever connected to the sleeve has also moved toward the pump. Movement at the upper end is limited by the stop plate and therefore the fulcrum lever pivots about the droop screw with the movement being absorbed by the rotation of the fulcrum lever bracket against the torsion spring.

PART 4   SMOKE LIMITER (ANEROID)

The fuel aneroid is mounted on top of the governor housing. It is also called a smoke limiter. It reduces smoke anytime the engine goes from a light load to a full load condition. During light load conditions, fuel delivery is cut by the governor to meet engine requirements. If the engine load is suddenly increased, the governor reacts and increases fuel delivery. In the case of a naturally-aspirated engine, fuel delivery will be limited by the governor to en-

sure that a specific air-fuel ratio will not be exceeded. Looking at a naturally-aspirated engine as an air pump, at a given speed it will only pump so much air. If the air-fuel ratio gets too rich, smoke will occur as the result of incomplete combustion. If a turbocharger is added to the engine, fuel delivery can be increased and the turbocharger forces the engine to breath more air, so the air-fuel ratio is not exceeded; however, during sudden load changes, the turbocharger will not come up to speed quickly enough, and the engine smokes, acting momentarily like an overfueled naturally-aspirated engine. The fuel aneroid limits fuel delivery to approximate naturally-aspirated engine levels until boost pressure from the turbocharger working against a spring loaded diaphragm moves the stop plate in the direction of increased fuel delivery. This momentary lag in fuel delivery does not affect engine performance; all it does is eliminate the puff of smoke often associated with turbocharged engines.

PART 5   EXCESS FUEL STARTING DEVICE

In the normal start position, the spring holds the piston back in the housing, thereby allowing more fuel to be metered to the engine. The lube oil that is supplied to the device is normally taken from the rear camshaft bushing area and piped to the underside of the excess fuel starting device housing. With the engine stopped and no lube oil pressure within the device, the spring force moves the piston and stop plate back to the excess fuel starting position.

With the cranking motor engaged and the accelerator pedal depressed, an additional pressure is exerted on the stop plate by the fulcrum lever and cam. When the engine has started, lube oil pressure overcomes the spring force and the cam pressure, and the stop plate is literally pushed into "run" position. The stop plate will remain in the "run" position until the engine is shut down.

PART 6   AMERICAN BOSCH MODEL 75 ELECTRONICALLY CONTROLLED DISTRIBUTOR PUMP*

The present series of pumps includes 3, 4, 5 and 6 cylinder versions, driven at half engine speed and applicable to engines ranging in power from 50 to

*(Courtesy of United Technologies-American Bosch)

180 horsepower (134 kW). An internal cam automatically actuates 2, 3 or 4 plungers for uniformity of fuel delivery.

Description of Operation

The supply pump transfers the fuel from the fuel tank through a combination water separator and filter and produces a hydraulic pressure which varies as a function of pump rpm. This is controlled by the pressure regulator. (See Fig. 26-14 and 26-15.)

Fuel at supply pressure flows to the charging sump and is conducted through holes in the porting sleeves and spill to move the plungers out to the base circle of the internal cam ring.

As the distributor shaft rotates, the angled slots admitting fuel to fill the pumping chamber close off the ports in the spill sleeve. At a selected angular position of the distributor shaft, the plungers are forced inward by the cam. At this angle, or a slightly later angle, the angled slots close off the ports in the porting sleeve. This is called "port closing." The plungers now create pressure in the pumping chamber. This pressure is directed by the distributor shaft to one of the outlet ports, through the snubber or delivery valve, thus opening the nozzle.

At a further angle of the distributor shaft, the metering sleeve angled slot opens the axial passage of the distributor shaft to spill the remainder of the pumping stroke, thus ending the injection.

The electronic control module controls the fuel quantity by adjusting the axial position of the metering sleeve. It also programs the start of injection over the speed range by driving a servo valve, which

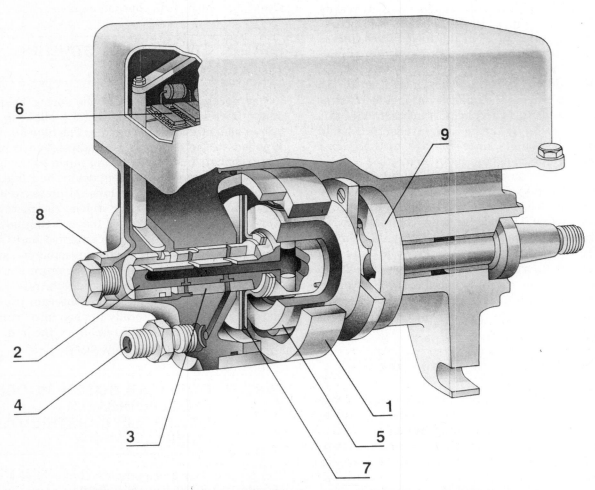

FIGURE 26-14. United Technologies Model 75 electronically controlled distributor injection pump cutaway view. *(Courtesy of American Bosch, subsidiary of United Technologies Corporation)*

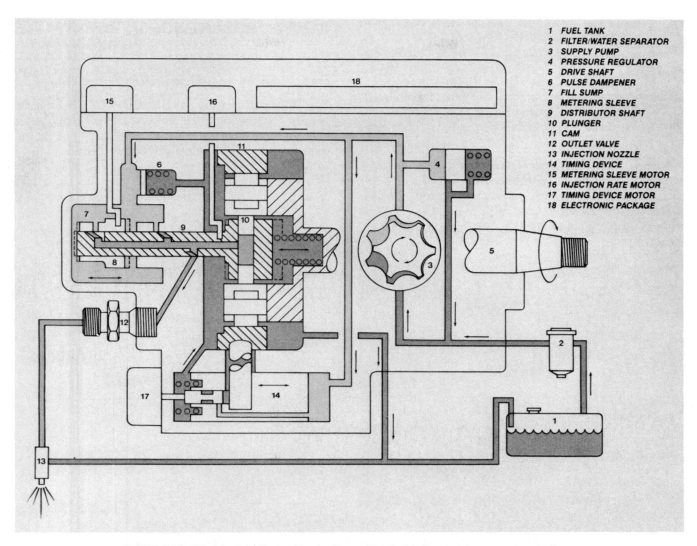

FIGURE 26–15. Model 75 electronically controlled injection pump schematic.
(Courtesy of American Bosch, subsidiary of United Technologies Corporation)

in turn specifies the correct hydraulic pressure for the advance mechanism function.

This servo valve controls the angular location of the injection cam and also, through ramped adjusted plates, controls the axial location of the distributor shaft. The angled port closing slot in the distributor shaft maintains the proper alignment of port closing with the cam.

The axial location of the distributor shaft can also be adjusted by the electronic control, independently of the cam, moving port closing to a different slope on the cam profile and thereby changing the injection rate.

A pulse dampener is built into the pressure gallery to minimize spill pulsations and aid in uniform filling.

Microprocessor based control device

Control capabilities for governing, timing and rate of injection varied as required during engine operation for more precise operation

Sleeve metering with constant beginning of injection

Integral supply pump

Fuel lubrication

Pressure capability for direct-injection and indirect-injection engines up to 850 bar (12,500 psi)

Pump displacement rate capability up to 3.75 mm^3 per crank angle degree

System self diagnostic feature with visual display on vehicle dashboard

On-engine service diagnostic access port

Starting and Warm-Up. With microprocessor controls, engine starting and warm-up is achieved by modulating fuel flow and injection timing based upon sensed engine parameters at no extra cost with no extra components.

Hydraulic/Electronic Timing Advance. The hydraulic timing advance mechanism operates on supply pump pressure acting upon a hydraulic piston to rotate the cam ring. Electronic modulation of cam ring rotation is controlled by sensing speed, load or other inputs.

PART 7   ROBERT BOSCH ROTARY DISTRIBUTOR PUMPS

Description and Operation*

The Robert Bosch model BR, CR, and VE Fuel Injection Pumps are single plunger, rotary distributor type pumps incorporating a hydraulic spill piston governor or a mechanical governor. (See Fig. 26-16 to 26-21.)

A single plunger, actuated laterally by a face cam, pressurizes the fuel in two separate circuits.

*Information in this section is based on information provided courtesy of International Harvester Company.

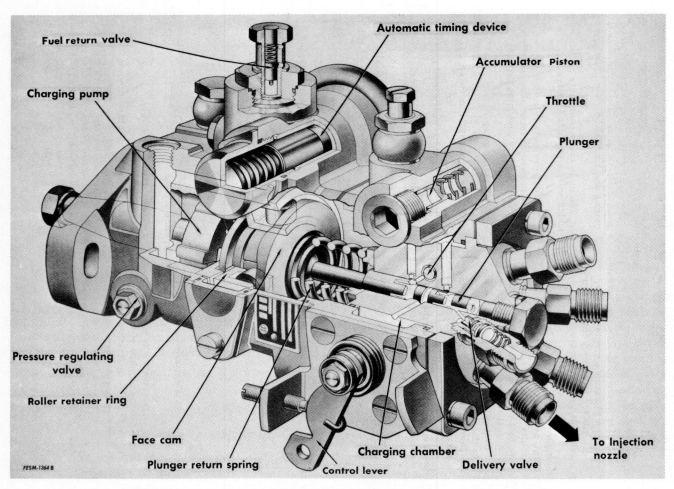

FIGURE 26-16. Cutaway of Robert Bosch injection pump. *(Courtesy of International Harvester Company)*

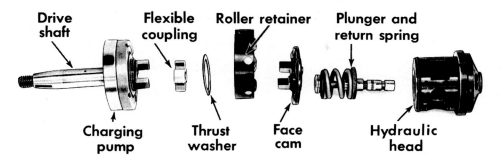

FIGURE 26-17. High-pressure pump components. *(Courtesy of International Harvester Company)*

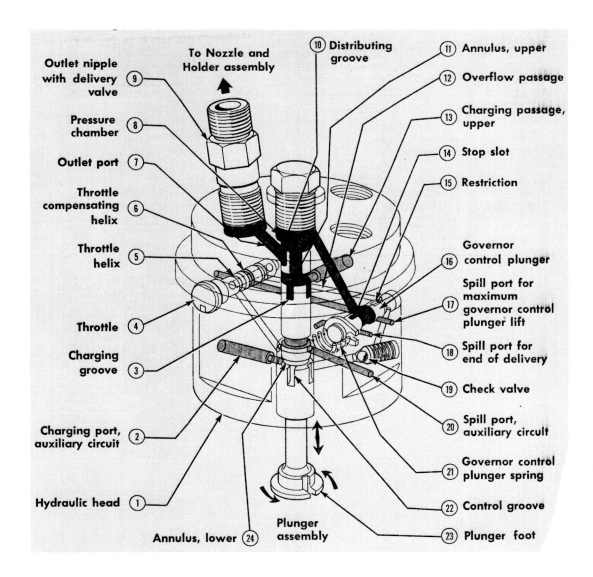

Outlet nipple with delivery valve ⑨

Pressure chamber ⑧

Outlet port ⑦

Throttle compensating helix ⑥

Throttle helix ⑤

Throttle ④

Charging groove ③

Charging port, auxiliary circuit ②

Hydraulic head ①

To Nozzle and Holder assembly

⑩ Distributing groove

⑪ Annulus, upper

⑫ Overflow passage

⑬ Charging passage, upper

⑭ Stop slot

⑮ Restriction

⑯ Governor control plunger

⑰ Spill port for maximum governor control plunger lift

⑱ Spill port for end of delivery

⑲ Check valve

⑳ Spill port, auxiliary circuit

㉑ Governor control plunger spring

㉒ Control groove

㉓ Plunger foot

Annulus, lower ㉔

Plunger assembly

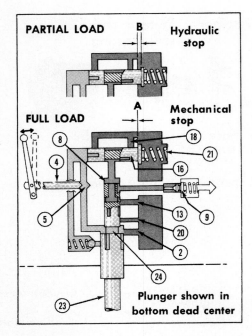

PARTIAL LOAD — Hydraulic stop

FULL LOAD — Mechanical stop

Plunger shown in bottom dead center

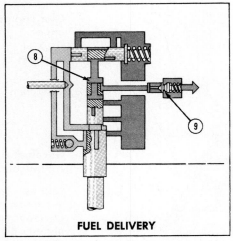

FUEL DELIVERY

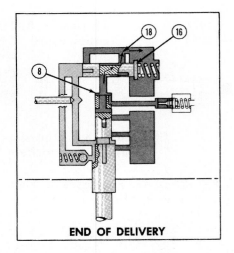

END OF DELIVERY

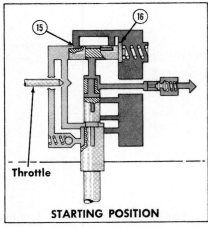

Throttle

STARTING POSITION

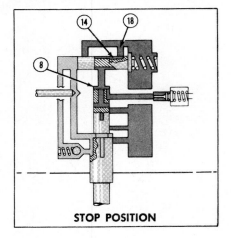

STOP POSITION

FIGURE 26-19. Various stages of fuel delivery and governor operation. *(Courtesy of International Harvester Company)*

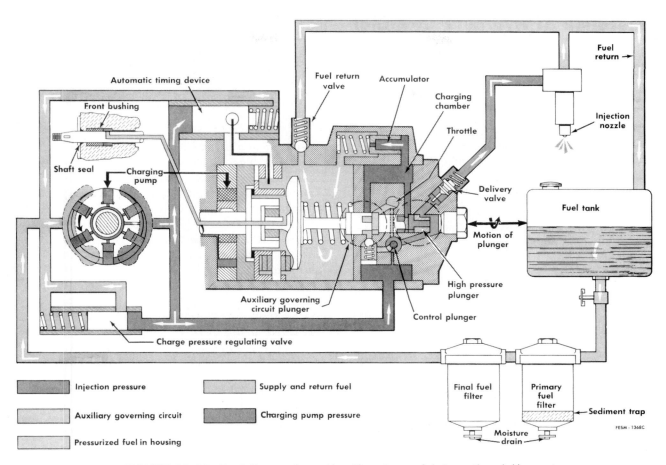

FIGURE 26-20. Fuel flow schematic. *(Courtesy of International Harvester Company)*

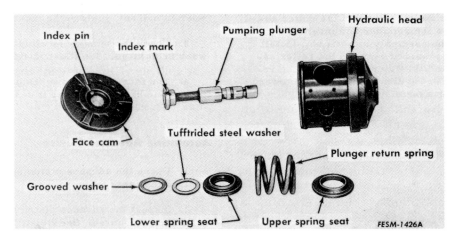

FIGURE 26-21. High pressure injection plunger and spring assembly. *(Courtesy of International Harvester Company)*

Primarily, it supplies high pressure fuel to the nozzles, and secondly a lower pressure fuel to an auxiliary circuit for controlling fuel delivery to the engine. At the same time, rotation of the plunger distributes pressurized fuel from the primary circuit to the individual nozzles.

The accessory units, such as the governor, charging pump and automatic timing device are located inside the pressurized fuel filled housing. Fuel is circulated within the pump housing to provide cooling and lubrication.

The main component parts of the pump, (explained separately) are:

1. High pressure injection plunger with hydraulic head (plunger also pressurizes fuel for auxiliary governing circuit).
2. Charging pump.
3. Pressure regulator valve.
4. Hydraulic variable speed governor.
5. Automatic timing device.
6. Accumulator.

The plunger is actuated by a face cam which has the same number of lobes as the engine has cylinders. A flexible coupling connects the face cam to the drive shaft. The cam lobes ride on four rollers in the roller retainer. This stationary assembly can be rotated for timing advance.

The plunger is indexed to the face cam. A plunger return spring holds the plunger and face cam against the rollers in the roller retainer assembly.

Referring to Figure 26-18, it can be seen that the plunger has two different diameters. The smaller diameter pressurizes and delivers metered fuel to the individual nozzles. It has an annulus (11) and charging grooves (3) with the pressure chamber (8). The plunger has a distributing groove (10) which connects the pressure chamber (8), with one outlet port (7) at a time, as the plunger rotates.

The larger diameter of the plunger supplies fuel under lower pressure and increased volume to an auxiliary governing circuit. This portion of the plunger also has an annulus (24) and six control grooves (22). On the injection stroke of the plunger, auxiliary fuel is delivered to both the throttle (4) and the check valve (19).

The displacement volume of the auxiliary fuel is constant for a given engine speed. Delivery starts as soon as the lower edge of the annulus (24) closes the charging port (2) on the injection stroke of the plunger foot (23). Delivery stops when the upper edge of the annulus opens the spill port (20) on the injection stroke of the plunger foot (23).

The hydraulic head (1) functions as the cylinder for the plunger foot (23), houses the delivery valves (9), the throttle (4) and the governor control plunger (16).

Fuel Delivery

The drive shaft, face cam and plunger rotate as one assembly. The face cam, which is held against the rollers by the plunger return spring, rotating against the cam rollers results in a reciprocating action of the plunger. This results in the charging and injection stroke of the plunger. Fuel enters the primary circuit from the upper charging passage (13) through the charging grooves (3), the upper annulus (11) and the central bore to the pressure chamber (8). The filling stops at bottom dead center of the plunger. On the injection stroke, when a filling groove on the rotating plunger has passed the passage (13), fuel is pressurized and directed by the distributing groove (10) through the various outlet ports (7) to the respective delivery valves (9) and to the nozzles.

End of Delivery

This injection pump, under all conditions of speed and fuel quantity, is described as having constant beginning of delivery. The pump controls the amount of fuel delivery by the spill port metering method, that is, variable end of delivery. The start of delivery, as related to crankshaft position, is varied by the automatic advance mechanism.

The governor control plunger (16) has a wide annulus in the middle, connecting with the pressure chamber (8). During the injection stroke, auxiliary fuel is pumped through the check valve (19) and exposed to one end of the control plunger, forcing it against the governor control spring (21) until the helix on the annulus of the control plunger aligns with the spill port (18). This ends the delivery as the main fuel now is permitted to flow back from the pressure chamber (8) to the charging chamber of the pump.

Delivery Valves

The delivery valves (Fig. 26-22) are designed to furnish line retraction and torque rise as needed for good performance. The delivery valve and its seat are selective fit and must not be interchanged.

All valves are identified with a number imprinted on the outside of the valve seat or on the flutes of the valve. Some have only the last six numerals of the complete number.

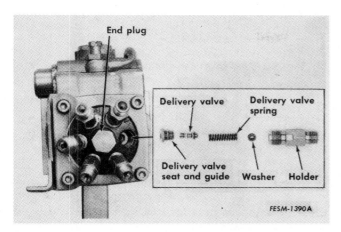

FIGURE 26-22. Delivery valve assembly. *(Courtesy of International Harvester Company)*

All valves have a retraction land, and some also have flats on the lands. The size of the flats, plus engine speed and fuel rate are considerations in determining the injection line "leak back" or line retraction.

Some valves have a torque land while others do not. If the valve has a torque land, it will have at least one flat. The presence or absence of a torque land, the number and size of the flats, plus fuel rate and engine speed are design considerations in providing the fuel delivery differential for overload torque rise.

Charging Pump

The vane type charging pump, having six steel vanes, is built around the drive shaft. The capacity of the charging pump is greater than the requirements of the engine. The excess fuel is used to cool and lubricate the pump and then returned to the fuel tank by an external line. Charging pump pressure is controlled by a pressure regulating valve. Pump output is directed to the automatic timing device as well as to the charging chamber around the hydraulic head. Charging pump pressure will vary with the speed of the engine.

PART 8   HYDRAULIC VARIABLE SPEED GOVERNOR

Note: The operation of the hydraulic governor will be covered by its three separate functions:

1. Fuel control at stop and starting positions
2. Fuel control at variable speed
3. Excess fuel for starting

Fuel Control at Stop and Starting Positions

The parts that control the fuel delivery are the throttle (4) (Figure 26-19), the governor control plunger (16), the governor control plunger spring (21), the check valve (19), and large diameter of the plunger and hydraulic head bore supplying fuel to the auxiliary governing circuit.

The throttle and governor control plunger are located in the hydraulic head with the same close tolerances as the plunger. The governor control plunger, operated by fuel from the auxiliary circuit, moves laterally to end the injection stroke. It can also be rotated by the operator for excess starting fuel or to shut-off the engine.

Rotation of the governor control plunger, for example in one direction, brings the stop slot (14) into alignment with the spill port (18), permitting all of the fuel from the pressure chamber (8) to flow back to the charging chamber.

The auxiliary fuel reaches one end of the governor control plunger (16) through the check valve (19) on the injection stroke of the main plunger. Another passage is provided for the auxiliary fuel leading to the throttle (4), then to the governor control plunger. On the charging stroke of the plunger, fuel cannot return from the governor control plunger past the check valve, but some of it is permitted to return past the throttle. The amount of fuel returned is controlled by the helix (5) on the throttle.

Fuel Control at Variable Speed

Engine speed is controlled by rotating the throttle (4). Helixes (5) and (6) open or restrict passages in accordance with the throttle position as established by the operator.

The controlling action of the hydraulic governor is based on the time taken for the return stroke of the governor control plunger, which is determined by the throttle setting.

The speed that the plunger travels, against the governor control plunger spring, depends on the reciprocating speed of the main plunger. The length of the stroke is determined by the amount of fuel supplied to the auxiliary circuit as controlled by the adjustment of the maximum fuel stop screw.

The speed of the return stroke is determined by the force of the governor control plunger spring and the restriction of fuel escaping past the throttle helices.

When the load is reduced, the engine tends to build up speed. Fuel is trapped beneath the plunger

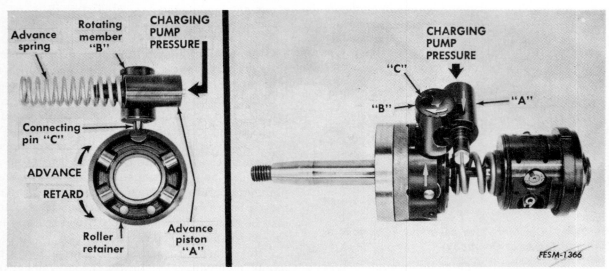

FIGURE 26-23. Automatic injection timing advance unit. *(Courtesy of International Harvester Company)*

and it, therefore, cannot return to its mechanical end of retraction. This hydraulically limited travel is called the hydraulic stop. This is more clearly shown in Figure 26-19 where the dimension "A" designates the full mechanical stroke of the control plunger. The inset dimension "B" shows the reduced travel of the governor control plunger as determined by the hydraulic stop.

On the next stroke of the main plunger, the governor control plunger has a shorter distance to travel until it opens the spill port from the pressure chamber. This results in a decreased fuel delivery. This reduction in fuel delivery is in proportion to the height of the hydraulic stop.

The governor control plunger does not reach the mechanical stop at any normal or overload speeds. At rated load speed the control plunger returns to a hydraulic stop. When the engine is overloaded the reduced speed allows an increased time for the return stroke of the governor control plunger which allows it to reach a point *closer* to the mechanical stop. Therefore, there is an increase of fuel delivery per injection stroke. This may not be apparent in the flow bench testing, but it is a fact.

Starting Position for Excess Fuel

Fuel delivery is at its maximum when the governor control plunger is rotated so that it does not open the high pressure circuit, that is, when there is no reciprocating movement of the control plunger. To accomplish this, a restriction (5) on the control plunger is positioned across the passage between the plunger section for the auxiliary fuel circuit and the charging chamber. This prevents the control plunger from moving toward the end of delivery position. For starting, the operator manually holds the control plunger in excess fuel position.

When the engine starts and attains sustaining speed, the auxiliary fuel quantity is increased to a point where it no longer can escape through the restriction. Pressure builds up behind the control plunger and it starts moving, thus reducing the fuel quantity by spill port action, even if the control lever is not immediately returned to the normal operating position. Although the restriction is still in operation, the engine speed will rise to approximately 1000 rpm.

PART 9   ACCUMULATOR (SURGE PISTON)

The movement of the face cam within the pump housing, the spill action of the control plunger and the displacement of fuel by the main plunger as it reciprocates create pressure fluctuations which are not desirable in the charging chamber. To dampen these fluctuations, a hydraulic accumulator is provided between the charging chamber and pump housing. Pressure surges in the charging chamber cause the accumulator piston to move against the force of the spring. A decrease in pressure in the charging chamber will cause the spring to force the piston back.

The design of the accumulator piston allows a constant flow of fuel to enter the pump housing for cooling and lubrication. This is accomplished in one of three ways depending on the version of the pump and its application.

1. An orifice in the accumulator piston
2. A flat on the side of the piston
3. The clearance between the piston and its bore

PART 10   AUTOMATIC TIMING DEVICE

Automatic timing advance (Figure 26-23) is accomplished by the rotation of the cam roller retainer. Charging pump fuel pressure is exposed to one end of the spring loaded advance piston "A." As the pressure increases, due to an increase in pump speed, the piston moves over against the pressure of the advance spring. This moves the rotating member "B" in the direction shown. Connecting pin "C," eccentrically located in the rotating member, rotates the roller retainer in the advance direction. The piston will continue the advance until it contacts the end plug.

As the pump speed decreases, the charging pump pressure likewise decreases, and the spring forces the advance piston "A" back, rotating the roller retainer in the retard direction.

The amount of advance is controlled by the travel of the advance piston. On older version pumps this is regulated by the amount of shims on the end of the piston. On the newer pumps it is controlled by the long end cap. The rotating member "B" is also used in determining the amount of advance.

End caps are marked with a number on the outside end. See the Specifications list for the correct end cap number for the application. The rotating member is also identified with a number. The Specifications list shows the number of the rotating member for each application.

The start of advance is controlled by the charge pump pressure and advance spring. Start of injection timing can be varied by the shims under the advance spring.

PART 11   ELECTRIC FUEL SHUT-OFF

Referring to Figure 26-24, with the electromagnet (1) engaged, the throttle (3) rests under spring pressure against the stop pin (2). This is the normal condition while the engine is running.

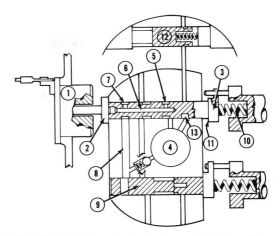

FIGURE 26-24. Electric fuel shut-off device. *(Courtesy of International Harvester Company)*

Fuel is directed from the auxiliary governing circuit (4) via the control edge (5) through the center port (6) into the rear annulus (7) and via the connecting port (8) toward the control plunger (9). The pump is shut-off by switching off the current supply to the electromagnet. With the electromagnet de-energized, the spring (10) pushes the throttle back until the shoulder of the throttle (11) rests against the hydraulic head. In this position of the throttle, the control edge (5) has closed the connection to the auxiliary governing circuit (4), i.e., the connection between the governing circuit and the control plunger is interrupted. This means that the control plunger is lifted during the upward stroke of the pump plunger via the check valve but the plunger still rests against the hydraulic stop. The control plunger has in this position opened the high pressure circuit via the helical control edge, thus the charging pump cannot build up pressure and no fuel delivery is possible.

For injection pumps with automatic excess fuel, the auxiliary governing circuit can be opened towards the charging chamber of the charging pump via starting piston (12). (Refer to the description of the automatic excess fuel starting device.) At speeds in excess of the low idle speed, the starting piston closes this connection and the excess fuel is cut-off. When shutting off via the electromagnet, blocking of the auxiliary governing circuit alone is not sufficient, the so-called starting circuit has likewise to be shut-off. This is done by establishing the connection between the auxiliary governing circuit and the starting piston, not directly, but via the throttle at the groove (13). During the lateral movement of the throttle this connection is then closed.

PART 12 FUEL INJECTION SYSTEM DIAGNOSIS (AMERICAN BOSCH)

TROUBLE - SHOOTING GUIDE

Cause	EXCESSIVE ENGINE SMOKE - BLACK	ROUGH IDLE, ENGINE VIBRATING	POOR PERFORMANCE, SURGE, ERRATIC ACTION	LOW POWER	EXCESSIVE ENGINE SMOKE - WHITE - BLUE	LUBE OIL DILUTION	ENGINE OVERSPEEDS	HARD STARTING	Recommended Remedy
1. FUEL SUPPLY PRESSURE									
a. Primary Filter Clogged or Restricted.	x	x							Remove and Clean or Replace.
b. Final Stage Filter Clogged or Restricted.	x	x							Remove and Replace.
c. Air Leaks in Fuel Suction System.	x	x	x					x	Inspect and Correct. Replace Parts where Required.
d. Reducers or Restrictions in Fuel Suction Lines.	x	x	x						Clean and Repair or Replace Parts as Required.
e. Fuel Tank Improperly Vented.	x	x	x					x	Clean or Replace Parts as Required.
f. Supply Pump Worn.	x	x	x					x	Remove Supply Pump. Repair by Authorized Personnel.
g. Supply Pump Regulating Valve Worn or Stuck.	x	x	x					x	Remove Supply Pump. Repair by Authorized Personnel.
h. Check Valve Leaking or Blocked.	x	x	x					x	Clean or Replace
i. Fuel Return Line Restricted.		x	x						Clean and Flush. Replace if Necessary.
2. TIMING									
a. Injection Pump to Engine Timing Incorrect.	x	x	x	x				x	Re-Time to Engine Manual Specifications.
b. Pump Drive Worn - Drive Gear, Keyway.	x	x	x						Remove, Inspect and Replace Worn Parts.
c. Plunger Drive Broken.		x	x	x				x	Remove Injection Pump. Repair by Authorized Personnel.
3. GOVERNOR									
a. Throttle Linkage Mis-Adjusted.	x	x	x						Adjust to Engine Manual Specifications.
b. Throttle Linkage Sticking or Binding.	x	x	x						Check for Binding, Worn or Loose Parts, Foreign Particles.
c. Shut-Off Lever Restricting Governor Operation.	x	x	x						Adjust to Engine Manual Specifications.
d. High Idle Screw Adjustment Incorrect.	x						x		Re-Adjust to Manual Specifications.
e. Low Idle Screw Adjustment Incorrect.		x							Re-Adjust to Manual Specifications.
f. Incorrect Governor Spring or Setting.	x	x	x				x		Re-Adjust to American Bosch Specifications.
g. Incorrect Friction Clutch Slippage Torque.	x	x					x		Re-Adjust by Authorized Personnel.
h. Control Rod Binding or Sticking.	x	x	x						Repair by Authorized Personnel.
4. CALIBRATION									
a. Fuel Delivery Incorrect.	x	x		x				x	Re-Calibrate by Authorized Personnel.
b. High Pressure Tubings Clogged or Restricted.	x	x	x						Remove, Check Inside Diameters, Drill or Ream, Clean and Flush.
c. Inadequate Lube Oil Pressure to Excess Fuel Device.		x	x	x					Clean Lines, Repair Injection Pump or Lube Pump as Required.
d. Excess Fuel Device Sticking in "Start" Position.		x		x					Remove, Repair or Replace, and Re-Set by Authorized Personnel.
e. Excess Fuel Device Sticking in "Run" Position.								x	Remove, Repair or Replace, and Re-Set by Authorized Personnel.
5. HYDRAULIC HEAD									
a. Injection Pump Plunger Worn or Scored.								x	Remove Pump. Repair by Authorized Personnel.
b. Injection Pump Plunger Sleeve Sticking or Binding.	x	x	x					x	Remove Pump. Repair by Authorized Personnel.
c. Injection Pump Control Unit Sticking or Binding.	x	x	x					x	Remove Pump. Repair by Authorized Personnel.
d. Delivery Valve Sticking or Leaking.	x	x	x					x	Remove, Clean, Repair or Replace as Required.
e. Delivery Valve Spring Broken.	x	x	x	x				x	Replace and Re-Set Opening Pressure.
6. NOZZLES									
a. Nozzles Defective - Leaking - Worn.	x	x	x	x				x	Remove, Replace or Repair, Reassemble, Test, Set Opening Pressure.
b. Incorrect Nozzle Opening Pressure.	x	x	x	x				x	Re-Set to American Bosch Specifications.
c. Nozzle Cap Nut Incorrectly Torqued.	x	x	x	x				x	Remove, Retighten Cap Nut, Replace Copper Gasket, Clean Engine Recess and Re-Install in Engine.
d. Nozzle Incorrectly Torqued in Engine.	x	x	x	x				x	Remove, Replace Copper Gasket, Clean Recess, Reassemble to Engine (Tighten Evenly to Required Torque).
e. Nozzle Valve Sticking.	x	x	x	x				x	Remove, Clean, Repair or Replace as Required.
f. Nozzle Spray Holes Plugged or Partially Plugged.	x	x	x						Remove Nozzle. Clean Holes or Replace Nozzle as Required.
7. OIL LEAKAGE									
a. Defective Supply Pump Seal.						x			Replace by Authorized Personnel.
b. Defective Hydraulic Head "O" Ring.						x			Replace by Authorized Personnel.
c. Defective Control Unit "O" Ring.						x			Replace by Authorized Personnel.
d. Loose Control Unit Fit.						x			Remove Pump. Repair by Authorized Personnel.
8. GENERAL									
a. Air Cleaner Restriction.	x	x	x	x					Remove and Clean.
b. Excessive Lube Oil in Air Cleaner.					x	x			Drain, Clean and Refill per Engine Manual Specifications.
c. Defective Tachometer.	x	x	x			x			Recheck with Master Tachometer. Replace if Necessary.
d. Cranking R.P.M. too Low (Cold or Hot).								x	Check Battery and Starter. Replace if Necessary.
e. Low Engine Compression.	x		x					x	Check Compression and Correct as Required.
f. Incorrect Engine Valve Lash.	x		x	x					Check and Correct as Required.
g. Water in Fuel.	x	x				x		x	Drain Water from Filters and Fuel Tank. If Necessary, Drain Fuel Tank and Refill.

Courtesy of United Technologies, American Bosch

PART 13   INJECTOR SERVICE

Cleaning Procedure

1. Wash nozzle and holder assemblies in Varsol or equivalent cleaning agent. Before washing, be sure to cap the nozzle holder inlets to prevent foreign matter in the cleaning solution from entering the nozzle and holder assembly.

2. If necessary, nozzle and holder assemblies can be wire brushed; however, tip (spray holes) of nozzle and holder threads must not come in contact with the wire brush.

3. Remove nozzle from holder, then remove nozzle valve from body.

4. Probe nozzle spray holes with correct size cleaning needle. Wash body and valve thoroughly. Then dip in clean test oil and reassemble.

5. Re-install nozzle in holder, using proper centering sleeve. Tighten cap nut to specified torque.

Testing

1. Connect each nozzle and holder assembly to the test stand.

2. Check to be sure that the valves are seating correctly and that there is no seat leakage. Also check to be sure that the nozzle produces good chatter.

3. Check to make sure that the nozzle opening pressures are within factory specifications. Readjust nozzle opening pressure if necessary.

PART 14   INJECTION PUMP REMOVAL AND INSTALLATION (MODEL 75-TYPICAL 6 CYL.)

Fuel Injection Pump Removal

1. Clean area around pump particularly around fittings, inspection cover.

2. Remove all lines, high pressure, inlet oil supply and return. Cap all open fittings to prevent dirt from getting in.

3. Remove inspection cover in front of pump drive gear and remove gear.

4. Remove flange bolts and remove pump.

Fuel Injection Pump Installation and Static Timing Procedure

1. Rotate engine until exhaust valve is just closing and intake valve is just opening on No. 6 cylinder and "TDC MARK" on pulley is in line with pointer. No. 1 and No. 2 push rods should be free.

2. Turn crank pulley counterclockwise past T.D.C. to 45 degrees BTDC on pulley then clockwise to degree setting per engine specification; do this firmly without "bumping" to avoid gear bounce which will create a backlash error.

3. Install F.I.P. with capscrews, flat washers, lock washers and nuts and tighten to specs. Verify that timing pointer and mark on F.I.P. gear drive hub are aligned.

Note: Service replacement F.I.P.'s may be friction locked in the No. 1 cylinder position. Do not turn pump shaft except to correct minor errors (less than .12 inch) then rotate hub so that the mark on the hub and pointer are exactly aligned.

4. Carefully, install pump drive gear onto drive hub in the approximate mid position of adjustment slots. Remember gear will rotate slightly while engaging teeth. Assure timing marks are still in alignment.

5. Install capscrews and flat washer through gear into the pump drive hub, tighten one at a time finger tight, checking alignment after each. Tighten all capscrews to specs. Assure correct alignment of pointer and F.I.P. drive hub visible through observation hole.

6. Remove friction locking screw with aluminum pin from F.I.P. housing which is located directly behind the F.I.P. mounting flange on outboard side of the F.I.P. (Throttle Lever Side) and replace with 1/4" pipe plug. Assure gear backlash is within specifications.

7. As a final timing confirmation, turn crank pulley counterclockwise past specified timing to 45 degrees BTDC then return (clockwise rotation) to specified timing and verify that timing marks are aligned. If not aligned, repeat steps 4 & 5 until alignment is reached.

8. Install gear cover and observation hole cover with capscrews and lock washers and tighten to specs.

9. Adjust throttle lines and idle speed to specifications.

If the replacement pump is not locked in No. 1 position, No. 1 position will have to be determined by one of the following methods.

Flow Timing Method of Positioning Pump on No. 1 Cylinder

1. Remove nozzle line from No. 1 outlet on pump.

2. Fabricate a port closing adapter. Install adapter on No. 1 outlet.

3. Remove delivery valve caps, remove delivery valve and spring, and replace cap.

4. Attach a clean fuel supply to hydraulic head inlet and plug return outlet.

5. Place throttle in 1/2 to wide open position.

6. Turn pump drive hub (or engine) in clockwise (viewing drive end) direction until a bubbleless stream of fuel is flowing from the adapter. Continue turning til flow stops. This is port closing. The pump timing pointer and mark on drive hub should now be in alignment. The pump is now in No. 1 position.

7. Replace delivery valve and torque per the torque chart.

Note: If there is no flow prior to coming up on mark the pump is probably coming up on No. 6.

Note: Pump drive hub turns at engine speed.

Face Gear Timing Mark Method of Positioning Pump on No. 1

Note: Flow timing method should be used if the service man has no method of resealing the pump.

1. Cut seal wire between control cover screw and top governor cover screw.

2. Remove control cover.

3. Turn pump until line on face gear comes into opening.

4. Align pointer and mark and pump drive hub while maintaining eye contact with line on face gear.

Note: The line only has to be visible in the window (usually near the front). It does not have to align with any mark in the housing.

The pump is now in the No. 1 position.

5. Install the pump drive gear while maintaining hub mark to pointer alignment.

6. Replace control cover and seal wire in appropriate attaching screws.

PART 15   INJECTION PUMP SERVICE (AMERICAN BOSCH)

Pump-Cleaning and Flushing

Cleaning Procedure

1. Thoroughly wash external surfaces of the injection pump with cleaning fluid. Make certain that all foreign matter is removed.

2. Remove inspection window cover or timing window cover. If applicable, remove or open drain plugs or fittings.

3. Remove governor top cover.

4. Flush lube oil compartments in pump and governor with approved cleaning fluid, such as Varsol, to remove all traces of dirt, grease, carbon and other foreign matter.

Note: After removing inspection covers, check components of governor for damage or binding of parts. Make certain fulcrum lever operates freely. Also check pump components for faulty or damaged parts, including broken spring, broken or worn control unit or sleeves, binding control rod or rack, broken plunger(s) or camshaft, cracked or damaged housing, etc.

5. Re-install covers.

Flushing Procedure

1. Mount pump on approved test stand.

Note: It may be necessary to replace the original pump drive hub and/or coupling with a coupling and/or hub that will fit the particular test stand being used.

2. Lubricate pump as follows: PSB, PSJ & PSM type—use a pressurized lube oil system (system and ducted adapters and fixtures available from Bacharach Instrument Co., Pittsburgh, Pa.) filled with SAE 30 oil, or equivalent, to properly lubricate the pump. The lube oil pressure must be 20 psi (1.38 Bars) minimum or, if pump includes lube oil pressure adjusting devices, a pressure of 35–40 psi (2.41–2.76 Bars) is required.

Note: Do not splash-lubricate PSB, PSJ or PSM type pumps, as this provides insufficient lubrication to certain components.

Important: Do not use calibrating oil or fuel oil as a lubricant.

Caution: Do not allow lube oil to mix with or contaminate calibrating or fuel oil.

3. Connect a set of high pressure tubings to the pump outlets and direct the opposite ends of the tubings into a pail, thereby bypassing the nozzle and holder assemblies and test stand tank.

4. To flush pump sump, proceed as follows: PSB, PSJ and PSM types—operate at a speed of about 1000 rpm with the operating lever in the full load position for approximately three minutes.

This procedure removes any fuel oil and foreign material from the pump sump and eliminates the possibility of contaminating the test oil or damaging the nozzles during the calibration check.

5. Stop test stand and remove the high pressure tubing.

Calibration Check

Test Set-Up

1. Install a good, matched set of calibration nozzle and holder assemblies or the engine nozzle and holder assemblies where specified on the test stand. Refer to calibration data for applicable pump to determine which type nozzle and holder assemblies are to be used.

2. Install required high pressure tubings. Refer to applicable calibration data for required tubing dimensions. Use a drill of the proper size to check the ends of each tubing. Drill and flush tubings that have crimped ends.

3. Install a transparent, plastic suction hose from the test oil tank to the inlet side of the fuel supply pump.

Calibrating

1. Observe fuel (test) oil entering supply pump. This must be free of air bubbles. Be sure to eliminate suction air leaks.

2. Check to be sure that supply pressure is adequate.

3. Check pump calibration in accordance with the data for applicable pump. Record the fuel delivery as received and also the final fuel deliveries. Use calibration form to provide a record of test performance for customer's information and agency's files. Identify the pump by serial number.

PART 16   SELF-CHECK

---

1. American Bosch distributor pumps are all flange mounted. True or False.

2. What is the speed ratio between the pump plunger and the pump camshaft in the American Bosch distributor pump?

3. How many cam lobes are there on the American Bosch distributor pump for a six-cylinder engine?

4. The quantity of fuel delivered per stroke in this pump is determined by the position of the _____ in relation to the _____.

5. What is the purpose of the delivery valve?

6. What two types of advance mechanisms are used with the PS series of pumps?

7. The governor controls engine _____ and _____ speed.

8. What is the purpose of the fuel aneroid?

9. The Robert Bosch rotary distributor pump plunger has two diameters, each serving a separate function. What are they?

10. The Robert Bosch rotary pump face cam has one lobe for each engine cylinder. True or False.

11. What are the three functions of the Robert Bosch rotary pump hydraulic variable speed governor?

12. During injection pump installation, static timing requires that number one piston be at the _____ position.

Chapter 27

Stanadyne Diesel Systems (Roosa Master) and CAV Distributor Pumps

Performance Objectives

After thorough study of this chapter, the appropriate training models and service manuals, you should be able to do the following with respect to diesel engines equipped with Stanadyne (Roosa Master) or CAV distributor injection pumps:

1. Complete the self-check questions with at least 80 percent accuracy.

2. Describe the basic construction and operation of the Stanadyne (Roosa Master) and CAV distributor injection pumps including the following:

    (a) method of pumping and metering

    (b) delivery valve action

    (c) lubrication

    (d) injection timing control

    (e) governor types

3. Diagnose basic fuel injection system problems and determine the required corrections.

4. Remove, clean, inspect, test, measure, repair, replace, and adjust fuel injection system components as needed to meet manufacturer's specifications.

This chapter is based on information courtesy of Diesel Systems Group, Stanadyne Inc., and CAV Limited.

PART 1 STANADYNE (ROOSA MASTER) DISTRIBUTOR INJECTION PUMPS

The following description of Stanadyne distributor injection pumps is provided courtesy of Stanadyne Diesel Systems and includes the microprocessor-controlled pump.

These pumps use a helix type metering valve. The radial position of the valve is controlled by throttle position and the governor. Cam-ring-operated pumping plungers pump high-pressure fuel for injection to the distributor head, where fuel is distributed to individual discharge ports for each engine cylinder.

The microprocessor-controlled injection pump uses a number of engine sensors to provide information to the microprocessor. This information is processed to provide the best possible fuel delivery for every engine operating condition and mode.

PART 2 PUMP OPERATION

The main rotating components of the D series of Stanadyne distributor pump are the drive shaft, transfer pump blades, distributor rotor, and governor.

With reference to Figure 27–1, the drive shaft (1) engages the distributor rotor (2) in the hydraulic head (6). The drive end of the rotor incorporates two or four pumping plungers (4) depending on pump model.

The plungers are actuated toward each other simultaneously by an internal cam ring (5) through rollers and shoes, which are carried in slots at the drive end of the rotor. The number of cam lobes normally equals the number of engine cylinders.

The transfer pump (3) at the rear of the rotor is of the positive displacement vane type and is enclosed in the end cap (7). The end cap also houses the fuel inlet strainer and transfer pump pressure regulator. The face of the regulator assembly is compressed against the liner and distributor rotor and forms an end seal for the transfer pump. The injection pump is designed so that the end thrust is against the face of the transfer pump pressure regulator. The distributor rotor incorporates two charging ports and a single axial bore with one discharge port to service all head outlets to the injection lines.

The hydraulic head contains the bore in which the rotor revolves, the metering valve bore, the charging ports, and the head discharge outlets. The

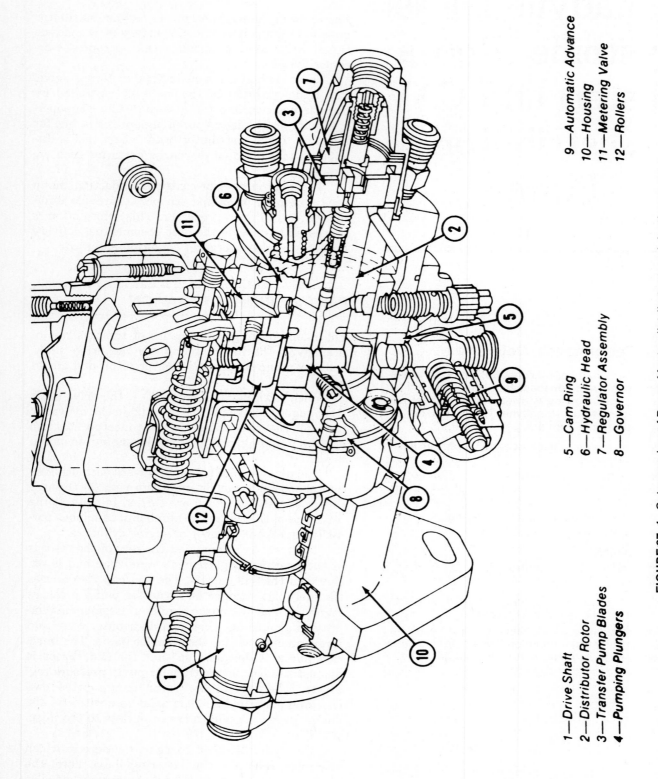

FIGURE 27-1. Cutaway view of Roosa Master distributor-type injection pump. Note pumping plungers and metering valve. *(Courtesy of Diesel Systems Group—Stanadyne, Inc.)*

1—Drive Shaft
2—Distributor Rotor
3—Transfer Pump Blades
4—Pumping Plungers

5—Cam Ring
6—Hydraulic Head
7—Regulator Assembly
8—Governor

9—Automatic Advance
10—Housing
11—Metering Valve
12—Rollers

high-pressure injection lines to the nozzles are fastened to these discharge outlets.

A high-pressure relief slot is incorporated in some regulators as part of the pressure-regulating slot to prevent excessively high transfer pump pressure, if the engine or pump is accidentally overspeeded.

The transfer pump works equally well with different grades of diesel fuel and varying temperatures, both of which affect fuel viscosity. A unique and simple feature of the regulating system offsets pressure changes caused by viscosity difference. Located in the spring adjusting plug is a thin plate incorporating a sharp-edged orifice. The orifice allows fuel leakage past the piston to return to the inlet side of the pump. Flow through a short orifice is virtually unaffected by viscosity changes. The pressure exerted against the back side of the piston is determined by the leakage through the clearance between the piston and the regulator bore and the pressure drop through the sharp-edged orifice. With cold or viscous fuels, very little leakage occurs past the piston. The additional force on the back side of the piston from the viscous fuel pressure is slight. With hot or light fuels, leakage past the piston increases. Fuel pressure in the spring cavity increases also, since the flow past the piston must equal the flow through the orifice. Pressure rises due to increased piston leakage and pressure rises to force more fuel through the orifice. This variation in piston position compensates for the leakage that would occur with thin fuels and design pressures are maintained over a broad range of viscosity changes.

Charging Cycle

As the rotor revolves, the inlet passages in the rotor register with the ports of the circular charging passage. Fuel under pressure from the transfer pump, controlled by the opening of the metering valve, flows into the pumping chamber, forcing the plungers apart. (See Fig. 27–2 to 27–4.)

The plungers move outward at a distance proportionate to the amount of fuel required for injection on the following stroke. If only a small quantity of fuel is admitted into the pumping chamber, as at idling, the plungers move out a short distance. Maximum plunger travel and, consequently, maximum fuel delivery are limited by the leaf spring, which contacts the edge of the roller shoes. Only when the engine is operating at full load will the plungers move to the most outward position. Note that while the angled inlet passages in the rotor are in registry

with the ports in the circular charging passage, the rotor discharge port is not in registry with a head outlet. Note also that the rollers are off the cam lobes. Compare their relative positions.

Discharge Cycle

As the rotor continues to revolve, the inlet passages move out of registry with the charging ports. The rotor discharge port opens to one of the head outlets. The rollers then contact the cam lobes, forcing the shoes in against the plungers, and high-pressure pumping begins. (See Fig. 27–5 to 27–7.)

The beginning of injection varies according to load (volume of charging fuel), even though rollers may always strike the cam at the same position. Further rotation of the rotor moves the rollers up the cam lobe ramps, pushing the plungers inward. During the discharge stroke the fuel trapped between the plungers flows through the axial passage of the rotor and discharge port to the injection line. Delivery to the injection line continues until the rollers pass the innermost point on the cam lobe and begin to move outward. The pressure in the axial passage is then reduced, allowing the nozzle to close. This is the end of delivery.

The delivery valve operates in a bore in the center of the distributor rotor. Note that the valve requires no seat—only a stop to limit travel. Sealing is accomplished by the close clearance between the valve and bore into which it fits. Since the same delivery valve performs the function of retraction for each injection line, the result is a smooth running engine at all loads and speeds.

PART 3   ROOSA MASTER INJECTION TIMING

Roosa Master basic injection timing is achieved by proper alignment of timing marks at the pump drive mounting. An automatic hydraulic advance mechanism controls timing in relation to engine speed. (See Fig. 27–8 and 27–9.)

The Roosa Master design permits the use of a simple hydraulic servomechanism, powered by oil from the transfer pump, to rotate the normally stationary cam ring to advance injection timing. Transfer pump pressure, increasing with speed, operates the servoadvance piston against spring pressure as required along a predetermined timing curve.

The automatic advance may also be used to retard injection, as an aid in cold starting, by starting the cam advance from a retard position.

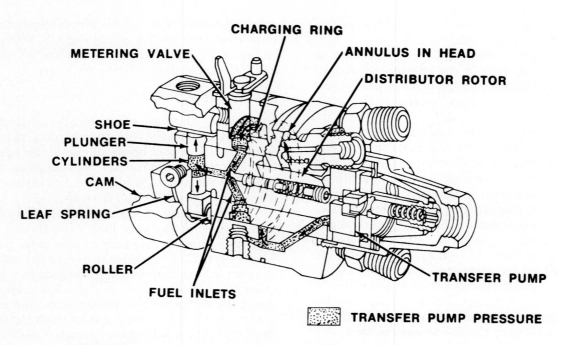

FIGURE 27-2. Roosa Master pump metering is achieved by three controlling factors: (1) Transfer pump fuel pressure, (2) metering valve position, and (3) leaf spring tension keeping the pumping plungers together. *(Courtesy of Diesel Systems Group—Stanadyne, Inc.)*

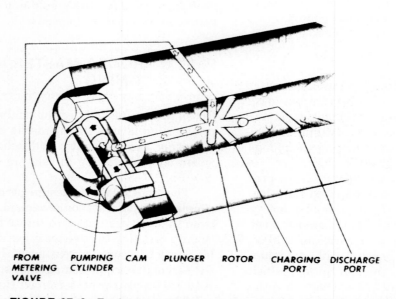

FIGURE 27-3. Fuel pressure force between plungers forces plungers apart against leaf-spring pressure during metering. (Leaf spring not shown here; see Figure 27-4). *(Courtesy of Diesel Systems Group—Stanadyne, Inc.)*

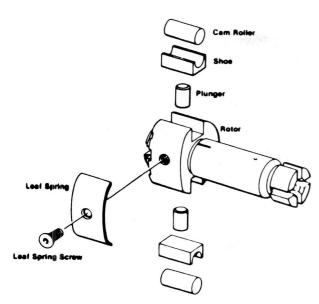

FIGURE 27-4. Exploded view of rotor assembly. *(Courtesy of Diesel Systems Group—Stanadyne, Inc.)*

Controlled movement of the cam in the pump housing is induced and limited by the action of the power piston and spring of the automatic advance against the cam advance screw.

During cranking the cam is in the retard position, since the force of the advance spring is greater than transfer pump pressure. As the engine rpm and transfer pump pressure increase, oil entering the advance housing behind the power piston moves the cam. Any amount of advance may be provided but the limit is 14 pump degrees. A ballcheck valve is provided to offset the normal tendency of the cam to return to the retard position during injection.

PART 4   MECHANICAL ALL-SPEED GOVERNOR

The governor serves the purpose of maintaining the desired engine speed within the operating range under various load settings. (See Fig. 27–10.)

In the mechanical governor the movement of the weights acting against the governor thrust sleeve rotates the metering valve by means of the governor arm and linkage hook. This rotation varies the registry of the metering valve opening to the passage

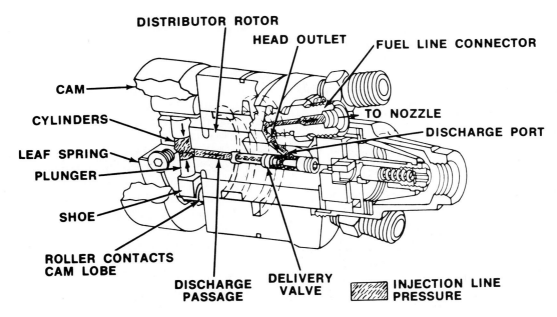

FIGURE 27-5. Continued rotation of the rotor causes plungers to be forced closer together, forcing fuel out of high-pressure discharge passage to injector. Inlet passage is closed when rotor is in this position. *(Courtesy of Diesel Systems Group—Stanadyne, Inc.)*

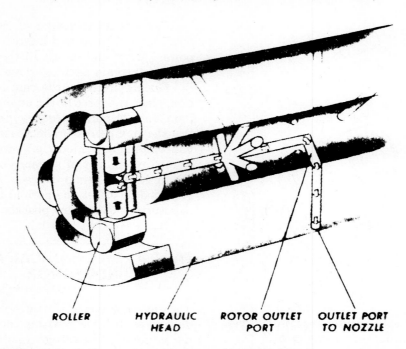

ROLLER   HYDRAULIC   ROTOR OUTLET   OUTLET PORT
HEAD   PORT   TO NOZZLE

FIGURE 27-6. High-pressure fuel passages. Rollers force plungers together every time they roll over the high parts of the cam ring. Cam ring has same number of cams as engine has cylinders. *(Courtesy of Diesel Systems Group— Stanadyne, Inc.)*

from the transfer pump, thereby controlling the quantity of fuel to the plungers. The governor derives its energy from weights pivoting in the weight retainer. Centrifugal force tips them outward, moving the governor thrust sleeve against the governor arm, which pivots on the knife edge of the pivot shaft and through a simple, positive linkage, rotates the metering valve. The force of the weights against

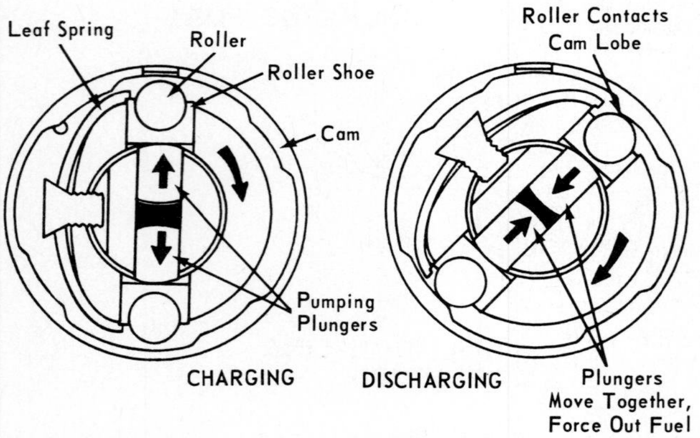

FIGURE 27-7. Cam roller and pumping plunger action during charging and discharging. *(Courtesy of Deere and Company)*

Leaf Spring   Roller   Roller Contacts
              Roller   Cam Lobe
                  Roller Shoe
                      Cam

CHARGING   DISCHARGING   Plungers
                           Move Together,
Pumping   Force Out Fuel
Plungers

the governor arm is balanced by the governor spring force, which is controlled by the manually positioned throttle lever and vehicle linkage for the desired engine speed.

In the event of a speed increase due to a load reduction, the resultant increase in centrifugal force of the weights rotates the metering valve clockwise to reduce fuel. This limits the speed increase (within the operating range) to a value determined by governor spring rate and setting of the throttle.

When the load on the engine is increased, the speed tends to reduce. The lower speed reduces the force generated by the weights permitting the spring force to rotate the metering valve in the counterclockwise direction to increase fuel. The speed of the engine at any point within the operating range is dependent upon the combination of load on the engine and the governor spring rate and setting as established by the throttle position. A light idle spring is provided for more sensitive regulation when weight energy is low in the low end of speed range. The limits of throttle travel are set by adjusting screws for proper low idle and high idle positions.

A light tension spring on the linkage assembly takes up any slack in the linkage joints and also al-

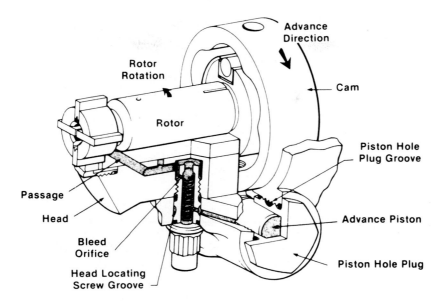

Transfer Pump Pressure

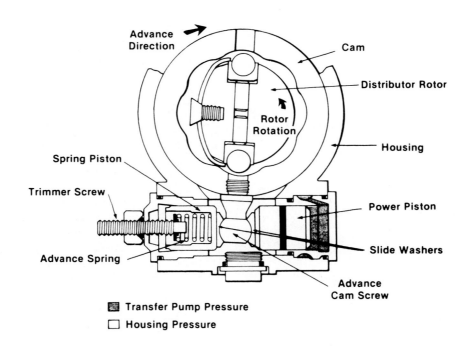

Transfer Pump Pressure

Housing Pressure

FIGURE 27-8. Roosa Master distributor pump automatic injection timing device. Fuel oil transfer pump pressure is the medium used to operate this device. *(Courtesy of Diesel Systems Group—Stanadyne, Inc.)*

lows the shutoff mechanism to close the metering valve without having to overcome the governor springing force. Only a very light force is required to rotate the metering valve to the closed position.

Electrical Shutoff

Electrical shutoff devices are available as an option in both Energized to Run (ETR) and Energized to

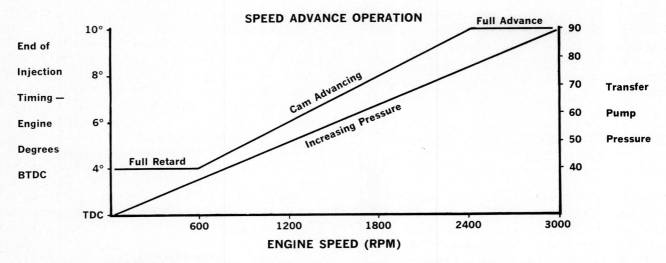

FIGURE 27-9. Chart showing amount of injection timing advance in relation to transfer pump pressure and engine speed. *(Courtesy of Diesel Systems Group—Stanadyne, Inc.)*

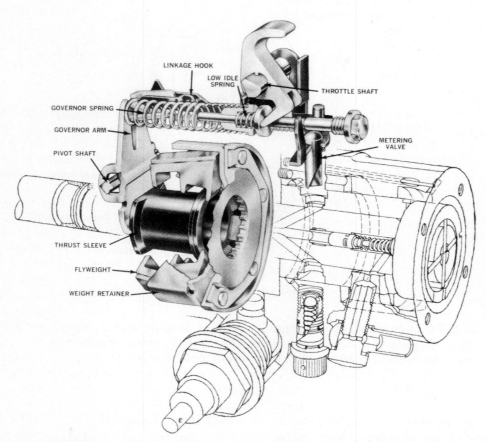

FIGURE 27-10. Centrifugal governor components. *(Courtesy of Diesel Systems Group—Stanadyne, Inc.)*

Shutoff (ETSO) models. These solenoids are included in various applications to control the run and stop functions of the engine. They accomplish this by positively stopping fuel flow to the plungers, thereby interrupting injection when the switch is turned off.

PART 5   TORQUE SCREW—TORQUE CONTROL

Torque is commonly defined as the turning moment or "lugging ability" of an engine. Maximum torque varies at each speed in the operating range for two reasons: (1) as engine speed increases, friction losses progressively increase and, (2) combustion chamber efficiency drops due to loss of volumetric efficiency (breathing ability of an engine), and due to reduction of time necessary to completely and cleanly burn the fuel in the cylinder. Since the torque increases with increased load conditions, a predetermined point at which maximum torque is desired may be selected for any engine. Thus, as engine rpm decreases, the torque generally increases toward this preselected point. This desirable feature is called "Torque Back-Up." In the Roosa Master pump three basic factors affect Torque Back-Up. These are:

1. Metering valve opening area
2. Time allowed for charging
3. Transfer pump pressure curve

Of these, the only control between engines for purposes of establishing a desired torque curve is the transfer pump pressure curve and metering valve opening, since the other factors involved are common to all engines. Torque control in Roosa Master fuel injection pumps is accomplished in the following manner:

The manufacturer determines at what speed for a specific application he wants his engine to develop its maximum torque. The maximum fuel setting is then adjusted for required delivery during dynamometer test. This delivery must provide acceptable fuel economy. The engine is then brought to full load governed speed. The fuel delivery is then reduced from that determined by the maximum fuel setting by turning in an adjustment or "torque screw" which moves the metering valve toward the closed position. The engine is now running at full load governed speed. When the engine is operating at high idle speed, no-load, the quantity of fuel delivered is controlled only by governor action through the metering valve. At this point, the torque screw and maximum fuel adjustment have no effect. As load is applied, the quantity of fuel delivered is controlled only by governor action and metering valve position until full-load governed speed is reached. At this point, further opening of the metering valve is prevented by its contact with the previously adjusted torque screw. Thus, the amount of fuel delivered at full-load governed speed is controlled by the torque screw and not by the roller-to-roller dimension. As additional load is applied and engine rpm decreases, a greater quantity of fuel is allowed to pass into the pumping chamber due to the increased time of registration of the charging ports. During this phase of operation the metering valve position remains unchanged, still being held from further rotation by the torque screw. As engine rpm continues to decrease under increasing load, the rotor charging ports remain in registry for a longer time period allowing a larger quantity of fuel into the pumping chamber. Fuel delivery increases until the predetermined point of maximum torque is reached.

PART 6   STANADYNE MODEL PCF ELECTRONIC FUEL INJECTION PUMP

The Model PCF electronic fuel control pump is a new pump concept. In place of the metering valve, a plunger control mechanism is used. Electronic governing eliminates the mechanical governing system from the pump, making it a smaller and lighter package. (See Fig. 27-11.)

Plunger control is achieved by physically limiting the radial displacement of the plungers. A yoke, located and guided by a slot at the driven end of the rotor, has a set of fingers which straddle the plungers. Ramps ground into the plungers make contact with these fingers during the charging sequence thereby limiting the amount of fuel available for the pumping sequence. Axial position of the yoke controls the outward radial displacement of the plungers. Disengagement of the plunger ramps from the yoke fingers begins when pumping commences.

Controlling plunger motion in this manner is advantageous because of lower dynamic loads on the control mechanisms.

Yoke positioning is accomplished by a cam follower which rotates against a cam profile ground into the pilot tube. The follower displacement is

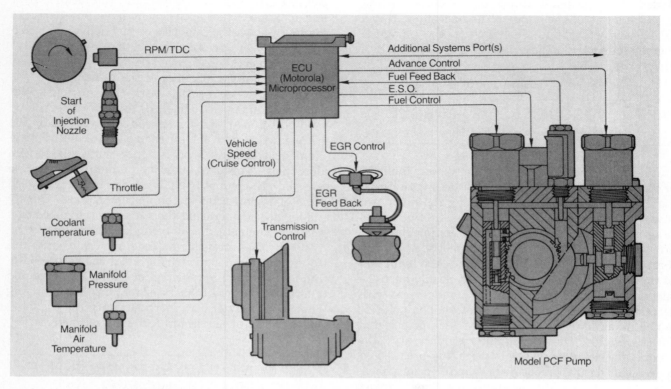

FIGURE 27-11. Electronic diesel fuel injection control system for Stanadyne Model PCF injection pump. *(Courtesy of Diesel Systems Group—Stanadyne, Inc.)*

transmitted by the thrust cup, cross pin and push rod to the yoke. Low force levels on the cam follower permit the use of actuators to rotate this part. A potentiometer measuring the position of the cam follower provides feedback to the ECU for fuel delivery accuracy.

The advance system is a servo type actuated by a stepper motor. A sensor in the nozzle provides a start of injection signal for timing feedback.

An eight-bit, single-chip microcomputer, engine-mounted sensors, and the Model PCF diesel fuel injection pump are the major components of the electronic system. Engine mounted sensors analyze engine speed, top dead center, start of injection, water temperature, manifold pressure and exhaust gas recirculation valve position. Start of injection is determined by a special nozzle that incorporates a Hall-effect needle lift sensor. The ECU, based on data received from the sensors, determines the desired advance and fuel delivery levels best suited to engine needs.

Electronic Control of Fuel Injection in the PCF Injection Pump Offers the Following Benefits:

1. Improved emission control, fuel economy and engine performance

2. Fueling control functions:

    • Maximum load fuel shaping

    • Altitude fuel compensation

    • Turbocharge boost compensation

    • Excess fuel at cranking speeds

    • Transient fuel trimming

    • Throttle progression tailoring

    • Minimum/Maximum or all-speed governing

3. Timing control functions

- Speed/Load advance

- Cold start advance

- Altitude timing compensation

- Pump installation error

- Drive shaft wear compensation

- Timing map flexibility

4. Optional control functions available

- Cruise control

- Optimum transmission control

- EGR-(Exhaust Gas Recirculation) control

PART 7   CAV-DPA DISTRIBUTOR PUMP

General Description

The pump is flange mounted to the engine (Fig. 27-12). It is oiltight, and during operation all moving parts are lubricated by fuel oil under pressure, so that no additional lubrication system is required. Pressure maintained within the pump housing prevents the entry of dust, water and other foreign matter.

Fuel injection is effected by a single pumping element having twin opposed plungers. These plungers are located within a transverse bore in a central rotating member which acts as a distributor and revolves in a stationary member known as the hydraulic head. The pump plungers are actuated by lobes on an internal cam ring. Fuel is accurately metered to the pumping element, and the high pressure charges are distributed to the engine cylinders at the required timing intervals through ports in the rotor and the hydraulic head. The single pumping element ensures equal delivery of fuel to each engine cylinder.

Most pumps are fitted with a device capable of automatically changing the timing of the fuel injection pulse.

The internally lobed cam ring, mounted in the pump housing, normally has as many lobes as there are engine cylinders and operates the opposed pump plungers through cam rollers carried in shoes sliding in the rotor body. The plungers are forced inward simultaneously as the rollers contact the diametrically opposed cam lobes. This is the injection stroke. The plungers are returned by pressure of the inflowing fuel and this forms the charging stroke.

The pump rotor is driven by the engine through a splined shaft, or other drive to suit the engine manufacturer's requirements. Pumps may be mounted horizontally, vertically or at any convenient angle.

The accurate spacing of cam lobes and delivery ports ensures the exact equality of the timing interval between injections, and the components which affect timing are designed with one assembly position only to ensure precision.

Fuel entering the pump through the main inlet connection is pressurized by a sliding vane transfer pump carried on the rotor inside the hydraulic head. The pressure rise is controlled by a regulating valve assembly located in the pump end plate. The fuel then flows through the passages to the pumping elements.

The outward travel of the opposed pumping plungers is determined by the quantity of fuel metered, which varies in accordance with the setting of the metering valve. The outward travel of the roller assemblies which are displaced centrifugally is always limited by the maximum fuel adjusting plates.

The maximum amount of fuel delivered is regulated by limiting the outward travel of the plungers by means of an adjustable stop.

As the rotor turns, the inlet port is cut off and the single distributor port in the rotor registers with an outlet port in the hydraulic head. At the same time the plungers are forced inward by the rollers contacting the cam lobes, and fuel under injection pressure passes to the central bore of the rotor through the aligned ports to one of the injectors. The rotor normally has as many inlet ports as the engine has cylinders, and a similar number of outlet ports in the hydraulic head.

The cam lobes are contoured to provide relief of pressure in the injector lines at the end of the injector cycle. This gives a sharp cut-off of fuel and prevents "dribble" at the nozzles.

The governor is contained in a small housing mounted on the pump body, the metering valve being operated by fuel at transfer pressure.

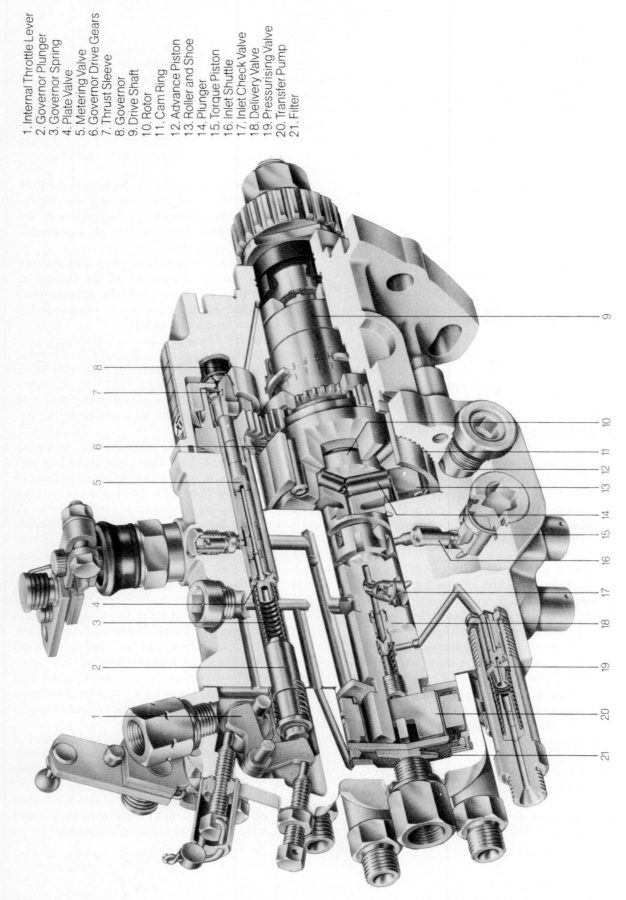

1. Internal Throttle Lever
2. Governor Plunger
3. Governor Spring
4. Plate Valve
5. Metering Valve
6. Governor Drive Gears
7. Thrust Sleeve
8. Governor
9. Drive Shaft
10. Rotor
11. Cam Ring
12. Advance Piston
13. Roller and Shoe
14. Plunger
15. Torque Piston
16. Inlet Shuttle
17. Inlet Check Valve
18. Delivery Valve
19. Pressurising Valve
20. Transfer Pump
21. Filter

FIGURE 27-12. CAV DP distributor injection pump. *(Courtesy of CAV Ltd.)*

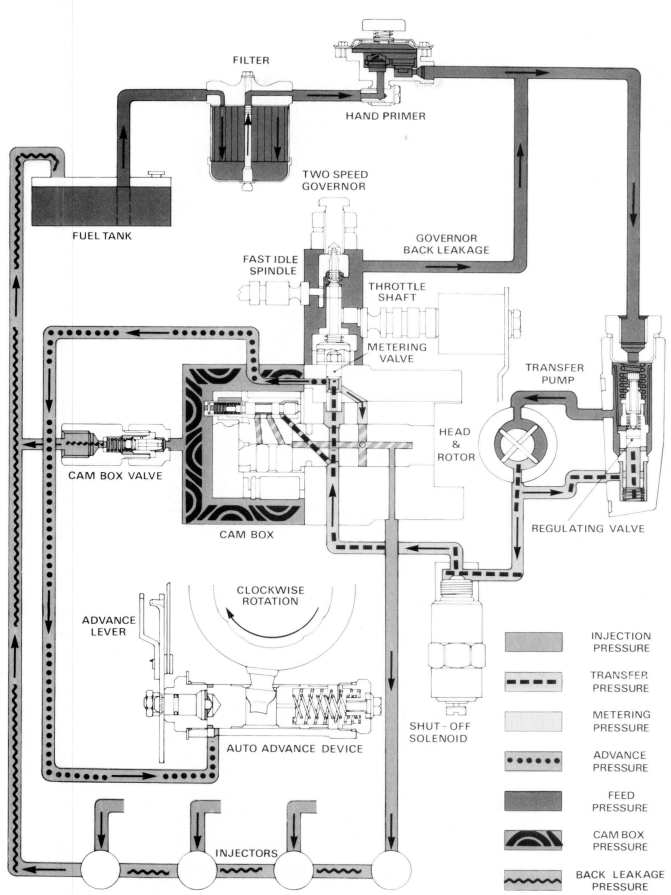

FILTER

HAND PRIMER

TWO SPEED
GOVERNOR

FAST IDLE
SPINDLE

GOVERNOR
BACK LEAKAGE

THROTTLE
SHAFT

METERING
VALVE

TRANSFER
PUMP

FUEL TANK

CAM BOX VALVE

HEAD
&
ROTOR

REGULATING VALVE

CAM BOX

CLOCKWISE
ROTATION

ADVANCE
LEVER

AUTO ADVANCE DEVICE

SHUT-OFF
SOLENOID

INJECTION
PRESSURE

TRANSFER
PRESSURE

METERING
PRESSURE

ADVANCE
PRESSURE

FEED
PRESSURE

CAM BOX
PRESSURE

BACK LEAKAGE
PRESSURE

INJECTORS

FIGURE 27–13. CAV DPA pump fuel system schematic with hydraulic governor. (Courtesy of CAV Ltd.)

PART 8 STANADYNE DISTRIBUTOR PUMP TROUBLESHOOTING

Problem columns:
- A. Fuel not reaching pump
- B. Fuel delivered from transfer pump but not to nozzles
- C. Fuel reaching nozzles but engine won't start
- D. Engine starts hard
- E. Engine starts and stops
- F. Erratic engine operation - surge, misfiring, poor governor regulation
- G. Engine idles imperfectly
- H. Engine does not develop full power or speed
- I. Engine smokes black
- J. Engine smokes blue or white

CAUSE — Numbers in "Problem" Check Chart indicate order in which to check possible "Causes" of Problem.

PROBLEM MAY OCCUR	CAUSE	A	B	C	D	E	F	G	H	I	J	CORRECTION
	Transfer Pump liner locating pin in wrong hole for correct rotation	7										Reinstall properly.
ON TEST STAND	Plunger missing.		9									Assemble new plunger.
	Cam backwards in housing.		8									Reassemble correctly.
FOLLOWING OVERHAUL	Metering Valve incorrectly assembled to metering valve arm		6									Reassemble correctly.
	Delivery Valve Sticking missing or assembled backwards				19		21	17	15			Remove, clean or replace as needed.
	Hydraulic head vent wires missing.		13						24			Install as indicated in reassembly instructions.
	Idling spring missing or incorrect.						15	8				Assemble as indicated in reassembly instructions.
FOLLOWING INSTALLATION ON ENGINE	Seizure of Distributor Rotor.	2										Check for cause of seizure. Replace hydraulic head and distributor rotor assembly.
	Failure of electrical shutoff		2		8							Remove, inspect and adjust parts, Replace parts as necessary.
	Fuel supply, lines clogged, restricted, wrong size or poorly located.	9	7	7	5	1	2	13	4			Blow out all fuel lines with filtered air. Replace if damaged. Remove and inspect all flexible lines.
	Air leaks on suction side of system.	11			6	7	8	3	5			Troubleshooting the system for air leaks. See Supplementary Inspections in manual.
	Transfer Pump Blades worn or broken.	8			12		18	5	13			Replace.
	Delivery Valve Retainer Screw loose and leaking or incorrectly installed.				25		20		21			Inspect Delivery Valve Stop seat for erosion, tighten Retainer Screw, or replace head and rotor assembly as needed.
DURING OPERATION	Transfer Pump Regulating Piston sticking.	10			13		19	12				Remove piston and regulator assembly and inspect for burrs, corrosion or varnishes Replace if necessary.
	Shut-off device at "stop" position		1									Move to "run" position.
	Plungers sticking.		10		21	10		18	17			Disassemble and inspect burrs, corrosion or varnishes.
	Metering Valve sticking or closed.		3		14	9	12	7	10			Check for governor linkage binding, foreign matter, burrs, missing metering valve shim, etc.
	Passage from Transfer Pump to Metering Valve clogged with foreign matter.		11									Disassemble and flush out Hydraulic Head.
	Tank valve closed.	1										Open valve.
	Fuel too heavy at low temperature.	6			8							Add kerosene as recommended for 0° F, -15° F and -30° F temperatures.
	Cranking speed too low.			1	2						7	Charge or replace batteries
	Lube oil too heavy at low temperature.			18	9							See engine manual.
	Engine engaged with load.			2	1							Disengage load.
	Nozzles faulty or sticking.			10	17		9	10		5		Replace or correct nozzles.
	Intake air temperature low.			5	3							Provide starting aids. See engine manual.
	Engine compression poor.			17	10			25	9	8		Correct compression. See engine manual.
	Pump timed incorrectly to engine.			3	4		4	4	7	4	3	Correct timing.
	Excessive fuel leakage past plungers (worn or badly scored).			15	22			16	18			Replace rotor and hydraulic head assembly.
	Filters or Inlet Strainer clogged.	5		7	6	3		6				Remove and replace clogged elements. Clean strainer.

(Courtesy of Diesel Systems Group—Stanadyne Inc.)

PROBLEM

CAUSE

Numbers in "Problem" Check Chart indicate order in which to check possible "Causes" of Problem.

CORRECTION

A. Fuel not reaching pump
B. Fuel not delivered from transfer pump but not to nozzles
C. Fuel reaching nozzles but engine won't start
D. Engine starts hard
E. Engine starts and stops
F. Erratic engine operation - surge, misfiring, poor governor regulation
G. Engine idles imperfectly
H. Engine does not develop full power
I. Engine smokes black
J. Engine smokes blue or white

PROBLEM MAY OCCUR

DURING OPERATION CONT.

CAUSE	A	B	C	D	E	F	G	H	I	J	CORRECTION
Cam, Shoes or Rollers worn.			14	20				16			Remove and replace.
Automatic advance faulty or not operating.			11	24		11	11	12	7	4	Remove, inspect, correct and reassemble.
Governor linkage out of adjustment or broken.				16		14	9	11			Adjust governor linkage hook.
Governor not operating: parts or linkage worn, sticking or binding, or incorrectly assembled.		4		26		13	6	9			Disassemble, inspect parts, replace if necessary and reassemble.
Maximum fuel setting too low.			12	18				14			Reset to pump specifications.
Engine valves faulty or out of adjustment.				28		10	15		6	9	Correct valves or valve adjustment as in engine manual.
Water in fuel.					2	5	1	23			Drain fuel system and pump housing, provide new fuel, prime system.
Return oil line or fittings restricted.				29	4	23	19	8			Remove line, blow clean with filtered air
Engine rotation wrong.	4										Check engine rotation. See engine manual.
Air intake restricted.					3			26	2		Check. See engine manual.
Wrong Governor spring.						17		27			Remove and replace with proper spring as in pump specifications.
Pump housing not full of fuel.						7	2				Operate engine for approximately 5 minutes until pump fills with fuel.
Low cetane fuel.			13	11		6	14	20	8		Provide fuel per engine specifications.
Fuel lines incorrect, leaking or connected to wrong cylinders.				6		1		28			Relocate fuel lines for correct engine firing sequence.
Tang Drive excessively worn.						22		19		5	Remove and install new head and rotor assembly and drive shaft as necessary.
Governor sleeve binding on drive shaft.						16					Remove, inspect for burrs, dirt, etc. Correct and reassemble.
Shutoff device interfering with Governor linkage.		8	15			2					Check and adjust governor linkage dimension.
Governor high-idle adjustment incorrect.							3				Adjust to pump specifications.
Torque Screw incorrectly adjusted.	5	9	23					22	11		Adjust to specification.
Throttle Arm travel not sufficient.		4						1			Check installation and adjust throttle linkage.
Rotor excessively worn.	12	16	27								Replace hydraulic head and rotor assembly.
Maximum fuel setting too high.									10		Reset to pump specifications.
Engine overheating.					5			3			Correct as in engine manual.
Exceeding rated load.									1		Reduce load on engine
Engine cold.				30						1	Check thermostats or shutter controls, warm to operating temperature. See engine manual.
Lube oil pumping past valve guides or piston rings in engine.										6	Correct as in engine manual.
Excess lube oil in engine air cleaner.										2	Correct as in engine manual.

PART 9   STANADYNE DISTRIBUTOR PUMP TESTING AND CALIBRATION

Any test is only as good as the testing equipment employed. Incorporation of quality test equipment and adherence to specifications and the following test procedures will reduce testing inaccuracies to a minimum.

Calibration Nozzles

Several different types of calibrating nozzles will be required for testing the various DB2 pump models. Be sure to use only the type of nozzle called for in the individual specification. Some of the permissible types are listed:

TYPE	OPENING PRESSURE—PSI
DN12SD12	2500–(170 ATS)
AMBAC PCU25D050.5	
Orifice plate (SAE Std.)	3000–(204 ATS)
AMBAC TSE77110 - 5/8	
.5 Orifice plate	1700–(116 ATS)

Use of the SAE/150 orifice plate nozzle is described in SAE recommended practices J968c and J969b.

Injection Lines

Several injection line sizes (length and inner diameter) have been released for service use. Refer to the individual specification for proper size, and see Roosa Master service bulletin 330 for preparation and maintenance instructions.

Calibrating Oil

Guidelines for calibrating oil are listed in SAE recommended practice J969d. Refer to service bulletin 201 for brands of calibrating oil approved for use with Roosa Master fuel injection equipment. Calibrating oil should be changed every three months or 200 pumps (whichever comes first).

Calibrating Oil Temperature

The temperature of the oil in the test bench must be maintained within 110° to 115°F while testing Roosa Master fuel injection equipment.

Note: This reading should be taken as close to inlet as possible. The test bench should be equipped with a heater and thermostatic control to maintain this temperature. Refer to SAE recommended practice J969b.

Test Bench

Mount and drive the DB2 pump models according to the test bench manufacturer's instructions. In addition, the test stand coupling should be of the self-aligning, "zero" backlash type, similar to the Thomas coupling types or Robert Bosch (SAE J969b).

General Test Procedure

1. Install applicable transfer pump inlet connector, using two wrenches so that the pump outlet fitting does not get moved at the same time. Install transfer pressure gauge connector 21900. Install a shut-off valve to isolate the gauge when not in use. Connect a pressure gauge to the 21900 adapter. Some automotive DB2 pumps require one-quarter of 1° setting accuracy. If so specified, remove the timing line cover and replace with the 21734 advance gauge. If the 21734 gauge is not required, replace the timing line cover with the 19918 advance window.

2. Determine the proper direction of pump rotation from the specification. Rotation is determined as viewed from the drive end of the pump.

3. If the pump is equipped with an energized-to-run electric shut-off device energize it at the lowest speed with the specified voltage (see service bulletin 108). Move the pump throttle lever to the full-load position. When the transfer pump is primed, allow fuel to bleed for several seconds from loosened injection line nuts at the nozzles. Tighten line nuts securely.

Note: Roosa Master pump specifications list fuel delivery in cubic millimeters/stroke. Some test benches measure fuel flows in cubic centimeters (milliliters). To convert from mm³/stroke to cubic centimeters, use the following formula:

$$\text{Delivery in cc} = \frac{\text{mm}^3/\text{stroke} \times \text{no. of strokes}}{1000}$$

Example: If the specification calls for 72 mm³/stroke and the test stand counter has been set for 500 strokes, simply substitute these numbers into the formula and calculate as follows:

$$\text{Delivery in cc} = \frac{72 \times 500}{1000} = 36 \text{ cc}$$

Bear in mind when testing Roosa Master pumps that the specifications refer to engine rpm (erpm) and that most test bench tachometers register pump rpm, which is one-half of engine speed for four-stroke-cycle engines.

4. Operate pump at 1000 erpm wide-open throttle (WOT) for 10 minutes. Dry pump off completely with compressed air. Observe for leaks and correct as necessary. Back out the high-idle, low-idle, and torque screw (if equipped).

Note: Refer to pump specification for correct sequence of test stand adjustments. Pressurize the transfer pump inlet to the amount indicated on specification or to a maximum of 5 psi, if not otherwise indicated.

5. Vacuum Check: Close the valve in the fuel supply line. At 400 erpm, the transfer pump must be capable of creating a vacuum of at least 18 inches of mercury. If it does not, check for air leaks between the pump inlet and shut-off valve or deficiency in transfer pump components.

6. Fill graduates to bleed air from the test stand and to wet graduates.

7. Check the return oil quantity by directing the return oil flow into an appropriately calibrated graduate for a given time. See individual specifications for allowable quantity and erpm at which to make the check.

8. Operate at the specified speeds with wide-open throttle and observe transfer pump pressure. Adjust pressure-regulating spring plug to raise or lower transfer pump pressure.

Caution: Under no circumstances should 130 psi be exceeded.

To adjust pressure, remove the line to the transfer pump inlet connector and use a 5/32-inch hex key wrench 13316 to adjust the plug. Clockwise adjustment increases pressure. Do not overadjust.

Note: The transfer pump pressure gauge must be isolated by the shut-off valve at the injection pump when checking fuel delivery and advance movement.

9. Check for minimum delivery at cranking speed.

10. Operate at high speed and adjust high-idle screw to obtain the specified delivery. Recheck transfer pump pressure upon completion of this adjustment.

11. Adjust the low-idle screw to the correct low-idle delivery.

12. Automatic Advance: Check the cam position at specified points in the speed range. Adjust trimmer screw, as required, to obtain proper advance operation. Each line on advance gauge 19918 equals two pump degrees. After setting the advance, check to see that the cam returns to its initial position at zero rpm. Recheck the transfer pump pressure after setting the advance and correct if necessary.

13. Record fuel delivery at checkpoints shown on the pump specification. *Roller settings should not be readjusted on the test bench.* Experience has proven that micrometer and dial indicator settings provide more consistent accurate results in performance. Variations in test bench drives, instrumentation, nozzle lines, and fuels in different areas sometimes result in nonconforming fuel flow readings.

14. While operating at full-load governed speed, set the torque screw (if employed) to specified delivery. Recheck the transfer pump pressure and advance movement upon completion of this adjustment.

15. Recheck delivery at the lowest speed checkpoint.

16. Check the governor cutoff at specified speed.

17. Check the electric shut-off (if equipped) at speeds indicated on specification.

18. Remove the pump from the test stand and assemble all sealing wires. The pump is then ready for installation to engine.

PART 10   CAV–DPA DISTRIBUTOR PUMP TESTING AND CALIBRATION*

Pressure Testing

All pumps must undergo a pressure test *BEFORE* and *AFTER* machine test, using the following method.

1. Drain all fuel from the pump and connect an air line to the pump inlet connection. Ensure the air supply is clean and free from water.

2. Seal off the low pressure outlet connection on the pump and completely immerse pump in a bath of clean test oil.

3. Raise the air pressure in the pump to 1.41 kg cm² (20 lb in²). Leave the pump immersed in oil for 10 minutes to allow any trapped air to escape.

4. Observe for leaks after pump has been immersed for 10 minutes; if the pump is not leaking reduce the air pressure to 0.14 kg cm² (2 lb in²) for

*(Courtesy of CAV Limited)

30 seconds; if there is still no leak increase the pressure to 1.41 kg cm² (20 lb in²) and if the pump is still leak free after 30 seconds it can be passed as satisfactory.

5. On pumps without a drive shaft oil seal it is necessary to stop the oil leaking past the drive shaft during pressure testing. Tool Part No. 7144-760 can be used but it is necessary to blank off the 12 × 1.5 mm threaded connection.

6. All leaks must be rectified before testing and setting the pump.

Test Procedure

General explanatory notes and an individual test plan, quoting the numbers of the range of pumps to which it may be applied, is published for each different model manufactured. The sequence of operations listed in the test data for the particular type of pump gives the test performance requirements at various pump speeds, the timing procedure and any special precautions necessary to safeguard the pump.

Note: All pump tests, calibrations and settings must be made using the specified test oils given in the CAV Test Plans. Where the word "fuel" is used in the following test instructions this means the approved test oil and *NOT* diesel fuel.

Test Machine

A typical test machine, adapted for use with DPA pumps, is shown in Figure 22–38. A suitable test machine must incorporate the following features.

1. A mounting bracket, for holding the pump.

2. A splined drive coupling for rotating the pump in either direction at all speeds specified in the test data.

Note: The pump *MUST* be rotated in the direction given in the test data and indicated on the pump nameplate. Incorrect rotation will cause serious damage.

3. A set of high pressure pipes, 863.6 mm long × 2 mm bore × 6 mm outside diameter, for coupling the pump outlet connections to a matched set of injectors, Type BDN 12SD12, set at 175 atm opening pressure.

4. An automatic trip mechanism that directs test oil from the injectors into graduated glasses during the period stated in the test data and then diverts the oil into a drain.

5. A set of graduated glasses for measuring the output from each injector, and one glass of larger capacity to measure the volume of back-leakage oil.

6. An oil feed system giving an adequate supply at constant pressure at the pump inlet. Required minimum fuel feed to the pump inlet is 1,000 cm³ (61 in³) per minute. If this figure is not obtainable with a gravity feed, a pressure feed of 0.14 kg cm² (2 lb in²) can be used.

7. One pressure gauge and one vacuum gauge for testing the output and efficiency of the transfer pump.

General Procedure

The following precautions must be observed.

1. The test machine must be set to run in the correct direction of rotation for the pump under test.

2. The pump must not be run with a low output for long periods at high speed.

3. The pump must not be run for long periods with shut-off control in the closed position.

4. The correct test machine adaptor plate must be used. A plate with a 50 mm hole must *never* be used for pump with 46 mm spigot.

5. Unless otherwise stated, standard radial high pressure connections must be fitted prior to testing. Information is given in the test data and explanatory notes.

6. Prime the pump thoroughly before testing also at all times indicated in the test plan.

Priming

Variations for certain pumps are explained in test data.

1. Slacken both the vent screw on the governor control casing and the head vent screw.

2. Connect the oil feed pipe to the pump inlet and connect the back leakage pipe.

3. Turn on the oil supply to feed pressure 0.14 kg cm² (2 lb in²) to fill the pump. Run pump at 100 rpm. When test oil free from air bubbles issues from the vents, retighten the vent valve and the head locking screw.

4. Slacken the connections at the injector end of the high pressure pipes or if fitted on the test machine open the bleeder valves at the injectors.

5. Run the pump at 100 rpm. When test oil free from air bubbles issues from all high pressure pipes, retighten the connections or close the bleeder valves.

6. Examine the pump after priming, for oil leaks at all jointing faces, connections and oil seals. Pumps must be free from leaks both when running and when stationary.

Pump Output

Fuel delivery is checked at full throttle setting at one or more speeds of rotation by measuring the volume passing through each injector during 200 pump cycles. The pump test data quotes the maximum fuel delivery, overall tolerance and the maximum permissible delivery variation between injectors.

Shut-Off Control

This is checked by running the pump at a specified speed, (see test plan) with the shut-off control closed. The maximum fuel delivery permitted on this setting is quoted.

Maximum Fuel Setting

1. The maximum fuel delivery is checked at a specified speed, with throttle and shut-off controls fully open.

If output is not within the specified limits, adjust as follows:

2. Remove the inspection cover.

3. Slacken the two drive plate screws.

4. Engage Tool Part No. 7144-875 with the slot in the periphery of the adjusting plate.

5. Adjust the plate by lightly tapping the knurled end of the tool. The direction in which the drive plate is turned to increase or to decrease fueling depends on the type of adjusting plates fitted.

When looking at the drive end of the pump, if the top adjusting plate has a shallow slot of 3 mm depth, the fueling can be increased by turning the adjusting plate in an anti-clockwise direction, and decreased by turning it in a clockwise direction.

If the top adjusting plate has a deeper slot of 5.5 mm depth, the fueling can be increased by turning the top adjusting plate in a clockwise direction, and decreased by turning it in an anti-clockwise direction.

6. Tighten the drive plate screws *evenly* to the listed torque value, using the Adaptor Tool Part No. 7144-482, Spanner Tool Part No. 7144-511A and a torque wrench.

7. Replace and secure the inspection cover, refill the pump, vent as necessary and re-check the maximum fuel delivery. Repeat until the fuel delivery is within the specified limits.

Governor Testing

Run the pump at more than half the maximum permissible speed of the engine to which it will be fitted, and adjust the maximum speed stop until the specified fuel delivery is obtained. This specified volume is less than at the maximum fuel setting. Reduce the speed of rotation, whereupon the fuel delivery should increase to a specified volume approximately equal to the maximum fuel delivery.

Note: Final governor setting must be carried out with the pump fitted to the engine, and in accordance with the engine manufacturer's instructions.

Transfer Pump Setting

Transfer pump vacuum is checked while the pump is running at a low speed, with the two-way cock in the fuel feed line turned to the position that cuts off the fuel supply and connects the pump inlet to the vacuum gauge. A given depression must be obtained in a specified time.

Note: The pump may need repriming after this test.

Transfer Pressure Adjustment

To adjust the transfer pressure the operation of the regulating valve is modified as follows.

On certain pumps with aluminum end plates, the transfer pressure can be adjusted within the limits of an individual specification in one or both of two ways: (a) by changing the end plate sleeve plug, and (b) by the adjustment of the transfer pressure adjusting screw passing through the plug.

Several plugs with different step thicknesses are available; the stepped portion in contact with the regulating spring determining the spring compression. The screwed transfer pressure adjuster limits the movement of the regulating piston and so controls the maximum uncovered area of the sleeve port. Method (a) modifies transfer pressure characteristics over the lower and middle speed ranges; method (b) controls the pressure rise over the upper speed range.

There is a third type of transfer pressure adjuster which varies the pre-loading of the regulating spring and maximum lift of the piston. It has a spring peg fitted between the adjusting screw and regulating spring. This has the same effect as method (a), but allows more variation and gives easier adjustment.

Additional information is given in the individual pump test data.

Pressurized CAM Boxes

On some pumps the cam box is pressurized during running by a spring loaded ball pressurizing valve located in the back leakage connection in the inspection cover.

The test pressure gauge is fitted to the governor housing vent screw hole.

Pressure limits are given in the appropriate test data, if the pressure is incorrect, check that the pump is not leaking and there is no restriction in the back leakage passages. If the pressurizing valve is faulty, a new inspection cover complete with valve must be fitted.

Testing of Advance Devices

The appropriate test plan for the individual pump specifies the type of advance device fitted and details the tests for the particular unit.

All advance devices are tested using special Tool Part No. 7244-50 which consists of a gauge with a scale covering 0 to 18° and a feeler pin Tool Part No. 7244-70. To fit this tool, proceed as follows:

1. Remove the small screw from the piston spring cap on the advance device.

2. Pass the threaded bushing of the feeler pin assembly through the hole in the tool bracket.

3. Insert the end of the plunger into the hole in the spring cap and screw the bushing into the spring cap hole. This will clamp the bracket between the spring cap and the shoulder on the threaded bushing.

4. Zero the gauge by moving the scale relative to the pointer.

Note: The pump must be reprimed after fitting the tool. After priming operate the throttle and press inward and release the advance gauge pin a few times with the pump running at 100 rpm.

Speed Advance Device

The tests outlined in the test data must be applied to ensure that the degree of advance obtained is within the stated limits at the speeds specified. To adjust the degree of advance, increase or decrease the thickness of the shims between the piston spring and the spring cap, see Test Data. When the tests are satisfactory, remove the special tool and prime the pump.

Combined Load and Speed Advance

These tests, at different speeds and fuel deliveries, check the movement of the outer piston in response to changes of speed, and of the inner piston to changes of load. Adjustment is made by altering the thickness of the shims beneath inner and outer piston springs. When the tests are satisfactory, remove special tool and prime the pump.

Timing

All pumps require timing (pump phasing).

After completion of tests remove the pump from the test machine and drain by slackening the inspection cover screws. Remove the inspection cover. For internal or external timing connect the stirrup pipe Part No. 7144-262A to the fuel outlet specified on the Test Plan and to the outlet diametrically opposite. Fit the relief valve Tool Part No. 7144 to the stirrup pipe, and connect the complete tool through a high pressure pipe to a nozzle testing unit.

For 3-cylinder pumps, connect one branch of the stirrup pipe to the specified outlet, and arrange the other branch to face away from the pump and seal it off with blanking plug.

Normally a pressure of 30 atm with a relief valve fitted in the system is specified in the test data but sometimes a higher pressure is quoted. To obtain the higher pressures adjust the relief valve appropriately or connect the stirrup pipe directly to the nozzle testing unit. *Do not exceed the specified pressure. Excess pressure can cause damage to the shoe assemblies and adjusting plates.*

Turn the pump drive shaft in the direction indicated on the nameplate until resistance to further movement is felt. At this point, the master spline on the drive shaft should be in the same plane as the outlet quoted in the test plan. This is the timing position. Access to the timing ring may be obtained through the inspection cover aperture. Move the timing ring until the straight edge of the timing circlip—or the line scribed on the ring in the case of old type clips—aligns with the mark on the drive plate as specified in the test plan. Circlips with two straight ears are only for spacing and the circlip ends are positioned remote from the inspection aperture. After carrying out this operation refit the inspection cover and tighten screws.

The test data gives specific information about the timing mark on the pump flange. A flange marking gauge Tool Part No. 7244-27 is available, and this consists of a cast aluminum body with a lock screw, around which slides a ring carrying the scribing plate. The ring is held in position by a scale plate on which direct readings in degrees are taken from the edge of the scribing plate.

Interchangeable spigot plates, one with a 46 mm bore and one with a 50 mm bore, are held in position by a cap-head screw, and these accommodate the different pump spigots. Four interchangeable inserts adapt the gauge to any type of pump drive. The inserts are held by two screws and positioned relative to the scale zero by a dowel pin.

To mark the flange, the pump is held in the timed position and the marking gauge, with the appropriate insert and spigot plate, is fitted to the pump drive and set to the indexing figure specified in the test data. Using the gauge as a template, scribe a line on the flange between the plates of the scribing gauge.

PART 11   SELF-CHECK

1. Name the main rotating components of the Stanadyne distributor pump.

2. How many pumping plungers does the DB2 pump have?

3. What actuates the DB2 pump plungers?

4. Automatic timing advance is achieved in the DB2 pump by a hydraulic _____ powered by oil from the _____ .

5. What three factors control DB2 pump metering?

6. How is torque control achieved in the DB2 pump?

7. What controls the functions of the Stanadyne PCF electronic fuel injection pump?

8. The PCF pump advance is controlled by a stepper motor. True or False.

9. What four areas of fuel injection pump functions are controlled electronically in the PCF pump?

10. The CAV–DPA distributor pump is _____ mounted.

11. The pumping plungers in the DPA pump are actuated by _____ .

12. What determines the accuracy of injection timing intervals between cylinders in the DPA pump?

Chapter 28

Diesel Engine Tune-Up

Performance Objectives

After thorough study of all the chapters in this text, the appropriate training models and shop manuals, and with the appropriate tools and equipment, you should be able to do the following:

1. Complete the self-check questions with at least 80 percent accuracy.
2. Perform the necessary tests to determine diesel engine performance.
3. Perform a diesel engine tune-up to meet the performance standards and specifications of the engine manufacturer.

PART 1   INTRODUCTION

A good diesel engine maintenance program always includes periodic engine tune-up procedures which are designed to ensure optimum performance and economy and prevent problems or failure during the periods between tune-ups. Significant operating and repair cost benefits are the obvious results of a good preventive maintenance program.

A distinction must be made between an engine tune-up and problem diagnosis or troubleshooting. The latter deals with identifying and correcting a specific problem, while a tune-up is a series of procedures and adjustments designed to both correct existing problems and prevent others from occurring.

A diesel engine tune-up should include the following basic steps, which are explained further in Part 2 of this chapter. Corrections and adjustments are made wherever required throughout the tune-up procedure to meet manufacturer's specifications.

1. Clean the engine exterior.
2. Change the engine oil and oil filters.
3. Service the cooling system.
4. Check for any loose fasteners, fittings, support brackets, and the like.
5. Check the air intake system.
6. Check the exhaust system.
7. Service the fuel system.
8. Adjust engine valve clearance.
9. Perform cylinder compression test if required.
10. Check throttle linkage.
11. Run engine and bring to operating temperature. Check operation of all engine monitoring gauges.
12. Perform dynamometer test of engine performance.

PART 2   TUNE-UP PROCEDURE

1. Clean the engine exterior. The importance of cleanliness when working on diesel engines and fuel injection systems must be strongly emphasized. It takes only a small particle of dirt to cause serious damage. It is therefore good practice to clean the exterior of the engine before any tune-up work is begun. Be sure to use a face mask and protective clothing when using a steam cleaner or high-pressure water cleaner.

Both the steam cleaner and the high pressure water cleaner work well; however some precautions must be observed. Be sure that water or steam do not enter the engine, induction system, or fuel system. Cover any vents or openings to prevent moisture contamination.

The steam cleaner presents another hazard that must be considered. Do not steam clean any fuel injection components while the engine is running. If fuel injection components are steam cleaned with the engine stopped, do not start the engine until you are sure that the temperature of the injection system components has completely stabilized. Steam cleaning fuel injection components causes consid-

TESTS			

SUFFICIENT CLEAN FUEL

1 FREE OF WATER – ICING – CLOUDING – CORRECT GRADE

INSTRUMENT		1st CHECK	2nd CHECK
VISUAL CHECK ☐			

EXTERNAL LEAKAGE †

2 ☐ FUEL ☐ OIL ☐ AIR ☐ WATER
† RECORD LOCATION OF LEAKS IN "COMMENTS"

INSTRUMENT		1st CHECK	2nd CHECK
VISUAL CHECK ☐			

ACCELERATOR LINKAGE

3
A. ADJUSTED TO OBTAIN OVERRIDE AT FULL THROTTLE (DT-466).
B. THROTTLE LEVER CONTAINS HI-IDLE STOP AT FULL THROTTLE (D-SERIES & V-800).
See Illustration – Reverse Side

INSTRUMENT		1st CHECK	2nd CHECK
VISUAL CHECK ☐			

SHUT OFF CABLE

4 ADJUSTED TO ALLOW FULL RUN POSITION

INSTRUMENT		1st CHECK	2nd CHECK
VISUAL CHECK ☐			

LOW IDLE (RPM) *

5
A. MANUAL TRANSMISSION - NEUTRAL
B. AUTOMATIC TRANSMISSION - DRIVE
C. MIXER - DRUM ENGAGED

INSTRUMENT	GUIDELINE DATA	1st CHECK	2nd CHECK
MASTER TACHOMETER			

HIGH IDLE RPM (NO LOAD) *

6 THROTTLE IN OVERRIDE POSITION OR AT MAXIMUM (HIGH IDLE) STOP

INSTRUMENT	GUIDELINE DATA	1st CHECK	2nd CHECK
MASTER TACHOMETER			

AIR CLEANER MAXIMUM RESTRICTION

7
A. TURBO-ENGINES MEASURE AT FULL LOAD AND RATED SPEED
B. N/A ENGINES MEASURE AT HIGH IDLE RPM.
See Illustration – Reverse Side

INSTRUMENT	GUIDELINE DATA	1st CHECK	2nd CHECK
WATER MANOMETER OR MAGNAHELIC GAUGE			

*ENGINE MUST BE AT NORMAL OPERATING TEMPERATURE

Courtesy of International Harvester Company

FIGURE 28-1. Tests. *(Courtesy of International Harvester Company)*

TESTS

8	**IN-TANK SUPPLY PUMP PRESSURE (D-SERIES ONLY)**			
	A. MEASURE AT HIGH IDLE RPM.			
	B. MUST HAVE ONE P.S.I. MINIMUM AFTER FINAL FUEL FILTER			
	INSTRUMENT	GUIDELINE DATA	1st CHECK	2nd CHECK
	PRESSURE GAUGE (15 PSI RANGE)			

9	**TRANSFER PUMP PRESSURE**			
	A. DT-466 AND V-800: TEST AT FINAL FILTER VENT PLUG.			
	B. D-SERIES ENGINE: TEST AT INJECTION PUMP BLEEDER FITTING.			
	See Illustration — Reverse Side			
	INSTRUMENT	GUIDELINE DATA	1st CHECK	2nd CHECK
	PRESSURE GAUGE (100 PSI RANGE)			

IF PRESSURE REMAINS LOW AFTER REPLACING FUEL FILTERS, PERFORM STEP C.

	C. TEST TRANSFER PUMP INLET RESTRICTION: 6" MAXIMUM HG VACUUM.			
	See Illustration — Reverse Side			
	INSTRUMENT	GUIDELINE DATA	1st CHECK	2nd CHECK
	VACUUM GAUGE			

10	**INJECTION PUMP INITIAL TIMING***			
	A. AMBAC — TIMING LIGHT OR FLOW TIME			
	B. ROBERT BOSCH — TIMING LIGHT (V-800) OR FLOW TIME (V-800 & D-SERIES)			
	INSTRUMENT	GUIDELINE DATA	1st CHECK	2nd CHECK
	TIMING LIGHT OR FLOW TIME			

11	**INJECTION PUMP AUTOMATIC ADVANCE***			
	DT-466 & V-800 ENGINES			
	INSTRUMENT	GUIDELINE DATA	1st CHECK	2nd CHECK
	DIESEL TIMING LIGHT			

12	**INTAKE MANIFOLD PRESSURE*** (TURBO-CHARGED ENGINES)			
	A. MEASURE AT FULL LOAD AND AT SPECIFIED SPEEDS.			
	B. SELECT TRANSMISSION GEAR TO ACHIEVE FULL LOAD. (AUTOMATIC TRANSMISSION — STALL RPM & LOAD).			
	See Illustration — Reverse Side			
	INSTRUMENT	GUIDELINE DATA	1st CHECK	2nd CHECK
	PRESSURE GAUGE (30 PSI RANGE)			

13	**CRANKCASE PRESSURE***			
	a. MEASURE AT BREATHER TUBE WITH SPECIAL TOOL {PLT-554			
	b. MEASURE AT HIGH IDLE (NO LOAD) RPM. {FES-533-5			
	See Illustration — Reverse Side			
	INSTRUMENT	GUIDELINE DATA	1st CHECK	2nd CHECK
	WATER MANOMETER OR MAGNAHELIC GAUGE			

FIGURE 28-2. Tests.

TESTS

STOP! READ THIS INSTRUCTION:
IF GUIDELINE DATA WAS OBTAINED DURING ABOVE
TESTS, **ENGINE OPERATION IS SATISFACTORY &**
BELOW TESTS ARE <u>NOT</u> REQUIRED. — **STOP!**

14 TEST INJECTION NOZZLES

A. SPRAY CONDITION

B. VALVE OPENING PRESSURE

C. LEAKAGE

INSTRUMENT	GUIDELINE DATA	1st CHECK	2nd CHECK
NOZZLE TESTER			

15 ANEROID DIAPHRAGM

PRESSURE TEST DIAPHRAGM (15-20 P.S.I. — FOR 30 SECONDS)
NO LEAKAGE PERMISSIBLE
See Illustration — Reverse Side

INSTRUMENT	GUIDELINE DATA	1st CHECK	2nd CHECK
D-100 PRESSURE KIT			

16 EXHAUST BACK PRESSURE

a. Measure at point not more than 18″ downstream of Turbo in straight length of pipe.
b. Measure at full load & rated speed. *See Illustration — Reverse Side*

INSTRUMENT	GUIDELINE DATA	1st CHECK	2nd CHECK
WATER MANOMETER OR MAGNAHELIC GAUGE			

17 MEASURE SMOKE INTENSITY

MEASURE AT FULL LOAD & RATED SPEED (RPM)

INSTRUMENT	GUIDELINE DATA	1st CHECK	2nd CHECK
SMOKE METER			

18 INTAKE AND EXHAUST VALVE CLEARANCE

ENGINE OFF — HOT OR COLD

INSTRUMENT	GUIDELINE DATA	1st CHECK	2nd CHECK
FEELER GAUGE			

All operating data is obtained from "PERFORMANCE DATA GUIDE-
LINE" for specified application listed in "OPERATION, MAINTENANCE
AND DIAGNOSTIC MANUAL," 1 085 843 R2 (or later).

FIGURE 28-3. Tests.

erable temperature differences in injection system components, which can cause scoring of parts or can result in actual seizure if done while the engine is running or if the engine is started soon after steam cleaning. Due to the extremely close tolerances between certain injection system components, uneven temperatures between these parts can cause serious damage if the engine is operated under these conditions.

Make sure all areas of the engine are properly cleaned, particularly places where disassembly is required. These areas are especially prone to the entry of dirt when covers or plates are removed to gain access to other parts of the engine.

2. Change the engine oil and oil filters. It is important that the engine oil and oil filters be changed at the recommended time or mileage intervals. See Chapter 11 for details on service and viscosity ratings of engine lubricating oils and filter changing.

3. Service the cooling system. The cooling system must be in good operating condition for it to be able to keep the engine at its most efficient operating temperature. A good engine tune-up includes a careful inspection of the cooling system and the correction of any faults that may be discovered. See Chapter 12 for complete details of cooling system service.

Check the coolant level. Check coolant condition before adding antifreeze or water to ensure that coolant quality is maintained. Check for coolant leakage and correct if necessary. Check condition of all hoses and clamps. Check condition and tension of all belts. Check pulleys, fan, fan shroud, fan drive, and radiator shutter operation. Change the coolant filter and conditioner as required. Replace any parts and make any adjustments that inspection procedures indicate are needed.

4. Check for loose fasteners, lines, fittings, supports, and mounting brackets. If there are any loose bolts or nuts, make sure that both the male and female threads are in good condition before tightening them to specified torque. Faulty fasteners must be replaced. Make sure that any required lock washers, cotter pins, or flat locks are in place. Missing or damaged locking devices must be replaced. Make sure that all lines and tubing are routed properly and are in good condition. Lines and tubing should be routed properly to prevent chafing and have sufficient distance from hot exhaust system components. All support brackets should be in place and in good condition.

5. Check the air intake system. Check all induction system hoses, tubes, and connections. The entire induction system should be airtight to prevent the entry of dust and dirt. Check the turbocharger connections as well. Make sure that piping is not damaged or dented, which could cause air intake to be restricted. Service the air cleaner as required. See Chapter 17 for details on induction system service.

6. Check the exhaust system. Make sure there are no exhaust leaks or damaged pipes that could restrict the exhaust. Exhaust leaks are usually evidenced by streaked lines of black extending from the leak. Repair as needed. See Chapter 17 for details on exhaust system service.

7. Service the fuel system. An adequate supply of clean fuel of the recommended type is absolutely essential to good engine performance. Incorrect fuel, contaminated fuel, or restricted fuel flow can all cause serious engine performance problems. Check the fuel tank and filler cap (be sure the vent is open), the water trap, and the fuel lines. Drain any water from the water trap. Make sure fuel lines are not dented or kinked to restrict flow in both supply and return lines. Clean any fuel screens that may be in the fuel supply system. Check the fuel supply pump pressure if necessary.

Remove, test, and calibrate the injectors if required. Injectors must perform in a uniform manner for all engine cylinders to ensure proper power balance between cylinders. On engines with engine cam-operated injectors, the engine valve clearances must be adjusted before injectors are adjusted. Unit injectors also require timing and fuel rack control adjustments. If an engine compression test is to be performed it should be done before the injectors are installed.

8. Perform engine cylinder compression test if needed. Cylinder compression must be adequate and must be even between cylinders as specified for all cylinders to produce equal power. See Part 3 of this chapter for compression testing and cylinder leakage tests.

9. Adjust engine valve clearance. The valve bridges do not normally need adjustment during a tune-up; however they must be checked. If adjustment of any amount is required, it is an indication of valve and seat problems. Adjust valve clearance after checking valve bridge adjustment.

10. Check throttle linkage. Make sure throttle linkage is able to move through the entire range of travel and does not bind throughout that range. If

the throttle linkage is not able to move to the full fuel position, the engine will not be able to produce its full power. Bent linkage or obstructions can be the cause of interference with linkage movement. Make sure connecting pins and clevises are not worn excessively. Correct any condition that interferes with free and full throttle movement. Adjust engine low-idle and high-idle speed to specifications.

11. Check engine monitoring gauges mounted in the instrument panel. Check any abnormal condition to determine whether the monitored function is in error or whether the gauge or sensor are at fault. Correct as needed.

12. Perform dynamometer test. See Part 3 in this chapter and Chapter 18 for this procedure. Be sure to follow the engine and equipment manufacturer's instructions for safety and proper procedure.

Summary

A good tune-up on a mechanically sound engine will restore the power, efficiency, and performance it was designed to provide when it was new. This means that the engine is capable of handling allowable maximum loads, will operate smoothly, produce limited exhaust emissions and smoke, will not use excessive amounts of fuel, and will reduce the possibility of costly and time-consuming down time.

PART 3   ENGINE PERFORMANCE TESTING

The general topic of engine performance may be divided into several areas such as:

- Won't start
- Poor idle
- Stalls
- Misses
- Lacks power
- Uses too much oil
- Noise—knocks, pings, rattles, squeaks, squeals, etc.
- Smoke from exhaust
- Overheats
- Uses too much fuel

These are the kinds of problems that are perceivable by the owner, the operator, and the technician alike. The source or cause of these problems may be any one or more of a number of engine systems. Thorough knowledge of each system is essential to isolate such problems.

The appropriate diagnostic charts and procedures for performance problems may be found in this text as follows and will therefore not be repeated here.

- Lubrication system diagnosis—Chapter 11.
- Cooling system diagnosis—Chapter 12.
- Diesel engine diagnosis—Chapter 13.
- Air supply system diagnosis—Chapter 17.
- Fuel system diagnosis—Chapters 20 to 27.
- Exhaust system diagnosis—Chapter 17.
- Battery diagnosis—Chapter 29.
- Starting system diagnosis—Chapter 34.
- AC charging system diagnosis—Chapter 32.
- DC charging system diagnosis—Chapter 33.

The method selected and used for diagnosis will depend on the type of equipment available for the purpose; however, all shops are normally equipped with sufficient test equipment of one kind or another to enable the technician to perform an accurate diagnosis of the problem.

Some of the types of test equipment used are as follows:

Hand held test equipment (or mounted in portable stands) including:

- Voltmeter
- Ammeter
- Ohmmeter
- Manometer
- Tachometer
- Pressure gauge
- Compression tester
- Cylinder leak down test adapter
- Test light
- Jumper wire
- Test type circuit breaker

Dynamometer—used in conjunction with test equipment listed above.

Compression Testing

To perform a compression test on a diesel engine, proceed as follows with a tester capable of 500 psi (3450 kPa) (Fig. 28–4).

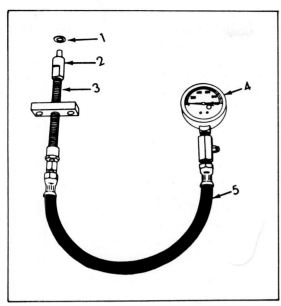

Tools for Checking Compression

1. **Nozzle Gasket**
2. **Adapter Tip**
3. **Adapter**
4. **Compression Gauge Assembly**

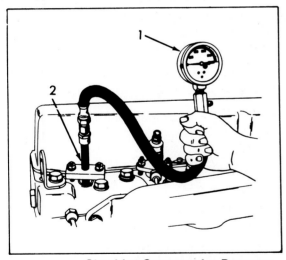

Checking Compression Pressure

1. **Compression Tester Gauge Assembly**
2. **Compression Tester Adapter**

FIGURE 28-4. Typical compression testing equipment. Different adapters are required for various engines. (*Courtesy of Allis-Chalmers*)

1. Engine should be at operating temperature.

2. Batteries and cranking system should be in good condition.

3. Remove air cleaner element.

4. Disable the fuel system and glow plug system as specified by the vehicle manufacturer. This may require disconnecting a fuel solenoid lead on some models, as well as glow plug connections. Refer to service manual for procedure to follow.

5. Remove either glow plugs or injectors to allow installation of compression tester. Some engines are compression tested through the glow plug holes, while others are tested through the injector holes. Compression testers may be equipped with screw-in or clamp-in adapters, depending on application. Be sure to follow engine and equipment manufacturer's instructions for use of compression tester.

6. Install compression tester into number one cylinder.

7. Crank the engine through at least six compression strokes—note the highest compression reading and the number of strokes required to obtain the reading. Repeat the test on each of the remaining cylinders, cranking the engine the same number of strokes as were required to obtain the highest reading for number one cylinder.

8. Record the results from all cylinders.

9. Analyze the test results as follows:

Normal—Compression builds up quickly and evenly to specified compression pressure on all cylinders.

Piston Ring Leakage—Compression low on first stroke but tends to build up on following strokes. Does not reach normal cylinder pressure.

Caution—Due to high compression ratio of diesel engine and very small combustion chamber volume, do not add oil to any cylinder for compression testing. Extensive engine damage can result from this procedure.

Comparison Between Cylinders

In order for cylinders to produce relatively even power output, compression pressures should be similar within certain limits. The comparison of compression pressures between cylinders is just as important as actual compression pressures. Since engine temperature, oil, viscosity, and engine cranking speed all have a bearing on test results, some variation from specified pressures may be expected. However, compression pressures should be comparable between cylinders. Some manufacturers allow a maximum difference in pressures between cylinders of 15 percent. A greater than 15 percent difference requires repair to the affected cylinders.

Cylinder Leakage Testing

Cylinder leakage tests are performed to determine whether compression pressures are able to leak past the rings into the crankcase, past the exhaust valves into the exhaust system, past the intake valves into the induction system, or past the head gasket into the engine coolant.

A simplified cylinder leakage test may be performed using a shop air line adapter made from a discarded injector of the appropriate type—a male shop air coupler is welded to the injector main body after the plunger etc. have been removed. This allows the adapter to be installed into the injector hole and shop air to be coupled to the adapter.

Shop air pressure of 150 psi (1000 kPa) is required for this test.

To perform a cylinder leakage test proceed as follows:

1. Engine at operating temperature.

2. Disconnect battery ground cable.

3. Remove air cleaner element.

4. Disable the fuel system and glow plug system and remove injectors as specified by the vehicle manufacturer. This may require disconnecting a fuel solenoid lead on some models, as well as glow plug connections. Refer to service manual for procedure to follow.

5. Install air line adapter into number one cylinder.

6. Remove crankcase oil dipstick, oil filler cap, and radiator cap.

7. Turn crankshaft to position number one piston at TDC position on the compression stroke. Make sure piston is exactly at TDC on the upstroke of the compression stroke. This is important for three reasons. First, the piston will be forced down by shop air if not exactly at the TDC position. Second, the piston rings should be at the bottom of their grooves for this test. Moving the piston up will do this. Third, valves are closed with the piston in this position.

8. Now connect shop air to the adapter in number one cylinder.

9. Listen for air leakage into the exhaust system at the exhaust pipe. If present, this indicates exhaust valve leakage. Listen for air leakage at the air intake. If present, this indicates intake valve leakage. Listen for air leakage at the oil filler cap or dipstick tube. If present, this indicates leakage past the rings. Listen for leakage at injector holes of cylinders adjacent to the one being tested. If present, this indicates cylinder head gasket leakage between cylinders. Watch for air bubbles in coolant in radiator. If present, this indicates cylinder head gasket leakage to cooling system.

10. Disconnect shop air line from cylinder adapter and repeat test procedure for all cylinders.

Dynamometer Testing

Diesel engine performance is determined by using a dynamometer designed to measure the turning effort the engine is able to produce. Several types are used, each designed for a particular application. These include the following:

1. Chassis dynamometer used to test highway vehicles.

2. Engine dynamometer used to test engines removed from the vehicle or equipment.

3. Power take-off dynamometer used to test agricultural and industrial tractors equipped with a power take-off shaft.

All three types use a power absorption unit to which engine power is applied for testing. The degree of load applied can be varied by a hand-held control device. The power absorption unit may be of the hydraulic or electrical type. See Chapter 6 for a description of dynamometer operation.

Dynamometer Test Instruments

A complete set of engine monitoring instruments is used to provide the operator with the necessary information during the test procedure. This includes the following:

1. A pressure gauge to determine engine lubricating oil pressure connected to the engine main oil gallery.

2. An oil temperature gauge inserted into the engine oil through the dipstick tube or hole.

3. A temperature gauge to determine engine coolant temperature, usually connected to the engine side of the thermostat housing.

4. A pressure gauge to measure cooling system pressure, usually connected to the water manifold.

5. A water manometer or low-reading pressure gauge to measure crankcase pressure, connected to the crankcase above the oil level.

6. A water manometer or low-reading vacuum gauge to determine negative inlet air pressure on

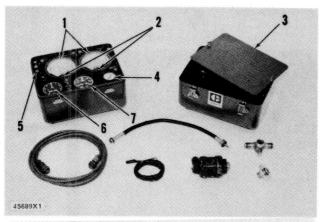

1. Differential pressure gauges. 2. Zero adjustment screw. 3. Lid. 4. Pressure gauge. 5. Pressure tap fitting. 6. Tachometer. 7. Manifold pressure gauge.

FIGURE 28-5. Test instruments for checking positive and negative pressures. *(Courtesy of Caterpillar Tractor Company)*

naturally aspirated engines, connected to the air inlet to the engine.

7. A mercury manometer or low-reading pressure gauge to measure boost pressure on turbocharged engines, connected to the intake air manifold.

FIGURE 28-6. Pyrometer used to test exhaust temperature. *(Courtesy of Caterpillar Tractor Company)*

8. A water manometer or low-reading pressure gauge to measure exhaust back pressure, connected to the exhaust manifold flange. (See Figure 28-5.)

9. A contact pyrometer to determine individual exhaust cylinder temperatures. Temperature is checked at the manifold port near the cylinder head at each cylinder to determine whether exhaust temperatures are within limits and are relatively even among cylinders. (See Figure 28-6.)

10. A tachometer to determine engine crankshaft speed in revolutions per minute.

11. A power gauge (horsepower and/or kilowatts) to indicate power produced by the engine.

12. A fuel consumption meter, connected to the fuel supply to the engine, to measure fuel consumption in pounds per brake horsepower hour (lbs./bhp./hr.).

Caution: Running recapped tires on a chassis dynamometer presents a risk of recap rubber flying off during testing. Chassis dynamometer tests should be conducted only with known good tires on the vehicle. If there is a heavy vibration, tire bounce, drive shaft whip, or the like, do not proceed with the testing until these problems have been corrected.

PART 4   SELF-CHECK

1. Why is an engine tune-up needed?
2. What must be done before proceeding with the tune-up?
3. A tune-up will restore the engine's power and efficiency only if it is _____.
4. List three reasons for engine performance testing.
5. What three types of dynamometers are used for engine performance testing?
6. Why is it dangerous to run recapped tires on a chassis dynamometer?
7. List ten engine functions that are monitored during performance tests.
8. Why is a compression test sometimes required?
9. What is the purpose of cylinder leakage tests?

Chapter 29

Batteries and Battery Service

Performance Objectives

After thorough study of this chapter and the appropriate training models and test equipment you should be able to:

1. Complete the self-check questions with at least 80 percent accuracy.
2. State the purpose of the lead acid storage battery.
3. Describe the basic construction features of the battery.
4. Describe the basic operation of the battery during charging and discharging cycles.
5. Clean and test any storage battery and interpret the test results to determine whether the battery is serviceable.
6. Diagnose basic battery problems.
7. Charge any storage battery needing a charge.
8. Replace a battery as required.

PART 1 PURPOSE AND CONSTRUCTION

The purpose of the lead-acid battery in the diesel-powered equipment is to provide electrical energy to the following:

• The electrical starting system, whenever needed;

• The lighting system

• The accessories (all power equipment, heating, air conditioning, radios, and tape players—whenever the charging system is not able to perform these functions)

The battery is also a sort of electrical shock absorber or voltage stabilizer for any abnormal stray voltages created by any of the various electrical systems.

Battery Cell

Each cell in the battery is made up of a number of positive plates and a number of negative plates separated by insulating separator plates. Negative and positive plates are arranged alternately in each cell. All the negative plates are connected to each other and so are the positive plates. This arrangement provides a positive cell connection and a negative cell connection. This assembly is submerged in a cell case full of battery electrolyte, which is 64 percent water and 36 percent sulfuric acid. (See Fig. 29–1 to 29–8.)

Each battery cell produces approximately 2 volts, regardless of the number or size of plates per cell. Six of these 2-volt cells arranged in a single battery case form a 12-volt battery. The battery case is usually made of a rubber composition or polypropylene. The case has built-in cell dividers and sediment traps. The six battery cells of a 12-volt battery are connected in series. This means that the positive side of a cell is connected to the negative side of the next cell throughout all six cells. If the cells were connected in parallel (positive to positive and negative to negative), the battery would have only a 2-volt potential.

Battery Polarity

The positive plate group in one end cell of a battery is connected to the positive battery external terminal. This terminal is identified in one of the following ways: POS, + sign, or red-colored terminal. The tapered positive post is also larger in diameter than the negative post.

The negative plate group at the other end of the battery is connected to the external negative battery terminal. It can be identified as follows: NEG or − sign on terminal. The tapered negative post is smaller in diameter than the positive post. Proper battery polarity must always be adhered to when working with electrical systems. With the exception of some types, most equipment is of the 24- or 32-volt negative-ground type. Be sure to check system voltage and polarity before performing any electrical system service.

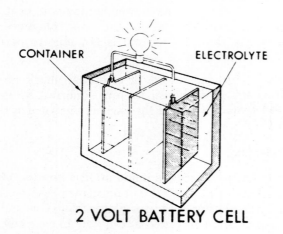

2 VOLT BATTERY CELL

FIGURE 29-1. A wet battery is a device that is able to convert chemical energy to electrical energy. The chemical action can be reversed to recharge the battery. *(Courtesy of Ford Motor Co. of Canada Ltd.)*

Battery Cables

Battery cables (Fig. 29-9 and 29-10) must be of sufficient current-carrying capacity to meet all electrical loads. Normal 12-volt cable size usually is 4 gauge (19 mm²) or 6 gauge (13 mm²). Various cable clamps and terminals are used to provide a good electrical connection at each end. All connections must be clean and tight to prevent arcing, corrosion, and high resistance.

PART 2   BATTERY TYPES

Wet and Dry Charged Batteries

Batteries are sometimes shipped without any electrolyte. These batteries are charged wet at the fac-

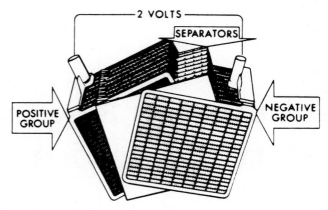

FIGURE 29-3. Lead acid battery cells have a number of positive and negative plates insulated from each other by separator plates. Increasing the number of plates increases the surface area exposed to chemical action, thereby increasing capacity. Regardless of the number of plates per cell, each cell is capable of producing only approximately 2 volts. *(Courtesy of Ford Motor Co. of Canada Ltd.)*

tory. The electrolyte is then removed and the cell plates dried out before shipment. Dry charged batteries can be stored longer than wet batteries. When a dry charged battery is put in use, it is filled to the correct level with electrolyte and activated by a short period of charging before installation.

Battery sales outlets may store wet batteries rather than dry batteries if their sales volume is high enough whereby batteries are not in storage for any significant length of time.

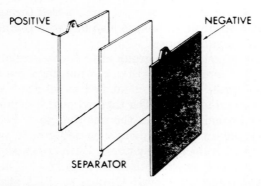

FIGURE 29-2. The basic components required in a battery cell. *(Courtesy of Ford Motor Co. of Canada Ltd.)*

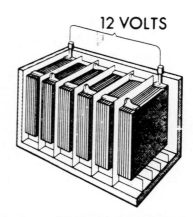

FIGURE 29-4. Assembling six 2-volt cells and connecting them in series inside a single battery case creates a 12-volt battery. *(Courtesy of Ford Motor Co. of Canada Ltd.)*

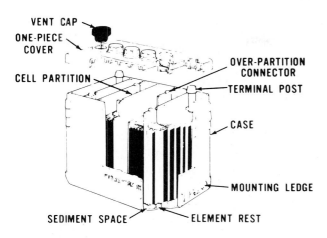

FIGURE 29-5. Battery case components. Note that partitions separate the cells from each other. *(Courtesy of Ford Motor Co. of Canada Ltd.)*

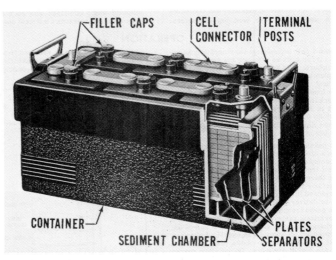

FIGURE 29-7. Typical heavy-duty battery construction. *(Courtesy of General Motors Terex)*

Maintenance-Free Batteries

The maintenance-free battery has cell plates made of a slightly different compound. This reduces the amount of vaporization that takes place during normal operation. A vapor recovery system is also used to reduce water loss through vaporization. This type of battery is also not as easily overcharged.

The battery is completely sealed except for a small vent so that acid and vapors cannot escape. An expansion chamber allows internal expansion and contraction to take place. Since vapors cannot

escape from this battery, it is not necessary to periodically add water to the battery. This also reduces the possibility of corrosion and surface discharge due to electrolyte on the surface of the battery.

Some maintenance-free batteries have a built-in hydrometer indicating the state of charge of the battery. Others cannot be tested with a hydrometer since they are sealed. A voltmeter is used in this case.

Heavy-Duty Batteries

Heavy-duty batteries are generally required for use in heavy equipment. Because of the severe operating angles and the higher levels of vibration and jarring that occur in many types of heavy equipment, batteries must be able to withstand these conditions.

Although construction is very similar to that of standard batteries, several special features are added, including the following:

1. Tilt reservoirs to prevent electrolyte spillage at up to 45-degree angles

2. Increased case strength

3. Plates bonded to case to prevent vibration damage

4. Heavier cell connectors

5. Higher density plate construction to increase charge and discharge cycle life.

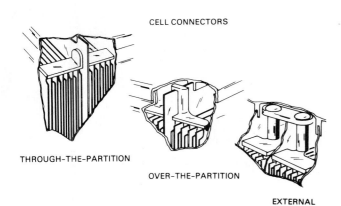

FIGURE 29-6. Different types of cell connectors. Sealed top batteries use the internal type of cell connectors. *(Courtesy of Ford Motor Co. of Canada Ltd.)*

FIGURE 29-8. Battery manufacturers do not always use the same types of battery terminals or terminal locations. *(Courtesy of General Motors Corporation)*

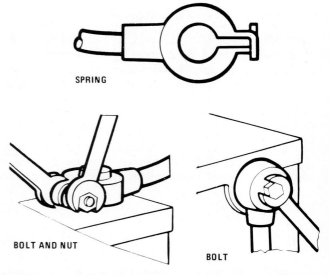

FIGURE 29-9. Comparison of battery cable diameters. *(Courtesy of Chrysler Corporation)*

FIGURE 29-10. Three types of battery cable connections. *(Courtesy of Chrysler Corporation)*

PART 3   CHEMICAL ACTION

Charging and Discharging

In operation, the battery is normally being partially discharged and recharged. This is actually a constant reversing of the chemical action taking place in the battery. The continual cycling of the charge and discharge modes slowly wears away the active materials on the battery cell plates. This eventually causes the battery positive plates to oxidize. When this oxidizing has reached the point of insufficient active plate area to charge the battery, the battery is worn out and must be replaced. (See Fig. 29-11 to 29-13.)

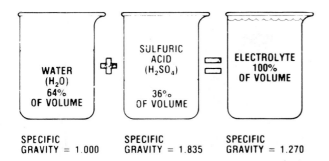

FIGURE 29-11. Composition of specific gravity of battery electrolyte. *(Courtesy of Chrysler Corporation)*

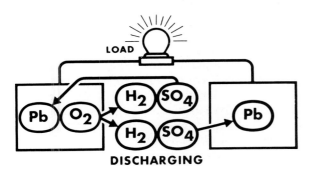

FIGURE 29-12. Chemical action inside battery as current is being used and battery is discharging. Pb is sponge lead, O_2 is oxygen; therefore, Pb O_2 is a lead oxide. H_2 is hydrogen, SO_4 is sulfate; therefore, H_2 SO_4 is sulfuric acid.

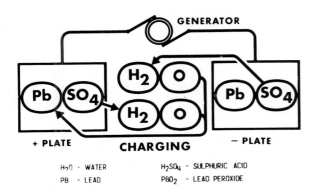

FIGURE 29-13. During the charging cycle, chemical action inside the battery is the reverse of that shown in Figure 29-12.

PART 4   BATTERY RATINGS

Battery capacity ratings are established by the Battery Council International (BCI) and the Society of Automotive Engineers. Commonly used ratings are as follows (Fig. 29–14):

- Cold cranking
- Reserve capacity
- Ampere-hour
- Watt-hour

Cold Cranking: The load in amperes a battery is able to deliver for 30 seconds at $0°F$ ($-17.7°C$) without falling below 7.2 volts for a 12-volt battery.

Reserve Capacity. The length of time in minutes a battery can be discharged under a specified load at $80°F$ ($26.6°C$) before battery cell voltage drops below 1.75 volts per cell.

Ampere Hour. Discharge the battery at one twentieth times the ampere hour rating for 20 hours at $80°F$ ($26.6°C$) without cell voltage falling below 1.75 volts per cell. This term was used for many years in the rating of storage batteries. The term and the method have been replaced by Cold Cranking and Reserve Capacity ratings described above.

Watt Hour. Ampere hours times battery voltage yields the watt-hour rating.

Factors that determine the battery rating required for a vehicle include engine size and type and climatic conditions under which it must operate. Battery power drops drastically as temperatures drop below freezing. As the temperature drops much below freezing, the engine is harder to crank owing to increased friction resulting from oil thickening.

AH Rating	Cranking Power at 0°F
48	325A
59	375A
70	440A
85	500A

FIGURE 29-14. Battery rating comparisons. Ampere hour rating is determined as follows: discharge the battery at 1.20 x AH rating of battery for 20 hours at $80°F$ ($26.6°C$). The battery voltage should not drop below 1.75 volts per cell under these conditions. Cranking power at $0°F$ is defined as follows: cranking load in amperes that a battery is able to deliver for 30 seconds at $0°F$ without falling below 7.2 volts for a 12-volt battery.

The correct shape, physical size, post location, and battery rating requirements must all be considered when replacing a battery.

PART 5   BATTERY DIAGNOSIS AND SERVICE

The battery is the heart of the various electrical systems in the diesel engine. Since this is so, it is important not to overlook the battery when servicing the electrical system.

General Precautions

To avoid personal injury and property damage, it is important to observe safe procedures when servicing or replacing lead-acid batteries. The following precautions, if observed properly, can prevent accidental injury or damage.

1. Battery acid is extremely corrosive. Avoid contact with skin, eyes, and clothing. If battery acid should accidentally get into your eyes, rinse thoroughly with clean water and see your doctor. Contacted skin should be thoroughly washed with clean water; some baking soda with the wash will neutralize the action of the acid. Painted surfaces and metal parts are also easily attacked by acid, and contact should be avoided.

2. When making connections to a battery in or out of the vehicle, always observe proper battery polarity: positive to positive and negative to negative.

3. Avoid any arcing (sparks) or open flame near battery. The battery produces highly explosive vapors that can cause serious damage if accidentally ignited.

4. When disconnecting battery cables always remove ground cable first; and when connecting cables, always connect ground cable last. This helps to avoid accidental arcing.

5. Observe equipment manufacturer's instructions when charging batteries. Never allow battery temperature to exceed 125°F (51.6°C).

6. Some battery manufacturers place restrictions on the use of a booster battery to jump start. Follow battery manufacturer's recommendations. One manufacturer says, "Do not charge, test, or jump start this battery when the built-in hydrometer is clear or yellow: battery must be replaced."

7. Use proper battery carrier to handle the battery; this avoids injury and possible battery damage.

8. Use proper protective clothing (apron, gloves, and face shield) when handling batteries to ensure safety.

9. Do not weld or smoke near a battery charging or storage area.

Emergency Jump Starting with Auxiliary (Booster) Battery

Notice: Do not push or tow the vehicle to start. Damage to the vehicle may result. Do not jump start the vehicle unless the manufacturer of the battery or the vehicle with a dead battery allows this procedure.

Both the booster and discharged battery should be treated carefully when using jumper cables. Follow the procedure outlined below, being careful not to cause sparks.

Caution: Departure from these conditions or the procedure below could result in (1) serious personal injury (particularly to eyes) or property damage from such causes as battery explosion, battery acid, or electrical burns; and/or (2) damage to electronic components of either vehicle.

Never expose a battery to an open flame or electric spark—batteries generate a gas that is flammable and explosive.

Remove rings, watches, and other jewelry. Wear approved eye protection. No smoking. Do not jump start a battery with a frozen electrolyte.

Do not allow battery fluid to contact eyes, skin, fabrics, or painted surfaces—its fluid is a corrosive acid. Flush any contacted area with water immediately and thoroughly. Be careful that metal tools or jumper cables do not contact the positive battery terminal (or metal in contact with it) or any other metal on the car because a short circuit could occur. Batteries should always be kept out of the reach of children.

1. Set the parking brake and place automatic transmission in *PARK* (*NEUTRAL* for manual transmission). Turn off the lights, and all other electrical loads.

2. Check the built-in hydrometer. If it is clear or light yellow, replace the battery.

3. Only 12-volt batteries can be used to start the engine equipped with a 12-volt starting system. For systems other than 12-volt, or other than negative ground, consult the appropriate service manual.

Notice: When jump starting a vehicle with special high rate charging equipment, be sure charging equipment is proper voltage and polarity.

4. Attach the end of one jumper cable to the positive terminal of the booster battery and the other end of the cable to the positive terminal of the discharged battery. Do not permit vehicles to touch each other because this could cause a ground connection and counteract the benefits of the procedure.

5. Attach one end of the remaining negative cable to the negative terminal of the booster battery, and the other end to a solid engine ground (such as A/C compressor bracket or generator mounting bracket) at least 18 inches from the battery of the vehicle started. (*DO NOT CONNECT DIRECTLY TO THE NEGATIVE TERMINAL OF THE DEAD BATTERY.*)

6. Start the engine that is providing the jump start and turn off electrical accessories. Then start the engine with the discharged battery.

7. Reverse these directions exactly when removing the jumper cables. The negative cable must be disconnected from the engine that was jump started first.

PART 6   STANDARD BATTERY SERVICE

Proper battery service includes the following (Fig. 29–15 to 29–33):

• Visual inspection of battery, battery cables, battery stand, and hold down for leakage, corrosion, and dirt.

• Check electrolyte level and specific gravity. If electrolyte is too low, add sufficient distilled water to raise electrolyte to correct level. After adding water, battery must be charged before a valid specific gravity test can be performed.

Battery Freezing Temperatures			
Specific Gravity	Freezing Temp.	Specific Gravity	Freezing Temp.
1.280	−90°F	1.150	+ 5°F
1.250	−62°F	1.100	+19°F
1.200	−16°F	1.050	+27°F

FIGURE 29–15. Specific gravity of battery electrolyte at which electrolyte will freeze is shown. *(Courtesy of Ford Motor Co. of Canada Ltd.)*

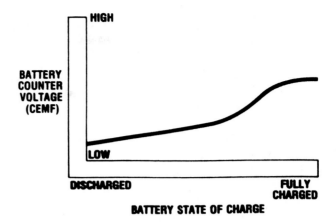

FIGURE 29–16. As the battery becomes fully charged, the battery's resistance to charging system output increases. Very little current is required to keep a battery in a fully charged condition if no current is being consumed. *(Courtesy of Chrysler Corporation)*

Determine battery state of charge with hydrometer or with expanded scale voltmeter. The battery must be stabilized (surface charge removed) before checking state of charge.

To stabilize a battery after charging, apply 15-ampere load for 15 seconds by turning the headlights on bright beam for 15 seconds; then proceed with testing. The following open circuit voltage figures are for batteries at 80°F (26.7°C).

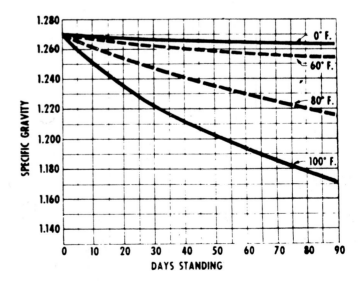

FIGURE 29–17. Self-discharge rates of batteries in storage at various temperatures and time periods. The least discharge occurs at the lower temperatures. *(Courtesy of Battery Council International)*

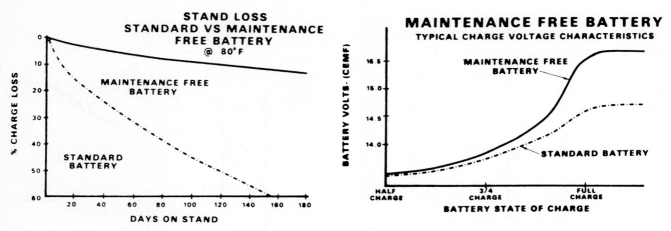

FIGURE 29-18. Comparison of resistance to charging (CEMF—Counter electromotive force) and percentage of charge loss over a given period (right) between standard battery and maintenance-free battery. *(Courtesy of Battery Council International)*

OPEN CIRCUIT VOLTS	PERCENT CHARGE
11.7 or less	0%
12.0	25%
12.2	50%
12.4	75%
12.6 or more	100%

Charging the Battery

Standard Battery Charging Guide (12-Volt and 6-Volt Batteries)

Caution: Do not use for maintenance-free batteries. Recommended rate and time for fully discharged condition.

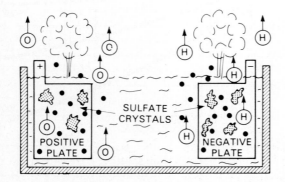

FIGURE 29-19. A battery becomes sulfated due to a discharged condition. Hardened sulfate crystals penetrate the pores of the plates. These crystals become insoluble. Prolonged charging at a low rate is required to restore it. Charging a sulfated battery at too high a rate can buckle the plate and destroy the battery. A battery that has been sulfated for too long cannot be restored.

FIGURE 29-20. A charging rate that is too high will cause gassing and will result in sulfate deposits at the battery terminals. Electrolyte on the terminal can also cause the creation of deposits. *(Courtesy of Chrysler Corporation)*

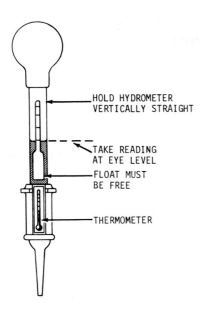

FIGURE 29-21. Battery hydrometer. *(Courtesy of Deere and Company)*

HOLD HYDROMETER
VERTICALLY STRAIGHT

TAKE READING
AT EYE LEVEL

FLOAT MUST
BE FREE

THERMOMETER

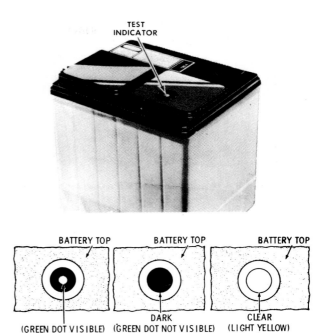

TEST
INDICATOR

BATTERY TOP   BATTERY TOP   BATTERY TOP

DARK   CLEAR
(GREEN DOT VISIBLE)   (GREEN DOT NOT VISIBLE)   (LIGHT YELLOW)

FIGURE 29-23. Maintenance-free battery with built-in color-coded hydrometer readings. Green = charged; dark, no dot = discharged; clear = replace battery. *(Courtesy of General Motors Corporation)*

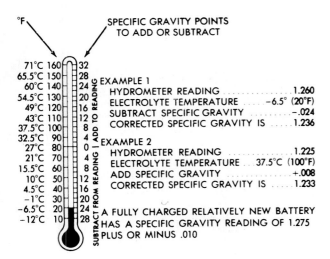

°F

SPECIFIC GRAVITY POINTS
TO ADD OR SUBTRACT

71°C	160	32
65.5°C	150	28
60°C	140	24
54.5°C	130	20
49°C	120	16
43°C	110	12
37.5°C	100	8
32.5°C	90	4
27°C	80	0
21°C	70	4
15.5°C	60	8
10°C	50	12
4.5°C	40	16
–1°C	30	20
–6.5°C	20	24
–12°C	10	28

ADD TO READING

SUBTRACT FROM READING

EXAMPLE 1
HYDROMETER READING 1.260
ELECTROLYTE TEMPERATURE . . . . . –6.5° (20°F)
SUBTRACT SPECIFIC GRAVITY . . . .–.024
CORRECTED SPECIFIC GRAVITY IS 1.236

EXAMPLE 2
HYDROMETER READING 1.225
ELECTROLYTE TEMPERATURE . . . 37.5°C (100°F)
ADD SPECIFIC GRAVITY . . . .+.008
CORRECTED SPECIFIC GRAVITY IS 1.233

A FULLY CHARGED RELATIVELY NEW BATTERY
HAS A SPECIFIC GRAVITY READING OF 1.275
PLUS OR MINUS .010

FIGURE 29-22. To accurately determine the state of charge of a battery, the reading must be temperature corrected as shown here. *(Courtesy of Chrysler Corporation)*

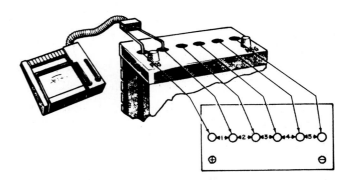

FIGURE 29-24. Individual battery cell voltage test.

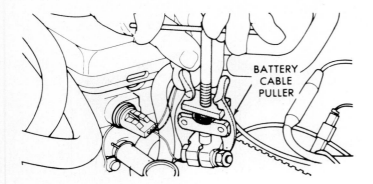

FIGURE 29-25. Since it is easy to damage battery posts, the correct tools should be used to remove battery cables. *(Courtesy of Chrysler Corporation)*

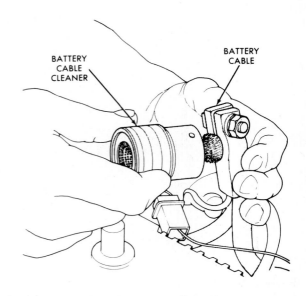

FIGURE 29-28. Cleaning battery cable clamp. *(Courtesy of Chrysler Corporation)*

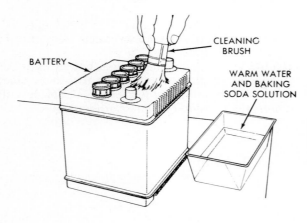

FIGURE 29-26. Cleaning the battery with a water and baking soda solution neutralizes battery electrolyte. Solution should not be allowed to enter battery. *(Courtesy of Chrysler Corporation)*

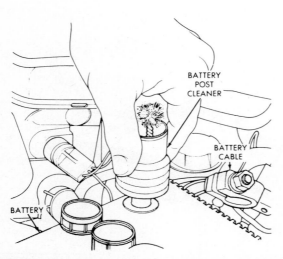

FIGURE 29-27. Battery post cleaner has internal wire brush. *(Courtesy of Chrysler Corporation)*

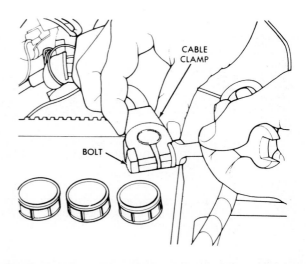

FIGURE 29-29. Battery connections must be clean and tight. Do not overtighten or break post loose. *(Courtesy of Chrysler Corporation)*

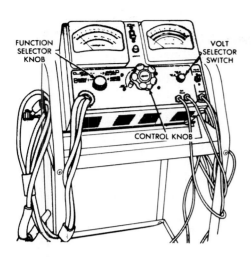

FIGURE 29-30. Equipment used to load test a battery. Unit includes voltmeter, ammeter, a variable resistor operated by the control knob, a function selector knob, a volt selector switch, and the required leads and connectors. *(Courtesy of Chrysler Corporation)*

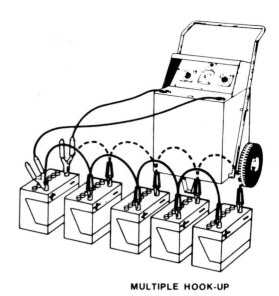

MULTIPLE HOOK-UP

FIGURE 29-32. Charging a number of batteries at the same time. Batteries are connected in parallel so combined battery voltage remains at 12 volts. *(Courtesy of Ford Motor Co. of Canada Ltd.)*

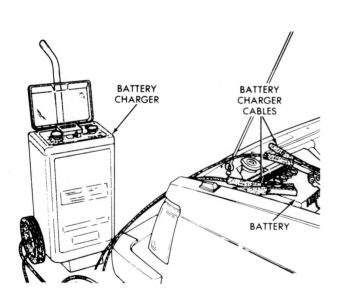

FIGURE 29-31. Charging battery in vehicles. Battery cables should be removed from battery posts for charging to prevent damage to vehicle electrical system. *(Courtesy of Chrysler Corporation)*

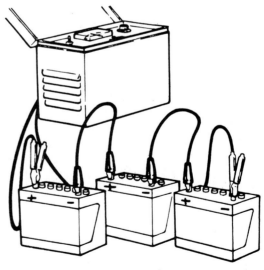

CONSTANT-CURRENT SLOW CHARGING

FIGURE 29-33. Many types of battery slow chargers are available. Always follow equipment manufacturer's instructions for hookup and charging rates. These batteries are connected in series. *(Courtesy of Ford Motor Co. of Canada Ltd.)*

RATED BATTERY CAPACITY (RESERVE MINUTES)	SLOW CHARGE	FAST CHARGE
80 minutes or less	14 hours @ 5 amperes	1-¾ hours @ 40 amperes
	7 hours @ 10 amperes	1 hours @ 60 amperes
Above 80 to 125 minutes	20 hours @ 5 amperes	2-½ hours @ 40 amperes
	10 hours @ 10 amperes	1-¾ hours @ 60 amperes
Above 125 to 170 minutes	28 hours @ 5 amperes	3-½ hours @ 40 amperes
	14 hours @ 10 amperes	2-½ hours @ 60 amperes
Above 170 to 250 minutes	42 hours @ 5 amperes	5 hours @ 40 amperes
	21 hours @ 10 amperes	3-½ hours @ 60 amperes
Above 250 minutes	33 hours @ 10 amperes	8 hours @ 40 amperes
		5-½ hours @ 60 amperes

If a battery is to be left on charge overnight, use only the slow charge rate. Charge maintenance-free batteries at rates given in Step 2, below. After charging, stabilize the battery before repeating state of charge tests.

If time is available, the lower charging rates in amperes are recommended; they must be used when a battery may have a problem (i.e., if it is sulfated or has a temperature below 15°F (−26.1°C).

• Check individual cell voltage on batteries where this applies. Minimum cell voltage should not vary more than 1/10 volt between cells. If it does, replace battery.

• Clean battery, cables, and battery carrier of vehicle. The cable terminals should be cleaned before connecting them to the battery. After the connections have been made, apply a thin coating of petroleum jelly on the post and cable terminals to retard corrosion. Never hammer cable terminals onto battery posts. The covers, undercover post connections, or post-to-cover connections could be severely damaged. Do not over-torque the terminal bolts of side terminal batteries. The threads may strip or the battery could be damaged.

Common Torque Valves

Tapered terminal posts (SAE): 50–70 lb. in.
  Side terminals 70 lb. in.
Use a baking soda and water solution after removing the battery. Clean the battery, battery stand,

and battery cables with this solution. Do not allow solution to enter battery. Rinse with clear water.

• Battery load test.

Follow the procedure in Step 3, below.

PART 7   MAINTENANCE-FREE BATTERY SERVICE

Step 1 Inspection. Visually inspect the battery and service the cables and battery carrier as in standard battery service above.

Step 2 Electrolyte Levels and State of Charge. Check the electrolyte level in the cells if possible. The level can be seen through translucent plastic cases. It can also be checked in batteries that are not sealed. If the electrolyte level is below the tops of the plates in any cell, add water if the vents are removable. If the battery is sealed and water cannot be added to it, replace the battery and check the charging system for a malfunction, such as a high voltage regulator setting. Follow instructions of manufacturer if the battery has a special indicating device.

BATTERY CAPACITY (RESERVE MINUTES)	SLOW CHARGE
80 minutes or less	10 hours @ 5 amps 5 hours @ 10 amps
Above 80 to 125 minutes	15 hours @ 5 amps 7-½ hours @ 10 amps
Above 125 to 170 minutes	20 hours @ 5 amps 10 hours @ 10 amps
Above 170 to 250 minutes	30 hours @ 5 amps 15 hours @ 10 amps
Above 250 minutes	20 hours @ 10 amps

If the level is O.K. and the stabilized open circuit voltage is below 12.4 volts, charge the battery using a constant potential taper charger. The initial charge rate should not exceed 30 amperes and should taper down to rates given here.

General Motors Delco Sealed Batteries

When it is necessary to charge the battery, the following basic rules must be followed:

1. Do not charge battery if hydrometer is clear or light yellow. Replace battery.

2. If the battery feels hot 125°F (52°C), or if violent gassing or spewing of electrolyte through the

vent holes occurs, discontinue charging or reduce charging rate.

Charge the battery until the green ball appears. Tipping or shaking the battery may be necessary to make the green ball appear. Temperature of the battery will affect the charging rate, and most charging equipment will not charge at a constant rate. For example, if the charger starts at 30 amperes and drops off to 10 amperes after 1 hour, the average current for that hour was 20 amperes. The actual boost charge was 20 ampere-hours. The sealed battery can be fast charged or slow charged with ordinary chargers in the same manner as conventional batteries. Either method will restore the battery to full charge.

Many chargers have special settings for sealed batteries. These settings reduce the charge voltage and limit the current. It is not necessary to use these settings with these sealed batteries.

Step 3 Load Test Procedure. The load test procedure is conducted to determine if the battery requires recharging or replacement. Test each battery separately.

1. Disconnect the battery cables (ground connection first) and connect the voltmeter and load test leads to the battery terminals, making sure the load switch on the tester is in the *OFF* position.

2. Apply a test load equal to 1/2 the cold cranking amperes rating @ 0°F (−18°C) of the battery for 15 seconds; for example, if a battery has a cold cranking rating @ 0°C of 350 amperes, use a test load of 175 amperes.

3. Read the voltage at 15 seconds and remove the load. If the voltage is less than the minimum specified below, replace the battery.

If the voltage meets or exceeds the specified minimum, clean the battery and return it to service.

If the battery tests well but fails, for no apparent reason, to perform satisfactorily in service, the following are some of the more important factors that may point to the cause of trouble:

1. Vehicle accessories left on overnight

2. Slow average driving speeds for short periods

MINIMUM VOLTS	ESTIMATED ELECTROLYTE TEMPERATURE	
9.6	70°F	(21°C) and above
9.5	60°F	(16°C)
9.4	50°F	(10°C)
9.3	40°F	(4°C)
9.1	30°F	(−1°C)
8.9	20°F	(−7°C)
8.7	10°F	(−12°C)
8.5	0°F	(−18°C)

3. The electrical load is more than the generator output, particularly with the addition of after market equipment

4. Defects in the charging system, such as electrical shorts, slipping fan belt, faulty generator, or voltage regulator

5. Battery abuse, including failure to keep the battery cable terminals clean and tight or loose battery hold-down

6. Mechanical problems in the electrical system, such as shorted or pinched wires

PART 8   SELF-CHECK

1. What is the purpose of the lead acid storage battery?

2. Battery electrolyte consists of _____percent water and _____percent sulfuric acid.

3. What is the specific gravity of pure water?

4. Battery voltage increases as the number of plates per cell is increased. True or false?

5. The two battery ratings in present use are _____ and _____.

6. The state of charge of a battery can be checked by using either a _____or a _____.

7. Describe how to perform a battery load test.

8. Battery vapors are highly explosive. True or false?

Chapter 30

Electrical Principles

Performance Objectives

After thorough study of this chapter, sufficient practice on adequate training models, and with the proper tools, equipment, and shop manuals, you should be able to accomplish the following:

1. Complete the self-check questions with at least 80 percent accuracy.

2. Follow the accepted general precautions as outlined in chapter 29.

3. Describe the principles of electricity and the electrical components covered in this chapter.

4. Properly use the voltmeter, ammeter, ohmmeter, and test light for basic circuit diagnosis.

A remarkable electric system is used by diesel engine powered-equipment. It produces electrical energy, stores it in chemical form, and delivers it on demand to any of the equipment's electrical systems.

Many components and accessories are operated by electricity: the starter motor that cranks the engine, the headlight and signal light systems that light the road and signal your intentions to other vehicles, the heater and the defroster, the radio, the gauges, and the air conditioning. These are but a few applications of electricity in diesel-powered equipment.

A thorough understanding of how electricity acts is necessary for the technician to be able to intelligently diagnose and service each system and its components.

PART 1   ELECTRICITY

Electricity behaves according to definite rules that produce predictable results and effects. (See Fig. 30–1 to 30–5.) The rules of electricity can best be explained by looking at the structure of an atom. An atom is composed of a complex arrangement of negatively charged electrons in orbit around a positively charged nucleus—much like our moon revolves around the earth. The nucleus consists of protons (positively charged particles) and neutrons (particles with no charge) tightly bound together. The nucleus exerts an attractive force on the electrons (due to its positive charge) holding them within fixed orbits around the nucleus.

The electrons are free to move within their orbits at fixed distances around the nucleus. When two electrons approach each other, they repel each other because they are negatively charged. The electrons try to stay as far away from each other as they can get without leaving their orbits.

Any substance composed of identical atoms (atoms all having the same number of electrons, protons, and neutrons) is called an element. When a substance is made of more than one element, each bound chemically together, it is called a compound. Electrically speaking, there are three types of substances: conductors, insulators, and semi-conductors. A conductor is capable of supporting the flow of electricity through it and an insulator is not. The difference is that the outermost electrons in a conductor's atoms are loosely held by the nucleus, whereas an insulator's atoms hold its outermost electrons very tightly. The electrons in the atoms of

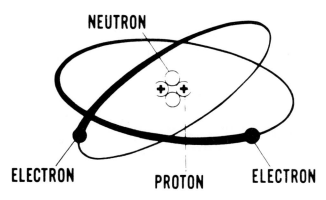

FIGURE 30–1. The principle of electricity can best be explained when the structure of an atom is understood.

BALANCED

FIGURE 30-2. The electrons in an electrical conductor remain balanced (no electrical flow) until a force is applied that will knock some of the electrons out of their orbit.

FIGURE 30-5. Current resulting from the pressure difference is measured in amperes.

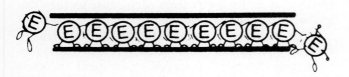

FIGURE 30-3. The continuation of an applied force causing electrons to be knocked out of their orbit results in a flow of electrons.

FIGURE 30-4. The rate of current is dependent on the strength of the applied force or voltage (pressure).

a conductor can be freed from their outer orbits by forces such as heat, light, pressure, friction, and magnetism. When electrons are knocked out of orbit, they can form an electrical current—under the proper conditions. When electrons in the atoms of a conductor are affected by a force, an unbalanced condition will occur in the atom between the negative charges of the electrons and the positive charges of the nucleus. The force tries to push the electron from its orbit. If an electron is freed from its orbit, the atom acquires a positive charge because it now has one more proton than electrons. The atom wants to return to an electrically balanced state. These unbalanced atoms attract electrons from the orbit of other balanced atoms to fill their empty orbit. This sets up a chain reaction of capturing and releasing electrons from one atom to another. These freed electrons try to move away from the force by transferring from one atom to the next through the conductor.

A stream of free electrons forms and an electrical current is born. The strength of the current depends on the strength of the affecting force.

An electrical current will continue to flow through the conductor as long as the electromotive force (EMF) is acting on the conductor's atoms and electrons. Stated in another way, current will continue to flow as long as a potential exists in the conductor, as long as there is a build up of excess electrons at the end of the conductor farthest from the EMF, and as long as there is a lack of electrons at the EMF end. This is called a voltage difference or *potential.*

PART 2   PRODUCING ELECTRICITY

An electrical current can be formed under certain conditions by the following forces: friction, chemical, heat (thermal), pressure (piezo-electric), and magnetism (induction).

Friction

When two materials are rubbed together, frictional contact between them will actually rub some of the electrons from their orbits. This causes a transfer of electrons to one of the materials and a lack of electrons in the other material. This gives a *static negative charge* to one material and a *static positive charge* to the other. When these materials are brought close to a grounded object, a spark will jump between the materials and the grounded object. A negatively charged object has a surplus of electrons. A positively charged object has a lack of electrons.

This accounts for the behavior of like charges repelling each other and unlike charges attracting each other.

Pressure

Perhaps the best known use of electricity produced by pressure is the crystal in the arm of a record player or in the diaphragm of a crystal microphone. Crystals of certain materials, such as quartz or Rochelle salt, develop a small electric charge when pressure is applied to them. When sound waves strike the diaphragm of a microphone, mechanical pressure is transferred to the crystal. This causes the crystal to flex and bend, producing a small voltage at its surface. This voltage is at the same frequency and amplitude as the incoming sound. This current is then amplified. This is called a piezo-electric effect.

Heat

A direct conversion of heat to electricity can be accomplished by heating a bimetallic junction of twisted wires made from two dissimilar metals like copper and iron. This type of junction is called a thermocouple.

Producing electricity in this manner is called *thermal conduction.* Since thermocouples cannot produce large amounts of electrical current, they are used as heat-sensing units. The amount of electrical current generated is dependent on the difference in temperature between the bimetallic junction of the

thermocouple and the opposite ends of the wire. The greater the temperature difference, the greater the flow of current and voltage produced. An example of this in the automobile is the oxygen (O_2) sensor.

Electricity can also be produced by magnetic induction as described in the next section.

PART 3   MAGNETS AND MAGNETISM

Magnetism produces nearly all the electricity used in our homes, offices, and industries. The alternator uses magnetism to produce the electrical current to run electrical systems and keep the battery fully charged. (See Fig. 30–6 to 30–13.)

There are two kinds of magnets—natural and artificial. All magnets have *polarity*—like the earth, they have a north and a south pole. *Like poles* (N-N, S-S) repel, while *unlike* or *opposite poles* (N-S) attract each other.

Natural magnets are found in the earth in the form of a black iron ore called *magnetite,* which will attract pieces of iron and steel. If you suspend these materials on a string, they will align themselves with the earth's magnetic north pole.

Artificial magnets can be produced by putting pieces of iron, steel, or certain alloys of aluminum and nickel (called *ALNICO*) in an intense magnetic

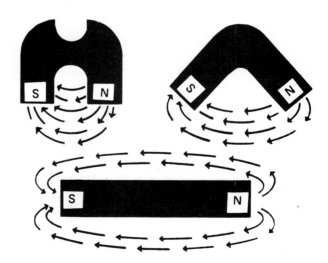

FIGURE 30-6. Permanent magnets have a north pole and a south pole regardless of physical shape. Invisible magnetic lines of force are present between the two poles. The direction of these lines of force is always from the north magnetic pole to the south magnetic pole.

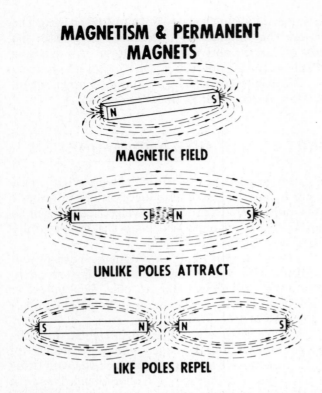

MAGNETISM & PERMANENT MAGNETS

MAGNETIC FIELD

UNLIKE POLES ATTRACT

LIKE POLES REPEL

FIGURE 30-7. Due to the direction of the magnetic lines of force, the opposite poles of two magnets will attract each other, while like poles of two magnets will repel each other.

CURRENT IN

FIGURE 30-9. Magnetic lines of force (magnetic field) also exist around a current-carrying conductor. In the conventional theory of current, current is said to be from positive to negative. Using the right hand as shown here, with the thumb pointing in the direction of current, the fingers indicate the direction of the magnetic field.

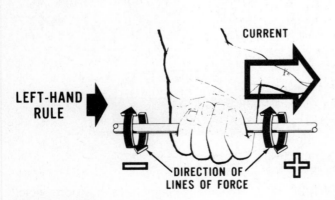

LEFT-HAND RULE

CURRENT

DIRECTION OF LINES OF FORCE

FIGURE 30-8. In the electron theory, current is from negative to positive. In this case, the left hand can be used with the thumb pointing in the direction of current. The fingers then point in the direction of the magnetic field. *(Courtesy of Chrysler Corporation)*

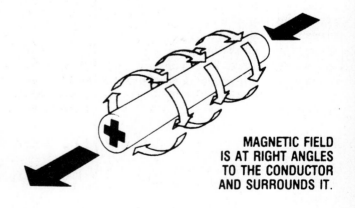

MAGNETIC FIELD IS AT RIGHT ANGLES TO THE CONDUCTOR AND SURROUNDS IT.

FIGURE 30-10. If the electron theory of current is used, current is from negative to positive. In this case, the left hand can be used with the thumb pointing in the direction of current. The fingers then point in the direction of the magnetic field. Note that whether you use the conventional theory or the electron theory (Figures 30-9 and 30-10), the direction of the magnetic field is the same. *(Courtesy of Chrysler Corporation)*

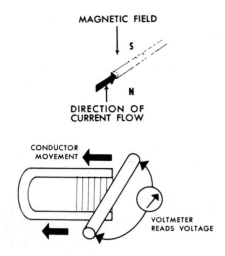

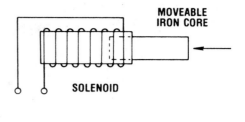

FIGURE 30-11. As a conductor is passed through a magnetic field, a current will be induced in the conductor. This principle is used to produce electricity in generators and alternators. *(Courtesy of General Motors Corporation)*

FIGURE 30-13. The magnetic lines of force can be further concentrated by the use of a soft iron core inside the coil. The soft iron core readily loses the induced magnetism when current in the coil is stopped. Examples of the use of this principle are the starter solenoid (movable iron core) and the relay (fixed iron core). *(Courtesy of Chrysler Corporation)*

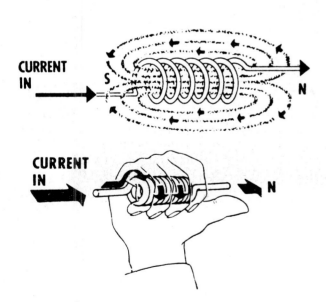

FIGURE 30-12. When a current-carrying conductor is wound in the form of a coil, the magnetic lines of force will be inside the coil and will be concentrated, making a stronger magnetic field. The polarity (direction) of these magnetic lines of force can be established by using the right hand with the fingers pointing in the direction of current in the coil winding. The thumb then points to the north pole. *(Courtesy of General Motors Corporation)*

field. These substances will acquire all the magnetic properties of magnetite, only stronger. The softer metal magnets (iron magnets) eventually lose their magnetic properties. The harder magnets made of steel and alnico tend to retain their magnetic properties indefinitely.

An electromagnet is made by wrapping a soft iron or steel core with a coil of insulated wire. When the ends of the wire are connected to an electric current, the steel or iron core acquires all the properties of a natural magnet but many times stronger. The iron core serves to concentrate the magnetism in an area surrounding the electromagnet.

Magnetic force is invisible. The only way we know it exists is by the effect it produces. A magnet's action can best be explained as having invisible lines of force (a *magnetic field*) leaving the magnet at one end or pole and reentering the magnet at the other end. These invisible lines of force are called *flux lines*. The shape and space they occupy is called a *flux pattern*. The number of lines of flux per inch is called the *flux density*. Flux lines are continuous and unbroken; they do not cross each other. The strength of the magnetic field is dependent on the flux density. The stronger the magnetic field, the greater the flux density.

Electricity and magnetism are very closely related. Magnetism can be used to produce electricity, and electricity can be used to produce magnetism. This is why generators and alternators produce electricity and motors use magnetism to produce mechanical energy. Whenever a magnetic field is passed through a conductor, a voltage is produced in that conductor—current begins to flow. When a lamp is connected to the ends of a conductor, a current will flow, lighting the lamp. This is called *electromagnetic induction*. This is the principle behind all generators and alternators.

Two types of magnets are used in diesel-powered equipment's electrical system—the permanent magnet and the electromagnet. Invisible lines of force (magnetic field) are present between the north and south poles of magnets. These lines of force have direction; in other words, they exert a force from the north pole to the south pole by convection. This force is utilized to induce an electrical current in a generator to produce electricity. It is also used in a starting motor to "push" a group of rotating conductors (armature) to start the engine, as well as for many other functions.

PART 4   HOW ELECTRICITY FLOWS THROUGH A CONDUCTOR

Keep in mind these six rules of electrical behavior:

1. All electrons repel each other.

2. All like charges repel each other. (Negatively charged objects repel other negatively charged objects; positively charged objects repel other positively charged objects.)

3. Unlike charges attract each other. (Positively charged objects attract negatively charged objects and vice versa.)

4. Electrons flow in a conductor only when affected by an electromotive force (EMF).

5. A voltage difference is created in the conductor when an EMF is acting on the conductor. Electrons flow only when a voltage difference exists between the two points in a conductor.

6. Current tends to flow to *ground* in an electrical circuit. Ground is defined as the area of lowest voltage. Electrical current moves through a conductor to ground in an attempt to reach a balance or equilibrium with the ground voltage (which is zero).

When electrons are set into motion, they display a variety of behaviors. The behavior of electrons moving through a conductor accounts for the many things electricity can do. By understanding how electricity behaves, you will be able to understand the function and operation of the various die-

sel engine electrical systems. This will aid you in diagnosing electrical problems.

Let us turn our attention to a single copper atom in a conductor. The copper atom has a single electron in its outermost orbit. This electron is not tightly held by the nucleus, and it can easily be freed by an electromotive force (EMF).

Once an electron escapes from its orbit, it is free to move—possibly colliding with other atoms in the conductor. As the free electron approaches the outer orbit of another copper atom, its electrostatic force starts to interact with the electron in orbit, repelling it. At the same time it is repelling this electron, the nucleus of that atom is attracting the free electron into its orbit. The free electron now enters the copper atom's orbit, replacing the ejected electron. As more and more electrons collide with other atoms in the conductor, an electrical current begins.

Once EMF is applied, it causes electrons to be freed from their orbits. This starts a chain reaction between the electrons and atoms in the conductor, causing the freed electrons to move away from the electromotive force. This effect is called *electron drift* and accounts for how electrons flow through a conductor. Whenever electrons flow or drift in mass, an electrical current is formed.

PART 5   BASIC TERMINOLOGY

Voltage

Voltage can be described as an electrical pressure. In diesel electrical systems, the battery or generator is used to apply this pressure. The amount of pressure applied to a circuit is stated in the number of volts. Another term for voltage is *electromotive force*. The symbol for electromotive force is E. The symbol for volts is V.

Current

Current can be described as the rate of electron flow. Current is measured in amperes. Current will increase as pressure or voltage is increased—provided that circuit resistance remains constant.

Another term for amperes is intensity of current. The symbol for current intensity is I. The symbol for amperes is A.

Resistance

Resistance in an electrical circuit is measured in ohms. The ohm is the unit of resistance (and of impedance) in the International System of Units (SI). The ohm is the resistance of a conductor such that a constant current of one ampere in it produces a voltage of one volt between its ends. The size and type of material used as a conductor, as well as its length and temperature, will determine the resistance of the conductor. The symbol for ohms is the Greek letter omega, Ω (See Fig. 30–14).

Conductors and Insulators

Conductors or wires of heavier cross-section are required for high-amperage current; smaller conductors can be used for low-amperage current. If too light a conductor is used, the conductor will overheat and actually melt the wire. Low-voltage conductors do not require heavy insulation. High-voltage conductors such as sparkplug wires require a heavy insulator to prevent current from jumping through the insulator to ground.

Ohm's Law

Ohm's law (Fig. 30–17) states that the current in an electric circuit is inversely proportional to the resistance of the circuit, and is directly proportional to the electromotive force (emf) in the circuit. Ohm's law applies only to linear constant-current circuits.

An electric circuit has a resistance of one ohm when an applied voltage (emf) of one volt causes current at the rate of one ampere. The equivalent equation would be:

$$R = \frac{E}{I}, \text{ or } R = \frac{V}{I},$$

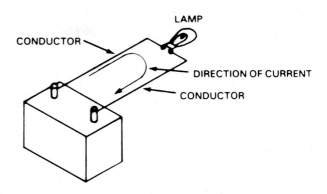

FIGURE 30-15. The lamp is brightly lit due to the current passing through it.

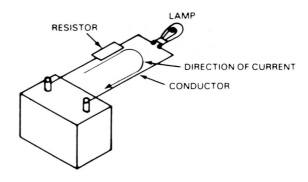

FIGURE 30-16. The lamp's light is reduced by the addition of a resistor (resistance) in the current path.

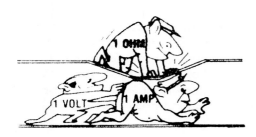

FIGURE 30-14. The flow of electrons encounters resistance in the form of lamp filaments, motor windings, resistors, and the conductors themselves. Resistance is measured in ohms (Ω).

OHM'S LAW

$$AMPERES = \frac{VOLTS}{OHMS}$$

$$OHMS = \frac{VOLTS}{AMPERES}$$

$$VOLTS = AMPERES \times OHMS$$

FIGURE 30-17. There is a definite relationship between volts, amperes, and ohms. Ohm's law can be used to calculate any one unknown factor when the other two are known.

where R is resistance, in ohms; E is the applied emf, in volts; V is the voltage drop across the resistance, in volts; and I is the resulting current through the resistance, in amperes. The use of E or V for voltage is determined by the circuit. E represents a voltage (emf or potential rise) of a power source such as a battery, a generator, or a battery charger. V represents the voltage drop (or potential fall) that occurs in the electric circuit connected to the power source. For normal operation, E (of the source) = V (of the circuit).

$I = E/R$ or $I = V/R$ with the same symbolic meanings as before. A further rearrangement of this formula allows one to find the voltage drop across a resistance; thus, $V = IR$.

To summarize:

$$\text{current, } I \text{ (in amperes)} = \frac{E \text{ (in volts)}}{R \text{ (in ohms)}} \text{ or}$$

$$I \text{ (in amperes)} = \frac{V \text{ (in volts)}}{R \text{ (in ohms)}}$$

$$\text{resistance, } R \text{ (in ohms)} = \frac{E \text{ (in volts)}}{I \text{ (in amperes)}} \text{ or}$$

$$R \text{ (in ohms)} = \frac{V \text{ (in volts)}}{I \text{ (in amperes)}}$$

In shorter form (without unit specifications), $V = IR$, $I = E/R$, or $I = V/R$, $R = E/I$, or $R = V/I$.

Here is an example of the application of one of the above equations. An engine with a 12-volt battery has a starting system in which the starting motor draws 150 amperes (A) of current. The resistance (R) of the starting motor circuit could be found by use of the equation:

$$R = \frac{E}{I} \cdot \text{ Therefore } R = \frac{12 \ (V)}{150 \ (A)} = 0.08 \ \Omega$$

The large battery cables and heavy starter windings offer little resistance to current, as shown by this example.

PART 6   ELECTRICAL POWER

Electricity is a medium for conveying energy, and the rate of work done by electricity is called *power*. In terms of other electrical units, power (P) = voltage (E or V) × current (I), or $P = EI$. Also $P = VI$,

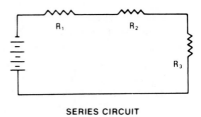

SERIES CIRCUIT

FIGURE 30-18. Three resistances (loads) are connected in series. *(Courtesy of Chrysler Corporation)*

where P is power in watts (W), E is an applied emf in volts, I is the current in amperes from the power source, and V is the voltage drop in volts, across the circuit resistance. Using the previous example of a 12-volt starting motor drawing 150 amperes, the power used by the motor would be: $P = EI = 12$ (V) × 150 (A) = 1800 W.

Another example would be the calculation of the current drawn by a 12-volt, 60-watt headlight. Rearranging the power equation $P = VI$ gives $I = P/V$. Therefore the headlight current is:

$$I = \frac{P}{V} = \frac{60 \ (W)}{12 \ (V)} = 5 \ A.$$

Horsepower ratings may be converted to electrical power ratings by use of the conversion factor: 1 horsepower (hp) = 746 watts (W) or 1 hp = 746 W.

PART 7   ELECTRICAL CIRCUITS

Every electrical system requires a *complete circuit* for it to function. A complete circuit is simply an uninterrupted path for electricity to flow from its source through all circuit components and back to the electrical source. Whenever the circuit is broken (interrupted), electricity will not flow. This interruption of current can be the result of a switch in the *off* position, a blown fuse, or an *open* (broken) wire. (See Fig. 30-19 to 30-22.)

There are basically three different types of electrical circuits: the series circuit, the parallel circuit, and the series-parallel circuit. An electrical system, however, may have one or more or a combination of these circuits.

Series Circuit

The series circuit provides only a single path for current from the electrical source through all the cir-

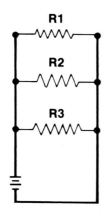

FIGURE 30-19. Parallel circuit showing three resistances (loads) connected in parallel. *(Courtesy of Chrysler Corporation)*

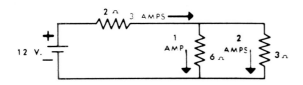

FIGURE 30-20. Series-parallel circuit. *(Courtesy of Chrysler Corporation)*

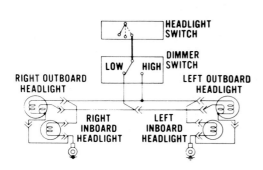

FIGURE 30-21. Headlamp circuit is an example of series-parallel circuit. *(Courtesy of Ford Motor Co. of Canada Ltd.)*

cuit's components and back to the electrical source. If any one component fails, the entire circuit will not function.

The total resistance in a series circuit is simply the sum of all the resistances in the circuit. For example, a series circuit (a light and two switches), would have a total resistance of 4 ohms if the light had a resistance of 2 ohms and the switches a resistance of 1 ohm each.

$$R = 2\,\Omega + 1\,\Omega + 1\,\Omega = 4\,\Omega$$

Parallel Circuit

A parallel circuit provides two or more paths for electricity to flow. Each path has separate resistances (or loads) and operates independently or in conjunction with the other paths in the circuit, depending on design. In this type of circuit, if one parallel path does not function, the other parallel sections of the circuit are not affected. An example of this is the headlight circuit; if one headlight is burned out the other headlight will still operate.

To calculate the total resistance in a parallel circuit, the following method must be used.

$$R = \frac{1}{\frac{1}{R1} + \frac{1}{R2} + \frac{1}{R3}} \text{ or } \frac{1}{R} = \frac{1}{R_1} + \frac{1}{R_2} + \frac{1}{R_3}$$

depending on the number of resistances and so on, that are in parallel.

If $R1$, $R2$, and $R3$ are 4, 6, and 8 ohms, respectively, total resistance would be calculated as follows:

$$R = \frac{1}{\frac{1}{4} + \frac{1}{6} + \frac{1}{8}}$$

$$= \frac{1}{\frac{6}{24} + \frac{4}{24} + \frac{3}{24}}$$

$$= \frac{1}{\frac{13}{24}} \text{ or } 1 \div \frac{13}{24}$$

$$= 1 \times \frac{24}{13} \text{ or } 1.85\,\Omega$$

In a parallel circuit the total resistance in the circuit will always be less than the lowest single device or resistance. The reason for this is that electricity has more than one path to follow.

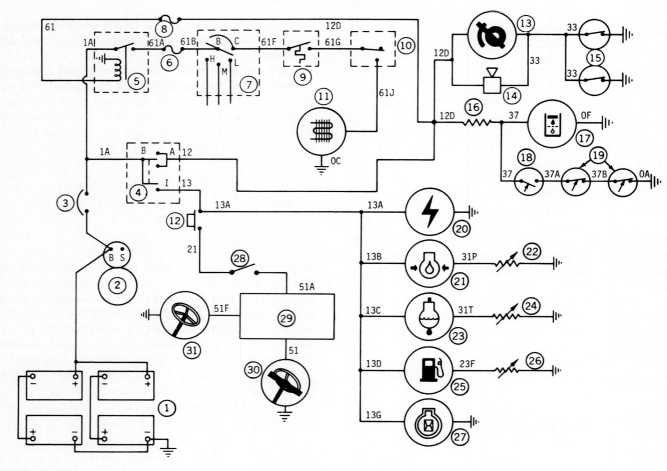

1—Batteries
2—Starting Motor
3—Main Circuit Breaker
4—Key Switch
5—Main A/C Relay
6—Pressurizer Fuse
7—Pressurizer Switch
8—Air-Heater Fuse
9—Thermostat
10—Thermal Switch
11—Plugged Condenser Indicator
12—Start Switch
13—Low Brake Pressure Indicator

14—Low Brake Pressure Buzzer
15—Low Brake Pressure Switches
16—Resistor
17—Filter Restriction Indicator
18—Test Switch
19—Filter Restrictor Indicator
        Sending Units
20—Voltmeter
21—Engine Oil Pressure Gauge
22—Engine Oil Pressure Sending
        Unit
23—Engine Coolant Temperature Gauge

24—Engine Coolant Temperature
        Sending Unit
25—Fuel Gauge
26—Fuel Sending Unit
27—Hourmeter
28—Neutral Start Switch
29—Emergency Steering
        Logic Board
30—Primary Steering
        Failed Indicator
31—Emergency Steering
        Failed Indicator

FIGURE 30-22. Typical loader instrument panel wiring diagram. *(Courtesy of Deere and Company)*

Series-Parallel Circuits

A series-parallel circuit combines the series and the parallel circuits. To calculate total resistance in a series-parallel circuit, calculate the series portion of the circuit as stated above. Then calculate the par-

allel portion of the circuit and add to the series resistance.

Examples of a series-parallel circuit are shown in Figures 30-20 and 30-21. In Figure 30-21 the headlight and dimmer switches are in series, while the headlights are in parallel with each other.

PART 8   VOLTAGE DROP

As current passes through a resistance, circuit voltage across it will drop. Total voltage drop in an electrical circuit will always equal available voltage at the source of electrical pressure.

Circuit resistance, if excessive at any point, will result in excessive voltage drop across that portion of the circuit. The voltage drop method is commonly used to determine circuit resistance. The voltage drop across a battery cable, for example, should not exceed 2/10 volt per 100 amps at 68°F (20°C).

The voltage drop method is the SAE recommended method for checking cable resistance. When checking voltage drop, the voltmeter is connected in parallel over the portion of the circuit being tested, and the results are compared to specifications.

Voltage drop can only be measured in a circuit when there is a flow of electricity present. With no electrical flow, the voltage (potential difference or pressure) remains the same anywhere in the circuit. Shop voltmeters are designed to indicate positive voltage potential on the upscale side of zero.

PART 9   OPENS, SHORTS, AND GROUNDS

Electrical systems may develop an *open* circuit, a *shorted* circuit, or a *grounded* circuit. Each of these conditions will render the circuit more or less ineffective.

Opens

An *open circuit* is a circuit in which there is a break in continuity. As stated earlier, for electricity to be able to flow there must be a complete and continuous path from the electrical source through the circuit back to the electrical source. If this path is broken, the condition is referred to as an open circuit. An open circuit, therefore, is no longer operational and acts the same as if it were switched off.

Shorts

A *shorted circuit* is a circuit that allows current to bypass part of the normal path. An example of this would be a shorted solenoid coil. Coil windings are normally insulated from each other; however, if this insulation breaks down and allows copper-to-copper contact between turns, part of the coil windings will be bypassed. In a starter pull-in winding, this condition would reduce the number of windings through which electricity will flow. If the short caused 50 windings of approximately 100 windings to be bypassed, this would reduce coil capacity by 50 percent and prevent starter engagement.

Grounds

A *grounded circuit* is a condition that allows current to return to ground before it has reached its intended destination. An example of this would be a grounded tail light circuit. If the wire leading to the tail light has an insulation breakdown allowing the wire to touch the frame or body of the vehicle, electricity will flow to ground at this point and return directly to the battery without reaching the tail light.

PART 10   ELECTRICAL CIRCUIT DIAGNOSIS AND SERVICE

A number of common tools and instruments are used to diagnose automotive electrical circuits. The most common are the test light, the jumper wire, the voltmeter, the ammeter, and the ohmmeter. The ammeter and voltmeter are often combined in a single piece of test equipment, which also may include a carbon pile resistor capable of applying varying loads to electrical circuits. The ohmmeter is often included in the multimeter, which is able to measure small amounts of current and voltage, as well as resistance. (See Fig. 30–23 to 30–36.)

Variations of equipment combinations are available from different manufacturers. Some of these types of multiuse test equipment will be dealt with

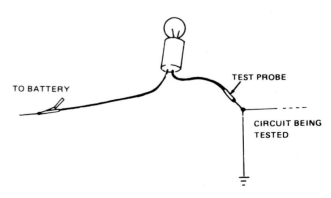

FIGURE 30–23. Using a jumper wire to bypass a switch that is suspected of being defective. *(Courtesy of Chrysler Corporation)*

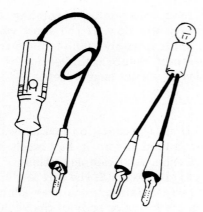

FIGURE 30-24. Two types of test lights for checking electrical circuits. *(Courtesy of Chrysler Corporation)*

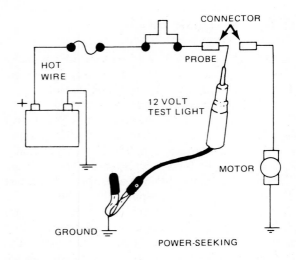

FIGURE 30-27. Isolating portions of a circuit to determine faulty section of the circuit. *(Courtesy of Chrysler Corporation)*

in the appropriate sections of this text. Examples include charging system and starting system testers.

In every case both the test equipment manufacturer's instructions and the engine manufacturers instructions should be followed for proper and accurate diagnosis. Test sequences given by manufacturers should be followed for systematic problem identification. Test results must be compared to manufacturer's specifications and recommended repair procedures followed. In many cases, component replacement is recommended rather than component repair particularly when solid-state components are involved.

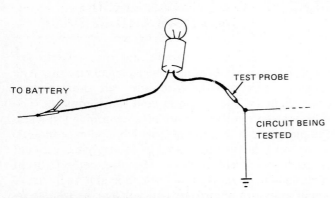

FIGURE 30-25. Testing continuity in a portion of a ground circuit. *(Courtesy of Chrysler Corporation)*

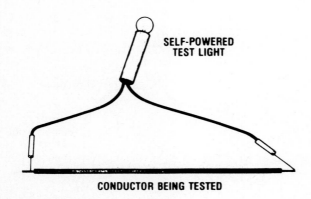

FIGURE 30-26. Battery-powered test light used to determine continuity in an electrical conductor. *(Courtesy of Chrysler Corporation)*

FIGURE 30-28. Using a circuit-breaker type of tester across a blown fuse allows the circuit being tested to remain live in order to locate faulty portion of circuit with other test equipment. *(Courtesy of Chrysler Corporation)*

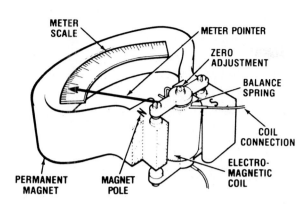

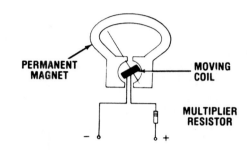

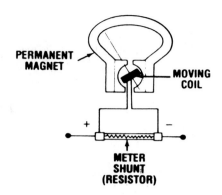

FIGURE 30-30. Ammeter construction is similar to voltmeter construction. The resistor here is connected in parallel, whereas in the voltmeter it is connected in series. The ammeter is used to measure the amount of current in a circuit and must always be connected to the circuit in series. Always connect voltmeter and the ammeter leads to a circuit by connecting positive leads to the positve side of circuit and negative leads to negative side of circuit. *(Courtesy of Chrysler Corporation)*

FIGURE 30-29. Typical voltmeter construction. The meter pointer is attached to a movable coil and is held at the 0 point on the scale by a coiled balance spring. As voltage is applied to the movable coil winding, it creates a magnetic field with opposite polarity to the permanent magnetic poles. Since opposite poles attract, the meter pointer will move up scale in direct proportion to the voltage applied. The voltmeter is used to test circuit voltage and voltage drop (resistance). The resistor determines the range of voltage that the voltmeter is able to measure. *(Courtesy of Chrysler Corporation)*

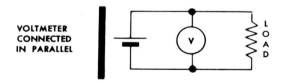

FIGURE 30-31. Method used to connect voltmeter properly. *(Courtesy of General Motors Corporation)*

Figures 30-23 to 30-36 illustrate the use of the basic equipment required for testing electrical circuits.

Basic circuit testing and the use of basic test equipment follows here.

Jumper Wire

The simplest electrical troubleshooting tool is also one of the most important—a jumper wire. Make it at least a meter in length and use alligator clips on the ends.

Connect one end to battery positive and you have an excellent 12-volt power supply. Use it to check lamp bulbs, motors, or as a power feed to any 12-volt component. But be careful and don't drop the other end; any place you touch on the engine or body is *ground*—battery negative—big sparks and

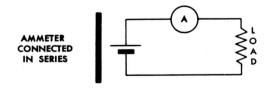

FIGURE 30-32. Proper method of connecting ammeter into circuit. *(Courtesy of General Motors Corporation)*

FIGURE 30-33. Typical volts/amp tester.

high current will result, possibly "cooking" the jumper wire!

Test Lamp

Sometimes you want to *look* for power, rather than *supply* it. That's when a *test lamp* is perfect. Just ground one side and you can go to most "hot" 12-volt points in the system and the lamp will light. But sometimes it won't light up fully with a hot circuit. For example, if you test the circuit *after* the voltage has "dropped" over a load.

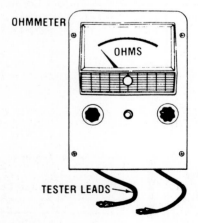

FIGURE 30-34. Typical ohmmeter used to test resistance and continuity in various electrical components. *(Courtesy of Chrysler Corporation)*

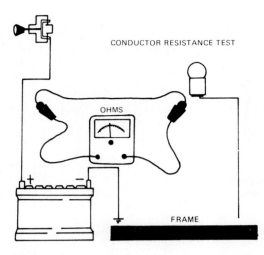

FIGURE 30-35. The ohmmeter is self-powered and should never be connected into a live circuit. Conductor has been removed from circuit to test resistance as shown here. *(Courtesy of Chrysler Corporation)*

Circuit Breaker

A circuit breaker reacts to excess current by heating, opening up, and cutting off the excess current. With no current flowing, the heating stops, and the breaker closes again and restores the circuit. If the high-current cause is still in the circuit, the breaker will open again. This cycling on and off will continue as long as the circuit is overloaded.

You can use a cycling breaker fitted with alligator clips in place of a fuse which keeps blowing. The circuit breaker will keep the circuit "alive" while you check the circuit for the cause of high current draw—usually a short.

Voltmeter

A voltmeter is connected in parallel with a circuit—it reads directly in volts. In parallel the meter draws only a small current—just enough to sample the voltage. That's why you can "short" right across the battery terminals with a voltmeter without damaging it.

However, never try to check voltage by putting the meter in *series*. The voltmeter hookup should always parallel the circuit being measured because you don't want the high resistance meter disrupting the circuit in a series connection.

Closed Circuit Voltage. In Figure 30-36 the voltage at *A* is 12 volts positive. There is a drop of six volts over the 1.0-ohm resistor and the reading

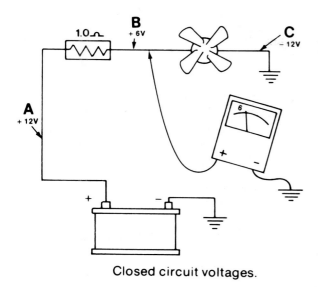

Closed circuit voltages.

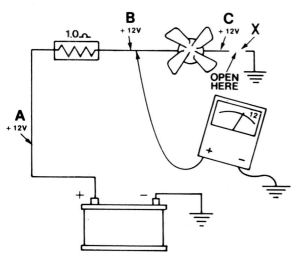

FIGURE 30-36. Using a voltmeter to isolate a problem in a motor circuit.

is 6 volts positive at *B*. The remaining voltage drops in the fan load and the voltmeter reads zero (12 volts negative) at *C* indicating normal motor circuit operation.

Open Circuit Voltage. Now we read the voltage in the same circuit as above but with *no* electricity flowing because there is an open circuit (broken wire or poor ground) at point *X*. The voltage at *A, B,* and *C* will be 12 volts positive indicating circuit continuity up to, but not through, point *X*. Remember, there is no voltage drop across a resistor or load if there is no electrical flow.

Ohmmeter

Another useful meter measures resistance. The ohmmeter is an excellent *continuity* checker, too. If you want to see if a wire is *continuous* or open inside a harness, clipping the ohmmeter leads to each end will show zero ohms (if the wire is good).

Just remember *never* to use an ohmmeter in a hot circuit. Always be sure there is no power (voltage or current) in the circuit when using an ohmmeter to avoid damaging it.

Ohmmeters have batteries for their power supply. For that reason, don't leave the ohmmeter connected for longer than necessary to read the scale or the batteries will run down.

Ammeter

You measure current draw with an ammeter. But unlike the parallel voltmeter, you must put the ammeter *in series* with the load to read the current draw. That means disconnecting the load and reconnecting with all the current going through the meter. Polarity must be followed—the red lead going to the positive side. The induction type of ammeter

WIRING TEST CHART

TYPE OF FAILURE	TEST UNIT AND EXPECTED RESULTS IF WIRING FAILED	
Open (Broken wire)	Ohmmeter —	Infinite resistance at other end of wire. Infinite to adjacent wire. Infinite to ground.
	Voltmeter —	Zero volts at other end of wire
Ground (bare wire touching frame)	Ohmmeter —	Zero resistance to ground. Infinite to adjacent wire. May or may not be infinite to other end of wire.
	Voltmeter —	Instead of testing, normally look for blown fuse or tripped circuit breaker.
Short (rubbing of two bare wires)	Ohmmeter —	Zero resistance to adjacent wire Infinite to ground. Zero to other end of wire.
	Voltmeter —	Voltage will be read on both wires.

with the clamp type of induction pickup is very convenient to use since the wire being tested is not disconnected.

Always use an ammeter that can handle the expected current since excessive current can damage a meter. Also, never connect an ammeter across a circuit (parallel) or you may damage the meter or the circuit.

PART 11   SELF-CHECK

1. Electron flow is caused by _____.
2. State Ohm's law.
3. What are the three common types of electrical circuits?
4. Like magnetic poles _____ and unlike poles _____.
5. All current-carrying conductors have a _____ surrounding the conductor.
6. Forming a conductor into a coil strengthens the _____ when the conductor carries current.
7. What causes voltage drop?
8. List four methods by which electricity can be produced.
9. What causes electron flow?
10. Define electrical opens, shorts, and grounds.

Chapter 31

Electrical and Electronic Devices

12-VOLT

FULL SIZE CABLES   UNDERSIZE

NO. 4 GAUGE   NO. 6 GAUGE   NO. 8 GAUGE

6-VOLT

FULL SIZE CABLES   UNDERSIZE CABLES

NO. 0 GAUGE   NO. 1 GAUGE   NO. 2 GAUGE   NO. 4 GAUGE

FIGURE 31-1. Comparison of battery cable diameters.

Performance Objectives

After sufficient study of this chapter, the appropriate training models, and shop manuals, you should be able to:

1. Complete the self-check questions with at least 80 percent accuracy.
2. Identify, on a vehicle or equipment specified by your instructor, any of the components covered in this chapter and describe their operation.

The diesel industry uses a wide variety of electric and electronic devices. These range from a very simple on-off switch to a computer capable of doing a multitude of jobs. The diesel technician should have a good basic understanding of how these devices operate in order to successfully diagnose and correct electrical system problems.

The basic operation of these devices is described here, and examples of their use are given as well.

PART 1   ELECTRICAL WIRING

Electrical wires may be one solid single strand, or a number of smaller wires twisted together to form a stranded wire. Wires are usually stranded for more flexibility in diesel equipment wiring. (See Fig. 31–1 to 31–5.)

Wire diameter is specified in gauge sizes or in millimeters (mm^2). Gauge sizes use numbers to indicate size. The smaller the gauge number is, the larger the wire diameter. The larger the metric number designation, the larger the wire cross section.

Large-diameter wires are required for circuits subject to high current (amperes). Smaller-diameter wires are used for low current.

Heavy insulation is required for high-voltage wires (e.g., spark plug wires) and lighter insulation is used for low-voltage circuits. The uninsulated side of electrical system circuits uses the equipment frame and body for a return path. The insulated side of the circuit is sometimes called the live side.

Microprocessor control system wiring may contain special design features not found in standard

CABLE CONVERSION CHART	
METRIC SIZE	**CURRENT GAGE**
.5mm²	20 GA.
.8mm²	18 GA.
1.0mm²	16 GA.
2.0mm²	14 GA.
3.0mm²	12 GA.
5.0mm²	10 GA.
8.0mm²	8 GA.
13.0mm²	6 GA.
19.0mm²	4 GA.

FIGURE 31-2. Wire size comparison chart. *(Courtesy of General Motors Corporation)*

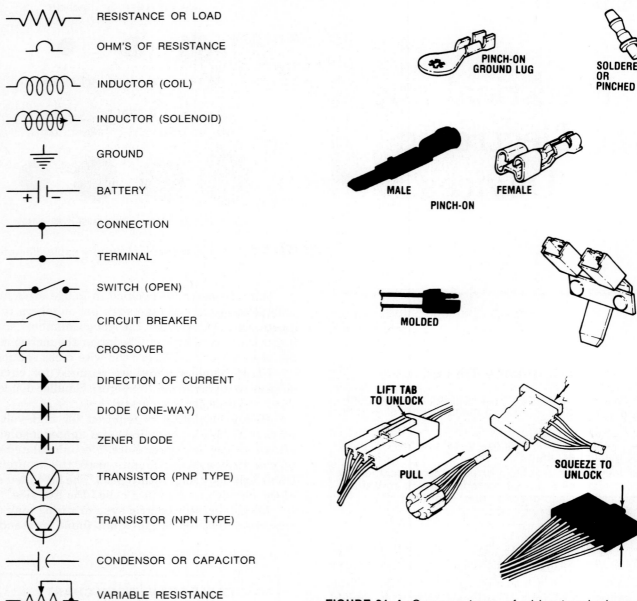

RESISTANCE OR LOAD

OHM'S OF RESISTANCE

INDUCTOR (COIL)

INDUCTOR (SOLENOID)

GROUND

BATTERY

CONNECTION

TERMINAL

SWITCH (OPEN)

CIRCUIT BREAKER

CROSSOVER

DIRECTION OF CURRENT

DIODE (ONE-WAY)

ZENER DIODE

TRANSISTOR (PNP TYPE)

TRANSISTOR (NPN TYPE)

CONDENSOR OR CAPACITOR

VARIABLE RESISTANCE

FIGURE 31-3. Wiring diagram symbols. *(Courtesy of Deere and Company)*

PINCH-ON GROUND LUG

SOLDERED OR PINCHED

MALE

FEMALE

PINCH-ON

MOLDED

LIFT TAB TO UNLOCK

PULL

SQUEEZE TO UNLOCK

FIGURE 31-4. Common types of wiring terminals and connectors.

wiring. Environmental protection is used extensively to protect electrical circuits. Special twisted and twisted/shielded cable is essential to system performance. Additionally, special high density connections are used in many of the wiring harnesses.

The current and voltage levels in this system are very low. There are three types of cable construction used: straight wire, twisted wires, and twisted/shielded wires. Unwanted induced voltages

are prevented from interfering with computer operation. Environmental deterioration by moisture, rust, and corrosion is prevented for the same reason.

Wire terminals and connectors include a great variety of types and sizes. Terminals and connectors are attached to wires by crimping, soldering the connection, or both. Terminals may be of the flat round hole, round male and female, flat male and female single, or multiple designs. Bulkhead or multi-connectors group two or more wires together in a two-piece plastic connector. Many of these use a locking

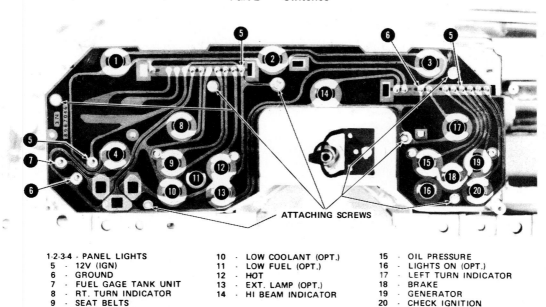

1-2-3-4 - PANEL LIGHTS	10 - LOW COOLANT (OPT.)	15 - OIL PRESSURE
5 - 12V (IGN)	11 - LOW FUEL (OPT.)	16 - LIGHTS ON (OPT.)
6 - GROUND	12 - HOT	17 - LEFT TURN INDICATOR
7 - FUEL GAGE TANK UNIT	13 - EXT. LAMP (OPT.)	18 - BRAKE
8 - RT. TURN INDICATOR	14 - HI BEAM INDICATOR	19 - GENERATOR
9 - SEAT BELTS		20 - CHECK IGNITION

FIGURE 31-5. Printed instrument panel circuit reduces number of wires required. *(Courtesy of General Motors Corporation)*

device to prevent the connector from separating due to vibration and the like. Various locking methods are used. The locking may be of the squeeze-to-unlock, or the lift-tab-to-unlock type. Be sure to unlock the connection before attempting to pull it apart, and be sure the connection is locked after completing the connection.

PART 2   SWITCHES

Electrical switches are used to open and close electrical circuits. Some of these are operated manually, whereas others operate automatically. Manually operated switches include the following types: push-pull, toggle, turn, and slider. Manually operated switches are used to operate: headlamps, radios, tape players, speakers, heaters, air conditioners, windshield wipers, speed control, power seats, power door locks, glow plugs, the starting system, trip and fuel calculators, and the like. (See Fig. 31-6 and 31-7.)

Some of these are switched on manually but switch off automatically; others must be switched on and off manually.

Automatic switches include those controlled by heat, pressure, vacuum, solenoids, and relays. Heat sensitive switches are used for coolant temperature indicators controlled by a thermal sending unit in contact with engine coolant. They may control cold

engine temperature indicator lights as well as hot engine indicators. An example of a pressure sensitive switch is the oil pressure indicator sending unit or switch screwed into a main engine oil gallery. With the engine off, there is no oil pressure: the switch is closed. When the start switch is turned on, the oil pressure indicator light goes on. When the engine is started and engine oil pressure rises above approximately 8 to 12 psi (55 to 82 kPa), the switch contacts separate, opening the circuit; then the dash indicator light goes out.

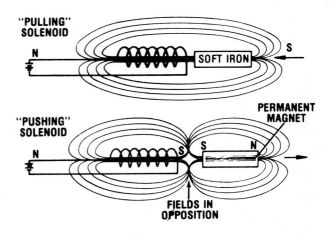

FIGURE 31-6. Principle of solenoid operation. *(Courtesy of Chrysler Corporation)*

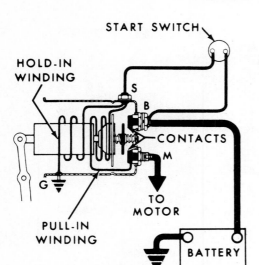

FIGURE 31-7. The starter solenoid is used to engage the starter pinion gear with the flywheel ring gear. *(Courtesy of General Motors Corporation)*

A relay is a switch that opens and closes an electrical circuit conducting relatively high current controlled by a circuit of relatively low current value. This reduces the need for long, heavy electrical wiring in many cases since only light wire is used to connect the relay to the actuating device or switch. Starter relays are good examples of the relay type of switch. When the control circuit switch is closed, the relay coil winding becomes an electromagnet which causes the armature of the relay to be drawn toward the coil, closing the relay points or contacts. One of the relay contacts is connected to the power source and the other to the device to be operated. Closing the relay points actuates the starting motor.

PART 3　CIRCUIT PROTECTION

Fuses and Fusible Links

A chain is only as strong as its weakest link; when a load is applied the weakest link will break.

In the same way, a fuse or a fusible link is the weakest point electrically in an electric circuit. It is needed in order to protect wiring and other components in the circuit from damage due to overloading of the circuit. Circuit overload can occur due to mechanical overload of the electrical device (i.e., windshield wiper motor, heater motor) or to shorts or grounds in the circuit. (See Fig. 31-8 to 31-11.)

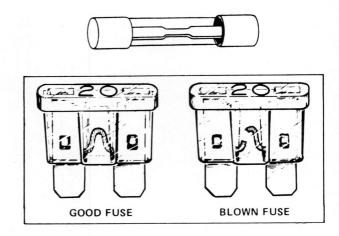

FIGURE 31-8. Typical fuse types.

Because of their lower current capacity, fuses and fusible links are designed to "blow" or "burn out" at a predetermined value, depending on the circuit capacity they are designed to protect. One type of fuse is the cylindrical glass type with the fusible link visible in the glass and connected at each end to a metal cap. The metal capped ends snap into place between two spring clip connectors in the fuse holder. Another type is enclosed in transparent plastic and has two blade terminals that plug into corresponding connectors in the fuse holder. Fuse capacity ranges anywhere from about 3 to 30 amperes. A failed fuse is easily identified by the gap in the wire visible in the fuse. The cause for fuse failure should be determined and corrected before fuse replacement. Replacement fuses should never exceed original fuse capacity.

A fusible link is a short piece of wire of smaller diameter than the wire the circuit is designed to protect. When the circuit is overloaded, the link burns in two before damage can occur to the rest of the circuit. Fusible links are identifiable in the wiring harness by color code or by a tag attached to it. Fusible links are insulated in the same way as the rest of the circuit. A failed fusible link can often be identified by heat-damaged insulation or exposed wire. They are used in such circuits as charging and lighting systems.

Circuit Breakers

Circuit breakers are designed for circuit protection as are the fuse and the fusible link. The circuit breaker is more costly but has the advantage of opening and closing the circuit intermittently. In a

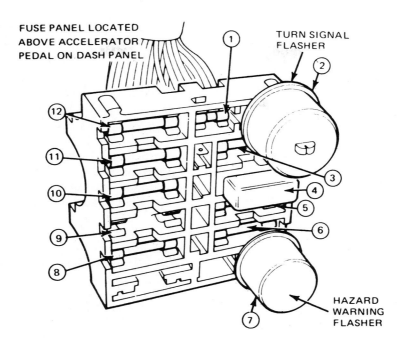

FUSE PANEL LOCATED
ABOVE ACCELERATOR
PEDAL ON DASH PANEL

TURN SIGNAL
FLASHER

HAZARD
WARNING
FLASHER

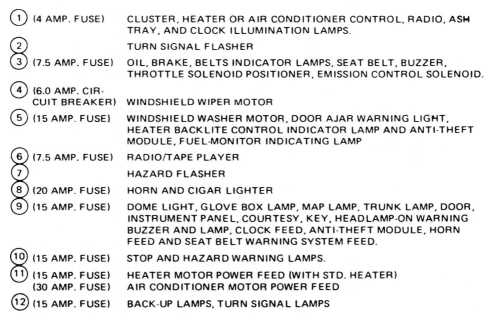

(1) (4 AMP. FUSE)   CLUSTER, HEATER OR AIR CONDITIONER CONTROL, RADIO, ASH TRAY, AND CLOCK ILLUMINATION LAMPS.

(2)   TURN SIGNAL FLASHER

(3) (7.5 AMP. FUSE) OIL, BRAKE, BELTS INDICATOR LAMPS, SEAT BELT, BUZZER, THROTTLE SOLENOID POSITIONER, EMISSION CONTROL SOLENOID.

(4) (6.0 AMP. CIRCUIT BREAKER) WINDSHIELD WIPER MOTOR

(5) (15 AMP. FUSE)   WINDSHIELD WASHER MOTOR, DOOR AJAR WARNING LIGHT, HEATER BACKLITE CONTROL INDICATOR LAMP AND ANTI-THEFT MODULE, FUEL-MONITOR INDICATING LAMP

(6) (7.5 AMP. FUSE) RADIO/TAPE PLAYER

(7)   HAZARD FLASHER

(8) (20 AMP. FUSE)   HORN AND CIGAR LIGHTER

(9) (15 AMP. FUSE)   DOME LIGHT, GLOVE BOX LAMP, MAP LAMP, TRUNK LAMP, DOOR, INSTRUMENT PANEL, COURTESY, KEY, HEADLAMP-ON WARNING BUZZER AND LAMP, CLOCK FEED, ANTI-THEFT MODULE, HORN FEED AND SEAT BELT WARNING SYSTEM FEED.

(10) (15 AMP. FUSE) STOP AND HAZARD WARNING LAMPS.

(11) (15 AMP. FUSE) HEATER MOTOR POWER FEED (WITH STD. HEATER)
     (30 AMP. FUSE) AIR CONDITIONER MOTOR POWER FEED

(12) (15 AMP. FUSE) BACK-UP LAMPS, TURN SIGNAL LAMPS

FIGURE 31-9. Individual circuit overload protection is provided by fuses or circuit breakers as shown in this fuse block. *(Courtesy of Ford Motor Co. of Canada Ltd.)*

headlight circuit, for example, the circuit breaker allows headlights to go on and off, which allows the driver to safely pull over to the side and stop. A fuse or fusible link failure, in this circuit, would cause the lights to go out completely, leaving the driver in the dark.

A circuit breaker has a pair of contact points, one of which is attached to a bimetal arm. The arm and contacts are connected in series in the circuit.

When circuit overload current heats the bimetal arm, the arm bends to open the contacts, stopping electrical flow in the circuit. When the arm cools, the

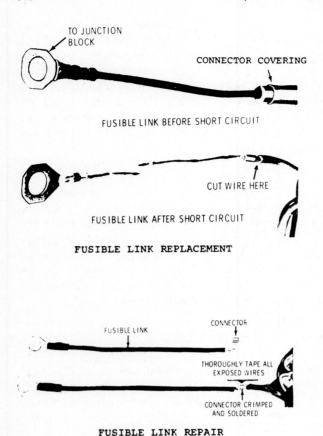

FIGURE 31-10. Typical fuse links and fuse link repair. *(Courtesy of General Motors Corporation)*

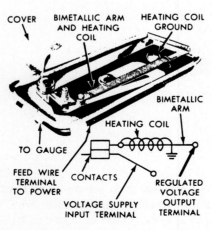

FIGURE 31-11. Typical instrument panel voltage limiter, components, and schematic. *(Courtesy of Ford Motor Co. of Canada Ltd.)*

contacts close again, energizing the circuit once more. This action continues until the circuit is switched off or repaired.

Voltage Limiter

The instrument voltage regulator is designed to limit voltage to the instrument panel gauges. Power to the voltage limiter is supplied when the start switch is in the "on" position. Voltage is limited to approximately five volts at the instrument gauges on some vehicles.

The voltage limiter consists of a bimetal arm, a heating coil, and a set of contact points enclosed in a housing. Two terminals provide connections in series into the circuit. When the switch is turned on, the heating coil heats the bimetal arm causing it to bend and open the contacts. This disconnects the voltage supply from the heating coil as well as from the circuit. When the bimetal arm cools sufficiently, the contacts close and the cycle repeats itself. The rapid opening and closing of the contacts results in a pulsating voltage at the output terminal averaging approximately five volts.

The voltage limiter protects the instrument gauges against high voltage surges and prevents erroneous gauge readings caused by voltage fluctuations.

PART 4   RESISTORS

Resistors are devices used in electrical circuits to reduce current and voltage levels from those supplied by the power source. Resistors are usually made from wire. They are used to protect devices or circuits designed to operate at lower current levels than that supplied by the battery or charging system. They are also used to control current and voltage levels produced by charging systems and to control light intensity. (See Fig. 31-12.)

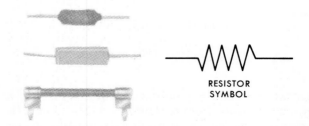

FIGURE 31-12. Some typical types of resistors used to control current and voltage values in the electrical system. Note resistor symbol used in wiring diagrams.

Resistors provide opposition to electron flow. This opposition causes the electrons to work harder to try to get through. The increased electron activity generates heat. Since some of the electrical energy is used up to produce heat, the voltage through the resistor is at a reduced level.

Several types of resistors are used in diesel equipment. These include fixed value resistors, variable resistors, ballast resistors, and thermistors.

The fixed resistors maintain a constant resistance value once operating temperature has been reached. A manually operated rheostat or variable resistor (as in the dash light control switch incorporated in the headlamp switch) inserts more or less resistance into the circuit to dim or brighten the dash lights as the switch knob is turned. A ballast resistor is a wire coil housed in a ceramic block to regulate temperature changes. The resistance of a ballast resistor increases with increased current and decreases with decreased current. Increased current causes the resistor to heat up which, in turn, increases resistance.

The resistance value of a thermistor (a type of resistor) varies with temperature. As the temperature of the thermistor increases, its resistance decreases. It is used in charging systems to vary charge voltage with ambient temperature change.

PART 5   CAPACITORS

A capacitor is a device that is used in an electrical circuit to temporarily store an electrical charge until it is needed to perform its job or until it can be safely dissipated if it is not to be used. (See Fig. 31-13.)

The typical condenser consists of several thin layers of electrically conductive material, such as metal foil, separated by thin insulating material known as dielectric material. Alternate layers of foil are connected to one terminal of the condenser. The

CAPACITOR
SYMBOLS

FIGURE 31-13. Typical capacitors used to prevent arcing, control radio interference, and provide circuit protection. Wiring diagram symbols used are also shown. Capacitors (condensers) are used to store undesirable electrical surges and eddy currents temporarily until they can be safely dissipated.

other layers of foil are connected to ground. The entire assembly is rolled up tightly and enclosed in a metal cylinder. The unit is completely sealed and moisture proof. The metal container is the ground connection, and a pigtail lead provides the other connection. The condenser is connected in parallel with the circuit. Any surge of current (excess electrons) enters the condenser and is stored on the condenser plates.

Capacitors of various types and sizes are used in electrical circuits to collect and dissipate stray or unwanted current. This prevents the unwanted current from interfering with other electrical functions. A radio-suppressor type of capacitor in the alternator is a typical example. The capacity or capacitance of a condenser is measured in units called *farads*. A farad is a charge of one ampere for one second producing a one-volt potential difference. A microfarad is 0.000001 (1/1,000,000) farad.

PART 6   TRANSDUCER

A transducer is a device that converts another form of energy to an electrical signal. MAP (manifold absolute pressure) sensors are used to provide a varying voltage signal depending on intake manifold pressure to the computer which controls injection timing, and fuel rate.

A throttle position switch or transducer is used to send a varying voltage signal, dependent on throttle position, to the electronic control unit or computer. The computer uses this information to increase or decrease fuel delivery accordingly.

PART 7   SEMICONDUCTORS, DIODES, AND TRANSISTORS

Semiconductors are neither good conductors nor good insulators. Semiconductor materials, such as silicon, are used in diodes and transistors. Silicon is an ingredient commonly found in beach sand. The capability of the silicon wafer *chip*, as used in computers, is astounding. (See Fig. 31-14 to 31-18)

A tiny silicon wafer one quarter the size of a man's fingernail can be manufactured to accommodate a million or more electronic components, diodes, and transistors, all interconnected with extremely thin films of metal. The silicon chip also has the ability to make decisions (logic) and to recall stored information (memory).

The amazing qualities of this material have revolutionized the electronics industry and have heav-

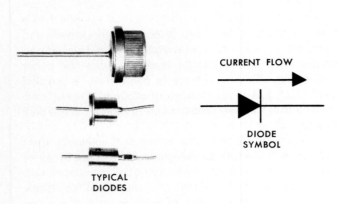

FIGURE 31-14. Typical diodes and diode symbol showing direction of current. *(Courtesy of General Motors Corporation)*

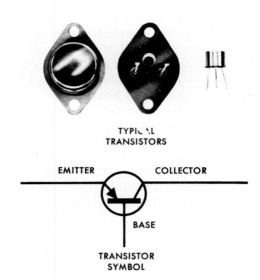

FIGURE 31-16. Typical transistor and transistor symbols. A small base current from the emitter to the base turns the transistor on, allowing a larger current from emitter to collector. *(Courtesy of General Motors Corporation)*

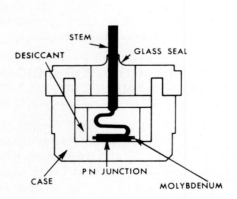

FIGURE 31-15. Cross-sectional view of diode construction. Materials used vary somewhat in different diodes. Both positive and negative diodes are used. A positive diode will allow current in one direction only from the stem, through the diode material (PN junction), to the diode case. A negative diode allows current in the opposite direction only. Diode polarity is established by the side of the PN junction that is connected to the stem. *(Courtesy of General Motors Corporation)*

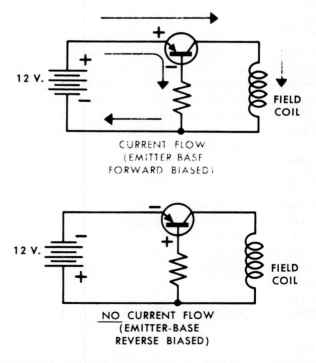

FIGURE 31-17. Diagrams showing transistor operation. *(Courtesy of General Motors Corporation)*

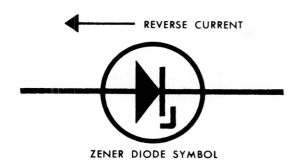

REVERSE CURRENT

ZENER DIODE SYMBOL

FIGURE 31-18. The Zener diode will allow current in the reverse direction when specified voltage is imposed. *(Courtesy of General Motors Corporation)*

ily impacted on the diesel industry—both in the factory through robotics, and in the service industry. Computers now control timing, fuel ratios, braking ratios, lighting systems, instrument panels, integral diagnostic systems, and other items.

It is not necessary for the diesel technician to fully understand the intricacies of the computer in order to diagnose and service diesel engines since these components are not serviced in the diesel shop. However, a basic understanding of the operating principles of these devices is needed to properly diagnose and repair systems using electronics.

Diodes

A diode is a solid state (completely static) device that allows current to pass through itself in one direction only (within its rated capacity). Acting as a one-way electrical check valve, it allows current to pass in one direction and blocks it in the other direction.

The silicon wafer is chemically treated to produce either a positive or negative diode. Diodes may be encased in non-corrosive heat conductive metal with the case acting as one lead and a metal wire connected to the opposite side of the wafer as the other lead. The unit is hermetically sealed to prevent the entry of moisture. This type of diode is used in AC charging system alternators. A minimum of six diodes is used—three positive diodes and three negative diodes to provide full wave rectification (changing alternating current to direct current). Many charging systems use more than six diodes.

Other diodes used in electronic systems are much smaller and may be sealed in epoxy resins with two leads for connection into the circuit. Diodes in computers may be very tiny in comparison to the more visible charging system diode.

Negative diodes are identified by a black paint mark, a part number in black, or a black negative sign. Positive diodes are similarly identified in red or with a red positive sign.

The manner in which the metallic disc is installed in the diode assembly determines whether the diode is negative or positive. (Inverting the disc in a positive diode would make it a negative assembly.) This disc is only .008″ to .010″ thick and approximately one-eighth of an inch square, depending upon current rating.

Some rectifier assemblies contain diodes that are exposed, while others have them built in. Those with built-in diodes contain only the wafer portion of the diode.

The silicon crystal material for diodes and transistors is processed or "doped" by adding other material to it. Phosphorus or antimony may be used to produce a negative or N type material. These materials have five electrons in the outer ring of their atoms. This results in the atoms of the N material having one extra or free electron. The free electron can be easily made to move through the material when voltage is applied. Electrons are considered to be negative current carriers.

Boron or indium may be used to treat silicon crystal to produce a positive or P type material. These elements have only three electrons in the outer ring of their atoms. This leaves a shortage of one electron in the atoms of P type material. This shortage or vacancy is called a *hole*. Holes are considered to be positive current carriers.

A diode consists of a very thin slice of each material, P type and N type placed together. The area where the two materials meet is called the *junction*. When the N material side of the diode is connected to a negative current supply, such as the battery negative terminal, and the P material side is connected to the positive battery terminal, the diode will conduct current. This happens because the negative battery terminal has an excess of electrons that repel the electrons in the diode toward the positive side. At the same time, the positive holes in the P material move toward the N side. This interchange of electrons and holes occurs at the junction of the N and P material in the diode. Connecting a diode in this manner is called *forward bias*.

When a diode is connected in the opposite manner (reverse bias) it will not conduct current. It cannot do so since the N material side of the diode is connected to the positive battery terminal and the P material side to the negative battery terminal. The electrons in the N material are attracted to the positive battery terminal side away from the diode junction. At the same time, the holes in the positive diode material are attracted to the negative battery terminal side of the diode away from the junction area. This in effect creates an open circuit which cannot conduct current.

Of course, these conditions apply only if normal diode design voltage is not exceeded. When applied in reverse bias, excessive current will cause the bond structure to break down and allow reverse current, which causes the diode to be damaged. Diodes are designed with the necessary current and voltage capacity for the circuit in which they are to be used.

Excessive reverse current will destroy a diode due to excessive heat. A "blown" diode will not conduct current, resulting in an open circuit. Blown diodes must be replaced. A shorted diode will conduct current in both directions and must be replaced.

Light emitting diodes (LED) may be used for digital display of instrument panel gauges on some vehicles.

Zener Diode

The zener diode is a specially designed diode that conducts current like a normal diode but will also safely conduct current in a reverse direction when reverse current reaches the specified design voltage. A zener diode can prevent reverse current if it is below design voltage, but when reverse current reaches and exceeds design voltage, the zener diode will conduct reverse current. This type of diode is used in control circuits such as in the field circuit in an alternator.

Transistors

A transistor is a solid state switching device used to control current in a circuit. It operates like a relay except that it has no moving parts. A relatively small current is used to control a larger current. The transistor either allows current to pass or stops it.

Transistors used in alternator applications are usually of the PNP type. This means that they are designed with a thin slice of N material sandwiched between two pieces of P material. The P material on one side is called the *emitter*, the N material in the middle is called the *base*, and the other P material is called the *collector*. NPN transistors are also produced but are not commonly used in diesel equipment applications.

The very thin slice of N type base material is attached to a surrounding metallic ring which provides the means for circuit connection. The emitter and collector material are also provided with circuit connections. The physical arrangement of the three pieces of material is such that the distance between the emitter and the collector is shorter than the distance between the emitter and the base. This feature results in the unique manner in which the transistor controls current.

A transistor is connected into a circuit in a manner that allows a low base-emitter current to control a larger collector-emitter current. A typical example of this is in the control module of an electronic voltage regulator.

When the base circuit is energized (by closing the ignition switch for example), a small base current is applied to the transistor emitter-base. This causes the electrons and holes in the emitter-base to act in a similar manner as in a diode described earlier. However, since the emitter is closer to the collector than it is to the base, most of the current is conducted by the emitter-collector section of the transistor. This is caused by the fact that electricity normally follows the path of least resistance.

The control current is called base current. The base circuit or current controls the emitter-collector current.

The same type of semiconductor material used in diodes is also used in transistors. The transistor, however, uses a second section of this material resulting in three terminals instead of two (as in the diode). If, for example, the base circuit of a transistor is energized with five amperes of current, the transistor divides this current into base current and emitter-collector current. This is known as the *current gain factor*. This factor varies with transistor design. The emitter-collector current may be 24 times that of the base current. In this example, therefore, base current would be 0.2 ampere and emitter-collector current would be 4.8 amperes.

Transistors are used in electronic voltage regulators to control charging system voltage and in computers.

PART 8 COMPUTERS AND MICROPROCESSORS

The computer has become a valuable component of the diesel injection system. The silicon chip and integrated circuitry are being used to control engine operation resulting in better performance, lower fuel consumption, and fewer emissions. The computer may even provide a correction factor for deviations from normal fuel ratios caused by production tolerances in the engine, the sensors, and the electronic interface with the fuel system, as well as drift from the norm, caused by system component aging.

Actually, the computer has taken over many of the functions formerly performed by electromechanical, or mechanical devices.

While it is not within the scope of this text to explain the electronic operation of the computer, it is worthwhile to consider its basic function. The computer consists of a power supply, a central processing unit (including the memory system), and an input and output system. The input signals received by the computer are processed in conjunction with stored information in the memory system. Based on input signal information and stored memory information, the computer makes the appropriate decision to provide the output signals needed to adjust or operate the various output devices.

Computer systems may include both RAM and ROM memory capability. Random access memory (RAM) temporarily stores bits of information from the various input signals before being acted upon by the program. Output data to be sent to output devices are also stored by RAM. The computer must compute this information and produce output signals based on input signals.

Another type of memory called ROM (read only memory) is permanent. This information need not be computed; it need only be consulted or looked up by the computer. The information remains in the system even when the start switch is turned off. An example of this type of data storage is injection timing advance that can be precomputed for an engine and the data stored in ROM.

Each memory location in the computer can store eight bits of information. Eight bits equal one *byte* according to computer programmers. Computer memory capacity is stated in Ks. One K stands for 1024 bits of information. A 64K RAM is, therefore, able to store 65,536 bits of information. New chips have been developed with many times this capacity—vastly in excess of what is needed for diesel engine applications but required for high technology industrial applications.

The diesel technician must be acquainted with electronics terminology which includes such terms as: diodes, transistors, capacitors, zener diodes, transducers, semiconductors, ECU (electronic control unit), ECM (electronic control module), CPU (central processing unit), microprocessor, computer, digital computer, electronic voltage regulator, data panel, digital display panel, MCU (microprocessor control unit), LED display or readout (light emitting diode), and others depending on make, year, and model of equipment.

PART 9   SELF-CHECK

1. Why are most diesel engine wires of the stranded type?
2. Wire size is specified by _____ or by _____.
3. Why do some wires require heavy insulation?
4. State the basic purpose of the following devices in electrical circuits.
   (a) switch
   (b) fuse
   (c) circuit breaker
   (d) voltage limiter
   (e) fusible link
   (f) resistor
   (g) capacitor
   (h) diode
   (i) transistor
5. What is the basic difference in function between RAM and ROM memory systems in a computer?

Chapter 32

AC Charging Systems

Performance Objectives

After thorough study of this chapter, sufficient practice on adequate training models, and with the necessary tools, equipment, and manuals, you should be able to do the following:

1. Complete the self-check questions with at least 80 percent accuracy.
2. Observe the accepted general precautions outlined in this chapter.
3. Describe the purpose, construction, and operation of AC charging systems and their components.
4. Accurately diagnose charging system faults.
5. Safely remove, test, recondition, replace, and adjust faulty charging system components according to manufacturer's specifications.
6. Properly performance test the reconditioned components and the entire charging system to determine the success of the repairs performed.

PART 1 ALTERNATING CURRENT (AC) GENERATORS

The AC generator charging system produces electrical energy. This energy is used to maintain the proper state of charge in the battery and supply current to all electrically powered equipment. It does this by converting mechanical energy into electricity. (See Fig. 32-1 to 32-7.)

In an AC generator an electromagnet is rotated inside a stationary conductor so that lines of force cut across the conductor. The magnet is called the *rotor* and the conductor is called a *stator*.

The rotor core is wound with wire and connected to an external source of current through two slip rings and brushes. This represents a simple form of externally excited field used in many alternators.

The conductor, represented by the single loop of wire, forms the stator part of the alternator. Because the rotor is externally excited by direct current, one pole will always be north and the other south. Current through the stator will reverse directions constantly.

Rotor Assembly

Actually, the alternator rotor has more than two poles. A commonly used rotor has 12 poles. The rotor consists of a core, coil, two pole pieces, and two slip rings. An external direct current is supplied to the rotor coil through slip rings and brushes. The field coil is located inside two interlocking pole pieces. Each of these pole pieces has six poles, providing a total of twelve poles for the complete rotor assembly.

Voltage Output

The following factors affect the magnitude of voltage generated:

1. Voltage will increase as the speed of rotor is increased.

2. Voltage will increase as the strength of the rotor magnetic field is increased.

3. The strength of the rotor magnetic field may be increased by the following:

(a) The number of turns and type of wire used in the rotor.

(b) The air gap between the rotor poles and stator. Reducing the air gap increases the strength of the field.

(c) The voltage applied to the rotor through the slip rings and brushes.

4. Voltage will increase as the number of turns of wire in the stator winding is increased. This is because more conductors will be cut by the lines of force from the rotor.

Single-Phase Stator

A simple alternator having a single loop of wire to represent the stator winding serves to illustrate how an alternating current is produced. An alternator of this type is called single-phase alternator, regard-

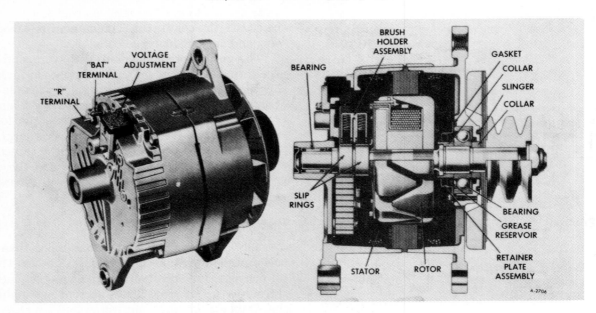

FIGURE 32-1. Typical brush-type alternator with internal regulator. *(Courtesy of General Motors Corporation)*

less of the number of turns of wire in the stator winding. When all the windings in the stator are connected in *series* to form one continuous circuit, the term *single-phase* applies.

Three-Phase Stator

The stator assembly in the alternator has three sets of windings. Each winding has one terminal or end that is independent of the others. The other end of each winding is connected to form an insulated junction called a Y connection.

The single-phase AC voltage is produced between any two of the open terminals. Combining these three single phases forms the three-phase connected stator. This means that it produces three overlapping sets of current. Some alternators use a *delta* connection of the three phases.

Six or More Diodes Used

Six silicone-diode rectifiers are used to rectify the AC output of the three-phase connected alternator to direct current. Three of these positive diodes are mounted on a heat sink that is called a *positive rectifier assembly*. This rectifier assembly is insulated from the alternator end shield and connected to the alternator output BATT terminal of the alternator.

The other three diodes have negative polarity and are mounted on a heat sink and are called a *negative rectifier assembly*. This rectifier assembly is mounted directly to the alternator end shield, which is a ground.

Both rectifier assemblies are mounted in the air stream to provide adequate cooling of the diodes.

Some alternator designs also use a diode trio (three diodes) connected to three stator windings to provide field current.

Rectifier Circuit

The diodes are connected into the alternator circuit between the stator winding and the battery and the vehicle electrical load. This arrangement provides a smooth direct current. The diodes also provide a blocking action to prevent the battery from discharging through the alternator. There are two diodes in each phase; one diode allows current in one direction, and the other diode in the same phase allows current in the opposite direction.

A capacitor is connected from the output BATT terminal to ground. It is used to absorb any peak voltages and thus protects the rectifiers and helps reduce radio interference.

Full-Wave Rectification

Alternators use full-wave rectification. Full-wave rectification of the stator output utilizes the total potential by redirecting the current from the stator windings so that all current is in one direction. All

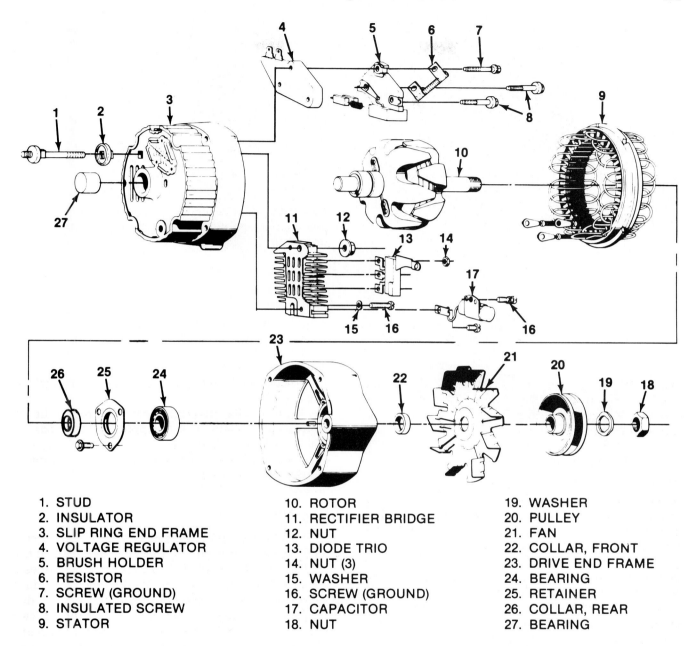

1. STUD
2. INSULATOR
3. SLIP RING END FRAME
4. VOLTAGE REGULATOR
5. BRUSH HOLDER
6. RESISTOR
7. SCREW (GROUND)
8. INSULATED SCREW
9. STATOR

10. ROTOR
11. RECTIFIER BRIDGE
12. NUT
13. DIODE TRIO
14. NUT (3)
15. WASHER
16. SCREW (GROUND)
17. CAPACITOR
18. NUT

19. WASHER
20. PULLEY
21. FAN
22. COLLAR, FRONT
23. DRIVE END FRAME
24. BEARING
25. RETAINER
26. COLLAR, REAR
27. BEARING

FIGURE 32-2. Components of brush-type alternator with integral regulator.
(Courtesy of General Motors Corporation)

current delivered to the output terminal is direct current.

Alternator Operation

This discussion of alternator operation (Fig. 32-8 and 32-9) follows the principle of positive voltage output. If desired, the electron theory of negative voltage output may be applied to the diagrams and

the discussion by reversing the direction of the arrows in the diagrams and by reversing the direction of current in the discussion.

For convenience, the three AC voltage curves provided by the "Y"-*connected stator* for each revolution of the rotor have been divided into six periods, 1 through 6. Each period represents one-sixth of a rotor revolution, or 60 degrees.

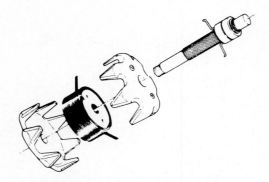

FIGURE 32-3. Exploded view of rotor. *(Courtesy of General Motors Corporation)*

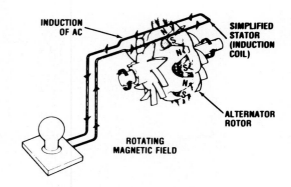

FIGURE 32-6. Rotating magnetic field induces current in stationary stator winding. *(Courtesy of Chrysler Corporation)*

FIGURE 32-4. Method used to energize rotor windings. *(Courtesy of General Motors Corporation)*

FIGURE 32-5. Magnetic field surrounding alternating N and S pole fingers of rotor. *(Courtesy of General Motors Corporation)*

An inspection of the voltage curves during period 1 reveals that the maximum voltage being produced appears across stator terminals *BA*. This means that the current will be from *B* to *A* in the stator winding during this period and through the diodes as illustrated.

In order to more clearly see why the current during period 1 is as illustrated, assume that the peak phase voltage developed from *B* to *A* is 16 volts. This means that the potential at *B* is zero volts, and the potential at *A* is 16 volts.

Between periods 1 and 2, the maximum voltage being impressed across the diodes changes or switches from phase *BA* to phase *CA*.

Taking the instant of time at which this voltage is 16 volts, the potential at *A* is 16 and at *C* is zero. Following the same procedure for periods 3–6, the current conditions can be determined, and they are shown in the illustrations. These are the six major current conditions for a three-phase "Y"-connected stator and rectifier combination.

The voltage obtained from the stator-rectifier combination when connected to a battery is not perfectly flat, but is so smooth that, for all practical purposes, the output may be considered to be a non-varying DC voltage. The voltage, of course, is obtained from the phase voltage curves and can be pictured as illustrated (Figure 32-8).

A delta-connected stator wound to provide the same output as a "Y"-connected stator also will provide a smooth voltage and current output when connected to a six-diode rectifier. For convenience, the three-phase AC voltage curves obtained from the basic delta connection for one rotor revolution are reproduced here and have been divided into six periods (Figure 32-9).

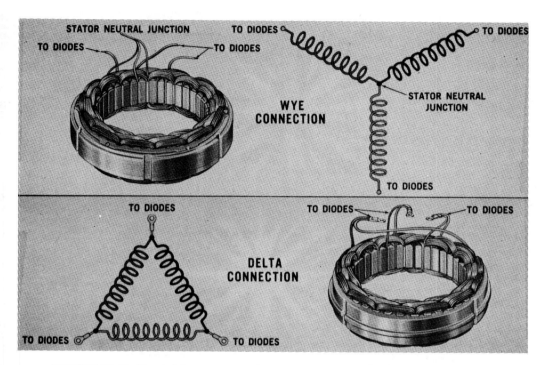

FIGURE 32-7. Most AC generators have a stator assembly with three sets of field windings (three phases) connected either in a Y (wye) arrangement or a delta arrangement. *(Courtesy of Ford Motor Co. of Canada Ltd.)*

During period 1, the maximum voltage being developed in the stator is in phase *BA*. To determine the direction of current, consider the instant at which the voltage during period 1 is at maximum, and assume this voltage to be 16 volts. The potential at *B* is zero and at *A* is 16.

An inspection of the delta stator, however, reveals a major difference from the "Y" stator. Whereas the "Y" stator conducts current through only two windings throughout period 1, the delta stator conducts current through all three. The reason for this is apparent, since phase *BA* is in parallel with phase *BC* plus *CA*. Note that, since the voltage from *B* to *A* is 16, the voltage from *B* to *C* to *A* must also be 16. This is true since 8 volts is developed in each of these two phases.

During period 2, the maximum voltage developed is in phase *CA*, and the voltage potentials are shown on the illustration (Figure 32-9) at the instant the voltage is maximum. Also shown are the other phase voltages, and again, the current through the rectifier is identical to that for a "Y" stator, since the voltages across the diodes are the same. However, as during period 1, all three delta phases conduct current as illustrated (Figure 32-9).

Following the same procedure for periods 3-6, the current directions are shown. These are the six major current conditions for a delta stator.

Open Rectifiers (Diode)

If the positive or negative diode is open circuited, there is no current in either direction. An open diode results in loss of current and an output less than normal.

Shorted Rectifiers (Diode)

When the positive or negative rectifier is shorted, (diode) current is allowed in either direction. When the positive diode is shorted, it may also drain the battery. When there is current through the heat sink, it can reverse through the shorted diode instead of to the battery. A shorted diode will reduce alternator current output.

Battery Polarity

If a battery were installed and connected with polarity reversed, a short circuit would be created through the diodes. Current would then be from the

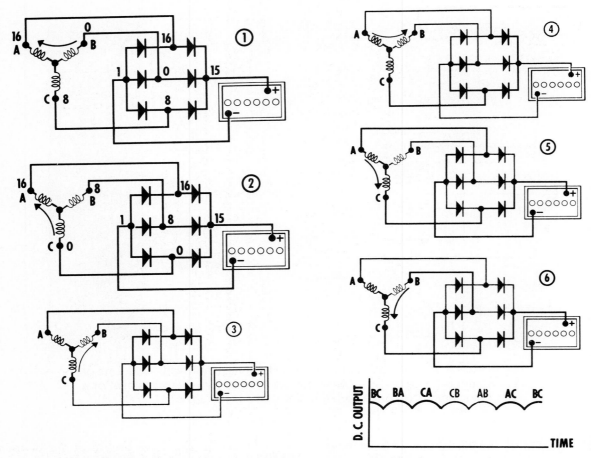

FIGURE 32.8. *(Courtesy of General Motors Corporation)*

positive and negative diodes and into the heat sink. From the heat sink, a completed circuit would exist back to the negative battery terminal. Full battery voltage would be impressed on the diodes. The resulting high-current would damage the diodes and/or damage the wiring harness. Some systems have a fusible link to protect the system in case of accidental reverse polarity.

Reverse Current Prevention

The positive post of the battery is connected to the alternator output terminal and the positive heat sink. However, the positive diode will not allow current into the stator windings. The diodes eliminate the need for a cutout relay used with DC generators.

Grounded-Brush Alternator

This unit has a single field terminal connected to a brush. The other brush is mounted to the alternator housing. In this system, the field circuit is grounded

at the alternator through the brush mounted to the alternator housing.

Insulated-Brush Alternator

This unit has both field brushes insulated from the alternator housing. The single terminal is connected to a brush; the other brush is connected to the heat sink instead of ground.

Isolated-Field Alternator

This unit has two insulated brushes, with each brush having a field terminal. Neither brush has a direct ground to the alternator housing or the heat sink. This means the internal field circuit of the alternator is isolated.

Grounding

The grounded-brush alternator's field circuit is grounded at the alternator housing. The insulated-

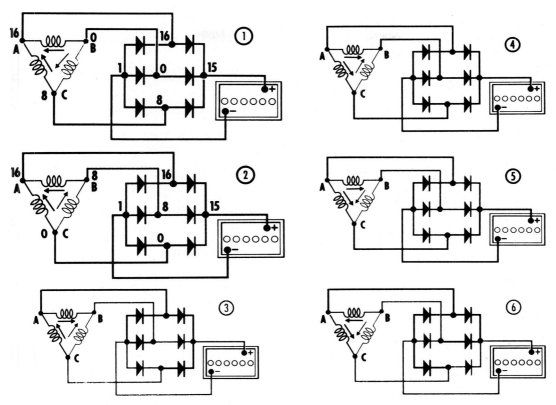

FIGURE 32-9. *(Courtesy of General Motors Corporation)*

brush and isolated-field alternator's field circuits are grounded in the electronic voltage regulator.

Summary

1. An alternator has two component circuits: one, the field circuit, is being fed current from a power source, the other is in the stator windings and terminal components.

2. The transfer of energy between the two circuits is effected by induction as the moving north and south rotor fingers alternately pass the stator windings. (The design gap between fingers collapses the magnetic field for each winding as rotation continues. This means that the magnetic fields build and collapse in typical alternating current cycles even though the phases overlap.)

3. The voltage induced by the field applies its current-driving force to the pair of stator windings being most strongly energized. Accordingly, it pushes current to the diodes.

4. The positive diodes pass current to the alternator output terminal. The negative diodes pass current returning through the ground circuit. (Each type of diode blocks current of opposite polarity.)

PART 2   BRUSHLESS ALTERNATOR

The following description applies to the Delcotron type 400 alternator, which is typical of this design. (See Fig. 32-11 to 32-15.)

The Integral Charging System is a self-rectifying, brushless unit featuring a built-in voltage regulator. The only movable part in the assembly is the rotor, which is mounted on a ball bearing at the drive end, and a roller bearing at the rectifier end. All current-carrying conductors are stationary. These conductors are the field winding, the stator windings, the six rectifying diodes, and the regulator circuit components. The regulator and diodes are enclosed in a sealed compartment.

A fan located on the drive end provides airflow for cooling. Extra large grease reservoirs contain an adequate supply of lubricant so that no periodic maintenance of any kind is required.

Only one wire is needed to connect the Integral Charging System to the battery, along with an adequate ground return. The specially designed output terminal is connected directly to the battery. A red output terminal is used on negative ground

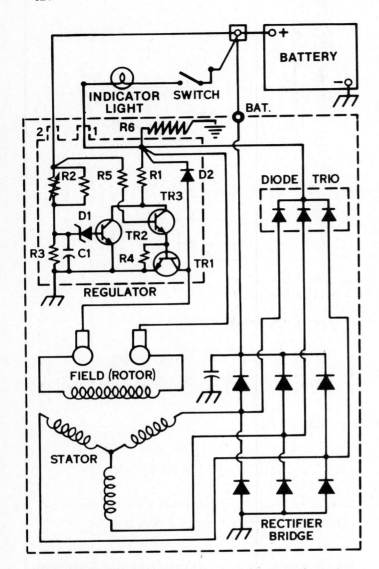

FIGURE 32-10. Wiring diagram of brush-type alternator with integral electronic regulator and Y-wound stator. *(Courtesy of General Motors Corporation)*

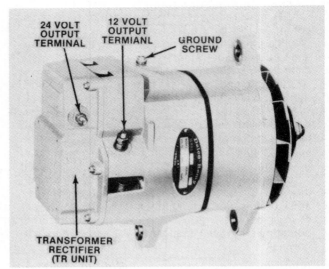

FIGURE 32-11. Brushless 12/24 volt alternator with integral regulator. *(Courtesy of General Motors Corporation)*

potentiometer mounted on the voltage regulator. The generator rear cover has to be removed for access to this adjustment.

The basic operating principles for the generators are explained as follows:

As the rotor begins to turn, the permanent magnetism therein induces voltages in the stator windings. The voltages across the six diodes cause current to flow to charge the battery.

Current from the stator flows through the three diodes to resistor R6 and the base-emitter of TR2

models, and is to be connected only to battery positive. A black output terminal is used on positive ground models and is to be connected only to battery negative. An "R" terminal is provided for use in some circuits to operate auxiliary equipment.

The hex head bolt on the output terminal is electrically insulated; no voltage reading can be obtained by connecting to the hex head.

Operating Principles

The wiring diagram Figure 32-14, applies to generator models with an internal voltage adjustment

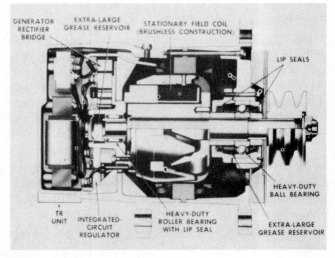

FIGURE 32-12. Brushless alternator with integral regulator. *(Courtesy of General Motors Corporation)*

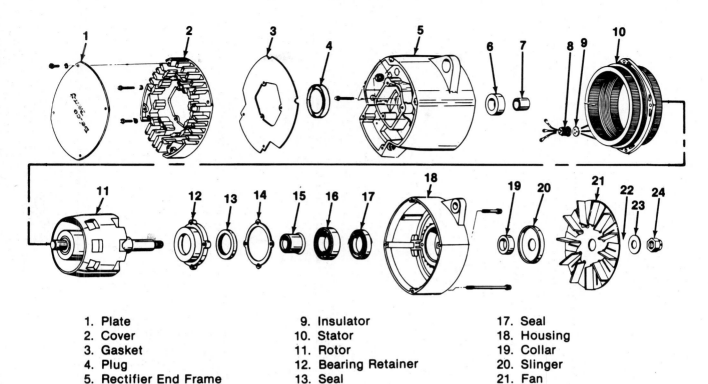

1. Plate
2. Cover
3. Gasket
4. Plug
5. Rectifier End Frame
6. Bearing—Outer Race
7. Bearing—Inner Race
8. Grommet
9. Insulator
10. Stator
11. Rotor
12. Bearing Retainer
13. Seal
14. Gasket
15. Collar
16. Bearing
17. Seal
18. Housing
19. Collar
20. Slinger
21. Fan
22. Pulley (Not Shown)
23. Washer
24. Nut

FIGURE 32-13. Components of brushless alternator with integral regulator. *(Courtesy of General Motors Corporation)*

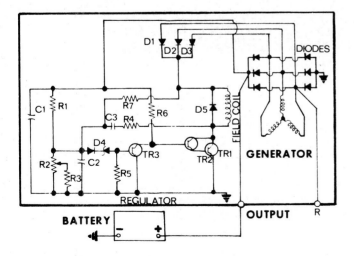

FIGURE 32-14. Wiring schematic of brushless integral charging system. *(Courtesy of General Motors Corporation)*

and TR1 to turn these transistors on. Current also flows from the stator through the diode trio D1, D2, and D3, the field coil and transistor TR1, returning to the stator through the other three diodes. All stator current, except through the diode trio D1, D2, and D3, flows through the six diodes connected to the stator.

Current flow through R1, R2, and R3 causes a voltage to appear at zener diode D4. When the voltage becomes high enough due to increasing generator speed, D4 and the base-emitter of TR3 conduct current and TR3 turns on. TR2 and TR1 then turn off, decreasing the field current and the system voltage decreases. The voltage at D4 decreases, D4 and TR3 turn off, TR2 and TR1 turn back on and the system voltage increases. This cycle then repeats many times per second to limit the system voltage as determined by the setting of the potentiometer R2, R3.

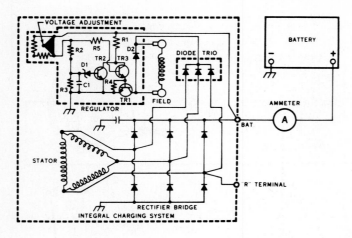

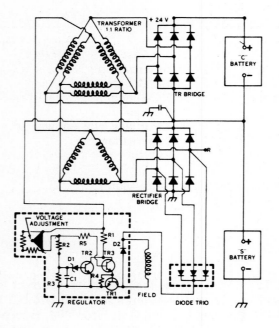

FIGURE 32-15. Wiring diagrams for 12 volt (top) and 12/ 24 volt (bottom) alternators with delta wound stators and integral regulators. *(Courtesy of General Motors Corporation)*

Capacitor C1 protects the generator diodes from high transient voltages and suppresses radio interference.

Resister R5 prevents current leakage through TR3 at high temperatures.

Diode D5 prevents high transient voltages in the field coil when the field current is decreasing.

Resistor R7, capacitor C3 and resistor R4 all act to cause transistors TR2 and TR1 to turn on and off more quickly.

PART 3   AC GENERATOR REGULATION

Current Control

The AC generator is "self limiting" in its maximum output. This occurs as the magnetic field produced by the current in the stator windings opposes in polarity and approaches in value the magnetic field provided by the rotor as the generator output increases. This causes the generator to limit its own output to a maximum value.

Voltage Regulator

The voltage delivered by an alternator must be regulated to protect the charging circuit. Circumstances or preference determine whether this is accomplished with a vibrating-type electromechanical regulator, with a transistorized regulator, or with an integral electronic regulator that incorporates transistors, diodes, and resistors in an arrangement which accomplishes the same end.

Voltage Limiter

The voltage limiter is a two-stage device. It is used to control field current and to extend the range of alternator rpm that can be controlled. When the alternator is operating at comparatively low speeds or when the system load is heavy, the voltage limiter armature will vibrate on the lower contacts and limit voltage by intermittently inserting resistance in the alternator field circuit. Regulation will continue on the lower contacts until the voltage rises to a value at which the resistor will no longer provide control.

At high alternator speeds, or when the battery is fully charged and the electrical load is light, the voltage attempts to rise; in so doing, the armature is pulled down and starts vibrating on the upper contacts. This provides additional control by intermittently inserting the resistance or bypassing the field to ground. When the contacts are closed, the current in the field circuit drops as the magnetic field in the rotor collapses. The result is a drop in alternator output voltage. When the voltage drops below the lower stage calibration point, the upper contacts open. This vibrating action takes place many times per second. Changes in the system load or changes in speed may cause the above sequence of operations to vary.

To summarize, the voltage limiter may be operating on either set of contacts (upper or lower) depending on load and speed conditions at the time, but never on both.

Electronic Voltage Regulator

The electronic voltage regulator is a solid-state unit with no moving parts or adjustment. It is serviced only by replacement. This regulator governs the electrical system voltage by limiting the output voltage that is generated by the alternator. This is accomplished by controlling the value of field current that is allowed to pass through the field windings. Basically, the electronic regulator operates as a voltage-sensitive switch as described earlier.

PART 4   AC CHARGING SYSTEM DIAGNOSIS AND SERVICE

AC CHARGING SYSTEM DIAGNOSIS

PROBLEM	CAUSE	CORRECTION
Battery low in charge or discharged	1. Loose or worn alternator drive belt 2. Defective battery not accepting or holding charge; electrolyte level low 3. Excessive resistance due to loose charging system connections 4. Defective battery temperature sensor (where fitted) 5. Defective voltage regulator 6. Defective alternator	1. Check and adjust tension or renew. 2. Check condition of battery and renew. Check, fill, and charge. 3. Check, clean, and tighten circuit connections. 4. Check and renew. 5. Check and renew. 6. Repair or replace as required.
Alternator charging at high rate (battery overheating)	1. Defective battery 2. Defective battery temperature sensor 3. Defective voltage regulator 4. Defective alternator	1. Check condition of battery and renew. 2. Check and renew. 3. Check and renew. 4. Repair or replace as required.
No output from alternator	1. Alternator drive belt broken 2. Loose connection or broken cable in charging system 3. Defective temperature sensor (where fitted) 4. Defective voltage regulator 5. Defective alternator	1. Renew and tension correctly. 2. Inspect system, tighten connections, and repair or renew faulty wiring. 3. Check and renew. 4. Check and renew. 5. Repair or replace as required.
Intermittent or low alternator output	1. Alternator drive belt slipping 2. Loose connection or broken cable in charging system 3. Defective temperature sensor (where fitted) 4. Defective voltage regulator 5. Defective alternator	1. Check and adjust tension or renew. 2. Inspect system, tighten connections, and repair or renew faulty wiring. 3. Check and renew. 4. Check and renew. 5. Repair or replace as required.
Warning light dimming and/or battery low	1. Faulty external charging circuit connections 2. Faulty rotor slip rings or brushes	1. Inspect system, clean and tighten connections. 2. Inspect and repair or renew.
Warning light going out or becoming brighter with increased speed	1. Faulty external charging circuit connections 2. Faulty rectifier or rectifying diodes	1. Inspect system, clean and tighten connections. 2. Check and renew.

Warning light normal but battery boiling	1. Defective voltage regulator 2. Faulty battery temperature sensor (where fitted)	1. Check and renew. 2. Check and renew.
Warning light normal but battery discharged	1. Defective voltage regulator 2. Faulty stator 3. Faulty rectifier or rectifying diodes	1. Check and renew. 2. Check and renew. 3. Check and renew.
Warning light illuminated continuously and/or flat battery	1. Loose or worn alternator drive belt 2. Defective surge protection diode (where fitted) 3. Defective isolation diodes (where fitted) 4. Faulty battery temperature sensor (where fitted) 5. Faulty rotor, slip rings, or brushes 6. Faulty voltage regulator 7. Defective stator 8. Defective rectifier or rectifying diodes	1. Check and adjust tension or renew. 2. Check and renew. 3. Check and renew. 4. Check and renew. 5. Inspect, repair, or renew. 6. Check and renew. 7. Inspect and renew. 8. Check and renew.
Warning light extinguished continuously and/or flat battery	1. Burned-out bulb 2. Alternator internal connections 3. Defective voltage regulator 4. Faulty rotor, slip rings, or brushes 5. Defective stator	1. Check and renew. 2. Inspect and test circuitry; repair or renew. 3. Check and renew. 4. Check, repair, or renew. 5. Check and renew.
Warning light flashing intermittently	1. Faulty external charging circuit 2. Alternator internal connections	1. Inspect circuit, clean and tighten connections, repair or renew faulty wiring. 2. Inspect and test circuitry repair or renew.
Warning light dimming continuously and/or flat battery	1. Defective rotor, slip rings, or brushes 2. Defective voltage regulator	1. Check, repair, or renew. 2. Check and renew.

PART 5   AC GENERATOR CHARGING SYSTEM SERVICE

Alternating-current generator charging system service includes a general overall visual inspection, system and component diagnosis, and component replacement or repair. A thorough understanding of the principles of operation is essential for accurate diagnosis and service. Continuing variations in design and application can be readily dealt with if operating principles are understood. (See Fig. 32–16 to 32–25.)

General practice is increasingly toward replacement of faulty components rather than repair. A good multiscale voltmeter, an ammeter, and an ohmmeter or test light are essential equipment for diagnosing AC charging systems. As with other electrical system service, battery inspection and testing as described in Chapter 29 should be included in charging system service.

FIGURE 32–16. Method of grounding field circuit to obtain maximum alternator output on one type of alternator. *(Courtesy of General Motors Corporation)*

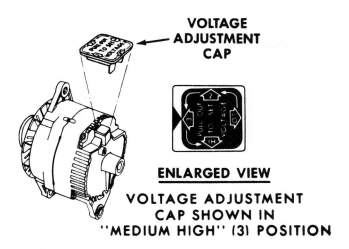

VOLTAGE
ADJUSTMENT
CAP

ENLARGED VIEW

VOLTAGE ADJUSTMENT
CAP SHOWN IN
"MEDIUM HIGH" (3) POSITION

FIGURE 32-17. Voltage adjustment cap can be installed in any of four positions to achieve specified voltage output on some alternators. *(Courtesy of General Motors Corporation)*

General Precautions

Refer to Chapter 29 for general precautions regarding battery service. In addition, the following precautions should be observed:

1. Follow test equipment manufacturer's instructions for testing.

2. Disconnect battery ground cable before removing alternator for service.

3. Always observe proper battery polarity.

4. Observe vehicle manufacturer's procedures and specifications.

5. Do not attempt to polarize the alternator.

6. Do not accidentally ground any alternator terminals.

7. Do not operate alternator on an open circuit.

8. Use only the induction pickup-type of ammeter on computer equipped vehicles—never the series connected ammeter.

The normal sequence of operations for servicing the AC charging system will depend on the nature of the problem. Charging system problems can be classified as follows:

1. Noisy alternator
2. Overcharging
3. Undercharging

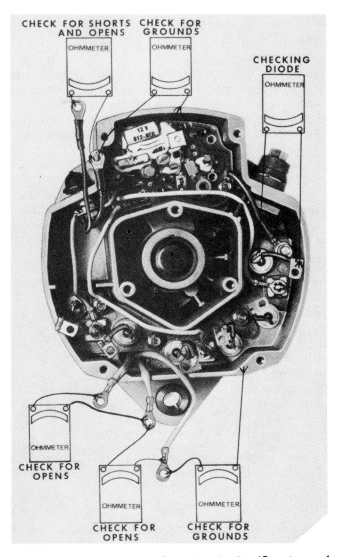

FIGURE 32-18. Typical ohmmeter tests. *(Courtesy of General Motors Corporation)*

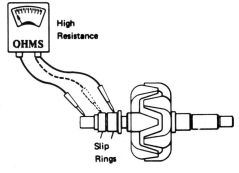

FIGURE 32-19. Checking field coil and slip rings for grounds with ohmmeter. *(Courtesy of General Motors Corporation)*

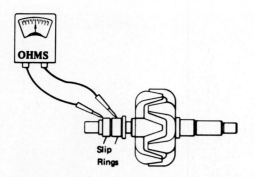

FIGURE 32-20. Checking field coil resistance with ohmmeter. *(Courtesy of General Motors Corporation)*

Noisy Alternator

A noisy water pump, power steering pump, fan, or other drive belt noise should not be confused with alternator noise. Once it has definitely been established that it is the alternator which is the source of the noise, follow the diagnostic chart to identify the possible reasons for alternator noise. Some of the problems listed can be corrected without alternator removal; others require alternator removal and repair or replacement.

Overcharging or Undercharging

Battery condition is a good indicator of charging system operation. Excessive gassing and the need for repeatedly adding water to the battery indicate that the charging system voltage may be too high.

A weak battery (poor cranking and low voltage) may indicate insufficient charging or a faulty battery. To determine whether the battery or the charging system is at fault, follow the diagnostic chart in this section.

Figures 32-18 to 32-25 show typical tests required to determine whether the alternator, the regulator, or connecting wiring are at fault. Faulty regulators are replaced. Faulty wiring is replaced or disassembled, tested, and repaired. After repairs or replacement have been completed, the system should again be tested for proper output.

AC Generator Removal

1. Disconnect negative battery cable at battery. Caution: Failure to observe this step may result in an injury from hot battery lead at generator.

2. Disconnect wiring leads from generator.

3. Loosen adjusting bolts and move generator to provide slack in belt.

4. Remove generator drive belt.

5. Remove thru bolts which retain generator.

6. Remove generator from engine.

Installation

1. If removed from engine, install generator to mounting bracket with bolts, washers, and nuts. Do not tighten.

2. Install generator drive belt.

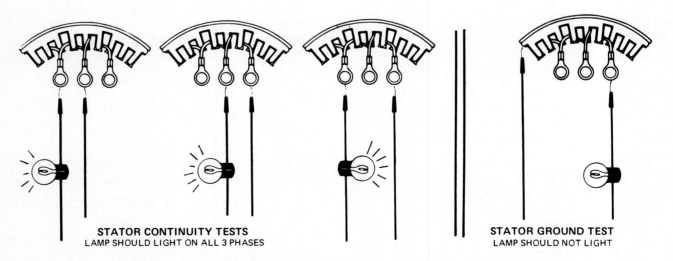

STATOR CONTINUITY TESTS
LAMP SHOULD LIGHT ON ALL 3 PHASES

STATOR GROUND TEST
LAMP SHOULD NOT LIGHT

FIGURE 32-21. Checking stator windings with a test lamp. Ohmmeter may be used for these tests. *(Courtesy of Ford Motor Company)*

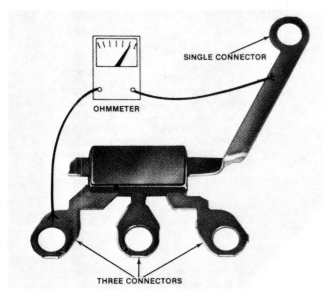

FIGURE 32-22. Checking the diode trio with an ohm-meter. *(Courtesy of General Motors Corporation)*

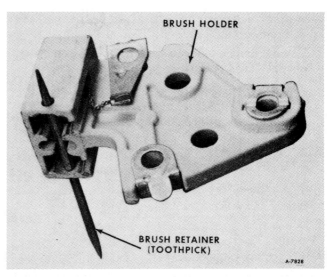

FIGURE 32-24. Using a toothpick to hold brushes back against spring pressure to aid in alternator assembly. *(Courtesy of General Motors Corporation)*

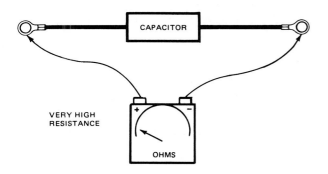

FIGURE 32-23. Checking the alternator capacitor with an ohmmeter. *(Courtesy of Ford Motor Company)*

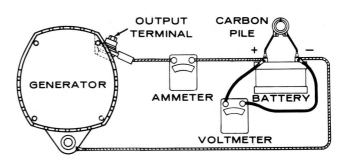

FIGURE 32-25. Typical test connections for bench testing alternator output. *(Courtesy of General Motors Corporation)*

3. Tighten belt to the specified belt tension. See the engine cooling section for proper belt tensioning procedures.

4. Tighten bolts.

5. Install generator terminal plug and battery leads to generator.

6. Connect negative battery cable.

PART 6   SELF-CHECK

1. The major parts of the AC generator are the _____.

2. How is the magnetic field in an AC generator increased?

3. The AC generator has a rotating electromagnet. True or false?

4. Without energizing the field, the AC generator will not produce current. True or false?

5. The AC generator brushes carry total generator output. True or false?

6. The AC generator requires three diodes for full-wave rectification. True or false?

7. Stator windings are connected to each other by the _____ or the _____ method.

8. Current in the AC generator stator windings is always in the same direction. True or false?

9. The alternator diode is a two-way electrical check valve. True or false?

10. What effect does a shorted diode have on an alternator output?

11. What is the purpose of the fusible link between the battery and the diodes on some AC charging systems?

12. The AC charging system regulator controls current and voltage output. True or false?

13. What four pieces of test equipment are used to check charging systems and components?

14. A noisy alternator may be the result of one or more shorted or open diodes. True or false?

15. Corroded battery posts or terminals can cause improper charging. True or false?

16. How should a diode be tested?

17. The rotor should be tested for

(a) _____,

(b) _____,

(c) _____.

18. The stator should be tested for (a) _____, (b) _____.

Chapter 33

DC Charging Systems

Performance Objectives

After thorough study of this chapter, sufficient practice on adequate training models, and with the necessary tools, equipment, and manuals, you should be able to do the following:

1. Complete the self-check questions with at least 80 percent accuracy.

2. Observe the accepted general precautions outlined in this chapter.

3. Describe the purpose, construction, and operation of DC charging systems and their components.

4. Accurately diagnose charging system faults.

5. Safely remove, test, recondition, replace, and adjust faulty charging system components according to manufacturer's specifications.

6. Properly performance test the reconditioned components and the entire charging system to determine the success of the repairs performed.

PART 1   DC GENERATOR PRINCIPLES

There is a great deal of similarity between a cranking motor and a generator. The cranking motor converts electrical energy into mechanical energy, whereas the generator converts mechanical energy into electrical energy. (See Fig. 33-1 and 33-2.)

The generator functions on the principle that any conductor connected into a completed circuit will have a current induced in it when it is moved through a magnetic field. Residual magnetism in the pole shoes provides enough field strength (magnetic lines of force) for the generator to start charging. The generator is made up of two major parts, the armature and the field winding assembly. The armature consists of a number of conductors placed in insulated slots around a soft iron core, which is assembled onto the armature shaft. The conductors are connected to each other and to the commutator. When brushes are placed on the commutator, any current induced in the conductors is carried away from the armature through the brushes. Part of this current is shunted through the field windings so that the magnetic field in which the conductors are moved is strengthened. This tends to produce a greater amount of current in the conductors.

Figure 33-2 shows a simple one-turn armature generator. If the armature loop is rotated in a clockwise direction, as indicated, current will be induced in the left-hand side of the loop away from the reader and in the right-hand side of the loop toward the reader. Use the right-hand rule to check the directions of current in the two sides of the loop. The current that is induced is to the right-hand segment of the commutator and through the right-hand brush. From there most of it is through the external circuit and the connected electrical load. Then it is to the left-hand brush and left-hand segment of the commutator and back into the loop. The remainder of the induced current is through the two field windings, which are assembled around the two magnetic pole shoes. This current is in the right direction to strengthen the magnetic field between the two pole shoes. In the complete generator assembly the pole shoes would be fastened into a field frame, and the field frame would form the return magnetic circuit for the lines of force from the north pole to the south pole. The magnetic lines of force pass from the magnetic north to the magnetic south pole through the external magnetic circuit.

The use of a commutator with copper segments separated by mica keeps the current in the external electrical circuit in the same direction, regardless of the relative positions of the two sides of the loop with respect to the north and south magnetic poles. For instance, when the left-hand side of the loop has rotated 180°, it will be moving in the opposite direction with respect to the magnetic field. However, since the commutator segment to which it is connected also has rotated 180°, this current will be fed to the right-hand brush. As far as the external circuit is concerned, the current still continues in the same direction.

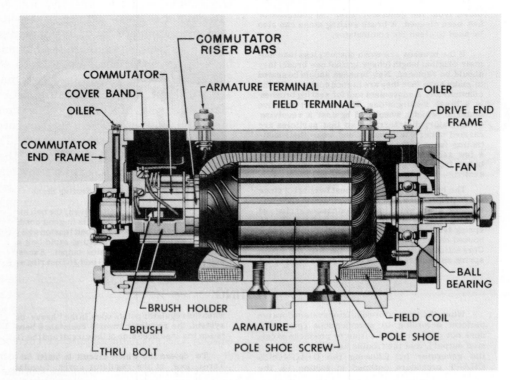

FIGURE 33-1. DC generator cross section. *(Courtesy of General Motors Terex)*

When the conductor is halfway between the two poles, it is in the neutral position where no current is generated and where the direction of current in the conductor reverses. This is also the position where the brush changes connections with the com-

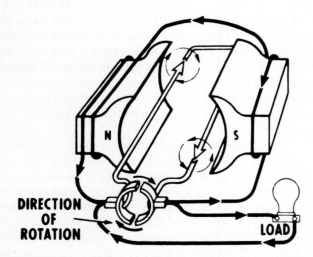

FIGURE 33-2. Simple dc generator with a single-loop armature and a two-segment commutator. Commutator rectifies generator current from ac to dc. Residual magnetism in the pole shoes provides sufficient magnetic field strength for the generator to start charging.

mutator bars. Actually, AC current is generated, but the commutator acts as a rectifier or rotating reversing switch that maintains DC current in the external circuit.

PART 2   GENERATOR CIRCUITS

Generators are connected internally in two different ways, depending on the type of generator regulator used. The two methods of connection are shown in Figure 33-3. It is important to bear these differences in mind in any check or adjustment of the generator since different checks are required for each type.

All shunt generators must have some form of external regulation. Regulation is normally provided by the use of spring-loaded contacts and a resistance. The contacts are opened magnetically by means of a winding or windings. When the points are opened, the resistance is inserted into the generator field circuit, so that the amount of field current is cut down. This reduces the strength of the generator magnetic field and lowers the generator output. When the contacts are closed, the resistance is removed, field current increases, and generator output rises.

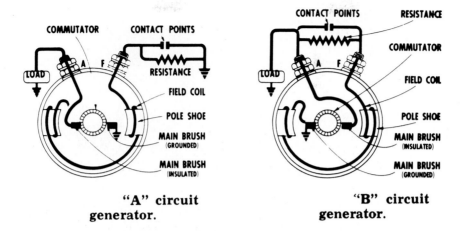

"A" circuit
generator.

"B" circuit
generator.

FIGURE 33-3. Two most common dc generator circuits. A-circuit generator has shunt-wound field coils, which are grounded externally through the regulator. B-circuit generator also has shunt-would field coils (through regulator), but field coils are grounded inside the generator through the grounded brush. *(Courtesy of General Motors Corporation)*

The difference between the two types of generator circuits lies mainly in that part of the field circuit in which the resistance is inserted. In the B circuit the generator regulator inserts resistance between the insulated side of the circuit and the field. In the A circuit system the resistance is inserted between the field and ground.

There is no particular advantage to either circuit, but because of the difference in the method of connecting the regulator into the field circuit, the regulators must not be interchanged.

Bucking Field

Some generators have additional turns on the armature to produce the necessary voltage to obtain a charging rate at very low speeds. When the range of operating speeds is great, voltage regulation at the higher speeds becomes a very important problem. As stated earlier, the voltage produced in a generator depends upon the strength of the magnetic field, the number of the armature conductors in series, and the speed of armature rotation. When the number of conductors and the speed are both great, then only a very weak field is required to produce the required voltage. On some generators, operating at very high speeds, it is possible to produce more than the required voltage even with an open-circuited field. The residual magnetism of the pole shoes supplies enough magnetic field to produce the voltage, but this voltage cannot be controlled. Even though the voltage winding of the regulator opens

the contacts and inserts a resistance in the field circuit, the voltage will continue to climb.

To control the voltage on this type of unit, it is necessary to use what is known as a bucking field. It is a shunt coil wound on one pole shoe and connected directly across the brushes of the armature. The winding is in the reverse direction from the normal field winding and has an opposing magnetic effect. At low speeds, when the normal field is strong, the opposing effect is not great, but at higher speeds, when the current in the regular field circuit is reduced by the voltage regulator, the opposing effect is greater than the residual magnetic field. Thus, the current through the regular field coils can be controlled by the regulator and a normal voltage maintained.

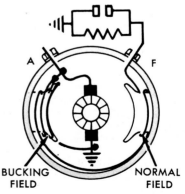

FIGURE 33-4. Bucking field generator. *(Courtesy of Deere and Company)*

PART 3   DC GENERATOR REGULATORS

For the generating system to operate, the armature wire from the generator must be connected to the generator (GEN) terminal of the regulator, the field terminal of the generator connected to the field (F) terminal of the regulator, and the wire from the battery through the ammeter to the battery (BAT) terminal of the regulator. (See Fig. 33-6 to 33-8.)

Circuit Breaker (Cutout Relay)

When the generator voltage builds up to a value great enough to charge the battery, the magnetism induced by the current through the shunt winding is sufficient to overcome the armature spring tension and pull the armature toward the core so that the contact points close. This completes the circuit between the generator and battery. The current from the generator to the battery passes through the series winding in the proper direction to add to the magnetism, holding the armature down and the points closed.

When the generator slows down or stops, current reverses from the battery to the generator. This reverse direction of current flows through the series winding, thus causing a reversal of the series winding magnetic field. The magnetic field of the shunt winding does not reverse. Therefore, instead of helping each other, the two windings now magnetically oppose each other so that the resultant magnetic field becomes insufficient to hold the armature down. The flat spring pulls the armature away from the core so that the points separate; this opens the circuit between the generator and battery.

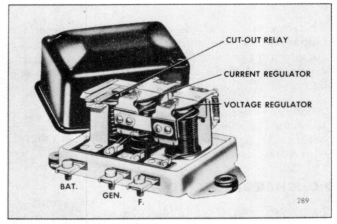

FIGURE 33-6. Three-unit generator regulator. *(Courtesy of General Motors Terex)*

Voltage Regulator

When the generator voltage reaches the value for which the voltage regulator is adjusted, the magnetic field produced by the winding overcomes the armature spring tension and pulls the armature down so that the contact points separate. This inserts resistance into the generator field circuit so that the generator field current and voltage are reduced. Reduction of the generator voltage opens the regulator series winding circuit so that its magnetic field collapses completely. This allows the spiral spring to pull the armature away from the core so that the contact points again close. This directly grounds the generator field circuit so that the generator voltage and output increase. The above cycle of action again takes place, and the cycle continues at a rate of 150 to 250 times a second, regulating the voltage to a constant value. By thus maintaining a constant voltage, the generator supplies varying amounts of current to meet the varying states of battery charge and electrical load.

Temperature Compensation

The voltage regulator is compensated for temperature by means of a bimetal thermostatic hinge on the armature. This causes the regulator to regulate for a higher voltage when cold, which partly compensates for the fact that a higher voltage is required to charge a cold battery.

Current Regulator

When electrical devices are turned on and the battery is in a discharged condition, the voltage may not increase to a value sufficient to cause the volt-

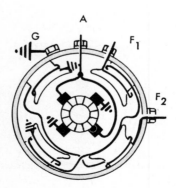

FIGURE 33-5. Split field generator. *(Courtesy of Deere and Company)*

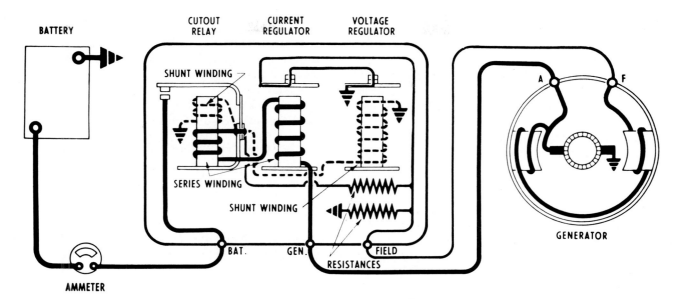

FIGURE 33-7. Schematic diagram of A-circuit generator and three-unit regulator. *(Courtesy of General Motors Corporation)*

age regulator to operate. Consequently, generator output will continue to increase until the generator reaches its rated maximum output. This is the current value for which the current regulator is set. Therefore, when the generator reaches its rated output, this output through the current regulator winding creates sufficient magnetism to pull the current regulator armature down and open the contact points. With the points open, resistance is inserted into the generator field circuit so that the generator output is reduced.

As generator output starts to fall, the magnetic field of the current regulator winding is reduced, the spiral spring tension pulls the armature up, and the contact points close, removing the resistance from the field circuit. Output increases and the above cycle is repeated. The cycle continues to take place while the current regulator is in operation 150 to 250 times a second, preventing the generator from exceeding its rated maximum.

When the electrical load is reduced, electrical devices are turned off, or battery comes up to charge, the voltage increases so that the voltage regulator begins to operate and tapers the generator output down. This prevents the current regulator from operating. Either the voltage regulator or the current regulator controls the generator at any one time; the two never operate at the same time.

Resistances

The current and voltage regulator circuits use a common resistance, which is inserted in the field circuit when either the current or voltage regulator operates. A second resistance is connected between the regulator field terminal and the relay frame, which places it in parallel with the generator field coils. The sudden reduction in field current occurring when either the current or voltage regulator contact points open is accompanied by a surge of induced voltage in the field coils as the strength of the magnetic field changes. These surges are partially dissipated by arcing at the contact points.

The cutout relay disconnects the battery from the charging system any time the engine is shut off. If the cutout points stick, the battery current will try to drive the generator. This will discharge the battery very quickly.

The voltage regulator controls maximum voltage output in order to prevent the battery from being overcharged and to protect the electrical components in the vehicle. Excessive voltage can cause lights to flare and to burn out prematurely and reduce the life of other electrical components.

The current regulator controls the maximum current output of the generator to prevent the generator from overheating and being burned out.

PART 4   DC GENERATOR CHARGING SYSTEM SERVICE

Direct-current generator charging systems have not been commonly used for many years. However, there are still some older units around equipped with this system.

In general, the practice has been to diagnose the system to determine whether the battery, generator, regulator, or related wiring are at fault, and then to replace or repair the faulty components. The following procedures will quickly lead to the source of the problem. The equipment required is the voltmeter and the ammeter. As with any other electrical problem, the battery must be properly checked as outlined in Chapter 18, Part 5. Follow the same General Precautions given in Chapter 22 when servicing DC charging systems.

DC Charging System Checks

In analyzing complaints of generator-regulator operation, any of several basic conditions may be found.

Fully Charged Battery and Low Charging Rate. This indicates normal generator-regulator operation.

Fully Charged Battery and a High Charging Rate. This usually indicates that the voltage regulator unit either is not limiting the generator voltage as it should or is set too high. A high charging rate to a fully charged battery will damage the battery, and the accompanying high voltage is very injurious to all electrical units. This operating condition may result from the following:

1. Improper voltage regulator setting.

2. Defective voltage regulator unit.

3. Grounded generator field circuit in either generator, regulator, or wiring; may be trouble in A circuit.

Low Battery and High Charging Rate. This is normal generator-regulator action.

Low Battery and Low or No Charging Rate. This condition could be due to the following:

1. Loose generator belt.

2. Loose connections, or damaged external wiring.

3. Defective battery.

4. High circuit resistance.

5. Low regulator setting.

6. Oxidized regulator contact points.

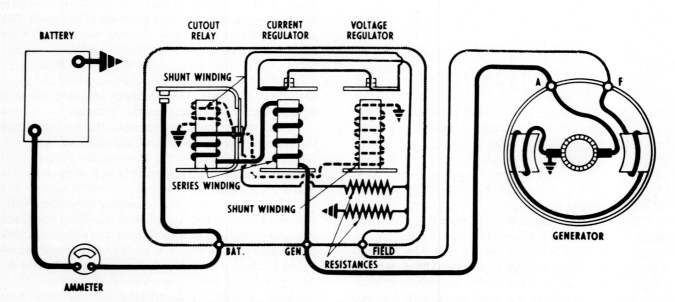

FIGURE 33-8. Schematic diagram of B-circuit generator and three-unit regulator. *(Courtesy of General Motors Corporation)*

7. Defects within the generator.

8. Cutout relay not closing.

9. Open series circuit within regulator.

10. Generator not properly polarized.

If the condition is not caused by loose connections or damaged wires, proceed as follows to locate cause of trouble.

A Circuit

To determine whether the generator or regulator is at fault, momentarily ground the F terminal of the regulator and increase generator speed. If output does not increase, the generator is probably at fault and it should be checked. Other causes for the output not increasing may be the relay not closing or an open series winding in the regulator. If the generator output increases, the trouble is due to the following:

1. A low voltage (or current) regulator setting.

2. Oxidized regulator contact points that insert excessive resistance into the generator field circuit so that output remains low.

3. Generator field circuit open within the regulator.

B Circuit

To determine whether the generator or regulator is at fault, momentarily place a jumper lead between the GEN and FLD terminals of the regulator and increase generator speed. If output does not increase, the generator is probably at fault and it should be checked. Other causes for the output not increasing may be the relay not closing or an open series winding in the regulator. If the generator output increases, the trouble is due to the following:

1. A low voltage (or current) regulator setting.

2. Oxidized regulator contact points that insert excessive resistance into the generator field circuit so that output remains low.

3. Generator field circuit opens within the regulator.

Burned Resistances, Windings, or Contacts. These result from open circuit operation, open resistance units, or loose or intermittent connections in the charging circuit. Where burned resistances, windings, or contacts are found, always check equipment wiring before installing a new regulator. Otherwise, the new regulator may also fail in the same way.

Burned Relay Contact Points. This may be due to reversed generator polarity. Generator polarity must be corrected after any checks of the regulator or generator or after disconnecting and reconnecting leads.

Polarizing Generator: A Circuit. After reconnecting leads, momentarily connect a jumper lead between the GEN and BAT terminals of the regulator. This allows a momentary surge of current through the generator, which correctly polarizes it. Failure to do this may result in severe damage to the equipment, since reversed polarity causes vibration, arcing, and burning of the relay contact points.

Polarizing Generator: B Circuit. To polarize circuit B generators, disconnect the lead from the FLD terminal of the regulator and momentarily touch this lead to the BAT terminal of the regulator. This allows a momentary surge of current through the generator, which correctly polarizes it. Failure to do this may result in severe damage to the equipment, since reversed polarity causes vibration, arcing, and burning of the relay contact points.

Generator Overhaul

If the generator is faulty, it must be removed and overhauled or be replaced with a new or reconditioned unit.

Disassemble the generator, but do not remove the field coils or the brush holders until tests indicate that this is necessary. Clean all the parts. Do not wash the brushes, armature, or field coils in solvent or other cleaning fluids. Compressed air and cloth wipers can be used for this purpose. Proceed with testing as follows.

Armature Testing

Inspect the armature for physical damage such as molten solder, rubbing on the pole shoes, burned commutator segments, and shaft wear or damage. A tester called a *growler* is used to test the armature electrically. Place the armature in the cradle of the growler and switch it on. Holding a thin, flat steel strip (long feeler gauge or hack saw blade) loosely parallel to the top of the armature, rotate the armature slowly for at least one turn. If the steel strip vibrates, there is an electrical short in the armature. A shorted armature must be replaced. Turn off the growler magnetic switch. (See Fig. 33-10 to 33-15.)

Using the growler test light leads, check the armature for grounds by placing one lead on the armature core and then touching each commutator segment in turn with the other lead. If the test lamp

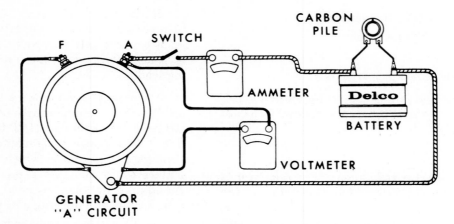

FIGURE 33-9. Method used to check A-circuit type of charging system output. Note jumper wire from generator field terminal to ground. This causes maximum field energization and therefore maximum generator output. *(Courtesy of General Motors Corporation)*

lights, the armature is grounded and must be replaced.

An armature with an open circuit must also be replaced. This problem is identified by visual inspection. The commutator segment connected to an open circuit will be burned due to arcing during operation.

An armature that passes these tests should be placed in an armature lathe and checked for runout. Runout should not exceed 0.005 inch (0.1270 mm). True the commutator in the lathe to remove scoring and runout. Undercut the mica insulators between the commutator bars. Since mica does not wear as quickly as copper, the insulators will protrude slightly as the commutator wears, causing brushes to bounce, arc, and wear rapidly.

Field Coil Tests

Using the growler test light, test the field coils for continuity (opens) and grounds. On B-type generators the grounded end of the field coil must first be disconnected for these tests. Place one lead of the test light at each end of the field coil winding. If the

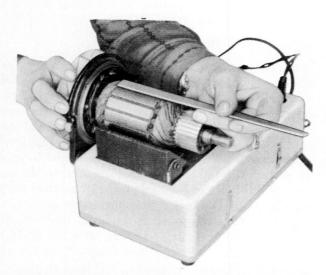

FIGURE 33-10. Testing armature for shorts on growler. *(Courtesy of Deere and Company)*

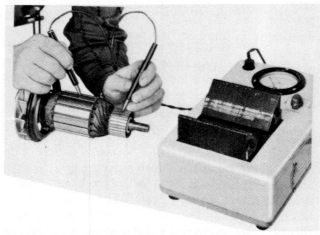

FIGURE 33-11. Testing armature for grounds with test light. *(Courtesy of Deere and Company)*

FIGURE 33-12. Testing armature for open circuit with growler tester. *(Courtesy of Deere and Company)*

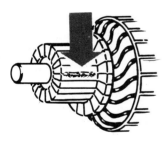

FIGURE 33-13. Evidence of open circuit in armature. *(Courtesy of Deere and Company)*

FIGURE 33-14. Turning down commutator to restore concentricity and full-brush contact. *(Courtesy of Deere and Company)*

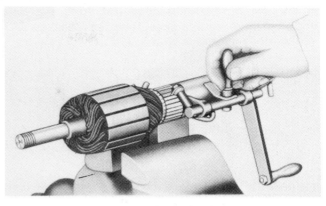

FIGURE 33-15. Undercutting mica between commutator bars. *(Courtesy of Deere and Company)*

test lamp lights, there are no opens. If it does not light, the coil has an open circuit and must be replaced. To test for grounds, place one lead on the generator frame and the other on the field coil terminal. The test light should not light. If it does, the field coil is grounded and must be replaced. Test each field coil in this manner.

Testing Brush Holders

Using the same test light, place one lead on the brush holder and the other on the field frame (Fig. 33-16). On insulated brushes, the lamp should not light. If it does, the insulation has broken down and must be replaced. On grounded brushes, the lamp should light. If it does not light, it is not properly grounded and must be removed, cleaned, and installed to restore electrical continuity. Brush holders should also be checked for damage and brush fit.

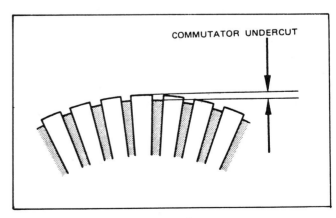

FIGURE 33-16. Appearance of properly undercut mica. *(Courtesy of Allis-Chalmers)*

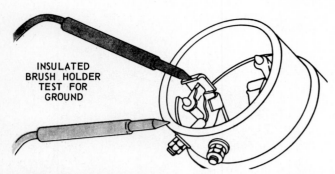

INSULATED
BRUSH HOLDER
TEST FOR
GROUND

FIGURE 33-17. Testing brush holder with test light. *(Courtesy of Deere and Company)*

Spring tension should be checked with a spring scale and, if below specifications, the brush holder and spring assembly should be replaced.

New brushes should be installed and positioned to ensure good brush-to-commutator contact across the width of the brush. Replace worn or rough bearings and bushings. Reassemble and install the generator as outlined earlier.

PART 5   SELF-CHECK

1. Name all the major components of a DC generator.

2. The DC generator operates on the principle of moving a _____through a _____to induce current.

3. The initial magnetic field in the DC generator is the result of residual magnetism. True or false?

4. What is the purpose of the commutator and brush assembly?

5. Explain the basic difference between A circuit and B circuit DC generators.

6. The DC regulator consists of *what* three major components?

7. What prevents battery current from reaching the generator when the engine is shut off?

8. What could cause battery overcharging?

9. List six causes of battery undercharging.

Chapter 34

Cranking Systems

Performance Objectives

After studying this chapter thoroughly and sufficient practical experience on adequate training models, and with the appropriate shop manuals, tools, and equipment, you should be able to do the following:

1. Complete the self-check questions with at least 80 percent accuracy.

2. Follow the accepted general precautions outlined in this chapter.

3. Describe the purpose, construction, and operation of the cranking system and each of its components.

4. Diagnose basic cranking system problems according to the diagnostic charts provided.

5. Safely remove, recondition, adjust, and replace any faulty cranking system components in accordance with manufacturer's specifications.

6. Properly performance test the reconditioned cranking system to determine the success of the repairs performed.

PART 1 ELECTRICAL CRANKING SYSTEM

The electrical cranking system is designed to change the electrical energy of the battery to mechanical energy by the cranking motor (starter). The system must crank the engine over at sufficient speed to allow the engine to run when the cylinders begin to fire. (See Fig. 34-1 and 34-2.)

The cranking system consists of the following units:

- storage battery and cables
- cranking motor and solenoid switch
- engine drive unit
- control circuit

Cranking systems are much the same in general design and operation. The motor consists of the drive mechanism, the frame, armature, brushes, and field windings. Some cranking motors also have a magnetically operated switch that closes and opens the circuit between the battery and cranking motor. On cranking motors that use overrunning clutch drive, the magnetically operated switch is referred to as a *solenoid switch*. In addition to closing and opening the circuit, the solenoid switch also shifts the drive pinion of the cranking motor into mesh with the teeth on the engine flywheel so that the engine can be cranked.

The *armature* is supported on bearings so that it can rotate freely. Current enters the motor through the field windings and then goes to the brushes, which ride the armature commutator. Current then passes through the armature windings, creating two strong magnetic fields. These fields oppose each other in such a way that the armature is forced to rotate.

Rotation of the armature causes the cranking motor drive pinion to rotate. When the drive pinion is meshed with the teeth of the flywheel ring gear, cranking of the engine takes place.

The cranking motor is designed to operate under great overload and to produce a high horsepower for its size. It can do this for a short period of time only, since to produce such power a high current must be used. If the cranking motor operation is continued for any length of time, heat will cause serious damage. For this reason, the cranking motor must never be used for more than 30 seconds at any one time, and cranking should not be repeated without a pause of at least 2 minutes to allow the heat to escape.

The *drive mechanism* is a vital part of the cranking motor, since it is through the drive that power is transmitted to the engine, cranking it as the cranking motor armature rotates. The drive mechanism has two functions. The first is to transmit the cranking torque to the engine flywheel when the cranking motor is operated and to disconnect the cranking motor from the flywheel after the engine has started. The second is to provide a gear reduction between the cranking motor and the engine so

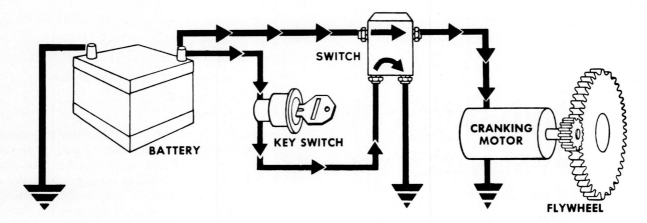

FIGURE 34-1. Cranking system schematic and gear reduction. *(Courtesy of General Motors Corporation)*

that there will be sufficient torque to turn the engine over at cranking speed.

If the cranking motor drive pinion remained with the flywheel ring gear at engine speeds above 1000 rpm and the pinion transmitted its rotation to the cranking motor armature, the armature would be spun at high speeds. Such speeds would cause the armature windings to be thrown from the armature slots and the segments to be thrown from the commutator. To avoid this, the drive mechanism must disengage the pinion from the flywheel ring gear as soon as the engine begins to operate.

Motor Principles

Magnetically, the cranking motor is made up of two parts, the armature and the field winding assembly. The armature consists of a number of low-resistance

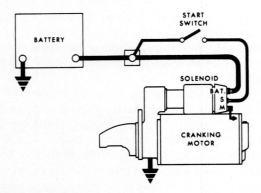

FIGURE 34-2. Basic electric control circuit for starter without relay-type solenoid. *(Courtesy of Mack Trucks Inc.)*

conductors placed in insulated slots around a soft iron core assembled onto the armature shaft. The commutator consists of a number of copper segments insulated from each other and from the armature shaft. The conductors are so connected to each other and to the commutator that electricity will flow through all the armature conductors when brushes are placed on the commutator and a source of current is connected to the brushes. This creates magnetic fields around each conductor. Current through the field windings creates a powerful magnetic field. Figure 34-3 shows the relationship between a magnetic field from a permanent magnet, a single current-carrying conductor, and the direction of the force that is exerted on the conductor. The magnetic lines of force pass from the north to the south pole as the arrows indicate. Current through the conductor is in the direction shown.

Using the right-hand rule, it can be seen that, when the direction of the current through the conductor is as shown, the magnetic lines of force around the conductor will be in a counterclockwise direction, as indicated by the circular arrow around the conductor. In Figure 34-4, when looking end on at the conductor it will be noted that to the right of the conductor the magnetic field from the permanent magnet and the circular magnetic field around the conductor oppose each other. To the left of the conductor they are in the same direction. When a current-carrying conductor is located in a magnetic field, the field of force is distorted, creating a strong field on one side of the conductor and a weak field on the opposite side. The conductor will be forced to move in the direction of the weak field; therefore, in this instance, the conductor will be pushed to the

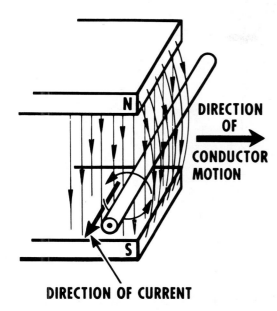

FIGURE 34-3. Magnetic field of magnet and magnetic field of current-carrying conductors are used in cranking motor operation.

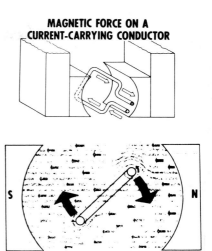

FIGURE 34-5. Action of magnetic fields causes looped conductor to rotate.

right. The more current through the conductor, the stronger is the force exerted on the conductors.

The application of this principle is shown in Figure 34-6, with a simple, one-turn armature electric motor. The magnetic field is created by means of current through the field coil windings, which are assembled around the two poles. The direction of current tends to increase the magnetic field between

the two poles. The U-shaped armature winding that is placed between the two poles is connected to a two-segment commutator. In the position shown, current is from the battery through the right-hand brush, through the right-hand segment of the commutator, and into the armature winding, where it flows first past the south pole and then, in returning, past the north pole then into the left-hand commutator segment and left-hand brush, through the north pole field winding, and back to the battery. The magnetic fields around the conductor will be in the two directions shown by the circular arrows. It can be seen that the left-hand side of the armature

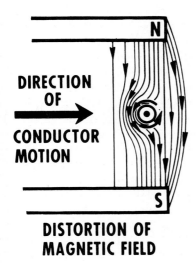

FIGURE 34-4. Strong magnetic field at left of current-carrying conductor forces conductor to move toward weak magnetic field at right of conductor.

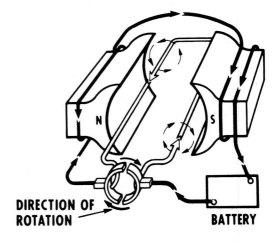

FIGURE 34-6. Armature loop and field windings are connected in series through brushes and commutator to create magnetic fields that cause armature to rotate.

winding will be pushed downward, thus imparting a clockwise rotation.

Since the armature winding and commutator are assembled together and must rotate together, movement of the winding causes the commutator also to turn. By the time the left-hand side of the winding has swung around toward the south pole, the commutator segments will have reversed their connections with respect to the brushes. Current is in the opposite direction with respect to the winding; since the winding has turned 180°, the force exerted on it still will tend to rotate it in a clockwise direction.

Cranking Motor Circuits

In cranking motors, the field windings and the armature normally are connected in such a way that all the current that enters the cranking motor passes through both field windings and the armature. In other words, the motor is series wound, meaning that the fields and armature are connected in series. All conductors are heavy ribbon-copper types that have a very low resistance and permit a high current. The more current, the higher is the power developed by the cranking motor. (See Fig. 34-7 and 34-8.)

Some cranking motors are four-pole units but have only two field windings, thus keeping resistance low. Notice the part of the current through this cranking motor. It will be noted that in operation the poles with field coil windings have a north polarity at the pole shoe face. The lines of force pass through the armature, enter the plain pole shoes, and pass through the frame and back to the original

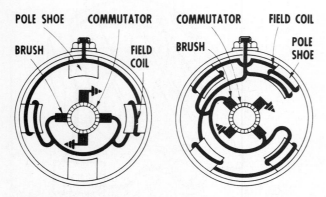

FIGURE 34-7. Schematic wiring diagram of two-field, four-pole, four-brush, series-wound cranking motor (at left). Schematic wiring diagram of four-brush, four-field, four-pole, series-wound cranking motor (at right). *(Courtesy of General Motors Corporation)*

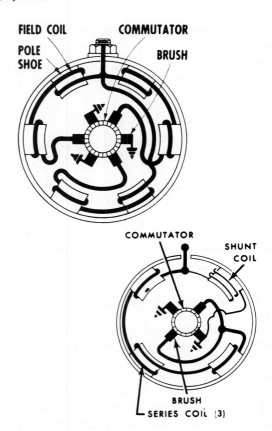

FIGURE 34-8. Schematic wiring diagram of a six-field, six-pole, six-brush, series-wound cranking motor (at left). Shunt coil controls excessive armature speed at light load, allowing heavier field coil winding for more torque (at right). *(Courtesy of General Motors Corporation)*

north shoe to complete the magnetic circle. In all cranking motors the adjacent pole shoes must be of opposite polarity, so that in a four-pole motor there is an N, S, N, S sequence around the frame.

Other cranking motors are four-pole, four-field winding, four-brush units. Here the field windings are paired off so that half the current is through one set of field windings to one of the insulated brushes, while the other half of the current is through the other set of field windings to the other insulated brush. With four-field coil windings, it is possible to create more ampere turns and consequently stronger magnetic fields, thus producing cranking motors with greater torque or cranking ability. By tracing the current from the terminal, it will again be noted that the poles alternate S, N, S, N, providing four magnetic paths through the armature core.

The cranking motor designed for heavy-duty service uses six poles and six brushes; 24V starters have twelve brushes. Here the current is split in three ways, one-third through each pair of field windings to one of the three insulated brushes.

Increasing the number of circuits through the cranking motor helps to keep the resistance low so that a high current and a high horsepower can be developed.

As a rule, all the insulated brushes are connected together by means of jumper leads or bars, so that the voltage is equalized at all brushes. Without these equalizing bars there may be conditions that cause arcing and burning of commutator bars, eventually insulating the brush contact and preventing cranking.

Cranking Motor Solenoid Circuits

The solenoid switch on a cranking motor not only closes the circuit between the battery and the cranking motor but also shifts the cranking motor pinion into mesh with the engine flywheel ring gear. This is accomplished by means of a linkage between the solenoid plunger and the shift lever on the cranking motor. Solenoids are energized directly from the battery through the switch or in conjunction with a solenoid relay. (See Fig. 34-9 to 34-13.)

When the circuit is completed to the solenoid, current from the battery is through two separate windings, designated as the *pull-in* and the *hold-in* windings. These windings produce a magnetic field that pulls in the plunger, so that the drive pinion is

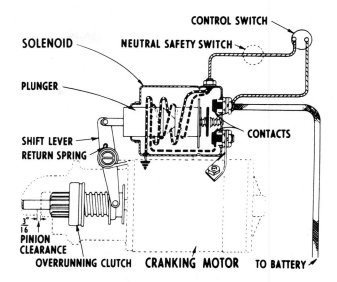

FIGURE 34-10. Shift mechanism on solenoid type of cranking motor. *(Courtesy of General Motors Corporation)*

shifted into mesh and the main contacts in the solenoid switch are closed, completing the cranking motor circuit.

Closing the main contacts in the solenoid switch at the same time shorts out the pull-in winding, since it is connected across the main contacts. The heavy current draw through the pull-in winding occurs only during the movement of the plunger.

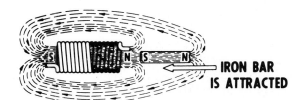

THE EFFECT ON THE FIELD OF FORCE CAUSED BY THE PRESENCE OF AN IRON BAR. THE BAR WILL BE DRAWN TO THE CENTER OF THE COIL.

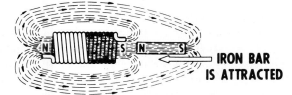

REVERSAL OF CURRENT FLOW CAUSES CHANGE OF POLARITY. THE BAR WILL BE DRAWN TO THE CENTER OF THE COIL WITH THE SAME FORCE.

FIGURE 34-9.

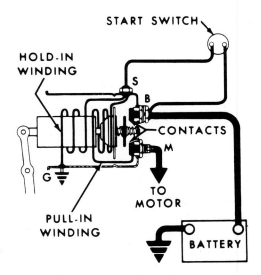

FIGURE 34-11. Solenoid wiring schematic showing pull-in and hold-in winding. Pull-in winding is shorted out when solenoid is engaged. *(Courtesy of General Motors Corporation)*

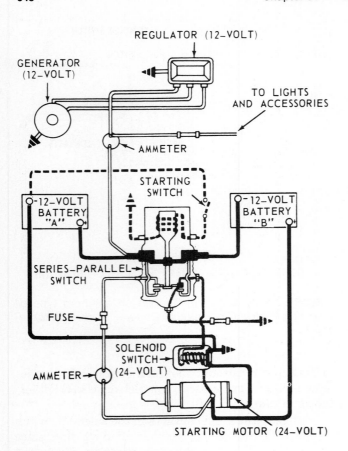

FIGURE 34-12. Series-parallel switch for a 24-volt starting circuit. The series connection between the two batteries and the starting motor is shown in solid line.

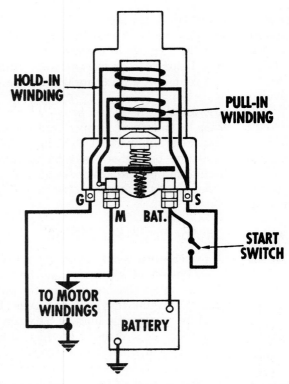

FIGURE 34-13. Solenoid switch circuit. *(Courtesy of Mack Trucks Inc.)*

When the control circuit is broken after the engine is started, current no longer reaches the hold-in winding. Tension of the return spring then causes the plunger to return to the at-rest position. Low system voltage or an open circuit in the hold-in winding will cause an oscillating action of the plunger. The pull-in winding has sufficient magnetic strength to close the main contacts, but when they are closed, the pull-in winding is shorted out, and there is no magnetic force to keep the contacts closed. Check for a complete circuit of the hold-in winding as well as the condition of the battery whenever chattering of the switch occurs.

Whenever a solenoid is replaced on a cranking motor, it is necessary to adjust the pinion travel, with the exact clearance and the method of adjustment varying between different motor designs.

Cranking Motor Drives

There are two common types of cranking motor drives in general use: the Bendix drive and the overrunning clutch.

Bendix Drive

The Bendix drive (Fig. 34-14 and 34-15) depends on inertia to provide meshing of the drive pinion with the engine flywheel ring gear.

The Bendix drive consists of a drive pinion, sleeve, spring, and spring-fastening screws. The drive pinion is normally unbalanced by a counterbalance on one side. It has screw threads on its inner bore. The Bendix sleeve, which is hollow, has screw threads cut on its outer diameter that match the screw threads of the pinion. The sleeve fits loosely on the armature shaft and is connected through the Bendix drive spring to the Bendix drive head, which is keyed to the armature shaft. Thus, the Bendix sleeve is free to turn on the armature shaft within the limits permitted by the flexing of the spring.

When the cranking motor switch is closed, the armature begins to revolve. The rotation is trans-

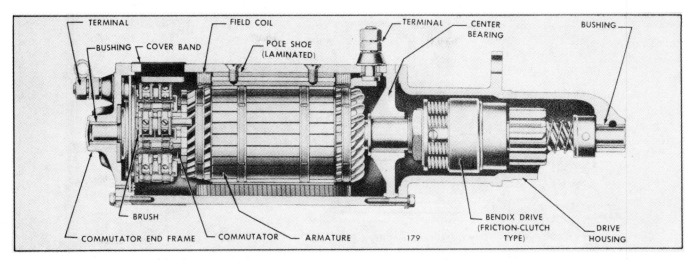

FIGURE 34-14. Starting motor with Bendix drive. *(Courtesy of Detroit Diesel Allison)*

mitted through the drive head and the spring to the sleeve, so that all these parts pick up speed with the armature. The pinion, however, being a loose fit on the sleeve screw thread, does not pick up speed along with the sleeve. In other words, the increased inertia of the drive pinion due to the effect of the counterbalance prevents it from rotating. The result is that the sleeve rotates within the pinion. This forces the drive pinion endwise along the armature shaft so that it goes into mesh with the flywheel teeth. As soon as the pinion reaches the pinion stop, it begins

to rotate along with the sleeve and armature. This rotation is transmitted to the flywheel. The Bendix spring takes up the shock of meshing.

When the engine begins to operate, it spins the pinion at a higher speed than that of the cranking motor armature. This causes the pinion to rotate relative to the sleeve, so that the pinion is driven back out of mesh from the flywheel teeth. Thus, the Bendix drive automatically meshes the pinion with the flywheel ring gear as soon as the engine starts. The spring-loaded antidrift pins prevent disengagement

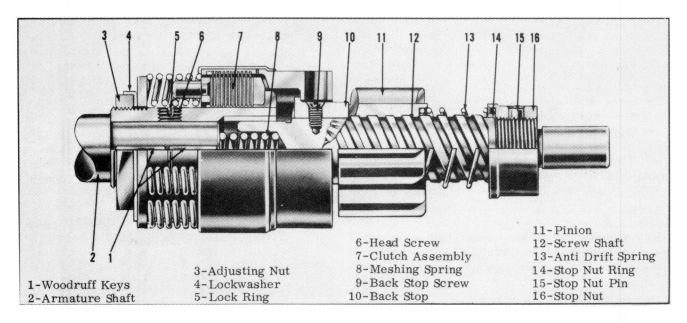

1-Woodruff Keys
2-Armature Shaft
3-Adjusting Nut
4-Lockwasher
5-Lock Ring
6-Head Screw
7-Clutch Assembly
8-Meshing Spring
9-Back Stop Screw
10-Back Stop
11-Pinion
12-Screw Shaft
13-Anti Drift Spring
14-Stop Nut Ring
15-Stop Nut Pin
16-Stop Nut

FIGURE 34-15. Bendix drive components. *(Courtesy of General Motors Terex)*

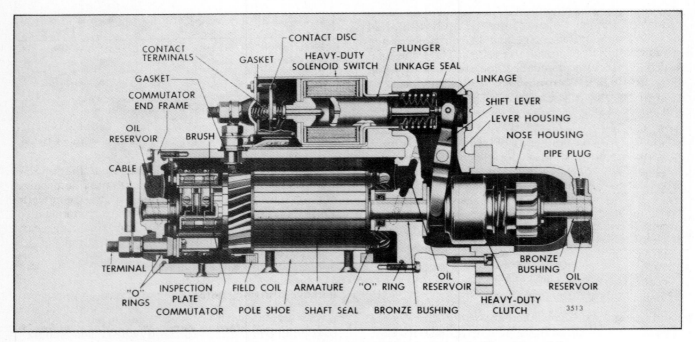

FIGURE 34-16. Starting motor with heavy-duty sprag clutch. *(Courtesy of Detroit Diesel Allison)*

until the engine speed reaches approximately 350 rpm.

Some heavy duty Bendix drives incorporate a friction clutch which is spring loaded and provides momentary slippage to cushion engagement.

Sprag Clutch

The sprag clutch is the device that has made the solenoid-actuated type of starter feasible. It is a roller-type or sprag-type clutch that transmits torque in only one direction, turning freely in the other. In this way, torque can be transmitted from the starting motor to the flywheel but not from the flywheel to the starting motor.

A typical sprag clutch is shown in Figure 34-17. The clutch housing is internally splined to the starting motor armature shaft. The drive pinion turns freely on the armature shaft within the clutch housing. When the clutch housing is driven by the armature, the spring-loaded rollers are forced into the small ends of their tapered slots and wedge tightly against the pinion barrel. This locks the pinion and clutch housing solidly together, permitting the pinion to turn the flywheel and thus crank the engine.

When the engine starts, the ring gear begins to drive the pinion faster than the starter motor because of the pinion-to-ring gear reduction ratio. This

action unloads and releases the clutch rollers, permitting the pinion to rotate freely around the armature shaft without stressing the starter motor.

The operator always should be careful not to reengage the cranking motor drive too soon after a false start. It is advisable to wait at least 5 seconds between attempts to crank. Burred teeth on the flywheel ring gear are an indication of attempted engagement while the engine is running.

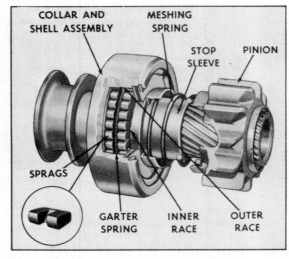

FIGURE 34-17. Heavy-duty sprag clutch. *(Courtesy of General Motors Terex)*

Dyer Drive

In this design, used on some larger diesel engines, the drive pinion meshes with the flywheel ring gear before the starter motor switch is closed and before the armature shaft starts to turn. (See Fig. 34–18.)

When the starter switch is engaged by the operator, the starter solenoid operates the shift lever to move the drive toward the flywheel ring gear. When the drive pinion is in mesh with the ring gear, it has moved against the pinion stop, which prevents further pinion movement along the armature shaft. At this point the starting motor switch contacts in the solenoid have closed and the armature shaft starts to turn and cranks the engine. As soon as the engine starts, the flywheel tries to turn the pinion faster than the armature shaft is turning. This causes the pinion and pinion guide to spin back and demesh. The pinion guide automatically locks the pinion in the disengaged position.

The dyer drive has been replaced by the newer Positork drive.

Positork Drive

This drive unit provides positive engagement and remains engaged until the electrical circuit is de-energized. The starter cannot be activated until the pinion is in mesh with the flywheel ring gear.

When the starter control circuit is energized, the solenoid is activated, causing the shift lever to move the drive toward the flywheel until the pinion is in mesh. When the pinion is fully meshed, the starting motor switch contacts in the solenoid assembly are closed, completing the electrical circuit to the starting motor. When the engine starts, the flywheel turns the pinion faster than the armature and drive parts, causing the drive collar teeth to demesh from the pinion teeth. A lockout device prevents these teeth from re-engaging as long as the pinion is meshed with the ring gear.

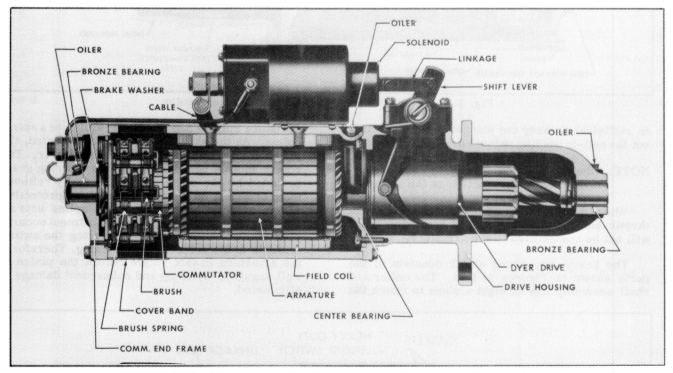

FIGURE 34–18. Starting motor with Dyer drive. *(Courtesy of General Motors Terex)*

PART 2   ELECTRIC CRANKING SYSTEM SERVICE

STARTING SYSTEM DIAGNOSTIC CHART

PROBLEM	POSSIBLE CAUSE	REMEDY
Engine will not crank and starting motor relay or solenoid does not engage	1. Battery discharged	1. Check battery and change or renew.
	2. Key start switch, safety start switch, relay or solenoid inoperative	2. Check circuitry and repair or renew faulty components.
	3. Starting circuit open or high resistance	3. Check circuit connections and repair or renew faulty wiring.
Engine will not crank but starting motor relay or solenoid engages	1. Engine seized	1. Check engine crankshaft free to turn.
	2. Battery discharged	2. Check battery and charge or renew.
	3. Defective starting motor connections or loose battery connections	3. Check, clean, and tighten connections.
	4. Starting motor faulty	4. Inspect, repair, or renew.
	5. Relay or solenoid contacts burned	5. Renew relay or solenoid.
Starting motor turns but does not crank engine	1. Defective starting motor drive assembly	1. Inspect and repair or renew.
	2. Defective solenoid or pinion engagement levers	2. Inspect and repair or renew.
	3. Defective flywheel ring gear	3. Inspect and renew.
Engine cranks slowly	1. Discharged battery	1. Check battery and charge or renew.
	2. Excessive resistance in starting circuit	2. Check circuit connections and repair or renew faulty wiring.
	3. Defective starting motor	3. Inspect and repair or renew.
	4. Tight engine	4. Investigate cause and effect repair.

General Precautions

When servicing the starter system the same general precautions should be followed as are outlined in Chapters 1 and 29. In addition, several other precautions should be observed.

• Always disconnect the battery ground cable before disconnecting wiring from the starting motor or removing the starting motor.

• Always have the unit properly positioned on a hoist or safely supported for any work underneath.

• For any cranking tests, make sure that transmission is in neutral or park with the parking brakes applied. Follow directions to prevent engine starting during cranking motor tests.

• Do not wash or immerse electrical components in solvent; clean with compressed air only.

In general, cranking system diagnosis and service includes checking the following:

1. engine mechanical condition

2. battery and battery cables

3. starting control circuit

4. cranking motor current draw

5. cranking motor removal, cleaning, inspection, testing, overhaul, and installation, or replacing motor with a new or rebuilt unit

Preliminary Inspection

Always begin with a quick visual check of the supply circuits to note any obvious trouble sources such as corroded or loose connections. The supply circuit consists of the battery, battery cables, clamps, and connectors.

Many slow-turning starters have been corrected by simply cleaning the battery terminal posts and cable clamps. Inspect starter and ground cables for corrosion or damage. In checking the supply circuit, always begin with a visual inspection of the battery post and cable clamps.

Test the battery to make sure that it is in good condition and has a minimum specific gravity reading of 1.220, temperature corrected, and see that the battery passes the high rate discharge test shown in the battery section.

Engine Won't Crank Properly

There are several possible causes for this condition. Assuming that the battery checks out and has been eliminated as a possible cause, either battery power is being prevented from reaching the starter motor or the motor is defective and must be repaired or replaced. First perform the resistance (voltage drop) tests.

Resistance (Voltage Drop) Tests

Excessive resistance in the circuit between the battery and starter motor will reduce cranking performance. The resistance can be checked by using a voltmeter to measure the voltage drop in the circuit while the starter motor is operated.

There are three checks to be made (Fig. 34–19):

1. The voltage drop between the positive battery terminal and the fastening device at the other end of the positive battery cable.

2. The drop between the unit frame and the starter motor field terminal (or the field frame if

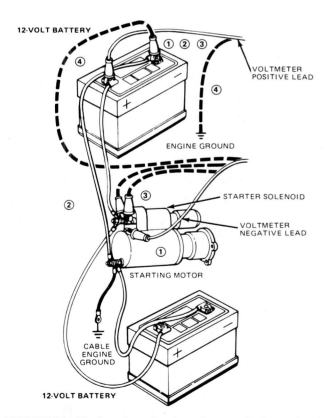

FIGURE 34–19. Starting circuit resistance (voltage drop) test connections. *(Courtesy of Ford Motor Company)*

there is no terminal). Disregard this test on starter motors used in 24 volt insulated systems.

3. The drop between the negative battery terminal and the fastening device at the other end of the negative battery cable.

Each of the checks should show no more than 0.1 volt drop with the starter motor cranking the engine.

If excessive voltage drop is found in any of these circuits, make correction by disconnecting the cables, cleaning the connections carefully, and then reconnecting the cables firmly in place. If this does not help, replace the cable. Broken strands in the cable which are not visible to the eye are probably creating excessive resistance.

Note: On some applications, extra long battery cables are required due to location of batteries and starter motor. This may result in somewhat higher voltage drops than the above recommended 0.1 volt. On such applications, the normal voltage should be established by checking several units. Then, when a voltage drop well above this normal figure is found, abnormal resistance will be indicated and correction can be made as already explained.

Starter Removal

1. Disconnect all cables and wiring coming into the starter motor.

2. Remove the capscrews which pass through the drive housing flange into the flywheel housing while supporting the starter.

3. Remove starter motor from engine.

Cranking Motor Tests

With the cranking motor removed from the engine, the armature should be checked for freedom of rotation by prying the pinion with a screwdriver. Tight bearings, a bent armature shaft, or a loose pole shoe screw will cause the armature to not turn freely. If the armature does not turn freely the motor should be disassembled immediately. However, if the armature does rotate freely, the motor should be given a no-load test before disassembly.

Free Speed Test

In the free speed test (Fig. 34–20), the starter motor is connected in series with a battery of the specified voltage and an ammeter capable of reading at least 700 amperes. A 1000 ampere capacity would be preferable. An rpm indicator also should be used to measure the armature revolutions per minute.

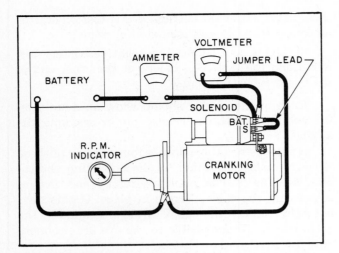

FIGURE 34-20. Test instruments connected to perform starter no-load test. *(Courtesy of Mack Trucks Inc.)*

Torque Test

The torque test requires such equipment as illustrated in Fig. 34-21. The starter motor is securely mounted and the brake arm hooked to the drive pinion. Then, when the specified voltage is applied, the torque can be computed from the reading on the scale. If the brake arm is one foot long as shown, the torque will be indicated directly on the scale in foot pounds. A high current carrying variable resistance should be used so that the specified voltage can be applied. Many torque testers indicate the developed pounds feet of torque on a dial.

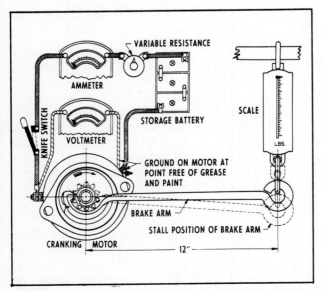

FIGURE 34-21. Starter torque test set-up. *(Courtesy of General Motors Terex)*

Warning: Stand clear of the apparatus when current is applied. A broken arm can result if the brake arm slips.

The specifications are normally given at low voltages so the torque and ammeter reading obtained will be within the range of the testing equipment available in the field.

Interpreting Results of Tests

1. Low free speed and high current draw with low developed torque may result from:

a. Tight, dirty or worn bearings, bent armature shaft or loose field pole screw which would allow armature to drag.

b. Shorted armature. Check further after armature is disassembled by revolving armature on growler with a steel strip such as a hacksaw blade held above it. The blade will vibrate above the area of the armature core in which the short circuit is located.

c. A grounded armature or field. Check by raising grounded brushes and insulating them from the commutator with cardboard, and then checking with a test lamp between the armature terminal and starter motor frame. If test lamp lights, there is a ground, so raise other brushes from commutator and check fields and commutator separately to determine whether it is the fields or armature that is grounded (See D Below). Grounds in the armature can be detected after disassembly by use of a test lamp and test points. If the lamp lights when one test point is placed on the commutator with the other point on the core or shaft, the armature is grounded.

d. Grounded field coil. Check after disassembly with a test lamp. If the starter has one or more coils normally connected to ground, the ground connection must be disconnected. Connect one lead of the test lamp to the field frame and the other lead to the field connector. If the lamp lights, at least one field coil is grounded which must be repaired or replaced.

2. Failure to operate with high current draw:
a. A direct ground in the switch, terminal, or fields.

b. Frozen shaft bearings which prevent armature from turning.

3. Failure to operate with no current draw:
a. Open field circuit. Connect test lamp leads to ends of field coils. If lamp does not light, the field coils are open and must be replaced.

b. Open armature coils. Inspect the commutator for badly burned bars. Running free speed, an open armature will show excessive arcing at the commutator bar which is open.

c. Broken or weakened brush springs, worn brushes, improperly seated brushes, high mica on the commutator, or other causes which would prevent good contact between the brushes and commutator. Any of these conditions will cause burned commutator bars.

4. Low no-speed with low torque and low current draw:

a. An open field winding. Raise and insulate ungrounded brushes from commutator and check fields with test lamp.

b. High internal resistance due to poor connections, defective leads, dirty commutator and causes listed under Step 3, above.

5. High free speed with low developed torque and high current draw indicates shorted fields. There is no easy way to detect shorted fields, since the field resistance is already low. If shorted fields are suspected, replace the fields and check for improvement in performance.

Solenoid Test

Proper operation of the solenoid switch depends on maintaining a definite balance between the magnetic strengths of the pull-in and hold-in windings. The two windings may be tested separately to check their efficiency. Test the pull-in winding by connecting between the solenoid switch terminal and the solenoid motor terminal with a source of variable voltage (battery and variable resistance in series), and an ammeter. Connect voltmeter between the same terminals. Adjust variable resistance to secure specified voltage and note current draw. Test hold-in coil in similar manner except making connections between solenoid switch terminal and other small solenoid terminal. If the solenoid does not come up to specifications and corrections cannot be made by wire brushing the control disc the solenoid should be replaced. (See Fig. 34–22.)

Magnetic Switch Test

Connect a test voltmeter between the switch control terminals and connect a source of variable voltage to these terminals. Where the magnetic switch has but one control terminal, make these connections between that terminal and the switch base. A battery and a variable resistance connected in series will be found to be a suitable source of variable voltage.

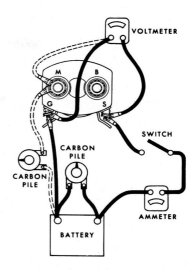

FIGURE 34-22. Checking solenoid hold-in and pull-in windings. *(Courtesy of Mack Trucks Inc.)*

Increase the voltage on the switch slowly and note the voltage required to close the contacts. Closing of the contacts is indicated by a click which can normally be heard. A more accurate way of checking the instant that the contacts close is to connect test lamp points between the two main switch terminals. The lamp will light as the contacts close.

With an ammeter connected in series with the magnetic switch windings, increase the voltage to the specified value and note the current draw. (See Fig. 34–23 and 34–24.)

Dyer Drive Check

There is one check on the Dyer drive mechanism which should be made while the starter motor is off

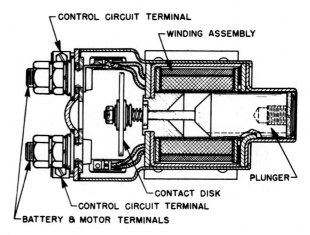

FIGURE 34-23. Sectional view of solenoid. *(Courtesy of General Motors Terex)*

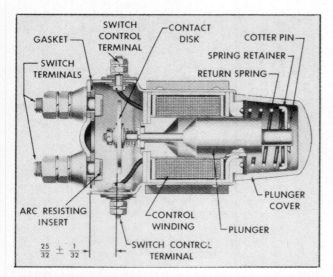

FIGURE 34-24. Sectional view of magnetic switch. *(Courtesy of General Motors Terex)*

the engine. This is the pinion travel against the spring with the motor in cranking position. Disconnect the lead between the solenoid and starter motor and connect a battery of the specified voltage to the two small solenoid terminals (or to the small terminal and the solenoid base if there is but one small terminal). If the plunger does not pull in, move it in by hand so the shift lever is in cranking position. Battery current will then maintain it in the operating position so that the pinion travel can be checked. It should be possible to push the pinion back against spring pressure 1/8 to 3/16 inch. Adjustment is made by turning the stud in the solenoid plunger in or out as required. A half turn of the stud allows 1/16 inch pinion travel adjustment. (See Fig. 34-25.)

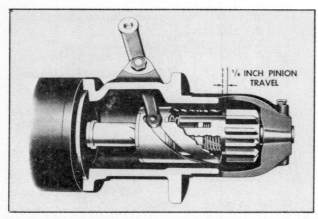

FIGURE 34-25. Checking Dyer drive pinion clearance. *(Courtesy of General Motors Terex)*

Bendix Drive Check

Hold the drive in your hands with the fingers extending over the front end of the clutch and squeeze it. By doing this, the meshing spring should be compressed inside the clutch housing. If the spring cannot be compressed, sufficient wear or damage has occurred to render the clutch inoperative and the clutch assembly should be replaced.

Note: Compress the meshing spring before removing the drive from the armature shaft. It is possible for the meshing spring to fall out once the drive assembly is removed. When this happens the check cannot be made.

Sprag Clutch Check

To check the pinion clearance, connect a battery from the solenoid switch terminal to the motor frame. Also to prevent motoring, connect a heavy jumper lead from the solenoid motor terminal to the motor frame. (See Fig. 34-26 to 34-28.)

With the solenoid energized and the clutch shifted toward the pinion, push the pinion back toward the commutator end as far as possible to take up any slack movement, then check the clearance between the pinion and housing. The clearance is adjusted by removing the plug on the lever housing and turning the nut on the plunger rod inside the housing. Turn the nut clockwise to decrease the clearance and counterclockwise to increase the clearance.

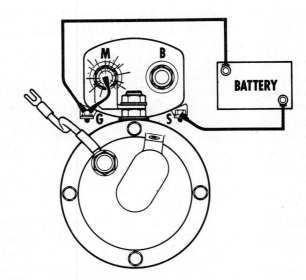

FIGURE 34-26. Hookup for checking pinion clearance on one type of sprag-type starter. *(Courtesy of Mack Trucks Inc.)*

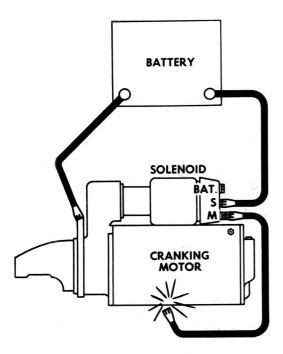

FIGURE 34-27. Hookup for another type of sprag drive starter for checking pinion clearance. *(Courtesy of Mack Trucks Inc.)*

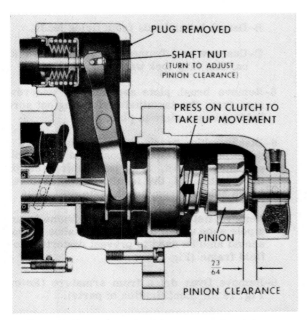

FIGURE 34-28. Checking sprag clutch pinion travel. *(Courtesy of General Motors Terex)*

A. Starter Drive Replacement (Fig. 34-29)

1. Clean the outside of the starter thoroughly.

2. Scribe a mark on the starter frame, drive end housing and lever housing for assembly.

3. Completely loosen the drive-end housing to lever housing screws, and separate the lever-end housing from the drive housing, drive assembly, and armature.

4. Loosen lever housing to frame screws and remove the drive gear and lever-housing assembly from the armature shaft and frame.

5. Separate the drive-gear assembly from the lever housing.

6. Position the new drive-gear assembly and engage the shift lever in the drive-assembly flange.

7. Position the lever housing on the armature shaft, align the scribe marks made during disassembly, and tighten the lever housing to the frame.

8. Position the drive-end housing assembly to the lever housing. Align the scribe marks made during disassembly, then adjust drive travel.

B. Brush Replacement

1. Remove the bolts securing the brush end plate, and rotate the end plate to expose the screws securing the field coil leads to the brush plate. Remove the screws.

2. Remove the brush end plate.

3. Remove the lever housing screws and remove the housing, drive and lever assembly.

4. Remove the starter armature. Inspect for an open or grounded circuit as in C below.

5. If the surface of the commutator is rough or more than 0.002 inch out of round, turn down the commutator. Do not undercut the mica. Polish with 00 or 000 sandpaper to remove all burrs left by the turning operation. Be sure that no copper particles remain on the insulation between the segments.

6. On 12-volt starter, remove the screws securing the brushes in the brush holders, and remove the eight brushes. On 24-volt starter, remove screws securing the 12 brushes in the brush holders.

7. Install new brushes in the brush holders. Tighten the screws securely. Be certain that the brush tension springs are functioning correctly.

8. Assemble the brush plate to the armature and then the armature to the frame.

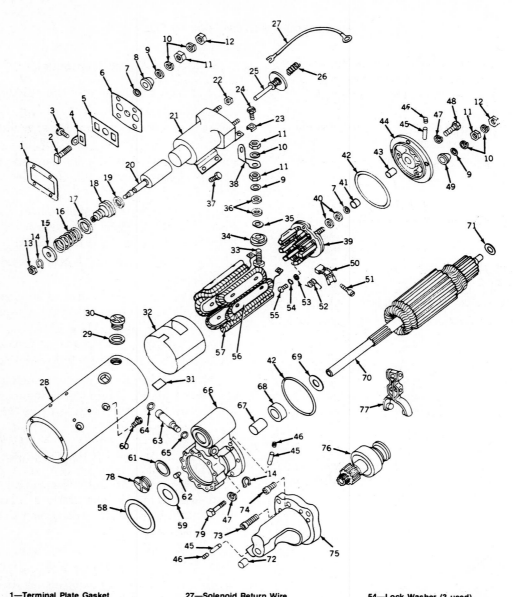

1—Terminal Plate Gasket
2—Terminal Stud (2 used)
3—Machine Screw
4—Solenoid Winding Terminal
5—Terminal Insulation
6—Terminal Plate
7—Packing (3 used)
8—Insulator (2 used)
9—Plain Washer (4 used)
10—Lock Washer (7 used)
11—Jam Nut (5 used)
12—Hex. Nut (3 used)
13—Self-Locking Nut
14—Snap Ring (2 used)
15—Spring Retainer
16—Solenoid Return Spring
17—Spring Retainer
18—Boot
19—Washer
20—Solenoid Plunger
21—Solenoid Winding and Case Assembly
22—Sealing Nut (4 used)
23—Lock Clip (2 used)
24—Machine Screw (2 used)
25—Contact Assembly
26—Contact Return Spring

27—Solenoid Return Wire
28—Field Frame
29—Gasket (4 used)
30—Brush Plug (3 used)
31—Field Coil Lead Insulator (2 used)
32—Field Coil and Brush Lead Insulator
33—Field Terminal Stud
34—Insulator
35—Special Packing
36—Insulating Washers (2 used)
37—Solenoid Mounting Screw (4 used)
38—Field Coil Connector
39—Brush Holder Assembly
40—Insulating Washers (2 used)
41—Insulator Bushing
42—O-Ring (2 used)
43—Commutator End Frame Bushing
44—Commutator End Frame
45—Lubricating Wick (3 used)
46—Plug (3 used)
47—Lock Washer (9 used)
48—Cap Screw (6 used)
49—Insulator
50—Brush (12 used)
51—Machine Screw (12 used)
52—Brush Spring (12 used)
53—Plain Washer (3 used)

54—Lock Washer (3 used)
55—Machine Screw (3 used)
56—Field Pole Shoe (6 used)
57—Field Coil Assembly
58—Gasket
59—Brake Washer
60—Pole Shoe Screw (12 used)
61—Adjusting Hole Plug Gasket
62—Seal Plug
63—Shift Lever Shaft
64—O-Ring
65—Small O-Ring
66—Lever Housing
67—Lever Housing Bushing
68—Oil Seal
69—Spacer Washer
70—Armature
71—Thrust Washer
72—Drive End Housing Bushing
73—Screw (5 used)
74—Special Screw
75—Drive End Housing
76—Sprag Clutch Assembly
77—Shift Lever
78—Adjusting Hole Plug
79—Cap Screw (7 used)

FIGURE 34-29. Exploded view of typical starter. *(Courtesy of Deere and Company)*

9. On a 12-volt starter, assemble the lever housing, drive and lever assembly, install the field coil attaching screws, and secure the field coil leads in place.

10. Install the bolts which attach the brush end plate to the starter frame.

11. Install the inspection plug and gasket.

C. Armature and Field Coil Replacement (Fig. 34–30 to 34–35)

1. Clean the outside of the starter thoroughly.

2. Scribe the starter frame, drive end, and brush end plate to ensure correct assembly.

3. Disassemble the starter as outlined above.

4. Wipe the field coils, armature, commutator and armature shaft with a clean cloth. Wash the springs, shims, thrust washer, trip collar, locking collar, and brush end plate assembly in solvent and dry the parts.

5. Inspect the armature for broken or burned insulation and unsoldered connections. Inspect the commutator surface for grooves, and check for run-out.

6. If the surface of the commutator is rough or more than 0.051 mm (0.002 inch) out of round, turn it down. Do not undercut the mica.

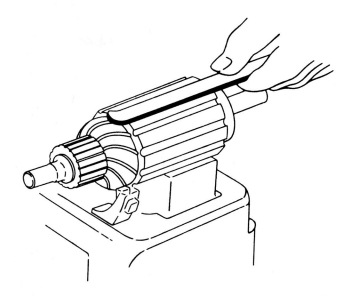

FIGURE 34-31. Checking the armature for short circuits. *(Courtesy of Deere and Company)*

7. Check the armature windings with a growler tester.

8. With the starter disassembled, check the field windings with a growler light for shorts to the housing or pole shoes. Connect one growler light lead to the starter housing; connect the other growler lead to each of the field coil brush leads in turn. A short circuit will light the growler light. Replace damaged coils.

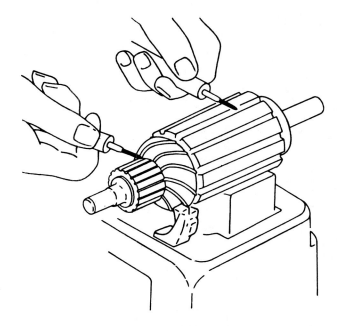

FIGURE 34-30. Checking the armature for grounds. *(Courtesy of Deere and Company)*

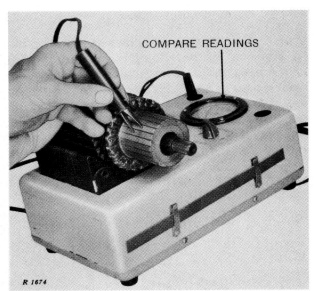

COMPARE READINGS

R 1674

FIGURE 34-32. Testing the armature for open circuits. *(Courtesy of Deere and Company)*

GROUNDED CIRCUIT TEST

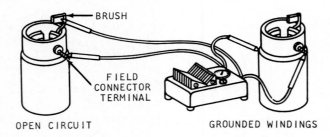

FIGURE 34-33. Checking the field coils for opens and grounds. *(Courtesy of Deere and Company)*

FIGURE 34-34. Installing snap ring on armature shaft. *(Courtesy of Mack Trucks Inc.)*

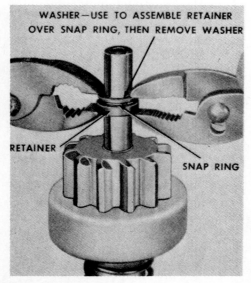

FIGURE 34-35. Installing retainer over snap ring. *(Courtesy of Mack Trucks Inc.)*

9. Examine the starter bushings for excessive wear. Replace worn or badly scored bushings.

10. Examine the contacts on the solenoid assembly, and if necessary, clean them with very fine sandpaper or crocus cloth.

11. The solenoid windings can be tested for internal shorts or open circuits with an ohmmeter. Connect the instrument across the windings and note the reading. Resistance of the pull-in winding and the hold-in winding should be as specified. A low reading will generally indicate an internal short circuit, while no reading at all indicates an open circuit.

If the solenoid contacts are badly burned, or the solenoid winding is shorted replace the solenoid assembly, then adjust drive travel.

12. Inspect the pinion teeth for excessive wear or damage from improper engagement. If a new drive is to be installed, be sure it has the same number of teeth as the replaced component. Install the new drive on the armature shaft and check for free movement. If necessary, the pinion and armature shaft can be lightly lapped together, then adjust drive travel.

13. Check the tension of the brush springs for conformance to specifications.

14. Check the brushes for free movement in the brush holders. Inspect the brushes for excessive wear. Where brush leads are insulated, be sure that the insulation is not burned or worn.

15. The brushes should be well seated to the commutator; that is, contacting the commutator over at least 60 percent of the brush contact area. If not, fit the brushes to the commutator with fine sandpaper until the desired seating area is obtained. Do only one brush at a time. Clean the brushes and commutator with compressed air. Be sure that no abrasive particles from the sandpaper are embedded in the contact area of the brushes.

New brushes are not ground to fit the commutator; therefore, they must be seated as described.

Mark each brush and brush holder with which it was seated to permit assembly in the same position.

Check the tension of each brush spring at the point of contact with the brush. Use a spring scale hooked under the brush spring lip. Brush spring tension must be within specifications.

16. Assemble the starter as outlined in A and B above.

Engagement Mechanism Check

1. Energize the solenoid with the proper voltage battery. Connect the battery between the solenoid terminal and the ground terminal; the pinion should move forward. Adjust clearance between pinion and pinion housing to specifications.

2. Disconnect the battery at the solenoid; the pinion must return to its normal position in one sharp movement.

PART 3   AIR CRANKING SYSTEM

The air-cranking motor consists of a five-vaned air motor with gear reduction, which drives the engine to be started through a conventional Bendix-type drive. (See Fig. 34-36 and 34-37.)

An air-starting reservoir provides air for the cranking motor only. The connection to the cranking motor is through flexible hose with a quick-acting control valve to permit operation of the motor.

Air from the compressor, which is adjusted to deliver 95 to 120 psi maximum pressure, flows through a check valve to the reservoir.

A trailer coupling connection or glad-hand is provided at the reservoir drain connection to permit charging of the reservoir from an external source, such as another vehicle or from the service shop air supply.

The energy source for an air-starting system is compressed air, which is usually stored in a separate receiver tank. The air starter has a rotor that is located eccentrically in a larger diameter bore and usually fitted with five vanes that can slide radially in and out within slots in the rotor. Between the

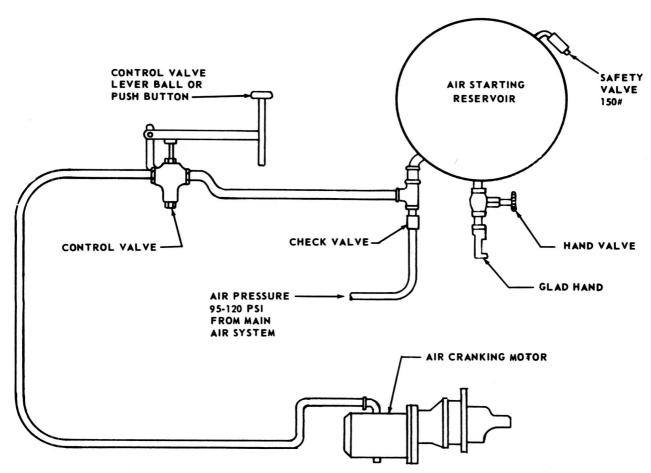

FIGURE 34-36. Typical air-starting system components and layout. *(Courtesy of Mack Trucks Inc.)*

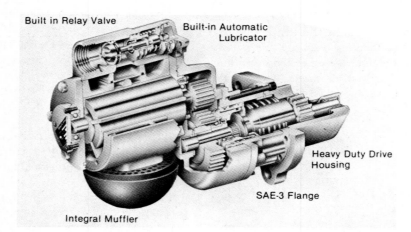

Built in Relay Valve   Built-in Automatic
                                      Lubricator

                                      Heavy Duty Drive
                                      Housing

                              SAE-3 Flange

Integral Muffler

FIGURE 34-37. Cutaway view of Startmaster 250 air-starting motor. *(Courtesy of Mack Trucks Inc.)*

starter and the compressed air tank is a control valve that holds the air in a ready condition. When the valve is opened, the compressor air is released, and the resultant force on the blades causes the rotation of the rotor—much like a paddle wheel.

In an inertia air-starting system, air is introduced into the air receiver tank by a one-way check valve from the air brake system. The compressed air within the receiver tank also serves as energy for the servo control. When the pushbutton is activated, it sends a servo signal to the main relay valve, which then allows a high-volume air flow to pass on to the starter and crank the engine. Some starters are equipped with a device that is mounted on the inlet of the starter, which automatically injects a measured quantity of lubricant into the air stream so that the moving parts within the starter are adequately lubricated during operation. Also to prevent an extremely loud discharging sound—made by exhaust air—from escaping into the environment, a muffler is used on most starters to bring the overall sound level of the air-starting system down near the 80 decibel range.

Today most starting systems used on over-the-road vehicular applications employ a preengaged cranking motor. A preengaged air-starting system is almost identical to the inertia-type system, except that it utilizes a starter that receives a servo signal from the push-button whereupon an internal actuator engages the starter pinion with the engine ring gear. If, and only if, the starter pinion is meshed with the ring gear will the servo signal come out of the starter and onto the relay valve. The life of the engine ring gear has been appreciably extended by the use of the pre-engaged type starters.

PART 4   AIR CRANKING SYSTEM SERVICE

The chart (page 663) lists likely problems and their possible causes and corrections is typical for air-operated starting systems.

PART 5   HYDRAULIC CRANKING SYSTEM

The hydrostarter system is a complete hydraulic system for cranking internal combustion engines. The system is automatically recharged after each engine start and can be manually recharged in an emergency. The starting potential does not deteriorate during long periods of inactivity; continuous exposure to hot or cold climates has no detrimental effect upon the hydrostarter system. Also, the hydrostarter torque for a given pressure remains substantially the same regardless of the ambient temperature. (See Fig. 34-38 and 34-39.)

The hydrostarter system consists of a reservoir, an engine-driven charging pump, a manually operated pump, a piston-type accumulator, a starting motor, and connecting hoses and fittings.

Operation

Hydraulic fluid flows by gravity or slight vacuum from the reservoir to either the engine-driven pump inlet or hand pump inlet. The hand pump is used only to supply the initial charge or to recharge the

AIR STARTER TROUBLESHOOTING CHART

PROBLEM	CAUSE	CORRECTION
Starting motor not operating	1. Low air pressure	1. Check air pressure in system for 95 to 120 psi. Correct as required.
	2. Inoperative or defective starting control valve	2. Check operation of starting control valve. Repair if needed.
	3. Loose starter	3. Check mounting of starter to flywheel housing. Correct if needed.
	4. Loose or leaking air line	4. Check for air leaks between starting control valve and starter when starting control valve is operated. Correct as necessary.
	5. Seized starter	5. Remove starter from flywheel and check for rotation. Repair or replace as required.
Slow starting motor speed	1. Low air pressure	1. Check air pressure in system for 95 to 120 psi. Correct as needed.
	2. Defective starting control valve	2. Check operation of starting control valve. Repair as needed.
	3. Loose starter	3. Check mounting of starter to flywheel housing. Correct if necesary.
	4. Restricted or loose air line	4. Check for dented, kinked, restricted, or loose air line and connections. Repair as required.
	5. Dirty air cleaner (when used)	5. Remove and clean air cleaner.
	6. Improperly or overlubricated starter	6. Remove starter, disassemble, clean, lubricate, reassemble, and reinstall on flywheel housing.
Engine not turning over, starting motor operating	1. Defective motor drive	1. Check for broken or stripped drive pinion or flywheel ring gear or broken drive spring. Repair as necessary.

system after servicing or overhaul. Fluid discharging from either pump outlet at high pressure flows into the accumulator and is stored at 3250 psi (22,408 kPa) under the pressure of compressed nitrogen gas. When the starter is engaged with the engine flywheel ring gear and the control valve is opened, high-pressure fluid is forced out of the accumulator (by the expanding nitrogen gas) and flows into the starting motor, which rapidly accelerates the engine to a high cranking speed. The used fluid returns from the starter directly to the reservoir.

The engine-driven hydrostarter charging pump runs continuously during engine operation, recharging the accumulator with fluid. When the proper amount of fluid has been returned to the accumulator, the pressure-operated unloading valve in the engine-driven pump opens and returns the pump discharge directly to the reservoir.

Reservoir

The reservoir is a cylindrical steel tank with a fine mesh screen at the outlet. The filler cap contains a filter to prevent dust and dirt from entering the reservoir.

Engine-Driven Charging Pump

The engine-driven charging pump is a single-piston positive displacement type and should run at approximately engine speed. It contains ball check valves and an unloading valve operated by the accumulator pressure. Its operation is entirely automatic and will operate in either direction of rotation.

Hand Pump

The hand pump is a single-piston, double-acting, positive displacement type. Flow through the pump is controlled by ball check valves. A manually operated relief valve is provided in this pump so that the accumulator pressure may be relieved when servicing of any components is required.

Accumulator

The piston-type accumulator is precharged with nitrogen through a small valve. A seal ring between the piston and the shell prevents the loss of gas into the hydraulic system. The accumulator is supplied with the proper precharge.

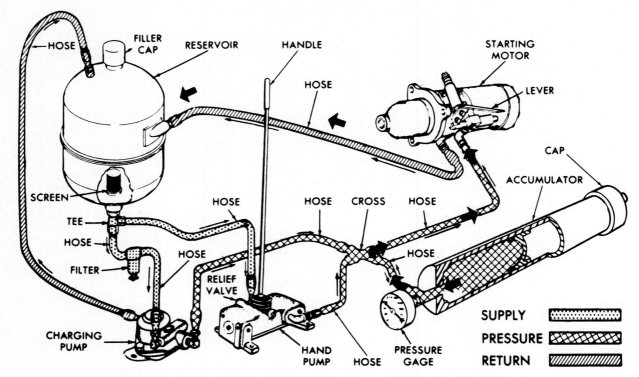

FIGURE 34-38. Schematic diagram of hydrostarter system, showing oil flows. *(Courtesy of Detroit Diesel Allison)*

SUPPLY

PRESSURE

RETURN

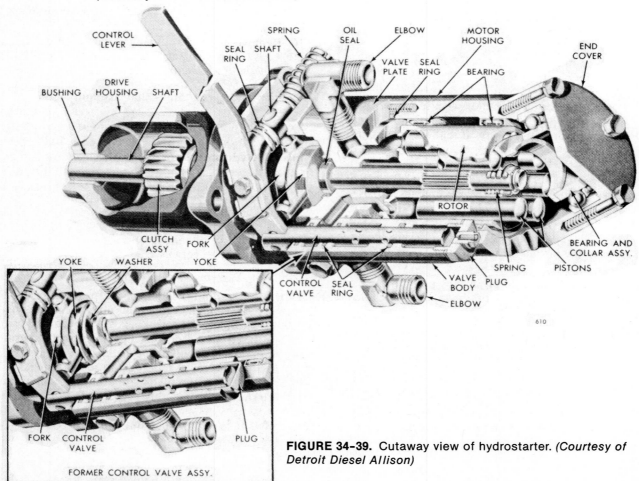

FIGURE 34-39. Cutaway view of hydrostarter. *(Courtesy of Detroit Diesel Allison)*

Starter

The starter mounts on the flywheel housing and has a pinion gear with an overrunning clutch for engaging the flywheel ring gear. Movement of the starter control lever engages the pinion and opens the control valve in the proper sequence. The motor is a multipiston, swash plate type. Provision is made so that if pinion tooth abutment occurs, the motor rotates slowly until the pinion snaps into full engagement. When the control lever is released, the pinion is disengaged and the valve is closed by spring action.

PART 6   HYDRAULIC CRANKING SYSTEM SERVICE

The following diagnostic charts and procedures are typical of hydraulic starting systems and are provided courtesy of Detroit Diesel Allison. They apply to the Detroit Diesel engine hydrostarter systems. Other systems are similar and use the same basic principles.

Chart 1

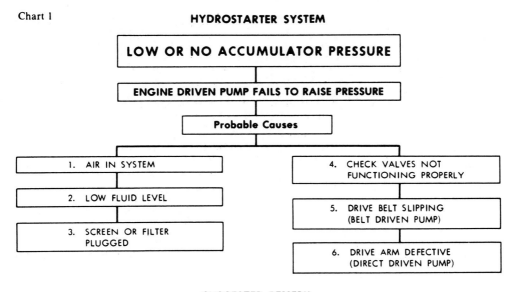

HYDROSTARTER SYSTEM

LOW OR NO ACCUMULATOR PRESSURE

ENGINE DRIVEN PUMP FAILS TO RAISE PRESSURE

Probable Causes

1. AIR IN SYSTEM
2. LOW FLUID LEVEL
3. SCREEN OR FILTER PLUGGED
4. CHECK VALVES NOT FUNCTIONING PROPERLY
5. DRIVE BELT SLIPPING (BELT DRIVEN PUMP)
6. DRIVE ARM DEFECTIVE (DIRECT DRIVEN PUMP)

─────────────── SUGGESTED REMEDY ───────────────

1. To purge the engine driven pump of air:

   a. Operate the engine at maximum no-load engine speed.

   b. Break the hose connection at the discharge side of the engine driven pump until a full stream of oil is discharged from the pump.

   c. Connect the hose to the pump and alternately loosen and tighten the swivel fitting on the discharge hose until oil leaking out, when the fitting is loose, appears free of air bubbles.

   d. Tighten the fitting securely and observe the pressure gage. The pressure must rise rapidly to the accumulator precharge pressure (1250 psi at 70°F.) then increase slowly to 2900 to 3300 psi in 6 to 10 minutes, depending upon the size of the particular accumulator. If the accumulator pressure does not rise, make certain that the hand pump relief valve is closed after the pressure is released and repeat the above purging procedure.

2. The fluid level in the reservoir must be sufficient to completely cover the screen at the bottom of the tank after the accumulator is charged and the engine driven pump is by-passing a full stream of fluid to the reservoir.

3. Remove and clean the reservoir screen and flush out the reservoir tank. Also clean the filter located in the supply hose between the reservoir and the engine driven pump.

4. Open the relief valve on the side of the hand pump, while the engine is running, to permit the engine driven pump to wash the check valves free from particles.

If the accumulator can be charged with the hand pump but not with the engine driven pump, then a check valve in the engine pump is defective. Replace the faulty check valve assembly.

5. Adjust or replace the drive belt if necessary.

6. Replace the pump drive arm.

Courtesy of Detroit Diesel Allison

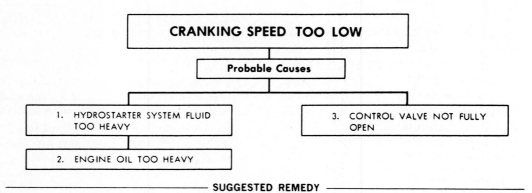

CRANKING SPEED TOO LOW

Probable Causes

1. HYDROSTARTER SYSTEM FLUID TOO HEAVY

2. ENGINE OIL TOO HEAVY

3. CONTROL VALVE NOT FULLY OPEN

———— SUGGESTED REMEDY ————

1. Check the mixture of fluid in the system. Use fluid consisting of 75% diesel fuel and 25% SAE 10 or 30 lubricating oil.

2. Replace the oil with the proper viscosity grade. Refer to the *Engine Lubrication Oil Specifications* in the *2-71 Engine Operators Manual.*

3. Check the travel of the control valve located on the side of the starter. Minimum travel is 1-1/16". Remove any obstruction preventing sufficient control valve or control lever handle travel.

Chart 3

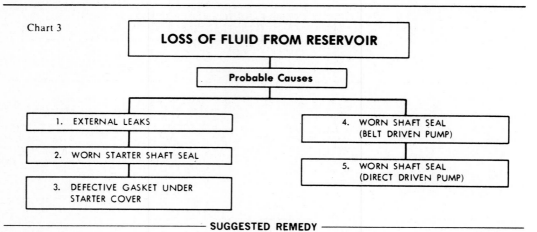

LOSS OF FLUID FROM RESERVOIR

Probable Causes

1. EXTERNAL LEAKS

2. WORN STARTER SHAFT SEAL

3. DEFECTIVE GASKET UNDER STARTER COVER

4. WORN SHAFT SEAL (BELT DRIVEN PUMP)

5. WORN SHAFT SEAL (DIRECT DRIVEN PUMP)

———— SUGGESTED REMEDY ————

1. With pressure in the system, check all hoses and fittings for leaks. Tighten or replace the fittings and any defective parts.

2. Remove the starter after releasing the system pressure and observe the inside of the clutch housing. If evidence of system fluid is found, replace the shaft seal.

3. Operate the starter. During the cranking cycle, watch closely for fluid leaking around the cover or any of the retaining bolts.

4. While the pump is by-passing at full system pressure, examine the shaft for evidence of leaks. Replace the seal if necessary.

5. After the pump has been by-passing at full system pressure, remove the pump from the flywheel housing and examine the back of the mounting plate near the seal for evidence of leaks. Replace the shaft seal if necessary.

Courtesy of Detroit Diesel Allison

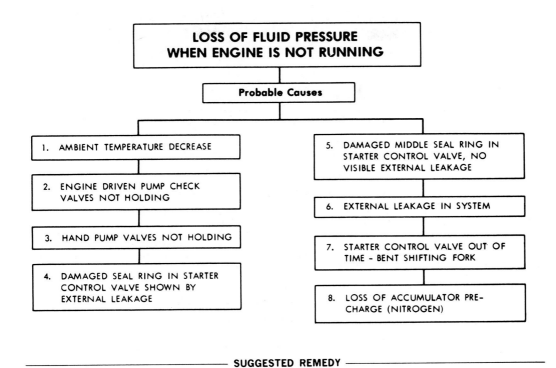

LOSS OF FLUID PRESSURE WHEN ENGINE IS NOT RUNNING

Probable Causes

1. AMBIENT TEMPERATURE DECREASE

2. ENGINE DRIVEN PUMP CHECK VALVES NOT HOLDING

3. HAND PUMP VALVES NOT HOLDING

4. DAMAGED SEAL RING IN STARTER CONTROL VALVE SHOWN BY EXTERNAL LEAKAGE

5. DAMAGED MIDDLE SEAL RING IN STARTER CONTROL VALVE, NO VISIBLE EXTERNAL LEAKAGE

6. EXTERNAL LEAKAGE IN SYSTEM

7. STARTER CONTROL VALVE OUT OF TIME - BENT SHIFTING FORK

8. LOSS OF ACCUMULATOR PRE-CHARGE (NITROGEN)

SUGGESTED REMEDY

---

1. A drop in temperature will decrease the nitrogen pressure. Adjust the pressure as needed for cranking requirements by use of the hand pump.

2. Disconnect the return hose and inlet hose from the engine driven pump. Leakage from the inlet fitting means that both check valves are defective. Leakage at the return fitting only means that the outlet check valve is defective. Replace the defective check valve assembly(s).

3. Disconnect the inlet hose from the hand pump. Leakage from the inlet fitting means that either the relief valve alone or both the inlet and outlet check valves are defective. Stone and clean the ball seats in the pump body and replace the balls and springs if necessary.

4. Remove the control valve from the starter and replace the seal ring.

5. Disconnect the return hose from the starter. Use the hand pump to raise the pressure if necessary. If fluid leaks from the return fitting when the control valve is closed, the middle seal ring is damaged. Remove the control valve and replace the seal ring.

6. Examine all hoses and fittings for leaks. Tighten or replace the fittings and any defective parts.

7. With the control valve closed, check the length of the piston protruding beyond the valve body. The correct length is 7/8" ± 1/32". If the length is incorrect, the shifting fork may be bent or the nylon yoke between the fork and the clutch collar may be damaged. Replace the faulty parts.

8. See Chart 7.

Courtesy of Detroit Diesel Allison

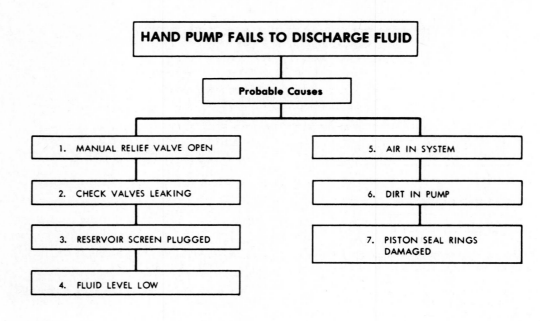

SUGGESTED REMEDY

1. Close the relief valve.

2. If caused by dirt, open the relief valve and operate the hand pump slowly for a few minutes to wash the particles out of the check valves. If this is unsuccessful, stone and clean the ball seats in the pump body and replace the balls and springs if necessary.

3. Remove and clean the reservoir screen, flush the reservoir tank and reassemble.

4. See Chart 1, Item 2.

5. To purge the hand pump of air:

a. Relieve any system pressure, then disconnect the outlet hose from the hand pump.

b. Close the manual relief valve and operate the pump until fluid is discharged when stroking in both directions.

c. Reconnect the outlet hose.

6. See Item 2.

7. Replace the seal rings.

Courtesy of Detroit Diesel Allison

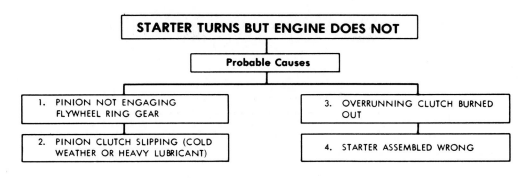

STARTER TURNS BUT ENGINE DOES NOT

Probable Causes

1. PINION NOT ENGAGING FLYWHEEL RING GEAR

2. PINION CLUTCH SLIPPING (COLD WEATHER OR HEAVY LUBRICANT)

3. OVERRUNNING CLUTCH BURNED OUT

4. STARTER ASSEMBLED WRONG

─────── SUGGESTED REMEDY ───────

1. Check the shifting fork. If the fork is bent, replace it.

2. Wash out the heavy lubricating oil and replace it with SAE 5W or SAE 10 oil.

3. Replace the clutch. If a mechanical linkage is attached to the control lever, add sufficient spring force to assure that the clutch is withdrawn from engagement, and that the control valve is returned to the shut-off position. If no mechanical linkage is used, disengage the starter as soon as the engine starts. Prolonging the period during which the clutch overruns will reduce clutch life.

4. The starter may be assembled for L.H. rotation but with a R.H. overrunning clutch. Remove the starter and assemble it correctly.

Chart 7

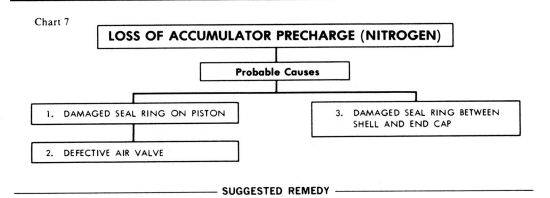

LOSS OF ACCUMULATOR PRECHARGE (NITROGEN)

Probable Causes

1. DAMAGED SEAL RING ON PISTON

2. DEFECTIVE AIR VALVE

3. DAMAGED SEAL RING BETWEEN SHELL AND END CAP

─────── SUGGESTED REMEDY ───────

1 With some nitrogen precharge but no fluid pressure in the system, bubbles and foaming in the reservoir indicate that the nitrogen is leaking past the seal ring on the accumulator piston. Overhaul the accumulator.

2. Release the pressure in the system by opening the relief valve on the side of the hand pump. Then, loosen the hex lock nut on the nitrogen valve approximately 3/4 turn to release the remaining precharge before attempting to remove the valve from the accumulator. Replace the air valve.

3. Apply light oil on the threaded end of the accumulator at the end of the cap. Bubbling of the oil indicates a leak past the end cap seal. Release the nitrogen precharge before removing the cap to replace the seals.

Courtesy of Detroit Diesel Allison

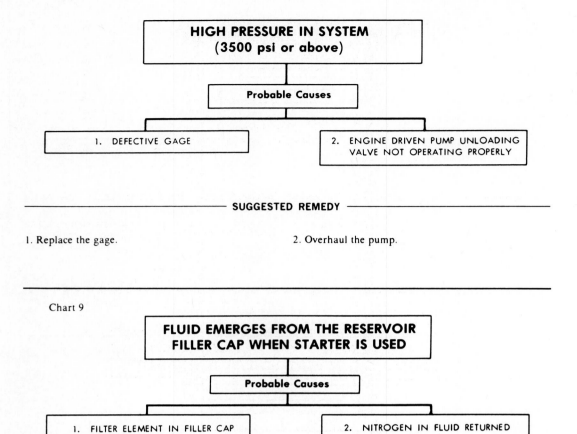

**HIGH PRESSURE IN SYSTEM
(3500 psi or above)**

Probable Causes

1. DEFECTIVE GAGE

2. ENGINE DRIVEN PUMP UNLOADING VALVE NOT OPERATING PROPERLY

───────────── SUGGESTED REMEDY ─────────────

1. Replace the gage.   2. Overhaul the pump.

Chart 9

**FLUID EMERGES FROM THE RESERVOIR
FILLER CAP WHEN STARTER IS USED**

Probable Causes

1. FILTER ELEMENT IN FILLER CAP LOADED WITH DIRT

2. NITROGEN IN FLUID RETURNED TO RESERVOIR

3. EXCESS FLUID IN RESERVOIR

───────────── SUGGESTED REMEDY ─────────────

1. Rinse the filler cap thoroughly in fuel oil and dry it with compressed air.

2. Overhaul the accumulator. See Chart 7, Item 1.

3. Check the fluid level after the accumulator is charged and the engine driven pump is by-passing a full stream of oil to the reservoir. The fluid level must be sufficient to completely cover the screen in the bottom of the tank.

Courtesy of Detroit Diesel Allison

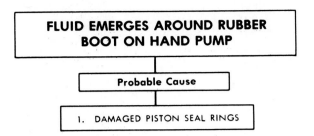

FLUID EMERGES AROUND RUBBER BOOT ON HAND PUMP

Probable Cause

1. DAMAGED PISTON SEAL RINGS

———————— SUGGESTED REMEDY ————————

1. Replace the seal rings and leather back-up rings on the pump piston.

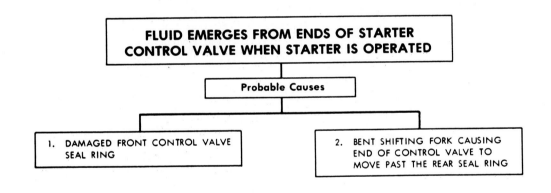

FLUID EMERGES FROM ENDS OF STARTER CONTROL VALVE WHEN STARTER IS OPERATED

Probable Causes

1. DAMAGED FRONT CONTROL VALVE SEAL RING

2. BENT SHIFTING FORK CAUSING END OF CONTROL VALVE TO MOVE PAST THE REAR SEAL RING

———————— SUGGESTED REMEDY ————————

1. Operate the starter. If fluid emerges around the front end of the control valve, the seal ring is damaged.

2. See Chart 4, Item 7. Also operate the starter. If fluid emerges from the cap on the rear of the control valve, the fork is bent and the seal ring may be damaged.

Courtesy of Detroit Diesel Allison

PART 7   SELF-CHECK

1. A solenoid switch is only used with a Bendix drive.   True or False

2. An overrunning clutch only provides for demeshing of the drive pinion.   True or False

3. Increasing the number of circuits in a cranking motor increases the resistance.   True or False

4. The ideal battery cable should be short and of heavy gauge.   True or False

5. Excessive heat is generated when a cranking motor is operated more than 30 seconds.   True or False

6. The armature core is made of iron to complete the magnetic circuit.   True or False

7. Cranking motor test specifications are given at high voltage to show errors more clearly.   True or False

8. The overrunning clutch uses a shift lever to actuate the drive pinion.   True or False

9. The armature windings and commutator rotate together.   True or False

10. All current through the field coils also goes through the armature.   True or False

11. A high-voltage drop is desirable in the cranking circuit.   True or False

12. Name the five major components of the hydraulic cranking system.

13. Fluid under high pressure is stored in the _____.

14. A continuously running (while engine is running) hydraulic pump is used to charge the accumulator.   True or False

15. For what is the hand-operated hydraulic pump used?

16. List four causes for low accumulator pressure.

17. Name three causes for slow hydraulic cranking motor cranking speed.

18. Name the four major components of the air-cranking system and state the purpose of each.

19. Cranking air is exhausted to atmosphere through a muffler.   True or False

20. List five causes of slow air cranking motor speed and the correction for each.

Appendix

Reference Charts

ENGLISH METRIC CONVERSION

Description	Multiply	By	For Metric Equivalent
ACCELERATION	Foot/sec²	0.304 8	metre/sec² (m/s²)
	Inch/sec²	0.025 4	metre/sec²
TORQUE	Pound-inch	0.112 98	newton-metres (N·m)
	Pound-foot	1.355 8	newton-metres
POWER	horsepower	0.746	kilowatts (kw)
PRESSURE or STRESS	inches of water	0.2488	kilopascals (kPa)
	pounds/sq. in.	6.895	kilopascals (kPa)
ENERGY or WORK	BTU	1 055.	joules (J)
	foot-pound	1.355 8	joules (J)
	kilowatt-hour	3 600 000. or 3.6 × 10⁶	joules (J = one W's)
LIGHT	foot candle	10.76	lumens/metre² (lm/m²)
FUEL PERFORMANCE	miles/gal	0.425 1	kilometres/litre (km/l)
	gal/mile	2.352 7	litres/kilometre (l/km)
VELOCITY	miles/hour	1.609 3	kilometres/hr. (km/h)
LENGTH	inch	25.4	millimetres (mm)
	foot	0.304 8	metres (m)
	yard	0.914 4	metres (m)
	mile	1.609	kilometres (km)
AREA	inch²	645.2	millimetres² (mm²)
		6.45	centimetres² (cm²)
	foot²	0.092 9	metres² (m²)
	yard²	0.836 1	metres²
VOLUME	inch³	16 387.	mm³
	inch³	16.387	cm³
	quart	0.016 4	litres (1)
	quart	0.946 4	litres
	gallon	3.785 4	litres
	yard³	0.764 6	metres³ (m³)
MASS	pound	0.453 6	kilograms (kg)
	ton	907.18	kilograms (kg)
	ton	0.90718	tonne
FORCE	kilogram	9.807	newtons (N)
	ounce	0.278 0	newtons
	pound	4.448	newtons
TEMPERATURE	degree farenheit	0.556 (°F −32)	degree Celsius (°C)

Left Column is units of 10, (0, 10, 20, 30 etc.);
Top Row is in units of one (0, 1, 2, 3, etc).

EXAMPLE: Feet to Inches Conversion Chart

feet	0	1	2	3	4	5	6	7	8	9	feet
	inches	inches	inches	inches	inches	inches	inches	inches	inches	inches	
..		12	24	36	48	60	72	84	96	108	..
10	120	132	144	156	168	180	192	204	216	228	10
20	240	252	264	276	288	300	312	324	336	348	20
30	360	372	384	396	408	420	432	444	456	468	30
40	480	492	504	516	528	540	552	564	576	588	40
50	600	612	624	636	648	660	672	684	696	708	50

12 feet equals 144 inches. Read across from 10 and down from 2.
6 feet equals 72 inches. Read down from 6.

FEET TO METERS

ft	0	1	2	3	4	5	6	7	8	9	ft
	m	m	m	m	m	m	m	m	m	m	
..		0.305	0.610	0.914	1.219	1.524	1.829	2.134	2.438	2.743	..
10	3.048	3.353	3.658	3.962	4.267	4.572	4.877	5.182	5.486	5.791	10
20	6.096	6.401	6.706	7.010	7.315	7.620	7.925	8.230	8.534	8.839	20
30	9.144	9.449	9.754	10.058	10.363	10.668	10.973	11.278	11.582	11.887	30
40	12.192	12.497	12.802	13.106	13.411	13.716	14.021	14.326	14.630	14.935	40
50	15.240	15.545	15.850	16.154	16.459	16.764	17.069	17.374	17.678	17.983	50
60	18.288	18.593	18.898	19.202	19.507	19.812	20.117	20.422	20.726	21.031	60
70	21.336	21.641	21.945	22.250	22.555	22.860	23.165	23.470	23.774	24.079	70
80	24.384	24.689	24.994	25.298	25.603	25.908	26.213	26.518	26.822	27.127	80
90	27.432	27.737	28.042	28.346	28.651	28.956	29.261	29.566	29.870	30.175	90
100	30.480	30.785	31.090	31.394	31.699	32.004	32.309	32.614	32.918	33.223	100

METERS TO FEET

m	0	1	2	3	4	5	6	7	8	9	m
	ft	ft	ft	ft	ft	ft	ft	ft	ft	ft	
..		3.2808	6.5617	9.8425	13.1234	16.4042	19.6850	22.9659	26.2467	29.5276	..
10	32.8084	36.0892	39.3701	42.6509	45.9318	49.2126	52.4934	55.7743	59.0551	62.3360	10
20	65.6168	68.8976	72.1785	75.4593	78.7402	82.0210	85.3018	88.5827	91.8635	95.1444	20
30	98.4252	101.7060	104.9869	108.2677	111.5486	114.8294	118.1102	121.3911	124.6719	127.9528	30
40	131.2336	134.5144	137.7953	141.0761	144.3570	147.6378	150.9186	154.1995	157.4803	160.7612	40
50	164.0420	167.3228	170.6037	173.8845	177.1654	180.4462	183.7270	187.0079	190.2887	193.5696	50
60	196.8504	200.1312	203.4121	206.6929	209.9738	213.2546	216.5354	219.8163	223.0971	226.3780	60
70	229.6588	232.9396	236.2205	239.5013	242.7822	246.0630	249.3438	252.6247	255.9055	259.1864	70
80	262.4672	265.7480	269.0289	272.3097	275.5906	278.8714	282.1522	285.4331	288.7139	291.9948	80
90	295.2756	298.5564	301.8373	305.1181	308.3990	311.6798	314.9606	318.2415	321.5223	324.8032	90
100	328.0840	331.3648	334.6457	337.9265	341.2074	344.4882	347.7690	351.0499	354.3307	357.6116	100

MILES TO KILOMETERS

mile	0	1	2	3	4	5	6	7	8	9	mile
	km	km	km	km	km	km	km	km	km	km	
..		1.609	3.219	4.828	6.437	8.047	9.656	11.265	12.875	14.484	..
10	16.093	17.703	19.312	20.921	22.531	24.140	25.750	27.359	28.968	30.578	10
20	32.187	33.796	35.406	37.015	38.624	40.234	41.843	43.452	45.062	46.671	20
30	48.280	49.890	51.499	53.108	54.718	56.327	57.936	59.546	61.155	62.764	30
40	64.374	65.983	67.593	69.202	70.811	72.421	74.030	75.639	77.249	78.858	40
50	80.467	82.077	83.686	85.295	86.905	88.514	90.123	91.733	93.342	94.951	50
60	96.561	98.170	99.779	101.39	103.00	104.61	106.22	107.83	109.44	111.04	60
70	112.65	114.26	115.87	117.48	119.09	120.70	122.31	123.92	125.53	127.14	70
80	128.75	130.36	131.97	133.58	135.19	136.79	138.40	140.01	141.62	143.23	80
90	144.84	146.45	148.06	149.67	151.28	152.89	154.50	156.11	157.72	159.33	90
100	160.93	162.54	164.15	165.76	167.37	168.98	170.59	172.20	173.81	175.42	100

KILOMETERS TO MILES

km	0	1	2	3	4	5	6	7	8	9	km
	mil	mil	mil	mil	mil	mil	mil	mil	mil	mil	
..		0.621	1.243	1.864	2.486	3.107	3.728	4.350	4.971	5.592	..
10	6.214	6.835	7.457	8.078	8.699	9.321	9.942	10.562	11.185	11.805	10
20	12.427	13.049	13.670	14.292	14.913	15.534	16.156	16.776	17.399	18.019	20
30	18.641	19.263	19.884	20.506	21.127	21.748	22.370	22.990	23.613	24.233	30
40	24.855	25.477	26.098	26.720	27.341	27.962	28.584	29.204	29.827	30.447	40
50	31.069	31.690	32.311	32.933	33.554	34.175	34.797	35.417	36.040	36.660	50
60	37.282	37.904	38.525	39.147	39.768	40.389	41.011	41.631	42.254	42.874	60
70	43.497	44.118	44.739	45.361	45.982	46.603	47.225	47.845	48.468	49.088	70
80	49.711	50.332	50.953	51.575	52.196	52.817	53.439	54.059	54.682	55.302	80
90	55.924	56.545	57.166	57.788	58.409	59.030	59.652	60.272	60.895	61.515	90
100	62.138	62.759	63.380	64.002	64.623	65.244	65.866	66.486	67.109	67.729	100

GALLONS (U.S.) TO LITERS

U.S. gal	0	1	2	3	4	5	6	7	8	9	U.S. gal
	L	L	L	L	L	L	L	L	L	L	
..		3.7854	7.5709	11.3563	15.1417	18.9271	22.7126	26.4980	30.2834	34.0638	..
10	37.8543	41.6397	45.4251	49.2105	52.9960	56.7814	60.5668	64.3523	68.1377	71.9231	10
20	75.7085	79.4940	83.2794	87.0648	90.8502	94.6357	98.4211	102.2065	105.9920	109.7774	20
30	113.5528	117.3482	121.1337	124.9191	128.7045	132.4899	136.2754	140.0608	143.8462	147.6316	30
40	151.4171	155.2025	158.9879	162.7734	166.5588	170.3442	174.1296	177.9151	181.7005	185.4859	40
50	189.2713	193.0568	196.8422	200.6276	204.4131	208.1985	211.9839	215.7693	219.5548	223.3402	50
60	227.1256	230.9110	234.6965	238.4819	242.2673	246.0527	249.8382	253.6236	257.4090	261.1945	60
70	264.9799	268.7653	272.5507	276.3362	280.1216	283.9070	287.6924	291.4779	295.2633	299.0487	70
80	302.8342	306.6196	310.4050	314.1904	317.9759	321.7613	325.5467	329.3321	333.1176	336.9030	80
90	340.6884	344.4738	348.2593	352.0447	355.8301	359.6156	363.4010	367.1864	370.9718	374.7573	90
100	378.5427	382.3281	386.1135	389.8990	393.6844	397.4698	401.2553	405.0407	408.8261	412.6115	100

LITERS TO GALLONS (U.S.)

L	0	1	2	3	4	5	6	7	8	9	L
	gal	gal	gal	gal	gal	gal	gal	gal	gal	gal	
..		0.2642	0.5283	0.7925	1.0567	1.3209	1.5850	1.8492	2.1134	2.3775	..
10	2.6417	2.9059	3.1701	3.4342	3.6984	3.9626	4.2267	4.4909	4.7551	5.0192	10
20	5.2834	5.5476	5.8118	6.0759	6.3401	6.6043	6.8684	7.1326	7.3968	7.6610	20
30	7.9251	8.1893	8.4535	8.7176	8.9818	9.2460	9.5102	9.7743	10.0385	10.3027	30
40	10.5668	10.8310	11.0952	11.3594	11.6235	11.8877	12.1519	12.4160	12.6802	12.9444	40
50	13.2086	13.4727	13.7369	14.0011	14.2652	14.5294	14.7936	15.0577	15.3219	15.5861	50
60	15.8503	16.1144	16.3786	16.6428	16.9069	17.1711	17.4353	17.6995	17.9636	18.2278	60
70	18.4920	18.7561	19.0203	19.2845	19.5487	19.8128	20.0770	20.3412	20.6053	20.8695	70
80	21.1337	21.3979	21.6620	21.9262	22.1904	22.4545	22.7187	22.9829	23.2470	23.5112	80
90	23.7754	24.0396	24.3037	24.5679	24.8321	25.0962	25.3604	25.6246	25.8888	26.1529	90
100	26.4171	26.6813	26.9454	27.2096	27.4738	27.7380	28.0021	28.2663	28.5305	28.7946	100

GALLONS (IMP.) TO LITERS

IMP gal	0	1	2	3	4	5	6	7	8	9	IMP gal
	L	L	L	L	L	L	L	L	L	L	
..		4.5460	9.0919	13.6379	18.1838	22.7298	27.2758	31.8217	36.3677	40.9136	..
10	45.4596	50.0056	54.5515	59.0975	63.6434	68.1894	72.2354	77.2813	81.8275	86.3732	10
20	90.9192	95.4652	100.0111	104.5571	109.1030	113.6490	118.1950	122.7409	127.2869	131.8328	20
30	136.3788	140.9248	145.4707	150.0167	154.5626	159.1086	163.6546	168.0005	172.7465	177.2924	30
40	181.8384	186.3844	190.9303	195.4763	200.0222	204.5682	209.1142	213.6601	218.2061	222.7520	40
50	227.2980	231.8440	236.3899	240.9359	245.4818	250.0278	254.5738	259.1197	263.6657	268.2116	50
60	272.7576	277.3036	281.8495	286.3955	290.9414	295.4874	300.0334	304.5793	309.1253	313.6712	60
70	318.2172	322.7632	327.3091	331.8551	336.4010	340.9470	345.4930	350.0389	354.5849	359.1308	70
80	363.6768	368.2223	372.7687	377.3147	381.8606	386.4066	390.9526	395.4985	400.0445	404.5904	80
90	409.1364	413.6824	418.2283	422.7743	427.3202	431.8662	436.4122	440.9581	445.9041	450.0500	90
100	454.5960	459.1420	463.6879	468.2339	472.7798	477.3258	481.8718	486.4177	490.9637	495.5096	100

LITERS TO GALLONS (IMP.)

L	0	1	2	3	4	5	6	7	8	9	L
	gal	gal	gal	gal	gal	gal	gal	gal	gal	gal	
..		0.2200	0.4400	0.6599	0.8799	1.0999	1.3199	1.5398	1.7598	1.9798	..
10	2.1998	2.4197	2.6397	2.8597	3.0797	3.2996	3.5196	3.7396	3.9596	4.1795	10
20	4.3995	4.6195	4.8395	5.0594	5.2794	5.4994	5.7194	5.9394	6.1593	6.3793	20
30	6.5593	6.8193	7.0392	7.2592	7.4792	7.6992	7.9191	8.1391	8.3591	8.5791	30
40	8.7990	9.0190	9.2390	9.4590	9.6789	9.8989	10.9189	10.3389	10.5588	10.7788	40
50	10.9988	11.2188	11.4388	11.6587	11.8787	12.0987	12.3187	12.5386	12.7586	12.9786	50
60	13.1986	13.4185	13.6385	13.8585	14.0785	14.2984	14.5184	14.7384	14.9584	15.1783	60
70	15.3983	15.6183	15.8383	16.0582	16.2782	16.4982	16.7182	16.9382	17.1581	17.3781	70
80	17.5981	17.8181	18.0380	18.2580	18.4780	18.6980	18.9179	19.1379	19.3579	19.5779	80
90	19.7978	20.0178	20.2378	20.4578	20.6777	20.8977	21.1177	21.3377	21.5576	21.7776	90
100	21.9976	22.2176	22.4376	22.6575	22.8775	23.0975	23.3175	23.5374	23.7574	23.9774	100

POUNDS TO KILOGRAMS

lb	0	.1	2	3	4	5	6	7	8	9	lb
	kg	kg	kg	kg	kg	kg	kg	kg	kg	kg	
..		0.454	0.907	1.361	1.814	2.268	2.722	3.175	3.629	4.082	..
10	4.536	4.990	5.443	5.897	6.350	6.804	7.257	7.711	8.165	8.618	10
20	9.072	9.525	9.979	10.433	10.886	11.340	11.793	12.247	12.701	13.154	20
30	13.608	14.061	14.515	14.969	15.422	15.876	16.329	16.783	17.237	17.690	30
40	18.144	18.597	19.051	19.504	19.958	20.412	20.865	21.319	21.772	22.226	40
50	22.680	23.133	23.587	24.040	24.494	24.948	25.401	25.855	26.308	26.762	50
60	27.216	27.669	28.123	28.576	29.030	29.484	29.937	30.391	30.844	31.298	60
70	31.751	32.205	32.659	33.112	33.566	34.019	34.473	34.927	35.380	35.834	70
80	36.287	36.741	37.195	37.648	38.102	38.555	39.009	39.463	39.916	40.370	80
90	40.823	41.277	41.730	42.184	42.638	43.092	43.545	43.998	44.453	44.906	90
100	45.359	45.813	46.266	46.720	47.174	47.627	48.081	48.534	48.988	49.442	100

KILOGRAMS TO POUNDS

kg	0	1	2	3	4	5	6	7	8	9	kg
	lb	lb	lb	lb	lb	lb	lb	lb	lb	lb	
..		2.205	4.409	6.614	8.818	11.023	13.228	15.432	17.637	19.842	..
10	22.046	24.251	26.455	28.660	30.865	33.069	35.274	37.479	39.683	41.888	10
20	44.092	46.297	48.502	50.706	52.911	55.116	57.320	59.525	61.729	63.934	20
30	66.139	68.343	70.548	72.752	74.957	77.162	79.366	81.571	83.776	85.980	30
40	88.185	90.389	92.594	94.799	97.003	99.208	101.41	103.62	105.82	108.03	40
50	110.23	112.44	114.64	116.84	119.05	121.25	123.46	125.66	127.87	130.07	50
60	132.28	134.48	136.69	138.89	141.10	143.30	145.51	147.71	149.91	152.12	60
70	154.32	156.53	158.73	160.94	163.14	165.35	167.55	169.76	171.96	174.17	70
80	176.37	178.57	180.78	182.98	185.19	187.39	189.60	191.80	194.01	196.21	80
90	198.42	200.62	202.83	205.03	207.23	209.44	211.64	213.85	216.05	218.26	90
100	220.46	222.67	224.87	227.08	229.28	231.49	233.69	235.89	238.10	240.30	100

POUNDS PER SQUARE INCHES TO KILOPASCALS

lb/in²	0	1	2	3	4	5	6	7	8	9	lb/in²
	kPa	kPa	kPa	kPa	kPa	kPa	kPa	kPa	kPa	kPa	
..	0.0000	6.8948	13.7895	20.6843	27.5790	34.4738	41.3685	48.2663	55.1581	62.0528	..
10	68.9476	75.8423	82.7371	89.6318	96.5266	103.4214	110.3161	117.2109	124.1056	131.0004	10
20	137.8951	144.7899	151.6847	158.5794	165.4742	172.3689	179.2637	186.1584	193.0532	199.9480	20
30	206.8427	213.7375	220.6322	227.5270	234.4217	241.3165	248.2113	255.1060	262.0008	268.8955	30
40	275.7903	282.6850	289.5798	296.4746	303.3693	310.2641	317.1588	324.0536	330.9483	337.8431	40
50	344.7379	351.6326	358.5274	365.4221	372.3169	379.2116	386.1064	393.0012	399.8959	406.7907	50
60	412.6854	420.5802	427.4749	434.3697	441.2645	448.1592	455.0540	461.9487	468.8435	475.7382	60
70	482.6330	489.5278	496.4225	503.3173	510.2120	517.1068	524.0015	530.8963	537.7911	544.6858	70
80	551.5806	558.4753	565.3701	572.2648	579.1596	586.0544	592.9491	599.8439	606.7386	613.6334	80
90	620.5281	627.4229	634.3177	641.2124	648.1072	655.0019	661.8967	668.7914	675.6862	682.5810	90
100	689.4757	696.3705	703.2653	710.1601	717.0549	723.9497	730.8445	737.7393	744.6341	751.5289	100

KILOPASCALS TO POUNDS PER SQUARE INCHES

kPa	0	1	2	3	4	5	6	7	8	9	kPa
	lb/in²	lb/in²	lb/in²	lb/in²	lb/in²	lb/in²	lb/in²	lb/in²	lb/in²	lb/in²	
..		.1450	.2901	.4351	.5801	.7252	.8702	1.0153	1.1603	1.3053	..
10	1.4504	1.5954	1.7404	1.8855	2.0305	2.1556	2.3206	2.4656	2.6107	2.7557	10
20	2.9007	3.0458	3.1908	3.3359	3.4809	3.6259	3.7710	3.9160	4.0610	4.2061	20
30	4.3511	4.4961	4.6412	4.7862	4.9313	5.0763	5.2213	5.3664	5.5114	5.6564	30
40	5.8015	5.9465	6.0916	6.2366	6.3816	6.5267	6.6717	6.8167	6.9618	7.1068	40
50	7.2518	7.3969	7.5419	7.6870	7.8320	7.9770	8.1221	8.2671	8.4121	8.5572	50
60	8.7022	8.8473	8.9923	9.1373	9.1824	9.4274	9.5724	9.7175	9.8625	10.0076	60
70	10.1526	10.2976	10.4427	10.5877	10.7327	10.8778	11.0228	11.1678	11.3129	11.4579	70
80	11.6030	11.7480	11.8930	12.0381	12.1831	12.3281	12.4732	12.6182	12.7633	12.9083	80
90	13.0533	13.1984	13.3434	13.4884	13.6335	13.7785	13.9236	14.0686	14.2136	14.3587	90
100	14.5037	14.6487	14.7938	14.9388	15.0838	15.2289	15.3739	15.5190	15.6640	15.8090	100

POUND FEET TO NEWTON-METERS

ft-lb	0	1	2	3	4	5	6	7	8	9	ft-lb
	N·m	N·m	N·m	N·m	N·m	N·m	N·m	N·m	N·m	N·m	
..		1.3558	2.7116	4.0675	5.4233	6.7791	8.1349	9.4907	10.8465	12.2024	..
10	13.5582	14.9140	16.2698	17.6256	18.9815	20.3373	21.6931	23.0489	24.4047	25.7605	10
20	27.1164	28.4722	29.8280	31.1838	32.5396	33.8954	35.2513	36.6071	37.9629	39.3187	20
30	40.6745	42.0304	43.3862	44.7420	46.0978	47.4536	48.8094	50.1653	51.5211	52.8769	30
40	54.2327	55.5885	56.9444	58.3002	59.6560	61.0118	62.3676	63.7234	65.0793	66.4351	40
50	67.7909	69.1467	70.5025	71.8584	73.2142	74.5700	75.9258	77.2816	78.6374	79.9933	50
60	81.3491	82.7049	84.0607	85.4165	86.7724	88.1282	89.4840	90.3898	92.1956	93.5514	60
70	94.9073	96.2631	97.6189	98.9747	100.3305	101.6863	103.0422	104.3980	105.7538	107.1096	70
80	108.4654	109.8213	111.1771	112.5329	113.8887	115.2445	116.6003	117.9562	119.3120	120.6678	80
90	122.0236	123.3794	124.7353	126.0911	127.4469	128.8027	130.1585	131.5143	132.8702	134.2260	90
100	135.5818	136.9376	138.2934	139.6493	141.0051	142.3609	143.7167	145.0725	146.4283	147.7842	100

NEWTON-METERS TO POUND FEET

N·m	0	1	2	3	4	5	6	7	8	9	N·m
	ft-lb	ft-lb	ft-lb	ft-lb	ft-lb	ft-lb	ft-lb	ft-lb	ft-lb	ft-lb	
..		.7376	1.4751	2.2127	2.9502	3.6878	4.4254	5.1692	5.9005	6.6381	..
10	7.3756	8.1132	8.8507	9.5883	10.3258	11.0634	11.8010	12.5385	13.2761	14.0136	10
20	14.7512	15.4888	16.2264	16.9639	17.7015	18.4390	19.1766	19.9142	20.6517	21.3893	20
30	22.1269	22.8644	23.6020	24.3395	25.0771	25.8147	26.5522	27.2898	28.0274	28.7649	30
40	29.5025	30.2400	30.9776	31.7152	32.4527	33.1903	33.9279	34.6654	35.4030	36.1405	40
50	36.8781	37.6157	38.3532	39.0908	39.8283	40.5659	41.3035	42.0410	42.7786	43.5162	50
60	44.2537	44.9913	45.7288	46.4664	47.2040	47.9415	48.6791	49.4167	50.1542	50.8918	60
70	51.6293	52.3669	53.1045	53.8420	54.5796	55.3171	56.0547	56.7923	57.5298	58.2674	70
80	59.0050	59.7425	60.4801	61.2176	61.9552	62.6928	63.4303	64.1679	64.9055	65.6430	80
90	66.3806	67.1181	67.8557	68.5933	69.3308	70.0684	70.8060	71.5435	72.2811	73.0186	90
100	73.7562	74.4938	75.2313	75.9689	76.7064	77.4440	78.1816	78.9191	79.6567	80.3943	100

DIMENSION AND TEMPERATURE CONVERSION CHART

Inches		Decimals	Milli-meters	Inches to millimeters		Millimeters to inches		Fahrenheit & Celsius			
				Inches	mm	mm	Inches	°F	°C	°C	°F
	1/64	.015625	.3969	.0001	.00254	0.001	.000039	-20	-28.9	-30	-22
1/32		.03125	.7937	.0002	.00508	0.002	.000079	-15	-26.1	-28	-18.4
	3/64	.046875	1.1906	.0003	.00762	0.003	.000118	-10	-23.3	-26	-14.8
1/16		.0625	1.5875	.0004	.01016	0.004	.000157	-5	-20.6	-24	-11.2
	5/64	.078125	1.9844	.0005	.01270	0.005	.000197	0	-17.8	-22	-7.6
3/32		.09375	2.3812	.0006	.01524	0.006	.000236	1	-17.2	-20	-4
	7/64	.109375	2.7781	.0007	.01778	0.007	.000276	2	-16.7	-18	-0.4
1/8		.125	3.1750	.0008	.02032	0.008	.000315	3	-16.1	-16	3.2
	9/64	.140625	3.5719	.0009	.02286	0.009	.000354	4	-15.6	-14	6.8
5/32		.15625	3.9687	.001	.0254	0.01	.00039	5	-15.0	-12	10.4
	11/64	.171875	4.3656	.002	.0508	0.02	.00079	10	-12.2	-10	14
3/16		.1875	4.7625	.003	.0762	0.03	.00118	15	-9.4	-8	17.6
	13/64	.203125	5.1594	.004	.1016	0.04	.00157	20	-6.7	-6	21.2
7/32		.21875	5.5562	.005	.1270	0.05	.00197	25	-3.9	-4	24.8
	15/64	.234375	5.9531	.006	.1524	0.06	.00236	30	-1.1	-2	28.4
1/4		.25	6.3500	.007	.1778	0.07	.00276	35	1.7	0	32
	17/64	.265625	6.7469	.008	.2032	0.08	.00315	40	4.4	2	35.6
9/32		.28125	7.1437	.009	.2286	0.09	.00354	45	7.2	4	39.2
	19/64	.296875	7.5406	.01	.254	0.1	.00394	50	10.0	6	42.8
5/16		.3125	7.9375	.02	.508	0.2	.00787	55	12.8	8	46.4
	21/64	.328125	8.3344	.03	.762	0.3	.01181	60	15.6	10	50
11/32		.34375	8.7312	.04	1.016	0.4	.01575	65	18.3	12	53.6
	23/64	.359375	9.1281	.05	1.270	0.5	.01969	70	21.1	14	57.2
3/8		.375	9.5250	.06	1.524	0.6	.02362	75	23.9	16	60.8
	25/64	.390625	9.9219	.07	1.778	0.7	.02756	80	26.7	18	64.4
13/32		.40625	10.3187	.08	2.032	0.8	.03150	85	29.4	20	68
	27/64	.421875	10.7156	.09	2.286	0.9	.03543	90	32.2	22	71.6
7/16		.4375	11.1125	.1	2.54	1	.03937	95	35.0	24	75.2
	29/64	.453125	11.5094	.2	5.08	2	.07874	100	37.8	26	78.8
15/32		.46875	11.9062	.3	7.62	3	.11811	105	40.6	28	82.4
	31/64	.484375	12.3031	.4	10.16	4	.15748	110	43.3	30	86
1/2		.5	12.7000	.5	12.70	5	.19685	115	46.1	32	89.6
	33/64	.515625	13.0969	.6	15.24	6	.23622	120	48.9	34	93.2
17/32		.53125	13.4937	.7	17.78	7	.27559	125	51.7	36	96.8
	35/64	.546875	13.8906	.8	20.32	8	.31496	130	54.4	38	100.4
9/16		.5625	14.2875	.9	22.86	9	.35433	135	57.2	40	104
	37/64	.578125	14.6844	1	25.4	10	.39370	140	60.0	42	107.6
19/32		.59375	15.0812	2	50.8	11	.43307	145	62.8	44	112.2
	39/64	.609375	15.4781	3	76.2	12	.47244	150	65.6	46	114.8
5/8		.625	15.8750	4	101.6	13	.51181	155	68.3	48	118.4
	41/64	.640625	16.2719	5	127.0	14	.55118	160	71.1	50	122
21/32		.65625	16.6687	6	152.4	15	.59055	165	73.9	52	125.6
	43/64	.671875	17.0656	7	177.8	16	.62992	170	76.7	54	129.2
11/16		.6875	17.4625	8	203.2	17	.66929	175	79.4	56	132.8
	45/64	.703125	17.8594	9	228.6	18	.70866	180	82.2	58	136.4
23/32		.71875	18.2562	10	254.0	19	.74803	185	85.0	60	140
	47/64	.734375	18.6531	11	279.4	20	.78740	190	87.8	62	143.6
3/4		.75	19.0500	12	304.8	21	.82677	195	90.6	64	147.2
	49/64	.765625	19.4469	13	330.2	22	.86614	200	93.3	66	150.8
25/32		.78125	19.8437	14	355.6	23	.90551	205	96.1	68	154.4
	51/64	.796875	20.2406	15	381.0	24	.94488	210	98.9	70	158
13/16		.8125	20.6375	16	406.4	25	.98425	212	100.0	75	167
	53/64	.828125	21.0344	17	431.8	26	1.02362	215	101.7	80	176
27/32		.84375	21.4312	18	457.2	27	1.06299	220	104.4	85	185
	55/64	.859375	21.8281	19	482.6	28	1.10236	225	107.2	90	194
7/8		.875	22.2250	20	508.0	29	1.14173	230	110.0	95	203
	57/64	.890625	22.6219	21	533.4	30	1.18110	235	112.8	100	212
29/32		.90625	23.0187	22	558.8	31	1.22047	240	115.6	105	221
	59/64	.921875	23.4156	23	584.2	32	1.25984	245	118.3	110	230
15/16		.9375	23.8125	24	609.6	33	1.29921	250	121.1	115	239
	61/64	.953125	24.2094	25	635.0	34	1.33858	255	123.9	120	248
31/32		.96875	24.6062	26	660.4	35	1.37795	260	126.6	125	257
	63/64	.984375	25.0031	27	690.6	36	1.41732	265	129.4	130	266

DECIMAL EQUIVALENTS AND TAP DRILL SIZES

DRILL SIZE	DECIMAL	TAP SIZE	DRILL SIZE	DECIMAL	TAP SIZE	DRILL SIZE	DECIMAL	TAP SIZE
1/64	.0156		17	.1730		Q	.3320	3/8-24
1/32	.0312		16	.1770	12-24	R	.3390	
60	.0400		15	.1800		11/32	.3437	
59	.0410		14	.1820	12-28	S	.3480	
58	.0420		13	.1850	12-32	T	.3580	
57	.0430		3/16	.1875		23/64	.3594	
56	.0465		12	.1890		U	.3680	7/16-14
3/64	.0469	0-80	11	.1910		3/8	.3750	
55	.0520		10	.1935		V	.3770	
54	.0550	1-56	9	.1960		W	.3860	
53	.0595	1-64, 72	8	.1990		25/64	.3906	7/16-20
1/16	.0625		7	.2010	1/4-20	X	.3970	
52	.0635		13/64	.2031		Y	.4040	
51	.0670		6	.2040		13/32	.4062	
50	.0700	2-56, 64	5	.2055		Z	.4130	
49	.0730		4	.2090		27/64	.4219	1/2-13
48	.0760		3	.2130	1/4-28	7/16	.4375	
5/64	.0781		7/32	.2187		29/64	.4531	1/2-20
47	.0785	3-48	2	.2210		15/32	.4687	
46	.0810		1	.2280		31/64	.4844	9/16-12
45	.0820	3-56,4-32	A	.2340		1/2	.5000	
44	.0860	4-36	15/64	.2344		33/64	.5156	9/16-18
43	.0890	4-40	B	.2380		17/32	.5312	5/8-11
42	.0935	4-48	C	.2420		35/64	.5469	
3/32	.0937		D	.2460		9/16	.5625	
41	.0960		E, 1/4	.2500		37/64	.5781	5/8-18
40	.0980		F	.2570	5/16-18	19/32	.5937	11/16-11
39	.0995		G	.2610		39/64	.6094	
38	.1015	5-40	17/64	.2656		5/8	.6250	11/16-16
37	.1040	5-44	H	.2660		41/64	.6406	
36	.1065	6-32	I	.2720	5/16-24	21/32	.6562	3/4-10
7/64	.1093		J	.2770		43/64	.6719	
35	.1100		K	.2810		11/16	.6875	3/4-16
34	.1110	6-36	9/32	.2812		45/64	.7031	
33	.1130	6-40	L	.2900		23/32	.7187	
32	.1160		M	.2950		47/64	.7344	
31	.1200		19/64	.2968		3/4	.7500	
1/8	.1250		N	.3020		49/64	.7656	7/8-9
30	.1285		5/16	.3125	3/8-16	25/32	.7812	
29	.1360	8-32, 36	O	.3160		51/64	.7969	
28	.1405	8-40	P	.3230		13/16	.8125	7/8-14
9/64	.1406		21/64	.3281		53/64	.8281	
27	.1440					27/32	.8437	
26	.1470					55/64	.8594	
25	.1495	10-24				7/8	.8750	1-8
24	.1520					57/64	.8906	
23	.1540					29/32	.9062	
5/32	.1562					59/64	.9219	
22	.1570	10-30				15/16	.9375	1-12,14
21	.1590	10-32				61/64	.9531	
20	.1610					31/32	.9687	
19	.1660					63/64	.9844	
18	.1695					1	1.000	
11/64	.1719							

PIPE THREAD SIZES

THREAD	DRILL	THREAD	DRILL
1/8-27	R	1 1/2-11 1/2	1 47/64
1/4-18	7/16	2-11 1/2	2 7/32
3/8-18	37/64	2 1/2-8	2 5/8
1/2-14	23/32	3-8	3 1/4
3/4-14	59/64	3 1/2-8	3 3/4
1-11 1/2	1 5/32	4-8	4 1/4
1 1/4-11 1/2	1 1/2		

Index